D0854884

The Antarctic Paleoenvironment: A Perspective on Global Change

Part One

American Geophysical Union

ANTARCTIC RESEARCH SERIES

Physical Sciences

ANTARCTIC OCEANOLOGY
   Joseph L. Reid, *Editor*
ANTARCTIC OCEANOLOGY II: THE AUSTRALIAN-
NEW ZEALAND SECTOR
   Dennis E. Hayes, *Editor*

ANTARCTIC SNOW AND ICE STUDIES
   Malcolm Mellor, *Editor*
ANTARCTIC SNOW AND ICE STUDIES II
   A. P. Crary, *Editor*

ANTARCTIC SOILS AND SOIL FORMING PROCESSES
   J. C. F. Tedrow, *Editor*
DRY VALLEY DRILLING PROJECT
   L. D. McGinnis, *Editor*
GEOLOGICAL INVESTIGATIONS IN NORTHERN
VICTORIA LAND
   Edmund Stump, *Editor*
GEOLOGY AND PALEONTOLOGY OF THE ANTARCTIC
   Jarvis B. Hadley, *Editor*
GEOLOGY OF THE CENTRAL TRANSANTARCTIC
MOUNTAINS
   Mort D. Turner and John F. Splettstoesser,
   Editors
GEOMAGNETISM AND AERONOMY
   A. H. Waynick, *Editor*
METEOROLOGICAL STUDIES AT PLATEAU STATION,
ANTARCTICA
   Joost A. Businger, *Editor*
OCEANOLOGY OF THE ANTARCTIC CONTINENTAL SHELF
   Stanley S. Jacobs, *Editor*
STUDIES IN ANTARCTIC METEOROLOGY
   Morton J. Rubin, *Editor*
UPPER ATMOSPHERE RESEARCH IN ANTARCTICA
   L. J. Lanzerotti and C. G. Park, *Editors*
THE ROSS ICE SHELF: GLACIOLOGY AND GEOPHYSICS
   C. R. Bentley and D. E. Hayes, *Editors*
VOLCANOES OF THE ANTARCTIC PLATE AND SOUTHERN
OCEANS
   W. E. LeMasurier and J. T. Thomson, *Editors*
MINERAL RESOURCES POTENTIAL OF ANTARCTICA
   John F. Splettstoesser and Gisela A. M. Dreschhoff,
   Editors
CONTRIBUTIONS TO ANTARCTIC RESEARCH I
   David H. Elliot, *Editor*
CONTRIBUTIONS TO ANTARCTIC RESEARCH II
   David H. Elliot, *Editor*
MARINE GEOLOGICAL AND GEOPHYSICAL ATLAS
OF THE CIRCUM-ANTARCTIC TO 30°S
   Dennis E. Hayes, *Editor*
MOLLUSCAN SYSTEMATICS AND BIOSTRATIGRAPHY
   Jeffery D. Stilwell and William J. Zinsmeister

American Geophysical Union

ANTARCTIC RESEARCH SERIES

Biological and Life Sciences

BIOLOGY OF THE ANTARCTIC SEAS
  Milton O. Lee, *Editor*
BIOLOGY OF THE ANTARCTIC SEAS II
  George A. Llano, *Editor*
BIOLOGY OF THE ANTARCTIC SEAS III
  George A. Llano and Waldo L. Schmitt, *Editors*
BIOLOGY OF THE ANTARCTIC SEAS IV
  George A. Llano and I. Eugene Wallen, *Editors*
BIOLOGY OF THE ANTARCTIC SEAS V
  David L. Pawson, *Editor*
BIOLOGY OF THE ANTARCTIC SEAS VI
  David L. Pawson, *Editor*
BIOLOGY OF THE ANTARCTIC SEAS VII
  David L. Pawson, *Editor*
BIOLOGY OF THE ANTARCTIC SEAS VIII
  David L. Pawson and Louis S. Kornicker, *Editors*
BIOLOGY OF THE ANTARCTIC SEAS IX
  Louis S. Kornicker, *Editor*
BIOLOGY OF THE ANTARCTIC SEAS X
  Louis S. Kornicker, *Editor*
BIOLOGY OF THE ANTARCTIC SEAS XI
  Louis S. Kornicker, *Editor*
BIOLOGY OF THE ANTARCTIC SEAS XII
  David L. Pawson, *Editor*
BIOLOGY OF THE ANTARCTIC SEAS XIII
  Louis S. Kornicker, *Editor*
BIOLOGY OF THE ANTARCTIC SEAS XIV
  Louis S. Kornicker, *Editor*
BIOLOGY OF THE ANTARCTIC SEAS XV
  Louis S. Kornicker, *Editor*
BIOLOGY OF THE ANTARCTIC SEAS XVI
  Louis S. Kornicker, *Editor*
BIOLOGY OF THE ANTARCTIC SEAS XVII
  Louis S. Kornicker, *Editor*
BIOLOGY OF THE ANTARCTIC SEAS XVIII
  Louis S. Kornicker, *Editor*
BIOLOGY OF THE ANTARCTIC SEAS XIX
  Louis S. Kornicker, *Editor*
BIOLOGY OF THE ANTARCTIC SEAS XX
  Louis S. Kornicker, *Editor*
BIOLOGY OF THE ANTARCTIC SEAS XXI
  Louis S. Kornicker, *Editor*

ANTARCTIC TERRESTRIAL BIOLOGY
  George A. Llano, *Editor*
TERRESTRIAL BIOLOGY II
  Bruce Parker, *Editor*
TERRESTRIAL BIOLOGY III
  Bruce Parker, *Editor*

ANTARCTIC ASCIDIACEA
  Patricia Kott
ANTARCTIC BIRD STUDIES
  Oliver L. Austin, Jr., *Editor*
ANTARCTIC PINNIPEDIA
  William Henry Burt, *Editor*
ANTARCTIC CIRRIPEDIA
  William A. Newman and Arnold Ross
BIRDS OF THE ANTARCTIC AND SUB-ANTARCTIC
  George E. Watson
ENTOMOLOGY OF ANTARCTICA
  J. Linsley Gressitt, *Editor*
HUMAN ADAPTABILITY TO ANTARCTIC CONDITIONS
  E. K. Eric Gunderson, *Editor*
POLYCHAETA ERRANTIA OF ANTARCTICA
  Olga Hartman
POLYCHAETA MYZOSTOMIDAE AND SEDENTIARIA OF ANTARCTICA
  Olga Hartman
RECENT ANTARCTIC AND SUBANTARCTIC BRACHIOPODS
  Merrill W. Foster

Volume 56

ANTARCTIC
RESEARCH
SERIES

The Antarctic Paleoenvironment:
A Perspective on Global Change

Part One

James P. Kennett
Detlef A. Warnke

Editors

American Geophysical Union
Washington, D.C.
1992

Volume 56 | **ANTARCTIC RESEARCH SERIES**

Library of Congress Cataloging-in-Publication Data

The Antarctic paleoenvironment.

(Antarctic research series ; v. 56)
Papers from a conference held at the University of California, Santa Barbara, Aug. 28–31, 1991.
Includes bibliographical references.
1. Paleogeography—Antarctic regions—Congresses.   2. Paleoecology—Antarctic regions—Congresses.
I. Kennett, James P.   II. Warnke, Detlef A.
QE501.4.P3A64   1992   560'.45'09989   92-37312
ISBN 0875908233 (pb. 1)   CIP

ISSN 0066-4634

Published by
American Geophysical Union
With the aid of grant DPP-89-15494 from the
National Science Foundation

Printed in the United States of America.

CONTENTS

The Antarctic Research Series: Statement of Objectives
Board of Associate Editors   xi

Preface
James P. Kennett and Detlef A. Warnke   xiii

Acknowledgments   xiv
James P. Kennett and Detlef A. Warnke

Introduction
James P. Kennett and John A. Barron   1

The Development of Paleoseaways Around Antarctica
Lawrence A. Lawver, Lisa M. Gahagan, and Millard F. Coffin   7

Biogeography of Campanian-Maastrichtian Calcareous Plankton in the Region
of the Southern Ocean: Paleogeographic and Paleoclimatic Implications
Brian T. Huber and David K. Watkins   31

Late Cretaceous–Early Tertiary Antarctic Outcrop Evidence for Past
Vegetation and Climates
Rosemary A. Askin   61

Paleogene Chronology of Southern Ocean Drill Holes: An Update
Wuchang Wei   75

Late Eocene–Early Oligocene Evolution of Climate and Marine Circulation:
Deep-Sea Clay Mineral Evidence
Christian Robert and Hervé Chamley   97

Evidence From Fossil Vertebrates for a Rich Eocene Antarctic Marine
Environment
Judd A. Case   119

Paleoecology of Eocene Antarctic Sharks
Douglas J. Long   131

Cenozoic Deep-Sea Circulation: Evidence From Deep-Sea Benthic
Foraminifera
Ellen Thomas   141

The Influence of the Tethys on the Bottom Waters of the Early Tertiary Ocean
Hedi Oberhänsli   167

Late Eocene–Oligocene Sedimentation in the Antarctic Ocean, Atlantic Sector
(Maud Rise, ODP Leg 113, Site 689): Development of Surface and Bottom
Water Circulation
Liselotte Diester-Haass   185

Geotechnical Stratigraphy of Neogene Sediments: Maud Rise and Kerguelen
Plateau
Frank R. Rack and Alan Pittenger   203

Cenozoic Glacial History of the Ross Sea Revealed by Intermediate Resolution
Seismic Reflection Data Combined With Drill Site Information
John B. Anderson and Louis R. Bartek   231

Toward a High-Resolution Stable Isotopic Record of the Southern Ocean
During the Pliocene-Pleistocene (4.8 to 0.8 Ma)
David A. Hodell and Kathryn Venz   265

Miocene-Pliocene Antarctic Glacial Evolution: A Synthesis of Ice-Rafted
Debris, Stable Isotope, and Planktonic Foraminiferal Indicators, ODP Leg 114
*Detlef A. Warnke, Carl P. Allen, Daniel W. Müller, David A. Hodell, and
Charlotte A. Brunner*   311

A Late Neogene Antarctic Glacio-eustatic Record, Victoria Land Basin
Margin, Antarctica
Scott E. Ishman and Hugh J. Rieck   327

Late Quaternary Climatic Cycles as Recorded in Sediments From the Antarctic
Continental Margin
Hannes Grobe and Andreas Mackensen   349

Paleoecological Implications of Radiolarian Distribution and Standing Stocks
Versus Accumulation Rates in the Weddell Sea
Demetrio Boltovskoy and Viviana A. Alder   377

List of Co-chief Scientists on DSDP and ODP Legs   385

The Antarctic Research Series:
STATEMENT OF OBJECTIVES

The Antarctic Research Series provides for the presentation of detailed scientific research results from Antarctica, particularly the results of the United States Antarctic Research Program, including monographs and long manuscripts.

The series is designed to make the results of Antarctic fieldwork available. The Antarctic Research Series encourages the collection of papers on specific geographic areas within Antarctica. In addition, many volumes focus on particular disciplines, including marine biology, oceanology, meteorology, upper atmosphere physics, terrestrial biology, geology, glaciology, human adaptability, engineering, and environmental protection.

Topical volumes in the series normally are devoted to papers in one or two disciplines. Multidisciplinary volumes, initiated in 1990 to enable more rapid publication, are open to papers from any discipline. The series can accommodate long manuscripts and utilize special formats, such as maps.

Priorities for publication are set by the Board of Associate Editors. Preference is given to research manuscripts from projects funded by U.S. agencies. Because the series serves to emphasize the U.S. Antarctic Research Program, it also performs a function similar to expedition reports of many other countries with national Antarctic research programs.

The standards of scientific excellence expected for the series are maintained by the review criteria established for the AGU publications program. Each paper is critically reviewed by two or more expert referees. A member of the Board of Associate Editors may serve as editor of a volume, or another person may be appointed. The Board works with the individual editors of each volume and with the AGU staff to assure that the objectives of the series are met, that the best possible papers are presented, and that publication is timely.

Proposals for volumes or papers offered should be sent to the Board of Associate Editors, Antarctic Research Series, at 2000 Florida Avenue, N.W., Washington, D.C. 20009. Publication of the series is partially supported by a grant from the National Science Foundation.

Board of Associate Editors
Antarctic Research Series

PREFACE

The Antarctic continent and the surrounding Southern Ocean represent one of the major climate engines of the Earth: coupled components critical in the Earth's environmental system. The contributions in this volume help with the understanding of the long-term evolution of Antarctica's environment and biota. The aim of this and the succeeding companion volume is to help place the modern system within a historical context.

The environment and biosphere of the Antarctic region have undergone dynamic changes through geologic time. These, in turn, have played a key role in long-term global paleoenvironmental evolution. The development of the Southern Ocean itself, resulting from plate tectonism, created first-order changes in the circulation of the global ocean, in turn affecting meridional heat transport and hence global climates. Biospheric changes responded to the changing oceanic climatic states. Comprehension of the climatic and oceanographic processes that have operated at various times in Antarctica's history are crucial to the understanding of the present-day global environmental system. This knowledge will become increasingly important in parallel with concerns about anthropogenically caused global change. How vulnerable is the Antarctic region, especially its ice sheets, to global warming? The question is not parochial, given the potential of sea level change resulting from any Antarctic cryospheric development. Conversely, how much of a role does the Antarctic region, this giant icebox, play in moderating global, including sea level, change?

This is the first of two volumes in the American Geophysical Union's Antarctic Research Series to present contributions that deal with the paleoenvironmental and biotic evolution of the Antarctic region. The papers are based on work presented at a conference held at the University of California, Santa Barbara, August 28–31, 1991, entitled "The Role of the Southern Ocean and Antarctica in Global Change: An Ocean Drilling Perspective." This conference, jointly sponsored by JOI/USSAC and the Division of Polar Programs, National Science Foundation, was attended by more than 100 scientists from around the world. The primary objectives of the meeting were successful in providing a forum (1) to summarize existing paleoenvi-

ronmental data from the Antarctic region; (2) to identify and debate major remaining questions, most of which are thematic in nature; (3) to assist in formulating plans for future Antarctic ocean drilling; and (4) to organize publication of a series of summary/synthesis papers leading to this, the first of two volumes. Although it has been the intention of the scientific community to produce summary or synthesis volumes of thematic or regional nature related to ocean drilling, few have yet been published. Therefore a major objective of this and the second volume is to help make the results of ocean drilling more widely available to the scientific community.

In addition to these volumes the conference also led to the production of a white paper, compiled by J. Kennett and J. Barron (available from JOI/USSAC, Washington, D.C.), that summarizes major remaining questions related to Southern Ocean paleoenvironmental evolution and outlines further ocean drilling required to assist in answering these questions. Selected material from the white paper has been modified and incorporated in the introduction to this volume.

This volume presents 18 papers of general and synthetic nature on a wide variety of topics related to the environmental and biotic evolution of the Antarctic. Following a contribution that provides plate tectonic reconstructions of the Antarctic region during the last 200 m.y. and related paleoenvironmental implications, the volume is organized so that the papers are presented in general order of geologic age, beginning with the Late Cretaceous and ending with the modern Antarctic ocean. This arrangement was selected to help emphasize the evolution of the Antarctic environmental and biotic system during the late Phanerozoic. Two of the contributions deal with the Late Cretaceous, seven emphasize the Paleogene, and seven the Neogene through modern Antarctic ocean. Of the proxies employed for interpretation of the paleoenvironmental record, eight of the contributions have used sediments, seven have used the fossil record, and one paper is an interpretation of the stable isotopic record.

James P. Kennett and Detlef A. Warnke

ACKNOWLEDGMENTS

A large number of workers have contributed much in providing the necessary reviews of the contributions published in this volume; we heartily thank you all: J. B. Anderson, J. H. Andrews, M.-P. Aubry, J. A. Barron, G. W. Brass, L. H. Burckle, C. Charles, A. K. Cooper, A. R. Edwards, D. K. Futterer, T. R. Janacek, M. Katz, L. D. Keigwin, L. A. Krissek, D. J. Long, B. P. Luyendyk, K. Moran, J. Morley, S. O'Connell, L. E. Osterman, J. T. Parrish, W. Sliter, R. Stein, J. D. Stewart, K. Takahashi, B. H. Tiffney, E. M. Truswell, W. Wei, J. K. Weissel, B. White, S. W. Wise, Jr., J. A. Wolfe, F. C. Woodruff, A. R. Wyss, J. C. Zachos, and A. M. Ziegler.

Publication of this volume was made possible by JOI/USSAC. We thank Ellen Kappel of JOI/USSAC for her unwavering support of this project, and also H. Zimmerman of the National Science Foundation for his support of the conference leading to this volume.

We also thank Diana M. Kennett, editorial assistant, for her major contributions toward the production of this volume and for her perseverance in keeping publication on schedule.

James P. Kennett and Detlef A. Warnke

INTRODUCTION

JAMES P. KENNETT

Marine Science Institute and Department of Geological Sciences, University of California, Santa Barbara
California 93106

JOHN A. BARRON

U.S. Geological Survey, Menlo Park, California 94025

Antarctica has played a key role in long-term global paleoenvironmental evolution. An understanding of the climatic, paleoceanographic, and cryospheric evolution of Antarctica is crucial to a broader understanding of global climate and oceanographic change on both short and long time scales.

Through time, major interactions between the hydrosphere, cryosphere, atmosphere, and biosphere have affected the Earth's environmental system through climatic feedback, biogeochemical cycles, deep ocean circulation, and sea level changes. The Antarctic region is almost certainly sensitive to externally imposed change and, conversely, has played a major role in affecting climatic change elsewhere on Earth during the Cenozoic. Processes linking the low-latitude regions with Antarctica as a heat sink continue to play a major role in driving global atmospheric and oceanic circulation.

The Antarctic region is a key repository of paleoenvironmental information important for understanding global climatic evolution and its causes, the development of the cryosphere, and sea level history. Studies of the Antarctic marine sedimentary record, in both shallow and deep waters and in uplifted sequences on land, are also essential for understanding the history of formation and supplies of deep and intermediate waters to the world ocean, oceanic upwelling and biological productivity, and the evolution of the marine biota. An integrated approach to studies of paleoenvironments has also become increasingly valuable in the understanding of modern oceanographic and climatic processes.

This integrated approach includes studies of drilled cores from the nearshore area, from the continental margin, and in the deep ocean surrounding Antarctica. It is also necessary to integrate stratigraphic information obtained from these sequences into the broader regional picture through the use of seismic and sequence stratigraphy. Valuable, indeed unique, information is also being obtained from marine and nonmarine sequences exposed on the continent itself. This volume and part 2 (to follow) include contributions about Antarctica's paleoenvironment and biota that use all of these various approaches. Certainly, the interpretations provided in these numerous contributions are not in full agreement; indeed, they are often in strong disagreement. This, in part, results from the character of the sequences that were examined. Sequences from the continent itself offer important advantages in location, especially with regard to the cryospheric evolution. However, these sequences can be complicated by processes that rework the sedimentary materials and cause hiatuses. This especially seems to be the case of Neogene sequences, the time when the Antarctic cryosphere was well developed. Conversely, deep-sea sequences are often continuous and relatively simple stratigraphically but are remote from the continent. Hence interpretations are often indirect. The greatest existing controversies relate to the history of the cryosphere, and the contributions in this and the companion volume express some of this disagreement. Although no consensus has yet been built on Antarctic cryospheric development, enormous progress has been made in the understanding of the broader paleoenvironmental and biotic evolution of Antarctica and the Southern Ocean.

Themes of special importance in the Antarctic region include the following:

1. The sequence and timing of the breakup and dispersal of the Gondwana continents.
2. Development of surface and deep-water circulation around Antarctica.
3. Climate evolution of continent and ocean.
4. Cryospheric development and evolution.
5. Sea level history.
6. Bottom water origin and circulation history.
7. Evolution of surface and deep-water chemistry.
8. Vertical water mass structure.
9. Evolution of oceanic paleoproductivity.

10. Evolution of the Antarctic biota in relation to environmental changes.

11. Paleobiogeographic development of Antarctic faunas and floras.

12. Cretaceous anoxic basins in the Antarctic.

Much of the new information that has been obtained during the last few years relative to Antarctica's paleoenvironmental and biotic evolution has resulted from ocean drilling. This volume and its companion reflect this activity. For various reasons, future advances in this field of endeavor will have to rely almost entirely on the drilling of sedimentary sequences in both shallow and deep seas around the Antarctic continent and on the integration of these data with seismic information. Because of this we have summarized as follows accomplishments of ocean drilling in the Antarctic region. Following this is a listing of important remaining questions and problems that need to be attacked using both ocean drilling in various water depths and locations and examination of sedimentary sequences exposed on the continent itself.

ACCOMPLISHMENTS OF OCEAN DRILLING PROGRAM IN ANTARCTICA

During the last two decades, studies of sedimentary sequences drilled in and around Antarctica have led to significant advances in the understanding of the evolution of climate, oceanography, and biota of the Antarctic continent and surrounding ocean. Four Ocean Drilling Project (ODP) expeditions carried out during 1987 and 1988 have contributed enormously toward the understanding of this evolution. These expeditions were Leg 113 in the Weddell Sea [*Barker et al.*, 1988, 1990], Leg 114 in the Subantarctic South Atlantic [*Ciesielski et al.*, 1988, 1991], Leg 119 in Prydz Bay and on Kerguelen Plateau [*Barron et al.*, 1989, 1991], and Leg 120 on Kerguelen Plateau [*Schlich et al.*, 1989, *Wise et al.*, 1992]. Important advances have also resulted from studies of sedimentary sequences on the continental shelf and continent [e.g., *Barrett*, 1989]. The recovery of numerous high-quality hydraulic piston cores and the successful application of detailed magnetostratigraphy have provided a stronger framework for paleoceanographic and biogeographic studies built upon the foundation of results from earlier drilling expeditions (Deep Sea Drilling Project (DSDP) to Antarctica [e.g., *Hayes et al.*, 1975; *Kennett et al.*, 1975; *Hollister et al.*, 1976; *Barker et al.*, 1977; *Ludwig et al.*, 1983]. For the first time, the scientific community now has a high southern latitude transect of sedimentary sequences with a working foundation of chemomagnetobiostratigraphy.

An extensive repository of sedimentologic, biotic, and isotopic evidence indicates that sequential cooling and cryospheric development of the Antarctic region during the Cenozoic profoundly affected the ocean/atmosphere circulation, sediments, and biota. The first

Cenozoic and Late Cretaceous oxygen isotopic records from near the continent now provide a stronger foundation for the understanding of climatic evolution of Antarctica and, indeed, the Earth. Important cooling steps occurred during the latest Cretaceous and the middle Eocene, near the Eocene/Oligocene boundary, and in the middle Oligocene, the middle Miocene, the early late Miocene, the latest Miocene, and the late Pliocene.

The analysis of a wide variety of proxy paleoenvironmental parameters from the latest drilled cores supports earlier concepts that the Cenozoic climatic development of Antarctica and the Southern Ocean resulted, in part, from the rearrangement of southern hemisphere land masses (see, for example, *Kennett and Barker* [1990]). Antarctica became increasingly isolated as Gondwana fragments dispersed northward and circumpolar circulation developed, allowing unrestricted latitudinal flow. Development of the Antarctic Circumpolar Current during the middle Cenozoic effectively isolated Antarctica thermally by decoupling the warmer subtropical gyres from the Antarctic continent. This increasing thermal isolation, in turn, assisted in the development of Antarctic glaciation and subsequently in the formation of major ice sheets. The climatic change included the cooling of waters surrounding the continent, extensive seasonal sea ice production, and wind-driven upwelling of nutrient-rich intermediate waters that have a profound effect on biogenic productivity in the Southern Ocean. Although a general consensus has emerged with regard to this general evolutionary progression, much debate remains about the scale of climatic and cryospheric change and its variability through much of the Cenozoic.

Discovery during ODP Leg 119 and Leg 120 of glaciomarine sediments of early Oligocene age in Prydz Bay, Antarctica, and a thin layer of ice-rafted detritus of earliest Oligocene age in the Kerguelen Plateau sequences far to the north of the continent indicates the presence of a major ice sheet on at least part of East Antarctica during the earliest Oligocene. Glacial sediments of possible late Eocene age were also cored. The drilling also provided new data supporting previous interpretations [*Drewry*, 1978; *Mercer*, 1978; *Ciesielski et al.*, 1982] that the East Antarctic ice sheet formed well before that of West Antarctica.

The discovery of a remarkable polar warming event at the end of the Paleocene is inferred to have been associated with unusual deep ocean circulation changes and massive extinctions in the deep sea. This event is of interest in global change considerations because it occurred very rapidly (in less than a few thousand years) and was clearly linked to changes in the Earth's environmental system. Oxygen isotopic evidence suggests the presence of warm saline deep water in the Weddell Sea at this time and during other Paleogene intervals. Antarctica may not have been the primary source of deep waters during parts of the Eocene and Paleocene.

At the beginning of the Oligocene, or perhaps during the late Eocene, cold deep waters began to form and strongly compete with lower-latitude sources. Evidence from the Weddell Sea sector suggests that cryospheric development began early in the Oligocene (~34 Ma), although the sites provide no evidence for major ice accumulation during the Oligocene. The combined evidence from different regions of Antarctica indicates that middle to late Paleogene climate change resulted dominantly, although not exclusively, in cooling, while latest Paleogene through Neogene change was dominated by ice accumulation.

The participants at a conference held at the University of California, Santa Barbara (The Role of the Southern Ocean and Antarctica in Global Change: An Ocean Drilling Perspective), in August 1991, assisted in compiling a list of specific major discoveries that have resulted from investigations of sequences drilled during DSDP and ODP legs 113, 114, 119 and 120 and during drilling in the Ross Sea region at sites referred to as MSSTS-1 and CIROS-1 [*Barrett*, 1986, 1989]. These discoveries are documented by many authors in the *Scientific Results of the Ocean Drilling Program* and include the following:

Discovery of middle Cretaceous anoxic sediments in the Weddell Sea.

Discovery of the oldest known (middle Cretaceous) diatoms and silicoflagellates, in pristine condition and of critical importance in the understanding of their early evolution.

Discovery of basement character and tectonic history of several oceanic rises such as Kerguelen Plateau and Maud Rise.

Discovery of middle Cretaceous nonmarine sediments on Kerguelen Plateau indicative of a relatively warm regional climate; documentation of colonization by land plants and their succession to a canopied forest.

Establishment of magnetobiostratigraphy for the Late Cretaceous to Quaternary for Antarctic and Subantarctic regions in the Atlantic and Indian oceans.

High-resolution stratigraphic studies across several Cretaceous/Tertiary boundary sections demonstrating important climatic changes preceding the iridium anomaly and inferred ocean paleoproductivity changes following the boundary.

Discovery of a brief global warming/carbon cycle/paleoceanographic event that caused latest Paleocene mass extinction in the deep sea.

Conclusion that there were warm, wet climates on the Antarctic continent during the Paleocene/Eocene transition.

Development of a weathering record (chemical versus physical) in Antarctica using clay mineralogy.

Confirmation of the seas around Antarctica as dominantly a carbonate ocean during the Paleogene. Evidence for initiation of biogenic silica production and deposition during the late Eocene and subsequent evolution leading to a silica-dominated ocean by the late Neogene.

Establishment of first oxygen isotopic temperature records for the Late Cretaceous and Paleogene of the Antarctic. Also, quality records from the Neogene and Paleogene of the Subantarctic.

First oxygen isotopic evidence for the existence of a warm saline deep water (WSDW) during the Paleogene. At times during the Paleogene there was a reduction in production of deep waters from Antarctica and a greater dominance of WSDW in the low-latitude regions (WSDW). Sources of deep ocean waters are much more complicated than previously believed.

Documentation of Eocene-Oligocene plant assemblages in Antarctica.

Discovery of a brief interval of ice-rafted debris at the Eocene/Oligocene boundary and of climate history in Kerguelen Plateau sequences. Implications for major East Antarctic ice sheet.

Discovery of earliest Oligocene and younger glaciomarine sediments in Prydz Bay. Implications for long-term glacial climates in East Antarctica.

Evidence for diachronism of development of East and West Antarctic glaciation and ice sheets.

Refinements of polar to equator Cenozoic meridional isotopic temperature gradients.

Documentation of late Cenozoic evolution of the polar frontal zone (PFZ).

Establishment of a high-resolution isotope and ice-rafting record from the latest Miocene (~6 Ma) in the Subantarctic region. Comparison with the northern hemisphere record.

Documentation that the Late Gauss (~2.7 Ma) represents a profound change in paleoceanographic conditions in northern Antarctic and Subantarctic regions, including northward migration of the PFZ, ice volume growth in Antarctica, and reduction of North Atlantic deep water (NADW) flux to the Southern Ocean.

Discovery of intervals of Quaternary carbonate ooze in Weddell Sea (Leg 113) sites that may correlate with "super-interglacial" conditions of stages 7, 9, and 11.

Extensive documentation of the evolution of the Cretaceous through Cenozoic marine biosphere including extinction, evolution, and migration.

IMPORTANT REMAINING QUESTIONS AND PROBLEMS

Recent investigations in the Antarctic region have resulted in many significant advances toward understanding of high-latitude paleoenvironmental evolution, including interrelations with the global environment. Nevertheless, major questions remain that can be answered using both future drilling in the Antarctic region and continued examination of deposits on the continent itself. For example, differences in interpretations have developed regarding the history of the Antarctic cryosphere and climate based upon the oceanic sedimentary record compared with continental and nearshore drilling sequences. The oceanic marine sequences have advantages because of their stratigraphic simplicity, continuity, and completeness, but criteria employed for estimating the extent and variability of the cryosphere can be open to question. In contrast, the land-based sequences have the advantage of being in close proximity to the Antarctic cryosphere but often suffer from stratigraphic incompleteness, poor age resolution, and significant reworking of sediments and microfossils. As a result, many significant questions remain concerning the variability of the Antarctic cryosphere during the middle and late Cenozoic and relations with global climate evolution.

Partial deglaciations and changes in the dynamics of the Antarctic ice sheet can affect the albedo at high latitudes and atmospheric circulation that, in turn, result in changes in the global heat budget. Northern hemisphere sequences, especially in the North Atlantic, represent a vital source of detailed information on global paleoceanographic and climatic changes for the last several million years. Knowledge of the evolution of late Neogene climate history in the Southern Ocean does not yet compare with that in high North Atlantic latitudes. It is clear that similar continuous, high-resolution climatic records can be developed for the Antarctic and Subantarctic regions in spite of a scarcity of carbonate-rich sedimentary sequences. High-resolution stratigraphic investigations are needed for comparison of climatic and glacial history between the northern and southern polar regions to better understand driving mechanisms for paleoenvironmental change. Better understanding of interhemispheric lead and lag relationships and coupling mechanisms will help with the development of predictive models and the estimation of response rates for short-duration events of global importance.

Participants at the conference at the University of California also helped with the compilation of a list of major questions related to the evolution of the Antarctic environmental and biotic system. The formulation of these questions is based upon the research of a large number of investigators presented in the Scientific Proceedings of the Ocean Drilling Program (listed above), in various other contributions in the literature, and in this volume. These questions, of relevance within a wide spectrum of geologic time, include the following:

1.   Development of the East Antarctic ice sheet: Consensus has emerged that major ice accumulation in East Antarctica began in the earliest Oligocene and that there has been at least some ice on the continent since that time. But was there appreciable ice in Antarctica before the early Oligocene? Preglacial marine sediments were not recovered on the Prydz Bay continental margin (Leg 119) or on the Weddell Sea continental margin (Leg 113) because of the presence of Cretaceous to

Oligocene hiatuses. Were there open seaways across the Antarctic continent during the Paleogene, and if so, at what times were they open?

2.   Stability of the Antarctic ice sheets: Once formed, were the ice sheets permanent? Did they fluctuate much in volume? Did partial deglaciation of the East Antarctic ice sheet occur at times during the Oligocene and early Miocene (the warmest interval of the Neogene)? What is the history of stability of the West Antarctic ice sheet under climatic conditions that were warmer than present-day conditions? Was there major deglaciation of the West or even the East Antarctic ice sheets during the early Pliocene, another episode of relative global warmth? (The West Antarctic ice sheet appears to have remained relatively stable during most Quaternary interglacial episodes; its history during particularly warm interglacial episodes is currently being debated.) If major deglaciation occurred during the early Pliocene, why is there no major signal in the deep-sea oxygen isotopic record? Is there a record of sea level change in response to such large-scale fluctuations in ice volume? What is the middle Cenozoic history of the terrestrial flora in Antarctica during the transformation from a greenhouse to an icebox Earth?

3.   Deep-water sources: How much change was there in the vertical water mass structure of the global oceans during the Cenozoic? There are a number of indications for the formation of warm saline deep water during Paleogene intervals, but conflicting evidence exists, including the nature of carbon isotopic gradients, and of development of benthic foraminiferal faunas, and a consensus has yet to emerge. If warm saline deep water was a major component of deep ocean waters during the Paleogene, at which depths did it largely reside? What were the spatial differences in deep-water masses between the major ocean basins, and what source areas were functional at various times? How do glacial-interglacial changes in Southern Ocean water mass paleochemistry affect the remaining world ocean? Additional depth transects of drilled sites are essential, including Maud Rise.

4.   Stable isotopic records: How much of the earliest Oligocene oxygen isotopic shift can be explained as ice volume, and how much as temperature effects? What complexities in the oxygen isotopic record need to be considered (e.g., salinity effects, precipitation effects)? Presently, discrepancies exist between temperature gradients based upon floral and faunal evidence and those based on oxygen isotopic evidence. Is it possible to resolve certain ambiguities in the oxygen isotopic records by comparison of carbonate and silicate records? This represents a promising new development, especially at the high latitudes where mixed carbonate-silicate sedimentary records are available.

5.   Influence of Antarctica on ocean geochemical evolution: Many of the estimates of accumulation rates for global Mesozoic and Cenozoic continental shelf and slope

sediments are based on available records from the other continents. The erosional history and sedimentary/geochemical contribution of Antarctica has been overlooked because of a lack of information on thicknesses and ages of shelf and slope sediments. This is a critical oversight considering the large amounts of debris that have been glacially eroded off the continent and deposited into the ocean since the early Oligocene.

6. Relations between oceanic and atmospheric development: There are no good records of eolian fluctuations from southern hemisphere regions such as the southwest Pacific to compare with those from the northern hemisphere. What was the east-west variation in oceanographic gradients in the southern Pacific, Indian, and Atlantic Ocean gyres? Is it possible to obtain proxy records on the presence and extent of sea ice during the pre-Pliocene?

7. Northern and southern hemisphere correlations: How does climatic development and variability compare between the northern and southern hemispheres during the Neogene, especially the latest Neogene? What were the leads and lags in the system? To answer these questions, more high-resolution records need to be produced from Antarctic and Subantarctic regions, especially in carbonate sequences, for comparison with low-latitude and northern hemisphere records.

8. Gondwana breakup: How and when did the breakup of eastern Gondwana occur? When and where did rifting begin, and what was the character of the early oceanic basins resulting from the breakup? Anoxic basins of early Cretaceous age existed in the Weddell Sea region, and biogeographic evidence suggests close connections with basins near Madagascar. The recent drilling was unable to dedicate sufficient time to the coring of this material, and yet it contains the earliest known records of evolution in the siliceous plankton. Use of micropaleontological, paleomagnetic, geophysical, and basement geochemical data is needed to help establish the timing of the opening of oceanic gateways around Antarctica, including those south of the South Tasman Rise, the Drake Passage, and the Scotia Sea.

9. Evolutionary processes: What events took place during intervals of major environmental change such as at the Cretaceous/Tertiary and Paleocene/Eocene boundaries and at the end of the Eocene? How did deep-water extinction events such as at the Paleocene/Eocene boundary relate to changes in the shallow-water and terrestrial environments? How did repopulation occur following these extinction events, both for planktonic and for benthic organisms? What effect did the cryospheric development of Antarctica have on the evolution of pelagic as well as benthic assemblages, including migrations to and from the Antarctic? What effects did paleoceanographic change during the late Phanerozoic have on endemism of Antarctic assemblages, including the benthics?

10. Biostratigraphy: Significant advances have been made in southern polar marine biostratigraphy and isotope stratigraphy as a result of the recent drilling. Nevertheless, important gaps remain to be filled that have significance in the understanding of biotic and paleoenvironmental evolution. Also needed are better correlations between siliceous and carbonate microfossil biostratigraphic schemes, preferably based upon sediment sequences that contain both elements. A search is also needed for distinct short-term "events" that are useful in assisting with high-precision correlations, such as the event that occurred during the latest Paleocene.

11. Evolution and nature of oceanic frontal systems: The history of these key areas needs examination in relation to their response to glacial-interglacial cycles. Particular attention is required on the effects of frontal change on gradients in temperature, productivity, silica versus carbonate deposition, and other parameters.

12. Continental shelf sequences: The erosional and depositional history of Antarctic continental shelves and slopes resulting from glaciation and sea level change needs to be established using a wide range of stratigraphic and geophysical approaches.

13. High-resolution Late Quaternary paleoclimate history: Marine sequences need to be recovered that can be correlated and compared with ice core data from the Antarctic continent.

14. Climate models: In the last instance, all of this new information is available for input into climate models, to delineate possible modes of operation of the ocean-atmosphere-biosphere system, and to use the past to help evaluate possible future changes in climate.

Acknowledgments. This summary is, in part, distilled from a white paper compiled by the authors on behalf of the participants in the International Conference on "The Role of the Southern Ocean and Antarctica in Global Change: An Ocean Drilling Perspective," held in Santa Barbara, California, during August 28–30, 1991. The white paper is available from JOI/USSAC, Washington, D.C. We acknowledge the efforts of the participants of the meeting and those who contributed notes for the white paper. We also thank E. Thomas, D. Hodell, S. Wise, Jr., J. Anderson, J. Zachos, A. Cooper, and J. Lipps for their valuable, thoughtful input to a draft of the white paper. This contribution benefited from a critical review by John Anderson. The conference and related publications were supported by JOI/USSAC and the Division of Polar Programs of NSF (grant DPP90-21690). Partial support for this research was from NSF grant DPP89-11554 (Division of Polar Programs).

REFERENCES

Barker, P. F., et al., Leg 36, *Initial Rep. Deep Sea Drill. Proj., 36,* 1977.

Barker, P. F., et al., Leg 113, *Proc. Ocean Drill. Program Initial Rep., 113,* 1988.

Barker, P. F., et al., Leg 113, *Proc. Ocean Drill. Program Sci. Results, 113,* 1033 pp., 1990.

Barrett, P. J. (Ed.), Antarctic Cenozoic history from MSSTS-1 drillhole, McMurdo Sound, *DSIR Bull. N. Z. 237,* 174 pp., 1986.

Barrett, P. J. (Ed.), Antarctic Cenozoic history from the CIROS-1 drillhole, McMurdo Sound, *DSIR Bull. N. Z.*, *245*, 254 pp., 1989.

Barron J., et al., Leg 119, *Proc. Ocean Drill. Program Initial Rep.*, *119*, 1989.

Barron, J., et al., Leg 119, *Proc. Ocean Drill. Program Sci. Results.*, *119*, 1991.

Ciesielski, P. F., M. T. Ledbetter, and B. B. Ellwood, The development of Antarctic glaciation and the Neogene paleoenvironment of the Maurice Ewing Bank, *Mar. Geol.*, *46*, 1–51, 1982.

Ciesielski, P. F., et al., Leg 114, *Proc. Ocean Drill. Program Initial Rep.*, *114*, 815 pp., 1988.

Ciesielski, P. F., et al., Leg 114, *Proc. Ocean Drill. Program Sci. Results*, *114*, 826 pp., 1991.

Drewry, D. J., Aspects of the early evolution of West Antarctic ice, in *Antarctic Glacial History and World Paleoenvironments*, edited by E. M. van Zinderen Bakker, pp. 25–32, Balkema, Rotterdam, 1978.

Hayes, D. E., et al., Leg 28, *Initial Rep. Deep Sea Drill. Proj.*, *28*, 1975.

Hollister, C. D., et al., Leg 35, *Initial Rep. Deep Sea Drill. Proj.*, *35*, 1976.

Kennett, J. P., and P. F. Barker, Latest Cretaceous to Cenozoic climate and oceanographic developments in the Weddell Sea, Antarctica: An ocean drilling perspective, *Proc. Ocean Drill. Program Sci. Results*, *113*, 937–960, 1990.

Kennett, J. P., et al., Leg 29, *Initial Rep. Deep Sea Drill. Proj.*, *29*, 1975.

Ludwig, W. J., et al., Leg 71, *Initial Rep. Deep Sea Drill. Project*, *71*, 1983.

Mercer, J. H., Glacial development and temperature trends in the Antarctic and in South America, in *Antarctic Glacial History and World Paleoenvironments*, edited by E. M. van Zinderen Bakker, pp. 73–93, Balkema, Rotterdam, 1978.

Schlich, R., et al., Leg 120, *Proc. Ocean Drill. Program Sci. Results*, *120*, 648 pp., 1989.

Wise, S. W., Jr., et al., Leg 120, *Proc. Ocean Drill. Program Sci. Results*, 120, 1155 pp., 1992.

(Received August 11, 1992;
accepted September 22, 1992.)

THE ANTARCTIC PALEOENVIRONMENT: A PERSPECTIVE ON GLOBAL CHANGE
ANTARCTIC RESEARCH SERIES, VOLUME 56, PAGES 7–30

THE DEVELOPMENT OF PALEOSEAWAYS AROUND ANTARCTICA

Lawrence A. Lawver, Lisa M. Gahagan, and Millard F. Coffin

Institute for Geophysics, University of Texas, Austin, Austin, Texas 78759-8397

Gondwana, with East Antarctica as its center, began to fragment during Late Triassic to Early Jurassic time. With the exception of the Permian or older convergent Pacific margin of the Antarctic Peninsula and Thurston Island, the margins of Antarctic are rifted or transform margins. The earliest seaway around the present-day margin of Antarctica developed during the Late Jurassic along the western part of Queen Maud Land of East Antarctica and may have included parts of the Weddell Sea margin of the Antarctic Peninsula. The next significant change occurred ~130 Ma, with the opening of the South Atlantic and rifting of India from East Antarctica. Circulation of deep water into the South Atlantic was blocked by the Falkland Plateau, and open, deepwater circulation between India and Antarctica was blocked by Sri Lanka. The Falkland Plateau cleared the southern tip of Africa ~100 Ma, but circulation from the South Atlantic northward into the Central Atlantic probably did not begin until Santonian time. Slow spreading between Australia and Antarctica, and a complicated spreading regime between various New Zealand plates, may have opened a seaway as early as 80 Ma. However, major circulation was probably delayed until as late as 40 Ma, when Tasmania and the South Tasman Rise finally cleared north Victoria Land of East Antarctica. The final barrier to circumpolar circulation was in the region of southern South America and the northern Antarctic Peninsula. There, various continental fragments may have partially blocked Drake Passage until the Miocene. General cooling of the world's oceans accompanied early Tertiary opening between Australia and Antarctica and the later opening of Drake Passage.

INTRODUCTION

According to *Kennett* [1982], paleoceanography requires the integration of sediment core analyses; biostratigraphy which has been chronologically calibrated and globally correlated; plate tectonic reconstructions; and paleontology, geochemistry, and mineralogy that can be applied to paleoenvironmental reconstructions. This paper deals with paleogeographic locations of the various plates around Antarctica and their motion through time starting at 160 Ma. In addition to the tectonic plates themselves, we show Ocean Drilling Program (ODP) and Deep Sea Drilling Project (DSDP) sites reconstructed to their paleopositions. The site locations have been assigned to their respective plates and appear on the reconstructions based on the age of the oldest sediments found in each of the holes.

Paleogeographic reconstructions are only as valid as the quality of the data employed. In some parts of the world, sufficient seafloor data are available so that it is unlikely that relevant relative plate rotations will change significantly from those used in this study (e.g., the central Atlantic data from *Klitgord and Schouten* [1986]). Our plate rotations are based on identified marine magnetic anomalies, other seafloor age determinations, seafloor bathymetric features, and tectonic lineations inferred from satellite altimetry data that

presumably indicate plate motion directions and assumed uniform seafloor spreading rates over reasonable periods of time. During the Cretaceous Normal Superchron, for example, which lasted from Chron M0 (118 Ma [*Kent and Gradstein*, 1986]) to Chron 34 (84 Ma), there are no identifiable magnetic reversals; so relative as well as absolute plate locations are not known precisely. In fact, in the central Atlantic, *Klitgord and Schouten* [1986] determined a spreading rate of 9 mm/yr (half rate) between Chron M4 (126 Ma) and 118 Ma, a spreading rate of 24 mm/yr between 118 Ma and Chron 33 (80 Ma), and a rate of 16 mm/yr until 67 Ma. It is not known how long the 9-mm/yr rate lasted after the start of the Cretaceous Normal Superchron, although any period of slow spreading after 118 Ma would require a spreading rate even faster than 24 mm/yr for some period between the end of the slow spreading and Chron 34. Until basement rocks are drilled in the central Atlantic for the Cretaceous Normal Superchron period and then accurately dated, precise relative positions of North America with respect to Africa for the period from 118 Ma to 84 Ma will remain unknown.

Other oceans are less well known than the central Atlantic. The southern oceans south of 60°S have an advantage in that the Geosat geodetic mission (GM) satellite altimetry data set has been declassified [*Sand-*

Fig. 1a.   Tight fit reconstruction of Gondwana at 200 Ma. Continents are indicated with coastlines
and 2000-m isobaths. Tick marks are a 5° grid based on the present-day latitude and longitude for the
continental fragments. Regions such as the northern extent of the Indian continent are devoid of tick
marks indicative of its having been subducted or distorted during collision with Eurasia. Gondwana is
in a paleolatitudinal framework based on the work of *Ziegler et al.* [1983]. AP, Antarctic Peninsula;
CP, Campbell Plateau; CR, Chatham Rise; KN, Kenya; LHR, Lord Howe Rise; LM, Lebombo
Monocline; MAD, Madagascar; MBL, Marie Byrd Land; MOZ, Mozambique; NNZ, north New
Zealand; SL, Sri Lanka; SNZ, south New Zealand; SP, Shillong Plateau; TI, Thurston Island.

well, 1992]; so seafloor tectonic lineations [*Gahagan et al.*, 1988] are quite prominent. Unfortunately, few magnetic anomalies have been reliably identified in the southern oceans; so exact locations of the continental masses with time are not precisely known. *Royer and*

Sandwell [1989] used 2500 magnetic anomaly identifications in the Indian Ocean to determine relative positions of the continents at 10 different times between 84 Ma and 9.8 Ma. Relative plate reconstructions can be checked by determining closure around triple junctions,

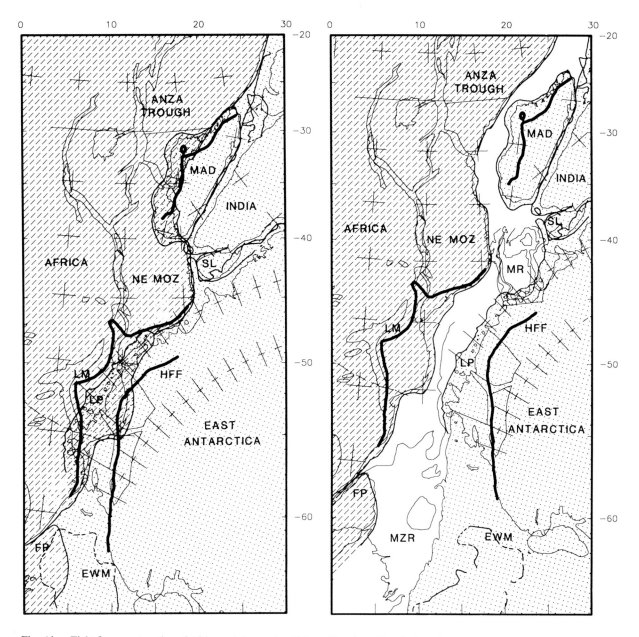

Fig. 1b. Tight fit reconstruction of Africa and Antarctica, Sri Lanka, India, and Madagascar at 200 Ma. Solid line on Africa (diagonal hachures) is edge of Lebombo Monocline taken from the edge of the Jurassic volcanic sequences of *Groenewald et al.* [1991]. Solid line on Antarctica (dotted continental region) is digitized edge of the Heimefrontfjella taken from the AVHRR map of Antarctica [*U.S. Geological Survey*, 1991]. EWM, Ellsworth-Whitmore mountains; FP, Falkland Plateau; GR, Gunnerus Ridge; HFF, Heimefrontfjella; LM, Lebombo Monocline; LP, Limpopo Plains or Mozambique Plains; MAD, Madagascar; NE MOZ, northeast Mozambique; SL, Sri Lanka.

Fig. 1c. Detail of continental regions at 160 Ma. Solid line on Africa (diagonal hachures) is edge of Lebombo Monocline taken from the edge of the Jurassic volcanic sequences of *Groenewald et al.* [1991]. Solid line on Antarctica (dotted continental region) is digitized edge of the Heimefrontfjella taken from the AVHRR map of Antarctica [*U.S. Geological Survey*, 1991]. Abbreviations as in Figure 1b. MZR, Mozambique Ridge; MR, Madagascar Ridge.

by ensuring there are no discrepancies when isochron charts are produced (i.e., the same seafloor was seemingly produced by two different spreading centers), and by comparing the geology of the conjugate margins of

the rotated plates. Absolute plate positions in a global framework depend on relative plate rotations and on paleomagnetic data from marine cores and terrestrial samples. Paleomagnetic data, combined with a datable hot spot reference frame, can be used to determine absolute plate positions through time if one assumes

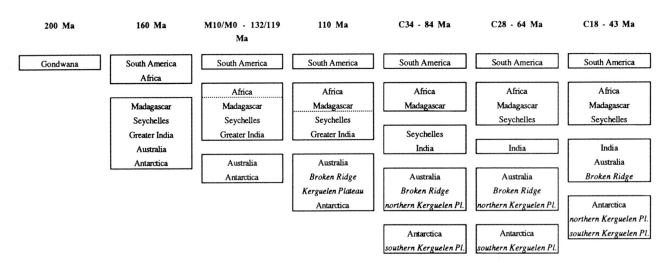

Fig. 2. Circum-Antarctic seafloor spreading chronology and plate configurations. Seaways developed between the major boxed coental assemblages shown. Dotted lines indicate uncertain configurations. Italics indicate crustal blocks of dominantly oceanic affinity.

that true polar wander is negligible. An absolute reference frame that neglects true polar wander has been devised for the last 130 m.y. [*Müller et al.*, 1991, 1992] but is considered most reliable for the last 84 Ma. The earlier absolute reference frame for Africa follows that of *Ziegler et al.* [1983]. The identified marine magnetic anomalies and tectonic lineations deduced from Geosat altimetry data used in our reconstructions are summarized by *Royer et al.* [1990].

Reconstructions of the major plates that bound the Southern Ocean are shown at 10-m.y. increments starting at 130 Ma, with two additional reconstructions at 160 Ma and approximately 200 Ma. The oldest reconstruction (Figure 1a) is a "tight fit" reconstruction that attempts to eliminate all possible prebreakup continental stretching. The 160 Ma reconstruction attempts to show the Gondwana continents just prior to the first identified seafloor spreading anomalies. These two reconstructions are partially based on the paleomagnetic work of *Grunow et al.* [1991].

In our reconstructions, continental boundaries are arbitrarily assumed to be the 2000-m contour. Postrifting subsidence and the rapid sedimentation that commonly occurs during the first phases of continental separations have not been considered in detail. The determination of prerift ocean-continent boundaries requires not only seismic reflection and refraction data, but also deep drilling data to determine the subsidence history of the continental margins. While some information is available and has been considered, it cannot be assumed that a uniform subsidence history applies to the entire length of a rifted margin. Consequently, it is difficult to ascertain if narrow gaps between reconstructed major plates were deepwater seaways at the time of initial rifting or if later thermal subsidence

deepened the seaways. Identified marine magnetic anomalies can generally be assumed to have been produced on true ocean floor and, with the exception of initial continental rifting, were probably produced at a water depth of ~2500 m [*Parsons and Sclater*, 1977]. Figure 2 indicates the ages for the development of most of the major seaways around Antarctica.

Ocean Drilling Program and Deep Sea Drilling Project sites drilled in the Southern Ocean region are listed in Table 1. The oldest sediments or basement rocks recovered at each site are given in Table 1 and are shown on the next youngest and on subsequent reconstructions. If *Kennett*'s [1982] four sources of paleoceanographic information are considered, it is obvious that plate reconstructions can only give approximate paleogeographic locations of the tectonic plates and the sediment cores taken on them. Precise information concerning the opening of major marine seaways will most likely be determined from the marine sedimentary record, if relevant sequences are drilled.

PREBREAKUP RECONSTRUCTION

A tight fit reconstruction of Gondwana in latest Triassic or earliest Jurassic (>200 Ma) time is shown in Figure 1a. The key to the tight fit reconstruction is the juxtaposition of Jurassic volcanic sequences of the Lebombo Monocline in southeast Africa with the Heimefrontfjella of East Antarctica [*Groenewald et al.*, 1991], shown in detail in Figure 1b. The tight fit superimposes Antarctic continental crust on top of what is now the Limpopo Plain of Mozambique. Our reconstruction is similar to the reconstruction of Africa and Antarctica of *Martin and Hartnady* [1986] but has been modified on the basis of a better definition of the

TABLE 1. Present-Day Location of DSDP and ODP Sites With the Age of the Oldest Sediments Found

Leg	Site	Lati-tude	Longi-tude	Oldest Sediment Age
21	206	−32.01	165.45	Middle Paleocene
21	207	−36.96	165.43	≤Upper Cretaceous (basement)
22	212	−19.19	99.30	Late Cretaceous? (basement)
22	214	−11.34	88.72	Paleocene (basement)
22	215	−8.12	86.79	mid-Paleocene (basement)
22	216	1.46	90.21	Maestrichtian (basement)
22	217	8.93	90.54	Campanian (basement)
25	243	−22.91	41.40	Early Cretaceous?
25	245	−31.53	52.30	Paleocene (basement)
25	246	−33.62	45.16	Early Eocene
25	248	−29.53	37.47	Paleocene
25	249	−29.95	36.08	Neocomian (basement)
26	250	−33.46	39.37	Coniacian (basement)
26	253	−24.88	87.37	mid-Eocene (basalt, basement)
26	254	−30.97	87.90	Eocene/Oligocene (basement)
26	255	−31.13	93.73	Santonian
26	256	−23.46	100.77	Albian (basement)
26	257	−30.99	108.35	Albian (basement)
26	258	−33.79	112.47	mid-Albian
27	259	−29.62	12.70	Aptian (basement)
27	260	−16.14	110.30	Albian (basement)
27	261	−12.95	117.89	Late Oxfordian (basement)
27	263	−23.32	110.98	Albian
28	264	−34.97	112.04	pre-Santonian (basement)
28	265	−53.54	109.95	mid-Miocene (basement)
28	266	−56.40	110.11	early Miocene (basement)
28	267	−59.26	104.49	mid-Oligocene (basement)
28	268	−63.95	105.16	mid-Oligocene
28	269	−61.68	140.07	≤mid-Oligocene
28	270	−77.44	−178.50	Oligocene (basement)
28	271	−76.72	−175.05	early Pliocene
28	272	−77.13	−176.76	early Miocene
28	273	−74.54	174.63	early Miocene
28	274	−69.00	173.43	early Oligocene (basement)
29	275	−50.44	176.32	Late Cretaceous
29	276	−50.80	176.81	Paleogene
29	277	−52.22	166.19	mid-Paleocene
29	278	−56.56	160.07	mid-Oligocene
29	279	−51.34	62.63	middle early Miocene
29	280	−48.96	147.23	early to mid-Eocene
29	281	−48.00	147.76	late Eocene
29	282	−42.25	143.49	late Eocene
29	283	−43.91	154.28	Paleocene
29	284	−40.51	167.68	late Miocene
35	322	−60.02	−79.42	Oligocene (basement)
35	323	−63.68	−97.99	late Cretaceous (basement)
35	324	−69.05	−98.75	Pliocene
35	325	−65.05	−73.67	Oligocene
36	326	−56.58	−65.30	Quaternary
36	327	−50.87	−46.78	Aptian
36	328	−49.81	−36.66	Upper Cretaceous
36	329	−50.66	−46.10	late Paleocene
36	330	−50.92	−46.88	Mid-Upper Jurassic (pre-Cretaceous)
39	356	−28.29	−41.09	Albian
39	358	−37.66	−35.96	Maestrichtian
40	361	−35.07	15.45	late Barremian (basement)
71	511	−51.00	−46.97	late Jurassic
71	512	−49.87	−40.85	mid-Eocene
71	513	−47.58	−24.64	early Oligocene (basement)
71	514	−46.05	−26.85	early Pliocene
72	516	−30.28	−35.28	Coniacian (basement)

TABLE 1. (continued)

Leg	Site	Lati-tude	Longi-tude	Oldest Sediment Age
73	524	−29.48	3.51	late Cretaceous
90	592	−36.47	165.44	Middle late Eocene
90	593	−40.51	167.67	late Eocene
90	594	−45.52	174.95	late early Miocene
113	689	−64.52	3.10	late Campanian
113	690	−65.16	1.20	late Campanian
113	691	−70.74	−13.80	Pliocene/Pleistocene
113	692	−70.72	−13.82	Valangian/Hauterivian
113	693	−70.83	−14.57	Albian
113	694	−66.85	−33.45	Middle middle Miocene
113	695	−62.39	−43.45	latest Miocene
113	696	−61.85	−42.93	Eocene/Oligocene
113	697	−61.81	−40.30	early Pliocene
114	698	−51.46	−33.10	≤Campanian (basement)
114	699	−51.54	−30.68	early Paleocene
114	700	−51.53	−30.28	Coniacian
114	701	−51.98	−23.21	mid-Eocene (basement)
114	702	−50.95	−26.37	late Paleocene
114	703	−47.05	7.89	mid-Eocene (basement)
114	704	−46.88	7.42	early Oligocene
119	736	−49.40	71.66	late early Pliocene
119	737	−50.23	73.03	mid-Eocene
119	738	−62.71	82.79	early Turonian (basement)
119	739	−67.28	75.08	Eocene–early Oligocene
119	740	−68.69	76.72	Permian?
119	741	−68.39	76.38	early Cretaceous
119	742	−67.55	75.40	Eocene?-Oligocene
119	743	−66.92	74.69	Quaternary
119	744	−61.58	80.59	late Eocene
119	745	−59.60	85.86	late Miocene
119	746	59.55	85.86	Late Miocene
120	747	−54.81	76.79	Santonian (basement)
120	748	−58.44	78.98	late Albian–Cenomanian
120	749	−58.72	76.41	early Cretaceous (basement)
120	750	−57.59	81.24	Albian (basement)
120	751	−57.73	79.81	early Miocene
121	752	−30.89	93.58	late Maestrichtian
121	753	−30.89	93.59	mid-Eocene
121	754	−30.94	93.57	early Maestrichtian
121	755	−31.03	93.55	Turonian
121	756	−27.35	87.60	Middle to late Eocene (basement)
121	757	−17.02	88.18	Paleocene (basement)
121	758	5.38	90.36	Santonian (basement)
122	759	−16.95	115.56	Carnian
122	760	−16.92	115.54	Carnian
122	761	−16.74	115.54	Norian
122	762	−19.89	112.25	Berriasian
122	763	−20.59	112.21	mid-Berriasian
122	764	−16.57	115.46	Rhaetian
123	765	−15.98	117.57	Berriasian (basement)
123	766	−19.93	110.45	late Valanginian (sills)
133	811	−16.52	148.16	early Eocene
133	824	−16.44	147.76	basement
133	825	−16.52	148.16	basement

References for DSDP and ODP legs are as follows: Leg 21 [*Burns et al.* 1973], Leg 22 [*von der Borch et al.*, 1974], Leg 25 [*Simpson et al.*, 1974], Leg 26 [*Davies et al.*, 1974], Leg 27 [*Veevers et al.*, 1974], Leg 28 [*Hayes et al.*, 1975], Leg 29 [*Kennett et al.*, 1974], Leg 35 [*Hollister et al.*, 1976], Leg 36 [*Barker et al.*, 1976], Leg 39 [*Supko et al.*, 1977], Leg 40 [*Bolli et al.*, 1978], Leg 71 [*Ludwig et al.*, 1983], Leg 72 [*Barker et al.*, 1983], Leg 73 [*Hsü et al.*, 1984], Leg 90 [*Kennett et al.*, 1986], Leg 113 [*Barker et al.*, 1988], Leg 114 [*Ciesielski et al.*, 1988], Leg 119 [*Barron et al.*, 1989], Leg 120 [*Schlich et al.*, 1989], Leg 121 [*Peirce et al.*, 1989], Leg 122 [*Haq et al.*, 1990], Leg 123 [*Gradstein et al.*, 1990], and Leg 133 [*Davies et al.*, 1991].

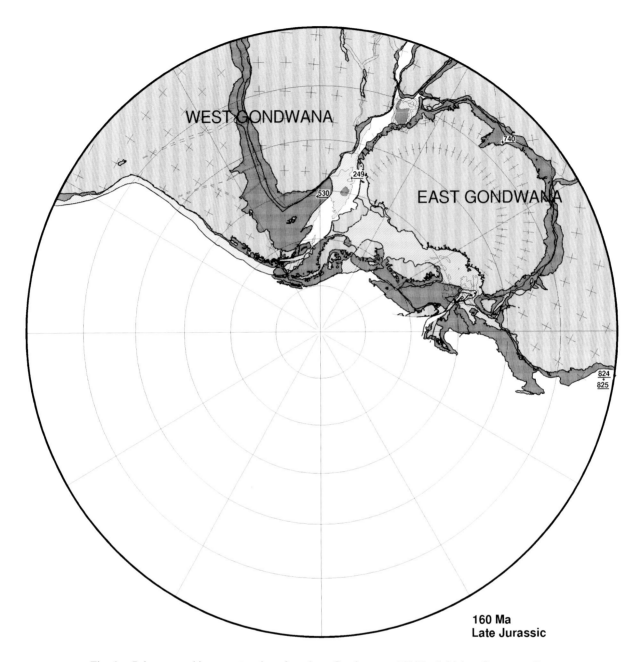

WEST GONDWANA

EAST GONDWANA

160 Ma
Late Jurassic

Fig. 3.   Paleogeographic reconstruction of southern Gondwana at 160 Ma. Initial seafloor spreading produced a shallow seaway that may have extended to the north of Madagascar and was discontinuous to the Rocas Verdes Basin of southern South America.

Figs. 3–17.   Polar stereographic reconstructions centered on the present south pole. Concentric circles are 10° of latitude to 30°S for times to 110 Ma, to 40°S for times 100 Ma to present. Longitude lines are every 30°. Medium shaded areas are continental. Dark shaded areas are continental margin or stretched continental crustal areas to the present-day 2000-m isobath. Lightly shaded areas are oceanic plateaus or oceanic crustal areas above the 4000-m isobath. Tick marks are a 5° grid based on present-day latitude and longitude marks of the continents. Numbers in the figures refer to DSDP and ODP holes that were drilled on crust within the areas of the figures. Underlined numbers indicate that the holes either bottomed in basement assumed older than the age of the reconstruction or that the oldest sediments found were older than the reconstruction indicated but not as old as the previous reconstruction. Regular numbers indicate that there are sediments older than the previous reconstruction found in the holes indicated. Age of the bottommost sediments is the assumed basement age indicated in Table 1. Isochrons are indicated only as approximations to the age of the seafloor but are not labeled.

Heimefrontfjella by the use of advanced very high resolution radiometry (AVHRR) data [*U.S. Geological Survey*, 1991]. While such a superposition of present-day continental crust would normally imply too tight a fit, there is no evidence for prebreakup continental material to the east of the Lebombo Monocline in southeast Africa [*de Wit et al.*, 1988]. An additional space difficulty is overlap of the western edge of Madagascar with the Kenyan region of east Africa. *Reeves et al.* [1987] discuss geophysical evidence for consideration of the Anza Trough as a failed third arm of a Jurassic-aged triradial rift in coastal Kenya. They conclude that Madagascar can be reassembled as part of Gondwana in a tighter fit than previously thought. If Permo-Triassic rocks of Madagascar are aligned along strike with the linear Permo-Triassic Karoo rocks of Zambia [*de Wit et al.*, 1988], then Jurassic-aged breakup rocks of western Madagascar fit with similar rocks exposed in Tanzania and southern Kenya. Precambrian rocks of coastal northern Mozambique and southern Tanzania fit with similar-aged rocks of Madagascar and define a coast that fits against Gunnerus Ridge, a submarine ridge off present-day East Antarctica at 32°E. This tight fit reconstruction also puts the Precambrian A to B rocks [*de Wit et al.*, 1988] of Mozambique in close proximity to rocks of similar age in Queen Maud Land, East Antarctica, and in the Lützow-Holm Bay region. *Coffin and Rabinowitz* [1988] summarize selected exploratory well data from coastal and offshore Madagascar and Tanzania that indicate some Upper Triassic and possibly even Permian rocks from the paleoseaway between Madagascar and Africa. These data suggest that our tight fit reconstruction may be even older than ~200 Ma.

The fit of Sri Lanka–India and Antarctica has been discussed previously [*Lawver and Scotese*, 1987]. The exact fit of Australia and Antarctica was significantly improved with release of Geosat exact repeat mission (ERM) data in 1987 [*Sandwell and McAdoo*, 1988]. The early breakup of Australia with respect to Antarctica was recently discussed by *Veevers et al.* [1991] and *Royer and Sandwell* [1989]. The southern margin of Australia fits along East Antarctica and ends to the west where the easternmost recognized continental rocks of the early Indian plate occur at the Shillong Plateau. Consequently, three continents alone, Africa, India, and Australia, encircled the present-day rifted margins of East Antarctica. With the obvious fit of South America and Africa [*Rabinowitz and LaBrecque*, 1979; *Nürnberg and Müller*, 1991], the tight fit reconstruction of Gondwana only lacks constraints along the trans-Antarctic margin of East Antarctica. The remaining southern Gondwana pieces included north and south New Zealand, the Campbell Plateau, Chatham Rise, Marie Byrd Land, the Ellsworth/Whitmore block, the Antarctic Peninsula with the South Orkney block, and Thurston Island. While there are no marine magnetic anomaly or Geosat data to constrain the fit of the West Antarctica blocks, paleomagnetic data and geometric and other considerations may be used to produce a reasonable representation of the entire Mesozoic prebreakup configuration of Gondwana (Figure 1a).

Marine seismic reflection and refraction data in the Ross Sea [*Cooper et al.*, 1991] support a 100% stretching of the continental crust. The ~200 Ma reconstruction of Marie Byrd Land with respect to East Antarctica used a 50% closure of the present-day Ross Sea Embayment in both north-south and east-west directions [*Lawver and Scotese*, 1987]. Since Campbell Plateau, Chatham Rise, and north and south New Zealand can be reconstructed to Marie Byrd Land [*Grindley and Davey*, 1982], this also constrains their positions in the prebreakup configuration of Gondwana. Consequently, if the tight fit reconstruction of Gondwana is valid, there is only a limited space in which to place the Antarctic Peninsula and Thurston Island. Paleomagnetic data [*Grunow et al.*, 1991] and geological considerations [*Dalziel and Elliot*, 1982] suggest that the Antarctic Peninsula can be rotated counterclockwise from its present-day position with respect to East Antarctica and can be reconstructed along the western margin of southernmost South America. Furthermore, paleomagnetic data [*Grunow et al.*, 1991] from Thurston Island indicate an almost 90° rotation with respect to the Antarctic Peninsula subsequent to breakup. Since the Thurston Island block is nearly circular, its exact rotational placement does not affect potential seaways.

By Late Permian to Early Triassic time, Madagascar and Africa were separated by a marine barrier [*Battail et al.*, 1987]. These authors state that "the Early Triassic ichthyofaunas from Madagascar are, in part, related to those of South Africa, but they display more affinities with Laurasian faunas from British Columbia, Greenland, and Spitsbergen." This suggests that a Permo-Triassic seaway between Madagascar and Africa extended to the northeast between India and Africa to connect with the Pacific Ocean (or Tethys) near the equator (Figure 3). The fauna could then have reached Greenland or Spitsbergen either across the Pacific and then southward or through a seaway cutting through Europe. The fact that ichthyofaunas show fewer affinities to those in south Africa suggests that the seaway did not extend much farther south than southern Madagascar and that any paleoseaway that reached as far north as south Africa came from the south and connected with the Pacific only near the south polar region. This evidence supports the fit of the Precambrian rocks of northeastern Mozambique against the Gunnerus Ridge of East Antarctica. Late Triassic fauna are similar in Madagascar and the western part of Laurasia. No connection between south Africa and Laurasia [*Battail et al.*, 1987] would indicate that no substantial stretching occurred between east and west Gondwana until at least latest Triassic–earliest Jurassic time. Conse-

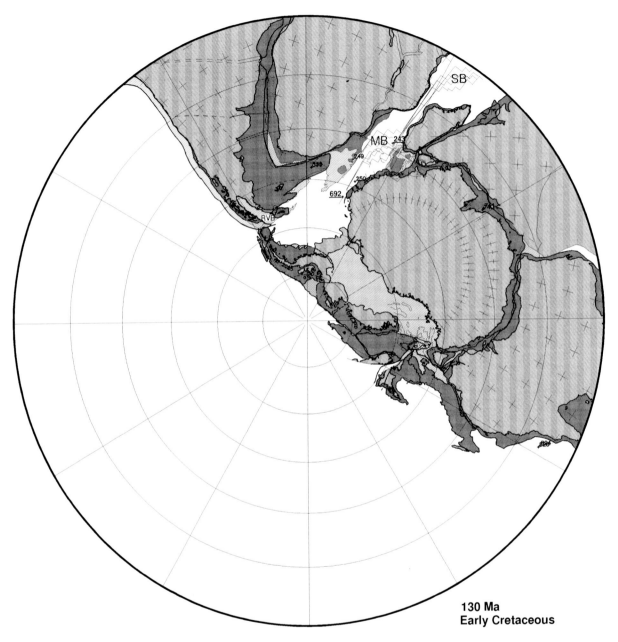

Fig. 4.   Paleogeographic reconstruction of southern Gondwana at 130 Ma. Initial rifting had begun between Australia and greater India. Stretching and possible initiation of seafloor spreading between India and East Antarctica and in the South Atlantic between South America and Africa. Paleolatitudinal framework for this and subsequent reconstructions based on *Müller et al.* [1992]. MB, Mozambique Basin; RVB, Rocas Verdes Basin; SB, Somali Basin.

quently, we estimate the tight fit reconstruction of Gondwana shown in Figure 1a as ~200 Ma.

EARLY JURASSIC TIME

Initial stretching between east and west Gondwana probably started in the north and progressed southward. *Cannon et al.* [1981] proposed a triradial rift system that included the Jurassic-aged Anza Trough. They show (Figure 1c) the approximate southern limit of marine Jurassic fauna of eastern coastal Africa and western Madagascar. An extremely tight fit between Madagascar and east Africa (Figure 1b) restores the southern limits of the Lower Jurassic fauna, aligns the Triassic and Permian-aged Karoo sediments, and may provide bounds on both the direction of initial stretching and timing of stretching. The southern limit of the marine Middle and Upper Jurassic are nearly in line for the poststretching phase, preseafloor spreading reconstruction of Gondwana. Stretching was certainly finished by the initiation of seafloor spreading during the earliest Late Jurassic in the Somali [*Ségoufin and Patriat*, 1980; *Rabinowitz et al.*, 1983; *Cochran*, 1988] and Mozambique [*Ségoufin*, 1978; *Simpson et al.*, 1979] basins and by inference in the southwest Weddell Sea. The stretching phase included movement between Madagascar and the Kenyan region of Africa, formation of the South Madagascar Ridge between the northeast Mozambique region of Africa and the Gunnerus Ridge off East Antarctica, and formation of the Mozambique Ridge and the Explora Wedge of the northwestern margin of Queen Maud Land [*Lawver et al.*, 1991]. While intrusion of the Ferrar Dolerites [*Elliot*, 1992] along the length of the present-day Transantarctic Mountains occurred during the time of stretching and was undoubtedly related to it, the main east-west Gondwana stretching probably extended through the Ross Sea Embayment region [*Lawver and Gahagan*, 1991]. Stretching involved right-lateral transtension between east and west Gondwana that included translation and rotation of the Ellsworth and Whitmore mountains block, as indicated by paleomagnetic results [*Grunow et al.*, 1991]. Related stretching also occurred in the Ross Sea and included extension between the Lord Howe Rise and the north New Zealand block. Stretching had to have occurred during this period; otherwise unacceptable overlap of north New Zealand on the Campbell Plateau results.

DEVELOPMENT OF PALEOSEAWAYS

160 Ma

When seafloor spreading commenced between east and west Gondwana (Figure 2), only small areas of what may be present-day seafloor existed. In the reconstruction for 160 Ma (Figure 3), a possible deep ocean basin lies south (in present-day coordinates) of the Falkland Plateau. Subsequently, that region would have been subducted at the South Sandwich Trench and hence no longer exists. It is possible, but unlikely, that the Ellsworth-Whitmore mountains fit into the space since *Grunow et al.* [1991] show them farther south by 175 Ma. The Mozambique Ridge to the 4000-m contour fills much of the rest of the space and was presumably produced during stretching that occurred sometime during the period ~240 Ma to 160 Ma (175 ± 10 Ma [*Kyle et al.*, 1981], if the Ferrar Dolerites are indicative of stretching). Very small basins are shown that later developed into the Mozambique and Somali basins. In fact, identified seafloor magnetic anomalies in both the Somali Basin [*Ségoufin and Patriat*, 1980; *Coffin and*

Rabinowitz, 1987; *Cochran*, 1988] and the Mozambique Basin [*Ségoufin*, 1978] support some creation of true ocean floor prior to 160 Ma.

130 Ma

Between 160 Ma and 130 Ma, the breakup of Gondwana was essentially a two-plate problem [*Lawver et al.*, 1991]. At 130 Ma, Gondwana seems to have divided into at least four or more fragments (Figure 2). Seafloor spreading commenced in the South Atlantic [*Rabinowitz and LaBrecque*, 1979; *Shaw and Cande*, 1990], off the western margin of Australia [*Veevers et al.*, 1985], and between India and Antarctica. While there are no identified Mesozoic magnetic anomalies that record breakup between India and Antarctica, timing is constrained by other means. One is the geometric fit of India, Madagascar, Sri Lanka, Antarctica, and Africa [*Lawver and Scotese*, 1987]. Unless there is transform motion between Madagascar and India that started about 130 Ma, then Madagascar would have overridden India. Also, there are dated rocks (114 Ma) from the Kerguelen Plateau that could only be that old if spreading between India and Antarctica had commenced by 130 Ma. Opening of the Gulf of Mannar between Sri Lanka and India probably occurred at about the same time or else slightly before rifting began between India and Antarctica. *Sastri et al.* [1977] found the earliest planktonic foraminifera in the Cauvery Basin of India to be Neocomian to Aptian in age, which fits with a breakup around 130 Ma. T. Munasinghe (personal communication, 1991) also estimates a breakup time of around 130 Ma, based on unpublished marine geophysical data around Sri Lanka.

At 130 Ma (Figure 4), the Weddell Sea appears to have been a substantial ocean basin connected to the Mozambique Basin and possibly to the Somali Basin through a seaway between Madagascar and Africa. The Weddell Sea extended north into the Rocas Verdes Basin of southern South America [*Dalziel*, 1981]. Another seaway started to develop between northern India and western Australia [*Powell et al.*, 1988]. There may have been a Permian seaway along the western margin of Australia, but unlike the Permian seaway between Madagascar and Africa which became the site of Late Triassic sedimentation, there is no evidence of pre-Cretaceous seafloor spreading activity on the western Australian margin [*Markl*, 1974; *Larson et al.*, 1979].

If the Ellsworth and Whitmore mountains were situated as *Grunow et al.* [1991] suggest, the Weddell Sea probably did not extend south of the Graham Land section of the Antarctic Peninsula. According to *Ryan et al.* [1978], fossil-bearing sedimentary successions from the deepest cores at sites 361 and 363 support an age of at least Aptian for the opening of the South Atlantic and suggest that the seafloor may be as old as M10N time (~130 Ma). More recent work [*Austin and Uchupi*,

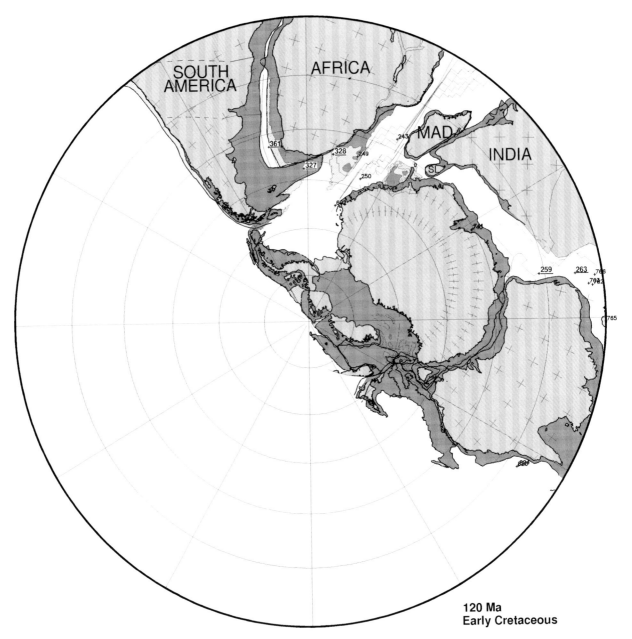

Fig. 5. Paleogeographic reconstruction of southern Gondwana at 120 Ma. Opening of deepwater seaways in South Atlantic and in Indian Ocean. MAD, Madagascar; SL, Sri Lanka.

1982] suggests that anomaly M10 found in the Cape Basin is actually on southwest Africa continental crust. Even if the oldest oceanic crust in the South Atlantic is Hauterivian, oxygen starvation in the Cape Basin of the South Atlantic lasted until well into Albian time, as evidenced at Site 361 [*Ryan et al.*, 1978] by sapropelic sediments rich in pyrite and plant debris. This would imply that there was no deepwater circulation into the South Atlantic until the Falkland Plateau cleared southern Africa.

120 Ma

By 120 Ma (Figure 5), seafloor spreading in the Somali Basin was virtually finished (spreading ceased shortly after anomaly M0 [*Cochran*, 1988]), and Madagascar had nearly reached its present-day position with respect to Africa. In order for 114 Ma rocks to exist on Kerguelen Plateau [*Leclaire et al.*, 1987], India would have to have moved a substantial distance away from Antarctica by 120 Ma. Even so, it appears that Sri

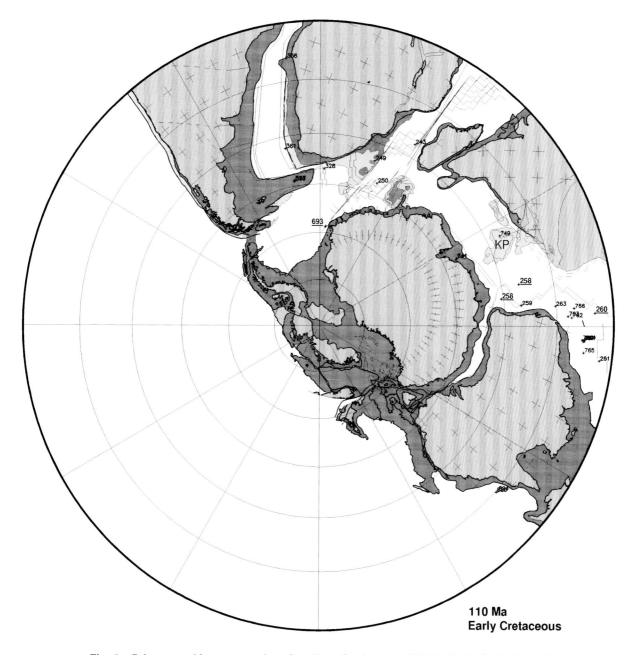

110 Ma
Early Cretaceous

Fig. 6.   Paleogeographic reconstruction of southern Gondwana at 110 Ma. India, Sri Lanka, and Madagascar clear East Antarctica and possible circulation of deep water into Weddell Sea region from south of India. KP, Kerguelen Plateau.

Lanka, Gunnerus Ridge, and Madagascar still formed a complete barrier to circulation from the east reaching to the Weddell Sea/Mozambique Basin region. The South Atlantic was slowly opening at this time while the Rocas Verdes Basin of southern South America was undergoing closure [*Dalziel*, 1981]. Deepwater circulation into the South Atlantic was still effectively blocked by the position of the Falkland Plateau with respect to southern Africa. Prerift stretching and rapid subsidence of the southern Australian margin began at 125 Ma, but there

is no evidence of marine fauna between Australia and Antarctica until after 100 Ma [*Hegarty et al.*, 1988], and actual seafloor spreading between Australia and Antarctica did not begin until ~96 Ma [*Cande and Mutter*, 1982; *Veevers et al.*, 1990].

110 Ma

At 110 Ma (Figure 6), west Africa and northeast South America were still together, and the Falkland Plateau

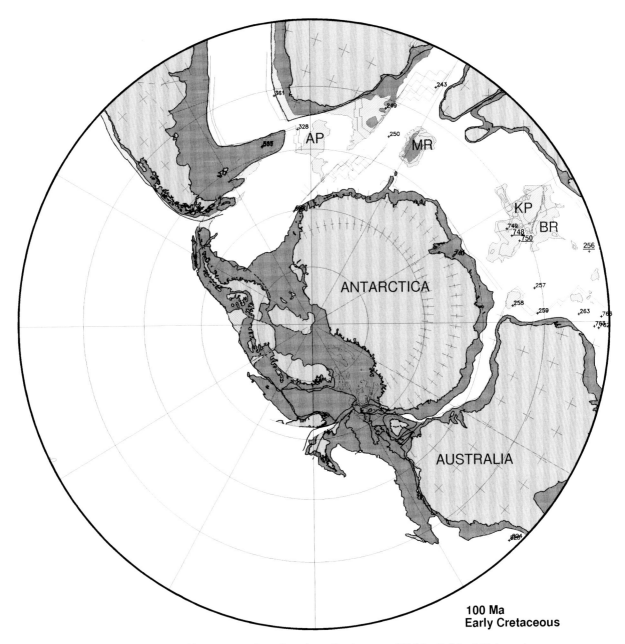

Fig. 7. Paleogeographic reconstruction of southern Gondwana at 100 Ma. Falkland Plateau clears South Africa and circulation into South Atlantic. Stretching and initiation of marine sedimentation south of Australia. Madagascar Rise and East Antarctica rifted. AP, Agulhas Plateau; BR, Broken Ridge; MR, Madagascar Ridge; KP, Kerguelen Plateau.

had not yet cleared the tip of south Africa; so the South Atlantic was a deep ocean basin that had no deepwater connection with other oceans. A deepwater passage between the Antarctic Peninsula and southern South America was presumably blocked by the South Orkney block. Major pieces of West Antarctica were nearly in their present-day positions with respect to East Antarctica, based on paleomagnetic data of *Grunow et al.* [1991], but some additional extension occurred later in

the Ross Sea Embayment, probably in the Late Cretaceous, with a lesser amount in the Cenozoic. Madagascar, India, and the southern Kerguelen Plateau had cleared Antarctica, and a wide, open seaway existed along that part of the present-day East Antarctic margin [*Leclaire et al.*, 1987]. While marine circulation between India and Antarctica was clearly possible, it is unlikely to have been vigorous if there was a land barrier that blocked Drake Passage, as we suggest. Excluding the

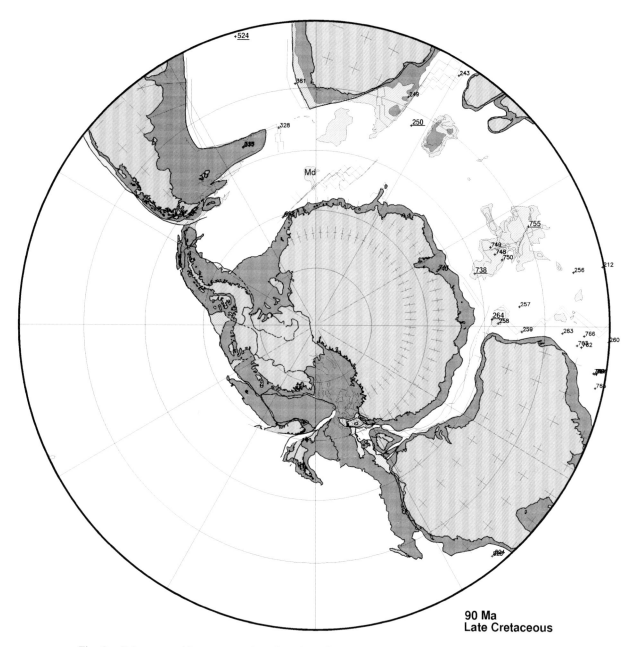

**90 Ma
Late Cretaceous**

Fig. 8. Paleogeographic reconstruction of southern Gondwana at 90 Ma. Initiation of slow seafloor spreading south of Australia. Final closure of the Rocas Verdes Basin in southern South America. Stretching and rifting in Bounty Trough between Chatham Rise and Campbell Plateau and in Great South Basin between south New Zealand and Campbell Plateau. South Atlantic finally open to deepwater circulation in the Angola Basin. Md, Maud Rise.

Tethyan part of the western Pacific Ocean, the Pacific Ocean at 110 Ma covered almost exactly half the surface of the Earth and extended nearly to the south pole. If there was equatorial circulation with a Tethyan Ocean fully open around the equator, then oceanic circulation would be dominated by the equatorial flow [*Luyendyk et al.*, 1972]. Otherwise, most of the oceanic circulation would have consisted of two large cyclonic gyres in the Pacific, one north of the equator and one south.

100 Ma

The seaway between West Africa and northeastern South America was blocked, although circulation into

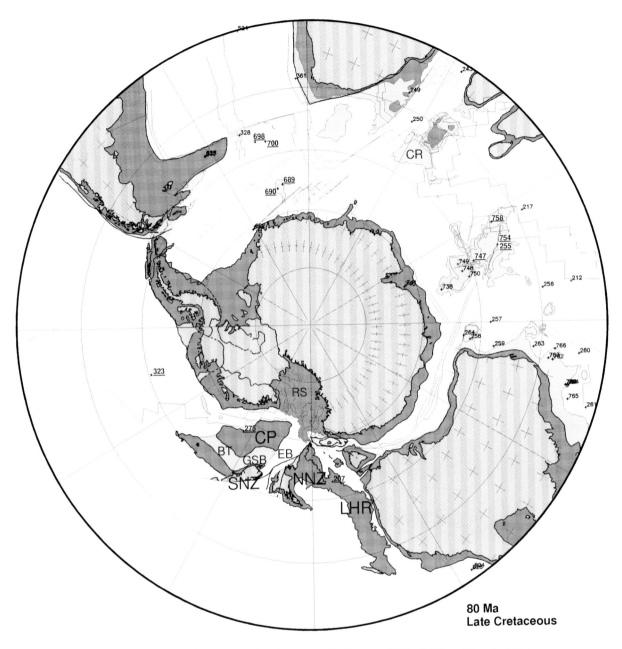

80 Ma
Late Cretaceous

Fig. 9. Paleogeographic reconstruction of southern Gondwana at 80 Ma. Rifting of Campbell Plateau from Marie Byrd Land with the extension of the Pacific-Antarctic spreading center to the west. Rifting in the Tasman Sea between Lord Howe Rise and eastern Australia. BT, Bounty Trough; CP, Campbell Plateau; CR, Conrad Rise; EB, Emerald Basin; GSB, Great South Basin; LHR, Lord Howe Rise; NNZ, north New Zealand; RS, Ross Sea Embayment; SNZ, south New Zealand.

the South Atlantic was possible between the eastern end of the Falkland Plateau and the Agulhas Rise (Figure 7). The Rocas Verdes Basin was nearly closed by 100 Ma [*Dalziel*, 1981]. Drake Passage did not yet exist because the Antarctic Peninsula and South Orkney group were still contiguous with southern South America. By this time, East Antarctica and West Antarctica can be considered to have been in their present-day positions with

respect to each other based on paleomagnetic data [*Grunow et al.*, 1987] and by closure around the plate circuit with respect to north and south New Zealand. A shallow seaway appears to have developed south of Australia, but actual seafloor spreading had probably not yet commenced [*Veevers et al.*, 1991; *Cande and Mutter*, 1982], and there is no evidence of 100 Ma marine sediments [*Hegarty et al.*, 1988]. The northernmost Kerguelen

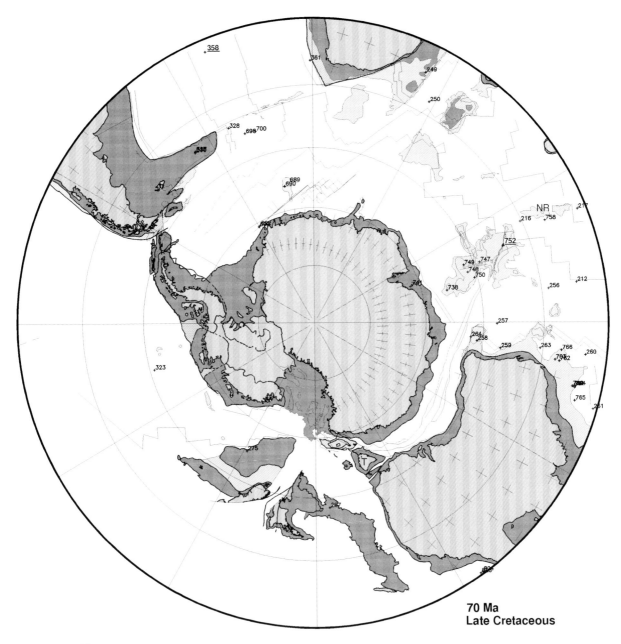

Fig. 10. Paleogeographic reconstruction of southern Gondwana at 70 Ma. NR, Ninety Degree East Ridge; T, Tasmania.

Plateau had probably just cleared the Indian continental margin. Deep circulation along the margins of East Antarctica was somewhat restricted by the submarine plateaus of the Southern Ocean, such as Agulhas Plateau, Maud Rise, Madagascar Rise, and Kerguelen Plateau.

The South Pacific margin of Gondwana had been an active subduction zone from Permian through Early Cretaceous based on evidence for magmatic arcs along South America [*Dalziel and Forsythe*, 1986], northern New Zealand [*Bradshaw*, 1989], and eastern Australia [*Cawood*, 1984]. The spreading center in the South Pacific began to be subducted along Australia and magmatism stopped. By 105 Ma, the tectonic regime in New Zealand changed significantly, and the spreading center was subducted off New Zealand and the Chatham Rise [*Bradshaw*, 1989]. Seafloor spreading and subduction ceased along that part of the margin of Gondwana. Subduction and arc-related volcanism continued along South America, the Antarctic Peninsula, and the Thurston Island region of West Antarctica [*Barker*, 1982].

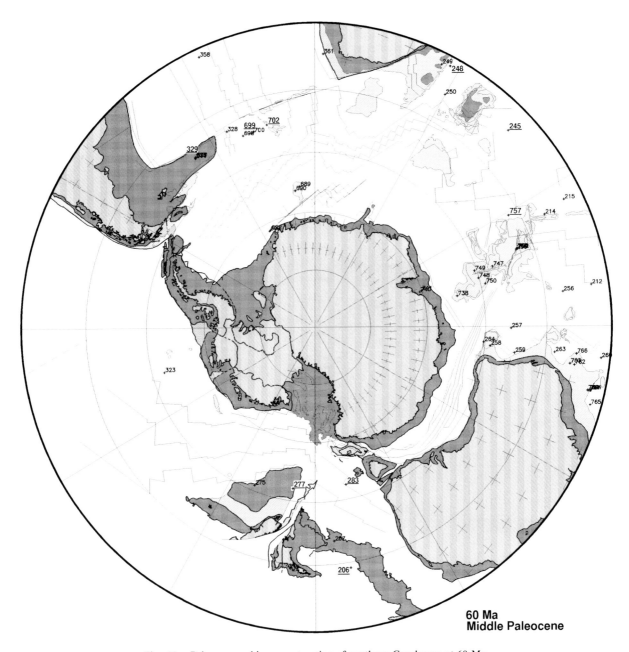

Fig. 11.   Paleogeographic reconstruction of southern Gondwana at 60 Ma.

90 Ma

The Indian Ocean and Weddell Sea regions (Figure 8) had open, deepwater circulation and marine sedimentation began south of Australia [*Hegarty et al.*, 1988]. Bounty Trough, between Chatham Rise and the Campbell Plateau, and Great South Basin [*Carter*, 1988] probably opened during this time (see Figure 9 for locations). Geosat data from this region show that both Bounty Trough and Great South Basin are gravimetri-

cally equivalent to oceanic crust. Rifting and subsidence of these features presumably postdate subduction of the Pacific spreading ridge to the north of New Zealand [*Bradshaw*, 1989]. They probably opened just prior to seafloor spreading that separated Campbell Plateau and north and south New Zealand from Marie Byrd Land of West Antarctica at about 84 Ma. Total subsidence for Great South Basin wells suggest initiation of rifting between 105 and 95 Ma [*Carter*, 1988], which correlates

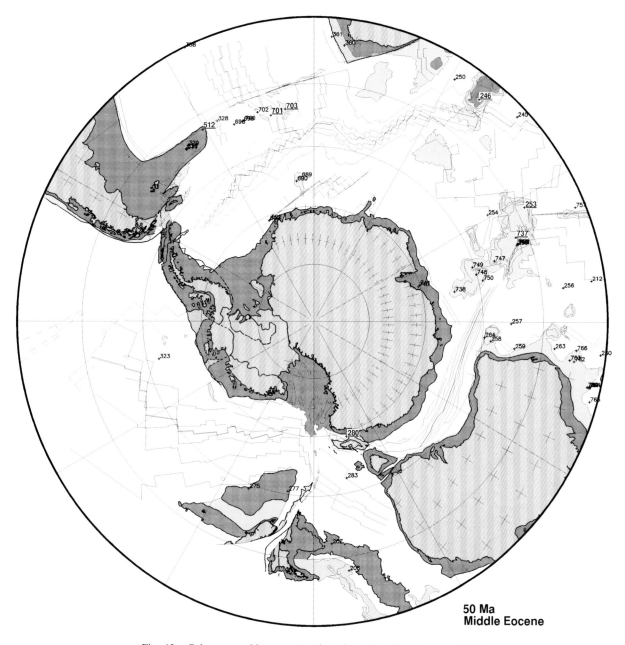

Fig. 12.   Paleogeographic reconstruction of southern Gondwana at 50 Ma.

closely with initiation of seafloor spreading between Australia and Antarctica [*Cande and Mutter*, 1982]. It might imply that as Australia rifted from East Antarctica, the South Tasman Rise rifted from Iselin Bank. This spreading center probably continued through the Emerald Basin/Solander Trough region between the Macquarie Ridge Complex and Campbell Plateau into Great South Basin [*Carter*, 1988] and finally connected with the Pacific-Antarctic spreading center via Bounty Trough.

Seafloor spreading between South America and Africa

and between Africa and Antarctica resulted in slow rotation of Antarctica away from South America about a pivot point located near the tip of South America. Drake Passage did not exist at this time. By 90 Ma, seafloor spreading may have opened a deepwater passage through the South Atlantic and into the equatorial Tethyan ocean, but Campanian sediments from the Angola Basin are anoxic [*Bolli et al.*, 1978]. This may indicate that both the Walvis Ridge to the south and the Romanche Fracture Zone ridge or other seafloor features to the north blocked open circulation in the Angola Basin.

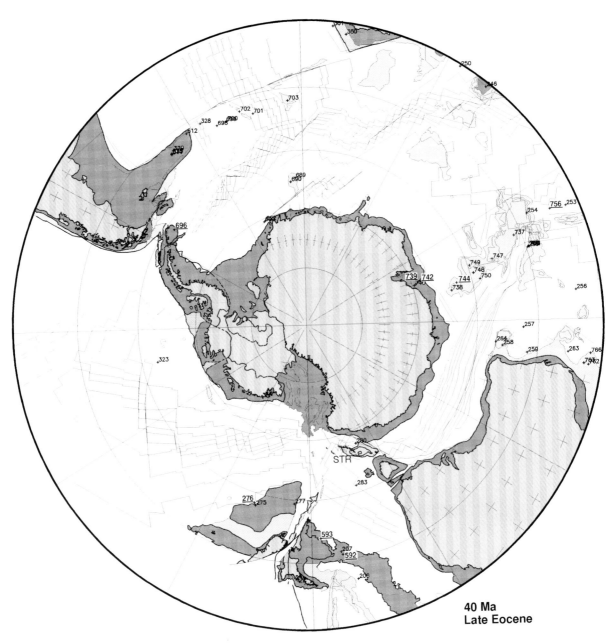

Fig. 13.   Paleogeographic reconstruction of southern Gondwana at 40 Ma. STR, South Tasman Ridge.

80 Ma

By 80 Ma (Figure 9), the magnetic anomaly 34 reversal (84 Ma) can be found in numerous places in the southern oceans. During Cretaceous Normal Superchron time (118 Ma to 84 Ma [*Kent and Gradstein*, 1986]), locations of major plates can only be approximated on the basis of drilling results, dated dredge samples, subsidence curves from continental margin wells, and other indirect methods. Intermittent euxinic conditions persisted at Site 364 in the Angola Basin into

the early Albian and recurred again in Albian to Santonian times [*Ryan et al.*, 1978]. By 84 Ma, deepwater circulation was established in the South Atlantic and persists to the present. The Indian Ocean was fully open to the Pacific Ocean and an equatorial Tethys. While there was marine deposition south of Australia, deepwater circulation between Tasmania and Antarctica was blocked. Seafloor spreading originated shortly before 84 Ma between Campbell Plateau [*Stock and Molnar*, 1987] and Marie Byrd Land and at about 80 Ma extended into the Tasman Sea

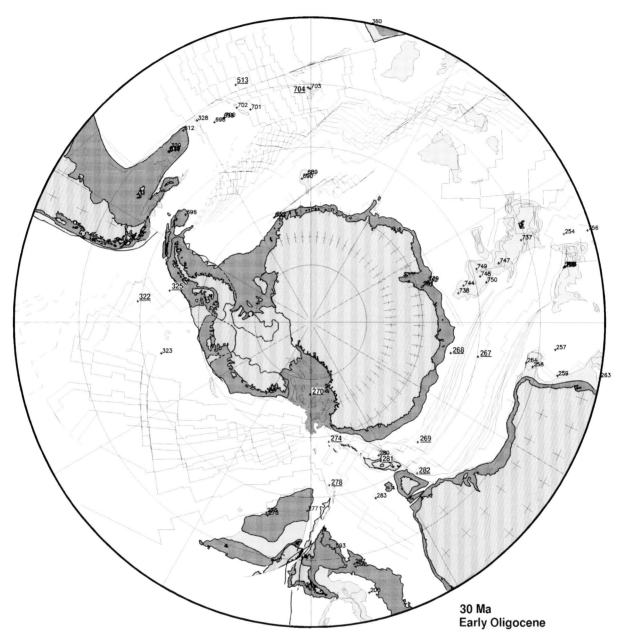

Fig. 14. Paleogeographic reconstruction of southern Gondwana at 30 Ma.

region [*Weissel and Hayes*, 1977]. Circulation south of Australia and between Campbell Plateau and Marie Byrd Land was not vigorous [*Kennett et al.*, 1974]. Subduction continued from about 120°W eastward along the Antarctic Peninsula and the western margin of South America. Drake Passage did not exist, and any high southern latitude circulation probably flowed north of Australia.

70–40 Ma

Australia initially moved very slowly northward (Figures 10–13) and did not undergo rapid northward motion until about 45 Ma, when there was a major reorientation of Indian Ocean plate motions [*Cande and Mutter*, 1982; *Royer and Sandwell*, 1989; *Veevers et al.*, 1991]. During this period, Tasmania and the South Tasman Rise

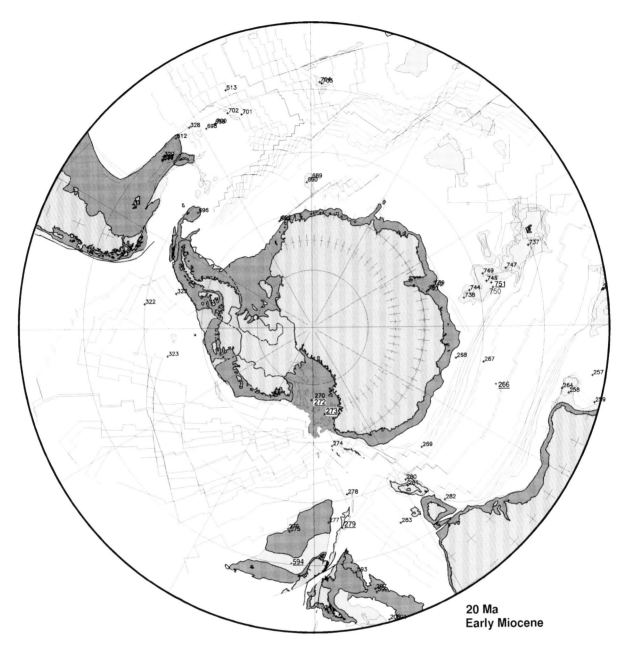

Fig. 15.   Paleogeographic reconstruction of southern Gondwana at 20 Ma.

formed an effective barrier to high southern latitude, deepwater circulation between Antarctica and Australia, although some deep water may have circulated between Campbell Plateau and Marie Byrd Land through the Tasman Sea. The Antarctic Peninsula moved eastward with respect to the southern tip of South America, but seafloor spreading had not commenced in the Scotia Sea according to the identified magnetic anomalies [*LaBrecque and Rabinowitz*, 1977]. There were numerous small continental fragments

[*Barker*, 1982] in the Drake Passage region that effectively blocked any circumpolar deepwater circulation through this region.

30 Ma to Present

As Australia and Antarctica separated at increasing rates (Figures 14 through 17), Tasmania and the South Tasman Rise cleared East Antarctica and vigorous high-latitude, deepwater circulation began. A comparison of

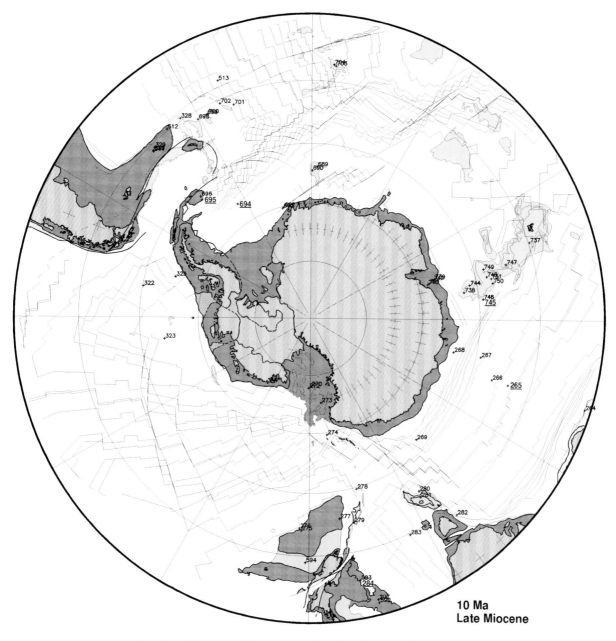

Fig. 16. Paleogeographic reconstruction of southern Gondwana at 10 Ma.

the Drake Passage region between Figures 13 and 14 indicates, on the basis of major plate motion, that the Antarctic Peninsula had already moved past the tip of southern South America prior to opening of the western Scotia Sea, which is dated at ~30 Ma [*Barker and Burrell*, 1977; *LaBrecque and Cande*, 1987]. The South Orkney block may have remained with the South American plate until the western Scotia Sea began to open, rather than being part of the Antarctic plate as indicated in the figures. By 20 Ma or shortly thereafter (Figures 15 through 17), a vigorous circumpolar current had undoubtedly developed.

SUMMARY

Paleogeographic reconstructions can be used to constrain the development of paleoseaways around Antarctica. Precise determination of the time of opening of a particular seaway is difficult even when dated marine magnetic anomalies are present because other information is lacking. At any rifted continental margin, it is important to estimate the amount and rate of thermal subsidence of the margin corrected for what is usually a rapid rate of sedimentation. In our figures, the 2000-m

Fig. 17.   Present-day map of the Southern Ocean.

isobath is arbitrarily taken as the edge of the continent. We assume that the outer continental margin continues to subside for up to 100 m.y. after rifting. Even if corrected for rapid sedimentation, vigorous circulation may erode older sediment. An additional unknown is the location of seafloor features that may have blocked the seaway after its apparent opening. The Walvis Ridge seems to have isolated the Angola Basin, even after the South Atlantic should have been fully open to the Weddell Sea and southwest Indian Ocean to the south and to the central Atlantic Ocean to the north. The best

way to determine the precise initiation of paleoseaways is to find an unequivocal change in the benthic sedimentation from an enclosed ocean basin to an open ocean basin.

Acknowledgments. This work was supported by the Division of Polar Programs, DPP 90-19247 to L.A.L., and by the PLATES Project of the Institute for Geophysics. The PLATES Project, supported by a consortium of industry sponsors, supports research in plate reconstructions and develops rotation software. This paper benefited from reviews by J. A.

Austin, B. Luyendyk, J. Weissel, and J. P. Kennett. This is contribution 0902 of the Institute for Geophysics.

REFERENCES

Austin, J. A., and E. Uchupi, Continental-oceanic crustal transition off southwest Africa, *Am. Assoc. Pet. Geol. Bull.*, *66*, 1328–1347, 1982.

Barker, P. F., The Cenozoic subduction history of the Pacific margin of the Antarctic Peninsula: Ridge crest-trench interaction, *J. Geol. Soc. London*, *139*, 787–801, 1982.

Barker, P. F., and J. Burrell, The opening of Drake Passage, *Mar. Geol.*, *25*, 15–34, 1977.

Barker, P. F., et al., Leg 36, *Initial Rep. Deep Sea Drill. Proj.*, *36*, 1080 pp., 1976.

Barker, P. F., et al., Leg 72, *Initial Rep. Deep Sea Drill. Proj.*, *72*, 1024 pp., 1983.

Barker, P. F., et al., Leg 113, *Proc. Ocean Drill. Program Initial Rep.*, *113*, 785 pp., 1988.

Barron, J., et al., Leg 119, *Proc. Ocean Drill. Program Initial Rep.*, *119*, 942 pp., 1989.

Battail, B., L. Beltan, and J.-M. Dutuit, Africa and Madagascar during Permo-Triassic time: The evidence of the vertebrate faunas, in *Gondwana Six: Stratigraphy, Sedimentology, and Paleontology, Geophys. Monogr. Ser.*, vol. 41, edited by G. D. McKenzie, pp. 147–156, AGU, Washington, D. C., 1987.

Bolli, H. M., et al., Leg 40, *Initial Rep. Deep Sea Drill. Proj.*, *40*, 1079 pp., 1978.

Bradshaw, J. D., Cretaceous geotectonic patterns in the New Zealand region, *Tectonics*, *8*, 803–820, 1989.

Burns, R. E., et al., Leg 21, *Initial Rep. Deep Sea Drill. Proj.*, *21*, 931 pp., 1973.

Cande, S. C., and J. C. Mutter, A revised identification of the oldest sea-floor spreading anomalies between Australia and Antarctica, *Earth Planet. Sci. Lett.*, *58*, 151–160, 1982.

Cannon, R. T., W. M. N. Simiyu Siambi, and F. M. Karanja, The proto–Indian Ocean and a probable Paleozoic/Mesozoic triradial rift system in east Africa, *Earth Planet. Sci. Lett.*, *52*, 419–426, 1981.

Carter, R. M., Post-breakup stratigraphy of the Kaikoura Synthem (Cretaceous-Cenozoic), continental margin, southeastern New Zealand, *N. Z. J. Geol. Geophys.*, *31*, 405–429, 1988.

Cawood, P. A., The development of the SW Pacific margin of Gondwana: Correlations between the Rangitata and New England orogens, *Tectonics*, *3*, 539–553, 1984.

Ciesielski, P. F., et al., Leg 114, *Proc. Ocean Drill. Program Initial Rep.*, *114*, 815 pp., 1988.

Cochran, J. R., The Somali Basin, Chain Ridge, and the origin of the northern Somali Basin gravity and geoid low, *J. Geophys. Res.*, *93*, 11,985–12,008, 1988.

Coffin, M. F., and P. D. Rabinowitz, Reconstruction of Madagascar and Africa: Evidence from the Davie Fracture Zone and western Somali Basin, *J. Geophys. Res.*, *92*, 9385–9406, 1987.

Coffin, M. F., and P. D. Rabinowitz, Evolution of the conjugate East African–Madagascan margins and the western Somali Basin, *Spec. Pap. Geol. Soc. Am.*, *226*, 78 pp., 1988.

Cooper, A. K., F. J. Davey, and K. Hinz, Crustal extension and origin of sedimentary basins beneath the Ross Sea and Ross Ice Shelf, Antarctica, in *Geological Evolution of Antarctica*, edited by M. R. A. Thomson, J. A. Crame, and J. Thomson, pp. 285–292, Cambridge University Press, New York, 1991.

Dalziel, I. W. D., Back-arc extension in the southern Andes: A review and critical reappraisal, *Philos. Trans. R. Soc. London, Ser. A*, *300*, 319–335, 1981.

Dalziel, I. W. D., and D. H. Elliot, West Antarctica: Problem child of Gondwanaland, *Tectonics*, *1*, 3–19, 1982.

Dalziel, I. W. D., and R. D. Forsythe, Andean evolution and the terrane concept, in *Tectonostratigraphic Terranes of the Circum-Pacific Region, CPCEMR ESS*, vol. 1, edited by D. G. Howell, pp. 565–581, American Association of Petroleum Geologists, Tulsa, Okla., 1986.

Davies, P. J., et al., Leg 133, *Proc. Ocean Drill. Program Initial Rep.*, *133*, 1496 pp., 1991.

Davies, T. A., et al., *Initial Rep. Deep Sea Drill. Proj.*, *26*, 1129 pp., 1974.

de Wit, M. J., M. Jeffery, H. Bergh, and L. O. Nicolaysen, Geological map of sectors of Gondwana reconstructed to their deposition—150 Ma, Am. Assoc. of Pet. Geol., Tulsa, Okla., 1988.

Elliot, D. H., Jurassic magmatism and tectonism associated with Gondwanaland breakup: An Antarctic perspective, in *Magmatism and the Causes of Continental Break-up, Geol. Soc. Spec. Publ. London*, in press, 1992.

Gahagan, L. M., et al., Tectonic fabric map of the ocean basins from satellite altimetry data, *Tectonophysics*, *155*, 1–26, 1988.

Gradstein, F. M., et al., Leg 123, *Proc. Ocean Drill. Program Initial Rep.*, *123*, 716 pp., 1990.

Grindley, G. W., and F. J. Davey, The reconstruction of New Zealand, Australia, and Antarctica, in *Antarctic Geoscience*, edited by C. Craddock, pp. 15–29, University of Wisconsin Press, Madison, 1982.

Groenewald, P. B., G. H. Grantham, and M. K. Watkeys, Geological evidence for a Proterozoic to Mesozoic link between southeastern Africa and Dronning Maud Land, Antarctica, *J. Geol. Soc. London*, *147*, 1115–1123, 1991.

Grunow, A. M., D. V. Kent, and I. W. D. Dalziel, Mesozoic evolution of West Antarctica and the Weddell Sea Basin: New paleomagnetic constraints, *Earth Planet. Sci. Lett.*, *86*, 16–26, 1987.

Grunow, A. M., D. V. Kent, and I. W. D. Dalziel, New paleomagnetic data from Thurston Island: Implications for the tectonics of West Antarctica and Weddell Sea opening, *J. Geophys. Res.*, *96*, 17,935–17,954, 1991.

Haq, B. U., et al., Leg 122, *Proc. Ocean Drill. Program Initial Rep.*, *122*, 826 pp., 1990.

Hayes, D. E., et al., Leg 28, *Initial Rep. Deep Sea Drill. Proj.*, *28*, 1017 pp., 1975.

Hegarty, K. A., J. K. Weissel, and J. C. Mutter, Subsidence history of Australia's southern margin: Constraints on basin models, *AAPG Bull.*, *74*, 615–633, 1988.

Hollister, C. D., et al., Leg 35, *Initial Rep. Deep Sea Drill. Proj.*, *35*, 929 pp., 1976.

Hsü, K. J., et al., Leg 73, *Initial Rep. Deep Sea Drill. Proj.*, *73*, 798 pp., 1984.

Kennett, J. P., *Marine Geology*, 813 pp., Prentice-Hall, Englewood Cliffs, N. J., 1982.

Kennett, J. P., et al., Leg 29, *Initial Rep. Deep Sea Drill. Proj.*, *29*, 1197 pp., 1974.

Kennett, J. P., et al., Leg 90, *Initial Rep. Deep Sea Drill. Proj.*, *90*, 1517 pp., 1986.

Kent, D. V., and F. M. Gradstein, A Jurassic to Recent geochronology, in *The Geology of North America*, vol. M, *The Western North Atlantic Region*, edited by P. R. Vogt and B. E. Tucholke, pp. 45–50, Geological Society of America, Boulder, Colo., 1986.

Klitgord, K. D., and H. Schouten, Plate kinematics of the central Atlantic, in *The Geology of North America*, vol. M, *The Western North Atlantic Region*, edited by P. R. Vogt and B. E. Tucholke, pp. 351–404, Geological Society of America, Boulder, Colo., 1986.

Kyle, P. R., D. H. Eliot, and J. F. Sutter, Jurassic Ferrar supergroup tholeiites from the Transantarctic Mountains,

Antarctica, and their relationship to the initial fragmentation of Gondwana, in *Gondwana Five*, edited by M. M. Cresswell and P. Vella, pp. 283–287, A. A. Balkema, Rotterdam, Netherlands, 1981.

LaBrecque, J. L., and S. C. Cande, Total intensity magnetic anomaly profiles, south, in South Atlantic Ocean and Adjacent Antarctic Continental Margin, atlas 13, Ocean Margin Drilling Program, *Regional Atlas Ser.*, sheet 9, edited by J. L. LaBrecque, Mar. Sci. Int., Woods Hole, Mass., 1987.

LaBrecque, J. L., and P. D. Rabinowitz, Magnetic anomalies bordering the continental margin of Argentina, *Map Ser. Cat. 826*, Am. Assoc. of Pet. Geol., Tulsa, Okla., 1977.

Larson, R. L., J. C. Mutter, J. B. Diebold, G. B. Carpenter, and P. Symonds, Cuvier Basin: A product of ocean crust formation by Early Cretaceous rifting off western Australia, *Earth Planet. Sci. Lett.*, *45*, 105–114, 1979.

Lawver, L. A., and L. M. Gahagan, Constraints on Mesozoic transtension in Antarctica (abstract), in *Abstracts, Sixth International Symposium on Antarctic Earth Sciences*, pp. 343–344, National Institute of Polar Research, Tokyo, 1991.

Lawver, L. A., and C. R. Scotese, A revised reconstruction of Gondwanaland, in *Gondwana Six: Structure, Tectonics, and Geophysics*, *Geophys. Monogr. Ser.*, vol. 40, edited by G. D. McKenzie, pp. 17–24, AGU, Washington, D. C., 1987.

Lawver, L. A., D. A. Sandwell, J.-Y. Royer, and C. R. Scotese, Evolution of the Antarctic continental margins, in *Geological Evolution of Antarctica*, edited by M. R. A. Thomson, J. A. Crame, and J. Thomson, pp. 533–539, Cambridge University Press, New York, 1991.

Leclaire, L., et al., Lower Cretaceous basalt and sediments from the Kerguelen Plateau, *Geo Mar. Lett.*, *7*, 169–176, 1987.

Ludwig, W. J., et al., Leg 71, *Initial Rep. Deep Sea Drill. Proj.*, *71*, 1187 pp., 1983.

Luyendyk, B. P., D. Forsyth, and J. D. Philips, Experimental approach to the paleocirculation of the oceanic surface waters, *Geol. Soc. Am. Bull.*, *83*, 2649–2664, 1972.

Markl, R. G., Evidence for the breakup of eastern Gondwanaland by the Early Cretaceous, *Nature*, *251*, 196–200, 1974.

Martin, P. K., and C. Hartnady, Plate tectonic development of the southwest Indian Ocean: A revised reconstruction of East Antarctica and Africa, *J. Geophys. Res.*, *91*, 4767–4785, 1986.

Müller, R. D., J.-Y. Royer, and L. A. Lawver, Evidence for hotspot group motion in the Late Cretaceous/early Tertiary (abstract), *Geol. Soc. Am. Abstr. Programs*, *23*(5), A318, 1991.

Müller, R. D., J.-Y. Royer, and L. A. Lawver, Revised plate motions relative to the hotspots from combined Atlantic and Indian ocean hotspot tracks, *Geology*, in press, 1992.

Nürnberg, D., and R. D. Müller, The tectonic evolution of the South Atlantic from Late Jurassic to present, *Tectonophysics*, *191*, 27–53, 1991.

Parsons, B., and J. G. Sclater, An analysis of the variation of ocean floor bathymetry and heat flow with age, *J. Geophys. Res.*, *82*, 803–827, 1977.

Peirce, J., et al., Leg 121, *Proc. Ocean Drill. Program Initial Rep.*, *121*, 1000 pp., 1989.

Powell, C. M., S. R. Roots, and J. J. Veevers, Pre-breakup continental extension in east Gondwanaland and the early opening of the eastern Indian Ocean, *Tectonophysics*, *155*, 261–283, 1988.

Rabinowitz, P. D., and J. L. LaBrecque, The Mesozoic South Atlantic Ocean and evolution of its continental margins, *J. Geophys. Res.*, *84*, 5973–6002, 1979.

Rabinowitz, P. D., M. F. Coffin, and D. A. Falvey, The separation of Madagascar and Africa, *Science*, *220*, 67–69, 1983.

Reeves, C. V., F. M. Karanja, and I. N. Macleod, Geophysical

evidence for a failed Jurassic rift and triple junction in Kenya, *Earth Planet. Sci. Lett.*, *81*, 299–311, 1987.

Royer, J.-Y., and D. T. Sandwell, Evolution of the eastern Indian Ocean since the Late Cretaceous: Constraints from Geosat altimetry, *J. Geophys. Res.*, *94*, 13,755–13,782, 1989.

Royer, J.-Y., et al., A tectonic chart for the Southern Ocean derived from Geosat altimetry data, *AAPG Stud. Geol.*, *31*, 89–100, 1990.

Ryan, W. B. F., et al., Objectives, principal results, operations, and explanatory notes of Leg 40, South Atlantic, *Initial Rep. Deep Sea Drill. Proj.*, *40*, 5–28, 1978.

Sandwell, D. T., Antarctic marine gravity field from high-density satellite altimetry, *Geophys. J. Int.*, *109*, 437–448, 1992.

Sandwell, D. T., and D. C. McAdoo, Marine gravity of the Southern Ocean and Antarctic margin from GEOSAT, *J. Geophys. Res.*, *93*, 10,389–10,394, 1988.

Sastri, V. V., A. T. R. Raju, R. N. Sinha, B. S. Venkatachala, and R. K. Banerji, Biostratigraphy and evolution of the Cauvery Basin, India, *J. Geol. Soc. India*, *48*, 355–377, 1977.

Schlich, R., et al., Leg 120, *Proc. Ocean Drill. Program Initial Rep.*, *120*, 648 pp., 1989.

Ségoufin, J., Anomalies magnétiques mesozoiques dans le bassin de Mozambique, *C. R. Seances Acad. Sci., Ser. B*, *287D*, 109–112, 1978.

Ségoufin, J., and P. Patriat, Existence d'anomalies mesozoïques dans le bassin de Somali, Implications pour les relations Afrique-Antarctique-Madagascar, *C. R. Seances Acad. Sci., Ser. B*, *291B*, 85–88, 1980.

Shaw, P. R., and S. C. Cande, High-resolution inversion for South Atlantic plate kinematics using joint altimeter and magnetic anomaly data, *J. Geophys. Res.*, *95*, 2625–2644, 1990.

Simpson, E. S. W., et al., Leg 25, *Initial Rep. Deep Sea Drill. Proj.*, *25*, 884 pp., 1974.

Simpson, E. S. W., J. G. Sclater, B. Parsons, I. O. Norton, and L. Meinke, Mesozoic magnetic lineations in the Mozambique Basin, *Earth Planet. Sci. Lett.*, *43*, 260–264, 1979.

Stock, J., and P. Molnar, Revised history of early Tertiary plate motion in the southwest Pacific, *Nature*, *325*, 495–499, 1987.

Supko, P. R., et al., Leg 39, *Initial Rep. Deep Sea Drill. Proj.*, *39*, 1139 pp., 1977.

U.S. Geological Survey, Satellite image map of Antarctica, *U.S. Geol. Surv. Misc. Invest. Map*, 1-2284, 1991.

Veevers, J. J., et al., Leg 27, *Initial Rep. Deep Sea Drill. Proj.*, *27*, 1060 pp., 1974.

Veevers, J. J., J. W. Tayton, B. D. Johnson, and L. Hansen, Magnetic expression of the continent-ocean boundary between the western margin of Australia and the eastern Indian Ocean, *J. Geophys.*, *56*, 106–120, 1985.

Veevers, J. J., H. M. J. Stagg, J. B. Willcox, and H. L. Davies, Pattern of slow seafloor spreading (<4 mm/year) from breakup (96 Ma) to A20 (44.5 Ma) off the southern margin of Australia, *BMR J. Aust. Geol. Geophys.*, *11*, 499–507, 1990.

Veevers, J. J., C. M. Powell, and S. R. Rotts, Review of seafloor spreading around Australia, I, Synthesis of the patterns of spreading, *Austr. J. Earth Sci.*, *38*, 373–389, 1991.

von der Borch, C. C., et al., Leg 22, *Initial Rep. Deep Sea Drill. Proj.*, *22*, 890 pp., 1974.

Weissel, J. K., and D. E. Hayes, Evolution of the Tasman Sea reappraised, *Earth Planet. Sci. Lett.*, *36*, 77–84, 1977.

Ziegler, A. M., C. R. Scotese, and S. F. Barrett, Mesozoic and Cenozoic paleogeographic maps, in *Tidal Friction and the Earth's Rotation II*, edited by P. Brosche and J. Sundermann, pp. 240–252, Springer-Verlag, New York, 1983.

(Received December 16, 1991;
accepted May 20, 1992.)

THE ANTARCTIC PALEOENVIRONMENT: A PERSPECTIVE ON GLOBAL CHANGE

ANTARCTIC RESEARCH SERIES, VOLUME 56, PAGES 31–60

BIOGEOGRAPHY OF CAMPANIAN-MAASTRICHTIAN CALCAREOUS PLANKTON IN THE REGION OF THE SOUTHERN OCEAN: PALEOGEOGRAPHIC AND PALEOCLIMATIC IMPLICATIONS

Brian T. Huber

Department of Paleobiology, Smithsonian Institution, Washington, D. C. 20560

David K. Watkins

Department of Geology, University of Nebraska, Lincoln, Nebraska 68588

Analysis of biogeographic distribution patterns among Campanian-Maastrichtian calcareous nannoplankton and planktonic foraminifera from the southern high latitudes provides insight to changes in circum-Antarctic climate and surface circulation surface routes. Both microfossil groups are similarly characterized in the early Campanian by low-diversity, cosmopolitan species with few or no austral provincial taxa. This changes by late Campanian–early Maastrichtian time as austral species diversified and began to dominate the high-latitude assemblages. Maximum diversity of austral provincial taxa occurs during the late Campanian among the planktonic foraminifera and in the early Maastrichtian among the calcareous nannoplankton. Climatic cooling is considered the cause for the decline from 53 nannofossil species during the early Maastrichtian to 20 species toward the end of the Maastrichtian as well as the equatorward shifts of the nannofossil *Nephrolithus frequens* and the planktonic foraminifer *Abathomphalus mayaroensis* during the late Maastrichtian. On the other hand, the poleward migrations of the planktonic foraminifer *Pseudotextularia elegans* and the nannofossil *Watznaueria barnesae* less than 500,000 years before the Cretaceous/Tertiary extinction event correspond with a negative $\delta^{18}O$ excursion observed at Maud Rise Site 690, suggesting that these species shifts were caused by a brief high-latitude warming event. The high degree of provinciality among the late Campanian–early Maastrichtian calcareous plankton reflects segregation of a cool, high-latitude water mass from warmer, subtropical surface waters. A long-term climatic cooling and paleogeographic changes related to the breakup of the southern Gondwana continents are considered the major factors that caused the paleocirculation and biogeographic changes. Seafloor spreading and subsidence between Antarctica, Australia, and New Zealand, northward drift of South America from the Antarctic Peninsula, and a global rise in sea level during the middle Campanian provided new routes for shallow marine communication between the Indian, Pacific, and South Atlantic ocean basins. Opening of these gateways may have also caused a widespread disconformity that separates lower Campanian from upper Campanian sediments in the Atlantic and Indian ocean sectors of the Southern Ocean. Reemergence of a South American–Antarctic Peninsula isthmus in the middle and late Maastrichtian is postulated to account for poleward migration of several keeled and nonkeeled planktonic foraminifera during a time of gradual climatic cooling of the polar oceans. Closure of this gateway could have been caused by a fall in sea level and renewed volcanism along the Antarctic Peninsula magmatic arc. This could have led to a diminished intensity of surface current flow between the southern South Atlantic and southern Indian ocean basins and enhanced vertical stratification and niche partitioning in the austral surface waters, thus enabling habitation by a greater diversity of depth-stratified planktonic foraminifera. A renewed terrestrial land bridge at this time would explain the selective dispersal of marsupials and terrestrial plants across the southern Gondwana continents that has been postulated in several paleobiogeographic studies.

INTRODUCTION

Changes in Late Cretaceous paleogeography of the circum-Antarctic region had an important influence on the prevailing patterns of ocean circulation, the evolution of Late Cretaceous climate, and the history of biotic interchange in the terrestrial and marine realms of the southern hemisphere. But direct geologic evidence bearing on the outline of the continental margin and interior seaways of Antarctica during Late Cretaceous time is largely inaccessible because of the presence of the thick Antarctic icecap.

One approach to getting around this problem is to compare the available fossil record from Antarctic and surrounding localities to determine former pathways of biotic dispersal in terrestrial and marine environments and to identify periods of increased provincialism versus enhanced biotic interchange. This approach has been effectively used to establish that a terrestrial connection linking South America, Antarctica, New Zealand, and Australia existed at various times during the Cretaceous [e.g., *Cranwell*, 1964; *Raven and Axelrod*, 1974; *Woodburne and Zinsmeister*, 1984; *Dettmann*, 1989; *Olivero et al.*, 1991]. Equally convincing evidence suggests that these areas were connected by shallow marine seaways along the Pacific margin of Antarctica and between some of the crustal blocks that comprise West Antarctica, leading to the development of a southern temperate biogeographic province [*Zinsmeister*, 1979, 1982; *Stevens*, 1980, 1989; *Macellari*, 1987; *Huber and Webb*, 1986; *Clarke and Crame*, 1989]. Unfortunately, the Cretaceous terrestrial and shallow marine paleontological records are too sparse and too poorly dated to accurately constrain the timing and duration of the trans-Antarctic biotic communication links.

We contend that insight to changes in Cretaceous geography within and around Antarctica can also be gained by study of the more complete and better dated deep-sea record of planktonic foraminifera and calcareous nannoplankton. It is well established that paleogeography plays an important role in determining the routes of surface circulation and the vertical structure of surface water masses. Physical oceanographic models have shown that the opening or closure of gateways between major ocean basins, and even the physiography of the ocean floor more than 2000 m deep, could influence the flow direction and velocity of surface water masses thousands of kilometers away [*Berggren and Hollister*, 1974, 1977; *Webb et al.*, 1991]. Disruption of marine communication across such gateways may lead to significant changes in the planktonic biota. For example, the early Pliocene closure of the Isthmus of Panama led to differences in stratigraphic ranges of planktonic foraminifera from Caribbean cores relative to those from Pacific cores and vicariant evolution of several species [*Parker*, 1973; *Keigwin*, 1982]. Closure during the late Miocene of the Indonesian Seaway is thought to have caused development of an easterly flowing Equatorial Undercurrent and a weakening of east-west differences among planktonic foraminifera in the equatorial Pacific [*Kennett et al.*, 1985]. This also led to intensification of the Kuroshio Current and a northward displacement of tropical planktonic foraminiferal assemblages [*Kennett et al.*, 1985].

The challenge posed by the Cretaceous circum-Antarctic record is how to discern biogeographic distribution patterns that were primarily influenced by paleogeographic changes from those patterns that were climatically controlled. Because of our limited understanding of the paleobiology of the Cretaceous plankton and the limited amount of data from the southern high latitudes, these signals may not be separable. Nonetheless, we will discuss implications of similarities and differences in the observed biogeographic distribution patterns and, using other available geologic information, propose hypothetical models for surface marine circulation and circum-Antarctic paleogeography for the Campanian-Maastrichtian time period.

STRATIGRAPHY OF THE SOUTHERN OCEAN SITES

The Cretaceous deep-sea data base for the circum-Antarctic region has vastly improved since the recent completion of a series of drilling legs in the region of the Southern Ocean by the Ocean Drilling Program (ODP). Prior to this phase of drilling, the only high-latitude sites yielding a Cretaceous record were Deep Sea Drilling Project (DSDP) sites 327 and 511, located at about 52°S on the Falkland Plateau (Figures 1 and 2). Although both sites bear calcareous microfossils that are remarkably well preserved, the sections are interrupted by stratigraphic gaps that span the middle through upper Maastrichtian, middle to upper Campanian, lower Turonian, and most of the Cenomanian [*Sliter*, 1977; *Wise and Wind*, 1977; *Wise*, 1983; *Wind and Wise*, 1983; *Krasheninnikov and Basov*, 1983].

A nearly complete Maastrichtian history of pelagic carbonate sedimentation was obtained from sites 689 and 690 on the Maud Rise (65°S). Both sites provide high-quality magnetostratigraphic records [*Hamilton*, 1990] and good microfossil preservation [*Huber*, 1990; *Thomas*, 1990; *Pospichal and Wise*, 1990], and they are regarded as the best biostratigraphic reference sections in the southern high latitudes [*Thomas et al.*, 1990]. The oldest sediments drilled at the Maud Rise were considered to be early Maastrichtian based on the calcareous nannofossil biostratigraphy of *Pospichal and Wise* [1990], but magnetobiostratigraphic correlation with the *Kent and Gradstein* [1985] geomagnetic polarity time scale suggests that these range into the late Campanian [*Huber*, 1991a].

Upper Cretaceous sediments were also cored at ODP sites 698 and 700 in the southern South Atlantic (Figures 1, 2). Microfossil preservation at both sites is moderate to good in the upper Maastrichtian sections but deteriorates in the lower Maastrichtian and older sediments [*Huber*, 1991a; *Crux*, 1991]. The uppermost Campanian-Maastrichtian sequences obtained from Site 698 is poorly represented because of incomplete core recovery and a hiatus in the uppermost Maastrichtian [*Ciesielski et al.*, 1988]. Recovery was much better in the upper Santonian–Maastrichtian limestone drilled at Site 700. A good magnetic polarity stratigraphy was obtained from this site [*Hailwood and Clement*, 1991],

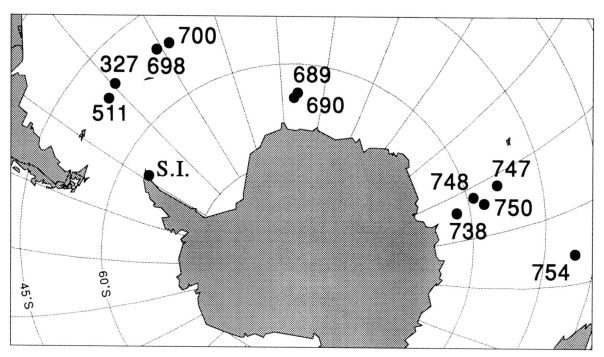

Fig. 1. Circum-Antarctic DSDP and ODP localities that have yielded Late Cretaceous microfossils.

but poor microfossil preservation limits the biostratigraphic accuracy. Hiatuses at Site 700 are recognized at the top of the Maastrichtian, in the lower upper Campanian, and in the upper part of the lower Campanian [*Crux*, 1991].

Four sites drilled on the Kerguelen Plateau between 52°S and 62°S recovered sediments of Late Cretaceous age. Microfossil preservation, however, is good only in the upper Maastrichtian of Site 738 and in the upper Campanian–Maastrichtian of sites 747 and 750 (Figure 2). Hiatuses occur at various levels in the upper Maastrichtian at sites 738, 747, and 748 and span most of the lower Campanian and the lower part of the upper Campanian at all four sites. Nevertheless, the composite upper Campanian–Maastrichtian record from the Kerguelen Plateau sites is nearly complete. A paleomagnetic reversal stratigraphy was obtained for part of the Maastrichtian at Site 738 [*Sakai and Keating*, 1991], but not from any of the other Kerguelen Plateau sections.

Although most of the stratigraphic gaps in the Southern Ocean sections appear to be only of local extent, a disconformity spanning the upper lower and lower upper Campanian is manifest at all localities. At Site 511 (Falkland Plateau), this disconformity occurs within a sequence of zeolitic claystones with the actual stratigraphic break marked by an interval barren of calcareous microplankton fossils. Benthic foraminifers indicate a dissolution facies associated with the disconformity, suggesting a paleodepth near the CCD calcite compensation depth (CCD) during the Campanian. This is

apparently the most complete mid Campanian section, as it is the only occurrence of the *G. diabolum* nannofossil subzone known from the Southern Ocean sites. At nearby Site 700, this disconformity (at or near Core 700B-50R) approximately corresponds to a lithologic change (ash-bearing zeolitic claystone horizons below, none above). Benthic foraminifers indicate 1500- to 2000-m water depth during this part of the Cretaceous [*Ciesielski et al.*, 1988]. At Site 747 (Kerguelen Plateau), this disconformity separates the condensed Turonian through lower Campanian section from the somewhat expanded upper Campanian through mid Maastrichtian pelagic section above. Pelagic deposition at Kerguelen sites 738 and 750 was also interrupted during the mid Campanian, although the duration of the hiatus is difficult to constrain at both sites because of poor nannofossil and foraminiferal preservation (Site 738) or coring gaps (Site 750). At Site 748, this disconformity interrupts the deposition of a sequence of glauconitic bryozoan grainstones and packstones. Benthic foraminifers indicate that the sequence was entirely neritic. Indeed, coralline red algae in grainstones overlying the disconformity indicate that this sequence was at least partially within the photic zone during the late Campanian [*Schlich et al.*, 1989]. The stratigraphic records from these sites demonstrate that this disconformity affected sites varying in paleodepth from neritic to bathyal (at CCD), suggesting that this event cannot be fully explained by an upward excursion of the CCD. This

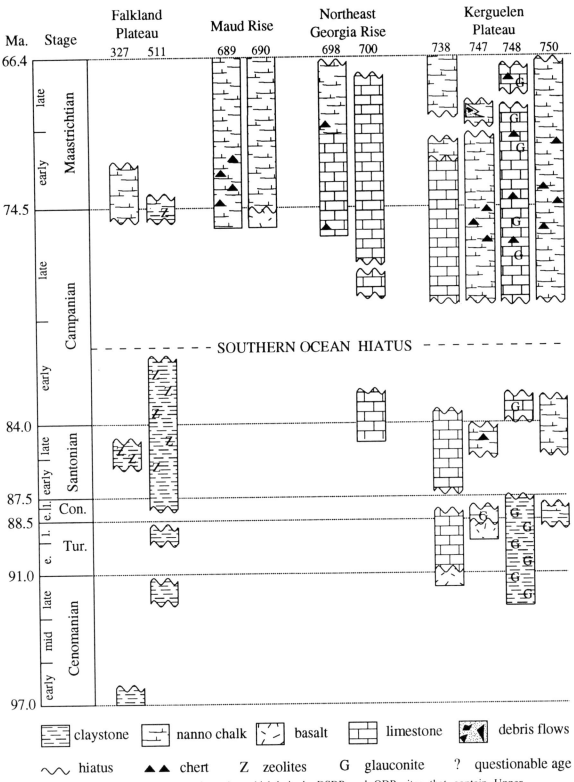

Fig. 2. Lithostratigraphy of southern high-latitude DSDP and ODP sites that contain Upper Cretaceous sediments.

pervasive stratigraphic gap is referred to as the Southern Ocean hiatus (Figure 2).

In summary, the Cretaceous record of deep-sea sedimentation in the circum-Antarctic region is most complete and best preserved in the upper Campanian–Maastrichtian interval. Good recovery and microfossil preservation and integration of detailed planktonic foraminiferal, calcareous nannoplankton, and paleomagnetic stratigraphies at Maud Rise sites 689 and 690 (Figure 3) have enabled development of a high-resolution age model that affords an unprecedented degree of accuracy in high- to low-latitude chronostratigraphic correlation. The Cenomanian through lower Campanian high-latitude record is not nearly as well constrained because of poor microfossil preservation at most sites, incomplete core recovery, and stratigraphic hiatuses. A middle Campanian disconformity occurs in all Upper Cretaceous cores in the South Atlantic and Indian ocean sectors of the Southern Ocean.

Calcareous Nannofossil Biostratigraphy

It has been clear since DSDP Leg 36 that the Upper Cretaceous nannofossils of the Southern Ocean were so significantly different from coeval tropical and subtropical assemblages that a separate high-latitude biostratigraphic zonation was necessary [e.g., *Wise and Wind*, 1977]. Several authors [*Wise and Wind*, 1977; *Wise*, 1983, 1988; *Pospichal and Wise*, 1990; *Crux*, 1991; *Watkins*, 1992] have proposed tentative zonal schemes as additional sections were collected. The zonation presented herein (Figure 4) is derived from a recent compilation of all available data by *Watkins et al.* [1992].

The distribution of the biozones used in this zonation is a strong reflection of the general paleobiogeographic pattern of the Southern Ocean. Only two polar species (*Thiersteinia ecclesiastica* and *Gephyrobiscutum diabolum*) are used in the zonation for the Cenomanian through Campanian interval. The other nine biohorizons are all recognized and used in the mid-latitude zonation of Sissingh (*Sissingh* [1977], as modified by *Perch-Nielsen* [1985]). This reflects the low degree of endemism and the largely cosmopolitan nature of Cenomanian through Campanian nannofossil assemblages. The Maastrichtian zonation, on the other hand, relies almost solely upon polar taxa, with only two cosmopolitan biohorizons and six polar ones.

The change from the cosmopolitan assemblages of the Cenomanian-Campanian to the strongly divided tropical/temperate and polar assemblages of the Maastrichtian appears to have begun in the late Campanian. The middle-latitude zonation is only marginally useful in the upper Campanian, as several important "cosmopolitan" species are absent or extremely rare and sporadic in the Southern Ocean sections. These biostratigraphically important species include *Bukryaster hayi*, *Cera-* *tolithoides aculeus*, *Quadrum sissinghii*, and *Quadrum trifidum*. *Lithastrinus grillii*, although common in Santonian and lower Campanian assemblages, is rare or absent near the end of its range, so that the last appearance datum of *L. grillii* is not a useful datum in the Southern Ocean. There are no polar species to replace these missing temperate species as biomarkers, because the proliferation of polar taxa did not commence in earnest until the early Maastrichtian. Thus the resolution of the Southern Ocean zonation is significantly less in the upper Campanian than that of the temperate zonation.

The Maastrichtian of the Southern Ocean is characterized by high rates of speciation and extinction in the Ahmuellerellaceae, Biscutaceae, and Podorhabdaceae (as discussed below). At least 12 Maastrichtian taxa from these families arose in the Southern Ocean. Although the stratigraphic ranges of several of these are still uncertain, four of them are sufficiently well documented to afford recognition of the isochronous first appearance datum (FAD) and last appearance datum (LAD) and are used to accurately correlate four biohorizons in the Southern Ocean zonation. These include *Nephrolithus corystus* (FAD), *Biscutum coronum* (LAD), *Neocrepidolithus watkinsii* (LAD), and *Biscutum magnum* (LAD). In addition, the zonation utilizes the acme of *Prediscosphaera stoverii*, an abundance event that seems to be restricted to the Southern Ocean [*Pospichal*, 1989]. Many of the species used in temperate Maastrichtian zonations are absent or sporadically rare in the Southern Ocean. Absent forms include *Quadrum trifidum*, *Lithraphidites praequadratus*, *Micula murus*, *Ceratolithoides kamptneri*, and *Micula prinsii*. *Tranolithus phacelosus* becomes very rare and sporadic in occurrence near the top of its stratigraphic range in the Southern Ocean, making it problematic as a useful biomarker. *Lithraphidites quadratus* is extremely rare in the Southern Ocean, rendering it useless for biostratigraphy. The first appearance of *Nephrolithus frequens*, a useful datum in temperate regions, has been shown to be diachronous in the Southern Ocean (as discussed below) and therefore inappropriate as a biohorizon. As a result, biostratigraphic correlation within the Southern Ocean must rely largely on high-latitude taxa.

Foraminiferal Biostratigraphy

Until ODP Leg 113, the only Upper Cretaceous zonal schemes proposed for southern high-latitude sections were developed from studies of nearshore sequences in New Zealand [*Webb*, 1971] and the Antarctic Peninsula [*Huber*, 1988]. The distributions of several of the defining taxa used in these zonations are strongly facies controlled and diachronous, and hence these zonations were primarily intended for local correlation. No biostratigraphic scheme was proposed for DSDP sites 327

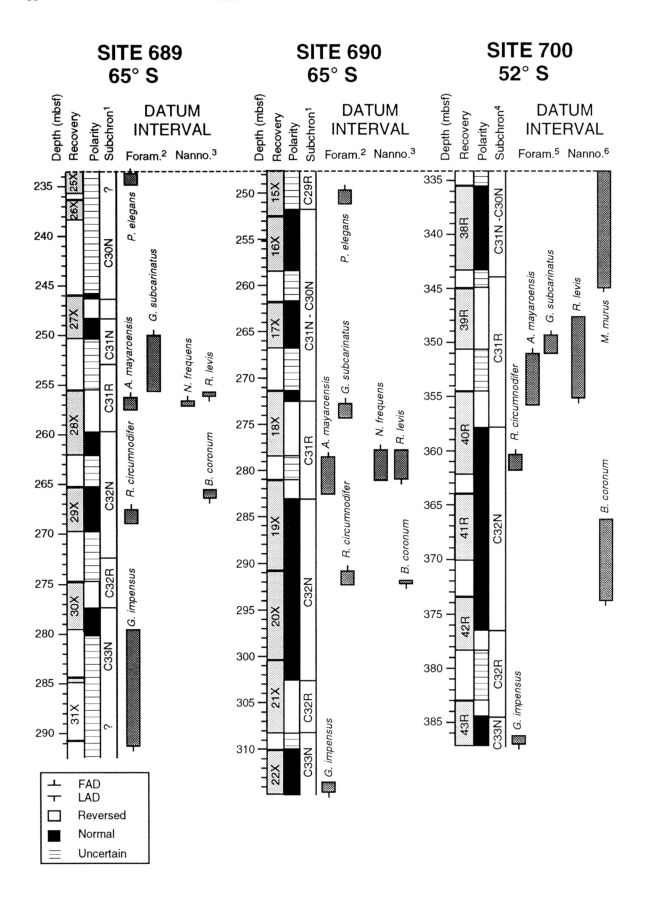

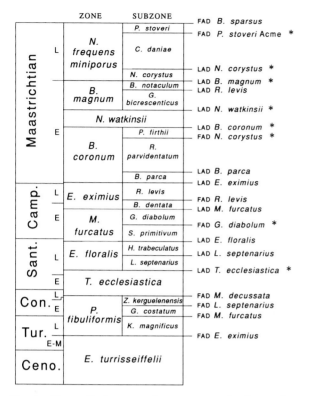

Fig. 4. Upper Cretaceous calcareous nannofossil zonation for the Southern Ocean [from *Watkins et al.*, 1992]. Austral endemics are denoted by asterisks.

ZONE	SUBZONE	
MAASTRICHTIAN L	Abathomphalus mayaroensis	P. elegans — FAD P. elegans
		G. subcarinatus — FAD G. subcarinatus
		G. petaloidea — FAD A. mayaroensis
MAASTRICHTIAN E	Globotruncanella havanensis	A. intermedius — FAD A. intermedius
		R. circumnodifer — FAD R. circumnodifer *
		A. australis — LAD G. impensus *
CAMP. L	Globigerinelloides impensus	FAD G. impensus *
CAMP. E	Archaeoglobigerina cretacea	LAD M. marginata
SANT. L/E	Marginotruncana marginata	
CON. L/E	Whiteinella baltica	FAD A. cretacea
TUR. L/E		LAD Praeglobotruncana spp.
CENO. L/E	Praeglobotruncana spp.	

Fig. 5. Upper Cretaceous planktonic foraminiferal zonation of the Southern Ocean [from *Huber*, 1992*b*]. Austral endemics are denoted by asterisks.

and 511 on the Falkland Plateau because of the absence of lower-latitude zonal marker taxa, cross-latitude correlation uncertainties, and incomplete stratigraphic recovery [*Sliter*, 1977; *Krasheninnikov and Basov*, 1983].

This situation considerably improved after completion of ODP drilling in the southern high latitudes. Comparison of planktonic foraminiferal assemblages from sites drilled during Leg 113 [*Huber*, 1990], Leg 114 [*Huber*, 1991*a*], Leg 119 [*Huber*, 1991*b*], and Leg 120 [*Quilty*, 1992] has revealed the following: (1) the taxonomic character and relative order of first and last occurrences are nearly identical at all circum-Antarctic sites; (2) species endemic to the Austral Biogeographic

Realm first appear during the late Campanian and range throughout the Maastrichtian; (3) a number of the identified cosmopolitan species have diachronous ranges relative to lower latitudes; (4) except for *A. mayaroensis*, marker taxa used in tropical and subtropical biozonations (e.g., *Rotalipora*, *Ticinella*, *Dicarinella*, *Radotruncana*, *Globotruncanita*, *Trinitella*, and *Racemiguembelina* and others) are completely absent from the high latitudes; and (5) keeled species are absent from uppermost Campanian through lower Maastrichtian sections, appear during the late early Maastrichtian, and then parallel an increase in total species richness during the late Maastrichtian [*Huber*, 1992*a*]. Observations of changes in taxonomic character and species richness among the austral assemblages suggest a transition from low-diversity, cool water assemblages of the upper Campanian to lower Maastrichtian to warmer water, higher-diversity assemblages of the upper Maastrichtian. These observations disagree with oxygen isotope paleotemperature results obtained by *Barrera and Huber* [1990] for the Maud Rise, which show a long-term cooling trend during this time period, as will be discussed in a later section.

Three planktonic foraminiferal zones were proposed for correlation of the Maud Rise holes [*Huber*, 1990] and were subsequently identified at all of the other upper Campanian–Maastrichtian sequences in the circum-Antarctic region. More recently, *Huber* [1992*b*] subdivided the two austral Maastrichtian zones into six subzones and proposed four additional zones for corre-

Fig. 3. (Opposite) Maastrichtian planktonic foraminiferal and calcareous nannoplankton magnetobiostratigraphic correlation of ODP sites 689, 690, and 700 in the southern South Atlantic. Datums are placed at the level of samples containing the marker taxa. Datum intervals represent the sampling uncertainty from the sample in which the marker species first or last occurs to the next underlying or overlying sample in which the species is absent. Light stipple in recovery columns represents the amount of sediment recovered for each core. Dashed line depicts the level of the Cretaceous/Tertiary boundary. Data sources are as follows: 1 [*Hamilton*, 1990], 2 [*Huber*, 1990]; 3 [*Pospichal and Wise*, 1990]; 4 [*Hailwood and Clement*, 1991]; 5 [*Huber*, 1991*a*]; 6 [*Crux*, 1991].

lation of Upper Cretaceous sediments in the Austral Realm (Figure 5). The greater number of datums recognized in the Maastrichtian interval reflects a sampling and taphonomic bias due to better core recovery and foraminiferal preservation in Maastrichtian sediments. Among the southern high-latitude sections drilled to date, 10 have recovered moderately to well-preserved foraminifera from the Maastrichtian Stage, whereas only three sites have moderate to good preservation in at least part of the Cenomanian-Campanian interval. Because the record of planktonic foraminifera in the Austral Realm is very spotty in the Cenomanian-Santonian interval, discussion of their biogeographic distribution patterns in this paper will be restricted to the Campanian and Maastrichtian time periods.

The oldest zone to be considered is the *Archaeoglobigerina cretacea* Zone, which extends from the LAD of marginotruncanids just above the base of the Campanian to the FAD of *Globigerinelloides impensus* in the lowermost upper Campanian. This zone is characterized by low-diversity assemblages of *Heterohelix, Globigerinelloides, Hedbergella,* and relatively common occurrences of the double-keeled species *A. cretacea.* Other double-keeled forms that occur more rarely are *Globotruncana linneiana* and *G. bulloides. Globotruncanita elevata* and other single-keeled taxa have never been observed in the high-latitude sections. The austral endemic taxon *Archaeoglobigerina australis* first appears near the top of the *A. cretacea* Zone.

The remainder of the upper Campanian is included in the *Globigerinelloides impensus* Zone. This zone has even lower species diversity than the *A. cretacea* Zone and is similarly dominated by biserial, planispiral, and low-trochospiral taxa. Ranging throughout the zone are the austral endemics *G. impensus* and *A. australis,* while two additional endemic species, *Hedbergella sliteri* and *Archaeoglobigerina mateola,* appear in the upper part of the zone. The only keeled taxon that has been reported in the *G. impensus* Zone is *G. linneiana,* which is very rare in the lower part of the zone and absent from the upper part. The LAD of *G. impensus* has been used to approximate the Campanian/Maastrichtian boundary, as this datum corresponds with the upper part of Subchron C33N at all sites where paleomagnetic data are available (Figure 3).

The extinction of *G. impensus* marks the base of the *G. havanensis* Zone, which was defined as a lower Maastrichtian partial range zone spanning from the LAD of *G. impensus* to the FAD of *A. mayaroensis* [*Huber,* 1990]. The endemic taxa *A. australis, A. mateola,* and *H. sliteri* are distinctive elements of the *G. havanensis* Zone, along with species of *Heterohelix* and *Globigerinelloides.* Three subzones are recognized in the *G. havanensis* Zone. Lowermost of these is the *A. australis* Subzone, which contains monotonous assemblages of endemic and long-ranging, cosmopolitan taxa but no keeled planktonic foraminifera. The FAD of the

austral taxon *Rugoglobigerina circumnodifer* marks the top of the *A. australis* Subzone and the base of the *R. circumnodifer* Subzone. This datum has been correlated with the middle of Subchron C32N at Sites 689 and 690, and the top of Subchron C32N at Site 700 [*Huber,* 1991*a*]. At the Maud Rise, two other double-keeled species, *Globotruncana subcircumnodifer* and juvenile forms of *G. arca,* occur near the top of the *R. circumnodifer* Subzone. These appearances are followed by the FAD of *Abathomphalus intermedius,* the nominate species of the *A. intermedius* Subzone. This datum occurs just below the FAD of *A. mayaroensis* in the middle of Subchron C31R [*Huber,* 1990].

The *A. mayaroensis* Zone ranges throughout the upper Maastrichtian, from the FAD of the nominate taxon to the FAD of *Eoglobigerina* spp. The relative frequency of endemic species declines toward the top of this zone, whereas keeled taxa remain conspicuously common. Included in the *A. mayaroensis* Zone are the *Globotruncanella petaloidea, Globigerinelloides subcarinatus,* and *Pseudotextularia elegans* subzones. The highest keeled species diversity occurs in the *G. petaloidea* Subzone. Marking the base of the *G. subcarinatus* Subzone is the FAD of *G. subcarinatus,* which has been correlated with lower Subchron C31N and the upper part of Subchron C31R in the austral paleomagnetic reference sections (Figure 3).

The last datum event recognized in the Maastrichtian is the FAD of *P. elegans,* which makes a very brief appearance in the southern high latitudes near the end of the Maastrichtian, within the upper part of Subchron C30N [*Huber,* 1990, 1991*b*]. This datum defines the base of the *P. elegans* Subzone.

BIOGEOGRAPHIC DISTRIBUTION PATTERNS

Austral Nannofossils

Nannofossil provincialism during the Maastrichtian was first noted by *Worsley and Martini* [1970], who documented the bipolar distribution of *Nephrolitus frequens* and the tropical-subtropical restriction of *Micula murus. Bukry* [1973] subsequently noted the absence of *Watznaueria barnesae* and the abnormally high abundance of *N. frequens* and *Kamptnerius magnificus* in assemblages from the Upper Cretaceous of DSDP Site 207 (Tasman Sea), New Zealand, and West Siberia. The first indication of provincialism in the early Late Cretaceous was the discovery of the high-latitude species *Seribiscutum primitivum* by *Thierstein* [1974]. In a later study, *Thierstein* [1981] identified 13 polar taxa based on the examination of 243 upper Campanian through upper Maastrichtian assemblages from outcrops and deep-sea drilling sites. These species included several (e.g., *Seribiscutum primitivum, Misceomarginatus pleniporus*) that were believed to be restricted to the Austral Realm during the later Cretaceous. Several other taxa (e.g., *Kamptnerius magnificus, Micula decussata*) were sig-

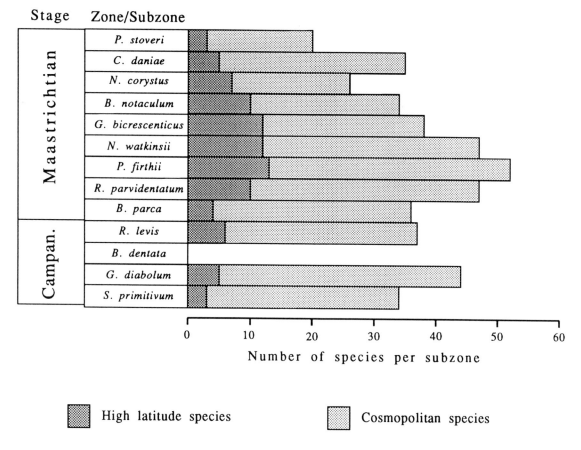

Fig. 6.   Distribution by subzone of maximized species richness in the Southern Ocean based on studies of sites 511, 689, 690, 747, and 750. The *Broinsonia dentata* Subzone is not included because it is represented only by poorly preserved (highly etched), sparse assemblages from Site 511. The near absence of this subzone characterizes the mid-Campanian Southern Ocean hiatus.

nificantly more abundant in high-latitude assemblages than in tropical-temperate ones. In addition, seven taxa were either absent or unusually rare in the Austral Realm during the Late Cretaceous according to *Thierstein* [1981].

The discovery and subsequent taxonomic description of many of the Upper Cretaceous austral taxa were the result of DSDP Leg 36 [*Wise and Wind*, 1977; *Wind*, 1979a, b]. *Wise* [1983] studied additional core material from the Falkland Plateau (DSDP Leg 71) and compiled a long list of "austral" taxa. These included species endemic to the Southern Ocean, species with bipolar distributions, and species that are significantly more abundant in the Southern Ocean relative to the tropical-temperate areas. *Pospichal and Wise* [1990], examining the Maastrichtian nannofossils from Maud Rise (Weddell Sea), found the assemblages similar to those on the Falkland Plateau in all respects except that the species richness of holococcoliths (Calyptrosphaeraceae) is much lower on Maud Rise than on the Falkland Plateau. This is most likely the result of the superior preservation

at the Falkland sites, as additional species have been found in the best preserved Maud Rise samples (J. J. Pospichal, personal communication, 1990). *Crux* [1991] documented the Upper Cretaceous succession on the northeast Georgia Rise (ODP Leg 114) and noted similarities with coeval boreal assemblages. *Wei and Thierstein* [1991] briefly described the Upper Cretaceous nannofossils from limestone recovered from Site 738 on the southern Kerguelen Plateau. *Watkins* [1992] documented the assemblages from the central Kerguelen Plateau (ODP Leg 120).

Examination of the nannofossil record of the Southern Ocean indicates a marked change in the degree of provincialism occurred during the Upper Cretaceous. The poor stratigraphic control in the Cenomanian through middle Turonian prevents analysis of this interval. The upper Turonian through Santonian is characterized by relatively few polar species (Figure 6). This interval has an average of approximately 4.3 polar species per zone. Of these, three species (*Repagalum parvidentatum*, *Biscutum dissimilis*, and *Seribiscutum*

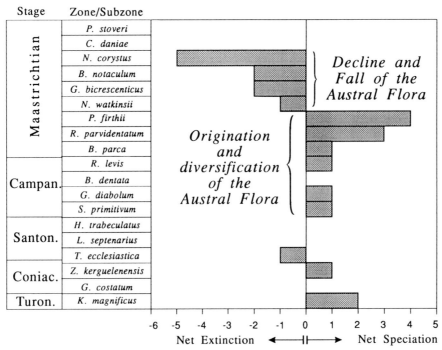

Fig. 7.   Speciation and extinction of the austral nannoflora by zone/subzone. The *Broinsonia dentata* Subzone is not included owing to lack of data.

primitivum) are known to occur in the northern hemisphere [*Burnett*, 1990; *Watkins and Liu*, 1992], indicating a bipolar distribution. The other taxa (*Thiersteinia ecclesiastica*, *Biscutum hattnerii*, and *Zeughrabdotus kerguelenensis*) have only been reported from Southern Ocean deep-sea sites and are assumed to be austral in distribution.

The number of polar nannofossils increases markedly beginning in the upper Campanian. Several taxonomic groups within the three nannofossil families Ahmuellerellaceae, Biscutaceae, and Podorhabdaceae gave rise to at least 14 species during the late Campanian through Maastrichtian. Within the Ahmuellerellaceae, four taxa arose within the *Monomarginatus-Misceomarginatus* complex. "*Neocrepidolithus*" *watkinsii* (which more probably should be assigned to the genus *Monomarginatus*; see *Watkins et al.* [1992]) was the root species for this lineage. Within the Biscutaceae, four species of *Biscutum* arose during the late Campanian and Maastrichtian. The extinctions of the two most common and distinctive forms, *B. coronum* and *B. magnum*, are used as biohorizons in the zonation. Five austral species evolve within the Podorhabdaceae during the Maastrichtian. *Teichorhabdus ethmos* is a podorhabdoidean species whose ancestral species is unknown at this time. Four species in the *Cribrosphaerella-Psyktosphaera-Nephrolithus* complex arise during the late Campanian to middle Maastrichtian. These species probably are derived from the australophilic *Cribrosphaerella ehren-*

bergii. *Nephrolithus frequens* was bipolar in distribution, at least during the upper part of its stratigraphic range. The others were apparently endemic to the Austral Realm throughout their duration.

Nephrolithus frequens is the last of the high-latitude species to appear (during the *Glaukolithus bicrescenticus* Subzone) in the Southern Ocean. Seven of the nine high-latitude species that were extant at the first appearance of *N. frequens* suffer extinction within the next two subzones. Only *N. frequens* and *Cribrosphaerella daniae* survive through the last two subzones prior to the Cretaceous-Tertiary boundary.

The evolutionary relationships within these lineages are still uncertain in some cases. Elucidation of these relationships must await more detailed ultrastructural work. However, it is clear that the rate of speciation within the *Cribrosphaerella-Psyktosphaera-Nephrolithis* complex was accelerated during the late Campanian to middle Maastrichtian. This was especially true during the early Maastrichtian, when the rate of speciation greatly exceeded the rate of extinction (Figure 7). Beginning in the *N. watkinsii* Zone, the rate of extinction exceeded the speciation rate, leading to the decline and fall of the austral nannoflora. By the *C. daniae* Subzone, the austral nannoflora had largely vanished. The nannofossil assemblages in the Southern Ocean reverted back to a depauperate subset of the coeval temperate assemblages, as they had been during the Turonian through early Campanian.

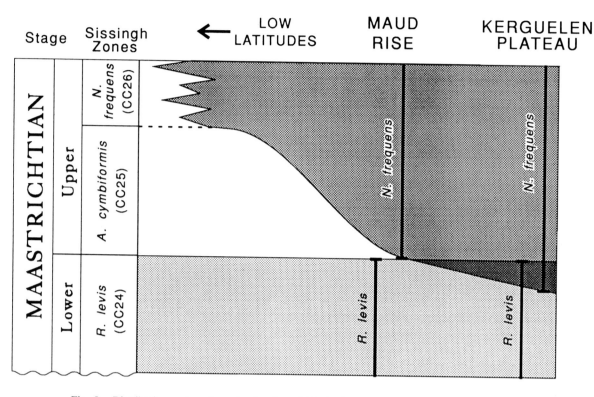

Fig. 8. Distribution and northward migration of *Nephrolithus frequens* during the Maastrichtian (modified from *Pospichal and Wise* [1990]).

Equatorward Migration. The diachroneity of *Nephrolithus frequens* has been known, in part, since *Worsley and Martini* [1970] first suggested significant nannofossil provincialism in the Upper Cretaceous ocean with their documentation of its bipolar distribution. *Wind* [1979a] found that *N. frequens* evolved from *N. corystus*, a form which was restricted to the Southern Ocean during its late Campanian to mid-Maastrichtian range. *Pospichal and Wise* [1990] documented the diachronous nature of the first occurrence of *N. frequens* relative to the LAD of *Reinhardtites levis*. They noted that the LAD of *R. levis* and the FAD of *N. frequens* occurred at approximately the same level at Maud Rise, indicating a significantly earlier first occurrence of *N. frequens* in the southern Atlantic. *Watkins* [1992] documented the overlap of these two species, indicating a greater degree of diachroneity for the FAD of *N. frequens* in the southern Indian Ocean (Figure 8). Unfortunately, preservation of the *Nephrolithus* complex at Site 700 is too poor to allow definition of the FAD of *N. frequens* in this critical area. It is clear, however, that the first occurrence of *N. frequens* is significantly younger in lower-latitude areas. The northward migration of *N. frequens* during the mid-Maastrichtian is probably related to the global cooling that was under way at that time.

Watznaueria barnesae is a nearly ubiquitous nanno-

fossil in Upper Cretaceous assemblages throughout the world. However, this species is generally rare or absent from Maastrichtian assemblages of the high latitudes. *Bukry* [1973] first recognized this in his analysis of high-latitude northern and southern assemblages. *Wind* [1979a, b] noted the inverse relationship of the distributions of *Micula decussata* versus *Watznaueria barnesae–Cyclagelosphaera margarelii*, the latter group being abundant in the tropics and absent or very rare in the austral ocean. Examination of several sections (Figure 9) indicates that there was a systematic pattern to the occurrence of this species in the Southern Ocean. The species became sporadic and rare in occurrence earlier at more southerly sites (e.g., *R. parvidentatum* Subzone at sites 747, 748, and 750) and later in more northerly sites (e.g., *M. watkinsii* Zone at sites 700 and 752). The species was absent for most of the late early and late Maastrichtian at the more southern sites. If this pattern of exclusion is interpreted as a temperature signal, it suggests that significant cooling of the Southern Ocean surface water mass began by at least the early Maastrichtian and was most pronounced during the late early Maastrichtian and the middle late Maastrichtian.

Poleward Migration. The only nannofossil species observed to have a poleward migration during the Late Cretaceous is *Watznaueria barnesae* (Figure 9). After disappearing from high-latitude sites during the early

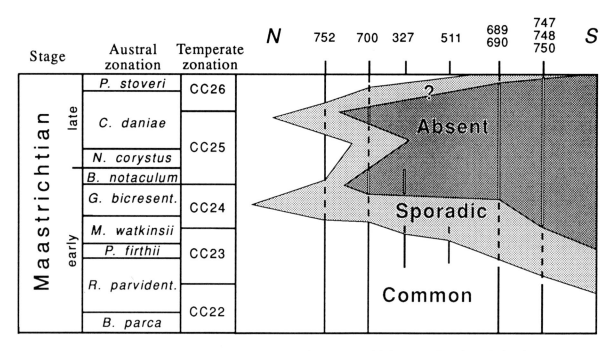

Fig. 9.   Distribution of *Watznaueria barnesae* during the Maastrichtian. Note the poleward migration at the end of the Maastrichtian after a prolonged period of absence from the southern high-latitude sites.

Maastrichtian, this species reappears within the *P. stoveri* Subzone of the latest Maastrichtian (uppermost Cretaceous sample at Site 690; uppermost meter of Cretaceous sediment at Site 750). This was about the same time that the planktonic foraminifer *Pseudotextularia elegans* migrated to the south polar region (see below). At the more northerly Site 700, *W. barnesae* is absent only from two intervals in the middle and upper Maastrichtian. Further to the north, *W. barnesae* was present throughout the Maastrichtian, although its occurrence was sporadic during parts of the late early Maastrichtian and late Maastrichtian. The southerly migration of *W. barnesae* apparently occurred during a brief warming event just prior to the Cretaceous-Tertiary boundary.

Austral Planktonic Foraminifera

General Trends. It is well understood that modern planktonic foraminiferal distributions generally parallel latitudinal climatic belts and the spatial configurations of surface water masses [*Bé*, 1977]. Five major faunal provinces are recognized in the modern oceans, including the tropical, subtropical, transitional, sub-Antarctic/sub-Arctic, and Antarctic/Arctic provinces. Maximum species diversity is in tropical waters, where a total of 36 species occur, whereas only six species have been found in polar waters [*Vincent and Berger*, 1981]. These biogeographic differences reflect latitudinal differences in the surface water habitat and the amount of vertical

niche space that could potentially be occupied by different groups of planktonic foraminifera.

Three biogeographic realms (originally called provinces) have been recognized in the Late Cretaceous southern hemisphere based on foraminiferal studies. The Austral Realm was first formally recognized by *Scheibnerova* [1971] based on the dominance of "cool water" agglutinated and calcareous benthic foraminifera and low species diversity of planktonic foraminiferal assemblages from Cretaceous nearshore marine facies in Australia, New Zealand, Madagascar, peninsular India, and southern South America. She considered the Austral Realm to be the southern hemisphere equivalent to the Boreal Realm. Scheibnerova's concept of the Austral Realm was based on the absence of Tethyan indicator taxa (e.g., the benthic Orbitoidacea and keeled planktonic taxa) rather than the presence of endemic forms. A Transitional Realm was similarly defined by *Scheibnerova* [1971] for foraminiferal assemblages "intermediate in species composition" between the Austral and Tethyan realms.

In his study of Cretaceous foraminifera from Site 327, *Sliter* [1977] more concisely defined the taxonomic character of the Tethyan, Transitional, and Austral realms for open ocean planktonic foraminiferal assemblages. Again, the Austral Realm was identified based on the absence of "thermophilic" taxa rather than the presence of endemic forms. Upper Cretaceous species considered to have been excluded from the Austral

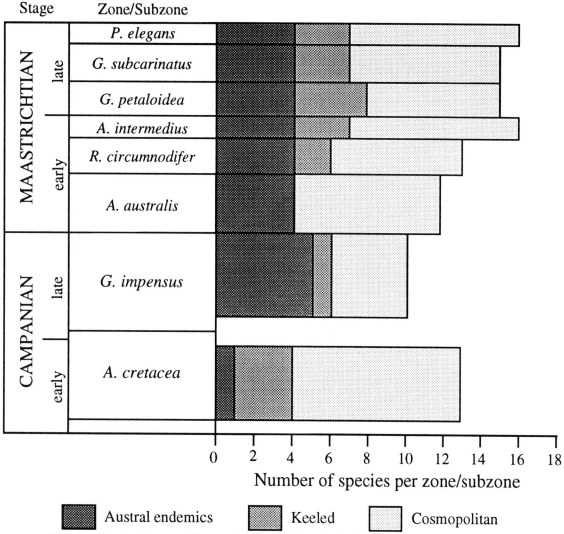

Fig. 10. Species richness of Campanian-Maastrichtian planktonic foraminifera for each Southern Ocean zone/subzone at sites 327, 511, 689, 690, 700, 747, and 750.

Realm include all species of *Globotruncanita, Gansserina, Contusotruncana, Plummerita, Trinitella, Racemiguembelina, Gublerina,* and *Pseudoguembelina,* as well as *Rugoglobigerina scotti, R. rugosa, Globotruncana subcircumnodifer,* and *Pseudotextularia elegans. Krasheninnikov and Basov*'s [1983, 1986] study of Site 511 basically echoed the conclusions of *Sliter* [1977], with the additional suggestion that the absence of keeled species from the upper Campanian–Maastrichtian cores resulted from a surface water cooling.

A much more refined understanding of Campanian-Maastrichtian biogeographic trends among the planktonic foraminifera has resulted from comparative study of the high-latitude DSDP and ODP sites [*Huber,* 1990, 1991*a, b,* 1992; *Quilty,* 1992]. Most importantly, five species were found to have a circumglobal distribution restricted to the southern high latitudes. This endemism

is first recognized in the upper lower Campanian by the FAD of *A. australis* and is amplified by the FADs of *G. impensus, H. sliteri,* and *A. mateola* in the upper Campanian and the FAD of *R. circumnodifer* in the upper lower Maastrichtian. The increased diversity and dominance of austral taxa within the *G. impensus* Zone and *A. australis* Subzone (Figure 10) indicate progressive biogeographic isolation of the southern high-latitude surface waters.

Latitudinal diversity gradients in the southern hemisphere were plotted by *Huber* [1992*a*] using keeled and total species diversity for the lower and upper intervals of the Campanian and Maastrichtian stages in an effort to delineate paleolatitudinal positions of the biogeographic boundaries. This revealed a gradual decrease in diversity at about 40°S during the early Campanian, sharp drops at about 38°S and 48°S during the late

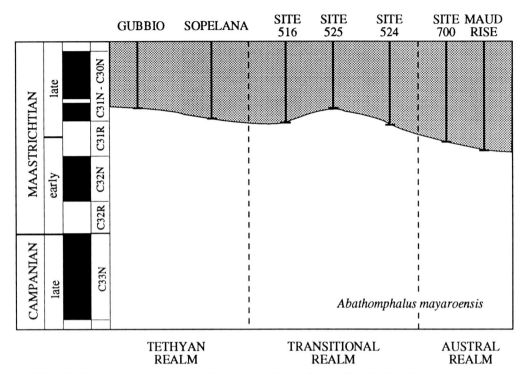

Fig. 11.   Equatorward migration of the keeled planktonic foraminifer *Abathomphalus mayaroensis* during the late Maastrichtian. See text for reference citations.

Campanian and early Maastrichtian, and sharp drops at about 43°S and 52°S during the late Maastrichtian. Huber concluded from this analysis that the Austral Realm is virtually indiscernible during early Campanian time but is well developed as a biogeographic entity during the late Campanian and Maastrichtian.

An additional observation from comparison of the high-latitude planktonic foraminiferal assemblages is that after an absence during most of late Campanian through early Maastrichtian time, keeled species reappear at the Southern Ocean sites during the late early Maastrichtian and become a significant component of the high-latitude assemblages during the late Maastrichtian [*Huber*, 1990, 1991*a*, *b*]. Keeled and total species diversity among the austral assemblages were highest at this time (Figure 10), largely as a result of poleward invasion of keeled and nonkeeled taxa from lower latitudes, as is discussed in the following section.

Equatorward Migration. Magnetobiostratigraphic correlation of the FAD of *Abathomphalus mayaroensis* demonstrates that this species underwent an equatorward migration during the late Maastrichtian (Figure 11). The earliest high-latitude occurrence of *A. mayaroensis* is within the middle of Subchron C31R at Maud Rise and Site 700 and is estimated at about 70.5 Ma [*Huber*, 1991*a*], whereas this datum was reported at the base of Subchron C31N at Hole 525A on the Walvis Ridge [*Boersma*, 1984], near the top of Subchron

C31R at Site 516 on the Rio Grande Rise [*Berggren et al.*, 1983], and in the middle of Subchron C31N in the equatorial Pacific [*Sliter*, 1989], Gubbio [*Premoli Silva*, 1977; *Monechi and Thierstein*, 1985], and Site 524 on the Walvis Ridge [*Poore et al.*, 1984]. More recently, *Mary et al.* [1991] identified the FAD of *A.* mayaroensis in the lower part of Subchron C31N in the Sopelana section of northern Spain.

Correlation between the magnetostratigraphically calibrated FADs of *A. mayaroensis* and the magnetic reversal time scale of *Kent and Gradstein* [1985] indicates an equatorward time-transgressive migration that took about 1.2 million years. On the other hand, the delayed first occurrence of this taxon in lower latitudes may be an artifact of the Signor-Lipps effect [*Signor and Lipps*, 1982], as this species is quite rare in tropical sections.

Poleward Migration. At least five species record migration from low to high latitudes during late early Maastrichtian time. The first of these are the keeled species *G. subcircumnodifer* and a dwarfed morphotype of *G. arca* (previously recorded as *G. bulloides* by *Huber* [1990, 1991*a*]), which both first appear within the *R. circumnodifer* Subzone in the southern high latitudes (Figures 12 and 13). The FAD of *G. subcircumnodifer* was recorded in the upper Campanian *Radotruncana calcarata* Zone at Site 465 [*Boersma*, 1981], at the base of the Maastrichtian at Site 305 [*Caron*, 1975]

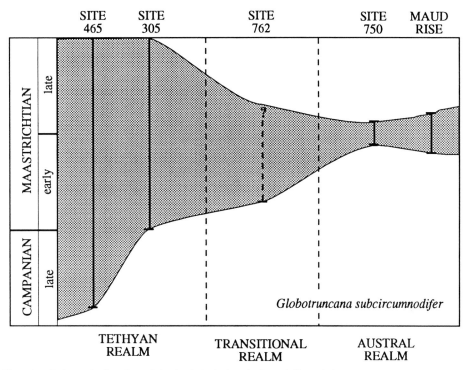

Fig. 12. Poleward migration of the keeled planktonic foraminifer *Globotruncana subcircumnodifer* during the late early Maastrichtian. See text for reference citations.

in the tropical Pacific, and in the early Maastrichtian at Site 762 [*Wonders*, 1992], which was at about 38°S in the Indian Ocean at about 74 Ma [*Scotese and Denham*, 1988]. *Globotruncana arca* first appeared in the latest Santonian or earliest Campanian in the Tethyan Realm of the Pacific Ocean [*Caron*, 1975; *Boersma*, 1981] and the Atlantic Ocean [*Weiss*, 1983] and in the Transitional Realm of the Indian Ocean [*Wonders*, 1992]. At Site 747 on the Kerguelen Plateau, *Quilty* [1992] recorded *G. arca* in one upper Campanian sample and again just below samples equivalent to the *R. circumnodifer* Subzone.

The delayed high-latitude appearances of the non-keeled species *G. petaloidea*, *G. subcarinatus*, and *P. elegans* have been found to be nearly synchronous within the Austral Realm [*Huber*, 1990, 1991*a*, *b*]. At all Southern Ocean sites, the FAD of *G. petaloidea* is at nearly the same level at the base of the upper Maastrichtian (Figure 14) and within Subchron C31R, whereas this datum has been recorded within the uppermost Campanian of the tropical Pacific [*Caron*, 1975; *Sliter*, 1989], the lowermost Maastrichtian in tropical Atlantic and European sections [*Caron*, 1985], and the lower Maastrichtian in the Transitional Realm at Site 524 [*Smith and Poore*, 1984]. The FAD of *G. subcarinatus* was found in the upper Campanian at Site 305 [*Caron*, 1975] and in samples correlated with Subchron

C32N in the lower Maastrichtian at Site 525 (B. T. Huber, unpublished data), but not until uppermost Subchron C31R, above the FAD of *G. petaloidea*, at the Maud Rise sites and Site 700 (Figure 15).

The most dramatic example of high-latitude diachroneity is demonstrated by *P. elegans* [sensu *Nederbragt*, 1989]. This species is reported to range into the lower Campanian [*Caron*, 1985; *Sliter*, 1989] and the Santonian [*Kassab*, 1978] in nearshore sequences of the Tethys, the upper Campanian at Tethyan sites 465 and 516 [*Boersma*, 1981; *Weiss*, 1983], and in the lower Maastrichtian at Transitional Realm sites in the Indian Ocean [*Hannah*, 1982] and South Atlantic [*Boersma*, 1984], whereas it does not appear in the circum-Antarctic region until near the end of the Maastrichtian (Figure 16), in the uppermost part of the *A. mayaroensis* Zone. It is important to note that *P. elegans* is also conspicuously absent from high northern latitude sections until the uppermost *A. mayaroensis* Zone [*Wicher*, 1953; *Berggren*, 1962; *Malmgren*, 1982].

The simultaneous poleward displacement of *P. elegans* and *W. barnsae* indicates that a brief warming event may have occurred in the south polar region during the latest Maastrichtian. This warming excursion was recorded in the oxygen isotope paleotemperature study of Site 690 by *Stott and Kennett* [1990] in samples estimated to be about 66.7 m.y. old.

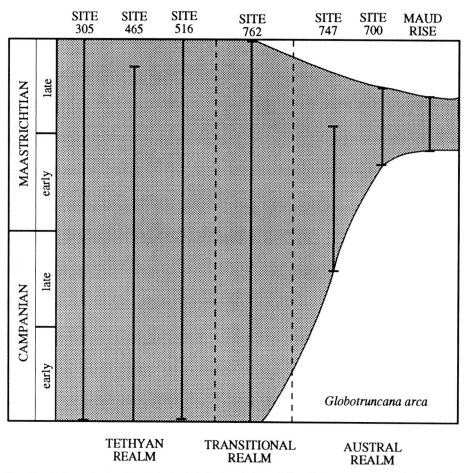

Fig. 13.   Poleward migration of the keeled planktonic foraminifer *Globotruncana arca* during the late early Maastrichtian. See text for reference citations.

PALEOGEOGRAPHIC AND PALEOCEANOGRAPHIC INFERENCES

Antarctic Paleogeography

Antarctica can be divided into two major physiographic features, including East Antarctica, which is a broad cratonic block extending from 35°W to 170°E, and West Antarctica, which is composed of a number of discrete continental blocks located between 50°W and 170°W (Figure 17). Geophysical studies indicate that the microcontinents in West Antarctica represent fragments of the Gondwana plate margin that were rearranged into their present configuration by translation and rotation by about 100 Ma [*Dalziel and Elliot*, 1982; *Watts et al.*, 1984]. Published paleogeographic reconstructions agree that Antarctica occupied a polar position throughout Late Cretaceous time [*Smith et al.*, 1981; *Barron*, 1987; *Lawver and Scotese*, 1987; *Lawver et al.*, 1991, this volume].

Our knowledge of the Cretaceous geography of Antarctica is poorly constrained, as most of the sedimentary and tectonic record is buried beneath the Antarctic ice sheet. In fact, the only outcrops of Upper Cretaceous sediments occur in the James Ross Island region of the northern Antarctic Peninsula (Figure 17). The thick succession of shallow marine sediments found in this area has yielded abundant and well-preserved fossil invertebrates, vertebrates, and plants ranging from Albian to early Tertiary in age. These were deposited in a retroarc basin (the James Ross Basin) on the eastern flank of an active magmatic arc that probably extended most of the length of the Antarctic Peninsula [*Farquharson*, 1982]. Subduction of Pacific crust beneath the magmatic arc proceeded throughout the Mesozoic and early Cenozoic history of the James Ross Basin, resulting in periodic phases of volcanic activity, plutonic emplacement, and uplift. Whereas plutonism and rates of uplift diminished during Late Cretaceous time, volcanic activity increased to a peak in the late Maastrichtian and Paleocene, with deposition of air fall debris in the James Ross Island region [*Macellari*, 1988; *Elliot*, 1988].

Despite the evidence for extensive tectonic activity

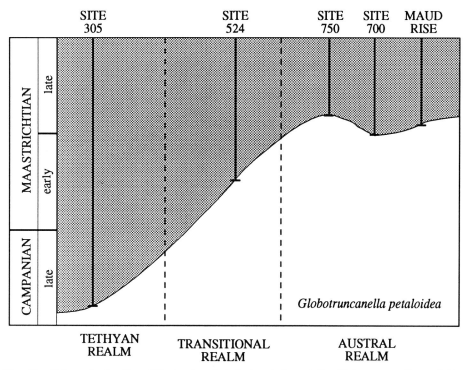

Fig. 14. Poleward migration of the trochospiral planktonic foraminifer *Globotruncanella petaloidea* during the late Maastrichtian. See text for reference citations.

during the late Mesozoic, it is not clear whether the Antarctic Peninsula existed as a continuous landmass connected with South America or as a series of emergent islands separated by broad, shallow seas. *Sliter* [1977, p. 537, Figure 15] suggested the presence of eastward flow through the Antarctic Peninsula and perhaps east of Ellsworth Land during the Late Cretaceous to explain the austral character of the sites 327 and 330 Cretaceous planktonic foraminifera. On the other hand, pollen [*Askin*, 1988, 1989] and fossil wood [*Francis*, 1986, 1991] found throughout the Campanian-Eocene sequence in the James Ross Basin indicate that at least some parts of the volcanic arc were above sea level and forested during that time.

A continuous land connection extending from southern South America to Australia was favored by *Woodburne and Zinsmeister* [1984] to explain dispersal of marsupials between these land areas during the Late Cretaceous and early Tertiary. This view was shared by several other authors in studying terrestrial fossils. Discovery of ankylosaurid dinosaur remains in Campanian sediments on James Ross Island led *Gasparini et al.* [1987] and *Olivero et al.* [1991] to suggest unimpeded terrestrial communication between South America and the Antarctic Peninsula, as these authors contend that it was not possible for ankylosaurs to cross water barriers. *Dettman* [1989] postulated that the southern beech *Nothofagus* and other floral elements may have origi-

nated in Antarctica during the Late Cretaceous and subsequently dispersed via terrestrial connections to South America, Australia, and New Zealand. *Case* [1988] and *Askin* [1989] further noted that *Nothofagus* dispersal required a continuous land connection across Antarctica since modern species of that genus are incapable of crossing all but the narrowest of water gaps.

Contrarily, *Briggs* [1987, p. 83] stated that terrestrial migrations from South America to Antarctica probably took place across a widely separated island chain rather than a continuous overland route. He pointed out that the presence of broad sea passages between South America and Australia would have effectively filtered out placental mammals, which are absent from Australia and New Zealand.

Evidence for the existence of trans-Antarctic seaways during the Cretaceous comes from several other sources. The most compelling among these is the discovery of reworked Late Cretaceous marine microfossils in glacial diamictites at a number of Antarctic localities in the Transantarctic Mountains and along the Antarctic continental margin [*Webb and Neall*, 1972; *Truswell*, 1983, 1987; *Webb et al.*, 1984; *Leckie and Webb*, 1985]. Occurrence of these microfossils implies that open marine conditions existed at least intermittently during the Late Cretaceous in sedimentary basins in and between East Antarctica and West Antarctica

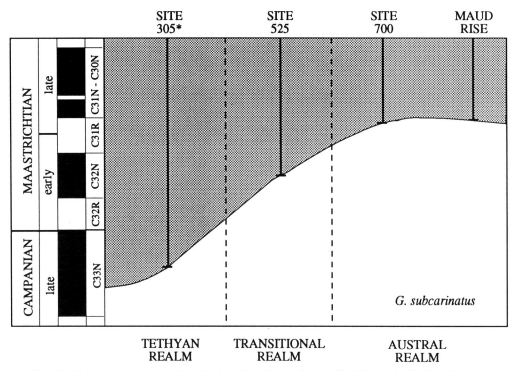

Fig. 15. Poleward migration of the planispiral planktonic foraminifer *Globigerinelloides subcarinatus* during the late Maastrichtian. The asterisk denotes the absence of paleomagnetic data at Site 305. See text for reference citations.

(Figure 17). These basins may have formed during an extensional phase of late Mesozoic rifting and graben downfaulting related to breakup of the Gondwana continents [*Bradshaw*, 1991; *Cooper et al.*, 1991; *Storey*, 1991].

A number of shallow marine invertebrate groups from southern South America, the Antarctic Peninsula, New Zealand, and southeast Australia shared strong taxonomic affinities during the Late Cretaceous [*Stevens*, 1980, 1989; *Zinsmeister*, 1979, 1982; *Macellari*, 1987; *Huber and Webb*, 1986; *Huber*, 1991*b*, 1992*a*; *Clarke and Crame*, 1989], This also indicates that shelfal marine communication existed between these regions, but the biogeographic record of shallow marine invertebrates is too imprecise to reveal the temporal duration and geographic extent of such shelfal seaways within Antarctica.

Circum-Antarctic Marine Gateways

Antarctic-Australian Margin. Analysis of seafloor magnetic anomalies between the rifted margins of Antarctica and Australia has revealed that spreading occurred in two distinct phases [*Cande and Mutter*, 1982]. Inception of a slow phase of rifting began during the early Cenomanian (about 95 Ma) and lasted until the middle Eocene (about 45 Ma), with a spreading rate averaging about 5 mm/yr [*Veevers*, 1987]. This was followed by

rapid northward drift of Australia and opening of a deep basin separating Tasmania and the Tasman Plateau from the north Victoria Land margin by about 35 Ma [*Veevers*, 1987]. Thus the final barrier to deep circulation between the southeast Indian and southwest Pacific oceans was removed by Eocene/Oligocene boundary time, enabling subsequent development of the Circum-Antarctic Current [*Kennett et al.*, 1975*b*].

Extensional tectonics during the Early Cretaceous phase of rifting between the Australia-Antarctic margin led to the formation of several large sedimentary basins, including the Otway, Bass, and Gippsland basins in southeast Australia (Figure 18), the Great Bight Basin in south central Australia, and the Bremer Basin in southwest Australia. Occurrence of in situ Aptian-Albian nonmarine siltstone on the continental rise near the George V Coast of East Antarctica [*Domack et al.*, 1980] indicates that at least one conjugate sedimentary basin had formed along the Antarctic margin during the earliest phase of rifting. Borehole studies and seismic profiling data indicate that a seaway extended from the Otway Basin westward along the southern margin of Australia and opened into the southeastern Indian Ocean by Cenomanian time [*Frakes et al.*, 1987]. This seaway progressively penetrated eastward toward the southwest Pacific as seafloor spreading continued between the Australian and Antarctic continents. By mid-

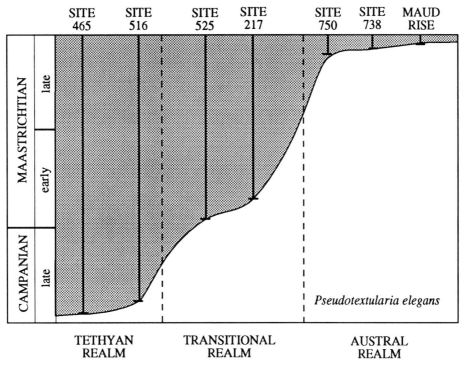

Fig. 16. Poleward migration of the biserial planktonic foraminifer *Pseudotextularia elegans* at the end of the Maastrichtian. See text for reference citations.

dle Campanian time, approximately 160 km of seafloor could have been generated between East Antarctica and southern Australia if a spreading rate of 5 mm/yr is assumed. But was there enough seafloor spreading and subsidence to allow unimpeded shallow marine commu-

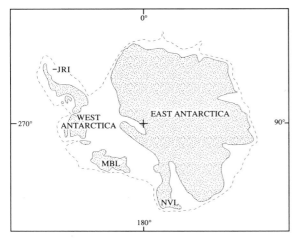

Fig. 17. Paleogeographic map of East Antarctica and West Antarctica showing areas inferred to have been flooded by shelfal seas during maximum highstands sea level during the Late Cretaceous. Modified from *Huber* [1992a]. JRI, James Ross Island; MBL, Marie Byrd Land; NVL, north Victoria Land.

nication along the entire Antarctic-Australian margin by this time?

A review of the patterns of sedimentation in the sedimentary basins of southeast Australia by *Deighton et al.* [1976] suggests that the marine influence was never strong in that region during the Late Cretaceous. Foraminiferal assemblages were reported only from the Belfast Mudstone in the Otway Basin, which was determined to range from Turonian to Santonian in age, and the maximum planktonic foraminiferal diversity was recorded as only three species [*Taylor*, 1964]. According to *Deighton et al.* [1976], open oceanic circulation in the region of the Otway Basin did not occur until middle to late Eocene time.

Although Australian plate motion was mostly perpendicular to the Antarctic margin, the Tasman Plateau and north Victoria Land margins separated more slowly because of a stronger strike-slip component in their relative plate motions. Determination of the spreading history and geology of this area is critical to paleoceanographic models for the Late Cretaceous. Two sites drilled in the vicinity of the Tasman Plateau during DSDP Leg 29, sites 280 and 281 (Figure 18), provide the oldest sediment record for this region. Unfortunately, Cretaceous sediments were not recovered, as the oldest sediments cored are middle to late Eocene in age [*Kennett et al.*, 1975a]. Site 280, which is the deeper of the two sites (4176-m depth), penetrated an uninter-

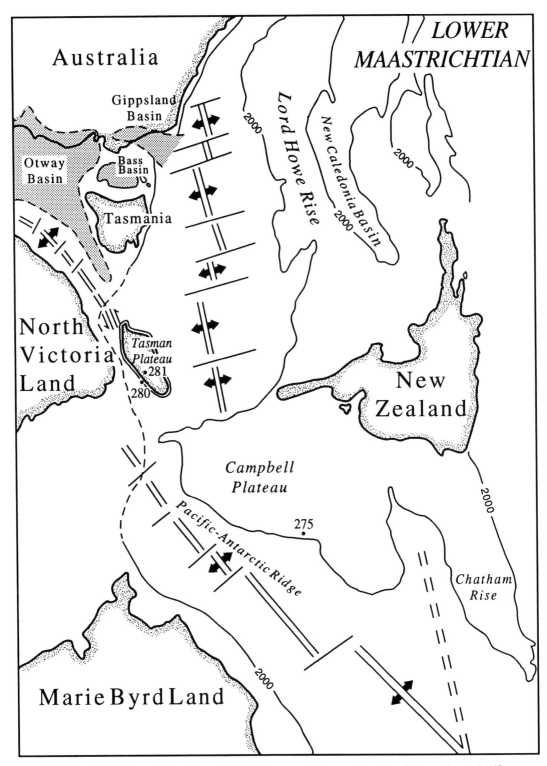

Fig. 18.   Early Maastrichtian (Anomaly 32) paleogeographic reconstruction of the southwest Pacific region showing the locations of sedimentary basins in southeast Australia and DSDP Leg 29 sites 280 and 281. Continental margins are taken to follow the 2000-m contour. Modified from *Kamp* [1986].

rupted sequence of late middle Eocene to early Oligocene terrigenous silts and clays underlain by oceanic basalt of uncertain age. The Eocene terrigenous clastic facies was probably deposited in a low-oxygen environment close to the CCD under the influence of sluggish bottom water circulation [*Kennett et al.*, 1975a]. If the CCD was no shallower than about 3200 m during the Eocene [*van Andel*, 1975], then the paleodepth of this site could have been about 2500–3000 m. Increase in the biogenic silica content near the Eocene/Oligocene boundary culminated with deposition of Oligocene siliceous oozes, reflecting intensification of open ocean circulation and development of upwelling as detrital deposition diminished [*Kennett et al.*, 1975b].

A fragmentary Paleogene record was recovered from Site 281 which was drilled in 1591 m below sea level on top of the Tasman Plateau. This sedimentary succession consists of upper Eocene glauconitic sandstone and biogenic glauconitic silty sands overlain by a short interval of lower Oligocene greensand and underlain by upper Paleozoic mica schist. Benthic foraminifera from the upper Eocene sediments indicate paleodepths no greater than 200–300 m [*Kennett et al.*, 1975b].

The shallow paleodepth of Site 281 during the Eocene indicates that marine inundation of the Tasman Plateau was unlikely during Late Cretaceous time. On the other hand, if the basalt underlying the much deeper Site 280 formed during the early phase of breakup, then this site, along with much of the continental slope along the southeast side of the Tasman Plateau and Tasmania, may well have been below sea level during the latest Cretaceous. Consistent with this view is the paleogeographic reconstruction for 70 Ma by *Lawver et al.* [this volume], who indicate that a shallow, narrow marine seaway had opened between Marie Byrd Land and the Tasman Plateau by early Maastrichtian time. According to these authors, the first complete marine connection separating East Antarctica and Australia was established at least 10 million years earlier between Tasmania and the Tasman Plateau. Lawver et al. state that deepwater circulation remained blocked in this region throughout the Late Cretaceous time.

The terrestrial fossil record on Australia and Antarctica is too poor to constrain when an epicontinental sea formed as a barrier to migration between Antarctica and Australia. The fossil record of marsupials only extends back to the middle Miocene in Australia [*Woodburne et al.*, 1985] and the Antarctic marsupial record is limited to Eocene sediments on Seymour Island [*Woodburne and Zinsmeister*, 1984]. The Cretaceous distribution of fossil *Nothofagus*, Protoaceae, and other plants in the circum-Antarctic region indicates that terrestrial migration routes between Tasmania and East Antarctica may have intermittently persisted until latest Cretaceous time [*Dettman*, 1989], but a more precise determination of when these routes disappeared is not possible based on present evidence.

Tasman Basin. Extension and rifting between New Zealand and Antarctica began during the Early Cretaceous at about 100 Ma [*Bradshaw*, 1991]. This marked the beginning of New Zealand's northward migration from 80°S at 95 Ma [*Oliver et al.*, 1979] to 62°S by 75 Ma [*Grindley et al.*, 1977]. Seafloor spreading between Australia and New Zealand was active from prior to anomaly 33 (76 Ma) in the Tasman Sea, causing the opening of another gateway for marine communication [*Stock and Molnar*, 1982]. The continental breakup history described by *Kamp* [1986] indicates that Late Cretaceous seafloor spreading in the Tasman Sea occurred at a faster rate than along the Pacific-Antarctic Ridge, leading to dextral motion along the Campbell Fault and a 25° counterclockwise rotation of the Campbell Plateau block (Figure 18).

Results from Site 275 of DSDP Leg 29 suggest that the southern Campbell Plateau was an area of open ocean circulation and significant bottom current activity during the late Campanian. Five cores containing well-preserved upper Campanian radiolarians and diatoms were recovered from this site in 2837-m water depth [*Kennett et al.*, 1975a]. Abundance of siliceous microfossils and absence of calcareous microfossils from those cores indicate sediment deposition in an area of high productivity below a relatively shallow CCD. Deepwater circulation apparently did not begin flowing between the Campbell Plateau and Marie Byrd Land and through the Tasman Sea until sometime after 70 Ma [*Lawver et al.*, this volume].

According to *Stevens* [1989], all land connections between New Zealand and Antarctica/Australia were severed by 85 Ma, but shallow marine seaways continued to link New Zealand to New Caledonia, Antarctica, and South America throughout Late Cretaceous time. *Dettman* [1989], on the other hand, suggested that terrestrial exchange across the Tasman Sea still existed during much of the Late Cretaceous since the *Nothofagus* pollen (*Nothofagidites fusca* group) appears at about the same time in New Zealand, South America, Antarctica, and Australia during the Maastrichtian. Nevertheless, published paleogeographic reconstructions of the southern hemisphere [e.g., *Smith et al.*, 1981; *Scotese and Denham*, 1988; *Lawver et al.*, 1991, this volume] suggest that New Zealand was completely isolated from the other southern hemisphere continents by early Campanian time.

South American–Antarctic Isthmus. A third gateway that opened as a result of breakup of the southern Gondwana continents was between South America and the Antarctic Peninsula. As was previously discussed, late Mesozoic tectonism in this region probably resulted in Early Cretaceous uplift of a magmatic arc and Late Cretaceous volcanism, resulting in an isthmus that may have been partly or wholly exposed above sea level. The Scotia Sea was closed and the South Orkney block and several smaller microcontinents remained juxta-

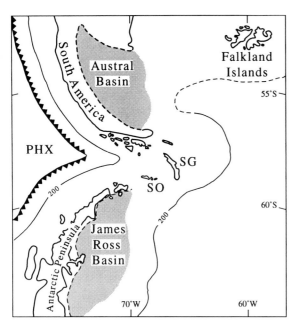

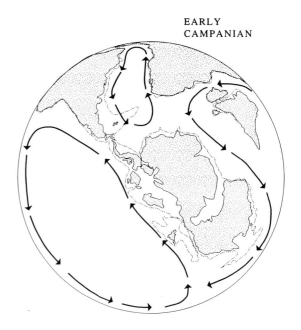

Fig. 19. Late Campanian paleogeographic reconstruction of the Antarctic Peninsula–southern South American region prior to opening of the Drake Passage and the Scotia Sea. PHX, Phoenix plate; SO, South Orkney block; SG, South Georgia block. Based on information from *Barker* [1982], *Riccardi* [1987], *Elliot* [1987], and *Toker et al.* [1991].

Fig. 20. Southern hemisphere surface circulation inferred for the early Campanian. Note that the oceanic communication gateways between Antarctica–South America, Antarctica, Australia, and New Zealand are closed and surface gyre configurations are broadly latitudinal. Paleogeographic reconstruction for 84 Ma from Terra Mobilis [*Scotese and Denham*, 1988], with modifications discussed in the text. Dashed contours represent shelf edge.

posed between southern South America and the Antarctic Peninsula prior to about 30 Ma [*Toker et al.*, 1991]. A shallow seaway probably developed in this region as South America began to slowly drift northward during the Late Cretaceous and early Tertiary [*Lawver and Scotese*, 1987]. It is unlikely, however, that a deepwater connection existed until the western Scotia Sea opened during the late Oligocene to Miocene [*Lawver et al.*, 1991, this volume].

Our late Campanian reconstruction (Figure 19) depicts eastward subduction of the Phoenix plate (PHX) beneath the Antarctic and South American plates [after *Barker*, 1982] and places microcontinents from the Scotia Arc in their inferred Late Cretaceous positions. Shallow marine communication between the South Atlantic and southeast Pacific Ocean basins may have occurred at this time. Elevation of a more continuous volcanic arc may have followed during a late Maastrichtian-Paleocene phase of intensified volcanic activity. This would have led to renewal of the terrestrial link between South America and Antarctica.

Surface Circulation

Accurate reconstruction of surface water gyres during the geologic past requires detailed knowledge of the relative positions of land and sea and the timing of the opening or closing of marine gateways that may have influenced changes in the circulation patterns. Unfortu-

nately, geologic information bearing on the evolution of the critical communication gateways between southern South America, Antarctica, and Australia during Late Cretaceous time is unavoidably incomplete because of more recent subduction and erosion episodes. Depiction of changes in the pattern of surface circulation that occurred during the Late Cretaceous phase of Gondwana breakup is therefore rather speculative. Nonetheless, we have modified the *Scotese and Denham* [1988] Terra Mobilis paleogeographic reconstruction maps of the southern hemisphere for 84 Ma, 79 Ma, and 70 Ma by incorporating information from several plate tectonic models [e.g., *Barker*, 1982; *Barker and Lawver*, 1988; *Kamp*, 1986; *Veevers*, 1986, 1987; *Lawver et al.*, this volume] and the Cretaceous sea level curve of *Haq et al.* [1987]. These reconstructions are shown in Figures 20–22 and discussed below.

Early Campanian. Although rifting between the continental margins of Australia and East Antarctica probably began during the Cenomanian, there is no sedimentologic evidence to accurately constrain when unimpeded oceanic circulation first separated the two continents. The history of this gateway opening is determined primarily from geophysical evidence of plate motion. The 90 Ma reconstruction of *Lawver et al.* [this volume] depicts an enclosed seaway penetrating eastward to Tasmania bordered by a continuous land

LATE CAMPANIAN-
EARLY MAASTRICHTIAN

LATE
MAASTRICHTIAN

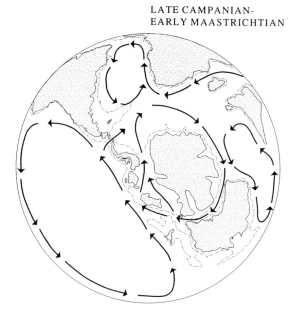

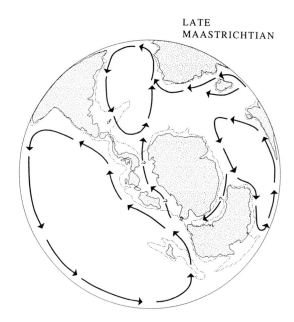

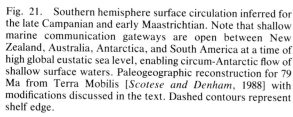

Fig. 21. Southern hemisphere surface circulation inferred for the late Campanian and early Maastrichtian. Note that shallow marine communication gateways are open between New Zealand, Australia, Antarctica, and South America at a time of high global eustatic sea level, enabling circum-Antarctic flow of shallow surface waters. Paleogeographic reconstruction for 79 Ma from Terra Mobilis [*Scotese and Denham*, 1988] with modifications discussed in the text. Dashed contours represent shelf edge.

Fig. 22. Southern hemisphere surface circulation inferred for the late Maastrichtian. Note that shallow seaways between Antarctica, Australia, and New Zealand continue to slowly widen, but uplift of a postulated South American–Antarctic Peninsula isthmus and a sea level fall have restricted surface communication between the South Atlantic and southwest Pacific Ocean basins. Paleogeographic reconstruction for 70 Ma from Terra Mobilis [*Scotese and Denham*, 1988] with modifications discussed in the text. Dashed contours represent shelf edge.

bridge connecting southeast Australia with East Antarctica. This land bridge was interrupted by a narrow passage between Tasmania and the Tasman Plateau by 80 Ma according to their model. A terrestrial connection between South America and the Antarctic Peninsula also probably existed during the early Campanian.

Earliest Campanian calcareous plankton distributions are consistent with evidence for relatively low latitudinal thermal gradients and longitudinal mixing of surface currents. During this time, the latitudinal species diversity gradients of calcareous nannoplankton and planktonic foraminifera were low, and provincialism among southern high-latitude assemblages was virtually nonexistent. Consequently, we suggest that early Campanian gyral circulation was weak and broadly latitudinal in the South Atlantic, poorly developed in the newly forming Indian Ocean, and very broad in the South Pacific (Figure 20).

Late Campanian–Early Maastrichtian. By middle Campanian time, seafloor spreading had opened shallow marine gateways between the Campbell Plateau, Marie Byrd Land, and the Tasman Sea, as well as between South America and Antarctica. We propose that this led to development of a shallow proto-circum-Antarctic current and caused changes in the southern high-latitude calcareous plankton distributions of this time. These

changes include progressive dominance by Austral endemic taxa (Figures 6, 10), impoverishment of keeled planktonic taxa, and increased latitudinal species diversity gradients with sharp drops in total and keeled planktonic foraminiferal diversity occurring between about 38°S and 52°S paleolatitude [*Huber*, 1992*a*]. We postulate that an oceanic front formed by circum-Antarctic flow of surface waters began to separate Austral Realm assemblages from subtropical assemblages of the Transitional Realm in all of the southern hemisphere ocean basins by late Campanian time. This water mass boundary has been previously postulated for late Campanian-Maastrichtian time in the southern South Atlantic by several authors [*Ciesielski et al.*, 1977; *Macellari*, 1985; *Huber*, 1992*a*]. The remarkably low species diversity and absence of deeper-dwelling taxa among the circum-Antarctic planktonic foraminiferal assemblages at this time suggest that surface waters within the Austral Realm were highly convective and poorly stratified [*Huber*, 1992*a*]. Another result of the opening of circum-Antarctic gateways may have been sediment erosion or nondeposition at shallow and intermediate sites under the influence of the proto-circum-Antarctic current. This would explain the presence of the Southern Ocean hiatus in widely separated deep-sea sites of the southern South Atlantic and southern Indian oceans.

According to *Haq et al.* [1987], the second highest stand of global eustacy during the Cretaceous was reached during the middle Campanian. This would have caused maximum flooding of shelfal basins within Antarctica, forming a continuous migration route for shallow marine taxa that originated in different regions of the remnant Gondwana continents. In addition, this would account for the strong taxonomic affinities shared among late Campanian–Maastrichtian nearshore benthic foraminifera [*Huber and Webb*, 1986; *Huber*, 1992a], ammonites [*Macellari*, 1987], and other molluscs [*Zinsmeister*, 1979, 1982] from southern South America, the Antarctic Peninsula, and New Zealand.

Late Maastrichtian. Slow widening of the Antarctic-Australian, Tasman, and South American–Antarctic gateways continued through the late Maastrichtian and into the Paleogene without the development of deep ocean passages [*Lawver et al.*, this volume]. Austral calcareous plankton assemblages, however, do not bear evidence of increased biogeographic isolation. Contrarily, austral provincial taxa become less common during the late early through late Maastrichtian, and several planktonic foraminifera and at least one calcareous nannoplankton species undergo poleward migrations. Are these assemblage changes the result of a climatic warming or some other factor or combination of factors that controlled the patterns of surface circulation?

Oxygen isotopic studies of monospecific assemblages of benthic and planktonic foraminifera have revealed that Antarctic surface and intermediate waters cooled by about 0.5 to 0.75‰ (2° to 3°C) from latest Campanian to late Maastrichtian time [*Barrera et al.*, 1987; *Barrera and Huber*, 1990]. This cooling trend has been substantiated by oxygen isotopic data obtained from pelagic carbonate sediments in the tropical Pacific [*Douglas and Savin*, 1975] and analyses of the upper Campanian–Maastrichtian record of terrestrial plants in polar regions [e.g., *Frederiksen*, 1989; *Askin*, 1989; *Francis*, 1986, 1991]. Furthermore, a global climatic cooling has been cited as an explanation for the time transgressive equatorward migrations observed for *Nephrolithus frequens* [*Worsely*, 1974; *Wise*, 1988; *Pospichal and Wise*, 1990] and *Abathomphalus mayaroensis* [*Huber*, 1990, 1992a]. The overall decrease in taxonomic diversity among austral calcareous nannoplankton (Figure 6) can also be attributed to high-latitude cooling. However, poleward migration of planktonic foraminifera during the late early Maastrichtian and late Maastrichtian and a significant increase in the number of keeled and non-keeled species at this time conflict with this for high-latitude cooling. Explanation for the apparent discrepancy between this biogeographic trends may be drawn from differences in the depth distribution of planktonic foraminifera and calcareous nannoplankton.

A number of studies have revealed that planktonic foraminifera undergo a depth migration during their life cycle, with different groups attaining different depth levels in the water column prior to gametogenesis [*Bé*, 1977, 1980; *Bé et al.*, 1985; *Fairbanks et al.*, 1980; *Deuser et al.*, 1981; *Deuser*, 1987; *Hemleben et al.*, 1989]. The maximum depths reached during the life of a planktonic foraminifera depend largely on the shell buoyancy, the duration of the reproductive cycle, and the vertical structure and physico-chemical characteristics of the water column. Because polar surface waters have a shallower photic zone, greater seasonal temperature and salinity variation, and are more poorly stratified than in tropical regions, polar planktonic foraminiferal assemblages are concentrated in a narrower zone of the upper water column than tropical assemblages.

The equator-to-pole reduction in planktonic species diversity and poleward decrease in the number of deeper dwelling keeled globotruncanids or other coarsely ornamented forms during the Late Cretaceous can be explained by a concomitant shallowing or habitable niche space in the surface waters of the Cretaceous oceans [*Huber*, 1992a]. Convective mixing of the circum-Antarctic surface layer may have been most pronounced during the late Campanian and early Maastrichtian when deeper dwelling morphotypes were virtually absent from the southern high latitudes. The subsequent poleward migration of keeled and nonkeeled taxa during the late early Maastrichtian and late Maastrichtian may have occurred as the circum-Antarctic surface waters became more depth stratified and the entire water column cooled; the increased vertical partitioning of the surface waters would have provided additional habitable niche space enabling reproduction of deeper dwelling taxa. Enhanced vertical stratification would not have affected the austral nannoflora distributions, since these organisms were probably concentrated in the uppermost part of the surface waters and their horizontal distributions were primarily influenced by temperature.

We postulate that the change in the vertical structure of the Antarctic surface waters resulted from a reemergence of the South America–Antarctic Peninsula isthmus during middle to late Maastrichtian time. This emergence would have resulted from a fall in the global eustatic sea level [*Haq et al.*, 1987] and/or a renewed phase of tectonism along the Antarctic Peninsula magmatic arc, as evidenced by increased deposition of volcanic air fall debris in upper Maastrichtian sediments on Seymour Island [*Macellari*, 1988; *Elliot*, 1988]. Short-term lowstands of eustatic sea level during the middle and late Maastrichtian [*Haq et al.*, 1987] may have also restricted marine communication along the southern margin of Australia, enabling selective dispersal of marsupials and terrestrial plants from southern South America, across West Antarctica and East Antarctica to Australia. The interruption in circum-Antarctic flow of surface waters would have caused a

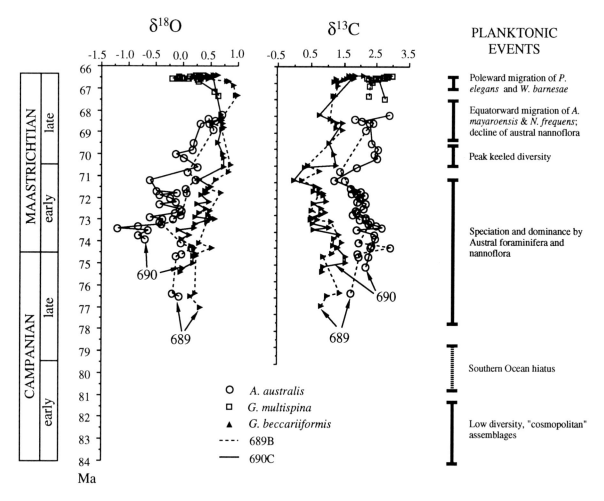

Fig. 23.   Oxygen and carbon isotopic data from Maud Rise sites 689 and 690 [from Barrera and Huber, 1990] and changes in Campanian-Maastrichtian nannofloral and planktonic foraminiferal assemblages at the Southern Ocean sites. Note that *A. australis* and *G. multispina* are planktonic species, whereas *G. beccariiformis* is a benthic species.

northward deflection of gyral flow in the southeast Pacific, more sluggish meridional flow in the southern Indian Ocean, and limited shelfal communication between East Antarctica and West Antarctica (Figure 22). Weakening of the Southern Ocean surface water gyres could have then led to enhancement of vertical stratification.

Oxygen and carbon isotopic analyses of planktonic and benthic foraminifera from Maud Rise sites 689 and 690 (Figure 23) do not support or repudiate the inferred enhancement of surface water stratification during the middle and late Maastrichtian. The difference between planktonic and benthic oxygen isotopic values would be expected to increase as surface waters become more stratified, while carbon isotopic values might decrease as surface water mixing and primary productivity diminish. However, the Maud Rise data do not show any clear trends in vertical isotopic gradients. The carbon

isotopic values of the uppermost surface water dweller *A. australis* are consistently 1.05‰ heavier than the benthic foraminifer *Gavelinella beccariiformis* throughout the Maastrichtian at Site 689, and this gradient only slightly increases at Site 690 from an average of 1.28‰ during the early Maastrichtian to a mean value of 1.38‰ during the late Maastrichtian [*Barrera and Huber*, 1990]. The difference in oxygen isotopic values between *A. australis* and *G. beccariiformis* increases by about 0.3‰ at both Maud Rise sites from the early to late Maastrichtian, whereas *G. multispina* shows a decrease by about 0.15‰ in the oxygen isotopic gradient relative to *G. beccariiformis* from early to late Maastrichtian time. Analyses of additional planktonic species are needed to determine whether isotopic gradients within the surface water mass show any significant changes from late Campanian to late Maastrichtian time.

Sedimentologic evidence for changes in the vertical

structure of Maastrichtian surface waters in the circum-Antarctic region may be inferred from the lithostratigraphic changes that occur at ODP sites 689, 698, 747, and 750. All of these sites have a greater concentration of chert in the upper Campanian–lower Maastrichtian interval than in the upper Maastrichtian sections. Assuming that the main source of silica for the chert was biogenic, we can speculate that the chert-rich levels could represent times of intensified surface water mixing leading to upwelling of nutrients and higher silicons microfossil productivity. However, this argument is weakened by the fact that a number of other ODP sites in the Southern Ocean do not show a parallel lithologic trend.

CONCLUSIONS

Our reconstructions of southern high-latitude paleogeography and paleoceanography during Campanian-Maastrichtian time are admittedly conjectural, as geologic information from the key oceanographic gateways around Antarctica are either buried beneath the Antarctic ice cap, subducted beneath the Pacific margin of Antarctica, or lost during the final breakup of the southern Gondwana continents. Nevertheless, we feel that these reconstructions provide the best explanation for the biogeographic distribution patterns that have been recorded from the terrestrial, shallow marine, and deep-sea realms in the circum-Antarctic region. As the time interval from the earliest Campanian to the latest Maastrichtian spans about 18 million years, a variety of tectonic, eustatic, and climatic changes could have operated at different rates and magnitudes to dramatically change corridors for terrestrial and marine dispersal.

It is clear from the terrestrial fossil record that an overland migration route connecting southern South America, West Antarctica and East Antarctica, and Australia must have existed during the Late Cretaceous. We propose that this migration route existed during the early Campanian but was severed during late Campanian and early Maastrichtian time as rifting progressed between the southern margin of Australia and East Antarctica and during a period of high global eustatic sea level. This phase of submergence of the continental margin of East Antarctica and much of West Antarctica could have led to development of a shallow to intermediate water proto-circumpolar current, causing biogeographic isolation of austral planktonic assemblages and, perhaps, erosion of sediments deposited at bathyal and shallower depths in the region of the Southern Ocean.

Renewed tectonic activity along the Antarctic Peninsula magmatic arc during the middle and late Maastrichtian and a global drop in sea level may have led to reemergence of a South American–Antarctic Peninsula isthmus and selective dispersal of terrestrial organisms across the southern Gondwana continents. Shelfal marine communication between the South Atlantic and Pacific and along the East Antarctic–Australian margins would have been restricted during maximum phases of island arc tectonism and marine regression. This could have resulted in weakened meridional surface flow between the South Atlantic and Indian ocean basins and enhancement of vertical stratification of the southern high-latitude surface waters. Consequently, deeper dwelling foraminiferal taxa previously living in lower latitudes could have migrated poleward to occupy newly available niches at deeper levels in the surface water mass.

The end of the Maastrichtian is marked by a brief but significant warming event in high latitudes, as evidenced by the poleward migration of *P. elegans* in both hemispheres, the poleward migration of *W. barnsae*, and an oxygen isotopic warming identified at about 66.7 Ma in Hole 690C on Maud Rise.

Acknowledgments. Our thanks are extended to Bill Sliter and Wuchang Wei for carefully reviewing this paper and providing constructive advice, Jim Kennett for his additional helpful comments and suggestions, L. Lawver for sending us a preprint of his contribution to this volume, and Chris Hamilton for his help and adeptness with the computer graphics. We would also like to thank JOI/USSAC and NSF Division of Polar Programs for sponsoring the Santa Barbara workshop and Jim Kennett for organizing a very enjoyable and stimulating meeting.

REFERENCES

Askin, R. A., Campanian to Eocene palynological succession of Seymor and adjacent islands, northeastern Antarctic Peninsula, in Geology and Paleontology of Seymour Island, Antarctic Peninsula, *Mem. Geol. Soc. Am.*, *169*, 131–153, 1988.

Askin, R. A., Endemism and heterochroneity in the Late Cretaceous (Campanian) to Paleocene palynofloras of Seymour Island, Antarctica: implications for origins, dispersal and palaeoclimates of southern floras, *Origins Evol. Antarct. Biota*, *47*, 107–119, 1989.

Barker, P. F., The Cenozoic subduction history of the Pacific margin of the Antarctic Peninsula: Ridge crest-trench interactions, *J. Geol. Soc. London*, *139*, 787–801, 1982.

Barker, P. F., and L. A. Lawver, South American–Antarctic plate motion over the past 50 Myr, and the evolution of the South American–Antarctic ridge, *Geophys. J.*, *94*, 377–386, 1988.

Barrera, E., and B. T. Huber, Evolution of Antarctic waters during the Maastrichtian: Foraminifer oxygen and carbon isotope ratios, ODP Leg 113, *Proc. Ocean Drill. Program Sci. Results*, *113*, 813–823, 1990.

Barrera, E., B. T. Huber, S. M. Savin, and P. N. Webb, Antarctic marine temperatures: Late Campanian through early Paleocene, *Paleoceanography*, *2*, 21–47, 1987.

Barron, E. J., Global Cretaceous paleogeography—International Geologic Correlation Program project 191, *Palaeogeogr. Palaeoclimatol. Palaeoecol.*, *59*, 207–216, 1987.

Bé, A. W. H., An ecological, zoogeographic and taxonomic review of recent planktonic foraminifera, in *Oceanic Micropaleontology*, vol. 1, edited by A. T. S. Ramsay, pp. 1–100, Academic, San Diego, Calif., 1977.

Bé, A. W. H., Gametogenic calcification in a spinose plank-tonic foraminifer *Globigerinoides sacculifer* (Brady), *Mar. Micropaleontol.*, *5*, 283–310, 1980.

Bé, A. W. H., J. K. B. Bishop, M. S. Sverdlove, and W. D. Gardner, Standing stock, vertical distribution and flux of planktonic foraminifera in the Panama Basin, *Mar. Micropaleontol.*, *9*, 307–333, 1985.

Berggren, W. A., Some planktonic foraminifera from the Maestrichtian and type Danian stages of southern Scandinavia, *Stockholm Contrib. Geol.*, *9*, 1–106, 1962.

Berggren, W. A., and C. D. Hollister, Paleogeography, paleobiogeography and the history of circulation in the Atlantic Ocean, *Stud. Paleoceanogr.*, *20*, 126–186, 1974.

Berggren, W. A., and C. D. Hollister, Plate tectonics and paleocirculation-commotion in the ocean, *Tectonophysics*, *11*, 11–48, 1977.

Berggren, W. A., N. Hamilton, D. A. Johnson, C. Pujol, W. Weiss, P. Cepek, and A. M. Gombos, Jr., Magnetobiostratigraphy of Deep Sea Drilling Project Leg 72, sites 515–518, Rio Grande Rise (South Atlantic), *Initial Rep. Deep Sea Drill. Proj.*, *72*, 939–947, 1983.

Boersma, A., Cretaceous-Tertiary foraminifers from Deep Sea Drilling Project Leg 62 sites in the central Pacific, *Initial Rep. Deep Sea Drill. Proj.*, *62*, 377–396, 1981.

Boersma, A., Cretaceous-Tertiary planktonic foraminifers from the southeastern Atlantic, Walvis Ridge area, Deep Sea Drilling Project Leg 74, *Initial Rep. Deep Sea Drill. Proj.*, *74*, 501–523, 1984.

Bradshaw, J. D., Cretaceous dispersion of Gondwana: Continental and oceanic spreading in the south-west Pacific-Antarctic sector, in *Geological Evolution of Antarctica*, edited by M. R. A. Thomson, J. A. Crame, and J. W. Thomson, pp. 581–585, Cambridge University Press, New York, 1991.

Briggs, J. C., *Biogeography and Plate Tectonics*, *Dev. in Palaeontol. and Stratigr.*, vol. 10, pp. 1–204, Elsevier, New York, 1987.

Bukry, D., Coccolith and silicoflagellate stratigraphy, Tasman Sea and southwestern Pacific Ocean, Deep Sea Drilling Project, Leg 21, *Initial Rep. Deep Sea Drill. Proj.*, *21*, 885–893, 1973.

Burnett, J., A new nannofossil zonation scheme for the Boreal Campanian, *Int. Nannoplankton Assoc. Newsl.*, *12*, 67–70, 1990.

Cande, S. C., and J. C. Mutter, A revised identification of the oldest seafloor spreading anomalies between Australia and Antarctica, *Earth Planet. Sci. Lett.*, *58*, 151–160, 1982.

Caron, M., Late Cretaceous planktonic foraminifera from the northwestern Pacific: Leg 32 of the Deep Sea Drilling Project, *Initial Rep. Deep Sea Drill. Proj.*, *32*, 719–724, 1975.

Caron, M., Cretaceous planktonic foraminifera, in *Plankton Stratigraphy*, edited by H. M. Bolli, J. B. Saunders, and K. Perch-Nielsen, pp. 17–86, Cambridge University Press, New York, 1985.

Case, J. A., Paleogene floras from Seymour Island, Antarctic Peninsula, in Geology and Paleontology of Seymour Island, Antarctic Peninsula, *Mem. Geol. Soc. Am.*, *169*, 523–530, 1988.

Ciesielski, P. F., W. V. Sliter, F. H. Wind, and S. W. Wise, Jr., Paleoenvironmental analysis and correlation of a Cretaceous *Islas Orcadas* core from the Falkland Plateau, southwest Atlantic, *Mar. Micropaleontol.*, *2*, 27–34, 1977.

Ciesielski, P. F., et al., Leg 114, *Proc. Ocean Drill. Program Initial Rep.*, *114*, 815 pp., 1988.

Clarke, A., and J. A. Crame, The origin of the Southern Ocean marine fauna, in *Origins and Evolution of the Antarctic Biota*, *Spec. Publ. 47*, edited by J. A. Crame, pp. 253–268, Geological Society of London, London, 1989.

Cooper, A. K., F. J. Davey, and K. Kinz, Crustal extension and origin of sedimentary basins beneath the Ross Sea and Ross Ice Shelf, Antarctica, in *Geological Evolution of Antarctica*, edited by M. R. A. Thomson, J. A. Crame, and J. W. Thomson, pp. 285–291, Cambridge University Press, New York, 1991.

Cranwell, L. M., Antarctica: Cradle or grave for its *Nothofagus*?, in *Ancient Pacific Floras, the Pollen Story*, edited by L. M. Cranwell, pp. 87–93, University of Hawaii Press, Honolulu, Hawaii, 1964.

Crux, J. A., Calcareous nannofossils recovered by Leg 114 in the subantarctic South Atlantic Ocean, *Proc. Ocean Drill. Program Sci. Results*, *114*, 155–177, 1991.

Dalziel, I. W. D., and D. H. Elliot, West Antarctica: Problem child of Gondwanaland, *Tectonics*, *1*, 3–19, 1982.

Deighton, I., D. A. Falvey, and D. J. Taylor, Depositional environments and geotectonic framework, southern Australian continental margin, *APEA J.*, *16*, 25–36, 1976.

Dettman, M. E., Antarctica: Cretaceous cradle of austral temperate rainforests?, in *Origins and Evolution of the Antarctic Biota*, *Spec. Publ. 47*, edited by J. A. Crame, pp. 89–105, Geological Society of London, London, 1989.

Deuser, W. G., Seasonal variations in isotopic composition and deep-water fluxes of the tests of perennially abundant planktonic foraminifera of the Sargasso Sea: Results from sediment-trap collections and their paleoceanographic significance, *J. Foraminiferal Res.*, *17*, 14–27, 1987.

Deuser, W. G., C. Hemleben, and M. Spindler, Seasonal changes in species composition, numbers, size, mass, and isotopic composition of planktonic foraminifera settling into the deep Sargasso Sea, *Palaeogeogr. Palaeoclimatol. Palaeoecol.*, *33*, 103–127, 1981.

Domack, E. M., W. W. Fairchild, and J. B. Anderson, Lower Cretaceous sediment from the East Antarctic continental shelf, *Nature*, *287*, 625–626, 1980.

Douglas, R. G., and S. M. Savin, Oxygen and carbon isotope analyses of Tertiary and Cretaceous microfossils from the Shatsky Rise and other sites in the North Pacific Ocean, *Initial Rep. Deep Sea Drill. Proj.*, *32*, 509–520, 1975.

Elliot, D. H., Tectonic setting and evolution of the James Ross Basin, northern Antarctic Peninsula, in Geology and Paleontology of Seymour Island, Antarctic Peninsula, *Mem. Geol. Soc. Am.*, *169*, 541–555, 1988.

Fairbanks, R. G., P. H. Wiebe, and A. W. H. Bé, Vertical distribution and isotopic composition of living planktonic foraminifera in the western North Atlantic, *Science*, *207*, 61–63, 1980.

Farquharson, G. W., Late Mesozoic sedimentation in the northern Antarctic Peninsula and its relationship to the southern Andes, *J. Geol. Soc. London*, *139*, 721–727, 1982.

Frakes, L. A., D. Burger, M. Apthorpe, J. Wiseman, M. Dettmann, N. Alley, R. Flint, D. Gravestock, N. Ludbrook, J. Backhouse, S. Skwarko, V. Scheibnerova, A. McMinn, P. S. Moore, B. R. Bolton, J. G. Douglas, R. Christ, M. Wade, R. E. Molnar, B. McGowran, B. E. Balme, and R. A. Day, Australian Cretaceous shorelines, stage by stage, *Palaeogeogr. Palaeoclimatol. Palaeoecol.*, *59*, 31–48, 1987.

Francis, J. E., Growth rings in Cretaceous and Tertiary wood from Antarctica and their paleoclimatic implications, *Palaeontology*, *29*, 665–684, 1986.

Francis, J. E., Palaeoclimatic significance of Cretaceous–early Tertiary fossil forests of the Antarctic Peninsula, in *Geological Evolution of Antarctica*, edited by M. R. A. Thomson, J. A. Crame, and J. W. Thomson, pp. 623–627, Cambridge University Press, New York, 1991.

Frederiksen, N. O., Changes in floral diversities, floral turnover rates, and climates in Campanian and Maastrichtian time, North Slope of Alaska, *Cretaceous Res.*, *10*, 249–266, 1989.

Gasparini, Z., E. Olivero, R. Scasso, and C. Rinaldi, Un

ankylosaurio (Reptilia, Ornithischia) campaniano en el continente antártico, *An. Congr. Bras. Paleontol. 10th, 1*, 131–141, 1987.

Grindley, G. W., C. J. D. Adams, J. T. Lumb, and W. A. Waters, Paleomagnetism, K-Ar dating and tectonic interpretation of Upper Cretaceous and Cenozoic volcanic rocks of the Chatham Islands, New Zealand, *N. Z. J. Geol. Geophys., 20*, 425–467, 1977.

Hailwood, E. A., and B. M. Clement, Magnetostratigraphy of sites 699 and 700, East Georgia Basin, *Proc. Ocean Drill. Program Sci. Results, 114*, 337–353, 1991.

Hamilton, N., Mesozoic magnetostratigraphy of Maud Rise, Antarctica, *Proc. Ocean Drill. Program Sci. Results, 113*, 255–260, 1990.

Hannah, M. J., Late Cretaceous foraminiferal biofacies of the northeastern Indian Ocean region, Ph.D. dissertation, Univ. of Adelaide, Adelaide, Australia, 1982.

Haq, B. U., J. Hardenbol, and P. R. Vail, The new chronostratigraphic basis of Cenozoic and Mesozoic sea level cycles, in *Timing and Depositional History of Eustatic Sequences: Constraints on Seismic Stratigraphy, Spec. Publ. Cushman Found. Foraminiferal Res., 24*, 7–13, 1987.

Hemleben, C., M. Spindler, and O. R. Anderson, *Modern Planktonic Foraminifera*, pp. 1–363, Springer-Verlag, New York, 1989.

Huber, B. T., Upper Campanian–Paleocene foraminifera from the James Ross Island region (Antarctic Peninsula), in Geology and Paleontology of Seymour Island, Antarctica, *Mem. Geol. Soc. Am., 169*, 163–251, 1988.

Huber, B. T., Maestrichtian planktonic foraminifer biostratigraphy of the Maud Rise (Weddell Sea, Antarctica): ODP Leg 113 holes 689B and 690C, *Proc. Ocean Drill. Program Sci. Results, 113*, 489–513, 1990.

Huber, B. T., Planktonic foraminifer biostratigraphy of Campanian-Maestrichtian sediments from ODP Leg 114, sites 698 and 700, southern South Atlantic, *Proc. Ocean Drill. Program Sci. Results, 114*, 281–297, 1991*a*.

Huber, B. T., Maestrichtian planktonic foraminifer biostratigraphy and the Cretaceous/Tertiary boundary at ODP Hole 738C (Kerguelen Plateau, southern Indian Ocean), *Proc. Ocean Drill. Program Sci. Results, 119*, 451–465, 1991*b*.

Huber, B. T., Paleobiogeography of Campanian-Maastrichtian foraminifers in the southern high latitudes, *Palaeogeogr. Palaeoclimatol. Palaeoecol., 92*, 325–360, 1992*a*.

Huber, B. T., Upper Cretaceous planktonic foraminiferal biozonation for the Austral Realm, *Micropaleontology*, in press, 1992*b*.

Huber, B. T., and P. N. Webb, Distribution of *Frondicularia rakauroana* (Finlay) in the southern high latitudes, *J. Foraminiferal Res., 16*, 135–140, 1986.

Kamp, P. J. J., Late Cretaceous–Cenozoic tectonic development of the southwest Pacific region, *Tectonophysics, 121*, 225–251, 1986.

Kassab, I. I. M., The genera *Pseudotextularia* and *Ventilabrella* (Foraminiferida) from northern Iraq, *Ann. Mines Geol. Tunis., 28*, 73–89, 1978.

Keigwin, L. D., Jr., Neogene planktonic foraminifers from Deep Sea Drilling Project sites 502 and 503, *Initial Rep. Deep Sea Drill. Proj., 68*, 269–288, 1982.

Kennett, J. P., et al., Leg 29, *Initial Rep. Deep Sea Drill. Proj., 29*, 1197 pp., 1975a.

Kennett, J. P., et al., Cenozoic paleoceanography in the southwest Pacific Ocean, Antarctic glaciation, and the development of the Circum-Antarctic Current, *Initial Rep. Deep Sea Drill. Proj., 29*, 1155–1169, 1975*b*.

Kennett, J. P., G. Keller, and M. S. Srinivasan, Miocene planktonic foraminiferal biogeography and paleoceanographic development of the Indo-Pacific region, in The

Miocene Ocean: Paleoceanography and Biogeography, *Mem. Geol. Soc. Am., 163*, 197–236, 1985.

Kent, D. V., and F. M. Gradstein, A Cretaceous and Jurassic geochronology, *Geol. Soc. Am. Bull., 96*, 1419–1427, 1985.

Krasheninnikov, V. A., and I. A. Basov, Stratigraphy of Cretaceous sediments of the Falkland Plateau based on planktonic foraminifers, Deep Sea Drilling Project, Leg 71, *Initial Rep. Deep Sea Drill. Proj., 71*, 789–820, 1983.

Krasheninnikov, V. A., and I. A. Basov, Late Mesozoic and Cenozoic stratigraphy and geological history of the South Atlantic high latitudes, *Palaeogeogr. Palaeoclimatol. Palaeoecol., 55*, 145–188, 1986.

Lawver, L. A., and C. R. Scotese, A revised reconstruction of Gondwanaland, in *Gondwana Six: Structure, Tectonics, and Geophysics, Geophys. Monogr. Ser.*, vol. 40, edited by G. D. McKenzie, pp. 17–24, AGU, Washington, D. C., 1987.

Lawver, L. A., J.-Y. Royer, D. T. Sandwell, and C. R. Scotese, Evolution of the Antarctic continental margins, in *Geological Evolution of Antarctica*, edited by M. R. A. Thomson, J. A. Crame, and J. W. Thomson, pp. 533–539, Cambridge University Press, New York, 1991.

Lawver, L. A., L. M. Gahagan, and M. F. Coffin, The development of paleoseaways around Antarctica, this volume.

Leckie, R. M., and P. N. Webb, Late Paleogene and early Neogene foraminifers of Deep Sea Drilling Project Site 270, Ross Sea, Antarctica, *Initial Rep. Deep Sea Drill. Proj., 90*, 1093–1142, 1985.

Macellari, C. E., Paleobiogeografía y edad de la fauna de *Maorites-Gunnarites* (Ammonoide a) del Cretácico Superior de la Antártida y Patagonia, *Ameghiniana, 21*, 223–242, 1985.

Macellari, C. E., Progressive endemism in the Late Cretaceous ammonite family Kossmaticeratidae and the breakup of Gondwanaland, in *Gondwana Six: Stratigraphy, Sedimentology, and Paleontology, Geophys. Monogr. Ser.*, vol. 41, edited by G. D. McKenzie, pp. 85–92, AGU, Washington, D. C., 1987.

Macellari, C. E., Stratigraphy, sedimentology and paleoecology of Late Cretaceous/Paleocene shelf-deltaic sediments of Seymour Island (Antarctic Peninsula), in Geology and Paleontology of Seymour Island, Antarctica, *Mem. Geol. Soc. Am., 169*, 25–53, 1988.

Malmgren, B. A., Biostratigraphy of planktonic foraminifera from the Maastrichtian white chalk of Sweden, *Geol. Foeren. Stockholm Foerh., 103*, 357–375, 1982.

Mary, C., M.-G. Moreau, X. Orue-Etxebarria, E. Apellaniz, and V. Courtillot, Biostratigraphy and magnetostratigraphy of the Cretaceous/Tertiary Sopelana section (Basque country), *Earth Planet. Sci. Lett., 106*, 133–150, 1991.

Monechi, S., and H. R. Thierstein, Late Cretaceous–Eocene nannofossil and magnetostratigraphic correlations near Gubbio, Italy, *Mar. Micropaleontol., 9*, 419–440, 1985.

Nederbragt, A. J., Maastrichtian Heterohelicidae (planktonic foraminifera) from the north west Atlantic, *J. Micropalaeontol., 8*, 183–206, 1989.

Oliver, P. J., T. C. Mumme, G. W. Grindley, and P. Vella, Paleomagnetism of the Upper Cretaceous Mt. Somers volcanics, Canterbury, New Zealand, *N. Z. J. Geol. Geophys., 22*, 199–212, 1979.

Olivero, E. B., Z. Gasparini, C. A. Rinaldi, and R. Scasso, First record of dinosaurs in Antarctica (Upper Cretaceous, James Ross Island): Palaeogeographic implications, in *Geological Evolution of Antarctica*, edited by M. R. A. Thomson, J. A. Crame, and J. W. Thomson, pp. 617–622, Cambridge University Press, New York, 1991.

Parker, F. L., Late Cenozoic biostratigraphy (planktonic foraminifera) of tropical Atlantic deep-sea sections, *Rev. Esp. Micropaleontol., 5*, 253–289, 1973.

Perch-Nielsen, K., Mesozoic calcareous nannofossils, in *Plankton Stratigraphy*, edited by H. M. Bolli, J. B. Saunders, and K. Perch-Nielsen, pp. 329–426, Cambridge University Press, New York, 1985.

Poore, R. Z., L. Tauxe, S. F. Percival, Jr., J. L. LaBrecque, R. Wright, N. P. Petersen, C. C. Smith, P. Tucker, and K. J. Hsü, Late Cretaceous–Cenozoic magnetostratigraphic and biostratigraphic correlations for the South Atlantic Ocean, Deep Sea Drilling Project Leg 73, *Initial Rep. Deep Sea Drill. Proj.*, *73*, 645–655, 1984.

Pospichal, J. J., Southern high latitude K/T boundary calcareous nannofossils from ODP sites 690 and 752, *Int. Nannoplankton Assoc. Newsl.*, *11*, 90–91, 1989.

Pospichal, J. J., and S. W. Wise, Jr., Maestrichtian calcareous nannofossil biostratigraphy of Maud Rise ODP Leg 113 sites 689 and 690, Weddell Sea, *Proc. Ocean Drill. Program Sci. Results*, *113*, 465–487, 1990.

Premoli Silva, I., Upper Cretaceous–Paleocene magnetic stratigraphy at Gubbio, Italy, II, Biostratigraphy, *Geol. Soc. Am. Bull.*, *88*, 371–374, 1977.

Quilty, P. G., Upper Cretaceous planktonic foraminifera and biostratigraphy, ODP Leg 120, southern Kerguelen Plateau, *Proc. Ocean Drill. Program Sci. Results*, *120*, 371–392, 1992.

Raven, P. H., and D. I. Axelrod, Angiosperm biogeography and past continental movements, *Ann. M. Bot. Gard.*, *61*, 539–673, 1974.

Riccardi, A. C., Cretaceous paleogeography of southern South America, *Palaeogeogr. Palaeoclimatol. Palaeoecol.*, *59*, 169–195, 1987.

Sakai, H., and B. Keating, Paleomagnetism of Leg 119—Holes 737A, 738C, 742A, 745B, and 746A, *Proc. Ocean Drill. Program Sci. Results*, *119*, 751–770, 1991.

Scheibnerova, V., Foraminifera and their Mesozoic biogeoprovinces, *Rec. Geol. Surv. N. S. W.*, *13*, 135–174, 1971.

Schlich, R., et al., Leg 120, *Proc. Ocean Drill. Program Initial Rep.*, *120*, 648 pp., 1989.

Scotese, C. R., and C. R. Denham, Terra Mobilis: Plate tectonics for the Macintosh, Earth in Motion Technologies, Austin, Tex., 1988.

Signor, P. W., and J. H. Lipps, Sampling bias, gradual extinction patterns and catastrophes in the fossil record, in Geological Implications of Impacts of Large Asteroids and Comets on the Earth, *Spec. Pap. Geol. Soc. Am.*, *190*, 291–296, 1982.

Sissingh, W., Biostratigraphy of Cretaceous calcareous nannoplankton, *Geol. Mijnbouw*, *56*, 37–50, 1977.

Sliter, W. V., Cretaceous foraminifera from the southwest Atlantic Ocean, Leg 36, Deep Sea Drilling Project, *Initial Rep. Deep Sea Drill. Proj.*, *36*, 591–573, 1977.

Sliter, W. V., Biostratigraphic zonation for Cretaceous planktonic foraminifers examined in thin section, *J. Foraminiferal Res.*, *19*, 1–19, 1989.

Smith, A. G., A. M. Hurley, and J. C. Briden, *Phanerozoic Paleocontinental World Maps*, 102 pp., Cambridge University Press, New York, 1981.

Smith, C. H., and R. Z. Poore, Upper Maastrichtian and Paleocene planktonic foraminiferal biostratigraphy of the northern Cape Basin, Deep Sea Drilling Project Hole 524, *Initial Rep. Deep Sea Drill. Proj.*, *73*, 449–457, 1984.

Stevens, G. R., Southwest Pacific faunal palaeobiogeography in Mesozoic and Cenozoic times: A review, *Palaeogeogr. Palaeoclimatol. Palaeoecol.*, *31*, 153–196, 1980.

Stevens, G. R., The nature and timing of biotic links between New Zealand and Antarctica in Mesozoic and early Cenozoic times, in *Origins and Evolution of the Antarctic Biota*, *Spec. Publ. 47*, edited by J. A. Crame, pp. 141–166, Geological Society of London, London, 1989.

Stock, J., and P. Molnar, Uncertainties in the relative positions of the Australia, Antarctica, Lord Howe, and Pacific plates since the Late Cretaceous, *J. Geophys. Res.*, *87*, 4697–4714, 1982.

Storey, B. C., The crustal blocks of West Antarctica within Gondwana: Reconstruction and break-up model, in *Geological Evolution of Antarctica*, edited by M. R. A. Thomson, J. A. Crame, and J. W. Thomson, pp. 587–592, Cambridge University Press, New York, 1991.

Stott, L. D., and J. P. Kennett, The paleoceanographic and paleoclimatic signature of the Cretaceous/Paleogene boundary in the Antarctic: Stable isotopic results from ODP Leg 113, *Proc. Ocean Drill. Program Sci. Results*, *113*, 829–848, 1990.

Taylor, D. J., Foraminifera and the stratigraphy of the western Victorian Cretaceous sediments, *Proc. R. Soc. Victoria*, *77*, 535–603, 1964.

Thierstein, H. R., Calcareous nannoplankton—Leg 26, Deep Sea Drilling Project, *Initial Rep. Deep Sea Drill. Proj.*, *26*, 619–667, 1974.

Thierstein, H. R., Late Cretaceous nannoplankton and the change at the Cretaceous-Tertiary boundary, *Spec. Publ. Soc. Econ. Paleontol. Mineral.*, *32*, 355–394, 1981.

Thomas, E., Late Cretaceous through Neogene deep-sea benthic foraminifers (Maud Rise, Weddell Sea, Antarctica), *Proc. Ocean Drill. Program Sci. Results*, *113*, 571–594, 1990.

Thomas, E., E. Barrera, N. Hamilton, B. T. Huber, J. P. Kennett, S. B. O'Connell, J. J. Pospichal, V. Spiess, L. D. Stott, W. Wei, and S. W. Wise, Jr., Upper Cretaceous-Paleogene stratigraphy of sites 689 and 690, Maud Rise (Antarctica), *Proc. Ocean Drill. Program Sci. Results*, *113*, 901–914, 1990.

Toker, V., P. F. Barker, and S. W. Wise, Jr., Middle Eocene carbonate-bearing marine sediments from Bruce Bank off northern Antarctic Peninsula, in *Geological Evolution of Antarctica*, edited by M. R. A. Thomson, J. A. Crame, and J. W. Thomson, pp. 639–644, Cambridge University Press, New York, 1991.

Truswell, E. M., Recycled Cretaceous and Tertiary pollen and spores in Antarctic marine sediments: A catalogue, *Palaeontographica, B*, *186*, 121–174, 1983.

Truswell, E. M., The palynology of core samples from the *S. P. Lee* Wilkes Land cruise, in *The Antarctic Continental Margin Geology and Geophysics of Offshore Wilkes Land, Earth Sci. Ser.*, vol. 5A, edited by S. L. Eittreim and M. A. Hampton, pp. 215–221, Circum-Pacific Council for Energy and Mineral Resources, Houston, Tex., 1987.

van Andel, T. H., Mesozoic/Cenozoic calcite compensation depth and the global distribution of calcareous sediments, *Earth Planet. Sci. Lett.*, *26*, 187–195, 1975.

Veevers, J. J., Break-up of Australia and Antarctica estimated at mid-Cretaceous (95 ± 5 Ma) from magnetic and seismic data at the continental margin, *Earth Planet. Sci. Lett.*, *77*, 91–99, 1986.

Veevers, J. J., Earth history of the southeast Indian Ocean and the conjugate margins of Australia and Antarctica, *J. Proc. R. Soc. N. S. W.*, *120*, 57–70, 1987.

Vincent, E., and W. H. Berger, Planktonic foraminifera and their use in paleoceanography, in *The Oceanic Lithosphere: The Sea*, 7, edited by E. C. Emiliani, pp. 1025–1119, Wiley-Interscience, New York, 1981.

Watkins, D. K., Upper Cretaceous nannofossils from Leg 120, Kerguelen Plateau, Southern Ocean, *Proc. Ocean Drill. Program Sci. Results*, *120*, 343–370, 1992.

Watkins, D. K., and H. Liu, Calcareous nannofossils from the Niobrara of western Kansas and eastern South Dakota, *Bull. Kans. Geol. Surv.*, in press, 1992.

Watkins, D. K., S. W. Wise, Jr., J. J. Pospichal, and J. A. Crux, Upper Cretaceous calcareous nannofossil biostratigraphy of the Southern Ocean, *Mar. Micropaleontol.*, in press, 1992.

Watts, D. R., G. C. Watts, and A. M. Bramall, Cretaceous and early Tertiary paleomagnetic results from the Antarctic Peninsula, *Tectonics*, *3*, 333–346, 1984.

Webb, D. J., P. D. Killworth, A. C. Coward, and S. R. Thompson, *The FRAM Atlas of the Southern Ocean*, 67 pp., National Environmental Research Council (Great Britain), Swindon, England, 1991.

Webb, P. N., New Zealand Late Cretaceous (Haumurian) foraminifera and stratigraphy: A summary, *N. Z. J. Geol. Geophys.*, *14*, 795–828, 1971.

Webb, P. N., and V. E. Neall, Cretaceous foraminifera in Quaternary deposits from Taylor Valley, Victoria Land, in *Antarctic Geology and Geophysics*, edited by R. J. Adie, pp. 653–657, Universitetsforlaget, Oslo, 1972.

Webb, P. N., D. M. Harwood, B. C. McKelvey, J. H. Mercer, and L. D. Stott, Cenozoic marine sedimentation and ice-volume variation on the East Antarctic craton, *Geology*, *12*, 287–291, 1984.

Wei, W., and H. R. Thierstein, Upper Cretaceous and Cenozoic calcareous nannofossils of the Kerguelen Plateau (southern Indian Ocean) and Prydz Bay (East Antarctica), *Proc. Ocean Drill. Program Sci. Results*, *119*, 467–493, 1991.

Weiss, W., Upper Cretaceous planktonic foraminiferal biostratigraphy from the Rio Grande Rise: Site 516 of Leg 72, Deep Sea Drilling Project, *Initial Rep. Deep Sea Drill. Proj.*, *72*, 715–721, 1983.

Wichter, C. A., Mikropalaontologische XX Beobachtungen in der hoheren XX borealen Oberkreide, besonders im Maastricht, *Geol. Jahrb.*, *Reihe B*, *68*, 1–25, 1953.

Wind, F. H., Late Campanian and Maestrichtian calcareous nannoplankton biogeography and high-latitude biostratigraphy, Ph.D. dissertation, Fla. State Univ., Tallahassee, 1979a.

Wind, F. H., Maestrichtian-Campanian nannoflora provinces of the southern South Atlantic and Indian oceans, in *Deep Drilling Results in the Atlantic Ocean: Continental Margins and Paleoenvironment*, *Maurice Ewing Ser.*, vol. 3, edited by M. Talwani, W. W. Hay, and W. B. F. Ryan, pp. 123–137, AGU, Washington, D. C., 1979b.

Wind, F. H., and S. W. Wise, Jr., Correlation of upper Campanian–lower Maestrichtian calcareous nannofossil assemblages in drill and lower piston cores from the Falkland Plateau, southwest Atlantic Ocean, *Initial Rep. Deep Sea Drill. Proj.*, *71*, 551–563, 1983.

Wise, S. W., Jr., Mesozoic and Cenozoic nannofossils recovered by Deep Sea Drilling Project Leg 71 in the Falkland Plateau region, southwest Atlantic Ocean, *Initial Rep. Deep Sea Drill. Proj.*, *71*, 481–550, 1983.

Wise, S. W., Jr., Mesozoic-Cenozoic history of calcareous nannofossils in the region of the Southern Ocean, *Palaeogeogr. Palaeoclimatol. Palaeoecol.*, *67*, 157–179, 1988.

Wise, S. W., Jr., and F. H. Wind, Mesozoic and Cenozoic nannofossils recovered by DSDP Leg 36 drilling on the Falkland Plateau, southwest Atlantic sector of the Southern Ocean, *Initial Rep. Deep Sea Drill. Proj.*, *36*, 269–491, 1977.

Wonders, A. A. H., Cretaceous planktonic foraminiferal biostratigraphy, Leg 122, Exmouth Plateau, Australia, *Proc. Ocean Drill. Program Sci. Results*, *122*, 587–599, 1992.

Woodburne, M. O., and W. J. Zinsmeister, The first land mammal from Antarctica and its biogeographic implications, *J. Paleontol.*, *58*, 913–948, 1984.

Woodburne, M. O., R. H. Tedford, M. Archer, W. D. Turnbull, M. D. Plane, and E. L. Lunedius, Biochronology of the continental mammal record of Australia and New Guinea, *Spec. Publ. South Aust. Dep. Mines Energy*, *5*, 347–363, 1985.

Worsley, T. R., The Cretaceous/Tertiary boundary event in the ocean, in Studies in Paleo-oceanography, *Spec. Publ. Soc. Econ. Paleontol. Mineral.*, *20*, 94–125, 1974.

Worsley, T. R., and E. Martini, Late Maestrichtian nannoplankton provinces, *Nature*, *225*, 1242–1243, 1970.

Zinsmeister, W. J., Biogeographic significance of the late Mesozoic and early Tertiary molluscan faunas of Seymour Island (Antarctic Peninsula) to the final breakup of Gondwanaland, in *Historical Biogeography, Plate Tectonics, and the Changing Environment*, edited by J. Gray and A. J. Boucot, pp. 349–355, Oregon State University Press, Corvallis, 1979.

Zinsmeister, W. J., Late Cretaceous–early Tertiary molluscan biogeography of the southern circum-Pacific, *J. Paleontol.*, *56*, 84–102, 1982.

(Received January 13, 1992;
accepted April 30, 1992.)

LATE CRETACEOUS–EARLY TERTIARY ANTARCTIC OUTCROP EVIDENCE FOR PAST VEGETATION AND CLIMATES

Rosemary A. Askin

Department of Earth Sciences, University of California, Riverside, California 92521

Fossil plant remains from the Late Cretaceous–early Tertiary Antarctic vegetation occur in exposed outcrops in the South Shetland Islands and James Ross Basin, Antarctic Peninsula area. These remains include foliage, wood, and palynomorphs, all of which provide useful information on the past high-latitude floras and their ambient environments. Qualitative and sometimes quantitative assessments of the paleoclimates are derived from overall floral composition, foliar physiognomy (leaf margin morphology, size, and shape), plant cuticles, wood anatomy and annual rings, and spore and pollen assemblages. During the Late Cretaceous and Paleocene, conifer-dominated (podocarp and araucarian) rain forest prevailed. Angiosperm diversification reached its peak during the Campanian-Maastrichtian, at which time mixed conifer-Proteaceae-*Nothofagus* rain forest floras are evident. Humid conditions continued throughout the Late Cretaceous–early Tertiary in the Antarctic Peninsula region, with climates varying between cool and warm temperate. Plant evidence suggests cooling through the Maastrichtian into the early Paleocene, with a brief warm interval in the latest Maastrichtian. Diverse angiosperm-rich Paleocene-Eocene floras grew in relatively warm conditions. There was a shift to *Nothofagus*-dominated vegetation in the Eocene. Reduction in vegetational diversity in the late Eocene/Oligocene and through the Oligocene is consistent with cooling through the late Paleogene.

INTRODUCTION

During the last decade, much new information has become available on the Antarctic Late Cretaceous–early Tertiary land vegetation and inferred ambient climates. This results in part from the increased need to understand past climatic change, particularly in the high southern latitudes over this important time interval. Palynologists and paleobotanists now have a better data base and greatly improved interpretive methods for fossil floras. *Truswell* [1990, 1991] reviewed Antarctic Cretaceous and Tertiary floras and their implied paleoclimates, *Dettmann and Thomson* [1987] and *Dettmann* [1989] discussed the development of these floras through the Cretaceous, and *Askin and Spicer* [1992] compared the northern and southern high-latitude records.

This review first outlines the different types of plant fossils and information that may be inferred from them, then summarizes and updates available data from the geologic record on the Antarctic continent. Exposed plant-bearing sedimentary rocks of Late Cretaceous and early Tertiary age occur in the Antarctic Peninsula area. This outcrop record provides a different and more complete perspective on land plants than the record recovered from drill hole, piston core, or dredge samples stressed in the rest of this symposium. The latter sources of information, such as piston core and dredge samples from around the continental margin [e.g., *Truswell*, 1983], drill holes on the coastal margin (e.g.,

Mildenhall [1989] and CIROS 1), and Deep Sea Drilling Project (DSDP) and Ocean Drilling Program (ODP) drill holes in Antarctic and Subantarctic waters [e.g., *Mohr*, 1990], reveal important clues to the largely unknown vegetative cover across the vast expanse of Antarctica.

TYPES OF MATERIAL AND INFORMATION FROM THE PLANT RECORD

We can interpret past vegetation and climates from four major types of plant fossil material: leaf (foliar) compressions and impressions, wood, palynomorphs (spores and pollen), and dispersed cuticular material. Sometimes roots and reproductive organs such as flowers, cones, fruits and seeds, plus symbionts and parasites (e.g., epiphyllous fungi), supply useful information.

1. Leaf fossils provide both quantitative and qualitative information. Wolfe [*Wolfe*, 1971, 1979, 1985, 1991; *Wolfe and Upchurch*, 1987] has devised an elegant technique for making realistic quantitative estimates of past ambient climatic conditions from foliar physiognomy (leaf margin morphology, size, and shape), as well as overall assemblage composition. These methods extend early observations by *Bailey and Sinnott* [1915] on living and fossil leaf assemblages and also rely on characterization and classification of extant floras [e.g., *Webb*, 1959]. Leaf margin morphologies (entire versus dentate) provide the most useful parameters. The pro-

portion of entire-margined leaves within a flora shows a linear relationship to mean annual temperature (MAT), with entire-margined leaves being more common in warmer climates, while leaves with dentate margins predominate in colder MATs. *Wolfe* [1971] suggests that assemblages containing 30 or more species may provide a temperature with ±5% accuracy, that assemblages containing between 20 and 29 species provide a temperature with ±10% accuracy, and that values for assemblages with fewer than 20 species "should be regarded as highly tentative." This is an important consideration for the often small and therefore species-poor Antarctic collections. Wolfe also cautions that the method only holds true for climax vegetation. Thus streamside collections that represent disturbed vegetation yield misleading results. Foliar physiognomic-climatic relationships also change in high-latitude/altitude (Arctic/subalpine) environments, where the small-leaved populations exhibit a relatively high proportion of entire-margined species [*Bailey and Sinnott*, 1915; J. A. Wolfe, personal communication, 1992]. The complex relationships of climatic parameters and foliar physiognomy are most accurately and precisely defined by the new Climate-Leaf Analysis Multivariate Program (CLAMP) of *Wolfe* [1991].

2. Wood anatomy and tree ring analysis provide useful qualitative climatic information. Width and variability of annual rings, cell size, and relative proportions of earlywood and latewood reflect availability of water supply, disruptions in normal growth (from water fluctuations, frost, etc.), productivity, and information about ambient temperatures and light [e.g., *Carlquist*, 1977]. Identification of fossilized tree taxa may also be possible from the wood anatomy, although relating the wood to extant families and genera is often very difficult. For well-preserved woody remains with distinctive anatomical characters, comparisons may be made to living families or sometimes genera [e.g., *Francis*, 1991], or even occasionally to species (e.g., *Carlquist* [1987] for Antarctic Pliocene material).

3. Palynomorphs, namely, spores of mosses, liverworts, lycophytes, articulates and ferns (cryptogams), and pollen of gymnosperms and angiosperms, provide indirect evidence of past vegetation and climate. This is a two-step process: first, affinities with the parent plants must be established; then, habitat and climatic preference of presumed modern counterparts (or nearest living relatives (NLR)) may be invoked with caution. This method has obvious drawbacks. Botanical affinities are not always certain, and the analogy to modern plants assumes evolutionary stasis. These limitations also hold for similar NLR interpretations of leaf assemblages. Inferences from palynomorph assemblages have sometimes produced inaccurate climatic parameters, or typically a rather broad qualitative view of past vegetation and climate. In Australia, however, great progress is being made toward acquiring more realistic quantitative

estimates of paleoclimate from pollen assemblages. Bioclimatic profiles of living tree taxa have been generated, and the method is applied using these well-defined climatic parameters for as many taxa as possible (BIOCLIM program, summarized by *Nix* [1984]).

Another factor applies for the majority of terrestrial palynomorph assemblages. Although spores and pollen representing the local vegetation are sometimes preserved in sediments of swamps and small lakes, palynomorphs are typically transported by wind and water from many different habitats, each with its own local environmental conditions. Thus most spore and pollen assemblages, especially those transported to shallow marine sediments, represent a broad-brush view of the regional vegetation and various ambient climates.

4. Dispersed plant cuticles, the resistant outer layer of leaves, stems and reproductive parts of plants, often enable recognition of the parent plant taxa and also reflect physiology of the plant [*Upchurch*, 1989]. For example, thin leaf cuticles are generally found in deciduous plants, while thick cuticles, like thick (coriaceous) leaves, characterize evergreen plants. Smooth cuticles are more common in humid environments, while hairy leaves predominate under more arid conditions.

A useful summary of physiological characteristics of land plants and their implications for the fossil record is provided by *Spicer* [1989]. *Wolfe and Upchurch* [1987] and *Upchurch and Wolfe* [1987] discuss the different types of plant evidence and apply them to the Late Cretaceous and Tertiary North American fossil record.

Antarctica was situated in polar latitudes throughout the Cretaceous and Cenozoic. Consequently, its vegetation has had to contend with the peculiar stringencies of life in high latitudes. Restrictions of and adaptations to these special conditions, in both "greenhouse" and present "icehouse" worlds, are discussed by many workers [e.g., *Wolfe*, 1980; *Spicer*, 1987, 1990; *Spicer and Parrish*, 1990; *Spicer and Chapman*, 1990; *Creber and Chaloner*, 1985; *Creber*, 1990; *Truswell*, 1991; *Askin and Spicer*, 1992].

THE ANTARCTIC PENINSULA PLANT RECORD

General

Antarctic outcrops of Late Cretaceous–early Tertiary age are restricted to the Antarctic Peninsula region. Fossil plant material occurs on various islands in the James Ross Basin; in the South Shetland Islands on Livingston, King George, and adjacent small islands; and on Alexander, Adelaide, and Brabant islands (Figure 1).

In the James Ross Basin, most of the Late Cretaceous, Paleocene, and Eocene interval is represented by well-exposed, siliciclastic, sedimentary strata (Figure 2), deposited in a back arc setting. Foliar material is rare in this predominantly shallow marine succession. Wood, some of it permineralized (typically carbonate)

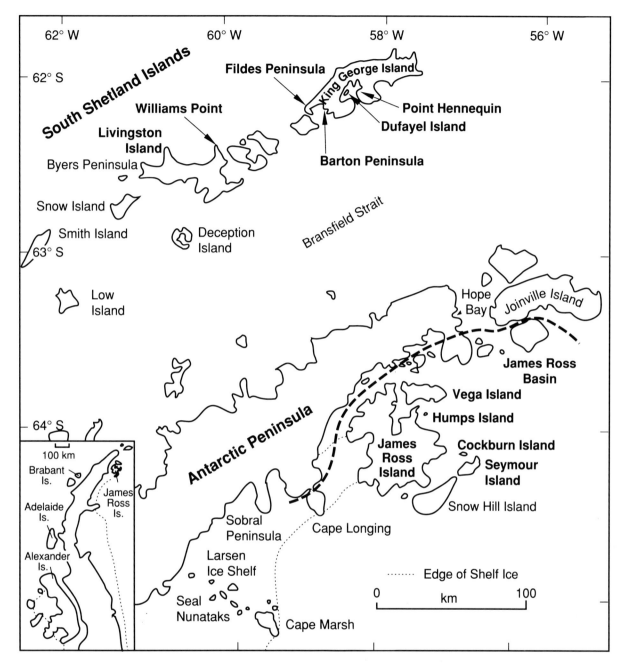

Fig. 1. Locality map for the Antarctic Peninsula area.

and some coalified, is relatively common and abundant in parts of the section. Marine and nonmarine palynomorphs provide the most continuous record through this richly fossiliferous succession. Dispersed plant cuticle is frequently encountered in palynological preparations, though it is best studied using specially prepared bulk samples.

In the South Shetland Islands, sediments were deposited in a volcanic island arc setting, preserving a somewhat different plant record. The strata exposed on King George Island (Figure 2) include interbedded volcaniclastic nonmarine and marine sediments, tuffs, and lavas. The sedimentary succession is discontinuous, although the lavas are datable by radiometric methods. Scattered Late Cretaceous and early Tertiary plant-bearing strata (from Senonian through the Oligocene) are known. Plant material includes leaves, wood (silicified), and palynomorphs in some localities.

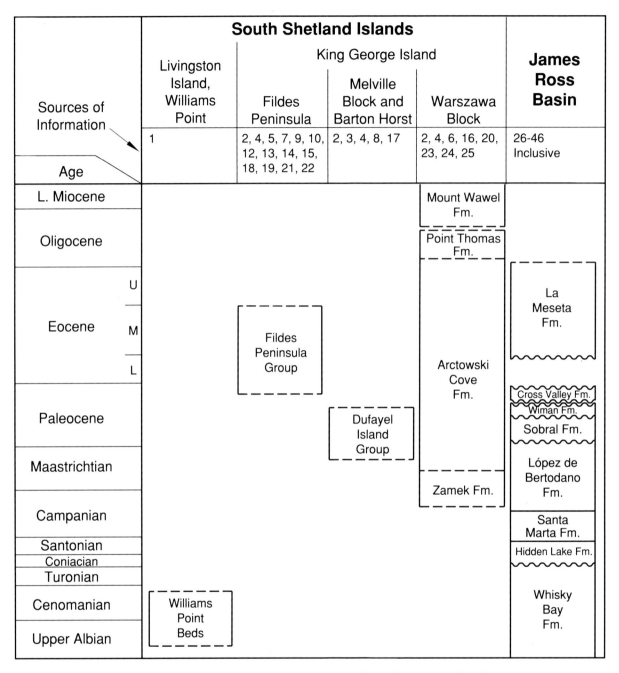

Fig. 2.   Stratigraphic distribution of plant-bearing beds in the South Shetland Islands and James Ross Basin, with sources of plant fossil information listed in Table 1. King George Island stratigraphy is after *Birkenmajer and Zastawniak* [1989a]. James Ross Basin stratigraphy is based on many sources, including *Olivero et al.* [1986], *Ineson et al.* [1986], *Askin* [1988a], *Elliot* [1988], *Elliot and Hoffman* [1989], and *Crame et al.* [1991].

A summary of the plant evidence from the South Shetland Islands and James Ross Basin, with paleoclimatic implications, is given below. Figures 2 and 3 show the stratigraphic distribution of plant remains, with sources cited in Table 1.

Late Cretaceous

At the beginning of the Late Cretaceous, angiosperms had already invaded the late Jurassic–Early Cretaceous vegetation of conifers (including araucarians and

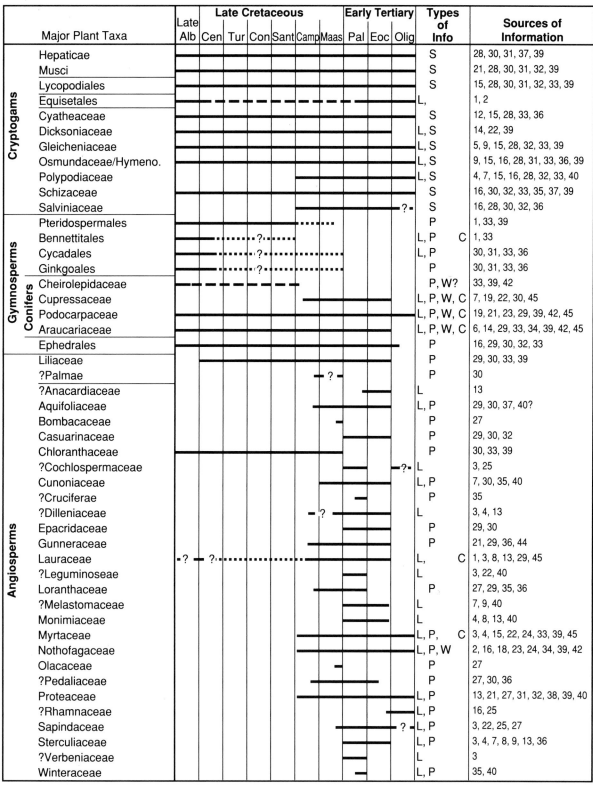

Fig. 3. Stratigraphic distribution of major plant taxa in the South Shetland Islands and James Ross Basin. Types of information are coded: L, leaf; S/P, spore/pollen; W, wood; C, cuticle. Sources of information are listed in Table 1. Note that for plant taxa with large numbers of possible sources (e.g., Podocarpaceae, Nothofagaceae), a few representative references have been selected.

TABLE 1. Sources of Late Cretaceous–Early Tertiary
Plant Information Cited in Figures 2 and 3

Number in Figure	Reference
	*South Shetland Islands**
1	*Rees and Smellie* [1989]
2	*Barton* [1964]
3	*Birkenmajer and Zastawniak* [1986]
4	*Birkenmajer and Zastawniak* [1989a]
5	*Cao* [1989]
6	*Cortemiglia et al.* [1981]
7	*Czajkowski and Rösler* [1986]
8	*Del Valle et al.* [1984]
9	*Li and Shen* [1989]
10	*Li and Song* [1988]
11	*Lucas and Lacey* [1981]
12	*Lyra* [1986]
13	*Orlando* [1964]
14	*Palma-Heldt* [1987]
15	*Shen* [1989]
16	*Stuchlik* [1981]
17	*Tokarski et al.* [1987]
18	*Torres* [1984]
19	*Torres et al.* [1984]
20	*Torres and LeMoigne* [1988]
21	*Torres and Meon* [1990]
22	*Troncoso* [1986]
23	*Zastawniak* [1981]
24	*Zastawniak* [1990]
25	*Zastawniak et al.* [1985]
	James Ross Basin
26	*Askin* [1988a]
27	*Askin* [1989]
28	*Askin* [1990a]
29	*Askin* [1990b]
30	R. A. Askin (unpublished data)
31	*Askin et al.* [1991]
32	*Baldoni and Barreda* [1986]
33	*Baldoni and Medina* [1989]
34	*Case* [1988]
35	*Cranwell* [1959]
36	*Cranwell* [1969]
37	*Dettmann* [1989]
38	*Dettmann and Jarzen* [1988]
39	*Dettmann and Thomson* [1987]
40	*Dusén* [1908]
41	*Francis* [1986]
42	*Francis* [1991]
43	*Gothan* [1908]
44	*Jarzen and Dettmann* [1991]
45	*Upchurch and Askin* [1989]
46	*Zamaloa et al.* [1987]

*Reference 1 [*Rees and Smellie*, 1989] is for Williams Point, Livingston Island. References 2–25 are all for King George Island.

podocarps), ferns and other cryptogams (mosses, liverworts, lycophytes, articulates), pteridosperms, cycadophytes (bennettitaleans and cycads), and ginkgophytes.

Leaf assemblages of possible Albian/Cenomanian age occur on Williams Point, northeast Livingston Island. The Williams Point Beds had previously yielded poorly preserved, fragmentary material assigned to various cryptogam and gymnosperm taxa and considered Triassic in age [*Orlando*, 1968; *Lacey and Lucas*, 1981; *Banerji and LeMoigne*, 1987; *Banerji et al.*, 1987; *LeMoigne*, 1987]. Recollecting better preserved and more complete specimens from more productive outcrops in this area reveals common angiosperm leaves. These can be grouped into six distinctive forms. One form, also the most abundant angiosperm leaf type, was compared with *Cinnamomoides*, while the others were unassignable to known taxa but probably represent distinct species [*Rees and Smellie*, 1989]. Other foliage includes various conifers, cycads, a bennettitalean (cf. *Pterophyllum*), ferns, and *Equisetites*. Although the age might be anywhere between the Barremian and Campanian (taking into account the floral composition and limiting radiometric dates), an Albian-Cenomanian age was considered the most likely, based on the presence of cf. *Cinnamomoides* and the primitive form and venation type of the angiosperms [*Rees and Smellie*, 1989]. The angiosperm leaves are almost all microphyllous (that is, small, <8 cm [*Wolfe*, 1985]), and entire-margined forms predominate over toothed types. Thus Rees and Smellie suggested a tentative MAT between 13°C and 20°C for this riparian collection (note well that six recorded taxa fall well below the 20+ species needed for reliable estimates; such temperature-physiognomic relationships may not have been fully established at this early stage of angiosperm evolution; note other limitations listed earlier). Fossil wood was also recently described from Williams Point and includes podocarp and araucarian conifer, bennettitalean, and four unassigned dicotyledonous angiosperm taxa, of Late Cretaceous age [*Torres and LeMoigne*, 1989].

Albian/Cenomanian palynological evidence is available from James Ross Island in the Whisky Bay Formation. *Dettmann and Thomson* [1987] described two samples (D.3006.2, D.3057.3) and *Baldoni and Medina* [1989] described three samples (M4, M5, M6) of late Albian and Cenomanian age. Vegetation types inferred from these spore and pollen assemblages are consistent with the preceding floral evidence. They are similar to Early Cretaceous associations with their abundant gymnosperms (particularly podocarp and araucarian conifers, plus other conifers, pteridosperms, and possible cycadophytes or ginkgophytes), and abundant, diverse cryptogams. Important differences from slightly older Albian assemblages are fewer cheirolepidacean conifers (*Classopollis* pollen), the last significant occurrences of conifers with *Brachyphyllum* foliage (represented by *Balmeiopsis* and *Cyclusphaera* pollen), and increasing diversity and frequency of angiosperm pollen. *Dettmann and Thomson* [1987] record eight pollen taxa representing monocotyledonous (*Liliacidites*) and dicotyledonous angiosperms (including *Clavatipollenites* (Chloranthaceae) and trichotomosulcate, trilcolpate, tricolpoidate, and tricolporate pollen). Some of these pollen represent "higher" or nonmagnoliid angiosperms, such as the tricolporate forms *Tricolporites* (now *Tubulifloridites*) *lilliei* and *Nyssapollenites* spp.

Published data on land-derived palynomorphs in the Turonian to lower Santonian part of the section are presently scarce. *Baldoni and Medina* [1989] outlined palynological results from three Coniacian–lower Santonian samples (M7, M8, M9) of the upper Hidden Lake Formation, James Ross Island. Plentiful information is potentially available in the James Ross Basin from abundant podocarp, araucarian, and angiosperm pollen and cryptogam spores. In rocks of this age, the focus has been largely on the biostratigraphically significant dinocyst floras [e.g., *Crame et al.*, 1991, appendix; *Riding et al.*, 1992].

In contrast, a wealth of plant fossil material of all four of the major types is known from the James Ross Basin and South Shetland Islands from upper Santonian, Campanian, and Maastrichtian rocks. This material has been discussed by *Cranwell* [1959, 1969], *Baldoni and Barreda* [1986], *Dettmann* [1986, 1989], *Dettmann and Thomson* [1987], *Dettmann and Hedlund* [1988], *Dettmann and Jarzen* [1988], *Jarzen and Dettmann* [1989], *Askin* [1988a, 1989, 1990a, b], *Askin et al.* [1991], *Askin and Spicer* [1992], *Birkenmajer and Zastawniak* [1989a, b], *Zastawniak* [1990], *Cao* [1989], and *Shen* [1989] and thus will not be repeated here. Familial (or higher) diversity for the latest Cretaceous is indicated in Figure 3, although the diagram only includes forms with known modern plant affinities. Many pollen types of unknown or uncertain affinities occur in these Maastrichtian rocks [e.g., *Askin*, 1990b].

Podocarp and araucarian rain forest vegetation with fern and a developing angiosperm understory grew in the northern Antarctic Peninsula area during the Late Cretaceous. In the James Ross Basin, ample moisture and favorable growing conditions with no frost (for at least some of this interval) are indicated by uniform annual rings in conifer wood, with no false rings, and large earlywood cell size [*Francis*, 1986, 1991]. Rings are well defined with narrow latewood, consistent with relatively sudden short-day/dark-induced dormancy at these paleolatitudes (~65°S).

Species diversity among the angiosperms greatly increased during the latest Cretaceous, with angiosperms replacing many of the cryptogams and gymnosperms that characterized earlier Cretaceous vegetation. The flora included immigrants from more northern latitudes and many newly evolved forms. By the end of the Cretaceous, almost half of the angiosperm taxa were endemic [*Askin*, 1989]. Antarctica was a locus of evolutionary innovation for many plant groups during the Senonian. The distinctive southern flora evolved in the southern high latitudes in the area encompassing much of Antarctica, southern-southeastern Australia, greater New Zealand, and southernmost South America. This area had already been the center of evolution for certain fern taxa [*Dettmann*, 1986, 1989] in the Early Creta-

ceous. Notable taxa that evolved in this region (the Weddellian Province) during the Senonian include podocarpaceous conifers such as *Lagarostrobus* and *Dacrydium* [*Dettmann*, 1989; *Dettmann and Jarzen*, 1990], *Nothofagus* [*Cranwell*, 1964; *Dettmann et al.*, 1990], *Ilex* [*Dettmann*, 1989], and Proteaceae [*Pocknall and Crosbie*, 1988; *Dettmann and Jarzen*, 1988, 1990; *Jarzen and Dettmann*, 1991]. Patterns of evolution and dispersal and possible causes for this impressive high-latitude diversification are discussed by *Dettmann* [1989], *Dettmann and Jarzen* [1990], *Askin* [1989], and *Askin and Spicer* [1992].

There is still much to be learned from the Senonian foliar material and palynomorphs on King George Island, the many hundreds of palynomorph samples from the James Ross Basin, and from the abundant wood. The ambient climate during the closing stages of the Cretaceous was apparently mild (frost free, at least on the coastal fringes), wet, and warm to cool temperate. Podocarp conifer-Proteaceae-*Nothofagus* rain forests flourished in these conditions. At paleolatitudes of 60° to 65°S, these rain forests grew close to the paleo-Antarctic circle. They included both evergreen and deciduous elements [*Birkenmajer and Zastawniak*, 1989a; *Upchurch and Askin*, 1989; *Askin and Spicer*, 1992].

The plant record provides us with a means of distinguishing climatic fluctuations that is independent of the deep-sea and isotopic record. Fossil wood indicates cooling through the Maastrichtian into the Paleocene [*Francis*, 1986, 1991], and palynomorph evidence suggests that a brief warm interval in the latest Maastrichtian is superimposed on this cooling trend [*Askin*, 1989]. Declining Maastrichtian temperatures are evident in the oxygen isotope record from James Ross Basin [*Barrera et al.*, 1987; *Pirrie and Marshall*, 1990] and from nearby Maud Rise ODP Leg 113, Site 689 and Site 690 [*Barrera and Huber*, 1990], with a well-defined latest Maastrichtian short warming episode, followed by marked cooling at the end of the Cretaceous [*Stott and Kennett*, 1990]. Plant evidence also indicates that the northern Antarctic Peninsula region experienced high rainfall [*Francis*, 1986, 1991; *Askin*, 1988a, 1989; *Dettmann*, 1989], possibly becoming wetter through the Maastrichtian [*Askin*, 1990a]. Preliminary dispersed plant cuticle evidence from the latest Maastrichtian on Seymour Island led G. R. Upchurch (personal communication, 1990) to suggest MAT of approximately 8° to 15°C, mean annual temperature range (MAR) of probably <16°C, and coldest month mean temperature (CMM) of >1°C. These parameters were derived from those established for the various extant vegetation types and plotted by Wolfe [e.g., *Wolfe*, 1985, Figure 1], the presumed Seymour vegetation being a broadleaved evergreen and coniferous type.

Cretaceous-Tertiary Transition

The only known Antarctic Cretaceous-Tertiary (K-T) boundary outcrop exposure is on Seymour Island. Terrestrial palynomorphs from sections across this boundary, as in other southern mid- to high-latitude sections, do not record major ecological trauma, although they do reflect some change during the K-T transition. Dominant components of the palynomorph assemblages continue unchanged across the boundary [Askin, 1988b, 1990b]; however, a few diagnostic pollen species disappeared at or near the K-T boundary throughout the Weddellian Province, reflecting some paleoenvironmental change in the southern high latitudes. These forms are only a rare component of the palynofloras, and the cause of their parent plants' demise is as yet unknown. Various spore and pollen taxa first appear throughout the latest Maastrichtian–earliest Danian interval [Askin, 1990b], coincident with significant sea level changes. Although long-term Maastrichtian to Paleocene cooling is evident from plant and other evidence, the consistent occurrence of the frost-sensitive epiphyllous fungi Trichopeltinites from Maastrichtian into the Danian on Seymour Island suggests that there was no sudden, devastating cold temperature event at the K-T boundary [Upchurch and Askin, 1989]. Any sudden temperature change might have been partly ameliorated, however, in the mild coastal conditions where most of the preserved Seymour floras grew. Furthermore, greatly increased local volcanism during the K-T transition and the Danian undoubtedly disrupted the Antarctic Peninsula vegetation (as well as the "normal" climatic patterns) and complicates interpretations of climatic change.

Early Tertiary

As in the Late Cretaceous, there is great potential in the Paleocene and Eocene for paleobotanical and palynological data collection and paleoclimatic interpretations. Much of this information remains untapped, although general floral content is known (Figure 3). Oligocene material is rare, and its occurrence is presently restricted to the South Shetland Islands.

Some wood and palynomorph assemblages and common foliar remains occur on King George Island. In the James Ross Basin, marginal marine and occasional nonmarine deposits contain abundant palynomorphs, wood, and rare leaf assemblages. Unfortunately, significant portions of the upper Paleocene and basal Eocene are probably missing in the otherwise relatively complete upper Campanian–Eocene section on Seymour Island, and the complex lensoid nature of the Eocene sediments makes it difficult to correlate individual sample sets.

Birkenmajer and Zastawniak [1989a, b] reviewed the King George Island floras. Leaves, wood, and palynomorphs have been described by Barton [1964], Orlando [1964], Stuchlik [1981], Cortemiglia et al. [1981], Za-

stawniak [1981], Zastawniak et al. [1985], Del Valle et al. [1984], Torres [1984], Torres et al. [1984], Lucas and Lacey [1981], Birkenmajer and Zastawniak [1986], Czajkowski and Rösler [1986], Lyra [1986], Troncoso [1986], Palma-Heldt [1987], Tokarski et al. [1987], Torres and LeMoigne [1988], Torres and Meon [1990], Li and Song [1988], Li and Shen [1989], and Shen [1989].

The Dufayel Island and Barton Peninsula collections of possible Paleocene age (ages are equivocal) are dominated by dicotyledonous angiosperm leaves, including Nothofagus. Other leaves from Dufayel Island represent Myrtaceae, various laurophyllous types, and possible Cochlospermaceae, Dilleniaceae, Leguminoseae, Sapindaceae, Sterculiaceae, and Verbeniaceae, plus an unknown monocotyledonous type and possible fern fragments [Birkenmajer and Zastawniak, 1986]. Birkenmajer and Zastawniak [1989a] compared the Dufayel flora with extant temperate South American broadleaved forests growing in MATs of 10°–12°C and annual rainfall of 1000–4000 mm.

Many leaf, wood, and palynomorph assemblages are known from Fildes Peninsula at the western end of King George Island. These come from several different localities and may represent a range of ages and paleoenvironments. The main locality (variously known as Mount Flora, Fossil Hill, or Leaves Hill) is now considered late Paleocene–early Eocene [Troncoso, 1986; Birkenmajer and Zastawniak, 1989a]. The fossils reflect a diverse vegetation of mixed broadleaved angiosperms, with podocarp, araucarian, and cupressacean conifers, and ferns. Details of assemblage composition, foliar physiognomy, and relationships between the various Fildes Peninsula collections need to be resolved for more accurate paleoenvironmental interpretations. In some assemblages, evergreen foliage predominates over deciduous types, and angiosperm leaves include microphyllous (<8 cm) and notophyllous (8–12 cm) size classes with entire and dentate margins (proportions not known). The addition to the typical angiosperm assemblage (Nothofagus, Proteaceae, Gunneraceae, Myrtaceae, etc.) of possible Anacardiaceae, Dilleniaceae, Icacinaceae, Monimiaceae, Sapindaceae, and Sterculiaceae led some authors (Figure 2, Table 1) to suggest warm, wet conditions for the Fossil Hill assemblages, based on modern subtropical distribution of at least some members of these latter families. Troncoso [1986] notes the apparent mix of subtropical elements with more cool temperate forms (including Nothofagus, Proteaceae, Laurelia-type Monimiaceae, Myricaceae, and conifers). Birkenmajer and Zastawniak [1989a] suggested, for their assemblages, drier conditions than for the Late Cretaceous Zamek flora.

Birkenmajer and Zastawniak [1989a, b] also summarize views expressed earlier [e.g., Birkenmajer, 1985] that the King George Island Eocene to Miocene sediments reflect a succession of glacial and interglacial events, the earliest (Krakow) glaciation occurring in the early Eocene (post-Fildes Fossil Hill flora). Plant fossils on King George Island were believed to be remains of vegetation that grew during intervening warm phases or interglacials. Early Eocene glaciation, however, seems incompatible with a plethora of other evidence (e.g., see *Stott et al.* [1990]) for a thermal maximum in the early Eocene. *Birkenmajer* [1991] later noted that the Krakow Glaciation was probably minor and related to local Antarctic Peninsula mountain glaciation.

In the James Ross Basin, outcrops of early Tertiary age are known from Seymour and Cockburn islands. Leaf fossil assemblages were described by *Dusén* [1908] and more recently by *Case* [1988], wood specimens were described by *Gothan* [1908] and *Francis* [1986, 1991], and spore and pollen assemblages were described by *Cranwell* [1959], *Fleming and Askin* [1982], *Askin* [1988a, 1990b], *Askin et al.* [1991], *Baldoni and Barreda* [1986], and *Zamaloa et al.* [1987].

Fossil foliage described by *Dusén* [1908] and *Case* [1988] from the Paleocene Cross Valley Formation includes diverse ferns and angiosperms plus some conifers. Dusén noted the presence of cool temperate taxa (*Nothofagus*, Proteaceae, Cunoniaceae, Winteraceae), plus some with more subtropical affinities (including forms assigned to Monimiaceae and Melanostomaceae; see Zastawniak, in the appendix of *Birkenmajer and Zastawniak* [1989b], for a differing opinion). Calcified wood from the Cross Valley Formation is from conifer and *Nothofagus* trees, and *Francis* [1991] notes that the widths of annual rings in this and Eocene La Meseta woods indicate warm, favorable growing conditions. This contrasts with the very narrow rings in woods from the late Maastrichtian–Danian cooler phase.

Leaves in the La Meseta Formation are mainly those of *Nothofagus*, which are relatively large (notophyllous) in the lower part (?upper lower/middle Eocene) and smaller in the upper part (upper Eocene) of the formation [*Case*, 1988]. Palynomorph assemblages suggest that many of the plant families represented in the Cretaceous vegetation also occur in the Paleocene and Eocene (Figure 3, Table 1). Some new pollen taxa appeared in middle and upper Eocene sediments on Seymour Island, including forms with unknown modern affinities (R. A. Askin, unpublished data).

There is a major difference between Eocene and earlier floras on Seymour Island. Conifers (and particularly *Lagarostrobus*) dominate Campanian to Paleocene floras, whereas Eocene palynofloras are noticeably much richer in *Nothofagus*. For Seymour floras, it is not known how rapidly this change occurred (data are unobtainable because section spanning this transition is missing), or why. Possibly, the proliferation of *Nothofagus* forests was a response to oceanic and atmospheric circulation changes causing a more seasonal rainfall pattern (with a more pronounced dry season) than the "ever-wet" climates of the Maastrichtian-Paleocene, or the change could have been temperature

related. In the South Shetlands, *Nothofagus* dominates ?middle/upper Eocene and younger floras, while Paleocene/early Eocene floras contain more mixed assemblages. To the northeast, on the South Orkney microcontinent, *Nothofagus* pollen dominate middle Eocene assemblages (ODP Leg 113, Site 696 [*Mohr*, 1990]). In Australia, a parallel shift to *Nothofagus*-dominated vegetation took place in the middle Eocene and is believed to be a response to cooling temperatures [*Macphail et al.*, 1992].

On the western side of the Antarctic Peninsula, poorly preserved angiosperm leaves, pollen (*Nothofagus* and other angiosperms), and wood of Tertiary age occur at the southern end of Adelaide Island [*Jefferson*, 1980]. Further south, also in Tertiary rocks, angiosperm leaf remains were found on the Elgar Uplands, northern Alexander Island (present latitude 70°S [*Thomson and Burn*, 1977]). In both localities the leaves may include *Nothofagus*, and the notophyllous size (leaves >10 cm at the more southern locality) indicated a possible warm temperate climate to *Thomson and Burn* [1977]. These occurrences may be pre-Oligocene, based on this climatic information. Podocarp and araucarian pollen were also reported from a single Tertiary sample from eastern Brabant Island [*Palma-Heldt*, 1987].

Oligocene plant material is presently scarce. The few Oligocene assemblages indicate vegetation that is greatly reduced in diversity (suggesting cooler temperatures?), when compared with older Eocene Fildes and La Meseta floras.

Leaf, wood, and palynomorph floras ranging in age from ?middle Eocene to early Oligocene are known from the southern part of King George Island (in Arctowski Cove and Point Thomas Formations, Ezcurra Inlet Group). *Nothofagus* and ferns are diverse and abundant in the Eocene/Oligocene palynomorph assemblages (uppermost Arctowski Cove Formation), suggesting a *Nothofagus* forest with fern understory, or a *Nothofagus* forest and fern-shrub communities resembling those growing today in warm (frost-free) moist lowlands [*Stuchlik*, 1981]. The diverse angiosperm floras characterizing Fildes and La Meseta assemblages are notably absent.

For the early Oligocene island arc floras with *Nothofagus* and ferns (Point Thomas Formation), *Birkenmajer and Zastawniak* [1989a] made comparisons with extant fern bush communities of southern oceanic islands with (quoting data for Gough and Auckland islands, respectively) MATs between 11.7° and 15°C and mean annual precipitation between 3225 and 1220 mm.

The youngest of the Paleogene floras is from Point Hennequin in the Mount Wawel Formation, deposited at about the Oligocene-Miocene transition. The Point Hennequin floras include foliage of several species of *Nothofagus* and Podocarpaceae, plus a few other angiosperms, *Equisetum*, and ferns. They were compared with extant southern, moist, cool temperate *Notho-*

faqus-podocarp communities [*Zastawniak*, 1981], perhaps growing in MATs of 5°–8°C and 600- to 4300-mm annual rainfall, as in Patagonian-Magellanian forests today [*Birkenmajer and Zastawniak*, 1989a]. *Barton* [1964] had earlier noted that the predominance of microphyllous leaves with dentate margins was consistent with cooler temperatures for the Hennequin floras, as compared with older assemblages.

More certain ages and relationships between the floras need to be established for the South Shetland early Tertiary plant remains and for the James Ross Basin assemblages, before their climatic signals can be fully appreciated. From what is presently known, however, there is an obvious trend of decreasing floral diversity in Antarctic Peninsula floras from the Eocene through the Oligocene (Figure 3; note that the apparent absence of some families may reflect, in part, the paucity of Oligocene data). A similar trend and change in vegetation took place in Australia [*Macphail et al.*, 1992]. *Stott et al.* [1990] recognized a series of cooling steps in their ODP Leg 113 stable isotope data for the Paleogene. Following the early Eocene thermal maximum, major cooling steps occurred at 43 Ma (middle middle Eocene), 40 Ma (latest Eocene), and ~36 Ma (middle Oligocene), with intervening warmer phases. It appears from the plant fossil record, although better stratigraphic control of floras is required to be tested, that the mixed broadleaved and conifer rain forests of the Paleocene-Eocene became progressively impoverished with each successive cooling phase. Magnetic anomaly data indicate spreading in the Drake Passage during the Oligocene, leading to deepwater circulation [e.g., *Barker and Burrell*, 1982]. The final land connections (easily passable by terrestrial biotas) between the Antarctic Peninsula and South America were probably lost by the end of the Eocene, if not somewhat earlier, from paleontologic evidence [*Woodburne and Zinsmeister*, 1984]. Henceforth, at each cooling phase a northward dispersal to warmer climes of plant taxa was not possible, nor a southward "restocking" when conditions ameliorated, except by taxa whose seeds could remain fertile after transport by birds, seawater, etc. Thus when conditions deteriorated, plant taxa that died out and did not survive in coastal refugia disappeared from the continent of Antarctica.

CONCLUSIONS

Presently known Cretaceous-Tertiary exposures are restricted to the Antarctic Peninsula region. Thus Late Cretaceous–early Tertiary vegetative cover for the vast majority of Antarctica remains unknown. Part of the flora can be inferred by indirect means: by determining likely dispersal paths between adjacent land masses, from drill hole records penetrating the coastal margins of the continent, and from palynomorph assemblages recycled into younger sediments.

Despite these constraints, the presently available "in situ" plant fossils provide us with a tantalizing glimpse of past vegetational types and the paleoclimatic information we can derive from them. For the Antarctic Peninsula area, some key points are as follows:

1. Albian/Cenomanian vegetation included araucarian and podocarp conifers, plus cheirolepidacean conifers, pteridosperms, bennettitaleans, cycads, ginkgos, abundant and diverse cryptogams, and some angiosperms.

2. Angiosperms diversified during the Late Cretaceous, replacing, by the end of the Cretaceous, many of the cryptogams and gymnosperms that characterized earlier Cretaceous vegetation.

3. By the Maastrichtian, the flora included immigrants from more northern latitudes and many newly evolved taxa. Antarctica was a locus of evolutionary innovation. The angiosperm flora contained many endemic species.

4. Campanian-Paleocene vegetation was primarily podocarp conifer-Proteaceae-*Nothofagus* rain forest.

5. Late Cretaceous-early Tertiary climates in the Antarctic Peninsula area were warm to cool temperate, with high rainfall. Cooling occurred through the Maastrichtian and into the early Paleocene, with a brief warm interval in the latest Maastrichtian.

6. Some change in the vegetation, but not major ecological trauma, is evident across the Cretaceous-Tertiary transition.

7. Late Paleocene–early/middle Eocene vegetation included mixed, diverse broadleaved angiosperms with conifers and ferns.

8. Humid, warm temperate conditions are likely for the diverse Paleocene/Eocene floras.

9. There was a shift from the Cretaceous-Paleocene conifer-dominated vegetation to *Nothofagus*-dominated vegetation in the Eocene.

10. Reduction in vegetational diversity is evident in the late Eocene/Oligocene and through the Oligocene. Sparse Oligocene floras suggest *Nothofagus*-fern communities.

11. Plant evidence is consistent with stepwise cooling during the late Paleogene.

Acknowledgments. This paper was greatly improved by helpful reviews from Stephen R. Jacobson, James P. Kennett, Elizabeth M. Truswell, Jack A. Wolfe, and one anonymous reviewer. I thank them for their valuable suggestions. Research was supported by National Science Foundation grant DPP-9019378.

REFERENCES

Askin, R. A., The Campanian to Paleocene palynological succession of Seymour and adjacent islands, northeastern Antarctic Peninsula, in Geology and Paleontology of Seymour Island, Antarctic Peninsula, *Mem. Geol. Soc. Am.*, *169*, 131–153, 1988a.

Askin, R. A., The palynological record across the Cretaceous/
Tertiary transition on Seymour Island, Antarctica, in Geology and Paleontology of Seymour Island, Antarctic Peninsula, *Mem. Geol. Soc. Am.*, *169*, 155–162, 1988*b*.

Askin, R. A., Endemism and heterochroneity in the Late
Cretaceous (Campanian) to Paleocene palynofloras of Seymour Island, Antarctica: Implications for origins, dispersal
and palaeoclimates of southern floras, in Origins and Evolution of the Antarctic Biota, *Spec. Publ. Geol. Soc. London*,
147, 107–119, 1989.

Askin, R. A., Cryptogam spores from the upper Campanian
and Maastrichtian of Seymour Island, Antarctica, *Micropaleontology*, *36*, 141–156, 1990*a*.

Askin, R. A., Campanian to Paleocene spore and pollen
assemblages of Seymour Island, Antarctica, in Proceedings
of the 7th International Palynological Congress, *Rev. Palaeobot. Palynol.*, *65*, 105–113, 1990*b*.

Askin, R. A., and R. A. Spicer, The Late Cretaceous and
Cenozoic history of vegetation and climate at northern and
southern high latitudes: A comparison, in *The Effects of Past
Global Change on Life*, Geophysics Study Committee, National Academy Press, Washington, D. C., in press, 1992.

Askin, R. A., D. H. Elliot, J. F. Stilwell, and W. J. Zinsmeister, Stratigraphy and paleontology of Cockburn Island, Antarctic Peninsula, *South Am. J. Earth Sci.*, *4*, 99–117, 1991.

Bailey, I. W., and E. W. Sinnott, A botanical index of
Cretaceous and Tertiary climates, *Science*, *41*, 831–834,
1915.

Baldoni, A. M., and V. Barreda, Estudio palinológico de las
Formaciones López de Bertodano y Sobral, Isla Vicecomo-
doro Marambio, Antártida, *Bol. Inst. Geol. Univ. Sao Paulo
Ser. Cient.*, *17*, 89–98, 1986.

Baldoni, A. M., and F. Medina, Fauna y microflora del
Cretácico, en bahía Brandy, isla James Ross, Antártida, *Ser.
Cient. INACH*, *39*, 43–58, 1989.

Banerji, J., and Y. LeMoigne, Significant additions to the
Upper Triassic of Williams Point, Livingston Island, South
Shetland Islands (Antarctica), *Geobios Jodhpur India*, *20*,
469–487, 1987.

Banerji, J., Y. LeMoigne, and T. Torres, Significant additions
to the Upper Triassic flora of Williams Point, Livingston
Island, South Shetland Islands (Antarctica), *Ser. Cient.
INACH*, *36*, 33–58, 1987.

Barker, P. F., and J. Burrell, The influence upon Southern
Ocean circulation, sedimentation, and climate of the opening
of Drake Passage, in *Antarctic Geoscience*, edited by C.
Craddock, pp. 377–385, University of Wisconsin Press,
Madison, Wis., 1982.

Barrera, E., and B. T. Huber, Evolution of Antarctic waters
during the Maastrichtian: Foraminifer oxygen and carbon
isotope ratios, ODP Leg 113, *Proc. Ocean Drill. Program
Sci. Results*, *113*, 813–827, 1990.

Barrera, E., B. T. Huber, S. M. Savin, and P. N. Webb,
Antarctic marine temperatures: Late Campanian through
early Paleocene, *Paleoceanography*, *2*, 21–47, 1987.

Barton, C. M., Significance of the Tertiary fossil floras of King
George Island, South Shetland Islands, in *Antarctic Geol-
ogy*, edited by R. J. Adie, pp. 603–609, North-Holland,
Amsterdam, 1964.

Birkenmajer, K., Onset of Tertiary continental glaciation in the
Antarctic Peninsula sector (West Antarctica), *Acta Geol.
Pol.*, *35*, 1–31, 1985.

Birkenmajer, K., Tertiary glaciation in the South Shetland
Islands, West Antarctica: Evaluation of data, in *Geological
Evolution of Antarctica*, edited by M. R. A. Thomson, J. A.
Crame, and J. W. Thomson, pp. 629–632, Cambridge Uni-
versity Press, New York, 1991.

Birkenmajer, K., and E. Zastawniak, Plant remains of the
Dufayel Island Group (early Tertiary?), King George Island,
South Shetland Islands (West Antarctica), *Acta Palaeobot.*,
26, 33–54, 1986.

Birkenmajer, K., and E. Zastawniak, Late Cretaceous–early
Tertiary floras of King George Island, West Antarctica:
Their stratigraphic distribution and palaeoclimatic signifi-
cance, in Origins and Evolution of the Antarctic Biota, *Spec.
Publ. Geol. Soc. London*, *147*, 227–240, 1989*a*.

Birkenmajer, K., and E. Zastawniak, Late Cretaceous–early
Neogene vegetation history of the Antarctic Peninsula sec-
tor, Gondwana break-up and Tertiary glaciations, *Bull. Pol.
Acad. Sci. Earth Sci.*, *37*, 63–88, 1989*b*.

Cao Liu, Late Cretaceous sporopollen flora from Half Three
Point on Fildes Peninsula of King George Island, Antarctica,
in *Proceedings of the International Symposium on Antarctic
Research*, pp. 151–156, China Ocean Press, Tianjin, China,
1989.

Carlquist, S., Ecological factors in wood evolution: A floristic
approach, *Am. J. Bot.*, *64*, 887–896, 1977.

Carlquist, S., Pliocene *Nothofagus* wood from the Transantarctic Mountains, *Aliso*, *11*, 571–583, 1987.

Case, J. A., Paleogene floras from Seymour Island, Antarctic
Peninsula, in Geology and Paleontology of Seymour Island,
Antarctic Peninsula, *Mem. Geol. Soc. Am.*, *169*, 523–530,
1988.

Cortemiglia, G. C., P. Gastaldo, and R. Terranova, Studio di
piante fossili trovate nella King George Island delle Isole
Shetland del Sud (Antartide), *Atti Soc. Ital. Sci. Nat. Mus.
Civ. Stor. Nat. Milano*, *122*, 37–61, 1981.

Crame, J. A., D. Pirrie, J. B. Riding, and M. R. A. Thomson,
Campanian-Maastrichtian (Cretaceous) stratigraphy of the
James Ross Island area, Antarctica, *J. Geol. Soc. London*,
148, 1125–1140, 1991.

Cranwell, L. M., Fossil pollen from Seymour Island, Antarctica, *Nature*, *184*, 1782–1785, 1959.

Cranwell, L. M., Antarctica: Cradle or grave for its *Nothofagus*, in *Ancient Pacific Floras, the Pollen Story*, edited by L.
M. Cranwell, pp. 87–93, University of Hawaii Press, Honolulu, Hawaii, 1964.

Cranwell, L. M., Palynological intimations of some pre-
Oligocene Antarctic climates, in *Palaeoecology of Africa*,
edited by van Zinderen Bakker, pp. 1–19, S. Balkema, Cape
Town, 1969.

Creber, G. T., The south polar forest ecosystem, in *Antarctic
Paleobiology*, edited by T. N. Taylor and E. L. Taylor, pp.
37–41, Springer-Verlag, New York, 1990.

Creber, G. T., and W. G. Chaloner, Tree growth in the
Mesozoic and early Tertiary and the reconstruction of palaeoclimates, *Palaeogeogr. Palaeoclimatol. Palaeoecol.*, *52*,
35–60, 1985.

Czajkowski, S., and O. Rösler, Plantas fósseis da Península
Fildes; Ilha Rei Jorge (Shetlands do Sul): Morfografia das
impressões foliares, *An. Acad. Bras. Cien.*, Suppl., *58*,
99–110, 1986.

Del Valle, R. A., M. T. Diaz, and E. J. Romero, Preliminary
report on the sedimentites of Barton Peninsula, 25 de Mayo
Island (King George Island), South Shetland Islands, Argen-
tine Antarctica, *Inst. Antart. Argen. Contrib.*, *308*, 1–19,
1984.

Dettmann, M. E., Significance of the Cretaceous-Tertiary
spore genus *Cyatheacidites* in tracing the origin and migration of *Lophosoria* (Filicopsida), *Spec. Pap. Palaeontol.*, *35*,
63–94, 1986.

Dettmann, M. E., Antarctica: Cretaceous cradle of austral
temperate rainforests?, in Origins and Evolution of the
Antarctic Biota, *Spec. Publ. Geol. Soc. London*, *147*, 89–
105, 1989.

Dettmann, M. E., and R. W. Hedlund, *Stellidiopollis*, a new
pollen genus from the Late Cretaceous of Antarctica and
southern Australia, *Pollen Spores*, *30*, 45–56, 1988.

Dettmann, M. E., and D. M. Jarzen, Angiosperm pollen from uppermost Cretaceous strata of southeastern Australia and the Antarctic Peninsula, *Mem. Assoc. Australas. Palaeontol.*, 5, 217–237, 1988.

Dettmann, M. E., and D. M. Jarzen, The Antarctic/Australian rift valley: Late Cretaceous cradle of northeastern Australasian relicts?, in Proceedings of the 7th International Palynological Congress, *Rev. Palaeobot. Palynol.*, 65, 131–144, 1990.

Dettmann, M. E., and M. R. A. Thomson, Cretaceous palynomorphs from the James Ross Island area, Antarctica—A pilot study, *Br. Antarct. Surv. Bull.*, 77, 13–59, 1987.

Dettmann, M. E., D. T. Pocknall, E. J. Romero, and M. del C. Zamaloa, *Nothofagidites* Erdtman ex Potonié, 1960; a catalogue of species with notes on the paleogeographic distribution of *Nothofagus* B1 (Southern Beech), *N. Z. Geol. Surv. Paleontol. Bull.*, 60, 79 pp., 1990.

Dusén, P., Über die tertiäre Flora der Seymour Insel, *Wiss. Ergeb. Schwed. Südpolarexped. 1901–1903*, 3(3), 127 pp., 1908.

Elliot, D. H., Tectonic setting and evolution of the James Ross Island basin, northern Antarctic Peninsula, in Geology and Paleontology of Seymour Island, Antarctic Peninsula, *Mem. Geol. Soc. Am.*, 169, 541–555, 1988.

Elliot, D. H., and S. Hoffman, Geologic studies on Seymour Island, *Antarct. J. U. S.*, 24(5), 3–5, 1989.

Fleming, R. F., and R. A. Askin, An early Tertiary coal bed on Seymour Island, Antarctic Peninsula, *Antarct. J. U. S.*, 17(5), 67, 1982.

Francis, J. E., Growth rings in Cretaceous and Tertiary wood from Antarctica and their palaeoclimatic implications, *Palaeontology*, 29, 665–684, 1986.

Francis, J. E., Palaeoclimatic significance of Cretaceous–early Tertiary fossil forests of the Antarctic Peninsula, in *Geological Evolution of Antarctica*, edited by M. R. A. Thomson, J. A. Crame, and J. W. Thomson, pp. 623–627, Cambridge University Press, New York, 1991.

Gothan, W., Die fossilien Holzer von der Seymour und Snow Hill Insel, *Wiss. Ergeb. Schwed. Südpolarexped. 1901–1903*, 3(8), 33 pp., 1908.

Ineson, J. R., J. A. Crame, and M. R. A. Thomson, Lithostratigraphy of the Cretaceous strata of west James Ross Island, Antarctica, *Cretaceous Res.*, 7, 141–159, 1986.

Jarzen, D. M., and M. E. Dettmann, Taxonomic revision of *Tricolpites reticulatus* Cookson ex Couper, 1953 with notes on the biogeography of *Gunnera* L., *Pollen Spores*, 31, 97–112, 1989.

Jarzen, D. M., and M. E. Dettmann, Pollen evidence for Late Cretaceous differentiation of Proteaceae in southern polar forests, *Can. J. Bot.*, 69, 901–906, 1991.

Jefferson, T. H., Angiosperm fossils in supposed Jurassic volcanogenic shales, Antarctica, *Nature*, 285, 157–158, 1980.

Lacey, W. S., and R. C. Lucas, The Triassic flora of Livingston Island, South Shetland Islands, *Br. Antarct. Surv. Bull.*, 53, 157–173, 1981.

LeMoigne, Y., Confirmation de l'existence d'une flore triasique dans l'île Livingston des Shetland du Sud (Oeust Antarctique), *C. R. Acad. Sci., Ser. 2*, 304, 543–546, 1987.

Li Haomin and Shen Yanbin, A preliminary study of the Eocene flora from the Fildes Peninsula of King George Island, Antarctica, in *Proceedings of the International Symposium on Antarctic Research*, pp. 128–135, China Ocean Press, Tianjin, China, 1989.

Li Haomin and Song Dekang, Fossil remains of some angiosperms from King George Island, Antarctica, *Acta Palaeontol. Sin.*, 27, 399–403, 1988.

Lucas, R. C., and W. S. Lacey, A permineralized wood flora of probable early Tertiary age from King George Island, South Shetland Islands, *Br. Antarct. Surv. Bull.*, 53, 147–151, 1981.

Lyra, C., Palinologia de sedimentos terciários da península Fildes, ilha Rei George, ilhas Shetland do Sul Antártica, e algumas considerações paleoambientais, *An Acad. Bras. Cien.*, Suppl., 58, 137–147, 1986.

Macphail, M. K., N. Alley, E. M. Truswell, and I. R. Sluiter, Early Tertiary vegetation: Evidence from spores and pollen, in *Australian Vegetation History: Cretaceous to Recent*, edited by R. S. Hill, in press, 1992.

Mildenhall, D. C., Terrestrial palynology, in Antarctic Cenozoic History From the CIROS-1 Drillhole, McMurdo Sound, edited by P. J. Barrett, *DSIR Bull. N. Z.*, 245, 119–127, 1989.

Mohr, B. A. R., Eocene and Oligocene sporomorphs and dinoflagellate cysts from Leg 113 drill sites, Weddell Sea, Antarctica, *Proc. Ocean Drill. Program Sci. Results*, 113, 595–612, 1990.

Nix, J. A., An environmental analysis of Australian rainforests, in *Australian National Rainforest Study Report*, vol. 1, edited by G. L. Werren and A. P. Kershaw, pp. 421–425, Department of Geography, Monash University, Clayton, Victoria, Australia, 1984.

Olivero, E., R. A. Scasso, and C. A. Rinaldi, Revision of the Marambio Group, James Ross Island, Antarctica, *Inst. Antart. Argen. Contrib.*, 351, 1–29, 1986.

Orlando, H. A., The fossil flora of the surroundings of Ardley Peninsula (Ardley Island), 25 de Mayo Island (King George Island), South Shetland Islands, in *Antarctic Geology*, edited by R. J. Adie, pp. 629–636, North-Holland, Amsterdam, 1964.

Orlando, H. A., A new Triassic flora from Livingston Island, South Shetland Islands, *Br. Antarct. Surv. Bull.*, 16, 1–13, 1968.

Palma-Heldt, S., Estudia palinológico en el terciario de las islas Rey Jorge y Brabante, territorio insular antártico, *Ser. Cient. INACH*, 36, 59–71, 1987.

Pirrie, D., and J. D. Marshall, High-paleolatitude Late Cretaceous paleotemperatures: New data from James Ross Island, Antarctica, *Geology*, 18, 31–34, 1990.

Pocknall, D. T., and Y. M. Crosbie, Pollen morphology of *Beauprea* (Proteaceae): Modern and fossil, *Rev. Palaeobot. Palynol.*, 53, 305–327, 1988.

Rees, P. M., and J. L. Smellie, Cretaceous angiosperms from an allegedly Triassic flora at Williams Point, Livingston Island, South Shetland Islands, *Antarct. Sci.*, 1, 239–248, 1989.

Riding, J. B., J. M. Keating, M. G. Snape, S. Newham, and D. Pirrie, Preliminary Jurassic and Cretaceous dinoflagellate cyst stratigraphy of the James Ross Island area, Antarctic Peninsula, *Newsl. Stratigr.*, 26, 19–39, 1992.

Shen Yanbin, Recent advances in research on the palaeontology of the Fildes Peninsula, King George Island, Antarctica, in *Proceedings of the International Symposium on Antarctic Research*, pp. 119–127, China Ocean Press, Tianjin, China, 1989.

Spicer, R. A., The significance of the Cretaceous flora of northern Alaska for the reconstruction of the climate of the Cretaceous, *Geol. Jahrb., Reihe A*, 96, 265–291, 1987.

Spicer, R. A., Physiological characteristics of land plants in relation to environment through time, *Proc. R. Soc. Edinburgh*, 80, 321–329, 1989.

Spicer, R. A., Reconstructing high-latitude Cretaceous vegetation and climate: Arctic and Antarctic compared, in *Antarctic Paleobiology*, edited by T. N. Taylor and E. L. Taylor, pp. 26–36, Springer-Verlag, New York, 1990.

Spicer, R. A., and J. L. Chapman, Climate change and the evolution of high latitude terrestrial vegetation and floras, *Trends Ecol. Evol.*, 5, 279–284, 1990.

Spicer, R. A., and J. T. Parrish, Late Cretaceous–early Tertiary palaeoclimates of northern high latitudes: A quantitative view, *J. Geol. Soc. London*, 147, 329–341, 1990.

Stott, L. D., and J. P. Kennett, The paleoceanographic and paleoclimatic signature of the Cretaceous/Paleogene boundary in the Antarctic: Stable isotope results from ODP Leg 113, *Proc. Ocean Drill. Program Sci. Results, 113*, 829–848, 1990.

Stott, L. D., J. P. Kennett, N. J. Shackleton, and R. M. Corfield, The evolution of Antarctic surface waters during the Paleogene: Inferences from the stable isotopic composition of planktonic foraminifers, ODP Leg 113, *Proc. Ocean Drill. Program Sci. Results, 113*, 849–863, 1990.

Stuchlik, L., Tertiary pollen spectra from the Ezcurra Inlet Group of Admiralty Bay, King George Island (South Shetland Islands, Antarctica), *Stud. Geol. Pol., 72*, 109–132, 1981.

Thomson, M. R. A., and R. W. Burn, Angiosperm fossils from latitude 70°S, *Nature, 269*, 139–141, 1977.

Tokarski, A. K., W. Danowski, and E. Zastawniak, On the age of fossil flora from Barton Peninsula, King George Island, West Antarctica, *Polish Polar Res., 8*, 293–302, 1987.

Torres, G. T., *Nothofagoxylon antarcticus*, n. sp., madera fósil del Terciario de la Isla Rey Jorge, Islas Shetland del Sur, Antártica, *Ser. Cient. INACH, 31*, 19–52, 1984.

Torres, T., and Y. LeMoigne, Maderas fósiles terciarias de la Formacion Caleta Arctowski, isla Rey Jorge, Antártica, *Ser. Cient. INACH, 37*, 69–107, 1988.

Torres, T., and Y. LeMoigne, Hallazgos de maderas fósiles de Angiospermas y Gimnospermas del Cretácico Superior en punta Williams, isla Livingston, islas Shetland del Sur, Antártica, *Ser. Cient. INACH, 39*, 9–29, 1989.

Torres, T., and H. Meon, Estudio palinologico preliminar de cerro Fosil, península Fildes, isla Rey Jorge, Antártica, *Ser. Cient. INACH, 40*, 21–39, 1990.

Torres, T., M. A. Hansen, and A. Linn, Flora fossil de aldredores de Punta Suffield, Isla Rey Jorge, islas Shetland del Sur, *Bol. Antart. Chil., 4*, 1–7, 1984.

Troncoso, A. A., Nuevas órgano-especies en la tafoflora terciaria inferior de Península Fildes, Isla Rei Jorge, Antártica, *Ser. Cient. INACH, 34*, 23–46, 1986.

Truswell, E. M., Recycled Cretaceous and Tertiary pollen and spores in Antarctic marine sediments: A catalogue, *Palaeontographica, 186B*, 121–174, 1983.

Truswell, E. M., Cretaceous and Tertiary vegetation of Antarctica: A palynological perspective, in *Antarctic Paleobiology*, edited by T. N. Taylor and E. L. Taylor, pp. 71–88, Springer-Verlag, New York, 1990.

Truswell, E. M., Antarctica: A history of terrestrial vegetation, in *The Geology of Antarctica*, edited by R. J. Tingey, pp. 499–528, Oxford University Press, New York, 1991.

Upchurch, G. R., Dispersed angiosperm cuticles, in *Phytodebris: Notes for a Workshop on the Study of Fragmentary Plant Remains*, edited by B. H. Tiffney, pp. 65–92, Botanical Society of America, Paleobotany Section, 1989.

Upchurch, G. R., and R. A. Askin, Latest Cretaceous and earliest Tertiary dispersed plant cuticles from Seymour Island, *Antarct. J. U. S., 24*(5), 7–10, 1989.

Upchurch, G. R., and J. A. Wolfe, Mid-Cretaceous to early Tertiary vegetation and climate: Evidence from fossil leaves and woods, in *The Origins of Angiosperms and Their Biological Consequences*, edited by E. M. Friis, W. G. Chaloner, and P. R. Crane, pp. 75–105, Cambridge University Press, New York, 1987.

Webb, L. J., Physiognomic classification of Australia rain forests, *J. Ecol., 47*, 551–570, 1959.

Wolfe, J. A., Tertiary climatic fluctuations and methods of analysis of Tertiary floras, *Palaeogeogr. Palaeoclimatol. Palaeoecol., 9*, 27–57, 1971.

Wolfe, J. A., Temperature parameters of humid to mesic forests of eastern Asia and relations to forests of other regions of the northern hemisphere and Australasia, *U.S. Geol. Surv. Prof. Pap., 1106*, 1–36, 1979.

Wolfe, J. A., Tertiary climates and floristic relationships at high latitudes in the northern hemisphere, *Palaeogeogr. Palaeoclimatol. Palaeoecol., 30*, 313–323, 1980.

Wolfe, J. A., Distribution of major vegetational types during the Tertiary, in *The Carbon Cycle and Atmospheric CO_2: Natural Variations Archean to Present, Geophys. Monogr. Ser.*, vol. 32, edited by K. E. T. Sundquist and W. S. Broecker, pp. 357–375, AGU, Washington, D. C., 1985.

Wolfe, J. A., CLAMP: A method of accurately estimating paleoclimatic parameters from leaf assemblages (abstract), *Geol. Soc. Am. Abstr. Programs, 23*, A179, 1991.

Wolfe, J. A., and G. R. Upchurch, North American nonmarine climates and vegetation during the Late Cretaceous, *Palaeogeogr. Palaeoclimatol. Palaeoecol., 61*, 33–77, 1987.

Woodburne, M. O., and W. J. Zinsmeister, The first land mammal from Antarctica and its biogeographic implications, *J. Paleontol., 58*, 913–948, 1984.

Zamaloa, M. C., E. J. Romero, and L. Stinco, Polen y esporas de la Formación La Meseta (Eoceno Superior-Oligoceno) de la isla Marambio (Seymour), Antártida, in VII Simposio Argentino de Paleobotanica y Palinologia Actas, *Actas Paleobot. Palin.*, 199–203, 1987.

Zastawniak, E., Tertiary leaf flora from the Point Hennequin Group of King George Island (South Shetland Islands, Antarctica), preliminary report, *Stud. Geol. Pol., 72*, 97–108, 1981.

Zastawniak, E., Late Cretaceous leaf flora of King George Island, West Antarctica, in *Proceedings of the Symposium "Paleofloristic and Paleoclimatic Changes in the Cretaceous and Tertiary,"* pp. 81–85, Prague, 1990.

Zastawniak, E., R. Wrona, A. Gazdzicki, and K. Birkenmajer, Plant remains from the top part of the Point Hennequin Group (upper Oligocene), King George Island (South Shetland Islands, Antarctica), *Stud. Geol. Pol., 81*, 143–164, 1985.

(Received January 2, 1992;
accepted April 13, 1992.)

PALEOGENE CHRONOLOGY OF SOUTHERN OCEAN DRILL HOLES: AN UPDATE

WUCHANG WEI

Scripps Institution of Oceanography, University of California, San Diego, La Jolla, California 92093-0215

The numerical ages of calcareous nannofossil and planktonic foraminiferal datums used in the Proceedings of the Ocean Drilling Program (ODP) volumes 113, 114, 119, and 120 are frequently those of Berggren et al. (1985). These ages were derived from calibration of biostratigraphic datums with magnetostratigraphy at mid-latitudes. As many of these microfossil datums are time transgressive from mid to high latitudes, interpretation of magnetostratigraphies and construction of age-depth curves for the Southern Ocean sites based on mid-latitude ages are prone to be in error. Fairly detailed magnetostratigraphies for different intervals of the Paleogene have recently become available for nine ODP sites (689, 690, 699, 700, 703, 744, 747, 748, and 752) in the Southern Ocean. This allows direct correlations of calcareous nannofossil and planktonic foraminifer datums with magnetostratigraphies and estimates of their numerical ages in the Southern Ocean. On the basis of these direct biomagnetostratigraphic correlations, published magnetostratigraphies for some intervals from Southern Ocean sites 690, 699, 700, and 703 are reinterpreted, and age-depth curves for these sites and sites 702, 737, 738, and 748 are constructed/reconstructed using the applicable published/reinterpreted magnetostratigraphic data and revised ages of biostratigraphic datums. The revised age models provide a more accurate time framework for these sediment sequences.

INTRODUCTION

The recent Ocean Drilling Program (ODP) legs 113, 114, 119, and 120 to the Atlantic Ocean and Indian Ocean sectors of the Southern Ocean have recovered a wealth of Paleogene sediment cores, which have yielded large amounts of paleoceanographic information. Undoubtedly, future studies of these cores will continue to provide important information on the developmental history of the Southern Ocean.

Accurate stratigraphy and chronology are important to virtually all such studies and quite often are critical to find causal links between events recognized in Southern Ocean cores and those in other areas. *Shackleton* [1986, p. 2] provided an excellent example to illustrate the importance of accurate stratigraphic correlation and precise chronology in advance of any attempt to find causal links:

> ... the Eocene-Oligocene boundary. At this time a major climatic change occurred, probably the most rapid cooling documented in the geological record: did it cause a major extinction in the ocean? No, all the extinctions used to recognize the boundary preceded the boundary. Did the cooling perhaps cause a big change in global erosion patterns that led to a change in the $^{87}Sr/^{86}Sr$ ratio in the ocean? No, the strontium isotope change preceded the cooling. Did the terminal Eocene tektite event cause the extinctions, or the cooling? No, it preceded in time, by at least a million years, any supposed effect. Until recently all these hypothetical links could be accommodated within the uncertainties of stratigraphic correlation.

With the importance of accurate stratigraphy and chronology in mind, I have recently examined some of the Southern Ocean cores and evaluated the biostratigraphies and magnetostratigraphies published in the work of *Barker et al.* [1990], *Ciesielski et al.* [1991], *Barron et al.* [1991*a*], and *Wise et al.* [1992]. On the basis of a comparison of results from these four Southern Ocean legs, I have reinterpreted some of the magnetostratigraphies, constructed new age-depth curves, or revised published age-depth curves based on reinterpreted magnetostratigraphies and revised ages of biostratigraphic datums. This paper provides an update on the Paleogene chronology of eight ODP sites (690, 699, 700, 702, 703, 737, 738, and 748) in the Southern Ocean (Figure 1).

METHODS

Biostratigraphic and magnetostratigraphic data are taken from *Barker et al.* [1990], *Ciesielski et al.* [1991], *Barron et al.* [1991*b*], and *Wise et al.* [1992]. Additional nannofossil data are derived from *Pospichal et al.* [1991], *Wei* [1991*a*], and this study. The ages of the biostratigraphic datums used in the ODP volumes to construct age-depth curves are frequently those of *Berggren et al.* [1985], which are based mostly on mid-latitude biomagnetostratigraphic correlations. Biomagnetostratigraphic correlations at several Southern Ocean sites, however, show considerably different ages

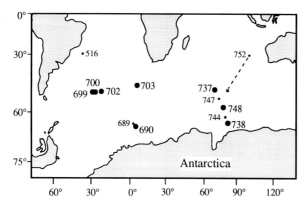

Fig. 1. Location map of Southern Ocean ODP sites for which chronology is updated in this paper. Other Deep Sea Drilling Project (DSDP)/ODP sites mentioned in the paper are shown by smaller dots. The paleogeographic position of Site 752 was substantially different from its present position and is shown as a cross.

in the high latitudes from those in the mid-latitudes for many nannofossil and planktonic foraminiferal datums (Figures 2–4, Tables 1–3). The details of the biomagnetostratigraphic correlations and datum ages summarized in Figures 2–4 and Tables 1–3 can be found in ODP volumes referenced above and in the work of *Wei and Wise* [1992]. The present paper will not further discuss these biomagnetostratigraphic correlations but will simply use this information in the reinterpretation of published magnetostratigraphies and in the construction of age-depth curves. The geomagnetic polarity time scale used in this study is that of *Berggren et al.* [1985]. As this polarity time scale is being revised [see *Berggren et al.*, 1992], the numerical ages given in this paper eventually should be adjusted accordingly, although most of the adjustments will be proportional.

REVISED CHRONOLOGY

Hole 690B

Biomagnetostratigraphic data in Hole 690B is summarized in Figure 5 with comparisons between the magnetostratigraphic interpretations of *Spiess* [1990], *Stott and Kennett* [1990], *Thomas et al.* [1990], and this study. *Thomas et al.* [1990] placed a major hiatus at about 102 meters below sea floor (mbsf) and interpreted the middle part of Core 12H as magnetic Chron C19, based primarily on the juxtaposition of the first occurrence (FO) of *Chiasmolithus oamaruensis* and the last occurrence (LO) of *Chiasmolithus solitus* at 102 mbsf, which eliminates nannofossil Subzone CP14b of *Okada and Bukry* [1980]. However, Subzone CP14b has a very short time span (≤0.2 m.y.) in the Southern Ocean

owing to the considerably older age of the FO of *C. oamaruensis* and the younger age of the LO of *C. solitus* relative to those in middle or low latitudes (Figure 2), and the juxtaposition of the two nannofossil datums does not need to indicate a hiatus. Furthermore, as the LO of *C. solitus* is at 102 mbsf, and this datum has been calibrated with the top of Chron C18 at sites 516, 689, and 748 (Figure 2), it is most appropriate to assign the middle part of Core 12 to Chron C18 rather than Chron C19 as indicated by *Thomas et al.* [1990].

The FOs of *Reticulofenestra reticulata* and *Reticulofenestra umbilica* are located in the lower part of Core 12 (Figure 5). These datums have been correlated with Chron C19 at a number of Southern Ocean sites, including the nearby Site 689. Consequently, the lower part of Core 12 is interpreted as Chron C19 rather than Chron C20 as suggested by *Spiess* [1990] and *Thomas et al.* [1990].

An unconformity slightly below the top of Chron C19 (Figure 5) is indicated by a scour mark at 106.8 mbsf [see *Barker et al.*, 1988, p. 193, Figure 8]. Probably not much sediment is missing based on a comparison of the thickness of chrons C19-C20 with chrons C17 and C18 above and Chron C21 below.

Stott and Kennett [1990] interpreted the interval between 130 mbsf and 137 mbsf as Chron C22. *Thomas et al.* [1990] gave a different interpretation but commented that "the interpretation of Stott and Kennett may be correct, because the nannofossil species might have diachronous first and last appearances" [*Thomas et al.*, 1990, p. 907]. The magnetostratigraphic interpretation of this study (Figure 5) is based on the FOs of *Discoaster lodoensis* and *Discoaster kuepperi* at about 136 mbsf. These two datums correlate with Subchron C23R at Site 752 (paleolatitude of ~50°S, Figure 1; Figure 3). In addition, the LO of *Tribrachiatus contortus* at 137 mbsf also suggests that the lower part of Core 15 is Subchron C23R or older. The magnetostratigraphic interpretation of *Stott and Kennett* [1990] would require that all these nannofossil datums are at least one chron younger at Site 690 than at Site 752 or at the mid-latitudes as summarized by *Berggren et al.* [1985]. This is considered unlikely.

An age-depth curve for the 95–120 mbsf interval is constructed in Figure 6 based on the reinterpreted magnetostratigraphy. The age-depth curve for other parts of the Paleogene remains the same as presented by *Spiess* [1990].

Hole 699A

Wei [1991a] recently worked out a detailed nannofossil biostratigraphy for the Eocene-Oligocene in Hole 699A. This results in a detailed age-depth curve for the 90–375 mbsf interval based on the nannofossil data and published magnetostratigraphic data of *Hailwood and*

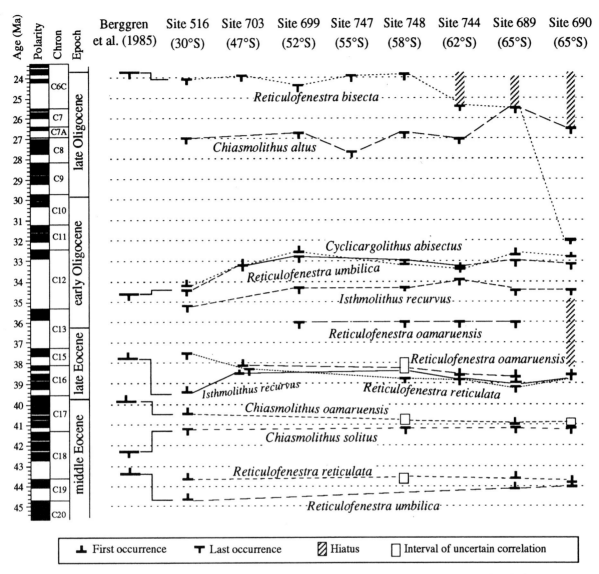

Fig. 2. Correlations of middle Eocene–Oligocene nannofossil datums with the geomagnetic polarity time scale of *Berggren et al.* [1985] at eight sites arranged from mid-latitudes to 65°S latitude. The datum placements of *Berggren et al.* [1985] are shown on the left for comparison. From *Wei and Wise* [1992].

Clement [1991a]. The main results from *Wei* [1991a] for Hole 699A are summarized here in Figure 7.

The magnetostratigraphic interpretation for the 85–215 mbsf interval is the same as that of *Hailwood and Clement* [1991a], who based it almost entirely on pattern matching when only limited biostratigraphic data were available then. The bottom of Subchron C12N, which could not be determined by *Hailwood and Clement* [1991a], has been identified at 249 mbsf with

the help of two nannofossil datums (the FO of *Cyclicargolithus abisectus* and the LO of *Reticulofenestra umbilica*) in Core 27. A number of nannofossil datums in the 240–375 mbsf interval have enabled the construction of a fairly detailed age-depth curve for the 215–273 mbsf interval. No age-depth curve was constructed by *Hailwood and Clement* [1991a] for this interval owing to the lack of paleomagnetic data and precise biochronologic data available then.

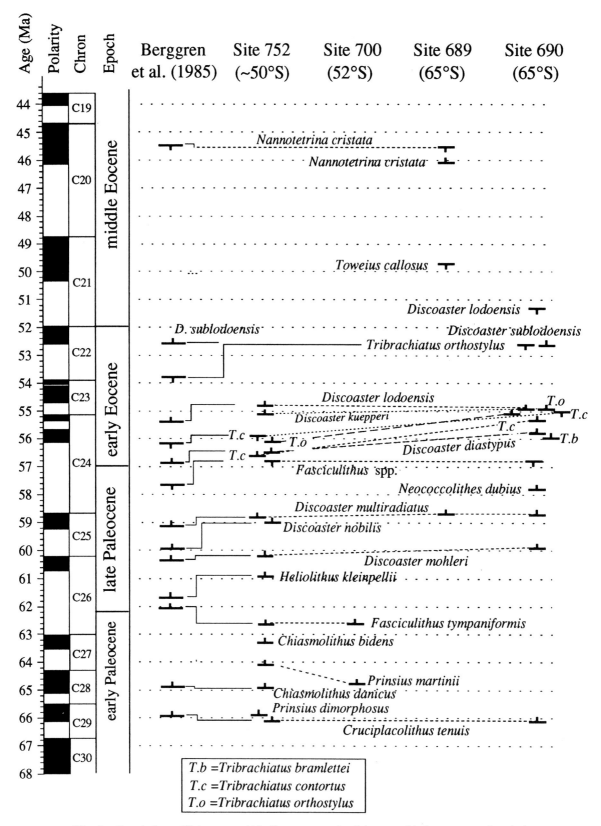

Fig. 3.   Correlations of Paleocene–middle Eocene nannofossil datums with the geomagnetic polarity time scale of *Berggren et al.* [1985] at four Southern Ocean sites.

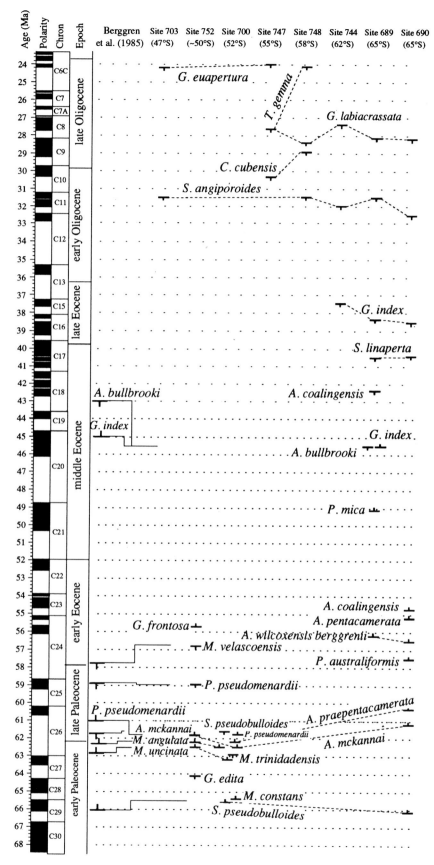

Fig. 4. Correlations of Paleogene planktonic foraminiferal datums with the geomagnetic polarity time scale of *Berggren et al.* [1985] at eight Southern Ocean sites.

TABLE 1. Estimated Ages of Middle Eocene–Oligocene Calcareous Nannofossil Datums as Calibrated With
Magnetostratigraphy at Different Southern Ocean ODP Sites (Higher Latitude Toward Right)

Nannofossil Datum	Site 703	Site 699	Site 747	Site 748	Site 744	Site 689	Site 690
LO *Reticulofenestra bisecta*	23.9	24.4	23.9	23.8	<25.5	<25.5	32.0
LO *Chiasmolithus altus*		26.7	27.7	26.7	27.0	<25.5	<26.6
FO *Cyclocargolithus abisectus*	33.2	32.6		33.1	33.5	32.7	32.8
LO *Reticulofenestra umbilica*	33.2	32.8		32.9	33.4	33.0	33.1
LO *Isthmolithus recurvus*				34.3	33.9	34.4	34.4
LO *Reticulofenestra oamaruensis*		36.0		36.0	36.0	36.0	
FO *Reticulofenestra oamaruensis*	38.2				38.7	38.7	
FO *Isthmolithus recurvus*	38.5				38.8	38.9	38.6
LO *Reticulofenestra reticulata*	38.3			38.7	38.8	39.1	38.6
FO *Chiasmolithus oamaruensis*						41.0	
LO *Chiasmolithus solithus*				41.2		41.2	41.2
FO *Reticulofenestra reticulata*						43.6	43.8
FO *Reticulofenestra umbilica*						44.1	44.0

Ages (Ma) are given according to the geomagnetic polarity time scale of *Berggren et al.* [1985]. FO, first occurrence; LO, last
occurrence. From *Wei and Wise* [1992].

TABLE 2. Estimated Ages of Paleocene–Middle Eocene Calcareous Nannofossil Datums as
Calibrated With Magnetostratigraphy

Nannofossil Datum	*Berggren et al.* [1985]	Site 752 (~50°S)	Site 700 (52°S)	Site 689 (65°S)	Site 690 (65°S)
LO *Nannotetrina cristata*	45.4			45.5	
FO *Nannotetrina cristata*	49.8			46.1	
LO *Toweius callosus*				49.7	46.2–49.5
LO *Discoaster lodoensis*					51.3
FO *D. sublodoensis*	52.6			51–52	52.6
LO *Tribrachiatus orthostylus*	53.7				52.6
FO *Discoaster lodoensis*	55.3	54.8			54.9
FO *Discoaster kuepperi*		55.1			54.9
LO *Tribrachiatus contortus*	56.8	55.9			55.0
FO *Tribrachiatus orthostylus*	56.6	56.1			55.1
FO *Discoaster diastypus*	56.5	56.5			55.8
FO *Tribrachiatus contortus*	56.8	56.6			55.3
FO *Tribrachiatus bramlettei*					55.9
LO *Fasciculithus* spp.	57.4 (57.6)	56.8			56.8
FO *Neococcolithes dubius*				57–58	57.8
FO *Discoaster multiradiatus*	59.2	58.8		58.7	58.7
FO *Discoaster nobilis*	59.4	59.0			
FO *Heliolithus riedellii*	60.0		?60.8		
FO *Discoaster mohleri*	60.4	60.2			59.9
FO *Heliolithus kleinpellii*	61.6	60.9	?61.1		
FO *Fasciculithus tympaniformis*	62.0	62.6	62.6		
FO *Fasciculithus* spp.		62.8			
FO *Chiasmolithus bidens*		63.3	?63.6		
FO *Prinsius martinii*		64.1	64.8		
FO *Chiasmolithus danicus*	64.8	64.9	?65		
FO *Prinsius dimorphosus*		65.9			
FO *Cruciplacolithus tenuis*	65.9	66.1			66.1

Ages (Ma) are given according to the geomagnetic polarity time scale of *Berggren et al.* [1985].

TABLE 3. Estimated Ages of Planktonic Foraminiferal Datums as Calibrated With Magnetostratigraphy

Foraminiferal Datum	Berggren et al. [1985]	Site 703	Site 752	Site 700	Site 747	Site 748	Site 744	Site 689	Site 690	Thomas et al. [1990]
LO *Globigerina euapertura*		24.2			24.0	22.9				
LO *Tenuitella gemma*					27.7	24.1				
LO *Globigerina labiacrassata*		22.5			27.7	28.4	27.4	28.1	28.2	28.2
LO *Chiloguemblina cubensis*					30.3	28.8				
LO *Subbotina angiporoides*		31.6				31.5	32.0	31.6	32.7	31.5
LO *Globigerinatheka index*		?38.0					37.5	38.3	38.6	38.4
LO *Subbotina linaperta*								40.5	40.4	40.0
LO *Acarinina coalingensis*								42.3		42.0
LO *Acarinina bullbrooki*	43.0							45.5		46.2
FO *Globigerinatheka index*	45.0							45.5		
FO *Pseudohastigerina mica*								49.1		47.5
FO *Acarinina coalingesis*									54.7	52.7
FO *Acarinina pentacamerata*									55.2	54.0
FO *Globigerina frontosa*			55.7							
FO *Acarinina wilcoxensis berggrenii*									56.5	56.6
FO *Planorotalites australiformis*								?56.2	57.5	57.5
LO *Morozovella velascoensis*	57.8		56.8							
LO *Planorotalites pseudomenardii*	58.8		58.9							
FO *Planorotalites pseudomenardii*	61.0			61.8						
LO *Subbotina pseudobulloides*	61.7			61.6						
FO *Morozovella pulilla*	62.0			62.1						
FO *Acarina mckannai*			61.8	62.5					61.2	62.4
FO *Acarina praepentacamerata*				62.2				?58.0	60.3	60.2
FO *Morozovella angulata*	62.3		62.2	62.5						
LO *Morozovella uncinata*				62.4						
FO *Planorotalites imitatus*									?62.8	63.4
LO *Morozovella trinidadensis*				62.9						
FO *Morozovella uncinata*	63.0		62.5	63.2						
FO *Planorotalites compressus*	64.5			?64.7						
FO *Subbotina trinidadensis*	64.5			?63.4						
FO *Subbotina inconstans*				65.3				?64.3	?64.8	65.1
FO *Subbotina pseudobulloides*	66.1			65.6					66.1	66.1

Ages (Ma) are given according to the geomagnetic polarity time scale of *Berggren et al.* [1985].

Hole 700B

Figure 8 shows the Paleocene biomagnetostratigraphic data for Hole 700B and a comparison of the magnetostratigraphic interpretation of this paper (left panel) with that of *Hailwood and Clement* [1991a]. The latter authors assigned the 327.3–333.3 mbsf interval to Subchron C29N, the 333.3–344.2 mbsf interval to Subchron C29R, and a major hiatus at 344.2 mbsf, which separates Chron C29 and Chron C31 (Figure 8). This magnetostratigraphy is reinterpreted here based on nannofossil [*Crux*, 1991; this study] and planktonic foraminiferal [*Nocchi et al.*, 1991] data (Figure 8).

Cruciplacolithus tenuis, *Prinsius dimorphosus*, and *Globoconusa daubjergensis* all first occur in Sample 36-CC (330.8 mbsf). The first occurrences of these species correlate with the middle part of Chron C29 in the Southern Ocean (Figures 2 and 3) and in the midlatitudes [*Berggren et al.*, 1985]. Below 330.8 mbsf, there are no Tertiary taxa but only Cretaceous forms.

That means the normally magnetized interval in the upper part of Core 37 should be assigned to Subchron C30N rather than Subchron C29N. Consequently, the normally magnetized interval in Core 38 is Subchron C31N, and there is no need to invoke a hiatus at 338 mbsf as suggested by *Hailwood and Clement* [1991a].

A detailed age-depth curve for the Paleogene section from Site 700 has not been published, although *Katz and Miller* [1991] utilized five biostratigraphic datums and two paleomagnetic events as their age model for their benthic foraminiferal and isotope study. A comparison of their age model parameters with those used in this study is given in Table 4 and graphically presented in Figure 9. The main differences between the two studies are as follows: (1) more biostratigraphic datums are used in the present study (total of 11) than in the previous one (total of 5); (2) the FO of *Globigerinatheka index* is placed at 71.42 mbsf in the present study based on the planktonic foraminiferal biostratigraphy of *Noc-*

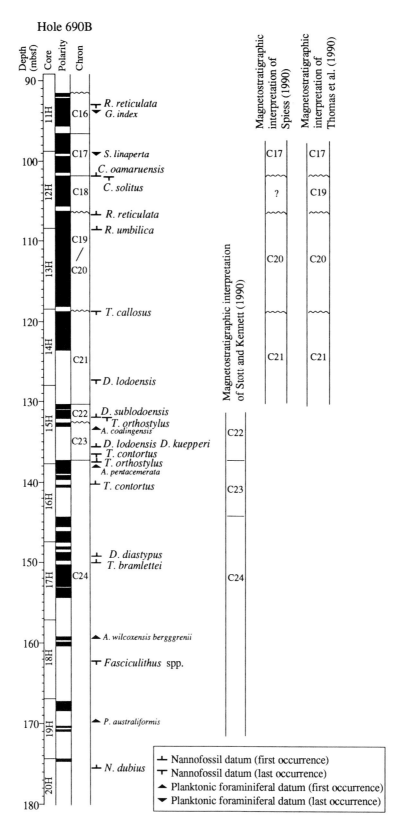

Fig. 5. Nannofossil stratigraphy [*Pospichal and Wise*, 1990; this study], planktonic foraminiferal stratigraphy [*Stott and Kennett*, 1990], and magnetostratigraphy [*Spiess*, 1990] of ODP Hole 690B. Magnetostratigraphic interpretation of this study (left column) is compared with that of *Stott and Kennett* [1990], *Spiess* [1990], and *Thomas et al.* [1990].

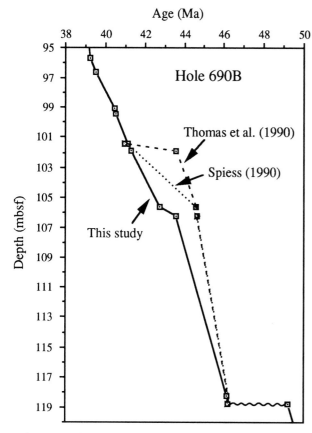

Fig. 6. Revised age-depth curve for ODP Hole 690B. Age-depth curves of *Spiess* [1990] and *Thomas et al.* [1990] are also shown.

chi et al. [1991] rather than at 44.50 mbsf in the work of *Katz and Miller* [1991], which is based on shipboard results published by *Ciesielski et al.* [1988]; and (3) age estimates used in this study are based on biostratigraphic and magnetostratigraphic correlations in the Southern Ocean, whereas *Katz and Miller* [1991] used age estimates from the mid-latitudes as compiled by *Berggren et al.* [1985].

Hole 702B

As in Site 700, no detailed age-depth curve has been published for Site 702. Five biostratigraphic datums were used in the age model of *Katz and Miller* [1991]. The present study has incorporated all the nannofossil and foraminiferal datums identified by *Crux* [1991] and *Nocchi et al.* [1991] in constructing a detailed age model for Site 702 (Table 5 and Figures 10a and 10b).

A major unconformity occurs around 20 mbsf, separating upper Eocene from upper Miocene sediments [*Nocchi et al.*, 1991]. *Isthmolithus recurvus* was found up to the unconformity [*Crux*, 1991], and the stratigraphic range of the species must have been truncated

by this unconformity. Consequently, the sediment below the unconformity must be older than 38.8 Ma, the age for the LO of *I. recurvus*. Extrapolation of sedimentation rates based on the LO of *C. solitus* and the Subchron C18N/C18R boundary suggests that the sediment subjacent to the unconformity is about 39.8 Ma (Figure 10a). *Katz and Miller* [1991] extrapolated sedimentation rates to ~29 mbsf (Figure 10a) based on the LO of *Acarinina primitiva* and the Subchron C18N/C18R boundary. Their age-depth curve for the 30–70 mbsf interval is thus considerably different from the present study. This is apparently due to the age uncertainty of the LO of *A. primitiva*, which has not been calibrated with magnetostratigraphy in the Southern Ocean.

The age model for the 200–280 mbsf interval is slightly different between the two studies (Figure 10b). This is mainly due to the fact that the present study uses age estimates based on biomagnetostratigraphic correlations in the Southern Ocean, whereas *Katz and Miller* [1991] adopted age estimates from *Berggren et al.* [1985].

Hole 703A

The recent study of *Wei* [1991a] has led to a reinterpretation of the paleomagnetic data of *Hailwood and Clement* [1991b] and a revised age model for the Eocene-Oligocene interval for Hole 703A. *Wei*'s [1991a] interpretation of the magnetostratigraphy is summarized here in Figure 11 along with those of *Hailwood and Clement* [1991b] and *Madile and Monechi* [1991].

Madile and Monechi [1991] interpreted the short magnetic interval in Core 13 as Subchron C13N. This interpretation is believed to be in error because the highest occurrence of *Reticulofenestra oamaruensis*, an abrupt increase in cool water taxa, and an abrupt shift in $\partial^{18}O$ values all occur above their interpreted Subchron C13N (see *Wei* [1991a] for details), and all these events have been shown to be synchronous in the Southern Ocean at the top of Subchron C13R [*Wei*, 1991b]. Furthermore, the short normal polarity in Core 13 is considered questionable by *Hailwood and Clement* [1991b] because it is represented by only one discrete sample measurement.

The normally magnetized interval in Core 15 is assigned to Subchron C16N (Figure 11) rather than Subchron C15N as indicated by *Hailwood and Clement* [1991b]. My reinterpretation is based on the three nannofossil datums (the FO of *R. oamaruensis*, the LO of *R. reticulata*, and the FO of *Isthmolithus recurvus*) in Core 15. These datums have been correlated consistently with Subchron C16N in several Southern Ocean sites [*Wei and Wise*, 1992], and it is unlikely that all three datums are one chron younger at Site 703 than at other Southern Ocean sites.

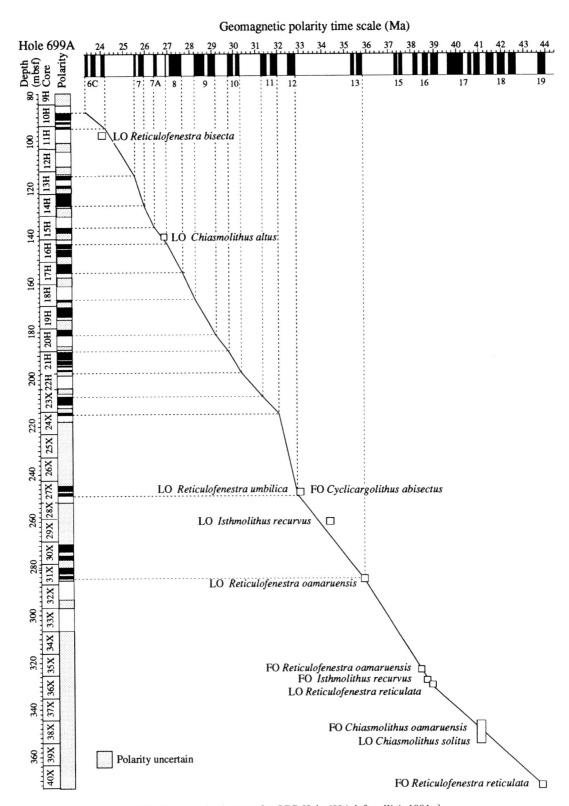

Fig. 7.   Age-depth curve for ODP Hole 699A [after *Wei*, 1991*a*].

Hole 700B

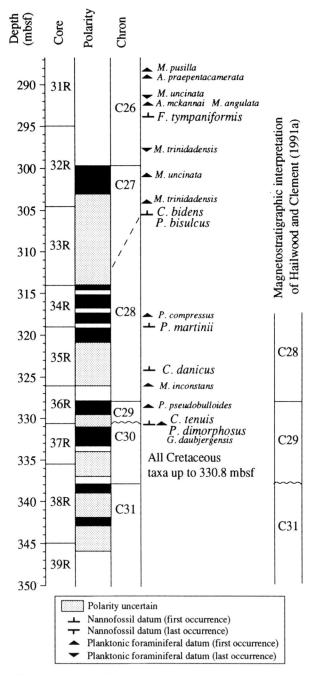

Fig. 8. Nannofossil stratigraphy [Crux, 1991; this study], planktonic foraminiferal stratigraphy [Nocchi et al., 1991], and magnetostratigraphy of Hailwood and Clement [1991a] of ODP Hole 700B. Magnetostratigraphic interpretation of this study (left column) is compared with that of Hailwood and Clement [1991a].

The age-depth curve of Hailwood and Clement [1991b] has been revised considerably (Figure 12). Hailwood and Clement [1991b] placed a major unconformity (representing more than 3 m.y.) at 53 mbsf. A major unconformity here is unlikely because the LO of Chiasmolithus altus is known to be about 26.7 Ma and the stratigraphic range of this species would have been truncated by the major unconformity if it existed. It is possible, however, that an unconformity occurs at 57 mbsf, where C. altus changes abruptly from abundant to absent. The interval from 57 mbsf to 66 mbsf is quite condensed and may encompass some additional hiatuses.

Another major unconformity (representing ~2 m.y.) indicated by Hailwood and Clement [1991b] at about 98 mbsf (Figure 12) was the result of using biostratigraphic datum ages derived from the mid-latitudes. This artificial unconformity simply disappears when Southern Ocean datum ages are applied (Figure 12).

On the other hand, Wei [1991a] identified a major unconformity at or slightly above 110 mbsf, where the highest occurrence of Reticulofenestra oamaruensis was found. An abrupt shift in $\partial^{18}O$ values of benthic foraminifers was also recorded at this level [Mead and Hodell, 1992], which coincides with an abrupt increase in cool water taxa [Wei, 1991b]. These three events are known to be at the top of Subchron C13N [Wei, 1991b]. Although detailed paleomagnetic data are available for the lower part of Core 12, no normally magnetized interval was identified in this core [Hailwood and Clement, 1991b]. This indicates that all of Subchron C13N and possibly the lower part of Subchron C12R and the upper part of Subchron C13R are missing owing to a major hiatus (Figure 12). The close association of the LOs of I. recurvus and R. oamaruensis also suggests a major hiatus between these two datums.

Hole 737B

Magnetostratigraphy is not available for this site, and the age-depth curve of Barron et al. [1991a] for Hole 737B is based exclusively on biostratigraphic data. Compilation of nannofossil and planktonic foraminiferal datum ages as calibrated with magnetostratigraphy in the Southern Ocean (see Figures 2–4, Tables 1–3) prompts the revision of some of the datum ages used by Barron et al. [1991b]. The ages applied in the work of Barron et al. [1991b] and the revised ages are presented in Table 6 for comparison. A new age-depth curve constructed using the revised ages is presented in Figure 13 along with the age-depth curve of Barron et al. [1991b].

Hole 738B

Similar to Hole 737B, no magnetostratigraphy could be established for Hole 738B, and the age model of

TABLE 4.   Comparison of Age-Depth Data Used in This Study With Those Used by *Katz and Miller* [1991] for ODP Hole 700B

Event	Age, Ma	Age of KM	Samples	Depth Range, mbsf	Mean, mbsf	± m	Depth of KM
LO *C. solitus*	41.2		2CC/3-1, 59	17.1/26.99	22.05	4.95	
FO *G. index*	45.5	45.00	7-5, 27/7-6, 27	70.67/72.17	71.42	0.75	44.5
FO *N. cristata*	48.0		11CC/12CC	102.9/112.3	107.55	4.65	
FO *D. sublodoensis*	52.6		16CC/18-1, 80	159.4/169.7	164.55	5.15	
LO *T. orthostylus*	53.0	53.70	18CC/20-2, 106	175/190.46	182.73	7.73	172.3
FO *A. pentacamerata*	55.2		21CC/22CC		199		
LO *F.* spp	56.8	57.40	24CC/26-1, 20	219.2/238.4	228.8	9.6	228.8
FO *P. australiformis*	57.0				219		
FO *D. multiradiatus*	58.8	59.20	26CC/27-1, 30	242.5/247.8	245.15	2.65	245.6
FO *H. kleinpellii*	60.9		29CC/30-2, 120	269.5/278.7	274.1	4.6	
FO *F. tympaniformis*	62.6	62.00	31-5, 115/31CC	292.65/295	293.83	1.18	293.9
C26R/C27N	63.03	63.03			299.13		299.13
C28R/C29N	65.50	65.50			327.31		327.31
C29R/C30N	66.74				330.8		
C30N/C30R	68.42				333.29		
C30R/C31N	68.52				338		

KM, *Katz and Miller* [1991].

TABLE 5.   Comparison of Age-Depth Data Used in This Study With Those Used by *Katz and Miller* [1991] for ODP Hole 702B

Event	Age, Ma	Age of KM	Depth, mbsf	± m	Depth of KM
Extrapololated sedimentation rate		39.19			29
LO *A. primitiva*		40.6			46.05
FO *I. recurvus*	38.8		20.55	4.75	
LO *C. solitus*	41.2		45.72	0.79	
C18N/C18R	42.73	42.73	71.9	0.06	71.9
C18R/C19N	43.60	43.60	84.26	0.01	84.26
C19R/C20N	44.66	44.66	98.7	0.05	98.7
C20N/C20R	46.17	46.17	113.3	0.16	113.3
C20R/C21N	48.75	48.75	157.3	0.05	157.3
Hiatus	50.41	50.41	181.29		181.29
Top C22N?	51.95	51.95	181.3		181.3
Base C22N	52.62	52.62	193.5		193.5
LO *D. lodoensis*	51.3		191.9	4.7	
FO *D. sublodoensis*	52.6		194.85	3.2	
LO *T. orthostylus*	53.0	53.7	200.45	2.5	200.23
FO *A pentacamerata*	55.2		212.0	4.9	
FO *A. wilcoxensis berggrenii*	56.5		225.6	0.7	
LO *F. tympaniformis*	56.8	57.4	239.78	4.1	239.63
FO *P. australiformis*	57.7		245.1	0.5	
FO *D. multiradiatus*	58.8	59.2	249.8	4.4	249.38
LO *S. pseudobulloides*	61.6		258.8	4.7	
FO *H. riedellii*	60.0	59.9	276.9	1.9	276.58
FO *H. kleinpellii*	60.9		283.35	4.6	
FO *A. mackannai*	61.8		287.7	0.3	
FO *M. pusilla*	62.0		283.1	4.4	
FO *A. praepentacamerata*	62.2		>288		
FO *M. angulata*	62.5		>288		
FO *F. tympaniformis*	62.6		>288		

KM, *Katz and Miller* [1991].

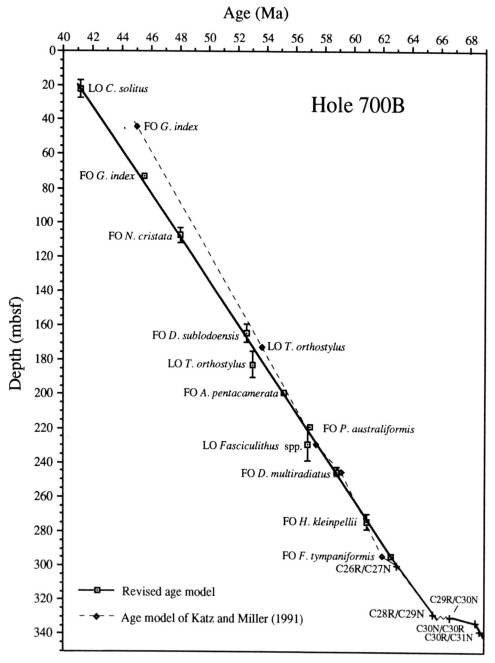

Fig. 9.   Revised age model and the age model of *Katz and Miller* [1991] for ODP Hole 700B.

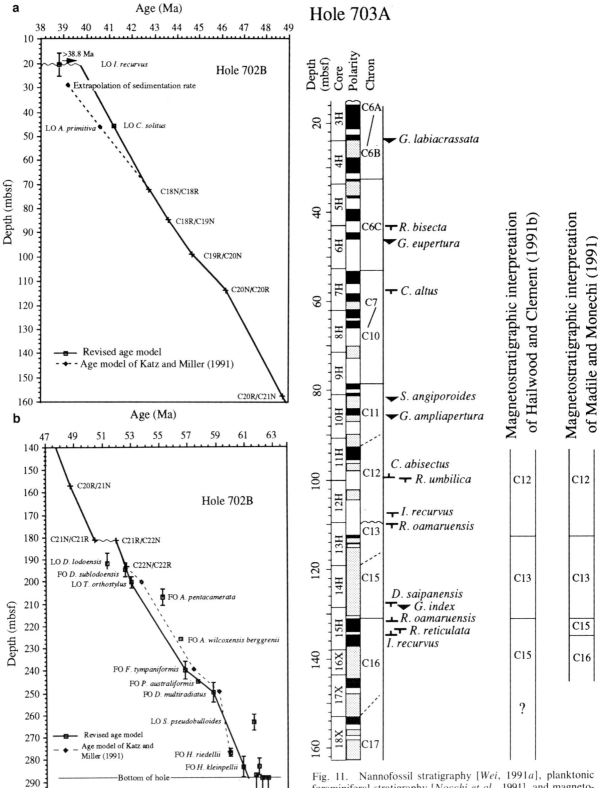

Fig. 10.   Revised age model and the age model of *Katz and Miller* [1991] for ODP Hole 702B.

Fig. 11.   Nannofossil stratigraphy [*Wei*, 1991*a*], planktonic foraminiferal stratigraphy [*Nocchi et al.*, 1991], and magnetostratigraphy of *Hailwood and Clement* [1991*b*] of ODP Hole 703A. Magnetostratigraphic interpretation of this study (left column) is compared with that of *Hailwood and Clement* [1991*b*] and *Madile and Monechi* [1991]. Symbols are the same as in Figure 8.

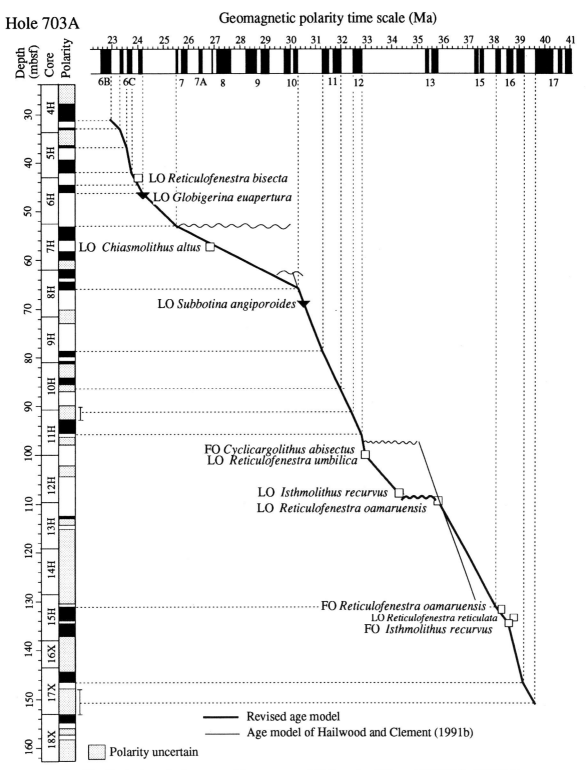

Fig. 12.   Revised age-depth curve and the age curve of *Hailwood and Clement* [1991b] for ODP Hole 703A [after *Wei*, 1991a].

TABLE 6.   Comparison of Age-Depth Data Used in This Study With Those Used by *Barron et al.*
[1991*b*] for ODP Hole 737B

Event	Age, Ma	Age of Barron	Top, mbsf	Bottom, mbsf	Mean, mbsf	± mbsf
LO *C. altus*	26.7	25.5	312.00	313.80	312.90	0.90
FO *B. veniamini*	27.7	27.7	330.80	?	>330.8	
LO *C. cubensis*	30.3	30.0	369.40	379.10	374.25	4.85
LO *S. angiporoides*	31.6	32.0	417.70	427.40	422.55	4.85
LO *R. umbilica*	32.9	33.2	543.30	552.90	548.10	4.80
LO *I. recurvus*	34.4	34.8	467.45	568.95	568.20	0.75
LO *D. saipanensis*	38.0	36.4	602.01	603.62	602.82	0.81
FO *R. oamaruensis*	38.7	38.0	605.42	605.99	605.71	0.29
FO *I recurvus*	38.8	38.3	649.30	650.20	649.75	0.45
FO *C. oamaruensis*	41.0	41.0	651.01	658.60	654.81	3.80
LO *C. solitus*	41.2	41.4	671.36	677.60	674.48	3.12
FO *R. reticulata*	43.7	42.1	686.50	687.01	686.76	0.26

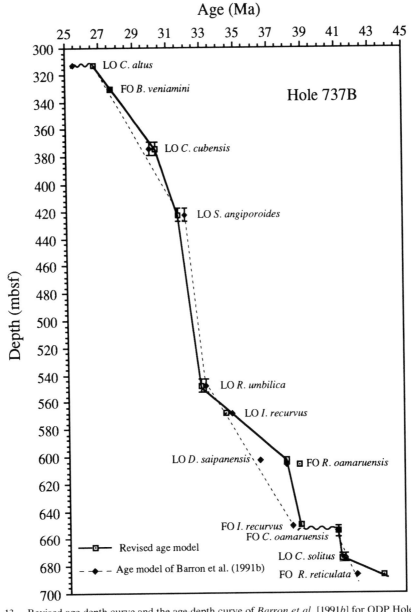

Fig. 13.   Revised age-depth curve and the age-depth curve of *Barron et al.* [1991*b*] for ODP Hole 737B.

TABLE 7. Comparison of Age-Depth Data Used in This Study With Those Used by *Barron et al.*
[1991*b*] for ODP Hole 738B

Event	Age, Ma	Age of Barron	Top, mbsf	Bottom, mbsf	Mean, mbsf	± mbsf
LO *R. oamaruensis*	36.0	36.0	23.66	25.16	24.41	0.75
LO *G. index*	38.3	37.0	34.90	36.40	35.65	0.75
FO *R. oamaruensis*	38.7	38.0	34.66	36.16	35.41	0.75
FO *I recurvus*	38.8	38.8	39.26	40.66	39.96	0.70
FO *C. oamaruensis*	41.0	41.0	69.70	71.16	70.43	0.73
LO *C. solitus*	41.2	41.4	69.70	71.16	70.43	0.73
LO *A. bullbrooki*		43.0	74.40	75.90	75.15	0.75
FO *R. reticulata*	43.7	42.1	96.66	98.16	97.41	0.75
FO *R. umbilica*	44.1	44.6	118.46	119.96	119.21	0.75
FO *G. index*	45.5	45.0	142.02	143.89	142.96	0.94
FO *A. bullbrooki*		47.5	179.70	180.26	179.98	0.28
FO *N. fulgens*	49.0	49.8	196.80	205.26	201.03	4.23

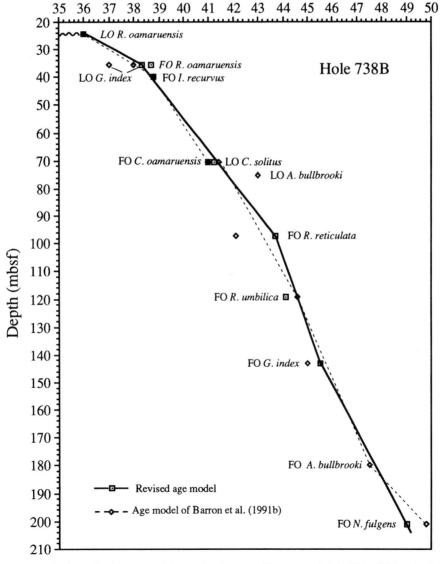

Fig. 14.   Revised age-depth curve and the age-depth curve of *Barron et al.* [1991*b*] for ODP Hole 738B.

TABLE 8.   Comparison of Age-Depth Data Used in This Study With Those Used by *Barron et al.*
[1991*b*] for ODP Hole 738C

Event	Age, Ma	Age of Barron	Top, mbsf	Bottom, mbsf	Mean, mbsf	± mbsf
FO *N. fulgens*	49.0	49.8	196.80	205.26	201.03	4.23
FO *D. sublodoensis*	52.6	52.6	226.26	227.79	227.03	0.77
FO *A. pentacamerata*	55.2	54.0	255.30	255.79	255.55	0.25
FO *D. lodoensis*	54.9	55.4	255.79	264.76	260.28	4.49
FO *A. wilcoxensis berggrenii*	56.5	56.6	278.90	284.55	281.73	2.83
FO *D. diastypus*	56.5	56.7	278.90	284.30	286.10	2.70
FO *P. australiformis*	57.5	57.5	286.86	302.83	294.85	7.99
FO *D. multiradiatus*	58.7	59.2	288.20	302.80	295.50	7.30
LO *S. pseudobulloides*	61.6	61.7	312.30	321.90	317.10	4.80
FO *D. mohleri*	60.2	60.5	312.44	322.00	317.22	4.78
FO *M. pusilla*	62.1	62.0	321.90	331.60	326.75	4.85
FO *A. praepentacamerata*	60.3	60.2	333.84	337.00	335.42	1.58
FO *H. kleinpellii*	60.9	61.6	322.07	338.30	330.19	8.12
FO *A. mackannai*	62.5	62.4	337.00	338.50	337.75	0.75
LO *G. daubjergensis*	64.0	64.0	349.40	350.90	350.15	0.75
FO *P. imitatus*	62.8	63.4	350.45	350.90	350.68	0.23
FO *C. bidens*	63.3	63.2	351.25	351.75	351.50	0.25
FO *P. martinii*	64.8	63.8	359.50	360.85	360.18	0.68
FO *C. danicus*	64.9	64.8	364.35	364.85	364.60	0.25
FO *P. compressus*	64.5	64.5	372.88	374.38	373.63	0.75
FO *S. inconstans*	65.4	65.1	372.88	374.88	373.88	1.00
FO *C. tenuis*	66.1	66.2	376.22	376.55	376.39	0.17
FO *S. pseudobulloides*	66.1	66.1	376.54	376.64	376.59	0.05
FO *G. daubjergensis*	66.3	66.4	376.76	376.85	376.81	0.05

Barron et al. [1991*b*] for Hole 738B is based entirely on nannofossil and planktonic foraminiferal data. Some of the datum ages used by *Barron et al.* [1991*b*] have been revised based on biomagnetostratigraphic correlations in the Southern Ocean, and an age-depth curve is constructed utilizing the revised ages (Table 7, Figure 14).

Hole 738C

Virtually no useful magnetostratigraphic information is available for the Paleogene section from Hole 738C. Some ages for the nannofossil and planktonic foraminiferal datums used by *Barron et al.* [1991*b*] have been revised (see Table 8), and the age-depth data are replotted in Figure 15. The age model for this hole is less well constrained than for other holes discussed, partly owing to poor core recovery.

Hole 748B

A Paleogene biomagnetostratigraphic synthesis for Hole 748B has not been published. It is thus useful to summarize the magnetostratigraphic [*Inokuchi and Heider*, 1992], nannofossil [*Wei et al.*, 1992], and foraminiferal [*Berggren*, 1992] data in Figure 16. The nannofossil data of *Aubry* [1992] are not incorporated here because a number of her datum placements are not consistent with levels identified in this study (e.g., the FO and LO of *Reticulofenestra oamaruensis* and the

LO of *Reticulofenestra bisecta*). Magnetostratigraphic interpretation for the 65–110 mbsf interval is straightforward and is virtually the same as that of *Schlich et al.* [1989]. However, the previous study interpreted the short normal interval in the middle of Core 14 as Subchron C13. *Wei et al.* [1992] noted that the LO of *R. oamaruensis* is above this interpreted Subchron C13N. Furthermore, an abrupt increase in foraminiferal $\partial^{18}O$ values and a sharp increase in cool water nannofossil taxa also coincide with the LO of *R. oamaruensis* at about 115.8 mbsf. These three events, as was mentioned before, correlate consistently with the top of Subchron C13R in the Southern Ocean [*Wei*, 1991*b*]. Consequently, Subchron C13N cannot be in the middle of Core 14 but should be near the top of Core 14 as interpreted by *Wei et al.* [1992]. *Inokuchi and Heider* [1992] concur with this interpretation.

An age model for the Eocene-Oligocene is presented in Figure 17 using magnetostratigraphic, nannofossil, and foraminiferal data given in Table 9. The age-depth curve accommodates the biostratigraphic and magnetostratigraphic data fairly well. Only three foraminiferal datums (LOs of *Globigerina euapertura*, *Globigerina labiacrassata*, and *Chiloguembilina cubensis*) are significantly off from the age-depth curve. The magnetostratigraphic ages of these foraminiferal datums are still not well established in the Southern Ocean, and they appear to be diachronous based on presently available data (see Figure 4).

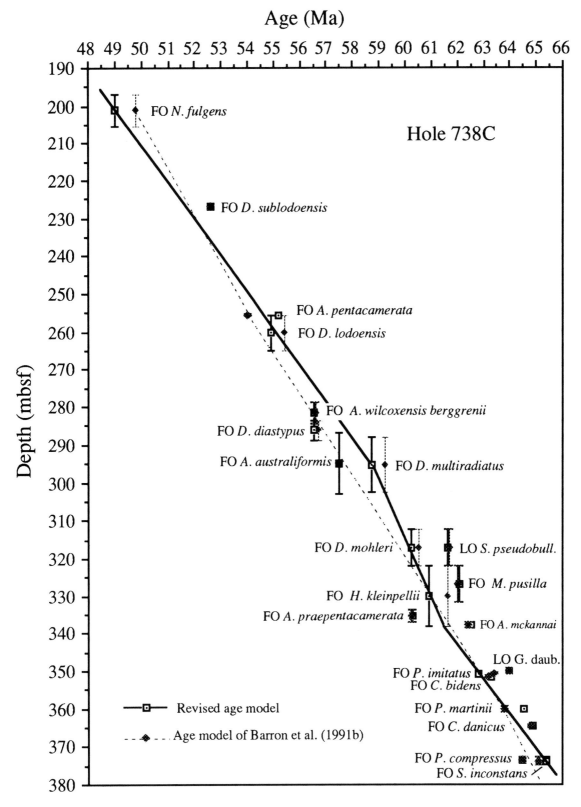

Fig. 15. Revised age-depth curve and the age-depth curve of *Barron et al.* [1991*b*] for ODP Hole 738C.

Hole 748B

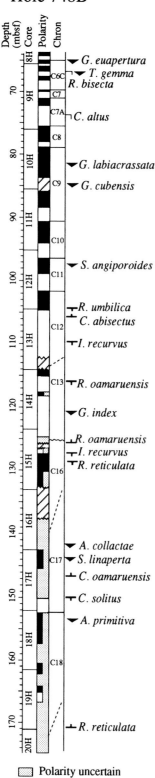

Fig. 16. Summary of nannofossil [*Wei et al.*, 1992], planktonic foraminiferal [*Berggren*, 1992], and paleomagnetic [*Inokuchi and Heider*, 1992] data for ODP Hole 748B. Symbols are the same as in Figure 5.

TABLE 9. Biomagnetostratigraphic Data Used to
Construct Age-Depth Curve for ODP Hole 748B

Event	Age, Ma	Depth, mbsf	± m
Nannofossil datum			
LO *Reticulofenestra bisecta*	24.0	66.68	0.1
LO *Chiasmolithus altus*	27.0	73.45	0.75
LO *reticulofenestra umbilica*	33.0	104.45	0.25
FO *Cyclicargolithus abisecutus*	33.0	105.95	0.75
LO *Ishmolithus recurvus*	34.6	109.95	0.75
LO *Reticulofenestra oamaruensis*	36.0	115.89	0.06
FO *Reticulofenestra oamaruensis*	38.5	125.95	0.75
FO *Isthmolithus recurvus*	38.8	127.45	0.75
LO *Chiasmolithus solitus*	41.2	149.45	0.75
FO *Reticulofenestra reticulata*	43.6	170.80	0.25
Planktonic foraminiferal datum			
LO *Globigerina euapertura*	24.2	64.25	
LO *Tenuitella gemma*		67.75	
LO *Globigerina labiacrassata*	27.7	81.75	
LO *Chiloguembelina cubensis*	30.3	84.75	
LO *Subbotina angiporoides*	31.6	97.95	
LO *Globigerinatheka index*	37.5	121.45	
LO *Acarinina collactea*		141.70	
LO *Subbotina linaperta*	40.5	144.15	
LO *Acarinina primitiva*		153.45	
Magnetostratigraphic boundary			
T C6CN	23.27	65.0	
B C6CN	24.21	68.5	
T C7N	25.50	70.0	
B C7N	25.97	70.5	
T C7AN	26.38	71.0	
B C7AN	26.56	72.5	
T C8N	26.86	75.5	
B C8N	27.74	78.0	
T C9N	28.15	79.0	
B C9N	29.21	88.5	
T C10N	29.73	90.5	
B C10N	30.33	94.0	
T C11N	31.23	96.5	
B C11N	32.06	99.0	
T C12N	32.46	101.5	
B C12N	32.90	105.0	
B C13N	35.87	115.0	
T C18N	41.29	152.5	

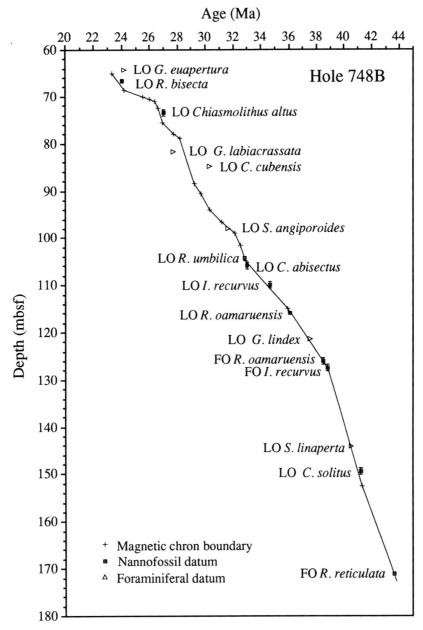

Age (Ma)

Hole 748B

+ Magnetic chron boundary
▣ Nannofossil datum
△ Foraminiferal datum

Fig. 17.   Age model for ODP Hole 748B.

SUMMARY

Based on biomagnetostratigraphic correlations at nine Southern Ocean sites, published magnetostratigraphies for some intervals from sites 690, 699, 700, and 703 are reinterpreted, and age-depth curves for these sites and sites 702, 737, 738, and 748 are constructed/reconstructed using published/reinterpreted magnetostratigraphic data and revised ages of biostratigraphic datums. The improved chronology should facilitate correlations of geologic events between these sites and those in other areas.

Acknowledgments. I am very grateful to J. P. Kennett for inviting me and providing funds for my participation in the International Conference on the Role of the Southern Ocean and Antarctica in Global Change: An Ocean Drilling Perspective, where this paper was presented. I sincerely thank J. Barron, A. R. Edwards and S. W. Wise for critical reviews and helpful suggestions, which greatly improved the quality of the paper. The major part of this work was carried out at Florida State University while I was a postdoc there and supported by National Science Foundation Grant DPP91-18480 (to S. W. Wise). I also acknowledge NSF Grant OCE91-15786 for support of my research at Scripps Institution of Oceanography.

REFERENCES

Aubry, M.-P., Paleogene calcareous nannofossils from the Kerguelen Plateau, Leg 120, *Proc. Ocean Drill. Program Sci. Results*, *120*, 471–491, 1992.

Barker, P. F., et al., Leg 113, *Proc. Ocean Drill. Program Sci. Results*, *113*, 1033 pp., 1990.

Barron, J., et al., Leg 119, *Proc. Ocean Drill. Program Sci. Results*, *119*, 1003 pp., 1991a.

Barron, J., J. G. Baldauf, E. Barrera, J.-P. Caulet, B. T. Huber, B. H. Keating, D. Lazarus, H. Sakai, H. R. Thierstein, and W. Wei, Biochronologic and magnetochronologic synthesis of Leg 119 sediments from the Kerguelen Plateau and Prydz Bay, Antarctica, *Proc. Ocean Drill. Program Sci. Results*, *119*, 813–847, 1991b.

Berggren, W. A., Paleogene planktonic foraminifer magneto-biostratigraphy of the southern Kerguelen Plateau (sites 747–749), *Proc. Ocean Drill. Program Sci. Results*, *120*, 551–568, 1992.

Berggren, W. A., D. V. Kent, and J. Flynn, Paleogene geochronology and chronostratigraphy, in *The Chronology of the Geological Record*, edited by N. J. Snelling, pp. 141–195, The Geological Society of London, London, 1985.

Berggren, W. A., D. V. Kent, J. D. Obradovic, and C. C. Swisher III, Toward a revised Paleogene geochronology, in *Eocene-Oligocene Climatic and Biotic Evolution*, edited by D. R. Prothero and W. A. Berggren, Princeton University Press, Princeton, N. J., in press, 1992.

Ciesielski, P. F., et al., Leg 114, *Proc. Ocean Drill. Program Initial Rep.*, *114*, 815 pp., 1988.

Ciesielski, P. F., et al., Leg 114, *Proc. Ocean Drill. Program Sci. Results*, *114*, 826 pp., 1991.

Crux, J. A., Calcareous nannofossils recovered by Leg 114 in the Subantarctic South Atlantic Ocean, *Proc. Ocean Drill. Program Sci. Results*, *114*, 155–177, 1991.

Hailwood, E. A., and B. M. Clement, Magnetostratigraphy of sites 699 and 700, East Georgia Basin, *Proc. Ocean Drill. Program Sci. Results*, *114*, 337–358, 1991a.

Hailwood, E. A., and B. M. Clement, Magnetostratigraphy of sites 703 and 704, Meteor Rise, southeastern South Atlantic, *Proc. Ocean Drill. Program Sci. Results*, *114*, 367–386, 1991b.

Inokuchi, H., and F. Heider, Magnetostratigraphy of sediments from sites 748 and 750, Leg 120, *Proc. Ocean Drill. Program Sci. Results*, *120*, 247–252, 1992.

Katz, M. E., and K. G. Miller, Early benthic foraminiferal assemblages and stable isotopes in the Southern Ocean, *Proc. Ocean Drill. Program Sci. Results*, *114*, 481–512, 1991.

Madile, M., and S. Monechi, Late Eocene to early Oligocene calcareous nannofossil assemblage from sites 699 and 703 of ODP Leg 114, Subantarctic South Atlantic Ocean, *Proc. Ocean Drill. Program Sci. Results*, *114*, 179–192, 1991.

Mead, G. A., and D. A. Hodell, Late Eocene to early Oligocene vertical oxygen isotopic gradients in the South Atlantic: Implications for warm saline deep water, *Palaeogeogr. Palaeoclimatol. Palaeoecol.*, in press, 1992.

Nocchi, M., E. Amici, and I. Premoli Silva, Planktonic foraminiferal biostratigraphy and paleoenvironmental interpretation of Paleogene faunas from the Subantarctic transect, Leg 114, *Proc. Ocean Drill. Program Sci. Results*, *114*, 233–279, 1991.

Okada, H., and D. Bukry, Supplementary modification and introduction of code numbers to the low-latitude coccolith biostratigraphic zonation (Bukry 1973, 1975), *Mar. Micropaleontol.*, *5*, 321–325, 1980.

Pospichal, J. J., and S. W. Wise, Jr., Paleocene to middle Eocene calcareous nannofossils of ODP sites 689 and 690, Maud Rise, Weddell Sea, *Proc. Ocean Drill. Program Sci. Results*, *113*, 613–638, 1990.

Pospichal, J. J., et al., Cretaceous-Paleogene biomagnetostratigraphy of sites 752–755, Broken Ridge: A synthesis, *Proc. Ocean Drill. Program Sci. Results*, *121*, 721–741, 1991.

Schlich, R., et al., Leg 120, *Proc. Ocean Drill. Program Initial Rep.*, *120*, 648 pp., 1989.

Shackleton, N. J., Preface, *Palaeogeogr. Palaeoclimatol. Palaeoecol.*, *57*, 1–2, 1986.

Spiess, V., Cenozoic magnetostratigraphy Leg 113 drill sites, Maud Rise, Weddell Sea, Antarctica, *Proc. Ocean Drill. Program Sci. Results*, *113*, 216–315, 1990.

Stott, L. D., and J. P. Kennett, Antarctic Paleogene planktonic foraminifer biostratigraphy: ODP Leg 113, sites 689 and 690, *Proc. Ocean Drill. Program Sci. Results*, *113*, 549–569, 1990.

Thomas, E., et al., Upper Cretaceous–Paleogene stratigraphy of sites 689 and 690, Maud Rise (Antarctica), *Proc. Ocean Drill. Program Sci. Results*, *113*, 901–914, 1990.

Wei, W., Middle Eocene–lower Miocene calcareous nannofossil magnetobiochronology of ODP holes 699A and 703A in the Subantarctic South Atlantic, *Mar. Micropaleontol.*, *18*, 143–165, 1991a.

Wei, W., Evidence for an earliest Oligocene abrupt cooling in the surface waters of the Southern Ocean, *Geology*, *19*, 780–783, 1991b.

Wei, W., and S. W. Wise, Jr., Eocene-Oligocene calcareous nannofossil magnetobiochronology of the Southern Ocean, *Newsl. Stratigr.*, *26*, 119–132, 1992.

Wei, W., G. Villa, and S. W. Wise, Jr., Paleoceanographic implications of Eocene-Oligocene calcareous nannofossils from ODP sites 711 and 748 in the Indian Ocean, *Proc. Ocean Drill. Program Sci. Results*, *120*, 979–999, 1992.

Wise, S. W., Jr., et al., Leg 120, *Proc. Ocean Drill. Program Sci. Results*, *120*, 1155 pp., 1992.

(Received October 20, 1991;
accepted May 6, 1992.)

LATE EOCENE–EARLY OLIGOCENE EVOLUTION OF CLIMATE AND MARINE CIRCULATION: DEEP-SEA CLAY MINERAL EVIDENCE

CHRISTIAN ROBERT

Géologie du Quaternaire, Centre National de la Recherche Scientifique—Luminy, Marseille, Cedex 9, France

HERVÉ CHAMLEY

Dynamique Sedimentaire et Structurale, Centre National de la Recherche Scientifique, Université de Lille, d'Ascq Cedex, France

Changes in clay mineral associations of marine sediments across the Eocene to Oligocene transition exhibit regional differences because large oceanic and associated continental regions acted as distinct thermal systems. The widespread abundance of pedogenic smectite in most Atlantic areas suggests globally warm climate and alternating wet and dry seasons on the adjacent landmasses. The appearance or slight increase of kaolinite content in the western Tethys area and from western Africa to northern Europe indicates increased transfer of heat and associated precipitation from ocean to continent. Dramatically increased illite abundance in and around East Antarctica resulted from the development of poorly weathered soils associated with the development of glacial conditions. The distinct dominance of smectite and the absence of any significant mineralogical change during the entire Paleogene in Central and North Atlantic areas indicate the persistence of climatic conditions on the adjacent continental margins. Therefore the Eocene/Oligocene transition events were not related to a global event of astronomical (solar input) or chemical (greenhouse gases content) origin. This would have resulted in climatic changes in the low latitudes, and associated redistribution of available heat at the surface of the Earth would have initiated similar and concomitant mineralogic variations in both hemispheres. Kaolinite in the Tethys region indicates increased precipitation, compensating for an increasing thermal deficit. That followed the restriction of the areas of shallow platforms likely to favor the formation and transport of warm, dense waters to the deep ocean. This restriction was a consequence of collision and tectonics along the continental margins of the Tethys. The tethyan warm water outflow also decreased through shoaling seaways. Dramatically decreased continental weathering in East Antarctica resulted from cooling due to oceanic development that thermally isolated the continent and reduced the energy budget at southern high latitudes. This activated meridional thermal exchanges (poleward surface circulation warmed the southwest Atlantic margins and equatorward currents cooled the southeast Atlantic margins), proto-Antarctic bottom waters formed along the East Antarctic margin, and the influence of these waters expanded through the Southern Ocean and southern South Atlantic, while deep circulation was reinforced locally from low-latitude areas other than the Tethys. Climatic and paleoceanographic development at the Eocene/Oligocene transition is principally the consequence of (1) expansion of the Southern Ocean which resulted in a dramatic cooling of East Antarctica and associated formation and circulation of cold water, beginning at about 37.5–38.0 Ma during the latest Eocene, and (2) restriction of the Tethys Ocean which resulted in decreased formation and circulation of Tethyan warm deep waters and associated transfer of heat, beginning at about 35.5–36 Ma during the earliest Oligocene. The Eocene/Oligocene transition reflects a threshold of climatic and oceanographic consequences of plate tectonic evolution.

INTRODUCTION

The Eocene/Oligocene boundary is commonly considered as a climatic threshold. Oxygen isotopes express a major event which has been interpreted as the result of climatic cooling, glacial increase, extension of sea ice, and development of cold deep waters [*Kennett and von der Borch*, 1986]. However, the relative impor-

tance of each of these changes toward the oxygen isotope signal has not yet been resolved [*Prentice and Matthews*, 1988; *Kennett and Barker*, 1990; *Wise et al.*, 1991]. Concomitant changes of planktonic foraminifer associations indicate an increased abundance of cool water species, but these are less marked than the oxygen isotope event or previous changes in planktonic foraminiferal associations of late middle and late Eocene age [*Keller*, 1983]. Coeval benthic faunas from the Atlantic Ocean indicate gradual updepth expansion of cool conditions from the latest Eocene to the early Oligocene [*Boersma*, 1986].

Clay mineral successions in oceanic basins have long been considered as reliable indicators of paleoenvironmental evolution on adjacent landmasses. This is especially the case in the Atlantic realm where clay diagenesis, volcanism, and hydrothermalism only locally influenced the composition of the clay mineral associations. In Atlantic sediments of Cenozoic age, the nature and association of clay particles especially reflects changes in climate and circulation. Both climate and circulation determine the nature and relative abundance of clay minerals in oceanic sediments, and in many instances discrimination between their respective influences have yet to be unravelled. Clay particles originate from continental weathering processes and soil formation, which are largely controlled by temperature and rainfall. Eroded by runoff and/or winds, sometimes reworked from ancient deposits, they have been transported to depositional areas by oceanic and atmospheric circulation. These points are detailed in the work of *Chamley* [1989]. Clay mineral variations in sedimentary sequences reflect changes in environmental conditions that prevailed on the continent as well as in the characters of circulation that controlled their transport and deposition. In most oceanic areas, clay mineral associations display only a very slight increase of physical weathering products near the Eocene/Oligocene boundary. This has been interpreted as a minor stage of the Cenozoic climate cooling which began at the end of the middle Eocene [*Chamley*, 1986]. However, recent deep-sea drilling at southern high latitudes provided critical new information about the intensity of the climatic change: in the Weddell Sea, a major modification in the composition of clay associations derived from East Antarctica occurred at the Eocene/Oligocene boundary. It was inferred that the climatic change created a strong decrease in chemical weathering (and soil formation) and increased erosion of parent rock in the source areas [*Robert and Maillot*, 1990].

The apparent importance of the Eocene/Oligocene boundary events varies both geographically and with the indicator examined. The purpose of this paper is to summarize information gained from studies of clay mineral associations, from the tropics to the high latitudes, and progress in the knowledge of both climate and circulation and their evolution near the Eocene/Oligocene boundary.

Clay mineral investigations have been conducted according to the method detailed by *Holtzapffel* [1985]. Each sample was sieved through a 63-μm mesh, and the fine fraction was leached in 0.2 N hydrochloric acid. The excess acid was removed by repeated centrifugations followed by homogenization. The <2-μm size fraction was separated by decantation (settling time based on Stoke's law), deposited on glass slides, and

dried at room temperature. Three X ray analyses were run at scan speeds of 1°2 θ/mn: (1) untreated sample, (2) glycolated sample, and (3) sample heated for 2 hours at 490°C. Semiquantitative evaluations are based on the peak heights and areas: Clay minerals were corrected by multiplying their peak height by a factor of 0.5 to 3.0, depending on their crystallinity. Data are given in percentages, the relative error being 5%. The relative abundance of smectite versus illite (S/I index) is obtained from the ratio of the 001 smectite (18 Å) and illite (10 Å) peaks of the glycolated sample. The relative abundance of kaolinite versus illite (K/I index) is obtained from the ratio of the 001 kaolinite (7 Å) and illite (10 Å) peaks of the glycolated sample. Then, clay mineral data have been compiled on the time scale of *Berggren et al.* [1985], and their geographical and chronological variations, as well as their relation to climatic and geodynamic events, have been examined. To meet this objective, 18 sedimentary sequences (mostly Deep Sea Drilling Project (DSDP) and Ocean Drilling Program (ODP) sites, but also land boreholes and outcrops) from the Southern and Atlantic oceans, western Tethys, and northern Europe have been used (Figure 1) and compared with data from the literature. Information on the clay mineral associations in these sequences, and on the origin of the clay particles as well, has been detailed in the work of *Chamley* [1979, 1986], *Deconinck et al.* [1986], *Diester-Haass and Chamley* [1980], *Leroy* [1981], *Maillot and Robert* [1984], *Mercier-Castiaux et al.* [1988], *Robert* [1982, 1987], *Robert and Kennett* [1992], and *Robert and Maillot* [1983, 1990].

CLAY MINERAL ASSOCIATIONS OF LATE EOCENE–EARLY OLIGOCENE AGE

Smectite largely dominated the clay fraction of oceanic sediments of late Eocene–early Oligocene age in most areas, as was also the case from the Cretaceous to the middle Eocene, with the exception of active tectonic areas [*Chamley*, 1979, 1989; *Robert*, 1987]. At locations considered in this study, smectite is mainly derived from continental weathering on subdued, low-lying relief where a restricted drainage favored its formation. Dominant pedogenic smectite in terrigenous deposits indicates that climate on adjacent continental areas was warm, with seasonally alternating wet and dry conditions. Widespread distribution of the mineral also indicates that these climatic conditions prevailed on most continental margins and remained fairly stable during the Late Cretaceous and early Paleogene. It is noteworthy that pedogenic smectite dominated in oceanic terrigenous fine fractions at times when warm saline deep waters may have dominated most oceanic basins [*Brass et al.*, 1982; *Kennett and Stott*, 1990]. It is probable that warm deep waters helped sustain the thermal equilibrium of the Earth, as well as relatively high temperatures along continental margins from the tropics to the

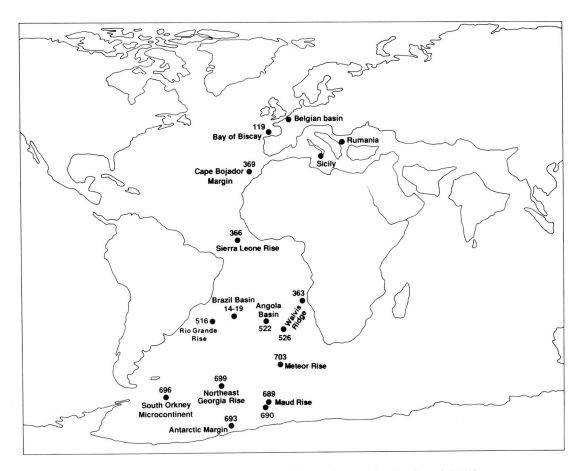

Fig. 1.   Location map. Paleoposition of the continents, after *Smith et al.* [1981].

high latitudes. In these conditions, it is likely that only slight thermal deficits occurred on continental margins, when lower incoming solar radiation in winter was compensated by heat transfer from the ocean and associated precipitation. Winter precipitation followed by summer dryness created conditions prone to favor smectite genesis [*Robert and Chamley*, 1991].

Kaolinite is generally only a minor component of clay mineral associations of Paleogene age in most oceanic sedimentary sequences. When climate is warm, this mineral characterizes weathering processes in rainy, well-drained continental areas, where running waters favor hydrolysis of the parent rock and removal of alkaline and alkaline-earth elements [*Millot*, 1970]. Kaolinite develops principally (1) in upstream areas of drainage basins; (2) during early stages of ocean opening and compressive episodes of active margins, associated with uplifted continental relief; (3) during sea level falls, when lower longitudinal profiles of the drainage basins result in steeper gradients; and (4) in warm and humid areas where high rainfall ensures an intense leaching of the parent rock and evacuation of the solutions by runoff [*Millot*, 1970; *Chamley*, 1989; *Weaver*, 1989]. During the Paleogene, kaolinite developed on the con-

tinental margins of the western Tethys submitted to compression and collision [*Deconinck et al.*, 1986; *Savostin et al.*, 1986]. During the same period, the subsiding passive margins of the Atlantic and Southern oceans were incompatible with the presence of significant relief and dominance of kaolinitic soils [*Robert*, 1987; *Robert and Maillot*, 1990]. However, the mineral formed abundantly in ferrallitic soils of continental interiors in West Africa and western Europe [*Millot*, 1970; *Mercier-Castiaux et al.*, 1988; *Thiry*, 1989]. In these areas, ascending warm air in summer (especially in uplifted areas) and compensation of continental thermal deficits increasing in winter could have locally caused important precipitation all the year round and favored the formation of kaolinite. Surprisingly, relatively abundant kaolinite also formed in high-latitude areas, where transfers of heat from low latitudes (involving warm surface oceanic currents and atmospheric circulation) likely compensated for the polar thermal deficit. This resulted in increased precipitation which in turn favored the formation of kaolinite in well-drained areas. Such a process has been especially effective at times of increased equator to pole thermal gradients [*Robert and Chamley*, 1991].

Illite (and chlorite) is derived from erosion of parent

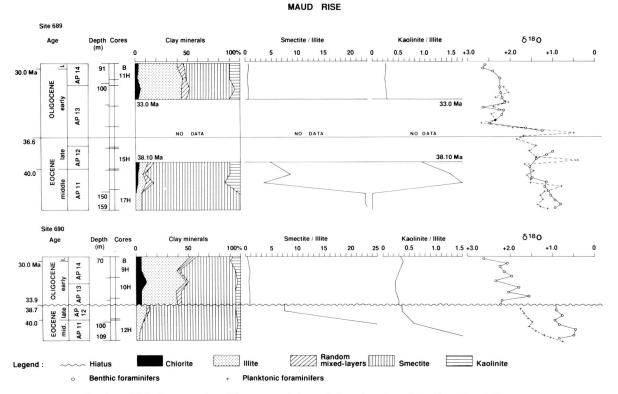

Fig. 2. Middle Eocene to late Oligocene variations of clay minerals at ODP Site 689 and Site 690 on Maud Rise, Weddell Sea. Comparison with oxygen isotope data from *Kennett and Stott* [1990] and *Stott et al.* [1990].

rock and/or poorly developed soils where they are some-times associated with random mixed-layer clays. The minerals principally originate from (1) areas of steep relief where active mechanical erosion decreases soil formation, especially during tectonically active periods, and (2) cold and/or desert areas where low temperatures and/or low rainfall reduce the rate of chemical weathering [*Millot*, 1970; *Chamley*, 1989]. In the western Tethys, illite and random mixed-layer clays were especially abundant in Paleogene deposits [*Deconinck et al.*, 1986] because inten-sive erosion probably removed continuously the surficial horizons, and the soils did not reach a state of equilibrium with local climate [*Chamley*, 1989]. In tectonically quies-cent Atlantic Ocean and Southern Ocean areas, contents of chlorite, illite, and random mixed-layer clays remain generally low. However, illite strongly increased in the Southern Ocean near the Eocene/Oligocene boundary, probably in relation to a drastic decrease of continental weathering on East Antarctica [*Robert and Maillot*, 1990].

CLAY MINERAL VARIATIONS IN THE SOUTH ATLANTIC AND SOUTHERN OCEAN

Antarctic Areas

During ODP Leg 113, sites 689 and 690 were drilled on Maud Rise, an aseismic structure isolated in the

eastern Weddell Sea, at 2080-m and 2914-m water depth, respectively. Deposits of late Eocene–early Oli-gocene age essentially consist of calcareous biogenics (locally associated with significant amounts of siliceous biogenics) and include minor amounts of terrigenous components [*Barker et al.*, 1988]. Clay mineral associ-ations of middle to late Eocene age (Figure 2) are largely dominated by smectite, and the variations of the clay mineral contents are especially expressed by S/I and K/I indices. They indicate that both smectite and kaolinite slightly decreased in relation to illite during the late Eocene, probably in relation to a minor decrease of continental weathering. Oxygen isotope display their lowest values in middle Eocene deposits. A slight in-crease of the $\delta^{18}O$ (Figure 2) is expressed by the planktonic foraminifers especially during the late Eocene, probably in relation to cooler water masses [*Kennett and Stott*, 1990; *Stott et al.*, 1990]. The corre-lation observed between clay mineral and oxygen iso-tope data suggests that the evolution of continental environments on East Antarctica during the late Eocene was triggered by cooling of the surrounding ocean [*Robert and Kennett*, 1992].

At Site 690 the 38.7 to 33.9 Ma time interval is marked by a hiatus, which possibly reflects an intensification of deep currents [*Barker et al.*, 1988]. At the shallower Site

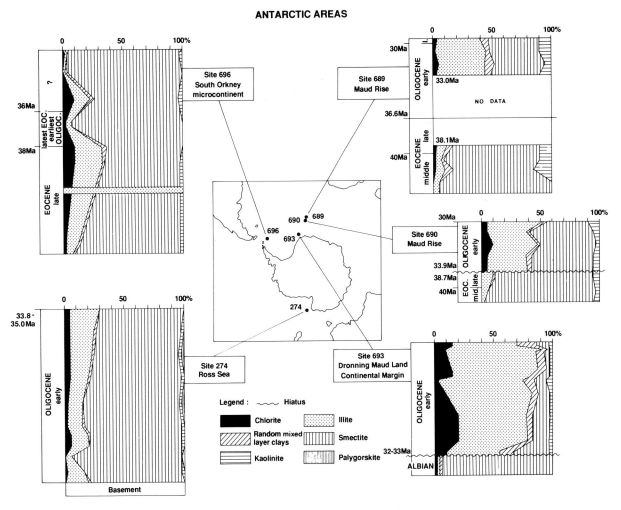

Fig. 3. Clay mineral associations near the Eocene/Oligocene boundary in Antarctic areas.

689 the sedimentary sequence is complete, and both planktonic and benthic foraminifers display a major increase of their $\delta^{18}O$ in the earliest Oligocene [*Kennett and Stott*, 1990; *Stott et al.*, 1990]. The isotope shift may record the beginning of important thermohaline circulation and initiation of the psychrosphere. It may also represent a cooling of Antarctic waters and/or the initial accumulation of continental ice [*Kennett and Barker*, 1990]. Very low terrigenous content at Site 689 from 38.1 to 33.0 Ma, where the sediment almost exclusively comprises biogenic elements, suggests intensive productivity and/or winnowing by intensified water circulation [*Robert and Maillot*, 1990]. When deposition of terrigenous elements resumed at both sites in the early Oligocene, the clay fraction was dominated by illite associated with chlorite and random mixed-layer clays. These minerals developed at the expense of smectite principally, as indicated by very low values of both S/I and K/I indices (Figure 2). On Maud Rise the strong correlation of clay mineral and oxygen isotope data

(i.e., higher contents of smectite associated with lower $\delta^{18}O$ and higher illite contents associated with higher $\delta^{18}O$) indicates climate and oceanographic control of clay mineral associations. It also suggests that a drastic decrease of chemical weathering and soil formation occurred on East Antarctica near the Eocene/Oligocene boundary, probably in association with cooler waters and intensified thermohaline circulation.

Clay mineral associations at ODP Site 693 [*Robert and Maillot*, 1990] drilled in the Weddell Sea on the East Antarctic continental margin of Queen Maud Land (Figure 3) contain the highest percentage of illite and chlorite. The abundance of these minerals decreases northward with increasing distance from East Antarctica. At DSDP Site 274 in the Ross Sea, the lower part of the sedimentary sequence is of early Oligocene age, older than 33.8–35.0 Ma. [*Lazarus and Caulet*, this volume]. The clay mineral association (Figure 3) is largely dominated by smectite, indicating warm conditions with alternating wet and dry seasons on the

adjacent continental margins. This clay mineral association, very similar to associations recognized in the
Weddell Sea during the middle and late Eocene, also
suggests the absence of any important modification of
climatic conditions and associated continental weathering near the Eocene/Oligocene transition in the Ross
Sea sector. Northeast of the Antarctic Peninsula, ODP
Site 696 was drilled at shallow water depth (650 m) on
the South Orkney microcontinent. Clay mineral associations at Site 696 [*Robert and Maillot*, 1990] are also
largely dominated by smectite. There, illite contents
progressively increased in the course of time and
reached a maximum at about 38.0 Ma (Figure 3). This
information is consistent with the cooling trend observed at ODP sites 689 and 690 on Maud Rise in late
Eocene deposits. However, at Site 696, this is followed
by a progression of smectite in two steps, separated by
temporary higher contents of illite. The first step occurred between 38.0 and 36.0 Ma, the second step being
younger than 36.0 Ma. It is probable that western
boundary circulation in the southwest Atlantic carried
southward clay particles originating from warmer regions in South America [*Robert and Maillot*, 1990].
Southward flowing currents also carried warm waters
toward high latitudes and probably helped sustain relatively warm temperatures and moisture along the Antarctic Peninsula. Clay mineral data finally suggest that
intensity of southward transfers of heat and fine particles increased after 38.0 and 36.0 Ma, near the Eocene/
Oligocene boundary.

To summarize, changes in clay mineral associations
in Antarctic areas near the Eocene/Oligocene boundary
result from climatic and oceanographic processes which
differ from region to region:

1. Near East Antarctica in the Weddell Sea sector,
dominant illite beginning in the earliest Oligocene indicates a drastic decrease of continental weathering and
soil formation, associated with cold climatic conditions.

2. In the Ross Sea, clay associations of early Oligocene age still dominated by smectite suggest much
less significant climatic changes on the Antarctic margins adjacent to the Pacific sector of the Southern
Ocean.

3. Northeast of the Antarctic Peninsula, higher contents of smectite near the Eocene/Oligocene boundary
suggest that enhanced warm, southward flowing surface
circulation favored the transport of fine particles from
lower latitudes and warmed the adjacent continental
margin to the south, creating conditions favorable to
smectite formation at higher latitudes.

Subantarctic Areas

ODP sites 699 and 703 have been drilled in the
western and eastern parts of the Subantarctic South
Atlantic, respectively (Figure 4). Late Eocene–early
Oligocene sediments at Site 699, drilled at 3705-m water

depth on the northeast Georgia Rise, consist of nannofossil and siliceous nannofossil oozes. At Site 703, drilled
at 1806-m water depth on Meteor Rise, deposits essentially consist of foraminifer-bearing nannofossil ooze
[*Ciesielski et al.*, 1988]. Clay mineral associations (Figure
4) are largely dominated by smectite and therefore indicate
predominantly warm climate and alternating wet and arid
conditions in the source areas. Increasing illite through
time indicates increasing contribution of terrigenous elements eroded from poorly weathered continental areas.

As observed at ODP Site 696 on the South Orkney
microcontinent, mineralogical variations near the
Eocene/Oligocene boundary can be divided into two
main stages. The first stage, of latest Eocene age, is
dated at about 37.5–38.0 Ma; mean abundances of illite
increased and mean values of the S/I index decreased at
both sites. Higher contents of illite on Northeast Georgia Rise than on Meteor Rise reflect a greater importance of detrital supply originating from poorly weathered continental areas (most probably East Antarctica)
in the eastern part of the ocean. The second stage, of
earliest Oligocene age, occurred at about 35.5–36.0 Ma;
the mean abundance of illite increased at Site 699 and
decreased at Site 703. Accordingly, the mean value of
the S/I index decreased on Northeast Georgia Rise and
increased on Meteor Rise. The evolution recorded at
Site 699 suggests that a further cooling favored genesis,
erosion, and transport of illite to the Subantarctic regions. At Site 703, however, these processes could have
interfered with increasing influence of oceanic circulation from low-latitude areas, which was needed to
reduce the increasing thermal deficit at southern high
latitudes. General circulation [*Wells*, 1986] suggests that
the currents flowed southward in the western Indian
Ocean rather than in the eastern South Atlantic. This is
corroborated by geological data, which indicate warm
conditions and development of evaporites on the African margin of the Indian Ocean during the early Oligocene and coeval development of dry temperate conditions on the Atlantic margin [*Dingle et al.*, 1983].

To summarize, clay particles eroded from East Antarctica extended at intermediate to deep water depth in
Subantarctic areas, and this was especially conspicuous
in the eastern part of the ocean. Their abundance was
raised in two steps of latest Eocene and earliest Oligocene age. During the second step (of earliest Oligocene age), the East Antarctic detrital supply (illite)
probably interfered locally with terrigenous components
from lower latitudes (smectite) carried southward by
reinforced warm oceanic circulation.

Southwest Atlantic

DSDP Site 516 was drilled during Leg 72 at relatively
shallow water depth (1313 m) on the western Rio
Grande Rise near the tropics (Figure 5). Sediments of
late Eocene–early Oligocene age mostly consist of nan-

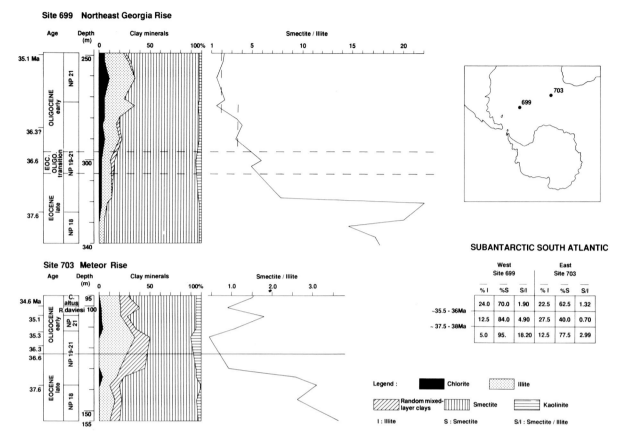

Fig. 4. Clay mineral associations near the Eocene/Oligocene boundary at ODP Site 699 on Northeast Georgia Rise and at ODP Site 703 on Meteor Rise, Subantarctic South Atlantic.

nofossil chalks [*Barker et al.*, 1983]. Clay mineral associations are largely dominated by smectite, and the clay mineral composition is very similar to that observed at Site 696, drilled at southern high latitudes. This similarity suggests that warm climates with alternating wet and dry seasons were probably widespread along the western margin of the South Atlantic, i.e., on the path of warm, southward flowing western boundary currents. The highest contents of illite, probably related to temporary cooler conditions, occurred during the late Eocene. At about 37.5 Ma, the mean abundance of smectite increased and the S/I index increased (Figure 5). After a short episode characterized by smaller amounts of smectite and higher abundances of illite, a further increase of smectite occurred at about 34.5–35.0 Ma. The Eocene/Oligocene boundary mineralogic event is therefore separated in two stages at shallow water depth in the tropical South Atlantic, as was also the case in Subantarctic areas (at intermediate to deep water depth) and at shallow water depth in the western Weddell Sea (Site 696). As observed at Site 696, both stages at Site 516 were characterized by higher contents of smectite. Progression of smectite at shallow water depth in the southwest South Atlantic confirms the increasing

importance of poleward transfer of heat through western boundary circulation.

DSDP sites 14 and 19 were drilled during Leg 3 at 4343-m and 4677-m water depth, respectively, in the southern Brazil Basin (Figure 5). Late Eocene–early Oligocene deposits consist of marly nannofossil and nannofossil oozes and chalks [*Maxwell et al.*, 1970]. Smectite dominated at these sites. Palygorskite is present in the lower part of the sedimentary sequence; considered to originate from the eastern Rio Grande Rise principally, it decreased during the Eocene with the progressive submergence of the rise [*Robert*, 1981]. In the abyssal Brazil Basin, the most important mineralogic change occurred during the earliest Oligocene at about 36.0 Ma; the mean abundance of smectite rose at the expense of illite, and the S/I index increased. It is probable that terrigenous supply from warm low latitudes intensified in areas under the influence of bottom water circulation. Further increase of illite from cooler areas is observed in early Oligocene deposits younger than 34.5–35.0 Ma.

To summarize, clay mineral data in the southwest South Atlantic off Brazil mostly indicate extension of climatic conditions favorable to smectite genesis on the

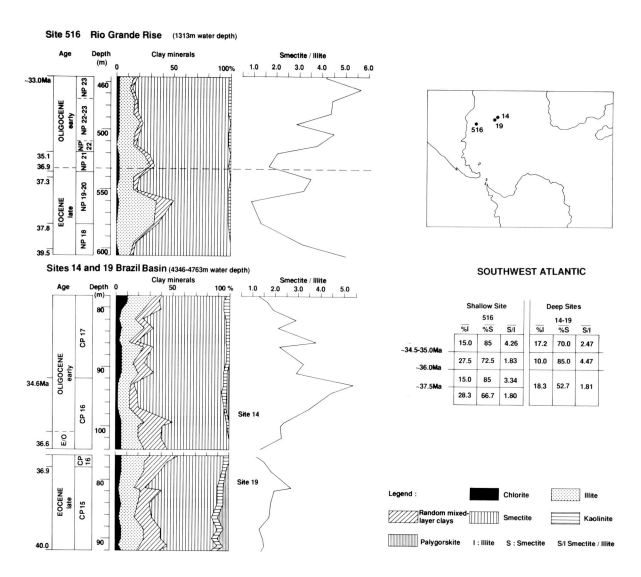

Fig. 5. Clay mineral associations near the Eocene/Oligocene boundary at DSDP Site 516 on Rio Grande Rise and at DSDP Site 14 and Site 19 in the Brazil Basin, western South Atlantic.

adjacent continental margin, as well as increasing influence of oceanic circulation from warm, low-latitude areas at shallow water depths and in the abyssal plain.

Southeast Atlantic

On the eastern Walvis Ridge, DSDP Site 363 was drilled during Leg 40 at 2248-m water depth (Figure 6). Sediment of late Eocene–early Oligocene age consists of foraminifer-nannofossil chalks [*Bolli et al.*, 1978]. The Eocene/Oligocene boundary there is missing by a hiatus. Clay mineral associations (Figure 6) are dominated by smectite. Palygorskite and sepiolite are present in late Eocene deposits only, indicating intensive evaporation on the adjacent coastal areas, as was also the case during the late Paleocene and early Eocene [*Robert and*

Chamley, 1991]. During the late Eocene, palygorskite at Site 363 was associated with warm temperatures of surface waters, expressed by the associations of planktonic foraminifers. This was followed by a retreat of warm waters at the end of the Eocene [*Boersma et al.*, 1987]. Palygorskite is absent above the hiatus in early Oligocene sediments, as a probable consequence of decreased evaporation in adjacent coastal areas. The retreat of warm waters probably slightly reduced temperatures in coastal areas, increasing the continental thermal deficit associated with lower insolation in winter. As a consequence, winter precipitation may have increased enough to compensate for the deficit and prevent the formation of palygorskite and, in turn, favor smectite genesis.

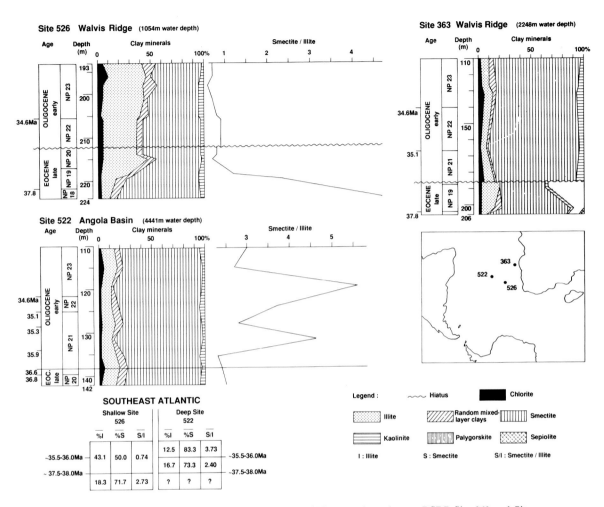

Fig. 6. Clay mineral associations near the Eocene/Oligocene boundary at DSDP Site 363 and Site 526 on Walvis Ridge and at DSDP Site 522 in the Angola Basin, eastern South Atlantic.

On the southwestern Walvis Ridge, DSDP Site 526 was drilled during Leg 74 at 1054-m water depth. Foraminifer oozes and nannofossil oozes were deposited from the late Eocene to the early Oligocene [*Moore et al.*, 1984]. Clay mineral associations (Figure 5) principally include smectite. By the end of the Eocene (Figure 6) at about 37.5–38.0 Ma, a major increase of illite contents occurred at the expense of smectite; mean abundances of illite rose while those of smectite decreased and the S/I index decreased. Abundant illite indicates higher terrigenous supply from poorly weathered continental areas. Illite could originate mostly from East Antarctica, where it formed abundantly, and be carried northward by cold surface current. Actually, the associations of planktonic foraminifers indicate the presence of cooler surface waters at Site 526 than at Site 363 [*Boersma et al.*, 1987]. Moreover, nannofossils originating from high latitudes have been found in Oligocene deposits at site 526 [*Manivit*, 1984]. But the abundance of illite is similar at Site 526 and at Site 689

and Site 690 on Maud Rise near East Antarctica, and its abundance is higher at Site 526 than at Site 699 and Site 703 in the Subantarctic South Atlantic. It is therefore possible that an additional source of illite contributed to the clay mineral association at Site 526 and could be located in southwestern Africa; northward flowing cold waters may have cooled the continental margin and locally favored aridity and illite formation. Such a hypothesis is supported by geological data, which point to the presence of dry temperate climatic conditions along the west coast of South Africa during the early Oligocene [*Dingle et al.*, 1983].

DSDP Site 522 was drilled during Leg 73, at 4441-m water depth in the deep Angola Basin (Figure 6). Sediment of late Eocene–early Oligocene age consists of nannofossil oozes [*Hsü et al.*, 1984]. Clay mineral associations are largely dominated by smectite, indicating an important fine terrigenous contribution from continental areas characterized by warm climate and alternating wet and arid seasons. A strengthening of

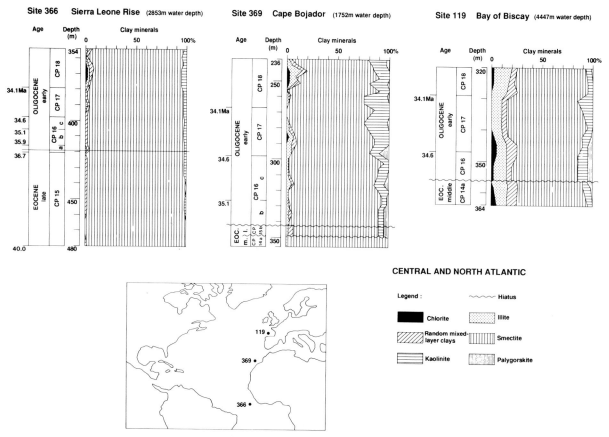

Fig. 7. Clay mineral associations near the Eocene/Oligocene boundary at DSDP Site 366 on Sierra Leone Rise, Central Atlantic, and at DSDP Site 369 on the continental margin off Cape Bojador (Morocco) and Site 119 on Cantabria Seamount in the Bay of Biscay, North Atlantic.

smectite abundance in the earliest Oligocene at about 35.5–36.0 Ma was associated with important fluctuations in the S/I index. Clay mineral data suggest that detrital particles eroded from low-latitude areas were carried to the deeper parts of the Angola Basin and that this terrigenous supply increased in the earliest Oligocene.

To summarize, clay mineral associations of latest Eocene age in the southeastern Atlantic mostly indicate cooler conditions and associated modification of the distribution of humidity in southwestern Africa, triggered by intensified circulation of cold surface oceanic waters. This was followed during the earliest Oligocene by a reinforcement of terrigenous components from low latitudes to the south, in the abyssal Angola Basin.

CLAY MINERAL VARIATIONS IN THE NORTHERN HEMISPHERE

Central and North Atlantic

DSDP Site 366 was drilled during Leg 41, at 2853-m water depth on Sierra Leone Rise (Figure 7). Sediments

of late Eocene–early Oligocene age consist of nannofossil chalks [Lancelot et al., 1978]. Clay mineral associations are almost exclusively smectite. No significant variation of the clay mineral assemblage is observed near the Eocene/Oligocene boundary. The age of the oldest occurrence of kaolinite, not well constrained, could have been sometime between 40 and 37 Ma. It most probably expresses an increase of precipitation in West Africa. Moreover, kaolinite at Site 366 may have been diluted by important smectitic terrigenous supply from the exposed continental margin; in the West African hinterland, locally important mineralogical changes including replacement of palygorskite by kaolinite in Mali [Millot, 1970] suggest that a drastic evolution of climatic conditions occurred, from intensive evaporation to abundant precipitation. A further, minor increase of kaolinite associated with small amounts of chlorite and illite is observed at Site 366 in early Oligocene sediments younger than 34.0 Ma.

DSDP Site 369 was drilled off Morocco during Leg 41, at 1752-m water depth on the Cape Bojador continental margin (Figure 7). Late Eocene–early Oligocene depos-

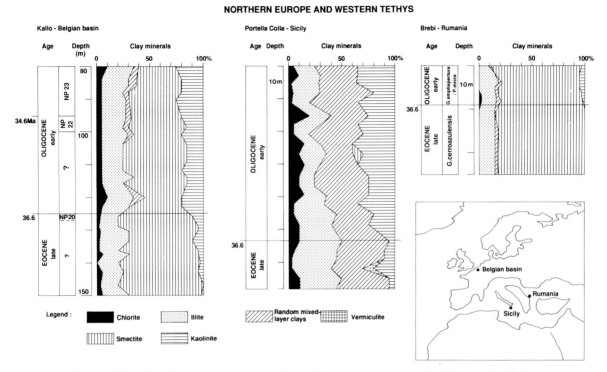

Fig. 8.   Clay mineral associations near the Eocene/Oligocene boundary in the Kallo well (Belgian Basin) in northern Europe and in the Portella-Colla (Sicily) and Brebi (Rumania) sections in the western Tethys.

its consist of marly limestones and diatomaceous and nannofossil marls [*Lancelot et al.*, 1978]. Sedimentation here has been interrupted by hiatuses of late middle–early late Eocene and latest Eocene–earliest Oligocene age. Clay mineral associations [*Diester-Haass and Chamley*, 1980; *Leroy*, 1981] are largely dominated by smectite. Absent from middle Eocene deposits, kaolinite occurs in clay associations of latest Eocene age (about 37.0–38.0 Ma) and slightly increases in earliest Oligocene sediments at about 35.0 Ma. A further, more important increase is evidenced in sediments younger than 34.5 Ma. As at Site 366, it is probable that kaolinite appeared at Site 369 at some time during the late Eocene, prior to the latest Eocene event. However, its increase of earliest Oligocene age is probably coeval with the earliest Oligocene event evidenced in the South Atlantic. Kaolinite at Site 369 indicates slightly higher terrigenous supply from the West African hinterland than at Site 366.

DSDP Site 119 was drilled during Leg 12 at 4447-m water depth on the Cantabria Seamount, in the deepest part of the Bay of Biscay where a hiatus spans the entire late Eocene. Middle Eocene and early Oligocene sediments consist of clays and nannofossil clays [*Laughton et al.*, 1972]. Clay mineral associations [*Leroy*, 1981] are dominated by smectite (Figure 7). On the West European continental margin, no significant variation of the

clay mineral association is observed from the middle Eocene to the early Oligocene. This indicates the presence of stable continental weathering (i.e., continental climatic conditions) and distribution of terrigenous particles in this sector of the Atlantic Ocean.

To summarize, no significant changes in clay mineral associations or related continental weathering were observed in Central and North Atlantic areas during the latest Eocene to earliest Oligocene. There were only slightly increased ocean-continent thermal exchanges and/or terrigenous supply from the West African hinterland. Of course, this does not preclude possible mineralogical and climatic exchanges in the African continental interior.

Northern Europe and Western Tethys

The Kallo well was drilled in the Belgian Basin (Figure 8). The Eocene/Oligocene boundary there is located at a depth of 125 m [*Steurbaut*, 1986], and late Eocene–early Oligocene sediments consist of clays and silty clays. Clay mineral associations [*Mercier-Castiaux et al.*, 1988] include smectite principally associated with illite and kaolinite. The abundance of kaolinite increased slightly at sometime during the late Eocene, the exact timing being unknown. Then a distinct augmentation of the abundance of kaolinite (associated with a minor increase of illite and chlorite) occurred just above the Eocene/Oligocene boundary. This is probably a consequence of increased erosion from the uplifted European hinterland (Ardennes especially) where kaolinite may have developed owing to intensified precipitation. Higher kaolinite contents (and associated precipitation) possibly resulted from intensified ocean-continent thermal exchanges needed to compensate slightly lower continental temperatures.

The Portella Colla section outcrops in southern Sicily (Figure 8). Late Eocene–early Oligocene sediments consist of clays with intercalated limestones and cherty limestones. Clay mineral associations [*Deconinck et al.*, 1986] include random mixed-layer clays, illite, kaolinite, and chlorite principally. This mostly results from active erosion of soils which did not reach a state of equilibrium with climate, in a tectonically active region. Just above the Eocene/Oligocene boundary, a significant increase of kaolinite (at the expense of illite) suggests either that increased precipitation favored a more rapid and/or intensive evolution of the soils or that decreased tectonic activity and mechanical erosion locally allowed evolution of the soils to maturity. As the western Tethys remained tectonically active (and tectonic activity even increased) during the late Eocene–early Oligocene [*Savostin et al.*, 1986], it is possible that higher kaolinite contents mostly express intensified precipitation.

The Brebi section crops out in Rumania (Figure 8). There, late Eocene–early Oligocene sediments mostly consist of marls, and the Eocene/Oligocene boundary has been placed at 35 m, at the top of the *G. cerroazulensis* biozone [*Bombita*, 1986]. Clay mineral associations [*Chamley*, 1986] are largely dominated by smectite. The late Eocene is totally devoid of kaolinite, the first occurrence of this mineral being recorded just above the Eocene/Oligocene boundary. It probably indicates slightly increased precipitation, associated with compensation of a minor continent-ocean thermal disequilibrium.

To summarize, no significant mineralogical change has been observed below the Eocene/Oligocene boundary in the western Tethys and northern Europe. Increase or appearance of kaolinite above the boundary indicates intensified transfer of heat and associated precipitation, involved in compensation of a thermal deficit in continental interiors. In these areas a single mineralogic event of earliest Oligocene age is observed.

GLOBAL EVOLUTION OF CLIMATE AND CIRCULATION AT THE EOCENE/OLIGOCENE TRANSITION

Geographical and chronological variations of late Eocene–early Oligocene clay mineral associations provide a combination of paleoclimatic and paleoceanographic information. Synchronous, but sometimes opposite, mineralogical trends near the Eocene/Oligocene

boundary may have resulted from a unique mechanism which mostly affected Antarctic climate, oceanic circulation, and their relationships. Frequent hiatuses probably resulted from rearrangement of the water masses and intensified circulation. Comprehensive sedimentary sequences allow a better resolution of the Eocene/Oligocene transition and recognition of two major stages:

1. The oldest stage is of latest Eocene age (37.5–38.0 Ma) and is locally associated with the extinction of some foraminifer species [*Boersma et al.*, 1987].

2. The youngest stage is of earliest Oligocene age (35.0–36.0 Ma) and is probably coeval with the major oxygen isotope shift [*Shackleton and Kennett*, 1975].

3. An additional, minor mineralogical event may have occurred locally at the Eocene/Oligocene boundary itself.

Climatic Evolution

The results detailed in this study are corroborated by our other investigations of late Eocene and early Oligocene Atlantic sediments, as well as by additional data from the Atlantic Ocean [*Froget*, 1981; *Latouche and Maillet*, 1984; *Chennaux et al.*, 1985; *Nielsen et al.*, 1989], Western Europe [*Sittler*, 1965; *Millot*, 1970], West Africa [*Millot*, 1970], and the Indian Ocean [*Cook et al.*, 1974; *Matti et al.*, 1974]. Essential information has been plotted on Figure 9.

No significant clay mineral change has been observed, either in low-latitude areas where pedogenic smectite largely dominates the clay associations and palygorskite still occurs locally in early Oligocene sediments [*Matti et al.*, 1974; *Debrabant et al.*, 1984] or in the North Atlantic where smectite is dominant at every location, including those at high latitudes: Labrador Sea [*Nielsen et al.*, 1989], Rockall Plateau [*Latouche and Maillet*, 1984], and Norwegian Sea [*Froget*, 1981]. The absence of any significant mineralogical evolution during the entire late Eocene–early Oligocene time interval indicates (1) the absence of any dramatic change of the energy budget in the northern hemisphere, (2) the steady state of climatic conditions on the North Atlantic continental margins, and (3) poleward transfer of heat (through surface and atmospheric circulation). This, in turn, suggests that mineralogical (and climatic) evolution in other areas did not predominantly result from variations of solar energy at the surface of the Earth. Therefore global processes of astronomical or chemical origin (resulting from variations of incoming solar radiation or from greenhouse gas contents) are not involved in the Eocene/Oligocene boundary events. This is corroborated by oxygen isotope data: planktonic foraminifers display only minor $\delta^{18}O$ variations in equatorial and tropical areas [*Keigwin*, 1980; *Rabussier-Lointier*, 1980; *Prentice and Matthews*, 1988]. Only minor increases of kaolinite are exhibited along the northwest

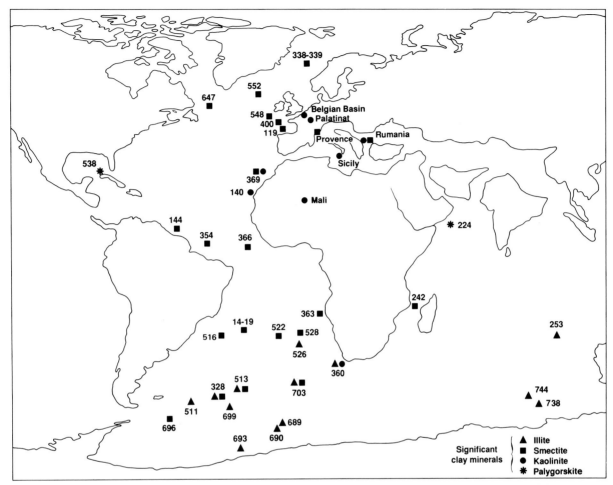

Fig. 9.   Distribution of major clay minerals near the Eocene/Oligocene boundary. Paleoposition of the continents after *Smith et al.* [1981]. Results from this study and also *Sittler* [1965], *Millot* [1970], *Cook et al.* [1974], *Matti et al.* [1974], *Diester-Haass and Chamley* [1980], *Froget* [1981], *Leroy* [1981], *Robert* [1982], *Debrabant et al.* [1984], *Latouche and Maillet* [1984], *Chennaux et al.* [1985], *Chamley* [1986, 1989], *Deconinck et al.* [1986], *Mercier-Castiaux et al.* [1988], *Nielsen et al.* [1989], *Robert and Maillot* [1990], and *Ehrmann* [1991].

African margin. Coincidence with the latest Eocene and/or earliest Oligocene events has not yet been established. Kaolinite probably originates from the African hinterland where drastic mineralogic changes have been observed [*Millot*, 1970]. Local disappearance of palygorskite and development of kaolinite suggest a replacement of evaporative conditions by intensive rainfall. Precipitation probably compensated for a thermal deficit in the continental interior, but this modification does not seem to be related to variations in the distribution of heat in the North Atlantic.

Generally minor, but distinct clay mineral variations have been observed in the western Tethys and northern Europe: kaolinite increased at most locations during the earliest Oligocene, i.e., at the time of the major oxygen isotope event. Although illite, chlorite, and random mixed-layer clays dominated in tectonically active regions and smectite dominated in tectonically quiescent areas, higher contents of kaolinite are observed at all of these sites. Intensified transfer of heat and associated precipitation probably compensated for a thermal deficit in continental interiors. No mineralogic change was observed in the latest Eocene, and the kaolinite increase occurred during the earliest Oligocene, a time of important modification of plate tectonic processes which began during magnetic anomaly 13, dated from 35.2 to 36.0 Ma. Changes in the poles of rotation of Africa and India relative to Eurasia [*Savostin et al.*, 1986] and associated collision and tectonic activity on the continental margin steepened continental relief. This led to the closure of the Para-Tethys [*Baldi*, 1984; *Zonenshain and Le Pichon*, 1986] and restriction of the volume of the warm Tethys Ocean as well as of the surface of shallow environments (platforms) especially favorable

SHALLOW SITES

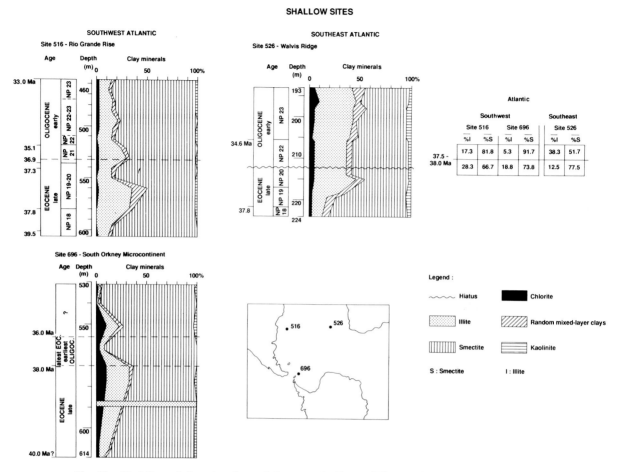

Fig. 10.   Variations of clay mineral associations near the Eocene/Oligocene boundary at shallow sites (650- to 1300-m water depth) in the South Atlantic and Southern oceans.

for the formation of warm, dense saline waters. The reservoir of heat in the Tethys Ocean decreased, disrupting thermal equilibrium on adjacent continental margins. This may have been enhanced by steep relief in uplifted areas. As a consequence, ocean-continent thermal exchanges and precipitation intensified. However, local presence of illite in mountain areas and of palygorskite in lacustrine basins [*Millot*, 1970] suggests that European climate was rather complex and that continental relief probably played an important role in climatic evolution. It is also probable that plate tectonic processes in the western Tethys influenced the evolution of climate on the adjacent African margin.

Major clay mineral variations occurred in the Southern Ocean [*Robert and Maillot*, 1990; *Ehrmann*, 1991] and influenced the composition of clay mineral associations in the South Atlantic; increased illite content near Antarctica resulted from a drastic decrease in continental weathering and soil formation. Association of increases in illite with higher planktonic and benthic $\delta^{18}O$ [*Kennett and Stott*, 1990; *Stott et al.*, 1990] suggests a

direct relation to cold conditions. Comprehensive Antarctic [*Ehrmann*, 1991] and Subantarctic sedimentary sequences show that the climatic change (and related mineralogic change) began during the latest Eocene and intensified during the earliest Oligocene. At shallow sites (drilled at 650- to 1300-m water depth) in the South Atlantic and Southern Ocean (Figure 10), the principal mineralogic event is of latest Eocene age (37.5–38.0 Ma). It follows increased content of illite at several localities, indicating local development of poorly weathered soils on the continental margins of the South Atlantic. At about 37.5–38.0 Ma, increased smectite (and associated influence from warm, low-latitude areas) in the western part of the ocean was concomitant with higher contents of illite (related to increased influence from cold, high-latitude areas) in the eastern part of the ocean. This probably resulted from intensified meridional thermal exchanges involving surface circulation principally. Increased poleward transfer of heat through the western boundary system was probably counterbalanced by equatorward extension of cold waters in the eastern part of the ocean. A further intensification of the mineralogic signal and increase of equator to pole thermal exchanges occurred during the earliest Oligocene at about 35.0–36.0 Ma. This is corroborated by the oxygen isotope composition of planktonic foraminifers; during the late Eocene–early Oligocene the $\delta^{18}O$ values on Rio Grande Rise in the southwest Atlantic [*Williams et al.*, 1985] were lower than those in the southern Angola Basin in the southeast Atlantic [*Oberhänsli and Toumarkine*, 1985], and $\delta^{18}O$ on the Falkland Plateau in the Subantarctic South Atlantic [*Muza et al.*, 1983] were similar to those in the southern Angola Basin.

The coincidence of increased poleward transfer of heat in the southern hemisphere and of increased cooling in East Antarctica during the latest Eocene indicates that the meridional heat transfer was largely insufficient to compensate for the polar thermal deficit. As a consequence, East Antarctic temperatures dropped, and weathering conditions became quite similar to those observed in modern deglaciated areas at southern high latitudes [*Campbell and Claridge*, 1982]. This change at high latitudes is observed in the southern hemisphere only and is not associated with any significant modification of continental weathering at low latitudes. Therefore it is concluded that the cooling of East Antarctica mostly resulted from plate tectonic processes. As the Southern Ocean expanded continuously and the South Tasman seaway deepened during the late Eocene [*Kennett et al.*, 1975], cold waters developed and the thermal deficit at southern high latitudes increased. Moreover, compensation of the southern thermal deficit was restricted by the consequences of the collision along the Eurasian margin of the Tethys [*Savostin et al.*, 1986] and associated shortening of the ocean and shoaling of the Arabian seaways, which reduced the Tethyan outflow and poleward transfer of heat through warm, deep waters [*Ricou et al.*, 1986]. Thus plate tectonic processes at low and southern high latitudes favored the development of cold conditions in and around East Antarctica near the Eocene/Oligocene boundary and largely influenced further climatic evolution of the Earth.

Evolution of Water Masses

During the late Eocene, smectite was widespread in all Atlantic Ocean and Southern Ocean basins. No significant evolution was observed from the northern high latitudes to the mid-latitudes in the southern hemisphere, near the Eocene/Oligocene boundary (Figure 11). At the end of the Eocene, reduction of the outflow of intermediate waters from the western Tethys to the Atlantic was compensated by development of waters originating in equatorial areas [*Boersma*, 1986]. This important event did not significantly alter the clay mineral associations, probably because in both cases

dense waters originated from areas favorable to the development of smectite on adjacent continental margins. At shallow water depth, warm waters from low latitudes expanded to the Antarctic areas in the southwest Atlantic, indicating reinforcement of the western boundary surface circulation system, to compensate for increasing polar thermal deficit.

Southward transfer of heat resulted in precipitation on the Antarctic margin, but climate in East Antarctica was probably too cold to favor further development of kaolinite. The paucity of continental weathering resulted in clay mineral associations dominated by illite. Release of heat also resulted in cold surface waters which flowed northward and entrained terrigenous particles (mostly illite) up to mid-latitude areas, together with high-latitude nannofossils [*Manivit*, 1984].

However, cold surface waters partly mixed along the Antarctic margin with warm deep waters [*Kennett and Stott*, 1990] to form proto-Antarctic waters. This was probably the beginning of important thermohaline circulation [*Kennett and Barker*, 1990]. These waters, enriched in illite from East Antarctica, also expanded northward, mostly at intermediate water depth. At deep sites (drilled below 1800-m water depth) in the Southern Ocean and the South Atlantic (Figure 12), clay mineral associations indicate a northward decrease of illite content and were still dominated by smectite in the Subantarctic South Atlantic. This indicates minor terrigenous contribution from East Antarctica (and associated dense waters) to the southern South Atlantic, where warm dense waters from low latitudes were probably still present during the early Oligocene. However, illite (i.e., Antarctic influences) increased in Subantarctic areas by the latest Eocene at 37.5–38.0 Ma. Further increased influence of Antarctic waters (enriched in illite) in the earliest Oligocene at 35.0–36.0 Ma locally interfered in the eastern South Atlantic with waters enriched in smectite, which probably flowed southward through the western Indian Ocean. Increased illite content was not found in the Brazil and Angola basins where smectite (and influences from warm, low-latitude areas) increased in the earliest Oligocene at 35.0–36.0 Ma. It is likely that the northern basins of the South Atlantic were still dominated by warm dense saline waters formed in low latitudes. Therefore it appears that either thermohaline circulation was still too weak to expand in the northern South Atlantic basins or the Rio Grande Rise–Walvis Ridge system still acted as a barrier against the deepwater circulation. Anyway, coeval increased influences from high latitudes (and thermohaline circulation) and from low latitudes (and halothermal circulation) occurred in the earliest Oligocene, simultaneously with a major oxygen isotope shift. This suggests synchronous reinforcement of both circulation systems, probably involved in intensified equator to pole thermal exchanges.

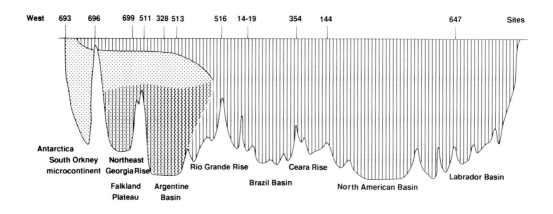

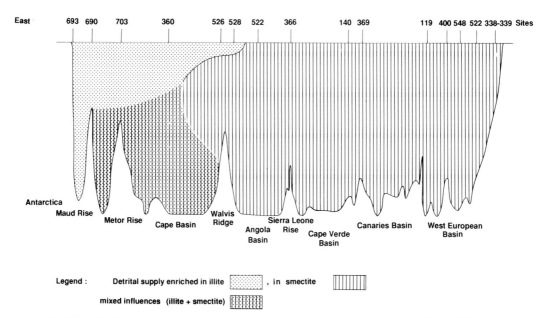

Fig. 11. Distribution of major clay minerals in the water column, near the Eocene/Oligocene boundary.

Variations of Oceanic Circulation in the Southern Hemisphere

Clay mineral associations at shallow sites indicate that surface western boundary circulation carried smectite and warm waters from low latitudes southward and warmed the South American margins (Figure 13). In winter, temperatures dropped more rapidly on the continent than in the ocean, creating a thermal deficit on the emerged continental margin of South America, likely to be compensated by transfer of heat from the ocean to the continent: warm air enriched in water vapor rose over cooler continental air masses, and condensation processes resulted in precipitation [Barry and Chorley, 1982; Wells, 1986]. The continental thermal deficit disappeared in summer, when temperatures rose faster on the continent than in the ocean. Wet winters and dry

summers favor smectite formation, and the dominance of this mineral on the South American margin, up to the Antarctic Peninsula, probably results in part from the warm influence of the western boundary circulation on local climate.

Release of heat in Antarctic areas resulted in precipitation and cooling of surface waters. Cold surface waters enriched in illite, eroded from East Antarctica, flowed northward in the southeast Atlantic, along the continental margin of southwest Africa (Figure 13). There, incoming solar radiation was probably high enough to sustain warm temperatures on the continent. The thermal deficit of the ocean may have been compensated by transfer of warm air from the continent [Barry and Chorley, 1982]. As continental air is poor in water vapor, this may have resulted in local develop-

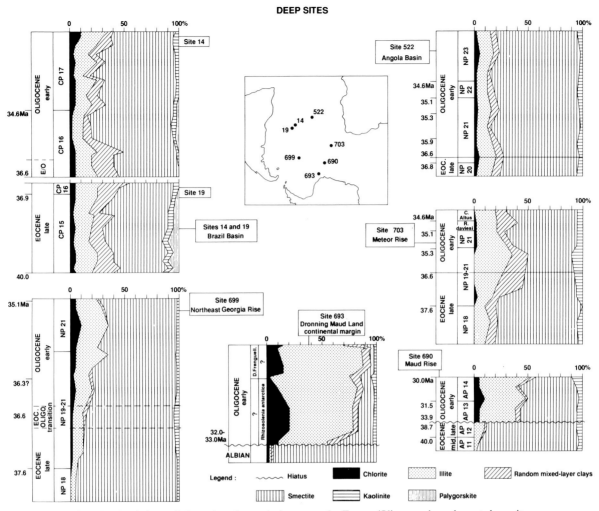

Fig. 12. Variations of clay mineral associations near the Eocene/Oligocene boundary at deep sites (below 1800-m water depth) in the South Atlantic and Southern oceans.

ment of dry environments, poor continental weathering, and development of illite, probably mostly eroded and transported to the ocean by reinforced atmospheric circulation. Cold surface waters progressively warmed as they approached low-latitude areas, while oceanward transfer of heat and inferred continental aridity declined.

Surface circulation in the South Atlantic probably formed a large anticyclonic gyre (Figure 13), similar to modern subtropical gyres [*Tchernia*, 1980]. However, owing to the absence of any seaway between South America and the Antarctica which opened later in the Oligocene [*Barker and Burrell*, 1977], warm waters were able to easily reach Antarctic areas. Activity of the anticyclonic gyre and its influence on climate increased significantly near the Eocene/Oligocene boundary. The interference of this gyre with cold surface waters at southern high latitudes from the earliest Oligocene and

its relation to a cyclonic proto-Weddell gyre [*Wei and Wise*, 1990] are not well constrained.

Proto-Antarctic waters, formed by mixing cold surface waters and warm deep waters along the East Antarctic margin [*Kennett and Stott*, 1990] flowed northward mainly at intermediate depth and carried illite eroded from East Antarctica. Near the Eocene/ Oligocene boundary, terrigenous supply enriched in illite is observed in both Antarctic and Subantarctic areas, at all locations in the western and the eastern part of the ocean. This suggests that proto-Antarctic intermediate waters followed a circulation pattern quite similar to that of modern Antarctic intermediate waters [*Tchernia*, 1980]. In southeastern Atlantic areas, relatively low amounts of illite and increased smectite contents suggest interference with warm, southward flowing intermediate circulation from the Indian Ocean. It is still unclear if proto-Antarctic waters at the Eocene/

DETRITAL SUPPLY AT SHALLOW WATER - DEPTH

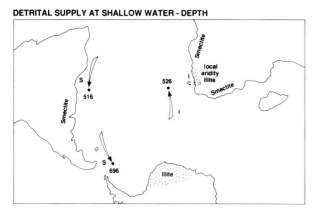

DETRITAL SUPPLY AT INTERMEDIATE WATER - DEPTH

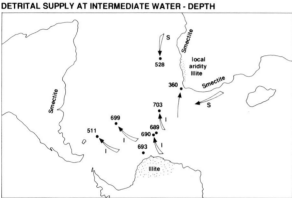

DETRITAL SUPPLY AT DEEP WATER - DEPTH

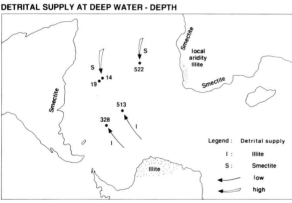

Fig. 13.   Clay mineral supply in the South Atlantic and Southern oceans near the Eocene/Oligocene boundary and possible relation to oceanic circulation and continental weathering.

Oligocene boundary reached the Rio Grande Rise–Walvis Ridge system or penetrated the northern South Atlantic basins. However, increased smectite contents at Site 528 on the northern flank of Walvis Ridge indicate, in this area, the intensified influence of warm intermediate waters from low latitudes.

In the abyssal plains, Antarctic influences as expressed by occurrences of illite are very weak in Subantarctic areas (Figure 13). These influences are observed at sites located on the path of modern Antarctic

Bottom Water circulation [*Tchernia*, 1980] and suggest the permanence of this pattern during the Cenozoic. However, dominant smectite at all locations suggests that the deepest parts of oceanic basins at southern high latitudes were still filled by warm waters from low latitudes. In the earliest Oligocene, minor increases of illite in Subantarctic areas occurred simultaneously with higher abundances of smectite in the Brazil and Angola basins. They express a reinforcement of both cold and warm bottom circulation systems, i.e., increased equator to pole thermal exchanges induced by a major cooling of East Antarctic areas.

CONCLUSION AND POSSIBLE SCENARIO FOR CLIMATIC AND OCEANOGRAPHIC DEVELOPMENT AT THE EOCENE/OLIGOCENE TRANSITION

Clay mineral evolution from the latest Eocene to the earliest Oligocene and its correlation with other indicators as well as geodynamic events are plotted in Figure 14. This provides information for a scenario likely to account for much of the climatic and oceanographic change at this time.

1.   The obvious dominance of smectite and the absence of any noticeable evolution of clay mineral associations in low-latitude areas (together with minor variations of the planktonic $\delta^{18}O$) and in the North Atlantic indicate that influences of astronomical (i.e., solar input) or chemical (i.e., greenhouse gas content) as well as inferred equator to pole thermal exchanges (which would have induced a similar clay mineral evolution in both hemispheres) did not change significantly near the Eocene/Oligocene boundary. Therefore mineralogical and climatic evolution evidenced in continent and oceanic areas of the southern hemisphere did not result from a redistribution of available heat at the surface of the Earth. Other processes and most probably plate tectonic evolution must have played a major role in the Eocene/Oligocene boundary events.

2.   Generally minor, but locally important, increases of kaolinite in the western Tethys as well as in northern Europe and in the African continental interior suggest increased thermal deficit on the emerged continental margins adjacent to the Tethys, compensated by intensified transfer of heat from the ocean to the continents. This deficit probably resulted from restriction of the areas of shallow environments favorable to warm, dense water formation and of the volume of the warm Tethys Ocean (i.e., reduction of a major reservoir of heat) due to plate tectonic collision along its margins and associated continental uplift. This also resulted in reduced output of Tethyan waters to the Atlantic and especially to the Indian Ocean.

3.   A major mineralogical event occurred in Antarctic areas where illite indicates a dramatic decrease of continental weathering associated with a temperature

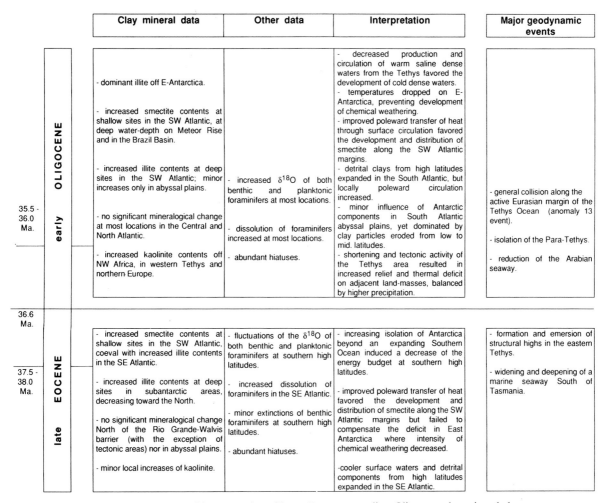

		Clay mineral data	Other data	Interpretation	Major geodynamic events
	OLIGOCENE early	- dominant illite off E-Antarctica. - increased smectite contents at shallow sites in the SW Atlantic, at deep water-depth on Meteor Rise and in the Brazil Basin. - increased illite contents at deep sites in the SW Atlantic; minor increases only in abyssal plains. - no significant mineralogical change at most locations in the Central and North Atlantic. - increased kaolinite contents off NW Africa, in western Tethys and northern Europe.	- increased δ¹⁸O of both benthic and planktonic foraminifers at most locations. - dissolution of foraminifers increased at most locations. - abundant hiatuses.	- decreased production and circulation of warm saline dense waters from the Tethys favored the development of cold dense waters. - temperatures dropped on E-Antarctica, preventing development of chemical weathering. - improved poleward transfer of heat through surface circulation favored the development and distribution of smectite along the SW Atlantic margins. - detrital clays from high latitudes expanded in the South Atlantic, but locally poleward circulation increased. - minor influence of Antarctic components in South Atlantic abyssal plains, yet dominated by clay particles eroded from low to mid. latitudes. - shortening and tectonic activity of the Tethys area resulted in increased relief and thermal deficit on adjacent land-masses, balanced by higher precipitation.	- general collision along the active Eurasian margin of the Tethys Ocean (anomaly 13 event). - isolation of the Para-Tethys. - reduction of the Arabian seaway.
35.5 - 36.0 Ma.					
36.6 Ma.					
37.5 - 38.0 Ma.	**EOCENE** late	- increased smectite contents at shallow sites in the SW Atlantic, coeval with increased illite contents in the SE Atlantic. - increased illite contents at deep sites in subantarctic areas, decreasing toward the North. - no significant mineralogical change North of the Rio Grande-Walvis barrier (with the exception of tectonic areas) nor in abyssal plains. - minor local increases of kaolinite.	- fluctuations of the δ¹⁸O of both benthic and planktonic foraminifers at southern high latitudes. - increased dissolution of foraminifers in the SE Atlantic. - minor extinctions of benthic foraminifers at southern high latitudes. - abundant hiatuses.	- increasing isolation of Antarctica beyond an expanding Southern Ocean induced a decrease of the energy budget at southern high latitudes. - improved poleward transfer of heat favored the development and distribution of smectite along the SW Atlantic margins but failed to compensate the deficit in East Antarctica where intensity of chemical weathering decreased. -cooler surface waters and detrital components from high latitudes expanded in the SE Atlantic.	- formation and emersion of structural highs in the eastern Tethys. - widening and deepening of a marine seaway South of Tasmania.

Fig. 14. Chronology and interpretation of latest Eocene to earliest Oligocene clay mineral changes and correlation with oceanic and geodynamic events.

drop particularly during the latest Eocene. It is likely that oceanic development at southern high latitudes (including widening and deepening of the Tasman seaway) resulted in the thermal isolation of the Antarctic continent and caused cooling. Increased thermal deficit activated equator to pole thermal exchanges through surface circulation; poleward transfer of heat warmed the southwest Atlantic margins (where smectite developed), and return circulation cooled the southeast Atlantic margins (where local aridity and illite erosion developed). Thermal exchange processes along the East Antarctic margin also favored the formation of proto-Antarctic waters (by mixing of cold surface waters and warm deep waters) which expanded at intermediate depth through the Southern Ocean and the South Atlantic.

4. The latest Eocene–earliest Oligocene events originated predominantly in the oceanic development at southern high latitudes; as the oceans need 5 times more energy than the continents to raise their temperature

and as incoming solar energy remains low at any time at high latitudes, this resulted in increased thermal deficit and cooling of both continental and oceanic Antarctic areas. Coeval restriction of the Tethys Ocean reduced the reservoir of heat at low latitudes and resulted in the lower influence of warm saline deep waters from the Tethys in distant oceanic basins. Both development of the Southern Ocean and restriction of the Tethys Ocean resulted in a dramatic increase of the southern polar thermal deficit. The development of warm deep waters originating from areas outside the Tethys Ocean and intensified meridional transfer of heat through surface circulation were insufficient to compensate for the Antarctic thermal deficit. This is due to the importance of the temperature drop at southern high latitudes, as well as to the absence of any increase of available solar energy at the surface of the Earth. As a consequence, cold conditions persisted in and around East Antarctica, and thermal exchanges resulted in expansion of cold waters and Antarctic influences to the oceans.

5. The latest Eocene–earliest Oligocene time interval represents a major step in climatic and oceanographic evolution as a consequence of plate tectonic processes. Progressive oceanographic evolution which began during the middle Eocene, in association with plate tectonic evolution, accelerated at the end of the Eocene. This was marked by the increased export of cold Antarctic deep water to the world ocean and consequent climatic cooling of the Earth.

Acknowledgments. We would like to express our thanks to the National Science Foundation for samples from the Deep Sea Drilling Project and the Ocean Drilling Program, to IFREMER and CNRS for financial support, to M. Decobert and F. Dujardin for technical assistance, to J. P. Kennett for stimulating discussions, and to G. Brass and R. Stein for the review of the original manuscript.

REFERENCES

Baldi, T., The terminal Eocene and early Oligocene events in Hungary and the separation of an anoxic, cool Paratethys, *Eclogae Geol. Helv.*, *77*, 1–27, 1984.

Barker, P. F., and J. Burrell, The opening of Drake Passage, *Mar. Geol.*, *25*, 15–34, 1977.

Barker, P. F., et al., Leg 72, *Initial Rep. Deep Sea Drill. Proj.*, *72*, 1024 pp., 1983.

Barker, P. F., et al., Leg 113, *Proc. Ocean Drill. Program Initial Rep.*, *113*, 785 pp., 1988.

Barry, R. G., and R. J. Chorley, *Atmosphere, Weather and Climate*, 407 pp., Holt, Rinehart, and Winston, New York, 1982.

Berggren, W. A., D. V. Kent, J. J. Flynn, and J. A. Van Couvering, Cenozoic geochronology, *Geol. Soc. Am. Bull.*, *96*, 1407–1418, 1985.

Boersma, A., Eocene/Oligocene Atlantic paleo-oceanography using benthic foraminifera, in *Terminal Eocene Events, Developments in Paleontology and Stratigraphy*, vol. 9, edited by C. Pomerol and I. Premoli-Silva, pp. 225–235, Elsevier, New York, 1986.

Boersma, A., I. Premoli-Silva, and N. Shackleton, Atlantic Eocene planktonic foraminiferal palaeohydrographic indicators and stable isotopic paleoceanography, *Paleoceanography*, *2*, 287–331, 1987.

Bolli, H. M., et al., Leg 40, *Initial Rep. Deep Sea Drill. Proj.*, *40*, 1067 pp., 1978.

Bombita, G., Eocene/Oligocene boundary in Romania, present-day state of investigation, in *Terminal Eocene Events, Developments in Paleontology and Stratigraphy*, vol. 9, edited by C. Pomerol and I. Premoli-Silva, pp. 121–127, Elsevier, New York, 1986.

Brass, G. W., J. R. Southam, and W. H. Peterson, Warm saline bottom water in the ancient ocean, *Nature*, *296*, 620–623, 1982.

Campbell, I. B., and G. G. C. Claridge, The influence of moisture on the development of soils of the cold deserts of Antarctica, *Geoderma*, *28*, 221–238, 1982.

Chamley, H., North Atlantic clay sedimentation and paleoenvironment since the late Jurassic, in *Deep Drilling Results in the Atlantic Ocean: Continental Margins and Paleoenvironment, Maurice Ewing Ser.*, vol. 3, edited by M. Talwani, W. Hay, and W. B. F. Ryan, pp. 342–361, AGU, Washington, D. C., 1979.

Chamley, H., Clay mineralogy at the Eocene/Oligocene boundary, in *Terminal Eocene Events, Developments in Paleontology and Stratigraphy*, vol. 9, edited by C. Pomerol and I. Premoli-Silva, pp. 381–386, Elsevier, New York, 1986.

Chamley, H., *Clay Sedimentology*, 623 pp., Springer-Verlag, New York, 1989.

Chennaux, G., J. Esquevin, A. Jourdan, C. Latouche, and N. Maillet, X-ray mineralogy and mineral geochemistry of Cenozoic strata (Leg 80) and petrographic study of associated pebbles, *Initial Rep. Deep Sea Drill. Proj.*, *80*, 1019–1046, 1985.

Ciesielski, P. F., et al., Leg 114, *Proc. Ocean Drill. Program Initial Rep.*, *114*, 815 pp., 1988.

Cook, H. E., I. Zemmels, and J. C. Matti, X-ray mineralogy data, southern Indian Ocean, *Initial Rep. Deep Sea Drill. Proj.*, *26*, 573–592, 1974.

Debrabant, P., H. Chamley, and J. Foulon, Paleoenvironmental implications of mineralogic and geochemical data in the western Florida Straits (D.S.D.P. Leg 77), *Initial Rep. Deep Sea Drill. Proj.*, *77*, 377–396, 1984.

Deconinck, J. F., P. Broquet, H. Chamley, F. Robaszynski, and F. Thiébault, Minéraux argileux de la zone de Sclafani (Madonies, Sicile): Diagenèse et paléoenvironnement du Permien au Miocène, *Geol. Mediterr.*, *13*, 3–11, 1986.

Diester-Haass, L., and H. Chamley, Oligocène climatic, tectonic and eustatic history off NW Africa (D.S.D.P. Leg 41, Site 369), *Oceanol. Acta*, *3*, 115–126, 1980.

Dingle, R. V., W. G. Siesser, and A. R. Newton, *Mesozoic and Tertiary geology of Southern Africa*, 375 pp., A. A. Balkema, Rotterdam, Netherlands, 1983.

Ehrmann, W. U., Implications of sediment composition on the southern Kerguelen Plateau for paleoclimate and depositional environment, *Proc. Ocean Drill. Program Sci. Results*, *119*, 185–210, 1991.

Froget, C., La sédimentation argileuse depuis l'Eocène sur le plateau Voring et à son voisinage, d'après le Leg 38 D.S.D.P. (Mer de Norvège), *Sedimentology*, *28*, 793–804, 1981.

Holtzapffel, T., Les minéraux argileux, Préparation, Analyse diffractométrique et détermination, *Publ. Soc. Geol. Nord*, *12*, 136 pp., 1985.

Hsü, K. J., et al., Leg 73, *Initial Rep. Deep Sea Drill. Proj.*, *73*, 798 pp., 1984.

Keigwin, L. D., Paleoceanographic change in the Pacific at the Eocene-Oligocene boundary, *Nature*, *287*, 722–725, 1980.

Keller, G., Biochronology and paleoclimatic implications of middle Eocene to Oligocene planktic foraminiferal faunas, *Mar. Micropaleontol.*, *7*, 463–486, 1983.

Kennett, J. P., and P. F. Barker, Latest Cretaceous to Cenozoic climate and oceanographic developments in the Weddell Sea, Antarctica: An ocean-drilling perspective, *Proc. Ocean Drill. Program Sci. Results*, *113*, 937–960, 1990.

Kennett, J. P., and L. D. Stott, Proteus and Proto-Oceanus: Ancestral paleogene oceans as revealed from Antarctic stable isotopic results, O.D.P. Leg 113, *Proc. Ocean Drill. Program Sci. Results*, *113*, 865–880, 1990.

Kennett, J. P., and C. Von der Borch, Southwest Pacific Cenozoic paleoceanography, *Initial Rep. Deep Sea Drill. Proj.*, *90*, 1493–1517, 1986.

Kennett, J. P., et al., Leg 29, *Initial Rep. Deep Sea Drill. Proj.*, *29*, 1197 pp., 1975.

Lancelot, Y., et al., Leg 41, *Initial Rep. Deep Sea Drill. Proj.*, *41*, 1259 pp., 1978.

Latouche, C., and N. Maillet, X-ray mineralogy study of Tertiary deposits, Leg 81, sites 552–555, *Initial Rep. Deep Sea Drill. Proj.*, *81*, 669–682, 1984.

Laughton, A. S., et al., Leg 12, *Initial Rep. Deep Sea Drill. Proj.*, *12*, 1243 pp., 1972.

Lazarus, D., and J. P. Caulet, Eocene to Recent radiolarian biostratigraphy, biogeography, diversity, and history of the Southern Ocean, this volume.

Leroy, P., Contribution à l'étude de la sédimentation argileuse sur les marges de l'océan Atlantique Nord depuis le Juras-sique supérieur, thesis, doctorat 3e cycle, 151 pp., Univ. de Lille, Lille, France, 1981.

Maillot, H., and C. Robert, Paleoenvironmental evolution of the Walvis Ridge deduced from inorganic geochemical and clay mineral data (DSDP Leg 74, southeast Atlantic), *Initial Rep. Deep Sea Drill. Proj.*, 74, 663–683, 1984.

Manivit, H., Paleogene and upper Cretaceous calcareous nan-nofossils from D.S.D.P. Leg 74, *Initial Rep. Deep Sea Drill. Proj.*, 74, 475–500, 1984.

Matti, J. C., I. Zemmels, and H. E. Cook, X-ray mineralogy data, western Indian Ocean, Leg 25 D.S.D.P., *Initial Rep. Deep Sea Drill. Proj.*, 25, 843–861, 1974.

Maxwell, A. E., et al., Leg 3, *Initial Rep. Deep Sea Drill. Proj.*, 3, 806 pp., 1970.

Mercier-Castiaux, M., H. Chamley, and C. Dupuis, La sédi-mentation argileuse tertiaire dans le bassin belge et ses approches occidentales, *Ann. Soc. Geol. Nord*, 107, 139–154, 1988.

Millot, G., *Geology of Clays*, 425 pp., Springer-Verlag, New York, 1970.

Moore, T. C., et al., *Initial Rep. Deep Sea Drill. Proj.*, 74, 894 pp., 1984.

Muza, J. P., D. F. Williams, and S. W. Wise, Paleogene oxygen isotope record from Deep Sea Drilling Project sites 511 and 512, Subantarctic South Atlantic Ocean: Paleotem-peratures, paleoceanographic changes, and the Eocene/ Oligocene boundary event, *Initial Rep. Deep Sea Drill. Proj.*, 71, 409–422, 1983.

Nielsen, O. B., M. Cremer, R. Stein, F. Thiébault, and H. Zimmerman, Analysis of sedimentary facies, clay mineral-ogy and geochemistry of the Paleogene sediments of Site 647, Labrador Sea, *Proc. Ocean Drill. Program Sci. Results*, 105, 101–110, 1989.

Oberhänsli, H., and M. Toumarkine, The Paleogene oxygen and carbon isotope history of sites 522, 523 and 524 from the central South Atlantic, in *South Atlantic Paleoceanography*, edited by K. J. Hsü and H. J. Weissert, pp. 125–148, Cambridge University Press, New York, 1985.

Prentice, M. L., and R. K. Matthews, Cenozoic ice-volume history: Development of a composite oxygen isotope record, *Geology*, 16, 963–966, 1988.

Rabussier-Lointier, D., Variations de composition isotopique de l'oxygène et du carbone en milieu marin et coupures stratigraphiques du Cénozoïque, thesis, doctorat 3e cycle, 182 pp., Univ. Pierre et Marie Curie, Paris, 1980.

Ricou, L. E., B. Mercier de Lépinay, and J. Marcoux, Evolu-tion of the Tethyan seaways and implications for the oceanic circulation around the Eocene/Oligocene boundary, in *Ter-minal Eocene Events, Developments in Paleontology and Stratigraphy*, vol. 9, edited by C. Pomerol and I. Premoli-Silva, pp. 387–394, Elsevier, New York, 1986.

Robert, C., Santonian to Eocene paleogeographic evolution of Rio Grande Rise (South Atlantic) deduced from clay miner-alogical data (D.S.D.P. legs 3 and 39), *Palaeogeogr. Palae-oclimatol. Palaeoecol.*, 33, 311–325, 1981.

Robert, C., Modalité de la sédimentation argileuse en relation avec l'histoire géologique de l'Atlantique Sud, thesis, doc-torat d'état—sciènces, 141 pp., Univ. d'Aix-Marseille, Marseille, France, 1982.

Robert, C., Clay mineral associations and structural evolution of the South Atlantic: Jurassic to Eocene, *Palaeogeogr. Palaeoclimatol. Palaeoecol.*, 58, 87–108, 1987.

Robert, C., and H. Chamley, Development of early Eocene warm climates, as inferred from clay mineral variations in oceanic sediments, *Global Planet. Change*, 89, 315–331, 1991.

Robert, C., and J. P. Kennett, Paleocene and Eocene kaolinite distribution in the South Atlantic and Southern Ocean: Antarctic climatic and paleoceanographic implications, *Mar. Geol.*, 103, 99–110, 1992.

Robert, C., and H. Maillot, Paleoenvironmental significance of clay mineralogical and geochemical data, southwest Atlan-tic, Deep Sea Drilling Project legs 36 and 71, *Initial Rep. Deep Sea Drill. Proj.*, 71, 317–343, 1983.

Robert, C., and H. Maillot, Paleoenvironments in the Weddell Sea area and Antarctic climates, as deduced from clay mineral associations and geochemical data, O.D.P. Leg 113, *Proc. Ocean Drill. Program Sci. Results*, 113, 51–70, 1990.

Savostin, L. A., J. C. Sibuet, L. P. Zonenshain, X. Le Pichon, and M. J. Roulet, Kinematic evolution of the Tethys Belt from the Atlantic Ocean to the Pamirs since the Triassic, *Tectonophysics*, 123, 1–35, 1986.

Shackleton, N. J., and J. P. Kennett, Paleotemperature history of the Cenozoic and the initiation of Antarctic glaciation: Oxygen and carbon isotope analyses in D.S.D.P. sites 277, 279, and 281, *Initial Rep. Deep Sea Drill. Proj.*, 29, 743–756, 1975.

Sittler, C., Le Paléogène des fossés rhénan et rhodanien, études sédimentologiques et paléoclimatiques, *Mem. Serv. Carte Geol. Alsace Lorraine*, 24, 392 pp., 1965.

Smith, A. G., A. M. Hurley, and J. C. Briden, *Phanerozoic Paleocontinental World Maps*, 102 pp., Cambridge Univer-sity Press, New York, 1981.

Steurbaut, E., The Kallo well and its key-position in establish-ing the Eo-Oligocene boundary in Belgium, in *Terminal Eocene Events, Developments in Paleontology and Stratig-raphy*, vol. 9, edited by C. Pomerol and I. Premoli-Silva, pp. 97–100, Elsevier, New York, 1986.

Stott, L. D., J. P. Kennett, N. J. Shackleton, and R. M. Corfield, The evolution of Antarctic surface waters during the Paleogene: Inferences from stable isotopic composition of planktonic foraminifers, O.D.P. Leg 113, *Proc. Ocean Drill. Program Sci. Results*, 113, 849–863, 1990.

Tchernia, P., *Descriptive Regional Oceanography, Pergamon Mar. Ser.*, vol. 3, 253 pp., Pergamon, New York, 1980.

Thiry, M., Geochemical evolution and paleoenvironments of the Eocene continental deposits in the Paris Basin, *Palaeo-geogr. Palaeoclimatol. Palaeoecol.*, 70, 153–163, 1989.

Weaver C. E., *Clays, Muds, and Shales, Developments in Sedimentology*, vol. 44, 819 pp., Elsevier, New York, 1989.

Wei, W., and S. W. Wise, Middle Eocene to Pleistocene calcareous nannofossils recovered by Ocean Drilling Pro-gram Leg 113 in the Weddell Sea, *Proc. Ocean Drill. Program Sci. Results*, 113, 639–666, 1990.

Wells, N., *The Atmosphere and Ocean: A physical introduc-tion*, 347 pp., Taylor and Francis, London, 1986.

Williams, D. F., R. C. Thunnell, D. A. Hodell, and C. Vergnaud-Grazzini, Synthesis of Late Cretaceous, Tertiary and Quaternary stable isotope records of the South Atlantic, based on Leg 72 D.S.D.P. core material, in *South Atlantic Paleoceanography*, edited by K. J. Hsü and H. J. Weissert, pp. 205–242, Cambridge University Press, New York, 1985.

Wise, S. W., J. R. Breza, D. M. Harwood, and W. Wei, Paleogene glacial history of Antarctica, in *Controversies in Modern Geology*, pp. 133–171, Academic, San Diego, Calif., 1991.

Zonenshain, L. P., and X. Le Pichon, Deep basins of the Black Sea and Caspian Sea as remnants of Mesozoic back-arc basins, *Tectonophysics*, 123, 181–211, 1986.

(Received January 2, 1992;
accepted May 26, 1992.)

EVIDENCE FROM FOSSIL VERTEBRATES FOR A RICH EOCENE ANTARCTIC MARINE ENVIRONMENT

JUDD A. CASE

Department of Biology, Saint Mary's College of California, Moraga, California 94575

The late Eocene fossil marine and terrestrial biota recovered from deposits on Seymour Island, Antarctic Peninsula, indicate that cool temperate climatic conditions prevailed. This interpretation agrees with oxygen isotope data from marine drill cores taken from the Weddell Sea (Ocean Drilling Program Leg 113). Sharks and penguins, the two major groups of fossil marine vertebrate taxa based on abundance of specimens recovered from the La Meseta Formation, Seymour Island, show a remarkable level of species diversity for a cool temperate marine environment. Fourteen species of sharks have been recorded from a single locality within Unit Telm 5 of the La Meseta Formation, while six species of penguins representing six distinct size classes have been noted in two localities in Unit Telm 7. The level of shark species diversity from this single locality ($H = 1.814$, based on the Shannon-Weiner diversity measure H) exceeds that obtained for present-day shark faunas sampled from cool temperate waters in the North Atlantic ($H = 0.365, 0.262$) and from warm temperate waters off of Durban, South Africa ($H = 0.872$). The shark paleofauna is nearly equal to an extant tropical shark fauna sampled from Baja California to Peru ($H = 1.920$). The number of penguin species (six) and the number of size classes (six) obtained from each of the late Eocene localities exceed the highest numbers of sympatric species (four) and size classes (four) found at present in regional breeding localities (primarily islands) in the cold Subantarctic to cool temperate waters south of the Subtropical Convergence. During the late Eocene, both sharks and penguins occupied positions at the top of the marine trophic structure. In order to support the high Eocene species diversity and large population sizes (the latter inferred from large sample sizes of fossil specimens) of the uppermost trophic level, an extensive biomass in the lower trophic levels must have been present. The amount of biomass and possible diversity suggested at all trophic levels implies a very rich late Eocene marine environment in the cool temperate waters east of the Antarctic Peninsula. Currently, high levels of productivity and biomass are generated in cool temperate marine environments only where coastal upwelling occurs. The waters off the east coast of the Antarctic Peninsula may have been such an area of upwelling during the late Eocene. This evidence indicates that the onset of upwelling conditions around Antarctica began in the late Eocene. Major upwelling conditions are predicted by paleoceanographic models to occur by the late Oligocene.

INTRODUCTION

The macroinvertebrate fauna recovered from the Eocene La Meseta Formation of Seymour Island, Antarctic Peninsula (Figure 1), is often described in the literature as being rich and diverse [*Zinsmeister and Camacho*, 1982; *Feldmann*, 1984; *Elliot*, 1988]. This fauna has been shown to be locally diverse (i.e., in the area surrounding Seymour Island, James Ross Basin) and to include abundant bivalves [*Zinsmeister*, 1984], gastropods [*Zinsmeister and Camacho*, 1980], and molluscs; asteroid [*Blake and Zinsmeister*, 1979, 1988], crinoid [*Doktor et al.*, 1988], echinoid [*McKinney et al.*, 1988], and ophiuroid (author's personal observations) echinoderms; brachiopods [*Wiedman et al.*, 1988]; and decapod crustaceans [*Feldmann and Zinsmeister*, 1984; *Feldmann and Wilson*, 1988]. This macroinvertebrate fauna indicates an apparent bountiful marine biota for that time period. However, in this sizable literature, there has been no attempt to document the degree of

richness of taxa or numbers of individuals, either quantitatively or qualitatively. The levels of abundance (number of specimens) or diversity (number of taxa) for any of the above groups have not been calculated or estimated, nor have comparisons been made to any modern fauna, to achieve some tangible measure of richness.

Levels of diversity and abundance for two marine vertebrate taxa, sharks and penguins, collected from the La Meseta Formation will be examined herein. These data allow for inferences about trophic level composition for vertebrates that permit interpretations of the richness (numbers of taxa and specimens) of the Antarctic marine environment in the region of the James Ross Basin during the late Eocene.

The late Eocene marine environment east of the Antarctic Peninsula has been viewed as cool temperate based on evidence from molluscan taxa [*Zinsmeister*, 1982, 1984] and other macroinvertebrates from the La

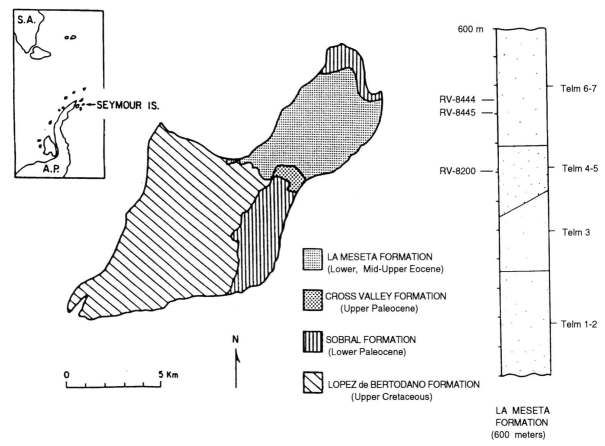

Fig. 1.   Map of Seymour Island and the Antarctic Peninsula (inset), plus the stratigraphic column of the La Meseta Formation. Fossil shark material recovered from locality RV-8200 [see *Long*, 1992] has served as the sample for measurements of shark faunal diversity. Fossil penguin material recovered from localities RV-8444 and RV-8445 has demonstrated the highest levels of species and body size class sympatry recorded from the late Eocene deposits.

Meseta Formation of Seymour Island [*Zinsmeister and Feldmann*, 1984]. Data from calcareous nannofossils [*Pospichal and Wise*, 1990; *Wei and Wise*, 1990] and planktonic and benthic foraminiferan taxa [*Stott and Kennett*, 1990; *Thomas*, 1990], as well as oxygen isotope data from marine drill cores taken during Ocean Drilling Program (ODP) Leg 113 in the Weddell Sea, east of Seymour Island [*Kennett and Stott*, 1990; *Stott et al.*, 1990; *Kennett and Barker*, 1990], also support the interpretation that the marine environment in this region was cool temperate during the late Eocene. In addition, the evidence of the terrestrial environment from fossil floras collected from localities within the late Eocene portion of the La Meseta Formation also indicates cool temperate conditions [*Case*, 1988]. Palynomorphs from Paleocene localities [*Askin*, 1988, 1989, 1990] and from the La Meseta Formation (R. A. Askin, personal communication, 1991) again indicate a cool temperate terrestrial environment.

SHARK DIVERSITY AND ABUNDANCE

Seventeen species of sharks have been recovered from Units Telm 3–5 of the La Meseta Formation (stratigraphic units as designated by *Sadler* [1988]; Figure 1). A single locality (RV-8200, the "mammal locality" [*Woodburne and Zinsmeister*, 1984]) in Unit Telm 5 is now known to have representatives of 14 of those 17 species [*Long*, 1992]. A plot of the relative abundance of each of the 14 species from RV-8200 is shown in Figure 2, where the relative abundance per species is calculated from the number of specimens/species per total number of specimens for the sample (440 specimens). From the relative abundance data, the Shannon-Weiner diversity measure ($H = \sum p_i \ln p_i$, where p_i is the proportion of species i) can be calculated, as well as a measure of evenness of relative abundance of the species within the fauna ($J = H/\ln S$, where S is the number of species) [*Levinton*, 1982]. The

RELATIVE ABUNDANCE OF SHARK SPECIES FROM LATE EOCENE, LOCALITY RV-8200, SEYMOUR ISLAND ANTARCTIC PENINSULA. (Data from Long, 1992)

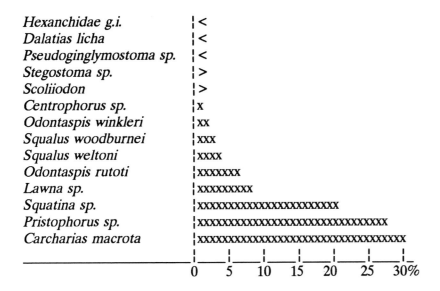

Percentage of total specimens (440 specimens)

$<$ = <0.5%; $>$ = >0.5%; x= values rounded to nearest whole %.
(Diversity Index - H = 1.814; Evenness Index - J = 0.687)

Fig. 2. Plot of relative abundance of shark species (i.e., percentage of the total number of specimens (= 440) recovered represented by each species) from the late Eocene La Meseta Formation, locality RV-8200.

values calculated from the late Eocene Seymour Island shark fauna for diversity (H = 1.814) and evenness (J = 0.687) are high, especially for a shark fauna which lived in cool temperate waters, as can be seen below.

Two examples of diversity of modern shark faunas from cool temperate waters are provided here (Figure 3). One example is from *Stevens* [1990], where a shark tagging project (i.e., a sample taken from an area in a short time span; but the sample may not represent the actual diversity, as biases could result owing to differing methods of capturing the animals) was performed in the northeast Atlantic. Only four shark species were recorded out of 2882 specimens tagged, with one species (*Prionace glauca*) composing 90% of the sample. The second example [*Daan et al.*, 1990] presents the mean values for tonnage figures recorded for sharks captured from the North Sea from 1977 to 1986. Here, five species of shark were recorded, with an even greater predominance of a single species (*Squalus acanthias*, 94.6%). Consequently, the two present-day cool temperate shark faunas exhibit low values for both diversity (H = 0.365, 0.262, respectively) and evenness (J = 0.263, 0.163, respectively).

Data from a shark faunal sample resulting from a tagging project at Durban, South Africa, represent a measure of extant shark diversity taken in warm temperate waters [*Davies and Joubert*, 1967]. The sample contained eight species and was dominated by a single species, *Carcharhinus obscurus*, which accounted for 73% of the sharks tagged. The diversity value was low (H = 0.869) in comparison with the sample from the late Eocene of Seymour Island (Figure 4). The fauna has a very low evenness value (J = 0.418), because of the predominance of *C. obscurus*.

A fourth sample is from a tagging project taken along a transect from the southern Pacific coast of Baja California to northern Peru (some 30° of latitude) and thus represents a sample from tropical waters [*Kato and Carvallo*, 1967]. Fifteen species were captured, with reasonable levels of relative abundance for six of the species (Figure 5). The diversity value of the sample is high (H = 1.920), as is the value for evenness (J = 0.709).

Thus the levels of diversity and evenness of the La Meseta shark fauna, which represents a fauna living in cool temperate waters, are comparable to those of a

A

RELATIVE ABUNDANCE OF SHARK SPECIES FROM THE NORTH-EAST ATLANTIC, SHARK TAGGING PROJECT 1970-1981. (Data from Stevens, 1990)

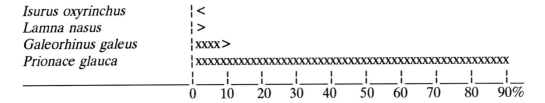

Percentage of total specimens (2882 specimens)

<= <1.0%; >= >1.0%; x= values rounded to nearest 2%.
(Diversity Index - H = 0.365; Evenness Index - J = 0.263)

B

RELATIVE ABUNDANCE OF SHARK SPECIES FROM THE NORTH SEA, ENGLISH GROUNDFISH SURVEYS, 1977-1986. (Data from Daan et al., 1990)

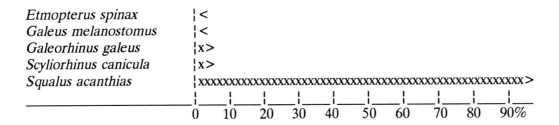

Percentage of total tonnage (185.5 tonnes)

<= <1.0%; >= >1.0%; x= values rounded to nearest 2%.
(Diversity Index - H = 0.262; Evenness Index - J = 0.163)

Fig. 3. Plot of relative abundance of cool temperate shark species from (a) a shark tagging project from the northeast Atlantic, 1970–1981 (percentage of total number of specimens captured (= 2882)) and (b) English groundfish surveys from the North Sea, 1977–1986 (percentage of mean of total tonnes of sharks captured per year (= 185.5)).

present-day shark fauna in tropical waters (a sample which covered a large range of latitude, unlike the single locality sample from the La Meseta Formation) and substantially above those of a currently existing shark fauna in temperate or cool temperate waters. The num-ber of shark specimens (n = 440) from locality RV-8200 also indicates a large biomass for this area during the late Eocene. Thus the late Eocene shark fauna from Seymour Island is rich and diverse despite the water temperature of the habitat.

RELATIVE ABUNDANCE OF SHARK SPECIES, DURBAN, SOUTH AFRICA (30°S).
SHARK TAGGING PROJECT 1964-1965. (Data from Davies & Joubert, 1967)

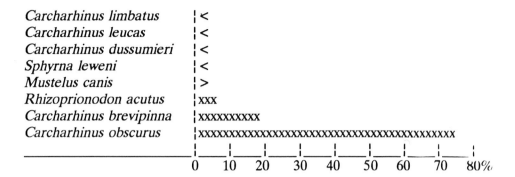

```
Carcharhinus limbatus      | <
Carcharhinus leucas        | <
Carcharhinus dussumieri    | <
Sphyrna leweni             | <
Mustelus canis             | >
Rhizoprionodon acutus      | xxx
Carcharhinus brevipinna    | xxxxxxxxxx
Carcharhinus obscurus      | xxxxxxxxxxxxxxxxxxxxxxxxxxxxxxxxxxxxxxxxxxxxxx
                           |__|__|__|__|__|__|__|__|__|
                           0  10  20  30  40  50  60  70  80%
```

Percentage of total specimens (984 specimens)

<= <1.0%; >= >1.0%; x= values rounded to nearest 2%.
(Diversity Index - H = 0.869; Evenness Index - J = 0.418)

Fig. 4.   Plot of relative abundance of warm temperate shark species from a shark tagging project from Durban, South Africa, 1964–1965 (percentage of total number of specimens captured (= 984)).

RELATIVE ABUNDANCE OF SHARK SPECIES FROM EASTERN PACIFIC
OCEAN, BAJA CALIFORNIA TO PERU (30°N-5°S), SHARK TAGGING
PROJECT 1962-1965. (Data from Kato & Carvallo, 1967)

```
Alopias vulpinus               | <
Carcharhinus velox             | <
Ginglymostoma cirratum         | <
Mustelus sp.                   | <
Carcharhinus longimanus        | <
Sphyrna zygaena                | >
Carcharhinus altimus           | >
Prionaca glauca                | >
Sphyrna leweni                 | xxx
Rhizoprionodon longurio        | xxxxxxxxxxx
Carcharhinus porosus           | xxxxxxxxxxx
Carcharhinus falciformis       | xxxxxxxxxxxxxxxx
Carcharhinus albimarginatus    | xxxxxxxxxxxxxxxxxx
Carcharhinus limbatus          | xxxxxxxxxxxxxxxxxxxxxxxx
Carcharhinus galapagensis      | xxxxxxxxxxxxxxxxxxxxxxxxxxxx
                               |__|__|__|__|__|__|
                               0  5  10  15  20  25  30%
```

Percentage of total specimens (860 specimens)

<= <0.5%; >= >0.5%; x= values rounded to nearest whole %.
(Diversity Index - H = 1.920; Evenness Index - J = 0.709)

Fig. 5.   Plot of relative abundance of tropical shark species from a shark tagging project in the eastern Pacific Ocean, Baja California to Peru, 1962–1965 (percentage of total number of specimens captured (= 860)).

PENGUIN SIZE CLASSES BASED ON FOSSILS FROM THE LATE EOCENE
UNITS OF THE LA MESETA FM., SEYMOUR IS., ANTARCTIC PENINSULA

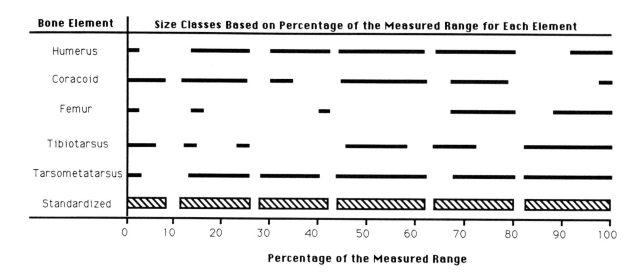

Standardized Size Classes: I=0-8%; II=12-26%; III=28-42%; IV=44-62%; V=64-80%; VI=82-100%.

Correlation of Species to Size Classes: I- *Archaeosphenicus wimani*;   II- *Delphinornis larseni*;
III- *Wimanornis seymourensis*;   IV- *Palaeeudyptes gunnari*
V- *Palaeeudyptes klekowskii*;   VI- *Anthropornis nordenskjoeldi*.

Fig. 6. Determination of the number of penguin size classes (six) based on a plot of the percentage of the measured range for each of the five bony elements sampled from the late Eocene units of the La Meseta Formation, Seymour Island, Antarctic Peninsula. A standardized range based on the compilation of the elemental ranges is shown in the bottom row. Percentage of the standardized range for each size class is given, as well as the correlation of those size classes to the species recovered.

PENGUINS: LEVELS OF SPECIES AND BODY SIZE SYMPATRY

The other frequently encountered fossil vertebrate taxon from the late Eocene La Meseta Formation on Seymour Island is the penguin family. Fossil penguin skeletal elements are found in low numbers in Unit Telm 3, with a slightly higher abundance of bones in Unit Telm 5 and a tremendous number of specimens in the uppermost Tertiary unit, Telm 7 (author's personal observations).

A variety of size classes of penguins existed in the late Eocene on Seymour Island [*Simpson*, 1975, 1976; *Myrcha et al.*, 1990]. It is possible to determine the number of penguin size classes present in this penguin paleofauna through measurements on five of the most frequently encountered skeletal elements (the humerus, coracoid, femur, tibiotarsus, and tarsometarsus) in the deposits of the La Meseta Formation. A single dimension is measured on a selected morphological structure for each element, that is most often preserved, in order to achieve as large a sample size as possible (J. A. Case and P. Shui, work in preparation). To equate differing

magnitudes of size among the five measurements (i.e., to compare each measured value to all other values on a dimensionless scale), the range of values for each measurement per element was determined, so that each value represents some percentage of the measured range for the particular bone, with the smallest value being set at the 0% mark and the highest value set at 100%. Distributions were plotted for the range percentage data for each element (summarized in Figure 6). To assist in standardizing ranges for all elements superimposed on one another, guidelines were determined from estimated standing height values for each of the Seymour Island species presented by *Simpson* [1975, 1976], where each species height range occupies about 15–20% of the total range of height. The standardized values at the bottom of Figure 6 were compiled from the ranges for each bone and then correlated to the late Eocene species known from Seymour Island.

Locality and stratigraphic unit data were recorded for each of the measured specimens, in order to compile data to determine the number of specimens/size class for each stratigraphic unit (Telm 3–7) and each penguin

RELATIVE ABUNDANCE OF PENGUIN SIZE CLASSES IN LATE EOCENE, UNIT Telm 7, LA MESETA FM. SEYMOUR ISLAND, ANTARCTIC PENINSULA

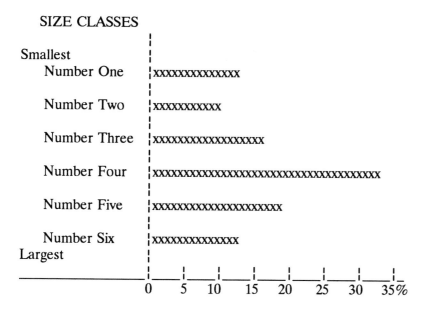

SIZE CLASSES

Percentage of total specimens (106 specimens)

x= values rounded to nearest whole %.
(Diversity Index - H = 1.619; Evenness Index - J = 0.904)

Fig. 7. Plot of relative abundance of cool temperate penguin species from stratigraphic Unit Telm 7, La Meseta Formation, Seymour Island, Antarctic Peninsula (percentage of total number of specimens recovered (= 106)).

locality (University of California, Riverside, localities, plus M. O. Woodburne (MOW) and R. E. Fordyce (REF) field localities). Relative abundances for each size class from Unit Telm 7 (the unit from which most of the specimens were recovered) were plotted, and values for both diversity and evenness were calculated (Figure 7).

Sample population data of the type necessary to produce relative abundance values are unavailable for extant penguins, as the population data from rookery sites for penguin species are given as crude estimates in ranges of thousands, to tens of thousands, to hundreds of thousands of individuals per species. Thus a different estimate of diversity, comparable for penguin species past and present, must be developed.

The Seymour Island penguin data from Unit Telm 7 indicate that all six size classes, and thus all six species, have been recovered from that stratigraphic unit. In fact, two localities (RV-8444 and RV-8445) have each yielded all six size classes (representing the six described species). Thirteen other localities (Figure 8) have produced a variety of combinations of size classes,

ranging from three to five size classes per locality. The above data indicate a high level of sympatry with the co-occurrence of as many as six species and six different size classes from the same locality.

The degree of sympatry among modern penguin species can be determined from penguin breeding colonies at discrete geographic localities, such as small southern oceanic islands or coastal areas of larger islands or continents (Figure 9). Data on breeding localities taken from *Stonehouse* [1975] show that there are 12 localities in the southern oceans where a maximum of four species occur sympatrically. No seasonal or temporal factor is considered here in regard to sympatry; only the presence of the species breeding at a locality is noted. Temporal differences in breeding between sympatric species at a locality have not been considered, because carcasses of dead individuals from different breeding times during the year would occur in the same stratigraphic horizon of possible, future fossil deposits. This would create a faunal sample similar to those taken from localities of late Eocene deposits on Seymour Island.

The distribution of modern penguin sizes based on

Localities	SIZE CLASSES					
	I	II	III	IV	V	VI
RV-8444	X	X	X	X	X	X
RV-8445	X	X	X	X	X	X
MOW-8656	X	X	X	X	X	
MOW-8679	X	X	X	X		
REF-13		X		X	X	X
REF-35		X	X	X	X	
REF-65	X		X	X	X	
REF-70			X	X	X	X
MOW-8655			X		X	X
MOW-8676	X		X		X	
MOW-8714			X	X	X	
REF-31	X			X	X	
REF-34			X	X	X	
REF-40			X	X	X	
REF-54				X	X	X
Number of Localities Per Size Class	7	6	12	13	14	6

(15 localities total)

Fig. 8. Distribution of penguin size classes present in 15 localities from Unit Telm 7, La Meseta Formation, Seymour Island, Antarctic Peninsula (MOW refers to field localities of M. O. Woodburne, J. A. Case, and D. S. Chaney, 1986–1987; REF refers to field localities of R. E. Fordyce, 1986–1987).

standing height reveals, coincidentally, six size classes, but spread over a different size range (40–115 cm) versus that estimated for the Seymour Island species (72–160 cm). The average size for the Eocene penguins is that of the largest modern penguin, the emperor, *Aptenodytes forsteri*.

Only three of the 12 sympatric localities, the Antarctic Peninsula, the South Shetland Islands, and the South Orkney Islands, can be considered as Antarctic, that is, located in latitudes greater than or equal to 60°S. All three localities have the three *Pygoscelis* species (i.e., *P. adeliae*, the adelie; *P. papua*, the gentoo; *P. antarctica*, the chinstrap), with the former locality also having the emperor penguin, while the latter two include the macaroni penguin (*Eudyptes chrysolophus*). The other nine localities are located north of 60°S, with most of the nine localities being north of the Antarctic Convergence as well. The associations here are predominantly com-

posed of king (*Aptenodytes patagonicus*), gentoo, macaroni, and rockhopper (*Eudyptes chrysocome*) penguins. The above species associations also represent penguins in four different body size categories at five localities (Figure 9). At a sixth locality, Macquarie Island, a royal penguin (*Eudyptes schlegeli*) replaces the macaroni penguin in the above association.

The maximum sympatric species and body size diversity among extant penguins are four species and four size classes, respectively. These numbers are lower than the values from the late Eocene deposits from Unit Telm 7 of the La Meseta Formation on Seymour Island (i.e., six sympatric species and size classes). In this, a more qualitative example, it is again shown that the marine vertebrate diversity levels in some groups were higher during the late Eocene in the cool temperate seas off the east coast of the Antarctic Peninsula than in areas within the high-latitude southern oceans today.

There has been an abundance of penguin material collected from the La Meseta Formation beginning with the Swedish South Polar Expedition at the turn of the century (material described by *Wiman* [1905]). Reports referencing the wealth of fossil penguin material indicate that the number of specimens collected is very large (with hundreds of specimens in each of a variety of institutional collections), especially from the uppermost unit of the La Meseta Formation [*Myrcha et al.*, 1990; *Woodburne and Zinsmeister*, 1984; J. A. Case, personal observations]. Thus not only are the diversity levels for penguins high as discussed above, but the population sizes, based on the number of fossil specimens collected, also appear to be large.

LATE EOCENE MARINE TROPHIC LEVELS

Penguins and sharks, the two predominant marine vertebrates (based on abundance) recovered from the late Eocene deposits of the La Meseta Formation on Seymour Island, have diversity levels which exceed the levels currently seen for those two taxa in most cool temperate waters (i.e., the same environment inferred from all fossil indicators for the late Eocene). The fossil penguins and sharks would represent the uppermost levels of the trophic pyramid (references in parentheses indicate taxa recovered from the La Meseta Formation on Seymour Island) for the marine region east of the Antarctic Peninsula during this time period (Figure 10; pyramid constructed after *Ryther* [1969], *Cushing* [1978], and *Levinton* [1982]). Along with large fish (which could also fall prey to sharks), these top marine predators would feed on both medium-sized fish (i.e., cod-sized fish, for example, codlike gadoids [*Eastman and Grande*, 1991] and hake (Merlucciidae) (D. J. Long, personal communication, 1991)) and schools of small shoaling planktivorous fish (e.g., clupeoids, or herring-like fish, which would also fall prey to the medium-sized fish [*Doktor et al.*, 1988]). The vertebrate trophic pyra-

MULTIPLE SPECIES SYMPATRY AMONG MODERN PENGUINS
(After Stonehouse, 1975)

SPECIES	Size Class	BREEDING LOCALITIES											
		1	2	3	4	5	6	7	8	9	10	11	12
Aptenodytes forsteri	6	(6)	-	-	-	-	-	-	-	-	-	-	-
Aptenodytes patagonicus	5	-	-	-	-	(5)	(5)	(5)	(5)	(5)	(5)	(5)	-
Pygoscelis papua	4	(4)	(4)	(4)	(4)	(4)	(4)	(4)	(4)	(4)	(4)	(4)	-
Pygoscelis adeliae	3	(3)	(3)	(3)	(3)	-	-	-	-	-	-	-	-
Eudyptes chrysolophus	3	-	(3)	(3)	(3)	(3)	(3)	(3)	(3)	(3)	(3)	-	-
Speniscus demersus	3	-	-	-	-	-	-	-	-	-	-	-	-
Speniscus magellanicus	3	-	-	-	-	-	-	-	-	-	-	-	-
Pygoscelis antarctica	3	(3)	(3)	(3)	(3)	(3)	-	-	-	-	-	-	-
Megadyptes antipodes	3	-	-	-	-	-	-	-	-	-	-	-	(3)
Spheniscus humboldti	3	-	-	-	-	-	-	-	-	-	-	-	-
Eudyptes schlegeli	3	-	-	-	-	-	-	-	-	-	-	(3)	-
Eudyptes sclateri	3	-	-	-	-	-	-	-	-	-	-	-	-
Eudyptes robustus	2	-	-	-	-	-	-	-	-	-	-	-	-
Eudyptes pachyrhynchus	2	-	-	-	-	-	-	-	-	-	-	-	(2)
Eudyptes chrysocome	2	-	-	-	-	-	(2)	(2)	(2)	(2)	(2)	(2)	-
Spheniscus mendiculus	2	-	-	-	-	-	-	-	-	-	-	-	-
Eudyptula albosignata	1	-	-	-	-	-	-	-	-	-	-	-	(1)
Eudyptula minor	1	-	-	-	-	-	-	-	-	-	-	-	(1)
TOTAL NUMBER OF SIZE CLASSES PER LOCALITY		3	2	2	2	3	4	4	4	4	4	4	3

Localities: 1) Antarctic Peninsula; 2) South Shetland Islands; 3) South Orkney Islands; 4) South Sandwich Islands; 5) South Georgia Is.; 6) Heard Is.; 7) Kerguelen Is.; 8) Falkland Islands; 9) Marion Is.; 10) Crozet Is.; 11) Macquarie Is.; 12) South Island, New Zealand.

Fig. 9.   Summary of locations where cases for modern penguin species of multiple species sympatry occur based on breeding localities for the 18 modern species. SC indicates the size class categories for modern penguins (does not correlate to the six size classes of late Eocene penguins). Numbers in parentheses designated the size class for that species at a particular locality. The summary only represents the highest levels of sympatry of four co-occurring species; the number of size classes co-occurring varies between localities.

mid would be supported by a large base of primary consumers, the megazooplankton, such as copepods. The broad underlying base to the entire overhead structure is supplied by macrophytoplankton, which includes such taxa as diatoms [*Harwood*, 1988] and dinoflagellates [*Wrenn and Hart*, 1988].

The values for shark diversity in the cool temperate late Eocene Antarctic waters are much higher than the diversity levels for present-day sharks in most cool temperate waters and are nearly equal to those for a tropical shark fauna. Penguin diversity levels for both species and size classes in the late Eocene also exceed values for modern cool temperate to cold temperate species or size classes. Fossil shark and penguin material has been collected in abundance from the La Meseta Formation, indicating high biomass as well as high diversity for the members of the uppermost trophic levels. In order to support the diversity and abundance

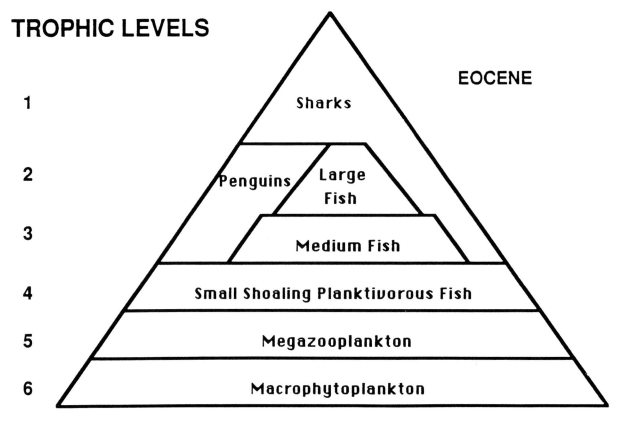

Fig. 10. Trophic structure for fossil vertebrates (levels 1–4) recovered from the late Eocene deposits of the La Meseta Formation, Seymour Island, Antarctic Peninsula. Levels 5 and 6 represent the foundation of the trophic structure with level 6 being the producer level and level 5 being the primary herbivore level.

of the top trophic levels, the mid to low trophic levels (vertebrate, invertebrate, protist) must themselves be very large in terms of abundance (they may be taxonomically diverse as well). This inferred abundance, in combination with a rich macroinvertebrate fauna, would certainly suggest a productive Antarctic marine biotic environment (cool temperate) during the late Eocene.

Today, rich and productive, cool temperate marine ecosystems only occur in areas of coastal upwelling. Here, cold bottom waters ascending to the surface transport nutrients to the photic zone, stimulating phytoplankton productivity and generating a large base of the food pyramid that supports large fish populations [*Levinton*, 1982]. Increased ocean upwelling can result from different processes; *Hay* [1988] and *Kennett and Stott* [1990] have hypothesized that in the early Oligocene, cold, dense Antarctic bottom waters were formed, which could help stimulate upwelling conditions around Antarctica by driving warm, saline deep water upward pushing nutrient-rich Antarctic intermediate waters to the surface. On the other hand, increased wind strengths resulting from high-latitude cooling can also induce upwelling. The biotic diversity and

abundance based on paleontological evidence presented here, plus that which has been inferred from the data for the cool temperate Antarctic waters during the late Eocene, imply a marine trophic structure that could only be supported by upwelling conditions and the nutrients which it provides. This suggests that upwelling conditions around Antarctica began during the late Eocene. Corroborating evidence for a late Eocene onset of upwelling conditions, as indicated by the marine vertebrate paleofauna from Seymour Island, comes from the first appearance of biosiliceous sediments in the sequence on Maud Rise, Weddell Sea, to the east of Seymour Island [*Kennett and Barker*, 1990].

SUMMARY

1.   A diverse and abundant macroinvertebrate fauna has been recorded from the late Eocene deposits of the La Meseta Formation, Seymour Island, Antarctic Peninsula, but to date, there have been no quantitative or qualitative comparisons to other invertebrate faunas in regard to levels of diversity and abundance. A cool

temperate marine environment during this interval is supported by numerous paleoenvironmental indicators.

2. Measures of diversity, evenness, and abundance are made for two groups of marine vertebrates, penguins and sharks.

3. The level of diversity of the late Eocene shark fauna from a single locality on Seymour Island is much higher than the level of diversity for most extant cool temperate shark faunas and is nearly equal to a present-day tropical shark fauna. The large number of specimens recovered for the paleofauna from this location indicates a high abundance of sharks at that time.

4. Late Eocene penguin diversity in the La Meseta Formation was compared with modern penguin diversity at different geographic localities through the number of sympatric species and size classes present. Two separate late Eocene localities from the La Meseta Formation produced six different penguin species and size classes. These numbers are higher than those reported for modern penguins, in which the highest levels of sympatry are represented by the occurrence of only four sympatric species and size classes. Extremely large numbers of fossil penguin specimens have been recovered from the La Meseta Formation, suggesting a high abundance for this group.

5. Sharks and penguins are representatives of the uppermost levels of the late Eocene marine trophic pyramid. The lower trophic levels must have been marked by a very large biomass in order to support the abundance and diversity of the uppermost trophic levels, which suggests a very rich marine environment at that time.

6. Today, high levels of biotic diversity and abundance in cool temperate waters occur only in areas of coastal upwelling. The vertebrate paleontological data suggest that the upwelling conditions may have begun near the end of the Eocene.

Acknowledgments. I would like to thank, Jim Kennett, Marine Science Institute, University of California, Santa Barbara (UCSB), for the invitation to the conference "The Role of the Southern Ocean and Antarctica in Global Change: An Ocean Drilling Perspective" at UCSB. I would like to offer my gratitude to Doug Long (Department of Integrative Biology, University of California, Berkeley) for his contributions to my knowledge of sharks and fish and a review of the manuscript and to Andy Wyss (Department of Geological Sciences, UCSB) and Philip Leitner (Department of Biology, Saint Mary's College) for helpful comments on the manuscript. Participation in the conference was supported by an award from the Faculty Development Fund, Saint Mary's College of California.

REFERENCES

Askin, R. A., Campanian to Paleocene palynological succession of Seymour and adjacent islands, northeastern Antarctic Peninsula, in Geology and Paleontology of Seymour Island, Antarctic Peninsula, *Mem. Geol. Soc. Am., 169,* 131–153, 1988.

Askin, R. A., Endemism and heterochroneity in the Late Cretaceous (Campanian) to Paleocene palynofloras of Seymour Island, Antarctica: Implications for the origins, dispersal and paleoclimates of southern floras, in Origin and Evolution of the Antarctic Biota, *Geol. Soc. Spec. Publ. London, 47,* 107–119, 1989.

Askin, R. A., Campanian to Paleocene spore and pollen assemblages, Seymour Island, Antarctica, *Rev. Palaeobot. Palynol., 65,* 105–113, 1990.

Blake, D. B., and W. J. Zinsmeister, Two early Cenozoic sea stars (class Asteroidea) from Seymour Island, Antarctic Peninsula, *J. Paleontol., 53,* 1145–1154, 1979.

Blake, D. B., and W. J. Zinsmeister, Eocene asteroids (Echinodermata) from Seymour Island, Antarctic Peninsula, in Geology and Paleontology of Seymour Island, Antarctic Peninsula, *Mem. Geol. Soc. Am., 169,* 489–498, 1988.

Case, J. A., Paleogene floras from Seymour Island, Antarctic Peninsula, in Geology and Paleontology of Seymour Island, Antarctic Peninsula, *Mem. Geol. Soc. Am., 169,* 523–530, 1988.

Cushing, D. H., Upper trophic levels in upwelling areas, in *Upwelling Ecosystems,* edited by R. Boje and M. Tomczak, pp. 101–110, Springer-Verlag, New York, 1978.

Daan, N., P. J. Bromley, J. R. G. Hislop, and N. A. Nielsen, Ecology of North Sea fish, *Neth. J. Sea Res., 26,* 343–386, 1990.

Davies, D. H., and L. S. Joubert, Tag evaluation and shark tagging in South African waters, 1964–65, in *Sharks, Skates, and Rays,* edited by P. W. Gilbert, R. F. Mathewson, and D. P. Rall, pp. 111–140, Johns Hopkins Press, Baltimore, Md., 1967.

Doktor, M., A. Gazdzicki, S. A. Marenssi, S. J. Porebski, S. N. Santillana, and A. V. Vrba, Argentine-Polish geological investigations on Seymour (Marambio) Island, Antarctica, 1988, *Pol. Polar Res., 9,* 521–541, 1988.

Eastman, J. T., and L. Grande, Late Eocene gadiform (Teleostei) skull from Seymour Island, Antarctic Peninsula, *Antarct. Sci., 3,* 87–95, 1991.

Elliot, D. H., Preface, in Geology and Paleontology of Seymour Island, Antarctic Peninsula, *Mem. Geol. Soc. Am., 169,* ix–x, 1988.

Feldmann, R. M., Seymour Island yields a rich fossil harvest, *Geotimes, 29,* 16–18, 1984.

Feldmann, R. M., and M. T. Wilson, Eocene decapod crustaceans from Antarctica, in Geology and Paleontology of Seymour Island, Antarctic Peninsula, *Mem. Geol. Soc. Am., 169,* 465–488, 1988.

Feldmann, R. M., and W. J. Zinsmeister, New fossil crabs (Decapoda; Brachyura) from the La Meseta Formation (Eocene) of Antarctica: Paleogeographic and biogeographic implications, *J. Paleontol., 58,* 1041–1061, 1984.

Harwood, D. M., Upper Cretaceous and lower Paleocene diatom and silicoflagellate biostratigraphy of Seymour Island, eastern Antarctic Peninsula, in Geology and Paleontology of Seymour Island, Antarctic Peninsula, *Mem. Geol. Soc. Am., 169,* 55–130, 1988.

Hay, W. W., Paleoceanography: A review for the GSA centennial, *Geol. Soc. Am. Bull., 100,* 1934–1956, 1988.

Kato, S., and A. H. Carvallo, Shark tagging in the eastern Pacific Ocean, 1962–65, in *Sharks, Skates, and Rays,* edited by P. W. Gilbert, R. F. Mathewson, and D. P. Rall, pp. 93–109, Johns Hopkins Press, Baltimore, Md., 1967.

Kennett, J. P., and P. F. Barker, Latest Cretaceous to Cenozoic climate and oceanographic developments in the Weddell Sea, Antarctica: An ocean-drilling perspective, *Proc. Ocean Drill. Program Sci. Results, 113,* 937–960, 1990.

Kennett, J. P., and L. D. Stott, Proteus and Proto-oceanus: Ancestral Paleogene oceans as revealed from Antarctic stable isotopic results, ODP Leg 113, *Proc. Ocean Drill. Program Sci. Results*, *113*, 865–880, 1990.

Levinton, J. S., *Marine Ecology*, Prentice-Hall, Englewood Cliffs, N. J., 1982.

Long, D. J., Sharks from the La Meseta Formation (Eocene), Seymour Island, Antarctic Peninsula, *J. Vertebr. Paleontol.*, *12*, 11–32, 1992.

McKinney, M. L., K. J. McNamara, and L. A. Wiedman, Echinoids from the La Meseta Formation (Eocene), Seymour Island, Antarctica, in Geology and Paleontology of Seymour Island, Antarctic Peninsula, *Mem. Geol. Soc. Am.*, *169*, 499–504, 1988.

Myrcha, A., A. Tatur, and R. Del Valle, A new species of fossil penguin from Seymour Island, West Antarctica, *Alcheringa*, *14*, 195–205, 1990.

Pospichal, J. J., and S. W. Wise, Paleocene to middle Eocene calcareous nannofossils of ODP sites 689 and 690, Maud Rise, Weddell Sea, *Proc. Ocean Drill. Program Sci. Results*, *113*, 613–638, 1990.

Ryther, J. H., Photosynthesis and fish production in the sea, *Science*, *166*, 72–76, 1969.

Sadler, P. M., Geometry and stratification of the uppermost Cretaceous and Paleogene units of Seymour Island, northern Antarctic Peninsula, in Geology and Paleontology of Seymour Island, Antarctic Peninsula, *Mem. Geol. Soc. Am.*, *169*, 303–320, 1988.

Simpson, G. G., Fossil penguins, in *The Biology of Penguins*, edited by B. Stonehouse, pp. 19–41, University Park Press, Baltimore, Md., 1975.

Simpson, G. G., *Penguins Past, Present, Here and There*, Yale University Press, New Haven, Conn., 1976.

Stevens, J. D., Further results from a tagging study of pelagic sharks in the north-east Atlantic, *J. Mar. Biol. Assoc. U.K.*, *70*, 707–720, 1990.

Stonehouse, B., Introduction: the Spheniscidae, in *The Biology of Penguins*, edited by B. Stonehouse, pp. 1–15, University Park Press, Baltimore, Md., 1975.

Stott, L. D., J. P. Kennett, N. J. Shackleton, and R. M. Corfield, The evolution of Antarctic surface waters during the Paleogene: Inferences from the stable isotopic composition of planktonic foraminifers, ODP Leg 113, *Proc. Ocean Drill. Program Sci. Results*, *113*, 849–863, 1990.

Thomas, E., Late Cretaceous through Neogene deep-sea benthic foraminifers (Maud Rise, Weddell Sea, Antarctica), *Proc. Ocean Drill. Program Sci. Results*, *113*, 571–594, 1990.

Wei, W., and S. W. Wise, Middle Eocene to Pleistocene calcareous nannofossils recovered by Ocean Drilling Program Leg 113 in the Weddell Sea, *Proc. Ocean Drill. Program Sci. Results*, *113*, 639–666, 1990.

Wiedman, L. A., R. M. Feldmann, D. E. Lee, and W. J. Zinsmeister, Brachiopoda from the La Meseta Formation (Eocene) Seymour Island, Antarctica, in Geology and Paleontology of Seymour Island, Antarctic Peninsula, *Mem. Geol. Soc. Am.*, *169*, 449–458, 1988.

Wiman, C., Uber die alttertiaren vertebraten der Seymour Insel, *Wiss. Ergeb. Schwed. Sudpolar Exped.*, *1901–1903*, *3*(1), 1–37, 1905.

Woodburne, M. O., and W. J. Zinsmeister, The first land mammal from Antarctica and its biogeographical implications, *J. Paleontol.*, *58*, 913–948, 1984.

Wrenn, J. H., and G. F. Hart, Paleogene dinoflagellate cyst biostratigraphy of Seymour Island, Antarctica, in Geology and Paleontology of Seymour Island, Antarctic Peninsula, *Mem. Geol. Soc. Am.*, *169*, 321–448, 1988.

Zinsmeister, W. J., Late Cretaceous–early Tertiary molluscan biogeography of the Circum-Pacific, *J. Paleontol.*, *56*, 84–102, 1982.

Zinsmeister, W. J., Late Eocene bivalves (Mollusca) from the La Meseta Formation, collected during the 1974–1975 joint Argentine-American expedition to Seymour Island, Antarctic Peninsula, *J. Paleontol.*, *58*, 1497–1527, 1984.

Zinsmeister, W. J., and H. H. Camacho, Late Eocene Struthiolariidae (molluscan: Gastropoda) from Seymour Island, Antarctic Peninsula and their significance to the biogeography of early Tertiary shallow-water faunas of the southern hemisphere, *J. Paleontol.*, *54*, 1–14, 1980.

Zinsmeister, W. J., and H. H. Camacho, Late Eocene (to possibly earliest Oligocene) molluscan fauna of the La Meseta Formation of Seymour Island, Antarctic Peninsula, in *Antarctic Geoscience*, edited by C. Craddock, pp. 299–304, University of Wisconsin Press, Madison, 1982.

Zinsmeister, W. J., and R. M. Feldmann, Cenozoic high latitude heterochroneity of southern hemisphere marine faunas, *Science*, *224*, 281–283, 1984.

(Received December 27, 1991;
accepted June 2, 1992.)

THE ANTARCTIC PALEOENVIRONMENT: A PERSPECTIVE ON GLOBAL CHANGE
ANTARCTIC RESEARCH SERIES, VOLUME 56, PAGES 131–139

PALEOECOLOGY OF EOCENE ANTARCTIC SHARKS

DOUGLAS J. LONG

Department of Integrative Biology and the Museum of Paleontology, University of California, Berkeley, California 94720

A diverse shark assemblage consisting of 17 taxa in 10 families has been collected from the middle to late Eocene La Meseta Formation on Seymour Island, northern Antarctic Peninsula. Paleoecological associations with the diverse shark assemblage suggest a temperate marine environment for the La Meseta Formation with four ecological components: (1) a resident shallow water community dominated by *Carcharias macrota*; (2) a deepwater fauna consisting mainly of squaloid sharks that occasionally entered the La Meseta area from deeper, offshore areas through daily or seasonal vertical migrations; (3) a transitional, eurybathic group of sharks that inhabited both inshore shallow and offshore pelagic and deep areas; and (4) shallow water tropical migrants that occasionally entered the La Meseta area from warmer northern areas.

INTRODUCTION

Hundreds of Cenozoic shark-bearing fossil localities are known throughout the world, and faunal assemblages from many of these localities are described in the literature [*Cappetta*, 1987]. In almost every case, the focus of these studies is on tooth identification and taxonomic designation and, to some degree, on stratigraphy and age correlation. Scant attention, if ever, is given on the ecological relationships of the sharks to other fossils recovered from the locality and between the different fossil shark taxa within. Sharks may have particular habitat or climatic requirements and can be separated into groups depending on the general habitats which they occupy (nearshore, outer shelf, slope, etc.) Thus they can supply ecological information regarding the depositional environment of the fossil locality. Conversely, analysis of invertebrate faunas and sedimentology can supply information about the sharks in a paleoecological context. This study will show the complexities, limitations, and importance in making a paleoecological interpretation of a diverse fossil shark fauna.

At least 17 taxa of sharks within 10 families (Table 1) occur in the middle to late Eocene La Meseta Formation on Seymour Island, Antarctic Peninsula (Figure 1). In contrast to the Eocene, the present-day seas around Antarctica are extremely cold and provide habitat for only a few species of sharks. The squaloid sharks *Somniosus microcephalus* and *Etmopterus lucifer* and the lamnoid shark *Lamna nasus* have been recorded within Antarctic waters; all occur infrequently within the Antarctic Convergence, and none are found in the waters around continental Antarctica [*Svetlov*, 1978; *Compagno*, 1990]. In the late Eocene, however, the waters adjacent to what is now Seymour Island were considerably warmer and had a much more diverse neoselachian fauna with several ecological components (Table 2, Figure 2).

GEOLOGY

Seymour Island lies in the Weddell Sea off the northeastern tip of the Antarctic Peninsula at approximately latitude 64°15'S and longitude 56°45'W, almost directly south of Tierra del Fuego, South America (Figure 1a). The stratigraphic sequence of marine sediments on Seymour Island (Figure 1b) extends from the Upper Cretaceous (Lopez de Bertodano Formation) through the lower Paleocene (Sobral Formation) and the upper Paleocene (Cross Valley Formation) to the middle and late Eocene (La Meseta Formation) [*Woodburne and Zinsmeister*, 1984; *Sadler*, 1988]. The Eocene beds of the La Meseta Formation are exposed on the northern portion of Seymour Island.

The La Meseta Formation is composed of poorly consolidated marine sandstones, siltstones, clays, and shell beds. Preserved within these strata are the remains of a diverse vertebrate and invertebrate nearshore fauna. The La Meseta Formation is divided into seven numbered units within three sections (Figure 3) [*Sadler*, 1988]. Unit I at the base of the La Meseta Formation is a sequence of unconsolidated fine and silty sands; Unit II is a sequence of highly fossiliferous fine laminated sands and conglomerates; Unit III consists of fine sands with intervals of clays and sandy gravels [*Welton and Zinsmeister*, 1980; *Woodburne and Zinsmeister*, 1984]. Of these seven units, Telm 2–5 contain the bulk of fossil shark localities, and the uppermost units, Telm 6–7, contain no fossil sharks [*Long*, 1992].

TABLE 1.   Sharks From the Eocene La Meseta Formation, Seymour Island, Antarctic Peninsula

	Primary Bathymetric Ecology
Hexanchidae	
Heptranchias howelli	deep water
Squalidae	
Squalus weltoni	deep water, transitional
Squalus woodburnei	deep water, transitional
Centrophorus sp.	deep water
Deania sp.	deep water
Dalatias licha	deep water
Pristiophoridae	
Pristiophorus lanceolatus	transitional
Squatinidae	
Squatina sp.	transitional
Stegostomatidae	
Stegostoma cf. *S. fasciatum*	shallow water tropical migrant
Ginglymostomatidae	
Pseudoginglymostoma cf. *P. brevicaudatum*	shallow water tropical migrant
Odontaspididae	
Carcharias macrota	shallow water resident
Odontaspis rutoti	transitional
Odontaspis winkleri	transitional
Mitsukurinidae	
Anomotodon multidenticulata	deep water
Lamnidae	
Carcharocles auriculatus	transitional
Lamna cf. *L. nasus*	transitional
Carcharhinidae	
Scoliodon sp.	shallow water tropical migrant

PALEOECOLOGY

The La Meseta Formation represents a shallow, nearshore, moderate to high-energy depositional environment. Precise depth has not been estimated, but the La Meseta Formation is believed to represent a littoral to shallow sublittoral inner shelf setting. Evidence supporting this interpretation comes from ichnofossils [*Wiedman and Feldmann*, 1988], decapods [*Feldmann and Wilson*, 1988], cirripeds [*Zullo et al.*, 1988], and molluscs [*Zinsmeister*, 1982] and from the sedimentology of the La Meseta Formation [*Sadler*, 1988].

Woodburne and Zinsmeister [1984], *Case* [1988], and *Mohr* [1990] suggested that the terrestrial environment around peninsular Antarctica during the late Eocene was heavily vegetated and had a humid, temperate climate similar to present-day New Zealand or southern South America. The marine environment was also temperate. This idea is supported primarily by the diverse molluscan and arthropod faunas found within the formation [*Zinsmeister*, 1982; *Feldmann and Wilson*, 1989]. Other lines of evidence for a temperate marine environment come from microfossils. The Weddell Sea, adjacent to present-day Seymour Island, began to cool after a period of relatively warm marine temperatures in the Paleocene and early Eocene; by the middle and late Eocene, the marine climate was temperate. Information from calcareous nannofossils [*Pospichal and Wise*, 1990; *Wei and Wise*, 1990] and planktonic and benthic

forams [*Stott and Kennett*, 1990; *Thomas*, 1990] confirms this. In a review of fossil penguins from Seymour Island, *Simpson* [1971] also implied that the marine climate at that time was temperate.

A comparison of the fossil shark taxa with those of their closest living relatives and their present distributions show that all of the sharks from the La Meseta Formation were adapted to a temperate marine environment or ranged well within such environments. *Lamna nasus* is widely dispersed in temperate oceans today, even ranging into subpolar regions, and prefers cool-temperate waters [*Svetlov*, 1978; *Castro*, 1983; *Compagno*, 1984, 1990]. Both the extinct *Carcharias macrota* and the living *C. taurus* are (and were) widespread in subtropical, temperate, and cool-temperate waters. Many species of *Squatina* live in tropical to temperate waters, often near the interface with sub-Arctic and Antarctic waters [*Compagno*, 1984]. The range of *Squalus* (*S. acanthias* in this case) in the northern hemisphere extends from subtropical to Arctic waters, including those of Scandinavia, Greenland, and the Bering Sea; *Squalus* in the southern hemisphere is also found in temperate waters [*Compagno*, 1984]. *Dalatias licha* is found in temperate waters off New Zealand, southern Australia, and South Africa, as well as in the cool-temperate waters on both sides of the North Atlantic [*Castro*, 1983; *Compagno*, 1984].

On the basis of the above distributions, as well as the

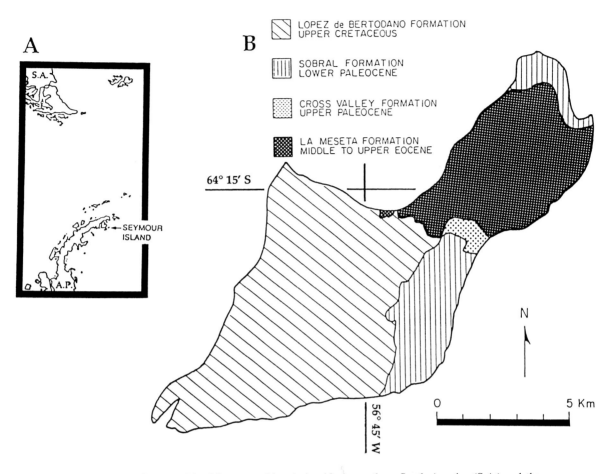

Fig. 1. (a) Seymour Island in geographic relationship to southern South America (S.A.) and the Antarctic Peninsula (A.P.). (b) Diagram of Seymour Island showing the four geological formations including the La Meseta Formation.

paucity of any Eocene taxa thought to have inhabited warm waters (such as the carcharhinids *Galeocerdo*, *Hemipristis*, *Sphyrna*, *Negaprion*, and *Carcharhinus*), the elasmobranch fauna of the La Meseta Formation seems to be characteristic of a temperate marine habitat. *Kennett and Barker* [1990] estimated the marine paleotemperature of the Seymour Island area, on the basis of oxygen isotope ratios of late Eocene fossil invertebrate shells, to be 8°C for the minimum winter temperature and 12°C for the minimum summer temperature. This temperature estimate is consistent with the water temperature inferred from the late Eocene shark fauna. However, several taxa of warm water sharks are found in the La Meseta Formation, but for reasons given below, these are considered anomalous occurrences.

As part of the paleobathymetric considerations of their geologic occurrences, fossil shark remains can be

TABLE 2. Habitats of Eocene Sharks, La Meseta Formation, Antarctic Peninsula

	Shallow Tropical Migrant	Shallow Temperate Resident	Transitional Temperate Resident	Deepwater Temperate Resident
No. of genera	3	1	6*	6*
No. of specimens	5	591	289	93
Percent of total	0.5	60	30	9.5

Squalus is included as both a transitional and a deepwater form, on the basis of other fossil taxa and on modern analogs.

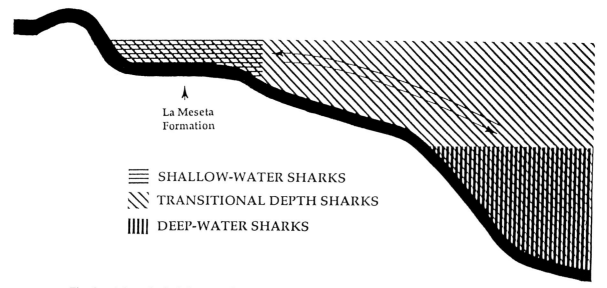

SHALLOW-WATER SHARKS

TRANSITIONAL DEPTH SHARKS

DEEP-WATER SHARKS

Fig. 2. A hypothetical diagram of a portion of the western Weddell Sea showing the three major shark habitats of the La Meseta depositional environment and adjacent areas. Arrows show the direction of occasional migration by deepwater sharks into the shallow depositional environment. Vertical and horizontal distances not to scale.

used as indicators of the general depth at which they once lived. Inasmuch as sharks can be large and highly mobile aquatic vertebrates, their application for paleobathymetry may not be as precise as those of sessile or limited-ranging invertebrates. However, extant sharks do live within certain depth ranges that are distinct, and these ranges can be compared and summarized with some reliability. In using such ranges to infer paleobathymetry, it is essential to compare the fossils to those of other known fossil sharks and their previously understood paleobathymetric ranges, as well as the depth ranges of living relatives. In doing so with the La Meseta shark fossils, a discontinuity becomes apparent. Although the La Meseta fossils are preserved within a single depositional setting, the sharks lived in different bathymetric and ecological zones. These bathymetric zones include an expected shallow water shark fauna dominated by *Carcharias* and an anomalous deepwater shark fauna characterized by a diversity of squaliform sharks. Complicating matters are several species of sharks that have ranges overlapping these zones. There are also several shark species that may not have been endemic to the La Meseta depositional paleoenvironment but may have strayed in from warmer northern waters. First, however, the shallow water forms will be discussed.

Temperate Shallow Water Residents

Zinsmeister [1979] and Woodburne and Zinsmeister [1984] state that the seas around western Antarctica and the Antarctic Peninsula were shallow and geographi-

cally extensive. This area, termed the Weddellian Province, extended from South America along both the east and west sides of the Antarctic Peninsula to eastern Australia and contained many small islands with shallow beaches and coastlines. Such habitats were ideally suited for *Carcharias*, whose modern depth range is normally no deeper than several meters. These sharks are most abundant along shallow coastal beaches and in bays, being found most often along the tideline and in the surf zone, although a few individuals may stray to deeper outer shelf waters [Bigelow and Schroeder, 1948; Castro, 1983; Compagno, 1984]. It is not surprising that the overwhelming majority of fossil shark teeth from the La Meseta Formation, 60% of the total sample, are those of *Carcharias* (Table 2). These coastal sharks most probably lived and foraged along the shorelines of the Seymour Island area in the late Eocene.

Temperate Deepwater Residents

A significant number of teeth from the La Meseta Formation (9.5%) are from sharks that are almost strictly deepwater inhabitants today but whose fossils are incorporated into the shallow La Meseta depositional environment (Table 2). These taxa were almost certainly specialized for a deepwater habitat in the Eocene as well; so their occurrence in the La Meseta Formation is noteworthy. These sharks include *Heptranchias*, *Centrophorus*, *Deania*, and *Dalatias*, which presently live on the outer shelf or upper slope at depths of 200 m or more, and *Squalus*, which commonly ranges from shallow water to well below 100 m [Bigelow and

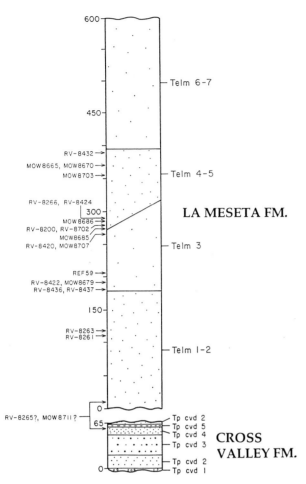

Fig. 3. Stratigraphic column of the La Meseta Formation showing the seven units within three sections and showing the shark-bearing localities. Locality abbreviations are as follows: RV, University of California Riverside vertebrate field locality; MOW, Michael O. Woodburne field locality; REF, R. Ewan Fordyce field locality [from *Long*, 1992].

early Cenozoic and radiated into deepwater habitats in the late Cenozoic. This scenario does not hold true for the deepwater sharks recovered from the La Meseta Formation. First, unlike the invertebrates, most of the shark taxa were not restricted to the La Meseta area in the Eocene, but had a cosmopolitan or bitemperate distribution by that time [*Long*, 1992]. Second, many of these deepwater shark species are known from strictly deepwater marine deposits, both Eocene and older, elsewhere on the globe [*Long*, 1992; *Welton*, 1979; *Keyes*, 1984; *Cappetta*, 1987]. This indicates they already lived primarily in deep water in the Eocene. In addition, remarkable whole-body preservation of deepwater sharks from the Cretaceous of Lebanon shows that present-day deepwater shark families (Hexanchidae, Squalidae, and Mitsukurinidae, among others) already possessed specialized morphological adaptations for a deepwater existence [*Cappetta*, 1980]. These lines of evidence, in addition to the known bathymetric ranges of their extant representatives, lend support to my contention that these Eocene taxa were inhabitants of deep water and were not shallow water taxa that radiated out into deeper waters in the late Cenozoic, as did many invertebrate groups.

The diversity of deepwater shark species found within the La Meseta Formation suggests the presence of local deepwater habitats. However, *Zinsmeister* [1979] and *Woodburne and Zinsmeister* [1984] claimed that western and peninsular Antarctica was a vast area of islands within a shallow sea. Such a large expanse of relatively shallow coastal water provided an ideal habitat for the shallow water sharks (especially *Carcharias*) and some bony fishes, but not for the deepwater squalomorph sharks. It seems that these deep-dwelling sharks must have lived in adjacent offshore deepwater trenches, canyons, rifts, and outer shelf and slope habitats better suited for them. If so, portions of the early Cenozoic seafloor in West Antarctica during the Eocene must have dropped steeply to sublittoral depths. In the Eocene Weddell Sea, many taxa of deepwater (upper abyssal to lower bathyal) foraminifera have been recovered, proving that deepwater habitats did exist around that area in the Eocene [*Thomas*, 1990]. According to *Woodburne and Zinsmeister* [1984], an extensive subduction zone ran from the western coast of South America down along the present-day Antarctic Peninsula from the mid-Mesozoic to the late Oligocene. This tectonic action could have created a series of deep and lengthy trenches that provided suitable habitat for the deepwater sharks and other deepwater fishes. Also, rifting between Antarctica-Australia and Antarctica-Africa produced many deepwater trenches and basins that would have provided suitable deepwater habitats.

These teeth could not have been reworked from older deposits because they do not exhibit any postdepositional wear or breakage. The teeth of *Centrophorus* and *Deania* are thin and delicate and especially prone to

Schroeder, 1948, 1957; *Compagno*, 1984]. *Cappetta* [1987] regards *Anomotodon* as a deepwater species because it is aligned with the presently deepwater family Mitsukurinidae which live at depths of 200 m to 700 m. The teeth of all of these deepwater species are not exceedingly abundant in the shallow water La Meseta Formation but are too common to be ignored as just anomalous occurrences. There are plausible explanations for their occurrence alongside shallow water coastal vertebrates and invertebrates.

Zinsmeister and Feldmann [1984], *Wiedman et al.* [1988], and *Blake and Zinsmeister* [1988] noted the occurrence of many species of endemic invertebrates from the shallow La Meseta Formation that are restricted to deepwater habitats today. Theoretically, these early forms evolved in south polar latitudes in the

breakage during erosion and reworking. Instead, these teeth were shed and deposited by the deepwater sharks while they were alive in the La Meseta environment. These deepwater sharks migrated into the La Meseta area from adjacent deepwater habitats. Many authors [*Bigelow and Schroeder*, 1948, 1957; *Bass et al.*, 1976; *Castro*, 1983; *Compagno*, 1984; *Compagno et al.*, 1990; *Lavenberg*, 1991] recorded the occurrences of primarily deepwater sharks in relatively shallow water (including a specimen of *Centrophorus* from a depth of 3 m) and suggested that some species may perform vertical migrations. Migration into the shallows could have been for one or several reasons.

Some deepwater sharks and fishes exhibit cyclical daily migrations from deep areas into shallow coastal areas and surface waters and back again. These migrations may be related to the diel movements of the deep scattering layers (DSL), stratified layers of fishes and invertebrates that ascend the water column at night to feed in surface waters and descend into deeper waters with the rise of the Sun [*Dietz*, 1962; *Holton*, 1969; *Lavenberg*, 1991]. Deep scattering layers are found in most of the world's oceans, including the coastal waters off Antarctica [*Dietz*, 1948, 1962]. The vertical depth at which these layers shift within the water column is dependent on the availability and intensity of direct or ambient light within the upper layers of water. The inhabitants of the DSL may protect themselves from predators by concealing themselves in the dark, unlit depths during the day [*Dietz*, 1962]. These organisms are found below the maximum depth that sunlight can penetrate. As the Sun sets and the intensity of sunlight decreases, the DSL rise to the surface, following the ascending depth of visible light, and diffuse into the inner shelf; on dark nights, DSL organisms can be found well into coastal areas [*Dietz*, 1962].

Sharks are a virtually unknown component of the DSL, and their direct relationship to the DSL is unclear. Sharks may follow these layers as they rise and fall through the course of the day, following behind and feeding on the migrating fishes and invertebrates [*Castro*, 1983; *Lavenberg*, 1991]. Large "blob fish" seen on the echograms of the DSL from sounding equipment are suspected to be sharks or large fishes; these "blob fish" are usually seen on the bottom of the DSL as it rises in the early evening [*Dietz*, 1962]. Little is known about sharks in the DSL because deepwater sharks are infrequently caught in trawls and tow sampling through the DSL. These relatively large and mobile organisms may avoid nets designed for smaller and slower deepwater organisms [*Merrett et al.*, 1991]. Like many fishes in the DSL [*Holton*, 1969; *Pearcey et al.*, 1977], sharks may make the vertical migration occasionally, not every day, and therefore would not frequently be captured.

If these deepwater sharks were not directly associated with the DSL, they may have performed diel vertical migrations in much the same way. These sharks

probably lived in deeper offshore areas during the day and moved into shallower coastal areas to hunt and feed at night [*Castro*, 1983; *Compagno et al.*, 1990; *Lavenberg*, 1991]. Capture data by *Bigelow and Schroeder* [1957] and others show that many of the shallow depth records are from sharks captured at night, suggesting nocturnal vertical migration. Like the DSL, some deepwater sharks also stay at depths below the level of sunlight penetration and may follow this level to shallower depths in the evening. Some species of deepwater sharks in the families Hexanchidae and Squalidae have an organ on the top of their head called a pineal window (connected to the pineal complex of the brain) that is believed to detect the amount of downwelling ambient light and may serve to mediate vertical migrations [*Clark and Kristof*, 1990, 1991]. It is interesting to note that the two families of deepwater sharks that have a pineal window are the most abundant families of deepwater sharks in the La Meseta Formation.

The proximity of these deepwater areas to the La Meseta Formation is uncertain, but the relative abundance of deepwater sharks and fishes [*Long*, 1991, 1992] suggests that these areas are in relatively close proximity. Vertically migrating sharks need not be fast swimmers, since they would be migrating onshore and offshore for at least 6 hours in each direction, allowing them enough time to move relatively long distances. For example, the DSL descends at speeds of up to 7.5 m/min [*Dietz*, 1962], and *Centrophorus* can swim at speeds of over 2 km per hour [*Yano and Tanaka*, 1986]; so they could travel up to 12 km or more inshore during one vertical migration. Larger squaloid hexanchid sharks (*Dalatias* and *Heptranchias*, respectively) can swim faster and hence travel further in a given time period. Additionally, since the depth of the DSL, and of sharks migrating independently of it, is determined by the depth of sunlight penetration, deepwater organisms would be found at shallower depths for longer periods of time during the almost continual darkness and twilight during the austral summer.

Deepwater sharks may have also entered shallow water on a seasonal basis. *Bigelow and Schroeder* [1948] and *Compagno* [1984] discuss the seasonal movements of *Squalus* and remark that their abundance in coastal areas is due largely to shifts in marine temperature. These sharks avoid water above 15°C, and so they are more numerous in coastal waters during cool seasonal periods. These sharks follow the temperature regime into deeper offshore waters as coastal waters warm with the onset of seasonal change. *Ketchen* [1986] and *Compagno* [1984] also note that pregnant females often migrate into shallow waters to give birth, and *Ketchen* [1986] also states that schools of *Squalus* will periodically swim into upper surface waters and into shallow coastal areas to feed on aggregations of prey.

Transitional Temperate Residents

Several other shark species were transitional between the shallow water and deepwater habitats, based on the ecologies of their living relatives and on the depositional environment of fossil material from other localities. These sharks either had individuals that roamed freely and sporadically between deepwater and shallow water areas without the influence of any particular diurnal or seasonal cycle or had individuals that inhabited particular areas between and including these areas. For example, *Squatina* lives at a depth between 3 m and 1300 meters, but a particular individual may move several hundred meters a day within a particular home range or upslope and downslope [*Compagno*, 1984; *Standora and Nelson*, 1977]. *Squatina* and *Pristiophorus* were primarily benthic in their habits. *Squalus* is regarded as both a deepwater and a transitional taxon, because some extant congeners live in both types of habitats. *Squalus* and both species of *Odontaspis* occupied lower water and midwater depths, and *Lamna* and *Carcharocles* primarily occupied upper water layers both inshore and offshore. All six of these taxa may have ranged freely between the deepwater and shallow water areas and exploited the resources of the habitats in these two zones. The common occurrence of teeth (30%) from these taxa in the La Meseta Formation agrees with this finding (Table 2).

Tropical Shallow Water Migrants

Several species of nonresident shallow water sharks, including *Pseudoginglymostoma*, *Stegostoma*, and *Scoliodon*, occur infrequently in the La Meseta Formation and probably were more common in warmer subtropical or warm-temperate waters to the north. The movements of these sharks into the La Meseta area during the Eocene could have been facilitated by fluctuations of ocean currents or periodic or prolonged warming of coastal waters; or these individuals may have strayed far outside of their normal range. Recent examples of this type of displacement can be seen off the west coast of North America, when warm El Niño currents periodically drive pelagic, inshore, and benthic sharks far into northern areas normally too cool for them. They may stay in these areas for as long as several years until the normal oceanographic condition returns [*Hubbs*, 1948; *Radovich*, 1961; *Karinen et al.*, 1985]. Similar phenomena in the southern hemisphere have been noted by *Velez et al.* [1984] off Peru and by *Compagno et al.* [1989] off southern Africa. These warm water sharks follow the advancing warm water masses into previously cooler areas, but since sharks continually shed their teeth, they will be deposited into their temporary environment and be preserved in the fossil record. The teeth of these shallow, warm water taxa are rare (0.5%, Table 2), and this rarity suggests that the required oceanographic phenomena were also infrequent.

SUMMARY

Most shark paleontologists refer to a collection of shark taxa from a certain locality as an "assemblage," merely a collection of teeth and usually not in an ecological or community context. The paleoecology of the sharks from a particular locality is little discussed aside from a gross generalization of the habitat or climate. As this study shows, a shark assemblage can be a very complex assortment of sharks from many different habitats converging on one specific locality. The La Meseta shark fauna supports many previous observations and interpretations on the paleoenvironment of the western Weddell Sea in the Eocene from invertebrates. More importantly, an understanding of the paleoecology, paleobiogeography, paleobathymetry, taphonomy, and sedimentology of the Eocene western Weddell Sea area is essential for an accurate and meaningful interpretation of the shark assemblage.

On the basis of their extant analogs, the Eocene shark fauna is indicative of a temperate marine climate influenced by periodic short-term warming. In the diverse La Meseta shark assemblage, four distinct ecological groups are represented at this single locality. The dominant ecological group is a shallow water fauna composed almost entirely of *Carcharias macrota* and accounts for 60% of the fossil specimens. The second most abundant ecological group is six genera that are transitional between shallow and deep waters. These account for 30% of the fossil specimens. There are also a significant number of sharks (9.5% of the fossil sample) that live almost entirely in deep water today but are found in the shallow La Meseta Formation in the Eocene. These sharks probably lived in adjacent deepwater areas and migrated into shallow waters on a cyclical diurnal or seasonal basis and deposited their teeth as they fed in the shallows. If this assumption is correct, it is the earliest indication of vertical migration in sharks. Several species of tropical, shallow water sharks (0.5% of the total sample) have also been recovered from the La Meseta Formation, and these are suspected to be migrants from warmer northern waters that entered the La Meseta area during periodic warm water fluctuations.

Acknowledgments. I would like to thank M. O. Woodburne, J. A. Case, and W. R. Daily for access to the La Meseta specimens and for help and encouragement on this project. I would also like to thank the following people who read and criticized this manuscript during various stages of development: G. W. Barlow, J. A. Case, J. Lipps, K. Padian, W.-E. Reif, J. D. Stewart, D. J. Ward, B. White, and M. O. Woodburne. This project was funded by National Science Foundation grants DPP-8215493 and DPP-8521368 to M. O. Woodburne.

REFERENCES

Bass, A. J., J. D. D'Aubrey, and N. Kistnasamy, Sharks of the east coast of southern Africa, VI, The families Oxynotidae,

Squalidae, Dalatiidae and Echinorhinidae, *Oceanogr. Res. Inst. Invest. Rep.*, *45*, 1–103, 1976.

Bigelow, H. B., and W. C. Schroeder, Fishes of the western North Atlantic, part 1, Lancelets, Cyclostomes, and sharks, *Mem. Sears Found. Mar. Res.*, *1*, 1–588, 1948.

Bigelow, H. B., and W. C. Schroeder, A study of the sharks of the suborder Squaloidea, *Bull. Mus. Comp. Zool.*, *117*, 1–150, 1957.

Blake, D. B., and W. J. Zinsmeister, Eocene asteroids (Echinodermata) from Seymour Island, Antarctic Peninsula, Geology and Paleontology of Seymour Island, Antarctic Peninsula, *Mem. Geol. Soc. Am.*, *169*, 489–498, 1988.

Cappetta, H., Les selaciens du Cretace Supérieur du Liban, I, Requins, *Paleontographica Abt. A*, *168*, 69–148, 1980.

Cappetta, H., *Handbook of Paleoichthyology*, vol. 3B, *Chondrichthyes II, Mesozoic and Cenozoic Elasmobranchii*, 193 pp., Gustav Fischer Verlag, New York, 1987.

Case, J. A., Paleogene floras from Seymour Island, Antarctic Peninsula, in Geology and Paleontology of Seymour Island, Antarctic Peninsula, *Mem. Geol. Soc. Am.*, *169*, 523–530, 1988.

Castro, J. I., *The Sharks of North American Waters*, 180 pp., Texas A&M University Press, College Station, 1983.

Clark, E., and E. Kristof, Deep-sea elasmobranchs observed from submersibles off Bermuda, Grand Cayman, and Freeport, Bahamas, in *Elasmobranchs as Living Resources*, H. L. Pratt jr., S. H. Gruber, and T. Taniuchi (eds.) NOAA *Tech. Rep. NMFS 90*, edited by H. L. Pratt, Jr., S. H. Gruber, and T. Taniuchi, pp. 269–284, National Oceanic and Atmospheric Administration, Boulder, Colo., 1990.

Clark, E., and E. Kristof, How deep do sharks go?, Reflections on deep sea sharks, in *Discovering Sharks*, *Spec. Publ. 14*, edited by S. H. Gruber, pp. 79–84, American Littoral Society, Highlands, N. J., 1991.

Compagno, L. V. J., *FAO Species Catalog*, vol. 4, *Sharks of the World: An Annotated and Illustrated Catalog of Shark Species Known to Date*, *FAO Fish. Synopsis 125*, part 1, Hexanchiformes to Lamniformes, 249 pp., Food and Agriculture Organization, United Nations, New York, 1984.

Compagno, L. V. J., Sharks, in *Fishes of the Southern Ocean*, edited by O. Gon and P. C. Heemstra, pp. 81–85, J. L. B. Smith Institute of Ichthyology, Grahamstown, South Africa, 1990.

Compagno, L. V. J., D. A. Ebert, and M. J. Smale, *Guide to the Sharks and Rays of Southern Africa*, 158 pp., New Holland Publishers, Cape Town, 1989.

Dietz, R. S., Deep scattering layer in the Pacific and Antarctic oceans, *J. Mar. Res.*, *7*, 430–442, 1948.

Dietz, R. S., The seas deep scattering layer, *Sci. Am.*, *207*(3), 44–50, 1962.

Feldmann, R. M., and M. T. Wilson, Eocene decapod crustaceans from Antarctica, in Geology and Paleontology of Seymour Island, Antarctic Peninsula, *Mem. Geol. Soc. Am.*, *169*, 465–488, 1988.

Holton, A. A., Feeding behavior of a vertically migrating lanternfish, *Pac. Sci.*, *23*, 325–331, 1969.

Hubbs, C. L., Changes in the fish fauna of western North America correlated with changes in ocean temperature, *J. Mar. Res.*, *7*, 459–482, 1948.

Karinen, J. F., B. L. Wing, and R. R. Straty, Records and sightings of fish and invertebrates in the eastern Gulf of Alaska and oceanic phenomena related to the 1983 El Niño event, in *El Niño North: El Niño Effects in the Eastern Subarctic Pacific Ocean*, edited by W. S. Wooster and D. L. Fluharty, pp. 253–267, Washington Sea Grant Program, University of Washington, Seattle, 1985.

Ketchen, K. S., The spiny dogfish (*Squalus acanthias*) in the northeast Pacific and a history of its utilization, *Spec. Publ.*

88, Fish. and Aquat. Sci., Can. Fish. and Oceans Sci. Inf. and Publ. Branch, Ottawa, Ont., 1986.

Keyes, I. W., New records of the fossil elasmobranch genera *Megascyliorhinus*, *Centrophorus*, and *Dalatias* (Order Selachii) in New Zealand, *N. Z. J. Geol. Geophys.*, *27*, 203–216, 1984.

Lavenberg, R. J., Megamania, the continuing saga of megamouth sharks, *Terra*, *30*, 30–39, 1991.

Long, D. J., Fossil cutlassfish (Perciformes: Trichiuridae) teeth from the La Meseta Formation (Eocene), Seymour Island, Antarctic Peninsula, *PaleoBios*, *13*(51), 3–6, 1991.

Long, D. J., Sharks from the La Meseta Formation (Eocene), Seymour Island, Antarctic Peninsula, *J. Vertebr. Paleontol.*, *12*, 11–32, 1992.

Merrett, N. R., J. D. M. Gordon, M. Stehmann, and R. L. Haedrich, Deep demersal fish assemblage structure in the Porcupine Seabight (eastern North Atlantic): Slope sampling by three different trawls compared, *J. Mar. Biol. Assoc. U.K.*, *71*, 329–358, 1991.

Mohr, B. A. R., Eocene and Oligocene sporomorphs and dinoflagellate cysts from Leg 113 drill sites, Weddell Sea, Antarctica, *Proc. Ocean Drill. Program Sci. Results*, *113*, 595–612, 1990.

Pearcy, W. G., E. E. Krygier, R. Mesecar, and F. Ramsey, Vertical distribution and migration of oceanic micronekton off Oregon, *Deep Sea Res.*, *24*, 223–245, 1977.

Pospichal, J. J., and S. W. Wise, Jr., Paleocene to middle Eocene calcareous nannofossils of ODP sites 689 and 690, Maud Rise, Weddell Sea, *Proc. Ocean Drill. Program Sci. Results*, *113*, 613–638, 1990.

Radovich, J., Relationships of some marine organisms of the northeast Pacific to water temperatures, particularly during 1957 through 1959, *Fish Bull. 112*, pp. 1–62, Calif. Dep. of Fish and Game Mar. Resour. Oper., Sacramento, 1961.

Sadler, P. M., Geometry and stratification of uppermost Cretaceous and Paleogene units on Seymour Island, northern Antarctic Peninsula, in Geology and Paleontology of Seymour Island, Antarctic Peninsula, *Mem. Geol. Soc. Am.*, *169*, 303–320, 1988.

Simpson, G. G., Review of fossil penguins from Seymour Island, *Proc. R. Soc. London, Ser. B*, *178*, 357–387, 1971.

Standora, E. A., and D. R. Nelson, A telemetric study of the behavior of free-swimming Pacific angel sharks, *Squatina californica*, *Bull. South. Calif. Acad. Sci.*, *76*, 193–201, 1977.

Stott, L. D., and J. P. Kennett, Antarctic Paleogene planktonic foraminifer biostratigraphy, ODP Leg 113, sites 689 and 690, *Proc. Ocean Drill. Program Sci. Results*, *113*, 549–570, 1990.

Svetlov, M. F., The porbeagle, *Lamna nasus*, in Antarctic waters, *J. Ichthyol.*, Engl. Transl., *18*, 850–851, 1978.

Thomas, E., Late Cretaceous through Neogene deep-sea benthic foraminifers (Maud Rise, Weddell Sea, Antarctica), *Proc. Ocean Drill. Program Sci. Results*, *113*, 571–594, 1990.

Velez, J., J. Zeballos, and M. Mendez, Effects of the 1982–83 El Niño on fishes and crustaceans off Peru, *Trop. Ocean Atmos. Newsl.*, *28*, 10–12, 1984.

Wei, W., and S. W. Wise, Jr., Middle Eocene to Pleistocene calcareous nannofossils recovered by Ocean Drilling Program Leg 113 in the Weddell Sea, *Proc. Ocean Drill. Program Sci. Results*, *113*, 639–666, 1990.

Welton, B. J., Late Cretaceous and Cenozoic squalomorphii of the northwest Pacific Ocean, Ph.D. thesis, 553 pp., Univ. of Calif., Berkeley, 1979.

Welton, B. J., and W. J. Zinsmeister, Eocene neoselachians from the La Meseta Formation, Seymour Island, Antarctic Peninsula, *Contrib. in Sci. 329*, pp. 1–10, Hist. Mus. of Los Angeles County, Los Angeles, Calif., 1980.

Wiedman, L. A., and R. M. Feldmann, Ichnofossils, tubiform body fossils, and depositional environment of the La Meseta Formation (Eocene) of Antarctica, in Geology and Paleontology of Seymour Island, Antarctic Peninsula, *Mem. Geol. Soc. Am.*, *169*, 531–539, 1988.

Wiedman, L. A., R. M. Feldmann, D. E. Lee, and W. J. Zinsmeister, Brachiopoda from the La Meseta Formation (Eocene), Seymour Island, Antarctica, in Geology and Paleontology of Seymour Island, Antarctic Peninsula, *Mem. Geol. Soc. Am.*, *169*, 449–457, 1988.

Woodburne, M. O., and W. J. Zinsmeister, The first land mammal from Antarctica and its biogeographical implications, *J. Paleontol.*, *58*, 913–948, 1984.

Yano, K., and S. Tanaka, A telemetric study on the movements of the deep sea squaloid shark *Centrophorus acus*, in *Indo-Pacific Fish Biology: Proceedings of the Second International Conference on Indo-Pacific Fishes*, edited by T. Uyeno, R. Arai, T. Taniuchi, and K. Matsuura, pp. 372–380, Ichthyological Society of Japan, Tokyo, 1986.

Zinsmeister, W. J., Biogeographic significance of the late Mesozoic and early Paleogene molluscan faunas of Seymour Island (Antarctic Peninsula) to the final breakup of Gond-wanaland, in *Historical Biogeography, Plate Tectonics, and the Changing Environment*, edited by J. Gray and A. J. Boucot, pp. 347–355, University of Oregon Press, Corvallis, 1979.

Zinsmeister, W. J., Late Cretaceous–early Tertiary molluscan biogeography of the southern Circum-Pacific, *J. Paleontol.*, *56*, 84–102, 1982.

Zinsmeister, W. J., and R. M. Feldmann, Cenozoic high latitude heterochroneity of southern hemisphere marine faunas, *Science*, *224*, 281–283, 1984.

Zullo, V. A., R. M. Feldmann, and L. A. Wiedmann, Balanomorph cirripedia from the Eocene La Meseta Formation, Seymour Island, Antarctica, in Geology and Paleontology of Seymour Island, Antarctic Peninsula, *Mem. Geol. Soc. Am.*, *169*, 459–464, 1988.

(Received January 3, 1992;
accepted March 11, 1992.)

CENOZOIC DEEP-SEA CIRCULATION: EVIDENCE FROM DEEP-SEA BENTHIC FORAMINIFERA

ELLEN THOMAS[1]

Department of Earth Sciences, University of Cambridge, Cambridge CB2 3EQ, United Kingdom

Deep-sea benthic foraminiferal faunas reflect the deep oceanic environment, the character of which is determined by interaction of deepwater circulation patterns, physicochemical parameters of the surface waters in the deepwater source areas, and nutrient influx from primary productivity in overlying surface waters. Three periods of turnover in deep-sea benthic foraminiferal assemblages can be recognized in Cenozoic sequences: (1) rapid ($<10^4$ yr), global extinction in the latest Paleocene, followed by migration and diversification; (2) gradual turnover in the late middle Eocene through early Oligocene, characterized by a decrease in diversity, a decrease in relative abundance of *Nuttallides truempyi* followed by its extinction, and a decreasing relative abundance or disappearance of *Bulimina* species in the lower bathyal to upper abyssal zones; and (3) gradual turnover in the late early through middle Miocene, characterized by the decrease in relative abundance or disappearance of uniserial species from the lower bathyal to abyssal reaches, the migration of miliolid species into these regions, and the evolution of *Cibicidoides wuellerstorfi*. The rapid mass extinction (35–50% of species) of deep-sea benthic foraminifera in the latest Paleocene was coeval with a transient 1–2‰ decrease in oxygen and carbon isotope ratios in benthic as well as planktonic foraminifera, superimposed on longer-term changes. The extinction could have resulted from a shift in dominant deepwater formation from high to low latitudes. Such a shift would change temperature and oxygen content of the intermediate to deep waters, but it would also change local nutrient input by changing global patterns of upwelling of nutrient-rich waters to the surface and thus of high-productivity areas. Faunal evidence suggests that this "reversed" pattern of oceanic circulation persisted no longer than the early Eocene, and possibly not more than about half a million years. The two periods of gradual benthic faunal changes overlap in time with two relatively rapid (of the order of 10^5 years) shifts toward heavier oxygen isotopic values of benthic foraminifera, in the earliest Oligocene and middle Miocene. Faunal changes started before the isotopic changes and were more gradual. The faunal changes might reflect periods of gradual change in the physicochemical character of surface waters in the source areas of deepwater formation (e.g., decrease in temperature), as well as changes in oceanic productivity. The more rapid changes in oxygen isotopic values are not directly reflected in benthic foraminiferal assemblage changes and might represent, at least in part, a rapid buildup of ice volume on land, a process that cannot be reflected in the benthic foraminifera faunas.

INTRODUCTION

The deep oceanic environment has been more stable through geologic time than the surface environments of the Earth, with less short-term variability, although spatial "patchyness" occurs [*Thiel et al.*, 1988; *Gooday and Lambshead*, 1989; *Lambshead and Gooday*, 1990]. This stability is reflected in the slow faunal turnover of benthic foraminiferal faunas as compared to planktonic microfossils. Benthic faunas have been a relatively neglected group and are commonly overlooked in reconstructions of deepwater circulation patterns of the past. Much information on Cenozoic deep-sea faunas has been obtained as an outcome of recent Ocean Drilling Program (ODP) drilling operations in the Indian and Southern oceans (legs 113, 114, 118, 119, and 121). The interpretation of these data is not straightforward, because no unequivocal relations between benthic foraminiferal faunal composition and environmental parameters have been discovered, and there exists no transfer function to translate benthic foraminiferal data into environmental parameters. The collection of long-term records, however, has helped to delineate benthic foraminiferal faunal history, making it possible to delineate periods of faunal change, alternating with long, relatively stable periods. Benthic foraminiferal faunal Cenozoic history proceeded stepwise, as did ocean history [e.g., *Berger et al.*, 1981; *Barron and Baldauf*, 1989].

Dramatic changes in the Cenozoic oceans are documented in stable isotopic and biotic records [e.g., *Savin et al.*, 1981; *Shackleton and Kennett*, 1975; *Kennett and Shackleton*, 1976; *Kennett*, 1977; *Savin*, 1977; *Berger*, 1981; *Berger et al.*, 1981; *Kemp*, 1978; *Shackleton and*

[1]Now at Center for the Study of Global Change, Department of Geology and Geophysics, Yale University, New Haven, Connecticut 06511.

Boersma, 1981; *Mercer*, 1982; *Shackleton*, 1986; *Miller et al.*, 1987*a*; *Kennett and Barker*, 1990; *Thomas*, 1989, 1990*a*; *Rea et al.*, 1990; *Webb*, 1990; *Barron et al.*, 1991; *Wise et al.*, 1992; *Kennett and Stott*, 1990, 1991; *Zachos et al.*, 1992*c*; *McGowran*, 1989, 1991]. Similarly dramatic changes were recorded in land floras [*Wolfe*, 1978; *Wolfe and Poore*, 1982; *Wing*, 1984; *Schmidt*, 1991; *Wing et al.*, 1991], as the Earth changed from essentially unglaciated to a world with very cold, glaciated polar regions.

Climatic change as reflected in the oxygen and carbon isotopic records did not proceed gradually, but stepwise [e.g., *Berger et al.*, 1981]. Deep waters in the world's oceans and surface waters at high latitudes cooled strongly after the very warm late Paleocene to earliest Eocene, the warmest period of the Cenozoic. Early Eocene surface water temperatures at high latitudes (65°S) are estimated to have been about 15°–17°C [*Stott et al.*, 1990; *Kennett and Stott*, 1991; *Wise et al.*, 1992; *Zachos et al.*, 1992*a*, *c*]. Climate as well as deepwater circulation and ocean chemistry during the Paleocene strongly resembled the Late Cretaceous situation [*Thomas*, 1990*b*; *Corfield et al.*, 1991]. At some time during the middle Eocene through early Oligocene the psychrosphere was established [*Benson*, 1975], as well as at least partial ice sheets on eastern Antarctica [*Savin et al.*, 1981; *Kennett and Shackleton*, 1976; *Keigwin and Keller*, 1984; *Miller and Thomas*, 1985; *Miller et al.*, 1987*a*; *Kennett and Barker*, 1990; *Barron et al.*, 1991; *Spezzaferri and Premoli-Silva*, 1991; *Wise et al.*, 1992]. The extent and nature of these ice sheets (true continental ice sheets, temperate ice sheets, or upland and coastal glaciers) are under intensive discussion [*Kennett and Barker*, 1990; *Barron et al.*, 1991; *Wise et al.*, 1992; *Zachos et al.*, 1992*a*, *b*, *c*]. Evidence from glaciomarine sediments in the Prydz Bay area (East Antarctica) suggests that some Antarctic ice sheets or large glaciers reached sea level during the earliest Oligocene, at least temporarily [*Barron et al.*, 1991], in agreement with the observation of ice-rafted material in lowermost Oligocene sediments on the Kerguelen Plateau [*Wise et al.*, 1992; *Zachos et al.*, 1992*a*, *b*, *c*].

There is no agreement on the interpretation of the Cenozoic oxygen isotopic record of benthic foraminifers, especially in how far it demonstrates establishment and growth of ice sheets (the ice volume effect) and in how far it represents cooling of deep waters in the oceans [e.g., *Matthews and Poore*, 1980; *Poore and Matthews*, 1984; *Keigwin and Corliss*, 1986; *Shackleton*, 1986; *Miller et al.*, 1987*a*; *Prentice and Matthews*, 1988; *Wise et al.*, 1992; *Zachos et al.*, 1992*a*, *c*; *Oberhänsli et al.*, 1991; *Spezzaferri and Premoli-Silva*, 1991]. Additional complications in the benthic record might result from storage of the heavier oxygen isotopes in deeper waters when these are formed at least partially by evaporation [*Railsback et al.*, 1989; *Railsback*, 1990].

It is not clear whether equatorial surface waters were cooler than at present during some globally warm periods [e.g., *Shackleton*, 1984, 1986] or remained essentially at the same temperature throughout the Cenozoic, as indicated by the distribution of tropical biota such as hermatypic corals, mangroves, and larger foraminifera [*Matthews and Poore*, 1980; *Adams et al.*, 1990; *McGowran*, 1989, 1991]. Especially during the Oligocene there is a discrepancy between macrofaunal and macrofloral, as well as microfaunal, data indicating fairly warm climates at high to middle latitudes and isotopic data suggesting cooler climates [*Adams et al.*, 1990; *McGowran*, 1991; *Spezzaferri and Premoli-Silva*, 1991]. There is no agreement on whether the flat Eocene-Oligocene latitudinal oxygen isotopic gradient in planktonic foraminifera truly reflects a very low temperature gradient in surface waters [*Shackleton and Boersma*, 1981; *Premoli-Silva and Boersma*, 1984; *Shackleton*, 1984; *Keigwin and Corliss*, 1986; *Boersma et al.*, 1987; *Boersma and Premoli-Silva*, 1991; *Zachos et al.*, 1992*a*]. The distribution of planktonic biota suggests a steeper temperature gradient than the oxygen isotopic data [*Wei*, 1991], and the planktonic oxygen isotope record may be complicated by local influx of fresh water into high-latitude surface waters [e.g., *Wise et al.*, 1991, 1992; *Zachos et al.*, 1992*c*].

There is similar discussion on modes of deepwater formation in the Cenozoic. The question is whether intermediate and bottom waters dominantly formed at high latitudes as a result of an increase in density because of cooling [e.g., *Manabe and Bryan*, 1985; *Barrera et al.*, 1987; *Katz and Miller*, 1991; *Thomas*, 1989, 1990*a*, *b*] or at low latitudes after an increase in density by evaporation [*Chamberlin*, 1906; *Matthews and Poore*, 1980; *Brass et al.*, 1982; *Hay*, 1989; *Woodruff and Savin*, 1989; *Kennett and Stott*, 1990]. Oceanic circulation dominated by deep and/or intermediate waters originating from low-latitude sources represents a reversal of the present deep-oceanic circulation [e.g., *Keith*, 1982; *Prentice and Matthews*, 1988; *Woodruff and Savin*, 1989; *Kennett and Stott*, 1990]. Oceans "running the reverse" from the modern circulation pattern might be required to model satisfactorily the high heat transfer from low to high latitudes required to maintain the warm Eocene climate at high latitudes [*Barron*, 1985, 1987; *Barron and Peterson*, 1991]. Carbon and oxygen isotopic data as well as deep-sea benthic foraminiferal data, however, may be seen as indicative of a Late Cretaceous–Paleocene Atlantic and Pacific circulation dominated by deep and intermediate waters formed at high latitudes, with possible exception of one or more short (<0.5 m.y.) periods of high-volume formation of warm, salty bottom waters during the early to early middle Eocene [*Barrera et al.*, 1987; *Miller et al.*, 1987*b*; *Thomas*, 1989, 1990*a*; *Katz and Miller*, 1991; *Zachos et al.*, 1992*c*; *Pak et al.*, 1991; *Barron and Peterson*, 1991; *Pak and Miller*, 1992]. The Indian

Ocean might have contained a relatively large volume of waters from a low latitude, as a result of outflow from the eastern end of the Tethys Ocean until its closure in the middle Miocene [*Woodruff and Savin*, 1989; *Nomura*, 1991; *Zachos et al.*, 1992c].

Deep waters formed at high latitudes during the periods in which the Antarctic was essentially ice free would have had a lower oxygen content than the present deep waters, because of the higher temperatures of the surface waters in the source areas, and thus lower oxygen content at the time of formation. The oxygen content would have been even lower in waters formed by evaporation at low latitudes, because of the nonlinear, reverse relation between solubility of oxygen and temperature. The ratio of the solubility of O_2 at 0°C to that at 24°C is 1.6 (solubilities taken per atmosphere pressure of the gas) [*Broecker and Peng*, 1982]. In recent oceans, most surface waters are supersaturated in oxygen by a few percent; the oxygen solubilities range from about 190 μmol/kg at 30°C, 225 mmol/kg at 20°C, 275 μmol/kg at 10°C, to 350 μmol/kg at 0°C. In the northern Pacific the oxygen content of the deep waters is the lowest, because these waters have been out of contact with the oxygen supply in the atmosphere for more than 1000 years. In this region the apparent oxygen utilization (AOU, the difference between saturated oxygen content and observed oxygen content) is about 190 μmol/kg [*Broecker and Peng*, 1982]. This value represents the amount of oxygen used during the travel of the deep waters from their source area, as a result of decay of organic material and respiration by the bottom-dwelling fauna. If the waters in the source area had been at a temperature of 30°C (and would have been dense enough to sink to the deep oceans at that temperature), they could not have taken up more than 190 μmol/kg. The oxygen in these waters would have been exhausted by the time that they had been out of contact with the atmosphere for about 1000 years (unless the oxygen content of the atmosphere was considerably higher at the time). Thus the warmer deep waters of the past possibly contained less dissolved oxygen than the present deep waters. A more sluggish deepwater ventilation would decrease the dissolved oxygen content even more [e.g., *Thierstein*, 1989]. Chemical and circulation modeling of the oceans suggests that dominance of warm saline deep waters in the oceans might likely drive the oceans to anoxia, which did not occur during the last 90 million years [*Herbert and Sarmiento*, 1991]. Benthic foraminiferal faunas have been interpreted as indicating good deep-ocean ventilation in Late Cretaceous through Paleocene and middle Eocene and later [*Thomas*, 1990a; *Kaiho*, 1991].

The ratio of the solubility of CO_2 at 0°C to that at 24°C is 2.2 (solubilities taken per atmosphere pressure of the gas [*Broecker and Peng*, 1982]). The amount of CO_2 that can be dissolved in the colder waters of the present oceans is thus much larger than the amount that could dissolve in the warmer oceans of the past. At lower temperatures the solubility of calcite is much higher than at high temperatures, so that the cold, present-day deep waters are potentially more corrosive to $CaCO_3$. More vigorous deepwater ventilation, however, could counteract this effect, because of the presence of overall "younger" deep waters (out of contact with the atmosphere for a shorter period) in an ocean with a faster turnover rate.

High-resolution carbon and oxygen isotopic records from many locations in the oceans and high-resolution, reliable stratigraphic data are necessary to develop models of deepwater sources, especially how many significant source areas there were, what their relative contribution to the total deepwater volume was, and where they were [e.g., *Woodruff and Savin*, 1989; *Mead and Hodell*, 1992; *Pak and Miller*, 1992]. We should not forget that all isotope information is derived from sites above the calcium carbonate compensation depth (CCD), and thus we do not know what characterized water masses in the middle and lower abyssal realms.

Recent deepwater benthic foraminiferal faunas reflect the complex interaction of deep oceanic circulation, the character of the surface waters in the source regions, and local primary productivity influx [e.g., *Lohmann*, 1978; *Douglas and Woodruff*, 1981; *Lutze and Coulbourn*, 1984; *Culver*, 1987; *Gooday*, 1988; *Gooday and Lambshead*, 1989]. Benthic foraminifera may form as much as 50% or more of the eukaryotic biomass in the deep sea [*Gooday et al.*, 1992]. They are the only fossil-providing organisms that live in large enough numbers on the nutrient-starved ocean floor environment to be represented in Deep Sea Drilling Project (DSDP)–ODP sized core samples in numbers large enough for statistically valid studies, at high time resolution. Therefore data on benthic foraminiferal faunal composition should be used as constraints in the reconstruction of deepwater formational processes as inferred from stable isotope or trace element studies. The large influence of nutrient influx on the deep-sea benthic foraminiferal faunas should be kept in mind, while assessing the influence of deepwater circulation on faunal patterns [*Gooday and Lambshead*, 1989; *Gooday and Turley*, 1990; *Lambshead and Gooday*, 1990].

In this paper I review information that has recently become available as a result of ODP drilling in the Southern and Indian oceans and interpret its significance for ocean circulation models for the Cenozoic, with emphasis on the Paleogene. The bathymetric division follows *Berggren and Miller* [1989] and *van Morkhoven* [1986]: neritic, <200 m; upper bathyal, 200–600 m; middle bathyal, 600–1000 m; lower bathyal, 1000–2000 m; upper abyssal, 2000–3000 m; lower abyssal, >3000 m. The information presented in this paper is dominantly based on calcareous taxa of benthic foraminifera. The calcium carbonate compensation depth during the Cenozoic fluctuated between 3500 and 5000

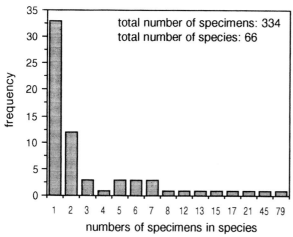

Fig. 1. Distribution of specimens within species for a highly diverse, Paleocene sample of benthic foraminifera (Site 690). Note the discontinuity in scale along the horizontal axis.

m [*van Andel*, 1975]; thus this information does not extend to the deeper parts of the ocean basins. The time scale used throughout is after *Berggren et al.* [1985], except where it has been modified by *Aubry et al.* [1988]. Major changes in numerical ages (especially in the Paleogene) will probably be proposed in the near future [*Montanari*, 1990; I. Premoli-Silva, personal communication, 1991; S. C. Cande and D. V. Kent, work in preparation] but are not yet available.

DEEP-SEA BENTHIC FORAMINIFERA AND PALEOCEANOGRAPHY

Deep-sea benthic foraminifera are not easy subjects of study. In most deep-sea samples, with the exception of those strongly affected by dissolution, they are outnumbered by several orders of magnitude by the tests of planktonic organisms, so that it is time consuming to extract enough specimens for a statistically valid study. This is aggravated by the fact that deep-sea benthic foraminifera form highly diverse assemblages, as do other deep-sea organisms, and consist of specimen-poor, species-rich assemblages [e.g., *Sanders*, 1968; *Douglas and Woodruff*, 1981]. Large numbers of specimens must be studied to obtain a valid representation of the total species richness, and many of the total number of species present are represented by only one or two specimens (Figure 1; *Douglas and Woodruff* [1981]). The minimum number needed to represent the species richness depends upon the diversity and can be determined by plotting rarefaction curves (*Sanders* [1968]; see also *Thomas* [1985]). The stratigraphic ranges of the rare species cannot be determined with precision [e.g., *Signor and Lipps*, 1982]. Biology of recent deep-sea benthic foraminifera is not well known; they feed at low trophic levels (consuming planktonic debris and bacte-

ria), and some species can respond quickly to the pulsed influx of detritus [*Gooday et al.*, 1992].

As a result of the high diversity and the presence of many rare species, results from quantitative studies are difficult to present. Data are commonly presented after *Q* mode or *R* mode multivariate analysis, or cluster analysis, on relative and/or absolute abundance data, and values of derived factors are plotted to supplement the information shown in simple species' relative or absolute abundances [e.g., *Lohmann*, 1978]. The derived factors, however, are commonly difficult to interpret unequivocally. In different studies, factors have been interpreted as representing water mass properties [*Lohmann*, 1978; *Bremer and Lohmann*, 1982], a combination of water mass properties, substrate and nutrient influx [*Lutze and Coulbourn*, 1984], or other combinations of environmental parameters. Nutrient influx in the shape of aggregates of phytodetritus appears to be an important factor in species composition as well as absolute abundance of recent deep-sea faunas [*Gooday*, 1988; *Gooday and Lambshead*, 1989; *Gooday et al.*, 1992]. Evaluation of the importance of this parameter for the past (for example, by using organic carbon content of the sediment as a proxy for nutrient influx) indicates that nutrient influx is indeed important [e.g., *Caralp*, 1984; *Lutze and Coulbourn*, 1984]. These studies, however, have not resulted in a transfer function relating absolute or relative abundances of species to nutrient influx.

In stratigraphic studies the factors commonly do not show a simple vertical succession, but several assemblages (as indicated by high loadings on a factor) alternate vertically in a drill hole or stratigraphic section. These assemblages thus probably represent not evolutionary and extinction events, but migratory events resulting from reversible environmental changes [e.g., *Mueller-Merz and Oberhänsli*, 1991; *Nomura*, 1991*a*; *Oberhänsli et al.*, 1991; *Mackensen*, 1992].

In few instances the relative abundance of species has been related to specific environmental properties. The recent species *Nuttallides umbonifera*, for instance, has been reported to be most abundant in waters that are highly corrosive to $CaCO_3$, thus with high concentrations of dissolved CO_2 [*Bremer and Lohmann*, 1982]. *Eilohedra weddellensis*, *Epistominella exigua*, *Cassidulina teretis*, and *Melonis barleeanus* appear to be dominant in faunas where there is a high influx of phytodetritus [*Caralp*, 1984; *Gooday*, 1988; *Gooday and Lambshead*, 1989; *Gooday et al.*, 1992]. Several biserial and triserial taxa such as *Bolivina* and *Uvigerina* spp. and thin-walled species of *Cassidulina* have been well described as abundant to dominant in low-O_2 environments, but their abundance may be primarily controlled by high nutrient as well as low O_2 conditions [e.g., *Douglas*, 1981; *Sen Gupta et al.*, 1981; *Caralp*, 1984; *Lutze and Coulbourn*, 1984; *Bernard*, 1986; *Corliss and*

Chen, 1988; *Niensted and Arnold*, 1988; *van der Zwaan et al.*, 1992].

Recently, attempts have been made to correlate not abundances of species, but general morphological types with environmental parameters, especially to an infaunal or epifaunal mode of life [*Corliss*, 1985; *Corliss and Chen*, 1988; *Corliss and Emerson*, 1990; *Rosoff and Corliss*, 1992]. These authors concluded that high relative abundance of infaunal morphotypes (including the biserial and triserial groups) reflects a relatively high flux of nutrients to the ocean floor. A similar approach for fossil material has been tried [*Keller*, 1988; *Thomas*, 1989, 1990*a*, *b*; *Kaiho*, 1991; *Oberhänsli et al.*, 1991], but the validity of the correlation between test morphology and life style is not beyond doubt.

The existing data base on the geological record of deep-sea benthic foraminiferal faunas is thus commonly difficult to access for nonmicropaleontologists; there is no simple numerical parameter to be plotted representing faunal composition, and there is no globally valid biostratigraphic zonation, or environmental zonation, or depth zonation. As a result, this data base is usually not consulted by nonmicropaleontologists while reconstructing circulation patterns and deepwater physicochemical properties. As an example, *Keith* [1982] argued that the end-Cretaceous extinction episode was caused by anoxia in the deep oceans, followed by catastrophic overturns and extinction of planktonic organisms as a result of poisoning. This theory conflicts with deep-sea benthic foraminiferal evidence, because it has long been known that extinction of deep-sea benthic foraminifera at the end of the Cretaceous was not catastrophic and hardly reached above background levels [*Cushman*, 1946; *Beckman*, 1960; *Webb*, 1973; *Dailey*, 1983; *Douglas and Woodruff*, 1981].

A biostratigraphic zonation for deep-sea benthic foraminifera (bathyal and abyssal) was proposed by *Berggren and Miller* [1989] (Figure 2), largely using material from Atlantic DSDP sites or land sections in the Caribbean region. Many of the zonal species, however, do not occur or are very rare at other sites or have different ranges. The timing of zonal boundaries, on the other hand, and thus of periods of faunal change, appears to be recognizable over wider areas, especially for the Paleogene. Faunal changes as recognized by the first and last appearances appeared to cluster around the times of zonal boundaries of *Berggren and Miller* [1989] for lower bathyal to uppermost abyssal faunas from Maud Rise, Antarctica [*Thomas*, 1990*a*] (Figure 2).

Informal benthic zones based upon cluster analysis of quantitative faunal data were proposed for parts of the Cenozoic for lower bathyal to upper abyssal faunas from the Walvis Ridge (DSDP Site 525), where an unconformity is present between middle Eocene and upper Oligocene [*Boltovskoy and Boltovskoy*, 1989] (Figure 2). For middle to upper bathyal sites on Ninetyeast Ridge and Broken Ridge (Indian Ocean), fewer

assemblages were recognized using multivariate analysis [*Nomura*, 1991*a*, *b*] (Figure 2), but periods of faunal change were coeval with zonal boundaries in the work of *Berggren and Miller* [1985]. In sections from Japan and New Zealand and material from DSDP sites in the North Pacific, the South Atlantic, and the Indian Ocean, *Kaiho* [1991, 1992] recognized four benthic zones, with boundaries overlapping in time with some of the zonal boundaries of *Berggren and Miller* [1989]. His data have low time resolution (less than one data point per million years) and are gathered at many different sites, so that records had to be spliced, with the resultant problems of precision in time correlation. He concludes that these faunal changes might reflect different oxygenation states of the oceanic bottom waters and that there was relatively low oxygenation of deep waters in the early to earliest middle Eocene (in agreement with *Thomas* [1989, 1990*a*]).

In the southernmost Atlantic east of the Falkland Plateau, *Katz and Miller* [1991] recognized the major faunal change close to the end of the Paleocene, which is prominent in all zonal schemes in Figure 2, in addition to faunal overturn over a wide depth range (1000–2500 m) at 54 Ma. At the southern tip of Kerguelen Plateau (Indian Ocean) faunal assemblages in the lower bathyal range changed in the late middle Eocene (44 Ma), in the middle late Eocene (39 Ma), shortly after the end of the Eocene (36 Ma), and in the early Oligocene (32 Ma) [*Schroeder-Adams*, 1991]. At middle bathyal to uppermost abyssal depths on the northern Kerguelen Plateau (Indian Ocean), major faunal change as demonstrated in multivariate analysis of quantitative faunal data occurred in the last part of the Paleocene and at the middle to late Eocene boundary [*Mackensen and Berggren*, 1992].

These results appear to be roughly in agreement, but in detail many differences in timing are seen (Figure 2), which can only partially be attributed to problems in stratigraphic correlation. Many of the data sets have fairly high resolution, and isotope, biostratigraphic, and magnetostratigraphic data allow reasonably reliable correlation. There is considerable agreement in the different evaluations of deep-sea benthic foraminiferal data if one realizes that some of the apparently conflicting data on the timing of zonal boundaries or assemblage ranges result from the difficulty of assigning an exact location to a boundary, when faunal change occurs gradually over one or several millions of years. Many authors have recognized that there was an extended period of faunal change over the full middle bathyal to abyssal depth range, starting in the late early Miocene through the middle to late middle Miocene [*Woodruff and Douglas*, 1981; *Thomas*, 1985, 1986*a*, *b*; *Woodruff*, 1985; *Schnitker*, 1979*b*, 1986; *Murray et al.*, 1986; *van Morkhoven et al.*, 1986; *Miller and Katz*, 1987; *Boltovskoy and Boltovskoy*, 1988, 1989; *Nomura et al.*, 1991*b*, 1992]. Another extended period of faunal

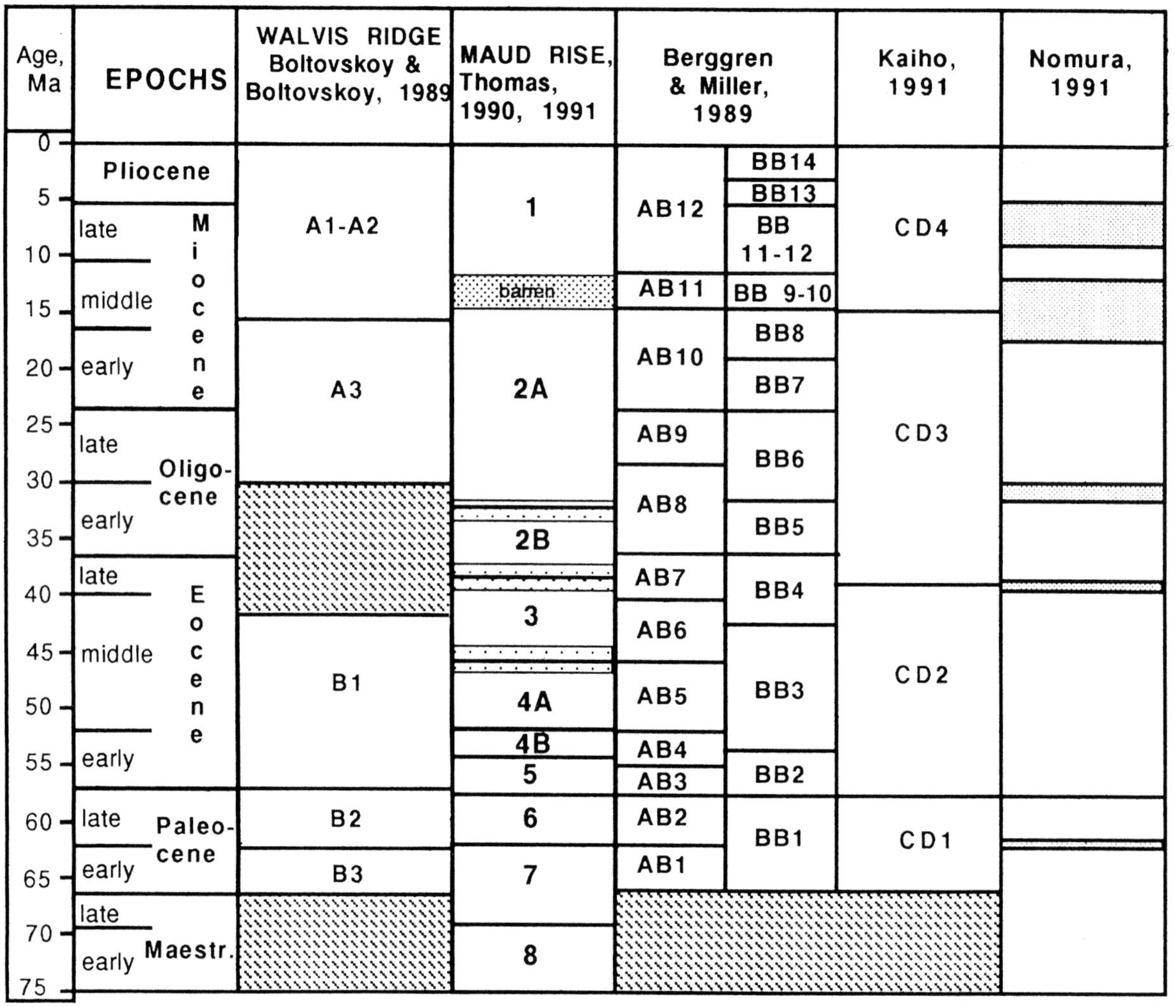

Age, Ma	EPOCHS	WALVIS RIDGE Boltovskoy & Boltovskoy, 1989	MAUD RISE, Thomas, 1990, 1991	Berggren & Miller, 1989		Kaiho, 1991	Nomura, 1991
0	Pliocene	A1-A2	1	AB12	BB14	CD4	
5	late M				BB13		
10	middle i o				BB 11-12		
15	c e n e		bahen	AB11	BB 9-10		
20	early	A3	2A	AB10	BB8		
25	late Oligo-			AB9	BB7	CD3	
30	cene			AB8	BB6		
35	early		2B		BB5		
40	late E			AB7	BB4		
45	middle o c	B1	3	AB6	BB3	CD2	
50	e n e		4A	AB5			
55	early		4B / 5	AB4 / AB3	BB2		
60	late Paleo-	B2	6	AB2	BB1	CD1	
65	early cene	B3	7	AB1			
70	late Maestr.		8				
75	early						

Fig. 2. Compilation of assemblages and informal and formal biostratigraphic zones of deep-sea benthic foraminifera. Data after *Boltovskoy and Boltovskoy* [1989], *Thomas* [1990a], *Berggren and Miller* [1989], *Kaiho* [1991], and *Nomura* [1991a, b]. Numerical ages after *Berggren et al.* [1985], except where amended by *Aubry et al.* [1988].

change started toward the end of the middle Eocene and ended close to the early/late Oligocene boundary [e.g., *Corliss*, 1981; *Tjalsma and Lohmann*, 1983; *Miller*, 1983; *Miller et al.*, 1984; *Boersma*, 1984, 1985, 1986; *Corliss and Keigwin*, 1986; *McGowran*, 1987; *Berggren and Miller*, 1989; *Boltovskoy*, 1980; *Boltovskoy and Boltovskoy*, 1988, 1989; *Mueller-Merz and Oberhänsli*, 1991; *Oberhänsli et al.*, 1991; *Thomas*, 1992] (see reviews in the work of *Douglas and Woodruff* [1981] and *Culver* [1987]).

In contrast with these extended periods of faunal change was the most profound benthic faunal change over the last 75 m.y. that occurred at the end of the Paleocene [*Cushman*, 1946; *Braga et al.*, 1975; *Schnitker*, 1979b; *Tjalsma and Lohmann*, 1983; *Miller et al.*, 1987b; *Berggren and Miller*, 1989; *Boltovskoy and Boltovskoy*, 1988, 1989; *Katz and Miller*, 1991; *Mack-*

ensen and Berggren, 1992; *Nomura*, 1991; *Kaiho*, 1988, 1991; *Thomas*, 1989, 1990a, b; *Katz and Miller*, 1991; *Berggren et al.*, 1992]. This was the only benthic foraminiferal mass extinction documented, resulting in a loss of diversity of 35 to 50% (Figure 2). The extinction occurred over 10,000 years or less [*Thomas*, 1990b, 1991; *Kennett and Stott*, 1991; E. Thomas and N. J. Shackleton, work in preparation]. At many sites, another, less prominent period of faunal change was observed at the boundary between upper and lower Paleocene (Figure 2); over this period, however, there are not many high-resolution data sets available. There was no major deep-sea benthic foraminiferal extinction at the end of the Cretaceous, and Paleocene faunas closely resemble upper Maestrichtian faunas [*Beckmann*, 1960; *Dailey*, 1983; *Keller*, 1988; *Thomas*, 1990b; *Widmark*, 1990; *Widmark and Malmgren*, 1992; *Kaiho*, 1992].

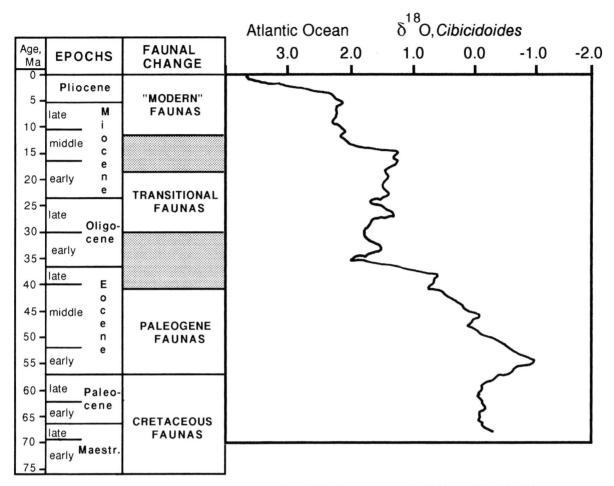

Fig. 3. Simplification of deep-sea benthic foraminiferal data, compared with the combined benthic foraminiferal oxygen isotopic curve for the Atlantic Ocean [*Miller et al.*, 1987a]. Numerical ages after *Berggren et al.* [1985], except where amended by *Aubry et al.* [1988].

Thus we can subdivide the Cenozoic into essentially three broad benthic foraminiferal zones (Figure 2), as was earlier recognized by *Berggren et al.* [1992]. These zones can be recognized worldwide and over a very large depth range (upper to middle bathyal to upper abyssal), suggesting that the periods of change reflect major changes in the deep oceanic environment. The first of the zones ended at the rapid extinction at the end of the Paleocene. This extinction occurred at the beginning of a long-term warming trend of deep waters as seen in oxygen isotopic records of deep-sea benthic foraminifera (Figure 3), and at the end of a strong decrease of $\delta^{13}C$ values in surface and deep waters (Figure 4; events X and A of *Shackleton* [1986] and event C of *McGowran* [1990]). The more gradual periods of faunal change in the Eocene/Oligocene and early/middle Miocene overlapped with the two large, fast shifts to heavier values of $\delta^{18}O$ in deep-sea benthic foraminifera in the earliest Oligocene (35.8 Ma) and the middle Miocene (14.6 Ma), but the faunal change was

more gradual, lasted longer, and started before the isotopic shifts.

THREE PERIODS OF BENTHIC FAUNAL CHANGE

Paleocene/Eocene Boundary Events

The Paleocene/Eocene boundary has not been generally recognized as a time of major biotic turnover, because generic extinction rates were low [e.g., *Raup and Sepkoski*, 1986]. These extinction patterns, however, largely show events in terrestrial and shallow marine environments and not in the deep ocean [*Thomas*, 1990b; *Kennett and Stott*, 1991]. The end of the Paleocene is within the lower reversed-polarity interval in Chron 24R [*Berggren et al.*, 1985; *Aubry et al.*, 1988], a period of profound changes in plate tectonic configuration [*Williams*, 1986; *McGowran*, 1989, 1989, 1991]. The northward motion of the Indian subcontinent slowed because of collision with Asia [*McGowran*,

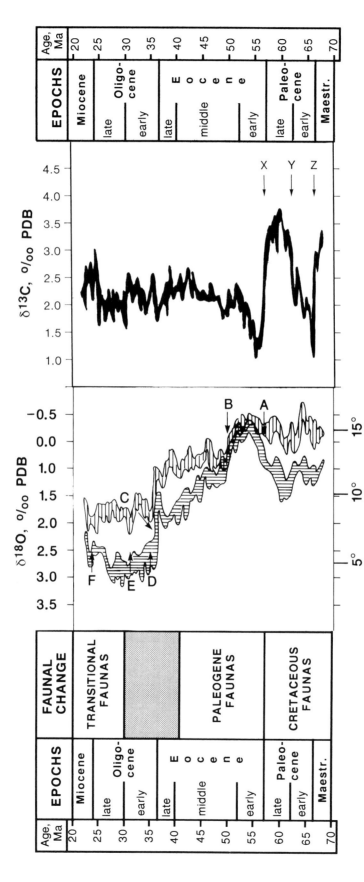

Fig. 4. Simplified benthic foraminiferal data for the Paleogene compared with oxygen isotopic data for bulk carbonate, reflecting surface water conditions (horizontal pattern) and benthic foraminifera (vertical pattern), and carbon isotopic data for bulk carbonate. Data and letters indicating events after *Shackleton* [1986]. Numerical ages after *Berggren et al.* [1985], except where amended by *Aubry et al.* [1989].

1991; *Klootwijk et al.*, 1991], the direction of subduction in the northern Pacific changed [*Goldfarb et al.*, 1991], and continental breakup started in the northern Atlantic. In this area, several millions of cubic kilometers of flood basalts erupted [*Roberts et al.*, 1984; *White and MacKenzie*, 1989; *White*, 1989; *Eldholm*, 1990] in unusually explosive eruptions [*Knox and Morton*, 1983, 1988; *Eldholm*, 1990]. The hydrothermal activity along the oceanic ridges in the Pacific was at its highest levels for the Cenozoic [*Leinen and Stakes*, 1979; *Owen and Rea*, 1985; *Leinen*, 1987; *Rea et al.*, 1990; *Kyte et al.*, 1992].

Rapid (10^3–10^4 years) environmental change occurred in the oceans, as observed in fluctuations in oxygen and carbon isotopic ratios of planktonic and benthic foraminifera. Carbon and oxygen isotopic values in benthic and planktonic foraminifera decreased by 1 to 2‰; the shift occurred within a few thousand years, and low values persisted for no longer than a few hundred thousand years [*Kennett and Stott*, 1991; *Pak and Miller*, 1992; *Lu and Keller*, 1992; E. Thomas and N. J. Shackleton, unpublished data]. The transient, large isotopic change has now been recognized in the Southern Ocean [*Kennett and Stott*, 1991; E. Thomas and N. J. Shackleton, unpublished data], the Atlantic Ocean [*Pak and Miller*, 1992; E. Thomas and N. J. Shackleton, unpublished data], the Atlantic Ocean [*Pak and Miller*, 1992; E. Thomas and N. J. Shackleton, unpublished data], the Indian Ocean [*Seto et al.*, 1991; *Barrera and Keller*, 1991; *Lu and Keller*, 1992; *Kennett*, 1991; *Thomas et al.*, 1992], and the equatorial Pacific [*Pak and Miller*, 1992] and is thus a global phenomenon. A coeval, large, transient shift to lower $\delta^{13}C$ values has also been observed in carbonate concretions and herbivore tooth enamel in North American land sections, suggesting that there was a major disturbance of the global carbon cycle, atmospheric as well as oceanic [*Koch et al.*, 1992]. The transient isotopic shift is at all oceanic sites coeval with the benthic foraminiferal extinction, suggesting that the extinction was coeval worldwide.

The short-term changes are superimposed on longer-term (10^6 years) changes [*Shackleton and Hall*, 1984, 1990; *Shackleton et al.*, 1984a, b, 1985; *Shackleton*, 1986, 1987; *Rea et al.*, 1990; *Thomas*, 1989, 1990b; *Corfield et al.*, 1991; *McGowran*, 1991; *Hovan and Rea*, 1992]; compare Figures 4–6. A reduction in the intensity of atmospheric circulation during this time was inferred from a sudden decrease of aeolian grain size, coeval with, or slightly before, the benthic foraminiferal extinction [*Rea et al.*, 1990; *Miller et at.*, 1987b; *Hovan and Rea*, 1992]. On land, there were major changes in mammalian faunas [*Butler et al.*, 1981, 1987; *Rea et al.*, 1990; *Koch et al.*, 1992]. Vegetation patterns suggest that the climate was warm and humid on the American continent [*Wolfe*, 1978; *Wing*, 1984; *Schmidt*, 1991; *Wing et al.*, 1991]. Increased abundance of the clay

mineral kaolinite in deep-sea sediment at high southern latitudes suggests a more humid climate over the Antarctic continent [*Robert and Maillot*, 1990; *Robert and Chamley*, 1991; *Robert and Kennett*, 1992].

The benthic foraminiferal extinction at the end of the Paleocene was the only catastrophically sudden, global extinction of bathyal and upper abyssal benthic foraminifera during the last 75 m.y. [*Cushman*, 1946; *Braga et al.*, 1975; *Schnitker*, 1979b; *Tjalsma and Lohmann*, 1983; *Miller et al.*, 1987; *Berggren and Miller*, 1989; *Boltovskoy and Boltovskoy*, 1988, 1989; *Katz and Miller*, 1991; *Mackensen and Berggren*, 1992; *Nomura*, 1991; *Kaiho*, 1988, 1991; *Thomas*, 1989, 1990a, b; *Katz and Miller*, 1991]. There was no bathyal to abyssal benthic foraminiferal mass extinction at the Cretaceous/ Tertiary boundary [*Beckmann*, 1960; *Dailey*, 1983; *Keller*, 1988; *Thomas*, 1990b; *Widmark and Malmgren*, 1992; *Kaiho*, 1992]. Paleocene deep-sea benthic foraminiferal faunas are so similar to Cretaceous faunas that early studies [e.g., *Cushman*, 1946] placed the Paleocene in the Cretaceous. The Paleocene $\delta^{13}C$ values of pelagic carbonates are more similar (heavy) to Cretaceous values than to values during the rest of the Cenozoic [*Shackleton*, 1987; *Corfield et al.*, 1991]. As to the development of the carbon cycle, the major break event in evolution toward the modern world seems to have occurred at the end of the Paleocene, and the major extinction at the end of the Cretaceous could be seen as a "freak accident," resulting from an impacting Apollo object [e.g., *Alvarez*, 1986]. The major environmental change in the latest Paleocene severely affected the deep oceans, which are volumetrically a very large part of the world's environment and of great importance to the heat balance of the ocean-atmosphere system.

At Maud Rise (Weddell Sea, Antarctica), Walvis Ridge (southern Atlantic Ocean), and Kerguelen Plateau (southern Indian Ocean), the extinction occurred slightly later than a decrease in $CaCO_3$ content of the sediments (Figure 6). At sites 525, 527 (Walvis Ridge), and 738 (Kerguelen), this decrease in $CaCO_3$ content from about 80–90% to 25–30% resulted in the presence of a dark brown clay layer across the extinction interval. The decrease in $CaCO_3$ content might reflect increased dissolution and a temporarily raised CCD. Directly after the extinction, both foraminiferal faunas and ostracode faunas are represented by small, thin-walled specimens, suggesting that the CCD did indeed move higher in the water column [*Thomas*, 1990a; P. Steineck, personal communication, 1991]. The decrease in $CaCO_3$ values, however, could also reflect a lowering of primary productivity, as is indicated to have occurred by a decrease in the $\delta^{13}C$ gradient between planktonic and benthic foraminifera [*Kennett and Stott*, 1991], and thus be similar in origin to the clay layers across the Cretaceous/ Tertiary boundary.

A general, oceanwide drop in productivity was thought to have occurred in the latest Paleocene through

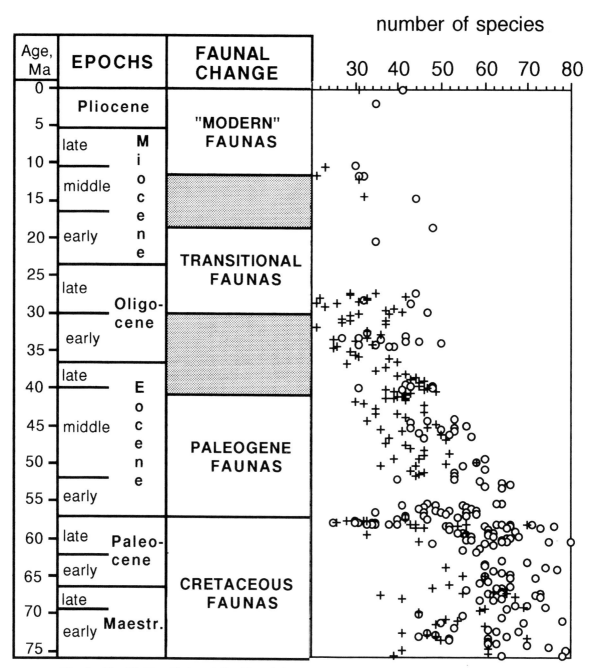

Fig. 5. Simplified benthic foraminiferal faunal changes compared with simple species richness at Site 689 (pluses) and Site 690 (circles) (Maud Rise, Antarctica). The number of species (normalized to 300 specimens) fluctuates strongly, but there is an overall decrease through the middle Eocene. Diversity is usually higher at the deeper site. Note the strong, sudden decrease in diversity at the end of the Paleocene.

earlier Eocene, from biogeographic patterns of planktonic foraminifera and carbon isotopes [*Shackleton et al.*, 1985; *Shackleton*, 1987; *Boersma and Premoli-Silva*, 1991; *Hallock et al.*, 1991]. Preliminary investigation of samples from Exmouth Plateau Site 762 and New Jersey margin Site 605, however, indicates that at these sites the productivity increased instead of decreased (E. Thomas, unpublished data), as inferred from absolute abundance of planktonic foraminifera, as well as from high relative abundance of chiloguembelinids.

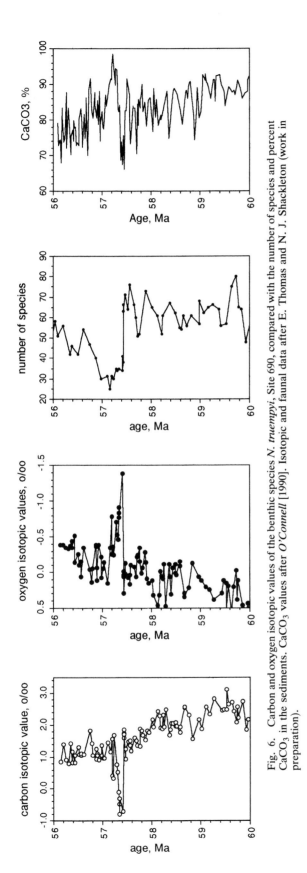

Fig. 6. Carbon and oxygen isotopic values of the benthic species *N. truempyi*, Site 690, compared with the number of species and percent CaCO₃ in the sediments. CaCO₃ values after *O'Connell* [1990]. Isotopic and faunal data after E. Thomas and N. J. Shackleton (work in preparation).

During the latest Paleocene deep-sea benthic foraminiferal extinction the diversity dropped sharply (by about 50%), and many long-lived, cosmopolitan and common species (such as *Gavelinella beccariiformis*) disappeared, as well as several typically Paleocene agglutinant species (e.g., *Tritaxia havanensis*, *Tritaxia paleocenica*, and *Dorothia oxycona*). After the extinction, low-diversity faunas were dominated by small, thin-walled specimens. Several species (e.g., *T. selmensis*) may have immigrated from shallower waters. Diversity never fully reached the levels of the Cretaceous and Paleogene, but it recovered in about 0.5 m.y. [*Berggren and Miller*, 1989; *Berggren et al.*, 1992; *Thomas*, 1989, 1990*a*, *b*, 1991; E. Thomas and N. J. Shackleton, work in preparation] (Figure 6). The faunal patterns after the extinction vary by site: at Atlantic sites the species *N. truempyi* is abundant after the extinction and is accompanied by small species such as *Abyssamina poagi*, *Quadrimorphina profunda*, and *Clinapertina planispira* [*Tjalsma and Lohmann*, 1983; *Miller et al.*, 1987*b*; *Berggren and Miller*, 1989; *Pak and Miller*, 1992; E. Thomas and N. J. Shackleton, work in preparation] (Figure 7). The faunal pattern at Site 762 in the eastern Indian Ocean resembles that at the Atlantic sites, but at Broken Ridge various *Anomalinoides* species dominated, together with *N. truempyi* [*Nomura*, 1991*a*]. At high-latitude sites on Maud Rise (Weddell Sea) and on the Kerguelen Plateau (Indian Ocean) the postextinction faunas are dominated by biserial species, most notably *Tappanina selmensis* [*Thomas*, 1989, 1990*a*, *b*, unpublished data]; at these sites *N. truempyi* disappeared temporarily, to return after about 150,000 years (Figure 7). The increase in the abundance of the biserial species at the high southern latitude sites could have resulted from lower oxygen content of the bottom waters and/or higher nutrient contents. It is difficult to see from where higher nutrient levels could have been derived at these sites, when there was a time of presumably low productivity (see above).

In land sections on Japan and New Zealand (deposited in the upper to middle bathyal zone) the typical Paleocene cosmopolitan taxa disappeared, and in the early Eocene more geographically limited faunas developed: the South Pacific–Atlantic–Tethyan Fauna (in New Zealand, Trinidad, and Italy) and the North Pacific Fauna (Japan and the Pacific coast of the United States [*Kaiho*, 1988].

The Paleocene deep oceans were populated by a cosmopolitan benthic fauna that occurred over a wide depth range, suggesting that a homogeneous water mass was present over large geographic and depth ranges [e.g., *Katz and Miller*, 1991]). After the extinction, more regional differences appeared, possibly indicating that deep and intermediate waters during the early to middle Eocene were more varied in physicochemical character and might have been derived from a number of source regions. This suggestion is in agreement with circulation

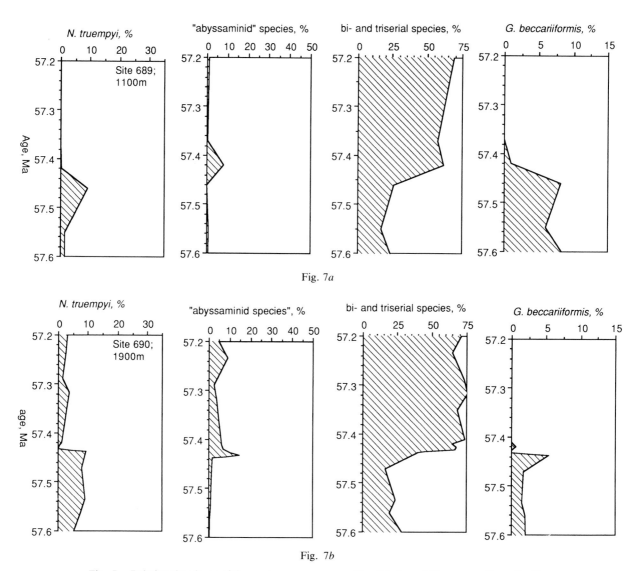

Fig. 7. Relative abundance of the most common groups of benthic foraminifera across the extinction at the end of the Paleocene, for two sites on Maud Rise (Weddell Sea: (*a*) Site 689, paleodepth of 1100 m; (*b*) Site 690, paleodepth of 1900 m) and two sites on Walvis Ridge (southern Atlantic Ocean: (*c*) Site 525, paleodepth of 1600 m; (*d*) Site 527, paleodepth of 3400 m).

models derived from carbon isotopic values [*Katz and Miller*, 1991; *Pak and Miller*, 1992; *Zachos et al.*, 1992*c*].

An explanation of the far-reaching, rapid climate change at the end of the Paleocene, and the benthic foraminiferal extinction, is not yet at hand [*McGowran*, 1991]. If the transient decrease in oxygen isotopic values at the time of extinction is explained wholly by increasing temperatures, there was a whole water column rise in temperature of 6°–7°C at Maud Rise [*Kennett and Stott*, 1990, 1991] (Figure 6). This increase was superimposed on a long-term increase in surface water temperatures at high latitudes, as indicated by the records of oxygen isotope values in bulk carbonate and

planktonic foraminifera [*Stott et al.*, 1991] (Figure 6) and by the penetration of tropical species of planktonic foraminifera and nannofossils to high southern latitudes [*Haq et al.*, 1977; *Kennett*, 1978; *Stott and Kennett*, 1990; *Pospichal and Wise*, 1990; *Boersma and Premoli-Silva*, 1991; *Hallock et al.*, 1991; *Lu and Keller*, 1992; E. Thomas and N. J. Shackleton, work in preparation]. No mass extinctions were documented for planktonic foraminifera and calcareous nannoplankton, but both groups show global high diversity and peak turnover rates [*Haq et al.*, 1977; *Backman*, 1986*a*, *b*; *Corfield and Shackleton*, 1988; *Corfield and Cartlidge*, 1992; *Boersma and Premoli-Silva*, 1991; *Hallock et al.*, 1991; *Ottens and Nederbragt*, 1992; *Lu and Keller*, 1992].

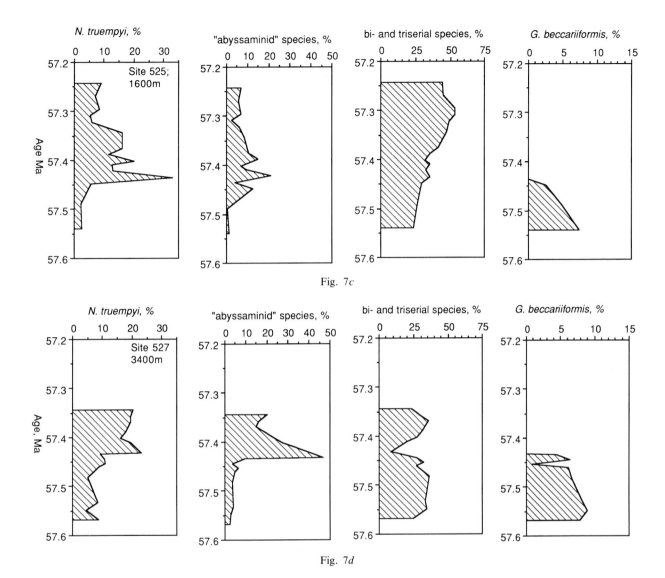

Fig. 7c

Fig. 7d

Calcareous nannoplankton shows extinction in *Fasciculithus* and *Rhomboaster* and radiations in *Discoaster* and *Cruciplacolithus* [*Romein*, 1979; *Backman*, 1986a].

Synchronous with the decrease in oxygen isotopic values there was a decrease of whole-ocean $\delta^{13}C$ values of dissolved carbonate by 1–2‰ (Figure 6; see also *Shackleton* [1987]). This change in carbon isotope value probably reflects at least partially an addition of isotopically light carbon to the oceans as well as the atmosphere (a so-called reservoir effect). It cannot purely be the result of productivity changes, because both surface and deep oceanic values are affected. The source of this isotopically light carbon was probably not continental weathering and erosion, because of the short time scale involved. It might have been the terrestrial biosphere, but there is no evidence for a collapse of terrestrial biota at the time [*Rea et al.*, 1990; *Wolfe*, 1978; *Wing*, 1984]. There is no indication of highly increased rate of erosion

of shelf sediments, because there is no evidence for a precipitous drop in sea level at the time [*Haq et al.*, 1987].

Isotopically light carbon might possibly have been derived from CO_2 emissions related to the massive flood basaltic activity related to the opening of the North Atlantic and the very explosive phase of subaerial volcanism in the latest Paleocene [*Backman et al.*, 1984; *Roberts et al.*, 1984; *Berggren et al.*, 1985; *Aubry et al.*, 1986; *White and MacKenzie*, 1989; *Thomas*, 1991; *Thomas and Varekamp*, 1992]. Additional volcanic input might have been derived from the Pacific region, where there was a strong increase in hydrothermal activity at the end of the Paleocene [*Owen and Rea*, 1985; *Leinen*, 1987; *Leinen and Stakes*, 1979; *Lenotre et al.*, 1985; *Olivarez and Owen*, 1989; *Rea et al.*, 1990].

Speculating about the causes of the benthic foraminiferal extinction, we could hypothesize that a very strong pulse volcanic activity in the northern Atlantic caused high CO_2 levels in atmosphere and ocean [*Eldholm*, 1990], leading to transient global warming, especially at high latitudes [*Manabe and Bryan*, 1985]. The high temperatures at high latitudes could have led to formation of a low-density surface water layer, effectively preventing the formation of deep to intermediate waters at these latitudes. Deepwater circulation could then have changed, to formation dominantly at low latitudes. Such a change in circulation would have caused an overall decrease in oxygen content of the deep to intermediate waters. High relative abundances of biserial and triserial species of benthic foraminifera at high latitudes suggest that the oxygen content of the waters at depths from 1000 to 2000 m decreased at the time of extinction [*Kaiho*, 1988; *Thomas*, 1990*a*, *b*]. This change in oxygen content of deep to intermediate waters specifically at high latitudes could be the result of a combination of (1) lower oxygen contents of the deep to intermediate waters in the source area, as a result of higher temperatures, and (2) the longer route from the source area to high latitudes, and thus a longer "aging" time of the waters, resulting in even more increased levels of CO_2 and nutrients and decreased levels of O_2.

There is no indication in the benthic foraminiferal faunas that oxygen levels in the deep oceans, and especially at the high-latitude sites at Maud Rise in the Weddell Sea, were very low before the latest Paleocene extinction or after the early to earliest middle Eocene [*Thomas*, 1990*a*; *Kaiho*, 1988, 1991]. At Site 689 (1100-m paleodepth) and Site 690 (1900-m paleodepth) on Maud Rise (Antarctica) benthic faunas were very similar during the Eocene, suggesting that a similar water mass bathed both sites. This suggestion conflicts with oxygen isotopic evidence [*Kennett and Stott*, 1990]. At the deeper site, benthic oxygen isotopic values were lower during the Eocene, which has been interpreted as a temperature inversion due to the presence of warm, salty bottom water at Site 690. At present, this discrepancy cannot be explained, but it has been suggested that alternative explanations of the oxygen isotopic records are possible [*Wright and Miller*, 1992].

There is thus no benthic foraminiferal evidence that poorly oxygenated warm salty bottom water persisted in the Atlantic or Pacific oceans after the earliest middle Eocene. Carbon isotopic patterns suggest that the postulated source of warm salty bottom water was active for less than several 100,000 years, at least for waters in the Pacific and Atlantic oceans [*Miller et al.*, 1987*a*; *Katz and Miller*, 1991; *Pak and Miller*, 1992; *Zachos et al.*, 1992*c*].

Benthic faunal changes at the boundary between early and late Paleocene were at a much smaller scale than events at the end of the Paleocene, but they have been recognized at many sites, by many different researchers (Figure 2). These faunal changes were approximately coeval with the strong increase in carbon isotopic values

of benthic and planktonic foraminifera and bulk carbonate (event Y of *Shackleton* [1987]; Figure 4). Presently, there is no clear explanation of this shift in carbon isotopes; it might at least result from increased productivity after recovery from the mass extinction at the end of the Cretaceous. The large isotopic shift, however, occurs in surface as well as deep waters [*Shackleton and Hall*, 1984; *Shackleton et al.*, 1984*a*, *b*, 1985; *Miller et al.*, 1987*b*; *Shackleton*, 1987; *Stott et al.*, 1990], and a reservoir effect must thus be involved. Possibly, the shift reflects the recovery of land biota after the end-Cretaceous extinction and thus storage of more light carbon in the terrestrial biosphere. More high-resolution records are needed over this interval that is commonly poorly recovered in DSDP and ODP drill holes.

The Middle Eocene Through Early Oligocene

The late Eocene was probably a period of cooling and growth of ice caps at high latitudes, as indicated by a relatively rapid increase in oxygen isotopic values of deep-sea benthic foraminifera [e.g., *Savin*, 1977; *Shackleton and Kennett*, 1975; *Kennett*, 1977; *Berger et al.*, 1981; *Corliss et al.*, 1984; *Shackleton*, 1984; *Corliss and Keigwin*, 1986; *Kennett and Stott*, 1990; *Zachos et al.*, 1992*c*]. It was also a period of high extinction rates in planktonic organisms as well as land faunas [*Raup and Sepkoski*, 1986; *Prothero and Berggren*, 1992]. Late Eocene extinctions have been suggested to have been caused by meteorite impact, because the tektites of the North American strewn field are upper Eocene [*Ganapathy*, 1982*a*, *b*; *Glass*, 1982; *Glass et al.*, 1983; *Glass and Zwart*, 1979; *Montanari*, 1990]. Overall, however, there is a poor correlation between the levels with tektites and those with extinctions. The tektites postdate the interval with most numerous extinctions in planktonic oceanic organisms and land fauna, the end of the middle Eocene, and predate the major shift in oxygen isotopic values of deep-sea benthic foraminifera [*Keller et al.*, 1983; *Keller*, 1983*a*, *b*, 1986; *MacLeod*, 1990; *Montanari*, 1990]. There is no short period of catastrophically sudden extinctions of deep-sea benthic foraminifera coeval with the short (100,000 years) oxygen isotopic shift in the earliest Oligocene [*Corliss*, 1981; *Thomas*, 1985, 1992; *Mackensen and Berggren*, 1992]. Benthic as well as planktonic foraminifera [*Keller*, 1983*a*, *b*; *Boersma and Premoli-Silva*, 1991; *Spezzaferri and Premoli-Silva*, 1991] show gradual extinction patterns from the middle middle Eocene on, without clusters of last appearances at the isotopic shift. The same pattern of gradual changes occurred in larger, neritic benthic foraminifera [*Hallock et al.*, 1991; *McGowran*, 1991] and smaller foraminifera in neritic sections [*McGowran*, 1987]. Extinctions occurred over a period of several millions of years from the middle Eocene into the early Oligocene in benthic foraminiferal faunas as well as in ostracode faunas [e.g., *Corliss*, 1981; *Tjalsma and Lohmann*, 1983; *Miller*, 1983; *Miller et al.*, 1984; *Boersma*, 1984, 1985, 1986; *Corliss and Keigwin*, 1986; *Kaiho*, 1988, 1991; *Berggren and Miller*, 1989; *Berggren et al.*, 1992; *Boltovskoy*, 1980; *Boltovskoy and Boltovskoy*, 1988, 1989; *Oberhänsli et al.*, 1991; *Thomas*, 1992] (see reviews in the work of *Douglas and Woodruff* [1981] and *Culver* [1987]).

Benthic foraminiferal faunal change started in the earliest part of the middle Eocene, between zones AB4 and AB5 [*Berggren and Miller*, 1989] and between assemblages 4B and 4A [*Thomas*, 1990a, 1992]; this change may not be worldwide and was certainly not noted by all investigators (Figure 2). It appears to be about coeval with oxygen isotopic event B of *Shackleton* [1986] (Figure 4), the first initiation of the gradual increase in $\delta^{18}O$ values in deep-sea benthic foraminiferal tests. Over this interval there are, however, few high-resolution data available. There is a parallel trend between the deep-sea benthic foraminiferal faunal events and the deep-sea oxygen isotopic record (Figure 5): the diversity of the faunas at Maud Rise sites 689 and 690, for instance, declines in parallel with the increase in oxygen isotopic values (compare Figure 5 with Figure 4).

At most sites the beginning of fast faunal overturn is placed somewhere near the end of the middle Eocene (Figure 2). At this time, there was a general decline in diversity at the high-latitude sites on Maud Rise (Antarctica; Figure 5). The cosmopolitan, very common species *Nuttallides truempyi* started to decline in abundance and migrated from bathyal to lower bathyal and abyssal depths [e.g., *Tjalsma and Lohmann*, 1983; *Miller*, 1983; *Miller et al.*, 1984; *Boersma*, 1984, 1985; *Corliss and Keigwin*, 1986; *Berggren and Miller*, 1989; *Mueller-Merz and Oberhänsli*, 1991; *Oberhänsli et al.*, 1991; *Thomas*, 1992]. The species had its last appearance at middle to lower bathyal depths at the end of the middle Eocene, at the end of the Eocene at lower bathyal to abyssal depths. A common phenomenon at many sites is the decline in relative abundance (to $<10\%$) of buliminid species at lower bathyal to abyssal depths, especially of the larger, heavily calcified species [*Miller*, 1983; *Miller et al.*, 1984; *Boersma*, 1984, 1985; *Thomas*, 1989, 1990a, 1992; *Mueller-Merz and Oberhänsli*, 1991; *Oberhänsli et al.*, 1991].

There does not appear to be a time correlation between these benthic faunal changes and fluctuation in large-scale features of the carbon isotopic record, but *Diester-Haass* [1991] and *Thomas* [1992] argued that there is evidence for increased surface productivity at the timing of the oxygen isotopic shift, as well as at the beginning of the benthic foraminiferal faunal change, at Site 689 (Maud Rise, Antarctica). *Zachos et al.* [1992c] recognized a change in carbon isotopes in Indian Ocean sites, which suggests increased productivity, starting shortly before the oxygen isotopic shift. *Thomas* [1992]

correlated the gradual faunal change at Maud Rise over this period primarily with the gradual decrease in temperatures of the deep waters and the concomitant increase in corrosivity of the waters as a result of increased solubility of CO_2. This increase in corrosivity is obvious at high-latitude sites, where the CCD decreased precipitously [*Barker et al.*, 1988], but at many sites in the Pacific and Atlantic oceans the CCD increased in the earliest Oligocene [*van Andel*, 1975]. Possibly the increased ventilation of the deep ocean, which started at the end of the Eocene [*Miller et al.*, 1987a], resulted in decreased levels of CO_2 in deep waters at middle to low latitudes and thus counteracted the increase in CO_2 levels resulting from a temperature drop at high latitudes. The decrease in the relative abundance of the *Bulimina* species could then be, at least partly, caused by increasing levels of oxygenation of the deep waters. This increased oxygenation could have resulted in more thorough oxidation of organic material, leaving less nutrients available for the benthic foraminiferal faunas.

The Early to Middle Miocene

Major changes occurred in the biosphere and the Earth's climate in the middle Miocene. The oxygen isotopic records indicate that some combination of rapid (100,000 years) ice growth and temperature decline at high latitudes occurred, as at the end of the Eocene [e.g., *Shackleton and Kennett*, 1975; *Kennett and Shackleton*, 1976; *Kennett*, 1977; *Savin*, 1977; *Berger et al.*, 1981; *Douglas and Woodruff*, 1981; *Woodruff and Douglas*, 1981; *Kemp*, 1983; *Savin et al.*, 1985; *Shackleton*, 1984; *Berger and Vincent*, 1985; *Vincent and Killingley*, 1985; *Webb*, 1990]. High-resolution studies have shown that during the Miocene isotopic changes, short-term high $\delta^{13}C$ values appear to occur at times of high $\delta^{18}O$ values, suggesting that periods of cooling and ice buildup were associated with rapid burial of organic material and lowered atmospheric CO_2 levels [*Berger and Vincent*, 1985; *Woodruff and Savin*, 1991]. The middle Miocene has been named as a period of faunal overturn in studies that argue for periodicity in extinction patterns [e.g., *Raup and Sepkoski*, 1986]. At some locations, iridium enrichments have been detected at middle Miocene levels, but no global pattern has been found [*Asaro et al.*, 1988].

Benthic faunal change clearly occurred during this time (about 17–13 Ma) [*Berggren and Miller*, 1989]) and started before the oxygen isotopic increase [*Thomas*, 1985, 1986; *Woodruff*, 1985; *Thomas and Vincent*, 1987, 1988; *Miller and Katz*, 1987; *Boltovskoy and Boltovskoy*, 1988, 1989; *Nomura*, 1991b; *Nomura et al.*, 1992]. The faunal change affected about 20% of the species [*Boltovskoy and Boltovskoy*, 1988; *Thomas*, 1986b]. The widespread species *Cibicidoides wuellerstorfi* and *Pyrgo murrhyna* evolved toward the end of the period of faunal change [e.g., *Thomas*, 1985; *Bolt-*

ovskoy, 1980, 1987; *Boltovskoy and Boltovskoy*, 1988; *Thomas and Vincent*, 1987, 1988; *Woodruff and Savin*, 1991]. Earlier in the period, uniserial species such as nodosariids and pleurostomellids, that had been numerous during the Cretaceous and the Paleogene, decreased in relative abundance, and miliolid species increased in relative abundance in the deep sea [*Boltovskoy and Boltovskoy*, 1988; *Thomas*, 1986]. The environmental significance of these faunal data is not clear.

The offset in timing between oxygen isotopic and benthic faunal change with the benthic faunal change leading cannot result from problems in correlation, because it has been noted at many sites where faunal and isotopic data were obtained from the same cores or even the same samples [*Woodruff*, 1985; *Miller and Katz*, 1987; *Thomas and Vincent*, 1987]. There might be a correlation in time between the initiation of benthic foraminiferal faunal change (about 17 Ma) and the early Miocene (Chron 16) carbon shift (the so-called Monterey event of *Berger and Vincent* [1985, 1986] *Miller and Fairbanks*, 1985; *Thomas and Vincent*, 1987]). The evolution of *C. wuellerstorfi* (15.0–15.6 Ma [*Thomas*, 1985] and 15.3 Ma [*Woodruff and Savin*, 1991]) occurred in the equatorial Pacific, from where the species spread out into the world's oceans fairly late in the period of faunal overturn.

It has been suggested that this faunal overturn was caused by changes in surface-ocean productivity or flux of organic carbon to the ocean floor [*Miller and Katz*, 1987; *Thomas and Vincent*, 1987; *Boltovskoy and Boltovskoy*, 1987]. *Thomas and Vincent* [1987] suggested that faunal overturn was influenced by changes in productivity as well as changes in corrosivity of the deep waters. *Woodruff and Savin* [1989], however, suggested that changing patterns of deepwater circulation, especially the volume of the outflow of salty, high-density waters at intermediate depths from the closing eastern end of the Tethys Ocean, might have been involved. Deep-sea benthic foraminiferal assemblages at DSDP sites 608 and 610 in the northeastern Atlantic suggest that just before the carbon isotopic shift there were episodes of at least local, sluggish circulation leading to poorly oxygenated basins in the North Atlantic [*Thomas*, 1986b]. For a period of about 1 m.y. (19–18 Ma), benthic faunas at these sites were strongly dominated by small, thin-walled bolivinids. Recently, these episodes of bolivinid-dominated faunas have also been recognized at other sites in the northern, southern, and equatorial Atlantic [*Smart*, 1991] but not in the equatorial Pacific [*Thomas*, 1985; *Woodruff*, 1985].

Presently, there is thus no clear, unequivocal correlation between faunal and oxygen or carbon isotopic events in the early to middle Miocene, but there appears to be some correlation between faunal change and changes in productivity.

DISCUSSION AND SPECULATION

In theory, we might expect global deep-sea benthic foraminiferal faunal change to occur as a result of change in source area of the deepwater masses and change in character of the waters in the source areas. This signal is expected to have become complicated by changes in productivity, which in turn might also have been influenced by changing oceanic circulation patterns.

Rapid Faunal Change

Commonly, benthic faunal change occurs rapidly at one site, as one assemblage is replaced by another. This type of faunal change is reversible, and several assemblages may alternate at one site in time [e.g., *Mueller-Merz and Oberhänsli*, 1991; *Nomura*, 1991b; *Oberhänsli et al.*, 1991]. There may appear to be a general correlation in the timing of these faunal changes from one site to another, but in the presence of high-resolution biostratigraphic and/or magnetostratigraphic data the timing proves to differ from site to site. In the absence of local effects such as tectonic changes in water depth at the site, we can interpret these changes most probably as indicating that the depth of boundaries between water masses at the site(s) fluctuated over time [e.g., *Oberhänsli et al.*, 1991], owing to changes in volume of the different water masses. Such an explanation has been proposed for the glacial/interglacial benthic foraminiferal faunal change in the northern Atlantic Ocean [*Streeter*, 1973; *Streeter and Shackleton*, 1979; *Schnitker*, 1974, 1979a].

To illustrate the pattern of faunal change expected from such a change in circulation, compare Figures 8a and 8b. The volume of deep water from the northern high latitude is greater in Figure 8a than in Figure 8b. If Figure 8b type circulation changes to Figure 8a type circulation, faunal change may be expected at site C, but not at other sites. The faunal changes will be rapid, because the water mass boundary passes quickly over each location. We need data on (at least) several sites at different depths, in the same general area, to decipher such motion of water mass boundaries [*Woodruff*, 1985; *Thomas*, 1986a; *Kurihara and Kennett*, 1988]. The faunal changes are diachronous from site to site, because the timing of the passage of water mass boundaries over different sites differs by site.

Gradual Faunal Change

Another type of faunal change occurs over millions of years and can be recognized globally. During the Cenozoic there were at least two times of such faunal change: (1) the late Eocene through early Oligocene and (2) the early middle Miocene through late middle Miocene [e.g., *Corliss*, 1981; *Tjalsma and Lohmann*, 1983; *Miller*, 1983; *Miller et al.*, 1984; *Boersma*, 1984, 1985;

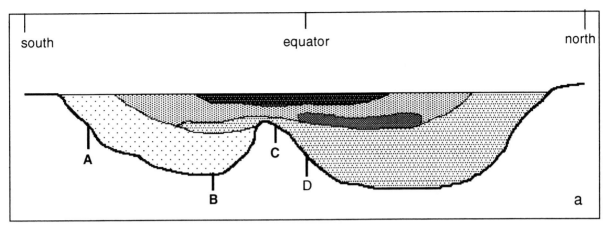

Fig. 8. North-south cross sections across an imaginary ocean to illustrate some of the many possible patterns of deepwater circulation. Letters A through D show possible locations from which deep-sea benthic foraminifera are studied and how these are affected by the changes in deepwater circulation.

Fig. 8a. Deep water formed dominantly at high northern as well as southern latitudes, with waters from the northern source filling the larger part of the ocean; the volume of warm salty deep water is very small (similar to the present Atlantic Ocean).

Corliss and Keigwin, 1986; Berggren and Miller, 1989; Boltovskoy, 1980; Boltovskoy and Boltovskoy, 1988, 1989; Oberhänsli et al., 1991; Douglas and Woodruff, 1981; Culver, 1987; Woodruff and Douglas, 1981; Thomas, 1985, 1986a, b, 1992; Woodruff, 1985; Murray et al., 1986; Schnitker, 1986; Thomas and Vincent, 1987, 1988; Miller and Katz, 1987]. Each of these periods of faunal overturn has been correlated, at least tentatively, with changes in productivity. These gradual changes might be influenced by gradual climate change in the areas of deep or intermediate water formation. During gradual climate change in the source region of the deep water, the surface waters change, and thus the deep

waters that form from these surface waters (e.g., temperature, O_2 content, CO_2 content, and preformed nutrient content) change.

Oxygen isotopic records suggest that deep waters cooled gradually from the middle Eocene on [e.g., *Miller et al.,* 1987a], while benthic faunas suffered gradual turnover at the same time (Figure 5). The benthic faunal overturn can thus be thought to reflect the gradual environmental change. This speculation suggests that at least part of the earliest Oligocene oxygen isotopic shifts must have been related to ice volume increase, because there was no rapid benthic extinction coeval with the isotopic shift. Benthic fora-

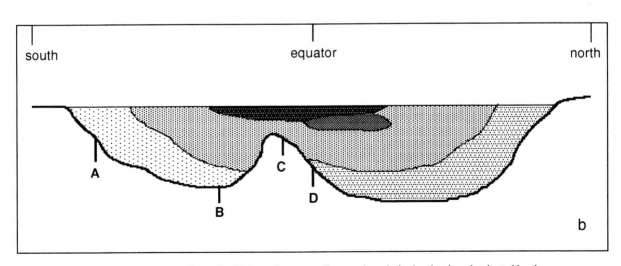

Fig. 8b. Deep water formed at high northern as well as southern latitudes, but less dominated by the northern source (possibly similar to the Atlantic during the last glacial). Note that after a change in circulation from Figure 8a to Figure 8b the faunas at site C are affected, but not those at sites A, B, and D.

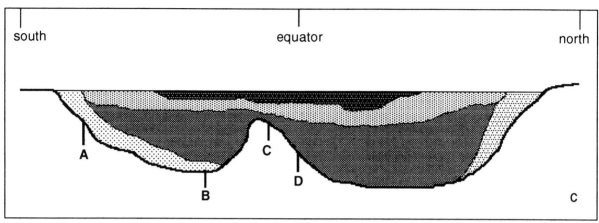

Fig. 8c. A hypothetical ocean in which the role of warm salty bottom water is much larger than that in Figures 8a and 8b, but deepest waters are formed at high latitudes. We can envisage this situation as well as the possibility that the high-latitude waters flow over the warm salty bottom waters (the Proteus Ocean [*Kennett and Stott*, 1990]). A change from Figure 8c to Figure 8a or 8b will be reflected in faunas at sites C and D, but not at sites A and B.

miniferal faunas would "see" the gradual decrease in temperatures leading up to the rapid isotopic shift, but not the increased ice volume on land. The correlation of benthic faunal change and decreasing high-latitude temperatures might be partially indirect and reflect changes in oceanic productivity resulting from high-latitude cooling and changes in the rate of oceanic turnover.

Global Mass Extinctions

This type of faunal change has been recorded only in the latest Paleocene, not at any other time in the Maestrichtian through Recent [*Schnitker*, 1979b; *Tjalsma and Lohmann*, 1983; *Miller et al.*, 1987; *Kaiho*, 1988; *Thomas*, 1989, 1990a, b; Nomura, 1991a]. The

extinction is rapid (10^2–10^3 years) and probably globally synchronous. An explanation for such an oceanwide event is hard to find. We need to speculate that there were either very fast changes in physicochemical character of the surface water in all deepwater source regions or a change in source area of deep waters affecting the circulation patterns of the whole ocean, not just one oceanic basin. After all, the Pleistocene glacial-interglacial circulation changes in the North Atlantic did not result in a mass extinction of deep-sea benthic foraminifera; assemblages appear to have migrated with the water masses [*Streeter*, 1974; *Streeter and Shackleton*, 1979; *Schnitker*, 1974]. The rarity of global extinction suggests that complete turnovers in deepwater

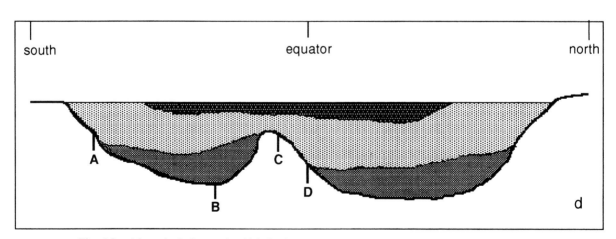

Fig. 8d. A hypothetical ocean in which the deepest parts of the basins are filled with warm salty deep water, but the largest parts of the oceans (including almost the entire depth range above the CCD) are filled with intermediate waters derived from high latitudes. In such an ocean the faunas over wide areas and depth ranges would be very similar (sites A, C, and D are in the same water mass), except for the deep regions of the basins (site B) from which we do not have information.

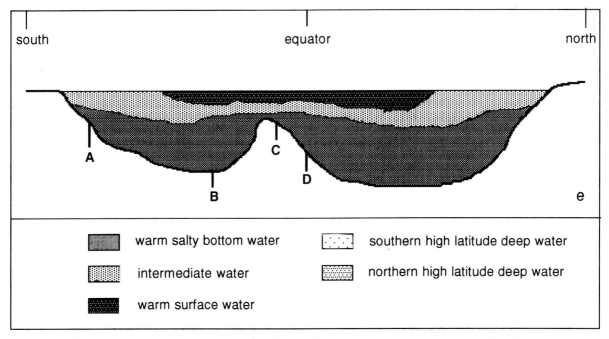

Fig. 8e.   Similar to the proposed ocean in Figure 8d, but the overall volume of the warm salty bottom water is much larger, and the intermediate waters are limited to a thin zone. Note that a change from the situation in Figure 8e to that in Figure 8d (and the reverse) will result in faunal change at sites A, C, and D from which we can obtain information.

circulation during Late Cretaceous through Cenozoic occurred only in the latest Paleocene.

We can envisage several possible configurations of "reversed" circulation in the deep oceans, as indicated in Figures 8c and 8d. A change from the type of oceanic circulation as shown in Figures 8a and 8b to one as shown in Figures 8d and 8e could occur, possibly through an intermediary stage similar to that in Figure 8c [see also *Kennett and Stott*, 1990]. Such large-scale circulation changes might affect faunas at all locations (hypothetical sites A through D). Faunal change might be exacerbated because of changes in locations of upwelling and in nutrient content of upwelled water and thus of the location of areas of high surface productivity.

Such a reversal in circulation of the deep to intermediate waters might have been triggered by the warming of surface waters at high latitudes (as indicated by oxygen isotopic as well as faunal data), resulting in high-latitude surface waters with a low density. This could have been aggravated by an increasingly humid climate and more precipitation at high latitudes, as indicated by changes in clay mineral assemblages at Maud Rise [*Robert and Maillot*, 1990; *Robert and Chamley*, 1991; *Robert and Kennett*, 1992]. Such a reversed circulation could have influenced upwelling patterns and nutrient availability to the plankton and thus plankton evolutionary patterns.

CONCLUSIONS

Benthic foraminiferal faunal data do not support the hypothesis that "warm salty bottom water" existed over large parts of the oceans during most of the Late Cretaceous and the Paleogene. They do support the possibility that such waters were dominant in the Atlantic and Pacific during one or a few such episodes during the early to earliest middle Eocene. The initiation of such an event of warm salty deepwater dominance in the oceans might have caused the extinction in the latest Paleocene. Benthic foraminiferal faunal data suggest that at least parts of the oxygen isotopic shifts to heavier values in the earliest Oligocene and the middle Miocene represent the buildup of ice volume.

Acknowledgments. I thank many colleagues for discussions over the last several years, including Enriqueta Barrera, Gerta Keller, Ken Miller, Mimi Katz, Edith Mueller-Merz, Hedi Oberhänsli, Nick Shackleton, and Jim Zachos. Constructive reviews by Fay Woodruff and Mimi Katz helped improve the paper. This research was partly supported by a grant from NSF-USSAC. This paper is a contribution to IUGS/IGCP project 308. This is Department of Earth Sciences contribution 2317.

REFERENCES

Adams, C. G., D. E. Lee, and B. R. Rosen, Conflicting isotopic and biotic evidence for tropical sea-surface temperature during the Tertiary, *Palaeogeogr. Palaeoclimatol. Palaeo-ecol.*, *77*, 289–313, 1990.

Alvarez, W., Toward a theory of impact crises, *Eos Trans. AGU*, *67*, 649–658, 1986.

Asaro, F., W. Alvarez, H. V. Michel, L. W. Alvarez, M. H. Anders, A. Montanari, and J. P. Kennett, Possible world-wide middle Miocene iridium anomaly and its relationship to periodicity of impacts and extinctions (abstract), in *Global Catastrophes in Earth History: Conference on Impacts, Volcanism and Mass Mortality, Abstracts Volume*, pp. 6–7, Lunar and Planetary Institute, Houston, Tex., 1988.

Aubry, M.-P., E. A. Hailwood, and H. A. Townsend, Magnetic and calcareous nannofossil stratigraphy of lower Paleogene formations of the Hampshire and London basins, *J. Geol. Soc. London*, *143*, 729–735, 1986.

Aubry, M.-P., W. A. Berggren, D. V. Kent, J. J. Flynn, K. D. Klitgord, J. D. Obradovich, and D. R. Prothero, Paleogene chronology: An integrated approach, *Paleoceanography*, *3*, 707–742, 1988.

Backman, J., Late Paleocene to middle Eocene calcareous nannofossil biochronology from the Shatsky Rise, Walvis Ridge and Italy, *Palaeogeogr. Palaeoclimatol. Palaeoecol.*, *57*, 43–59, 1986*a*.

Backman, J., Accumulation patterns of Tertiary calcareous nannofossils around extinctions, *Geol. Rundsch.*, *75*, 185–196, 1986*b*.

Backman, J., A. C. Morton, D. G. Roberts, S. Brown, K. Krumsieck, and R. M. MacIntyre, Geochronology of the lower Eocene and upper Paleocene sequences of Leg 81, *Initial Rep. Deep Sea Drill. Proj.*, *81*, 877–882, 1984.

Barker, P. F., et al., Leg 113, *Proc. Ocean Drill. Program Initial Rep.*, *113*, 785 pp., 1988.

Barrera, E., and G. Keller, Late Paleocene to early Eocene climatic and oceanographic events in the Antarctic Indian Ocean (abstract), *Geol. Soc. Am. Abstr. Programs*, *23*(5), A179, 1991.

Barrera, E., B. Huber, S. M. Savin, and P. N. Webb, Antarctic marine temperatures: Late Campanian through early Paleocene, *Paleoceanography*, *2*, 21–48, 1987.

Barron, E. J., Explanations of the Tertiary global cooling trend, *Palaeogeogr. Palaeoclimatol. Palaeoecol.*, *50*, 45–61, 1985.

Barron, E. J., Eocene equator-to-pole surface ocean temperatures: A significant climate problem?, *Paleoceanography*, *2*, 729–740, 1987.

Barron, E. J., and W. H. Peterson, The Cenozoic ocean circulation based on ocean general circulation model results, *Palaeogeogr. Palaeoclimatol. Palaeoecol.*, *83*, 1–18, 1991.

Barron, J. A., and J. G. Baldauf, Tertiary cooling steps and paleoproductivity as reflected by diatoms and biosiliceous sediments, in *Productivity of the Oceans: Past and Present*, edited by W. H. Berger, V. S. Smetacek, and G. Wefer, pp. 341–345, John Wiley, New York, 1989.

Barron, J. A., B. L. Larsen, and J. Baldauf, Evidence for late Eocene–early Oligocene Antarctic glaciation and observations on late Neogene glacial history of Antarctica: Results from ODP Leg 119, *Proc. Ocean Drill. Program Sci. Results*, *119*, 869–891, 1991.

Beckmann, J.-P., Distribution of benthonic foraminifera at the Cretaceous-Tertiary boundary of Trinidad (West Indies), in *Report of the 21st Session, Norden, Part 5: The Cretaceous-Tertiary Boundary*, pp. 57–69, International Geological Congress, Norden, Denmark, 1960.

Benson, R. H., The origin of the psychrosphere as recorded in changes of deep-sea ostracode assemblages, *Lethaia*, *8*, 69–83, 1975.

Berger, W. H., Paleoceanography: The deep sea record, in *The Sea*, vol. 7, *The Oceanic Lithosphere*, edited by C. Emíliani, pp. 1437–1519, John Wiley, New York, 1981.

Berger, W. H., and E. Vincent, Carbon dioxide and polar cooling in the Miocene: The Monterey hypothesis, in *The Carbon Cycle and Atmospheric CO_2: Natural Variations Archean to Present*, Geophys. Monogr. Ser., vol. 32, edited by E. T. Sundquist and W. S. Broecker, pp. 455–468, AGU, Washington, D. C., 1985.

Berger, W. H., and E. Vincent, Deep-sea carbonates: Reading the carbon isotope signal, *Geol. Rundsch.*, *75*, 249–269, 1986.

Berger, W. H., E. Vincent, and H. Thierstein, The deep-sea record: Major steps in Cenozoic ocean evolution, *Spec. Publ. Soc. Econ. Paleontol. Mineral.*, *32*, 489–504, 1981.

Berggren, W. A., and K. G. Miller, Cenozoic bathyal and abyssal calcareous benthic foraminiferal zonation, *Micropaleontology*, *35*, 308–320, 1989.

Berggren, W. A., D. V. Kent, J. J. Flynn, and J. A. Van Couvering, Cenozoic geochronology, *Geol. Soc. Am. Bull.*, *96*, 1407–1418, 1985.

Berggren, W. A., M. E. Katz, and K. G. Miller, Cenozoic deep-sea benthic foraminifera: A tale of three turnovers, in *Fourth International Symposium on Benthic Foraminifera*, edited by T. Saito, Yamagata University, Yamagata, Japan, in press, 1992.

Bernard, J. M., Characteristic assemblages and morphologies of benthic foraminifera, from anoxic, organic-rich deposits: Jurassic through Holocene, *J. Foraminiferal Res.*, *16*, 207–215, 1986.

Boersma, A., Oligocene and other Tertiary benthic foraminifera from a depth traverse down Walvis Ridge, Deep Sea Drilling Project Leg 74, *Initial Rep. Deep Sea Drill. Proj.*, *75*, 1273–1300, 1984.

Boersma, A., Oligocene benthic foraminifers from North Atlantic sites: Benthic foraminifers as water-mass indexes in the North and South Atlantic, *Initial Rep. Deep Sea Drill. Proj.*, *82*, 611–628, 1985.

Boersma, A., Eocene-Oligocene Atlantic paleo-oceanography, using benthic foraminifera, in *Terminal Eocene Events*, edited by C. Pomerol and I. Premoli-Silva, pp. 225–236, Elsevier, New York, 1986.

Boersma, A., and I. Premoli-Silva, Distribution of Paleogene planktonic foraminifera—Analogies with the Recent?, *Palaeogeogr. Palaeoclimatol. Palaeoecol.*, *83*, 29–48, 1991.

Boersma, A., I. Premoli-Silva, and N. J. Shackleton, Atlantic Eocene planktonic foraminiferal paleohydrographic indicators and stable isotope stratigraphy, *Paleoceanography*, *2*, 287–331, 1987.

Boltovskoy, E., On the benthonic bathyal-abyssal foraminifera as stratigraphic guide to fossils, *J. Foraminiferal Res.*, *10*, 163–172, 1980.

Boltovskoy, E., Tertiary benthic foraminifera in bathyal deposits of the Quaternary world ocean, *J. Foraminiferal Res.*, *17*, 279–285, 1987.

Boltovskoy, E., and D. Boltovskoy, Cenozoic deep-sea benthic foraminifera: Faunal turnovers and paleobiographic differences, *Rev. Micropaleontol.*, *31*, 67–84, 1988.

Boltovskoy, E., and D. Boltovskoy, Paleocene-Pleistocene benthic foraminiferal evidence of major paleoceanographic events in the eastern South Atlantic (DSDP Site 525, Walvis Ridge), *Mar. Micropaleontol.*, *14*, 283–316, 1989.

Braga, G., R. deBiase, A. Grunig, and F. Proto-Decima, Foraminiferi bentonici del Paleocene e dell'Eocene della Sezione Posagno, *Schweiz. Paleontol. Abh.*, *97*, 85–111, 1975.

Brass, G. W., J. R. Southam, and W. H. Peterson, Warm saline bottom water in the ancient ocean, *Nature*, *296*, 620–623, 1982.

Bremer, M., and G. P. Lohmann, Evidence for primary control of the distribution of certain Atlantic Ocean benthonic foraminifera by degree of carbonate saturation, *Deep Sea Res.*, *29*, 987–998, 1982.

Broecker, W. S., and T.-H. Peng, *Tracers in the Sea*, 690 pp., Eldigio Press, Palisades, N. Y., 1982.

Butler, R. F., P. Gingerich, and E. H. Lindsay, Magnetic polarity stratigraphy and biostratigraphy of Paleocene and lower Eocene continental deposits, Clark's Fork Basin, Wyoming, *J. Geol.*, *89*, 299–316, 1981.

Butler, R. F., D. Krause, and P. Gingerich, Magnetic polarity stratigraphy and biostratigraphy of middle-late Paleocene continental deposits of south-central Montana, *J. Geol.*, *95*, 647–657, 1987.

Caralp, M. H., Impact de la matière organique dans les zones de forte productivité sur certains foraminiferes benthiques, *Oceanol. Acta*, *7*, 509–515, 1984.

Chamberlin, T. C., On a possible reversal of deep-sea circulation and its influence on geologic climates, *J. Geol.*, *14*, 363–373, 1906.

Comiso, J. C., and A. L. Gordon, Recurring polynyas over the Cosmonaut Sea and the Maud Rise, *J. Geophys. Res.*, *92*, 2819–2833, 1987.

Corfield, R. M., and J. Cartlidge, Isotopic evidence for the depth stratification of fossil and recent Globigerinina: A review, *Hist. Biol.*, *5*, 37–63, 1992.

Corfield, R. M., and N. J. Shackleton, Productivity change as a control on planktonic foraminiferal evolution after the Cretaceous/Tertiary boundary, *Hist. Biol.*, *1*, 323–343, 1988.

Corfield, R. M., J. E. Cartlidge, I. Premoli-Silva, and R. A. Housley, Oxygen and carbon isotope stratigraphy of the Paleogene and Cretaceous limestones in the Bottacione Gorge and the Contessa Highway sections, Umbria, Italy, *Terra Nova*, *3*, 414–422, 1991.

Corliss, B. H., Deep-sea benthonic foraminiferal faunal turnover near the Eocene/Oligocene boundary, *Mar. Micropaleontol.*, *6*, 367–384, 1981.

Corliss, B. H., Microhabitats of benthic foraminifera within deep-sea sediments, *Nature*, *314*, 435–438, 1985.

Corliss, B. H., and C. Chen, Morphotype patterns of Norwegian deep-sea benthic foraminifera and ecological implications, *Geology*, *16*, 716–719, 1988.

Corliss, B. H., and S. Emerson, Distribution of Rose Bengal stained deep-sea benthic foraminifera from the Nova Scotian continental margin and Gulf of Maine, *Deep Sea Res.*, *37*, 381–400, 1990.

Corliss, B. H., and L. D. Keigwin, Jr., Eocene-Oligocene paleoceanography, in *Mesozoic and Cenozoic Oceans*, *Geodyn. Ser.*, vol. 15, edited by K. Hsu, pp. 101–118, AGU, Washington, D. C., 1986.

Corliss, B. H., M.-P. Aubry, W. A. Berggren, J. M. Fenner, L. D. Keigwin, Jr., and G. Keller, The Eocene/Oligocene boundary in the deep sea, *Science*, *226*, 806–810, 1984.

Culver, S. J., Foraminifera, in *Fossil Prokaryotes and Protists*, *Stud. in Geol.*, vol. 18, edited by J. R. Lipps, pp. 169–212, University of Tennessee, Knoxville, 1987.

Cushman, J. A., Upper Cretaceous foraminifera of the Gulf Coastal region of the United States and adjacent areas, *U.S. Geol. Surv. Prof. Pap.*, *206*, 241 pp., 1946.

Dailey, D. H., Late Cretaceous and Paleocene benthic foraminifera from DSDP Site 516, Rio Grande Rise, western South Atlantic, *Initial Rep. Deep Sea Drill. Proj.*, *74*, 757–782, 1983.

Diester-Haass, L., Eocene/Oligocene paleoceanography in the Antarctic Ocean, Atlantic sector (Maud Rise, ODP Leg 113, sites 689B and 690B), *Mar. Geol.*, *100*, 249–276, 1991.

Douglas, R. G., and F. Woodruff, Deep sea benthic foraminifera, in *The Sea*, vol. 7, *The Oceanic Lithosphere*, edited by C. Emiliani, pp. 1233–1327, John Wiley, New York, 1981.

Eldholm, O., Paleogene North Atlantic magmatic-tectonic

events: Environmental implications, *Mem. Soc. Geol. Ital.*, *44*, 13–28, 1990.

Ganapathy, R., Evidence for a major meteorite impact on the Earth 34 million years ago: Implication for Eocene extinctions, *Science*, *216*, 885–886, 1982*a*.

Ganapathy, R., Evidence for a major meteorite impact on the Earth 34 million years ago: Implication for the origin of North American tektites and Eocene extinctions, *Geol. Soc. Am. Spec. Publ.*, *190*, 513–516, 1982*b*.

Glass, B. P., Possible correlations between tektite events and climatic changes?, *Geol. Soc. Am. Spec. Publ.*, *190*, 251–256, 1982.

Glass, B. P., and M. J. Zwart, North American microtektites in Deep Sea Drilling Project cores from the Caribbean Sea and Gulf of Mexico, *Geol. Soc. Am. Bull.*, *90*, 595–602, 1979.

Glass, B. P., R. N. Baker, D. Storzer, and G. A. Wagner, North American microtektites from the Caribbean Sea and their fission track ages, *Earth Planet. Sci. Lett.*, *19*, 184–192, 1983.

Goldfarb, R. J., L. W. Snee, L. D. Miller, and R. J. Newberry, Rapid dewatering of the crust deduced from ages of mesothermal gold deposits, *Nature*, *354*, 296–298, 1991.

Gooday, A. J., A response by benthic foraminifera to the deposition of phytodetritus in the deep sea, *Nature*, *332*, 70–73, 1988.

Gooday, A. J., and P. J. D. Lambshead, Influence of seasonally deposited phytodetritus on benthic foraminiferal populations in the bathyal northeast Atlantic: The species response, *Mar. Ecol. Prog. Ser.*, *58*, 53–67, 1989.

Gooday, A. J., and C. M. Turley, Responses by benthic organisms to inputs of organic material to the ocean floor: A review, *Philos. Trans. R. Soc. London, Ser. A*, *331*, 119–138, 1990.

Gooday, A. J., L. A. Levin, P. Linke, and T. Heeger, The role of benthic foraminifera in deep-sea food webs and carbon cycling, in G. T. Rowe and V. Pariente, eds., *Deep-Sea Food Chains and the Global Carbon Cycle*, edited by G. T. Rowe and V. Pariente, pp. 63–91, Kluwer Academic Publishers, Dordrecht, Netherlands, 1992.

Hallock, P., I. Premoli-Silva, and A. Boersma, Similarities between planktonic and larger foraminiferal evolutionary trends through Paleogene paleoceanographic changes, *Palaeogeogr. Palaeoclimatol. Palaeoecol.*, *83*, 49–64, 1991.

Haq, B. U., I. Premoli-Silva, and G. P. Lohmann, Calcareous plankton biogeographic evidence for major climatic fluctuations in the early Cenozoic Atlantic Ocean, *J. Geophys. Res.*, *82*, 3861–3876, 1977.

Haq, B. U., J. Hardenbol, and P. R. Vail, The chronology of fluctuation of sealevel since the Triassic, *Science*, *235*, 1156–1167, 1987.

Hay, W. W., Paleoceanography: A review for GSA centennial, *Geol. Soc. Am. Bull.*, *100*, 1934–1956, 1989.

Herbert, T. D., and J. L. Sarmiento, Ocean nutrient distribution and oxygenation: Limits on the formation of warm saline bottom water over the past 91 m.y., *Geology*, *19*, 702–705, 1991.

Hovan, S. A., and D. K. Rea, Paleocene/Eocene boundary changes in atmospheric and oceanic circulation: A southern hemisphere record, *Geology*, *20*, 15–18, 1992.

Kaiho, K., Uppermost Cretaceous to Paleogene bathyal benthic foraminiferal biostratigraphy of Japan and New Zealand; latest Paleocene–middle Eocene benthic foraminiferal species turnover, *Rev. Paleobiol. Spec. Vol.*, *2*, 553–559, 1988.

Kaiho, K., Global changes of Paleogene aerobic/anaerobic benthic foraminifera and deep-sea circulation, *Palaeogeogr. Palaeoclimatol. Palaeoecol.*, *83*, 65–86, 1991.

Kaiho, K., A low extinction rate of intermediate-water benthic foraminifera at the Cretaceous/Tertiary boundary, *Mar. Micropaleontol.*, *18*, 229–259, 1992.

Katz, M. R., and K. G. Miller, Early Paleogene benthic foraminiferal assemblage and stable isotope composition in the Southern Ocean, ODP Leg 114, *Proc. Ocean Drill. Program Sci. Results*, *114*, 481–513, 1991.

Keigwin, L. D., Jr., and B. H. Corliss, Stable isotopes in late middle Eocene through Oligocene foraminifera, *Geol. Soc. Am. Bull.*, *97*, 335–345, 1986.

Keigwin, L. D., Jr., and G. Keller, Middle Oligocene cooling from equatorial Pacific DSDP Site 77B, *Geology*, *12*, 16–19, 1984.

Keith, M. L., Violent volcanism, stagnant oceans and some inferences regarding petroleum, strata-bound ores and mass extinctions, *Geochim. Cosmochim. Acta*, *46*, 2621–2637, 1982.

Keller, G., Biochronology and paleoclimatic implications on middle Eocene through Oligocene planktic foraminiferal faunas, *Mar. Micropaleontol.*, *7*, 474–486, 1983*a*.

Keller, G., Paleoclimatic analysis of middle Eocene through Oligocene planktic foraminiferal faunas, *Palaeogeogr. Palaeoclimatol. Palaeoecol.*, *49*, 73–94, 1983*b*.

Keller, G., Stepwise mass extinctions and impact events: Late Eocene to early Oligocene, *Mar. Micropaleontol.*, *10*, 267–293, 1986.

Keller, G., Biotic turnover among benthic foraminifera across the Cretaceous/Tertiary boundary, El Kef, Tunisia, *Palaeogeogr. Palaeoclimatol. Palaeoecol.*, *66*, 153–171, 1988.

Keller, G., S. L. D'Hondt, and T. Vallier, Multiple microtektite horizons in upper Eocene marine sediments: No evidence for mass extinctions, *Science*, *221*, 150–152, 1983.

Kemp. E. M., Tertiary climatic evolution and vegetation history in the southeast Indian Ocean region, *Palaeogeogr. Palaeoclimatol. Palaeoecol.*, *24*, 169–208, 1978.

Kennett, J. P., Cenozoic evolution of Antarctic glaciation, the circum–Antarctic Ocean, and their impact on global paleoceanography, *J. Geophys. Res.*, *82*, 3843–3860, 1977.

Kennett, J. P., The development of planktonic biogeography in the Southern Ocean during the Cenozoic, *Mar. Micropaleontol.*, *3*, 301–345, 1978.

Kennett, J. P., Paleoceanographic changes at the Paleocene/Eocene boundary in mid-latitude ODP Site 762C, Exmouth Plateau, NW Australia (abstract), *Geol. Soc. Am. Abstr. Programs*, *23*(5), A338, 1991.

Kennett, J. P., and P. F. Barker, Latest Cretaceous to Cenozoic climate and oceanographic developments in the Weddell Sea, Antarctica: An ocean-drilling perspective, *Proc. Ocean Drill. Program Sci. Results*, *113*, 937–962, 1990.

Kennett, J. P., and N. J. Shackleton, Oxygen isotope evidence for the development of the psychrosphere 38 myr. ago, *Nature*, *260*, 513–515, 1976.

Kennett, J. P., and L. D. Stott, Proteus and Proto-Oceanus: Ancestral Paleogene oceans as revealed from Antarctic stable isotopic results, *Proc. Ocean Drill. Program Sci. Results*, *113*, 865–880, 1990.

Kennett, J. P., and L. D. Stott, Terminal Paleocene deep-sea benthic crisis: Sharp deep sea warming and paleoceanographic changes in Antarctica, *Nature*, *353*, 225–229, 1991.

Klootwijk, C. T., J. S. Gee, J. W. Peirce, and G. M. Smith, Constraints on the India-Asia convergence: Paleomagnetic results from Ninetyeast Ridge, *Proc. Ocean Drill. Program Sci. Results*, *121*, 777–882, 1991.

Knox, R. W. O., and A. C. Morton, Stratigraphical distribution of early Palaeogene pyroclastic deposits in the North Sea Basin, *Proc. Yorks. Geol. Soc.*, *44*, 355–363, 1983.

Knox, R. W. O., and A. C. Morton, The record of early Tertiary N. Atlantic volcanism in sediments of the North Sea Basin, *Geol. Soc. Spec. Publ. London*, *39*, 407–420, 1988.

Koch, P. L., J. C. Zachos, and P. D. Gingerich, Coupled isotopic change in marine and continental carbon reservoirs at the Palaeocene/Eocene boundary, *Nature*, *358*, 319–322, 1992.

Kurihara, K., and J. P. Kennett, Bathymetric migration of deep-sea benthic foraminifera in the southwest Pacific during the Neogene, *J. Foraminiferal Res.*, *18*, 75–83, 1988.

Kyte, F. T., M. Leinen, G. Ross Heath, and L. Zhou, Elemental geochemistry of Core LL44-GPC3 and a model for the Cenozoic sedimentation history of the central North Pacific, *Geochim. Cosmochim. Acta*, in press, 1992.

Lambshead, P. J. D., and A. J. Gooday, The impact of seasonally deposited phytodetritus on epifaunal and shallow infaunal benthic foraminiferal populations in the bathyal northeast Atlantic: The assemblage response, *Deep Sea Res.*, *37*, 1263–1283, 1990.

Leinen, M., The origin of paleochemical signatures in North Pacific pelagic clays: Partitioning experiments, *Geochim. Cosmochim. Acta*, *51*, 305–319, 1987.

Leinen, M., and D. Stakes, Metal accumulation rates in the central equatorial Pacific during Cenozoic times, *Geol. Soc. Am. Bull.*, *90*, 357–375, 1979.

Lenotre, N., H. Chamley, and M. Hoffert, Clay stratigraphy at DSDP sites 576 and 578, DSDP Leg 86 (western North Pacific), *Initial Rep. Deep Sea Drill. Proj.*, *86*, 605–646, 1985.

Lohmann, G. P., Abyssal benthonic foraminifera as hydrographic indicators in the western South Atlantic Ocean, *J. Foraminiferal Res.*, *8*, 6–34, 1978.

Lu, G., and G. Keller, Climatic and oceanographic events across the Paleocene-Eocene transition in the Antarctic Indian Ocean: Inference from planktic foraminifera, *Mar. Micropaleontol.*, in press, 1992.

Lutze, G. F., and W. T. Coulbourn, Recent benthic foraminifera from the continental margin of northwest Africa: Community structure and distribution, *Mar. Micropaleontol.*, *8*, 361–401, 1984.

Mackensen, A., and W. A. Berggren, Paleogene benthic foraminifers from the southern Indian Ocean (Kerguelen Plateau): Biostratigraphy and paleoecology, *Proc. Ocean Drill. Program Sci. Results*, *120*, 603–630, 1992.

MacLeod, N., Effects of late Eocene impacts on planktonic foraminifera, *Geol. Soc. Am. Spec. Publ.*, *247*, 595–606, 1990.

Manabe, S., and K. Bryan, CO_2-induced change in a coupled ocean-atmosphere system and its paleoclimatic implications, *J. Geophys. Res.*, *90*, 11,689–11,707, 1985.

Matthews, R. K., and R. Z. Poore, Tertiary $d^{18}O$ record and glacioeustatic sea-level fluctuations, *Geology*, *8*, 501–504, 1980.

McGowran, B., Late Eocene perturbations: Foraminiferal biofacies and evolutionary overturn, southern Australia, *Paleoceanography*, *2*, 715–727, 1987.

McGowran, B., Silica burp in the Eocene ocean, *Geology*, *17*, 857–860, 1989.

McGowran, B., Fifty million years ago, *Am. Sci.*, *78*, 30–39, 1990.

McGowran, B., Evolution and environment in the early Paleogene, *Mem. Geol. Soc. India*, *20*, 21–53, 1991.

Mead, G. A., and D. A. Hodell, Late Eocene to early Oligocene vertical oxygen isotope gradients in the South Atlantic: Implications for warm saline deep water, *Paleoceanography*, in press, 1992.

Mercer, J. H., Cenozoic glaciation in the southern hemisphere, *Annu. Rev. Earth Planet. Sci.*, *11*, 99–132, 1982.

Miller, K. G., Eocene-Oligocene paleoceanography in the deep Bay of Biscay, *Mar. Micropaleontol.*, *7*, 403–440, 1983.

Miller, K. G., and R. G. Fairbanks, Oligocene-Miocene global carbon and abyssal circulation changes, in *The Carbon Cycle and Atmospheric CO_2: Natural Variations Archean to*

Present, Geophys. Monogr. Ser., vol. 32, edited by E. Sundquist and W. S. Broecker, pp. 469–486, AGU, Washington, D. C., 1985.

Miller, K. G., and M. Katz, Oligocene to Miocene benthic foraminiferal and abyssal circulation changes in the North Atlantic, *Micropaleontology*, *33*, 97–149, 1987.

Miller, K. G., and E. Thomas, Late Eocene to Oligocene benthic foraminiferal isotopic record, Site 574 Equatorial Pacific, *Initial Rep. Deep Sea Drill. Proj.*, *85*, 771–777, 1985.

Miller, K. G., W. B. Curry, and D. R. Ostermann, Late Paleogene (Eocene to Oligocene) benthic foraminiferal oceanography of the Goban Spur region DSDP Leg 80, *Initial Rep. Deep Sea Drill. Proj.*, *80*, 505–538, 1984.

Miller, K. G., R. G. Fairbanks, and G. S. Mountain, Tertiary isotope synthesis, sea level history, and continental margin erosion, *Paleoceanography*, *2*, 1–20, 1987a.

Miller, K. G., T. R. Janecek, M. R. Katz, and D. J. Keil, Abyssal circulation and benthic foraminiferal changes near the Paleocene/Eocene boundary, *Paleoceanography*, *2*, 741–761, 1987b.

Montanari, A., Geochronology of the terminal Eocene impacts: An update, *Geol. Soc. Am. Spec. Publ.*, *247*, 607–616, 1990.

Mueller-Merz, E., and H. Oberhaensli, Eocene bathyal and abyssal benthic foraminifera from a South Atlantic transect at 20°–30°S, *Palaeogeogr. Palaeoclimatol. Palaeoecol.*, *83*, 117–172, 1991.

Murray, J. W., J. F. Weston, C. A. Haddon, and A. D. J. Powell, Miocene to Recent bottom water masses of the northeast Atlantic: An analysis of benthic foraminifera, in *North Atlantic Palaeoceanography, Geol. Soc. Spec. Publ. London*, *21*, 219–230, 1986.

Niensted, J. C., and A. J. Arnold, The distribution of benthic foraminifera on seamounts near the East Pacific Rise, *J. Foraminiferal Res.*, *18*, 237–249, 1988.

Nomura, R., Paleoceanography of upper Maestrichtian to Eocene benthic foraminiferal assemblages at ODP sites 752, 753 and 754, eastern Indian Ocean, *Proc. Ocean Drill. Program Sci. Results*, *121*, 3–30, 1991a.

Nomura, R., Oligocene to Pleistocene benthic foraminifer assemblages at sites 754 and 756, eastern Indian Ocean, *Proc. Ocean Drill. Program Sci. Results*, *121*, 31–76, 1991b.

Nomura, R., K. Seto, and N. Niitsuma, Late Cenozoic deep-sea benthic foraminiferal changes and isotopic records in the eastern Indian Ocean, in *Fourth International Symposium on Benthic Foraminifera*, edited by T. Saito, Yamagata University, Yamagata, Japan, in press, 1992.

Oberhänsli, H., E. Mueller-Merz, and R. Oberhänsli, Eocene paleoceanographic evolution at 20°–30°S in the Atlantic Ocean, *Palaeogeogr. Palaeoclimatol. Palaeoecol.*, *83*, 173–216, 1991.

O'Connell, S., Variations in Upper Cretaceous and Cenozoic calcium carbonate percentages, Maud Rise, Weddell Sea, Antarctica, *Proc. Ocean Drill. Program Sci. Results*, *113*, 971–984, 1990.

Olivarez, A. M., and R. M. Owen, Plate tectonic reorganizations: Implications regarding the formation of hydrothermal ore deposits, *Mar. Min.*, *14*, 123–138, 1989.

Ottens, J. J., and A. J. Nederbragt, Planktic foraminiferal diversity as indicator of ocean environments, *Mar. Micropaleontol.*, *19*, 13–28, 1992.

Owen, R. M., and D. K. Rea, Sea floor hydrothermal activity links climate to tectonics: The Eocene CO_2 greenhouse, *Science*, *227*, 166–169, 1985.

Pak, D. K., and K. G. Miller, Late Paleocene to early Eocene benthic foraminiferal stable isotopes and assemblages: Implications for deepwater circulation, *Paleoceanography*, in press, 1992.

Pak, D. K., K. G. Miller, and J. D. Wright, Early Paleogene deep and intermediate water sources: Evidence from benthic foraminiferal and isotopic changes (abstract), *Geol. Soc. Am. Abstr. Programs*, *23*(5), A141–A142, 1991.

Poore, R. Z., and R. K. Matthews, Late Eocene–Oligocene oxygen and carbon isotopic record from the South Atlantic DSDP Site 522, *Initial Rep. Deep Sea Drill. Proj.*, *73*, 725–735, 1984.

Pospichal, J. J., and S. W. Wise, Jr., Paleocene to middle Eocene calcareous nannofossils of ODP sites 689 and 690, Maud Rise, Weddell Sea, *Proc. Ocean Drill. Program Sci. Results*, *113*, 613–638, 1990.

Premoli-Silva, I., and A. Boersma, Atlantic Eocene planktonic foraminiferal historical biogeographic and paleohydrologic indices, *Palaeogeogr. Palaeoclimatol. Palaeoecol.*, *67*, 315–356, 1984.

Prentice, M. L., and R. K. Matthews, Cenozoic ice-volume history: Development of a composite oxygen isotope record, *Geology*, *16*, 963–966, 1988.

Prothero, D. A., and W. A. Berggren (Eds.), *Eocene-Oligocene Climatic and Biotic Evolution*, 568 pp., Princeton University Press, Princeton, N. J., 1992.

Railsback, L. B., Influence of changing deep ocean circulation on the Phanerozoic oxygen isotopic record, *Geochim. Cosmochim. Acta*, *54*, 1501–1509, 1990.

Railsback, L. B., T. F. Anderson, S. C. Ackerly, and J. L. Cisne, Paleoceanographic modeling of temperature-salinity profiles from stable isotope data, *Paleoceanography*, *4*, 585–591, 1989.

Raup, D. M., and J. J. Sepkoski, Periodic extinction of families and genera, *Science*, *231*, 833–836, 1986.

Rea, D. K., J. C. Zachos, R. M. Owen, and D. Gingerich, Global change at the Paleocene-Eocene boundary: Climatic and evolutionary consequences of tectonic events, *Palaeogeogr. Palaeoclimatol. Palaeoecol.*, *79*, 117–128, 1990.

Robert, C., and H. Chamley, Development of early Eocene warm climates, as inferred from clay mineral variations in oceanic sediments, *Palaeogeogr. Palaeoclimatol. Palaeoecol.*, *89*, 315–332, 1991.

Robert, C., and J. P. Kennett, Paleocene and Eocene kaolinite distribution in the South Atlantic and Southern Ocean: Antarctic climatic and paleoceanographic implications, *Mar. Geol.*, *103*, 99–110, 1992.

Robert, C., and H. Maillot, Paleoenvironments in the Weddell Sea area and Antarctic climates as deduced from clay mineral associations and geochemical data, ODP Leg 113, *Proc. Ocean Drill. Program Sci. Results*, *113*, 51–70, 1990.

Roberts, D. G., A. C. Morton, and J. Backman, Late Paleocene-Eocene volcanic events in the northern North Atlantic Ocean, *Initial Rep. Deep Sea Drill. Proj.*, *81*, 913–923, 1984.

Romein, A. J. T., Lineages in early Paleogene calcareous nannoplankton, *Utrecht Micropaleontol. Bull.*, *22*, 231 pp., 1979.

Rosoff, D. B., and B. H. Corliss, An analysis of Recent deep-sea benthic foraminiferal morphotypes from the Norwegian and Greenland seas, *Palaeogeogr. Palaeoclimatol. Palaeoecol.*, *91*, 13–20, 1992.

Sanders, H. L., Marine benthic diversity: A comparative study, *Am. Nat.*, *102*, 243–282, 1968.

Savin, S. M., The history of the Earth's surface temperature during the past 100 million years, *Annu. Rev. Earth Planet. Sci.*, *5*, 319–344, 1977.

Savin, S. M., R. G. Douglas, G. Keller, J. S. Killingley, L. Shaughnessy, M. A. Sommer, E. Vincent, and F. Woodruff, Miocene benthic foraminiferal isotope records: A synthesis, *Mar. Micropaleontol.*, *6*, 423–450, 1981.

Schmidt, K.-H., Tertiary palaeoclimatic history of the southeastern Colorado Plateau, *Palaeogeogr. Palaeoclimatol. Palaeoecol.*, *86*, 283–296, 1991.

Schnitker, D., West Atlantic abyssal circulation during the past 120,000 years, *Nature*, *248*, 385–387, 1974.

Schnitker, D., The deep waters of the western North Atlantic during the past 24,000 years, and reinitiation of the Western Boundary Undercurrent, *Mar. Micropaleontol.*, *4*, 265–280, 1979*a*.

Schnitker, D., Cenozoic deep water benthic foraminifers, Bay of Biscay, *Initial Rep. Deep Sea Drill. Proj.*, *48*, 377–414, 1979*b*.

Schnitker, D., North-east Atlantic Neogene benthic foraminiferal faunas: Tracers of deepwater palaeoceanography, in North Atlantic Palaeoceanography, *Geol. Soc. Spec. Publ. London*, *21*, 191–204, 1986.

Schroeder-Adams, C. J., Eocene to Recent benthic foraminifer assemblages from the Kerguelen Plateau (southern Indian Ocean), *Proc. Ocean Drill. Program Sci. Results*, *119*, 611–630, 1991.

Sen Gupta, B. K., R. F. Lee, and S. M. Mallory, Upwelling and an unusual assemblage of benthic foraminifera on the northern Florida continental slope, *J. Paleontol.*, *55*, 853–857, 1981.

Shackleton, N. J., Oxygen isotopic evidence for Cenozoic climate change, in *Fossils and Climate*, edited by P. Brenchley, pp. 27–34, John Wiley, New York, 1984.

Shackleton, N. J., Paleogene stable isotope events, *Palaeogeogr. Palaeoclimatol. Palaeoecol.*, *57*, 91–102, 1986.

Shackleton, N. J., The carbon isotope history of the Cenozoic, in *Petroleum Source Rocks*, edited by J. Brooks and A. J. Fleet, pp. 427–438, Blackwell Scientific, Boston, 1987.

Shackleton, N. J., and A. Boersma, The climate of the Eocene ocean, *J. Geol. Soc. London*, *138*, 153–157, 1981.

Shackleton, N. J., and M. A. Hall, Carbon isotope data from Leg 74 sediments, *Initial Rep. Deep Sea Drill. Proj.*, *74*, 613–619, 1984.

Shackleton, N. J., and M. A. Hall, Carbon isotope stratigraphy of bulk sediments, ODP sites 689 and 690, Maud Rise, Antarctica, *Proc. Ocean Drill. Program Sci. Results*, *113*, 985–989, 1990.

Shackleton, N. J., and J. P. Kennett, Palaeotemperature history of the Cenozoic and the initiation of Antarctic glaciation: Oxygen and carbon isotope analyses in DSDP sites 277, 279 and 281, *Initial Rep. Deep Sea Drill. Proj.*, *29*, 743–755, 1975.

Shackleton, N. J., M. A. Hall, and A. Boersma, Oxygen and carbon isotope data from Leg 74 foraminifera, *Initial Rep. Deep Sea Drill. Proj.*, *74*, 599–612, 1984*a*.

Shackleton, N. J., et al., Accumulation rates in Leg 74 sediments, *Initial Rep. Deep Sea Drill. Proj.*, *74*, 621–637, 1984*b*.

Shackleton, N. J., M. A. Hall, and U. Bleil, Carbon isotope stratigraphy, Site 577, *Initial Rep. Deep Sea Drill. Proj.*, *86*, 503–511, 1985.

Signor, P., and J. Lipps, Sampling bias, gradual extinction patterns, and catastrophes in the fossil record, *Geol. Soc. Am. Spec. Publ.*, *19*, 291–296, 1982.

Smart, C. W., Ecological controls on patterns of speciation and extinction in deep-sea benthic foraminifera, M.Phil. report, Univ. of Southampton, Southampton, United Kingdom, 1991.

Spezzaferri, S., and I. Premoli-Silva, Oligocene planktonic foraminiferal biostratigraphy and paleoclimatic interpretation from Hole 538A, DSDP Leg 77, Gulf of Mexico, *Palaeogeogr. Palaeoclimatol. Palaeoecol.*, *83*, 217–263, 1991.

Stott, L. D., and J. P. Kennett, Antarctic Paleogene planktonic foraminiferal biostratigraphy: ODP Leg 113, sites 689 and 690, *Proc. Ocean Drill. Program Sci. Results*, *113*, 549–570, 1990.

Stott, L. D., J. P. Kennett, N. J. Shackleton, and R. M.

Corfield, The evolution of Antarctic surface waters during the Paleogene: Inferences from the stable isotopic composition of planktonic foraminifera, ODP Leg 113, *Proc. Ocean Drill. Program Sci. Results*, *113*, 849–864, 1990.

Streeter, S. S., Bottom water and benthonic foraminifera in the North Atlantic: Glacial-interglacial contrasts, *Quat. Res.*, *3*, 131–141, 1973.

Streeter, S. S., and N. J. Shackleton, Paleocirculation of the deep North Atlantic: 150,000 year record of benthic foraminifera and oxygen-18, *Science*, *203*, 168–171, 1979.

Thiel, H., O. Pfannkuche, G. Schriever, K. Lochte, A. J. Gooday, C. Hemleben, R. F. G. Mantoura, C. M. Turley, J. W. Patching, and F. Riemann, Phytodetritus on the deep-sea floor in a central oceanic region of the northeast Atlantic, *Biol. Oceanogr.*, *6*, 203–239, 1988.

Thierstein, H. R., Inventory of paleoproductivity records: The Mid-Cretaceous enigma, in *Productivity of the Oceans: Past and Present*, edited by W. H. Berger, V. S. Smetacek, and G. Wefer, pp. 355–375, John Wiley, New York, 1989.

Thomas, E., Late Eocene to Recent deep-sea benthic foraminifers from the central equatorial Pacific Ocean, *Initial Rep. Deep Sea Drill. Proj.*, *85*, 655–679, 1985.

Thomas, E., Changes in composition of Neogene benthic foraminiferal faunas in equatorial Pacific and North Atlantic, *Palaeogeogr. Palaeoclimatol. Palaeoecol.*, *53*, 47–61, 1986*a*.

Thomas, E., Early to middle Miocene benthic foraminiferal faunas from DSDP sites 608 and 610, North Atlantic, in North Atlantic, Palaeoceanography, *Geol. Soc. Spec. Publ. London*, *21*, 205–218, 1986*b*.

Thomas, E., Development of Cenozoic deep-sea benthic foraminiferal faunas in Antarctic waters, in Origins and Evolution of Antarctic Biota, *Geol. Soc. Spec. Publ. London*, *47*, 283–296, 1989.

Thomas, E., Late Cretaceous through Neogene deep-sea benthic foraminifers, Maud Rise, Weddell Sea, Antarctica, *Proc. Ocean Drill. Program Sci. Results*, *113*, 571–594, 1990*a*.

Thomas, E., Late Cretaceous–early Eocene mass extinctions in the deep sea, in *Global Catastrophes*, *Geol. Soc. Am. Spec. Publ.*, *247*, 481–496, 1990*b*.

Thomas, E., The latest Paleocene mass extinction of deep-sea benthic foraminifera: Result of global change (abstract), *Geol. Soc. Am. Abstr. Programs*, *23*(5), A141, 1991.

Thomas, E., Middle Eocene–late Oligocene bathyal benthic foraminifera (Weddell Sea): Faunal changes and implications for ocean circulation, in *Eocene-Oligocene Climatic and Biotic Evolution*, edited by D. A. Prothero and W. A. Berggren, pp. 245–271, Princeton University Press, Princeton, N. J., 1992.

Thomas, E., and J. C. Varekamp, Did volcanic CO_2 emissions cause the rapid global change at the end of the Paleocene?, paper presented at Chapman Conference on Climate, Volcanism and Global Change, AGU, Hilo, Hawaii, March 23–27, 1992.

Thomas, E., and E. Vincent, Major changes in benthic foraminifera in the equatorial Pacific before the middle Miocene polar cooling, *Geology*, *15*, 1035–1039, 1987.

Thomas, E., and E. Vincent, Early to middle Miocene deep-sea benthic foraminifera in the equatorial Pacific, *Rev. Paleobiol. Spec. Vol.*, *2*, 583–588, 1988.

Thomas, E., N. J. Shackleton, and M. A. Hall, Carbon isotope stratigraphy of Paleogene bulk sediments, Hole 762C (Exmouth Plateau, eastern Indian Ocean), *Proc. Ocean Drill. Program Sci. Results*, *122*, 897–901, 1992.

Tjalsma, R. C., and G. P. Lohmann, Paleocene-Eocene bathyal and abyssal benthic foraminifera from the Atlantic Ocean, *Micropaleontology Spec. Publ.*, *4*, 94 pp., 1983.

van Andel, T. H., Mesozoic-Cenozoic calcite compensation

depth and the global distribution of carbonate sediments, *Earth Planet. Sci. Lett.*, *26*, 187–194, 1975.

van der Zwaan, G. J., F. J. Jorissen, and W. J. Zachariasse (Eds.), Approaches to paleoproductivity reconstructions, *Mar. Micropaleontol.*, *19*, 1–180, 1992.

van Morkhoven, F. P. C., W. A. Berggren, and A. S. Edwards, *Cenozoic Cosmopolitan Deep-Water Benthic Foraminifera*, 421 pp., Elf-Aquitaine, Pau, France, 1986.

Vincent, E., and J. S. Killingley, Oxygen and carbon isotope record for the early and middle Miocene in the central equatorial Pacific (DSDP Leg 85), *Initial Rep. Deep Sea Drill. Proj.*, *85*, 749–770, 1985.

Webb, P.-N., Upper Cretaceous–Paleocene foraminifera from Site 208, Lord Howe Rise, Tasman Sea, DSDP Leg 21, *Initial Rep. Deep Sea Drill. Proj.*, *21*, 541–573, 1973.

Webb, P.-N., The Cenozoic history of Antarctica and its global impact, *Antarct. Sci.*, *2*, 3–21, 1990.

Wei, W., Evidence for an earliest Oligocene abrupt cooling in the surface waters of the Southern Ocean, *Geology*, *19*, 780–783, 1991.

White, R. S., Igneous outbursts and mass extinctions, *Eos Trans. AGU*, *70*, 1480, 1490–1491, 1989.

White, R. S., and D. Mackenzie, Magmatism at rift zones: The generation of volcanic continental margins and flood basalts, *J. Geophys. Res.*, *94*, 7685–7729, 1989.

Widmark, J. G. V., Upper Cretaceous–lower Tertiary deep-sea benthic foraminifera from the Walvis Ridge, South Atlantic Ocean: Taxonomy, paleobiogeography, and paleoecology, *Abstr. Uppsala Diss. Sci.*, *283*, 1–289, 1990.

Widmark, J. G. V., and B. Malmgren, Benthic foraminiferal changes across the Cretaceous-Tertiary boundary in the deep sea: DSDP sites 525, 527 and 465, *J. Foraminiferal Res.*, *22*, 81–113, 1992.

Williams, C. A., An oceanwide view of Palaeogene plate tectonics, *Palaeogeogr. Palaeoclimatol. Palaeoecol.*, *57*, 3–25, 1986.

Wing, S. L., A new basis for recognizing the Paleocene/Eocene boundary in western interior North America, *Science*, *226*, 439–441, 1984.

Wing, S. L., T. M. Bown, and J. D. Obradovich, Early Eocene biotic and climatic change in interior western America, *Geology*, *19*, 1189–1192, 1991.

Wise, S. W., J. R. Breza, D. M. Harwood, and W. Wei, Paleogene glacial history of Antarctica, in *Controversies in Modern Geology*, pp. 133–171, Academic, San Diego, Calif., 1991.

Wise, S. W., Jr., J. R. Breza, D. M. Harwood, W. Wei, and J. C. Zachos, Paleogene glacial history of Antarctica in the light of ODP Leg 120 drilling results, *Proc. Ocean Drill. Program Sci. Results*, *120*, 1001–1030, 1992.

Wolfe, J. A., A paleobotanical interpretation of Tertiary climates in the northern hemisphere, *Am. Sci.*, *66*, 694–703, 1978.

Wolfe, J. A., and R. Z. Poore, Tertiary marine and non-marine climatic trends, in *Climate in Earth History*, pp. 154–158, National Academy Press, Washington, D. C., 1982.

Woodruff, F., Changes in Miocene deep sea benthic foraminiferal distribution in the Pacific Ocean: Relationship to paleoceanography, *Mem. Geol. Soc. Am.*, *163*, 131–176, 1985.

Woodruff, F., and R. G. Douglas, Response of deep sea benthic foraminifera to Miocene paleoclimatic events, *Mar. Micropaleontol.*, *6*, 617–632, 1981.

Woodruff, F., and S. M. Savin, $d^{13}C$ values of Miocene Pacific benthic foraminifera: Correlations with sea level and productivity, *Geology*, *13*, 119–122, 1985.

Woodruff, F., and S. M. Savin, Miocene deep water oceanography, *Paleoceanography*, *4*, 87–140, 1989.

Woodruff, F., and S. M. Savin, Mid-Miocene isotope stratigraphy in the deep sea: High resolution correlations, paleoclimatic cycles, and sediment preservation, *Paleoceanography*, *6*, 755–806, 1991.

Wright, J. D., and K. G. Miller, Southern Ocean dominance of late Eocene to Oligocene deep-water circulation (abstract), *Eos Trans. AGU*, *73*, 171, 1992.

Zachos, J., W. A. Berggren, M.-P. Aubry, and A. Mackensen, Isotope and trace element geochemistry of Eocene and Oligocene foraminifers from Site 748, Kerguelen Plateau, *Proc. Ocean Drill. Program Sci. Results*, *120*, 839–854, 1992*a*.

Zachos, J., J. Breza, and S. W. Wise, Early Oligocene ice-sheet expansion on Antarctica: Sedimentological and isotopic evidence from Kerguelen Plateau, *Geology*, *20*, 569–573, 1992*b*.

Zachos, J., D. Rea, K. Seto, R. Nomura, and N. Niitsuma, Paleogene and early Neogene deepwater paleoceanography of the Indian Ocean as determined from benthic foraminifera stable isotope records, in *The Indian Ocean: A Synthesis of Results From the Ocean Drilling Program*, Geophys. Monogr. Ser., edited by R. A. Duncan et al., AGU, Washington, D. C., in press, 1992*c*.

(Received March 4, 1992;
accepted June 8, 1992.)

THE ANTARCTIC PALEOENVIRONMENT: A PERSPECTIVE ON GLOBAL CHANGE

ANTARCTIC RESEARCH SERIES, VOLUME 56, PAGES 167–184

THE INFLUENCE OF THE TETHYS ON THE BOTTOM WATERS OF THE EARLY TERTIARY OCEAN

Hedi Oberhänsli

Max-Planck-Institut für Chemie, D-6500 Mainz, Germany

In this paper the role of the Tethys as a bottom water source area is evaluated using the basin/shelf distribution and climatic reconstructions from coastal and adjacent continental settings. Palinspastic maps schematically document plate tectonic changes during Paleocene through Eocene. Lithofacies patterns of the Tethyan shelf and adjacent continental areas are compiled for four early Tertiary time slices (Paleocene, Ypresian, Lutetian, and Priabonian). For this type of study, spatial and temporal distributions of sediments with a high potential for paleoclimatic interpretation such as coal, bauxite, gypsum, and halite are particularly useful. Evaporite distribution indicates excess evaporation over a large area in the eastern Tethys during the Ypresian. As a result, surface waters became more saline with potential for transformation into bottom waters. Significant amounts of warm saline bottom water were exported during the Ypresian, probably preferentially in the direction of the Indian Ocean, which then successively filled the deep basins of the world oceans. At present, there is no indication that after the Ypresian significant amounts of warm saline water left the Tethys to become intermediate or deep water. However, it is very likely that before and after the Ypresian, proportional quantities of dense water were produced in the Tethys, much in the same way as dense water is produced in the modern Mediterranean Sea and ejected to the Atlantic. During the early Tertiary, this dense water probably left the Tethys in both directions (Atlantic and Indian oceans).

INTRODUCTION

In a stratified ocean, bottom water is denser than the overlying water. The density of seawater is controlled by its temperature and salinity. When sea ice forms, the density increases. Today bottom water generates in the perimeter of the Antarctic continent and in the Norwegian-Greenland Sea, which therefore have an important role in the global circulation pattern. Since the meridional temperature gradients are very high, the surface water temperature is low and sea ice forms regionally and seasonally. Both processes drive the surface water of the high latitudes toward high density. Depending on its density structure, the altered water sinks and fills the world's deep ocean basins. In the geologic past the situation might have been different.

Major points of the discussion of the deepwater history during the early Tertiary are the origin of the water masses and the evolution of the circulation pattern. Both source areas and flow tracts are related to global climatic history. Bottom water source areas and bottom water flow patterns are responses to the complex interplay between land/sea distribution and latitudinal extents of climatic belts because the position of continents controls the oceanic gateways and the global albedo pattern. Thus plate tectonic configuration is crucial too. A comparison of today's situation with the early Tertiary is not adequate, as the climate has changed considerably. Temperatures compared with the

present were generally warmer during the earlier Tertiary [*Krutzsch*, 1967; *Kemp*, 1978; *Wolfe*, 1978; *Hochuli*, 1984]. As a consequence, meridional temperature gradients were smaller during the early Tertiary. Therefore we may conclude that oceanic circulation patterns and therefore also the source area of bottom water were different at the beginning of the Tertiary.

Two potential source areas were possible before the first ice buildup on Antarctica. Besides high latitudes, formation of bottom water may also have occurred at mid- to low-latitude sites [*Brass et al.*, 1982]. It is undisputed that Late Cretaceous and earlier Paleogene oceans were partly filled with warm saline bottom water beneath shallow bathyal depths, as has already been suggested by *Chamberlin* [1906] and more recently by *Brass et al.* [1982], *Oberhänsli and Hsü* [1986], *Hay* [1988], and *Kennett and Stott* [1990]. The only controversy concerns the timing of the onset and termination and the intensity of the halokinetically driven circulation. The source areas of the dense, warm saline bottom water have so far not been identified.

In this paper, the influence of the Tethys as a potential low-latitude source area for bottom water formation is evaluated. As a basis for discussion, sedimentary facies sketches are used, which show environmental conditions mostly in shallow water and continental settings of the Tethyan and adjacent areas during four time slices (Paleocene (mainly Thanetian), Ypresian, Lutetian, and

Priabonian). These lithofacies patterns were used for qualitative interpretations of the evaporation/precipitation conditions in the Tethyan realm. The facies maps were designed for time slices rather than for particular datum planes considering that many of the stratigraphic intervals shown are poorly constrained. As was pointed out earlier, given the importance of the land/sea distribution on the climatic evolution and circulation pattern, the plate tectonic evolution will be presented for two time intervals: (1) the Late Cretaceous/early Paleocene and (2) the late Eocene/early Oligocene. These paleotectonic sketches exhibit the shelf/basin configuration and oceanic gateways during these time intervals.

THE PALEOTECTONIC EVOLUTION OF TETHYS DURING THE PALEOGENE

In this paper, two plate tectonic snapshots will be documented: (1) the Late Cretaceous–Paleocene and (2) the late Eocene–Oligocene (Figures 1 and 2). Additionally, Figure 3 shows a closeup of the western Tethys, exhibiting the tectonic status during the late Lutetian. Based on a kinematic solution proposed by *Savostin et al.* [1986], the evolution of the Tethys has been reconstructed by *Dercourt et al.* [1986]. According to *Dercourt et al.* [1986] and *Auboin et al.* [1986] the tectonic evolution of Tethys when related to the displacement of the Eurasian, Iberian, African/Arabian, and Indian continents came to its terminal stage during the earlier Tertiary.

The time interval shown in Figure 1 (80–65 Ma) documents the onset of a new kinematic regime. By the end of the Mesozoic the predominantly left-lateral motion between Africa and Eurasia is replaced by scissorslike movements along the African plate boundary after the rotation pole of Eurasia/Africa has shifted to the Gibraltar region [*Dercourt et al.*, 1986]. The convergence of the two plates was slowed considerably by this change in direction of movement, and the convergence rate decreased to near zero in the vicinity of Gibraltar and to about 2 cm/yr at the eastern rim of the Arabian plate. Main features of this phase are as follows: (1) the spreading in the western Tethys (Mesogea) ceased and Apulia was again linked to the African craton; as an African salient it caused folding in the sedimentary cover of the Alpine and Carpathian basins [*Tapponier*, 1977]; (2) along the western margin of Eurasia, the erosion of the Alpine and Dacidic reliefs resulted in the deposition of thick flysch sequences in many different troughs; (3) in the west, Iberia rotated and converged on the Pyrenees where deformation (Laramic phase) can be observed; and (4) a northward subduction of the eastern Tethys (Neo-Tethys) occurred along the southeastern margin of Eurasia.

Changes in the direction of motion caused major effects in the eastern Tethys. Between the western and eastern Tethys the deepwater connection was interrupted by the Campanian, although in the western Tethys a deepwater exchange continued to the Atlantic. Figure 1 indicates that by the latest Cretaceous, the oceanic basin off the Arabian northeast coast was still 200 to 500 km wide and widened to the southeast to approximately 1000 km. By 65 Ma, large oceanic basins were present in the western as well as in the eastern part of Tethys, with deepwater connections to the Atlantic and Indian oceans, respectively.

Figure 2 documents a palinspastic snapshot of the latest Eocene. The scissorslike movement between the African/Arabian and Eurasian plates exhibited slightly increased rates of motion between 65 and 35 Ma. Rates were approximately 3 cm/yr at the eastern rim of Arabia [*Dercourt et al.*, 1986]. The major reorganization of the eastern Tethyan realm was determined by the very high rate of motion of the Indian continent, which is estimated to have been 12 cm/yr during the earliest Paleogene (65–55 Ma) and later slowed to 8 cm/yr (54–35 Ma). This caused a decrease of 2500–3000 km in the length of the eastern Tethys between India and Eurasia during the Paleocene-Eocene time slice. As the continental collision progressed during the early Paleogene, the southern margin of Eurasia was reshaped. Previous major oceanic basins were fragmented and survived as small, occasionally isolated basins. The large Austro-Alpine nappe belt, which was constructed during the Late Cretaceous, emerged during the Paleogene and erosion began [*Trümpy*, 1980; *Dercourt et al.*, 1986]. As closure of the Tethys and its marginal seas proceeded, marginal volcanic belts (island arcs) developed. The Paleogene volcanic activity and the distribution of these belts are unique on the active margin of Eurasia and can be traced continuously from the Rhodope massif (Bulgaria) to Afghanistan [*Kazmin et al.*, 1986]. According to these authors the most active phase is the early-middle Eocene time interval.

Figure 3 gives detailed information on the tectonic evolution of the westernmost realm of the Tethys. The sketch portrays the palinspastic constellation during the middle/late Eocene transition. The situation is influenced by the Pyrenean compressional phase which took place in the late Lutetian and/or earliest Priabonian [*Wildi*, 1983]. Compression and shortening of the crust are widespread [*Wildi*, 1983]. However, at the moment, no estimates on crustal shortening for this area are known.

OCEANIC GATEWAYS IN THE TETHYAN REALM

As displayed in Figure 3, the seaway to the Atlantic Ocean is still functioning on both the northern and the southern edge of the Alboran block, although it is considerably narrower than at the Cretaceous/Tertiary transition. Sediment facies and their distribution indicate, however, that no deepwater (abyssal) connection between the western Tethyan and the Central Atlantic

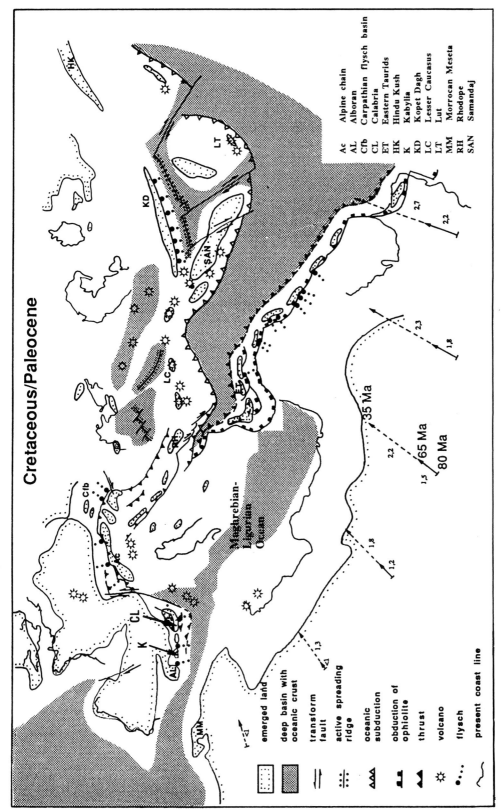

Fig. 1. The plate tectonic situation of the Tethyan realm during the latest Cretaceous (65 Ma) according to *Dercourt et al.* [1986]. The reconstruction represents the approximate position of the continents during anomaly 30 [*Savostin et al.*, 1986]. Arrows indicate the direction of the motion of the African/Arabian plate. The numbers indicate the motion rates (in centimeters per year) for the time slices shown [*Dercourt et al.*, 1986]. According to *Dercourt et al.* [1986], Mesogea includes the oceanic domain which mainly formed during the Cretaceous when Apulia was rotated. It extended from the Maghrebian-Ligurian Ocean (west) to the Neo-Tethys. Neo-Tethys circumscribes the oceanic domain formed during Triassic when northern Gondwana blocks separated from Africa/Arabia.

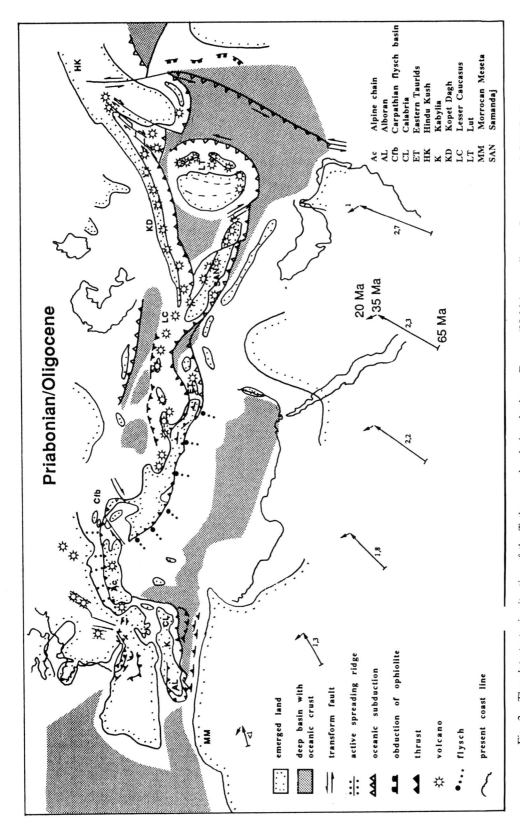

Fig. 2. The plate tectonic situation of the Tethyan realm during the latest Eocene (35 Ma) according to *Dercourt et al.* [1986]. The reconstruction represents the approximate position of the continents during anomaly 13 [*Savostin et al.*, 1986]. Arrows indicate the direction of the motion of the African/Arabian plate. The numbers indicate the motion rates (in centimeters per year) for the time slices shown [*Dercourt et al.*, 1986].

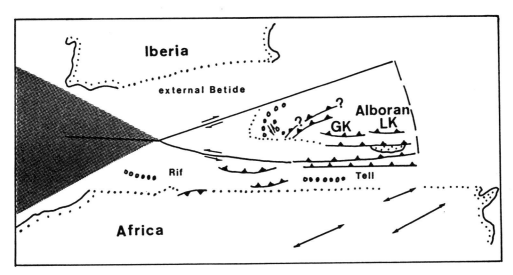

Fig. 3. Status of the tectonic evolution in the western Tethys during the upper middle Eocene (late Lutetian) after *Wildi* [1983]. Dotted lines contour the edges of the continents. Shaded area represents oceanic crust. GK, Greater Kabylia; LK, Lesser Kabylia.

basins existed until the late Eocene, although the sill depth might have been greater than it is today (*Wildi* [1983]; the current maximum depth is 450 m). I assume that this seaway was probably wider than it is today (15 km) because the paroxysm of the collision-related crustal shortening did not occur until the upper Eocene [*Dercourt et al.*, 1986].

To the south, water exchange between the Tethys and the South Atlantic existed across the Sahara, east of Al Hoggar. The Tethyan surface water stretched into the lower Niger valley, reaching the Gulf of Guinea. This shallow gateway existed until the upper Lutetian [*Ronov et al.*, 1989]. Between the Tethys and the North Sea and the Arctic Sea, interoceanic exchange occurred intermittently during the early Paleogene [*Ronov et al.*, 1989]. The shallow gateway between the Tethys and the Kara Sea existed east of the Ural during the Paleocene and most of the Eocene. This connection was interrupted by the Oligocene [*Ronov et al.*, 1989].

Major interoceanic gateways between the Tethys and the Atlantic and Indian oceans were intact throughout the Paleocene and the Eocene. Areas in which shallow water environments persisted were much larger in the eastern than in the western Tethys. In the latter, different small basins had formed progressively in the course of the early Paleogene and probably some of the basins were occasionally isolated [e.g., *Oberhänsli et al.*, 1983].

EVALUATION OF CLIMATIC PROXY DATA

Global climatic and geochemical changes in the ocean are recorded in planktic and benthic marine organisms. Analyses of fossil faunal assemblages and of the stable isotopic composition of fossils document all environ-mental modifications of a significant amplitude. We recognize an event when selected key species show their first appearance at a particular locality or when they become more abundant or less abundant or even disappear. The coeval isotopic signal may show a deviation in either direction.

Recently, *Adams et al.* [1990] discussed the conflicting interpretation of isotopic and biotic evidence and the discrepancy of temperature estimates from early Tertiary tropical isotopic records and those of today's tropical sea surface waters. A similar discrepancy is observed looking at the deepwater record, although the deepwater benthic fauna is in general thought to be less sensitive to temperature changes than the shallow water, low-latitude benthic assemblages [*Corliss*, 1981; *McGowran*, 1979]. The evolution of deep water during the Paleogene seems to be closely linked with overall global climatic evolution. However, the difference between tropical surface water temperature and bottom water temperature as deduced from oxygen isotope studies indicates that no significant changes occurred from the Late Cretaceous through the Oligocene. Vertical temperature gradients of 10°–15°C are suggested for the Late Cretaceous, Eocene, and Oligocene [*Shackleton and Boersma*, 1981; *Brass et al.*, 1982; *Hay*, 1988]. These temperature differences are considerably lower than the vertical temperature gradient observed today, which is approximately 25°C at low latitudes. These observations, suggesting relatively constant vertical temperature gradients through the Paleogene, conflict with faunal data in selected time intervals. Near the end of the Eocene, for instance, ostracode assemblages show drastic changes. *Benson* [1975] attributed this faunal reorganization to the establishment of the psychrosphere that created a cold, stable deepwater environment. According to Benson, bottom water began to form then on the perimeter of Antarctica.

In the geologic record, assignment of a specific isotope signal to a specific cause is sometimes equivocal. The $\delta^{18}O$ signal of benthic foraminifers is determined by the ambient temperature. However, it also includes information on the $\delta^{18}O$ of the seawater. The latter is influenced by two processes: (1) changes in evaporation/ precipitation rates due to climatic fluctuations in the bottom water source area and (2) formation of continental ice. Very little is known yet about the influence of a third process: hydrothermal activity on ocean floors. This process may be important when the residence time of the bottom water is prolonged for some reason.

The covariant $\delta^{18}O$ changes in bottom and surface water records at the Eocene/Oligocene transition have been interpreted by *Miller et al.* [1987] to represent stages in which polar ice existed. Before the buildup of polar ice caps, the calculation of temperature changes in surface and bottom water environments from $\delta^{18}O$ signals is not a direct task because it has to be assumed that the $\delta^{18}O$ of the water has not changed during the time interval studied. This assumption has been discussed recently with skepticism by several authors [e.g., *Horrell*, 1990; *Kennett and Stott*, 1990; *Mead et al.*, 1992]. These authors noted that before the first significant buildup of ice sheets on the Antarctic continent, not only water temperature but also salinity of the subsurface water masses may have changed. They suggested that warm water of higher salinity may have been present in bathyal and abyssal depths in high-latitude oceans. Thus in addition to temperature changes, changes in the global evaporation/precipitation conditions may be documented in the oxygen isotopic signal of benthic organisms. The enigma of sorting out the two different influences from isotopic records cannot easily be solved, since neither temperature nor water mass characteristics are depicted directly in the biogenic marine sediment particles.

Sedimentary facies patterns of shallow water and continental settings can also be used for interpretation of local and regional climatic evolution. However, not all facies are useful for this type of analysis. Direct correlations can be expected from evaporite and laterite deposits and, with some reservations, coal deposits [*Kozary et al.*, 1968; *Gordon*, 1975; *Parrish et al.*, 1982; *Ziegler et al.*, 1987]. They have an excellent paleoclimatic potential. Other lithofacies allow speculation on the climatic evolution: clay minerals, silicretes, calcretes, phosphorites, and iron pisolits [*Chamley*, 1979; *Robert*, 1980; *Baturin*, 1982; *Prasad*, 1983; *Valeton*, 1983a, b; *Bardossy and Aleva*, 1990].

In the context of this study, changes in evaporation/ precipitation conditions during the Paleocene and Eocene will be evaluated. For this purpose the Tethys and adjacent continents were studied because it is particularly promising to look at the regions where net evaporation occurs. Evaporite deposition commonly takes place where evaporation exceeds the total of precipitation plus the river discharge [*Gordon*, 1975]. The global distribution of modern evaporites is mainly controlled by the atmospheric circulation. Today net evaporation zones are located in the subtropical high-pressure belt, between 15°–35°S and 15°–35°N [*Gordon*, 1975]. In open ocean settings, surface water salinity is maximal at these latitudes [*Wüst*, 1954].

Estimates on excess precipitation are based on the laterite and coal distributions in the Tethyan realm. Laterites form when in situ pedogenic processes leach Si from bedrock in the peneplained hinterland or coastal plains. This process is limited to the warm, humid climate zone [*Brückner*, 1957; *Valeton*, 1983a; *Bardossy and Aleva*, 1990]. According to *Valeton* [1983a], "the early Eocene coastal plain laterites of India are accompanied by coeval terrestrial facies including red beds with kaolinite clays (Al-rich) or sepiolite/ attapulgite clays (Mg-rich), ferricretes or silicretes, peat or lignite beds with sulfides or sulfates (jarusite and alunite), and siderite." Chemical weathering on continents has its almost synchronous facies counterparts in the marine environment [*Prasad*, 1983; *Valeton*, 1983a; *Bardossy and Aleva*, 1990]. On the basis of field survey data, these authors suggest a marine facies succession related to the increased chemical weathering on continents. This succession is observed at many sites and is considered to represent the Late Cretaceous to early Tertiary climatic optimum. During this period the attapulgite/sepiolite facies (near shore) and siliceous chalks with phosphate nodules and smectite became more abundant in shelf environments. At the same time, the phosphate/glauconite facies associated with siderite was observed especially in marine shallow water environments. This deposition model proposed by Prasad, Valeton, and Bardossy and Aleva may explain why "bauxite episodes" (late Thanetian/early Ypresian and early Lutetian documenting intervals with increased chemical weathering) are accompanied by widespread, probably coeval deposition of siliceous limestones and marls with phosphorite nodules and marls with glauconite in shallow water environments of the vicinity. This model is not uncontested. Several investigators favor the "actualistic" model which relates chert and phosphorite deposits with upwelling activities. In the Tethyan realm, most of these facies assemblages and their spatial distribution as well (north and south Tethyan margin; see Figures 5a and 5b) point to temporally increased biological productivity in shallow water environments due to continental runoff rather than to upwelling processes.

The preservation of peat, which is considered to be a precursor of coal, is controlled by several factors, including temperature. However, according to *Ziegler et al.* [1987], the monthly continuity of rain is the most important factor for peat formation. In general, the preservation factor increases with decreasing temperatures. But the climate is not the only limiting factor for its preservation. The morphogenetic evolution of the deposition site is also important. The early Paleogene brown coal deposits of central and northern Europe had an excellent chance for preservation because northwestern Eurasia was structured by low-level swells and troughs formed by the alpidic tectonogenesis. This topography allowed the excessive accumulation of plant tissue, and the steady water table, maintained by the warm and humid climate, also favored preservation of coal deposits. Similar conditions existed for the coal accumulation in India and western Africa, although temperatures were probably considerably higher.

Climatic evolution, when reconstructed from geological data sets (flora, fauna, and geochemistry), may not be exact. The limitations are given because we do not have the necessary information on physical processes which determine the paleoclimate. Important information gaps concern the radiation budget and the troposphere circulation which are both linked to the land/sea distribution and the continental topography. In continental and epicontinental settings the evaporation/precipitation balance may be influenced by local factors such as topography. But the distribution of evaporites during the past 100 Ma suggests that zones where net evaporation has occurred remained mostly stationary [Kozary et al., 1968; Gordon, 1975]. Only during the latest Eocene can a slight shift, by approximately 10° toward north, be observed [Kozary et al., 1968]. Parrish et al. [1982], however, show that for climatic reconstructions, zonal models reflect less effectively the situation of the past than the more sophisticated areal climate models, which resemble those we observe today.

In the Tethys the reconstruction of sedimentary facies distribution on adjacent continents (bauxite, laterite, coal, redbed, silicified wood, duricrusts, and weathered clay beds) was often impeded by the patchy record and poor dating of continental deposits. Deciphering the facies patterns, I also faced another problem: local and regional sea level changes in shallow seas may have influenced the facies successions and their spatial distributions. The early Tertiary sea level history on the Tethyan margins and adjacent coastal plains was tightly linked with the tectono-eustatic evolution of the Alpine fold belt. Therefore many of the lithologic sections were partly biased by numerous erosional events related to the alpine tectonic evolution during the Eocene. Besides, we have to keep in mind that some facies changes can not directly be related to climatic changes. They were instead related to local subsidence events. The extent of the data bias due to sea level changes can be outlined by a comparison of two sections from the Helvetic facies belt (Figure 4). These sediments were deposited on the southern margin of the Eurasian con-

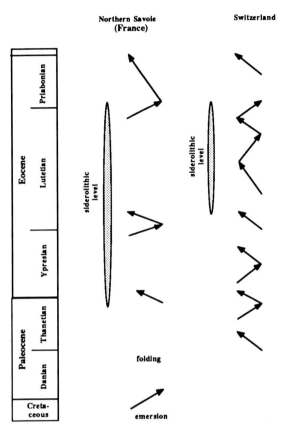

Fig. 4. Examples of regression/transgression cycles at the southwestern margin of the Eurasian plate. The onlap/offlap cycles of Savoie and Switzerland are shown after *Kerchhoven et al.* [1980] and *Herb* [1988]. In both areas, transgression proceeds progressively from southeast to northwest. "Siderolithic" describes continental residuals in fissures. They consist mostly of iron pisolites. At the eastern location, their occurrence is restricted in time and space. They are observed in western and northwestern Switzerland only during the middle Eocene. In southeastern France their occurrence in time is less restricted.

tinent. Comparing the onlap and offlap events of the two sections, the striking differences are evident: one section reveals several major hiatuses [*Herb*, 1988; *Kerckhoven et al.*, 1980]. Also, deposits from the Paris Basin sensitively record the Pyrenean and Alpine orogenic pulsations. *Cavalier and Pomerol* [1979] observe about 20 tectonic episodes during the Paleogene. The climatic record deduced from such sequences may remain fragmentary. Near the coastline, minimal sea level changes can cause an emersion and expose the marine deposits to pedogenetic processes. However, the processes which alter the unlithified, emerged sediments allow an interpretation of past climatic conditions. *Thiry* [1981] gave detailed descriptions of the climatic evolution for the Paris Basin based on the formation of duricrusts and "lateritic" pedogenesis both of which are common phenomena. His estimates on the evaporation/precipitation balance are promising.

SEDIMENTARY FACIES OF TETHYS AND ADJACENT CONTINENTS

Because of the imprecise age assignments and spotty recovery, only four time slices were considered here for reconstruction of the climatic evolution of the Tethyan realm. The generalized sediment facies patterns of four selected time intervals are shown in Figures 5a–5d. The information used for this compilation was extracted from the *Léxique Stratigraphique International*, volumes I–IV (1963) and the following authors: *Cerujevska* [1967], *Zaklinskaja* [1967], *Stöcklin* [1968], *Plaziat* [1970], *Murray and Wright* [1974], *Dercourt* [1980], *Pomerol* [1981], *Thiry* [1981], *Freytet and Plaziat* [1982], *Plint* [1983], *Wildi* [1983], *Hohl* [1985], *Mohr* [1986], *Herb* [1988], *Royden and Baldi* [1988], *Ronov et al.* [1989], *Bardossy and Aleva* [1990], and *Trappe* [1992]. During the earliest Tertiary the Tethyan realm covered the latitudes between 0° and 40°N. As the paleotectonic evolution of Tethys proceeded, the deposition area was reduced considerably. Toward the end of the Eocene the area of recorded sequences was reduced and moved to a latitudinal belt between 15°N and 50°N. This region is extended on either side by a further 10°–15° when the deposits from the Eurasian and African/Arabian continents are included. Thus for paleoclimatic reconstructions a latitudinal belt is represented that includes tropical/paratropical to temperate conditions.

Paleocene, Mostly Thanetian, 60.8–57.8 Ma (Figure 5a)

Along the northern margin of the African/Arabian plate, the shore facies consisted mostly of dolomite. From the littoral and inner shelf area, limestone with larger foraminifers and marly carbonates intercalated with siliceous limestones are described. In central and southern Morocco, silicretes, dolocretes, and stromatolite/chert associations, resembling the Lake Magadi deposits, are widespread on the coastal plain throughout the Paleocene to the Lutetian [*Trappe*, 1992]. Off Egypt and Sinai, marly limestones, calcareous claystones with phosphorites, and pisoliths were accumulated on the shelf. During the late Paleocene mostly dolomitic carbonates and thin gypsum beds formed on the southeastern Arabian platform. In southwest Iran, clays, gypsum, and dolomite with a few carbonates accumulated. Along the Afghanistan/Russian border area, paralic clayey sediments, gypsum, and dolomites were replaced by shallow water limestones with larger foraminifers. On the northwestern Indian continent, fine detrital fluviatile deposits with a few gypsum intercalations dominated at the base of this time interval. They passed into a thick marine limestone/shale/marl sequence rich in fossils.

In the Pyrenean-Provincal basin, fluviatile deposits and sediments of perennial lakes are observed. At the lake edges which are located along the slopes of the topographic elevations, smectite and attapulgite clays occurred commonly with dolomite and desiccation breccias.

Ypresian, 57.8–52 Ma (Figure 5b)

By this time the trans-Saharan seaway had been established and surface water exchange between the western Tethys and the South Atlantic had occurred. Along the northern margin of the African/Arabian continent, sedimentation started with nummulitic limestones (nearshore), silicified marls, and limestones. Also, in the trans-Saharan strait the silicified, marly limestone facies is widespread. Locally glauconite- and phosphorite-rich beds occurred.

In western India, lateritization of the Deccan Trap basalts took place at many sites along the shoreline [*Valeton*, 1983a, b]. Two laterite weathering horizons were preserved at the northwestern coast of India. Along the western margin of the Indian plate the earliest Eocene or latest Paleocene is documented by coal seams overlain by clayey marls and neritic limestones.

Facies distribution changed significantly during the middle Ypresian. Gypsum or anhydrite, dolomite accumulation and duricrusts formation were widespread along the northeastern margin of the African/Arabian continent (Rus Formation in Saudi Arabia). At the northern African margin, deposition of gypsum was restricted to southern Libya. Also in southwestern Iran (Amiran and Kashkan Formation) and in the Salt Range dolomite and gypsum were accumulated. Halites probably developed in central Iran and western Pakistan [*Stöcklin*, 1968] (S. H. Faruqi, personal communication, 1987).

In the Paris Basin, common silicretes and silicified clays, sand, and conglomerates alternated with "lateritic" horizons in soils and at lake shores. In the southern part of the Paris Basin, mainly kaolinitic clays and sand were accumulated, whereas in the northern part, lignitic clays prevailed [*Thiry*, 1981]. In southern France a marine incursion, triggered by tectonic activities of the Pyrenean Phase, occurred and drainage on shore had improved. The extensive discharge as traced from fluviatile sediments was probably not only influenced by the rainfall intensities, but the progressively evolving relief in the hinterland of the flood plains also modified the transport potential. Similar to the Thanetian, dolomites formed along the emerged lake shores. But locally mangrove-related molluscs and other faunas were observed [*Plaziat*, 1970; *Freytet and Plaziat*, 1982].

Lutetian, 52–43.6 Ma (Figure 5c)

In the eastern Tethys, dolomitic marls or dolomite appeared in many mid-latitude coastal sections. Sabkha-like conditions existed over vast areas along the eastern margin of the African/Arabian platform. Halite deposition was restricted to central Iran (Kulat Basin). In some areas, sporadic sulfatic intercalations in marly deposits are observed (Somalia, Yemen, southern margin of the Elborz, the northern margin of Hindu Kush, and Simla Himalayas). There is also lateritic weathering at several emerged areas during the early Lutetian (northwest India and Hindu Kush).

In the western Tethys, carbonate sedimentation was common in shallow marine environments. Locally, deposition of marls with chert nodules occurred. However, abundance of silicification phenomena was reduced in comparison with late Paleocene/early Ypresian time. Coal measures are restricted to northwestern and central Europe. Glauconite was a common mineral in shallow marine environments of Europe. There is also evidence for lateritic weathering at several emerged areas (western Africa, Hungary, Yugoslavia, Greece, and Spain). In the western Tethyan realm the northernmost occurrence of karst-bauxite was documented for Hungary and Yugoslavia [Grubic, 1972; Bardossy and Aleva, 1990]. The coeval "siderolithikum" of the Helvetic belt of the Alps (see Figure 4) and of the Paris Basin, considered to be a laterite equivalent, may prove that the warm, humid to semihumid climate had expanded by the middle Eocene to 40°N. According to Valeton [1983b], Eocene laterites may have formed over vast areas, but owing to their topographically vulnerable sites they were extremely sensitive to erosional processes. Only relict preservation of these horizons occurs [see Valeton, 1983b]. Nevertheless, even the relict occurrence of this climatically indicative facies allows us to trace its occurrence which reached far to the north during the early Tertiary.

Priabonian, 43.6–36.6 Ma (Figure 5d)

The Priabonian sedimentary record is less complete than that of the early and middle Eocene intervals. This may be attributed to the fact that tectonic evolution approached the paroxysm of the alpine orogeny by this time (Carpathians, Turkey, Iran, and Himalayas) and post-Eocene erosional events may have capped the latest Eocene record over vast areas. Volcanogenic and redepositional activities were widespread. At the western and eastern Tethys margins, this time was mostly represented by nummulitic limestone. The evaporitic facies was spatially very restricted. It was confined to northern Africa and western and southeastern Europe. Sulfates and halites are observed in coastal plain settings of southern Libya, northern Spain, and the Aquitan and Paris basins; in the upper Rhine valley; and in the Pontides.

THE EVAPORATION/PRECIPITATION BALANCE IN THE TETHYAN REALM

Based on the distribution of bauxite, coal, and evaporite deposits, and further on silicrets, dolocretes, and calcrits (Figures 5a–5d), estimates on the humidity and the aridity of four continental regions (India, Africa, Arabia/Iran, and Europe) are outlined. The curves displayed in Figure 6 delineate trends. In reality, the situation might be more complicated. Krutzsch [1967] mentioned that the Paleogene flora from central European localities does not show a linear climatic evolution but rather cyclic changes. This might also be valuable for the other localities.

The climatic reconstruction of the northwestern Indian province was based on the widely distributed sediment sequences of the Ranikot, Laki, and Charat formations. Maximum humidity is recorded for the latest Thanetian/earlier Ypresian, late Ypresian, and early Lutetian when bauxite and peat have been deposited. On the basis of the lithofacies patterns, I conclude that northwest India was mostly lying in the humid climate zone. Only occasionally, semihumid conditions influenced the depositional conditions during its movement from the equator (early Paleocene) to 30°N (Priabonian).

The climate curve of Arabia/Iran indicates that during the Paleocene and Eocene this area was mostly located within the arid zone. Maximum aridity occurred during the Ypresian when gypsum deposits formed over vast areas of the Arabian Peninsula (Rus Formation). In the Lut (Iran) and Kohat (western Pakistan) basins, halites were probably deposited coevally. The age assignments, however, of both the Lut and Kohat salt series are still a matter of severe disputes [Stöcklin, 1968] (S. H. Faruqi, personal communication, 1987).

Central west Africa shows a facies pattern similar to the one exhibited for northwestern India, although a slight time lag is evident. The climatic reconstruction is based on bauxite occurrences in Senegal and Ghana and on coal seams (early Ypresian) in Nigeria (Léxique Stratigraphique International). However, the age assignments of the bauxite are somewhat controversial: Valeton [1983a] gives an early Lutetian age, whereas Brückner [1957] determines a Lutetian or Bartonian age.

In central and southern Morocco, the early Paleogene sequences from the Atlas Mountains display a complete transgressive-regressive cycle. The climatically indicative facies document semiarid to arid conditions on the coastal plain throughout the Paleocene to the Lutetian, except during early Lutetian, when transgression reached a maximum [Trappe, 1992]. Then climate may have transiently changed in northwest Africa. Palynological studies from southern Morocco yield floras which represent a mangrove vegetation [Mohr, 1986], a flora which traces warm, humid to semihumid conditions at the Ypresian/Lutetian transition.

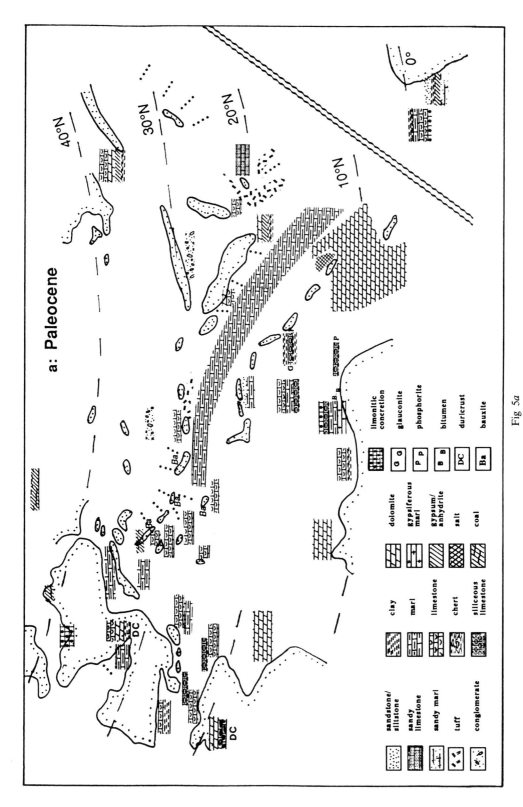

Fig 5a

Fig. 5.   Sedimentary facies in the Tethyan realm. Palinspastic sketches given by *Dercourt et al.* [1986] have been modified. (*a*) Paleocene (mostly Thanetian, 60.8–57.8 Ma). (*b*) Ypresian, 57.8–52 Ma. (*c*) Lutetian, 52–43.6 Ma. (*d*) Priabonian, 43.6–36.6 Ma. Age assignments are given according to *Berggren et al.* [1985].

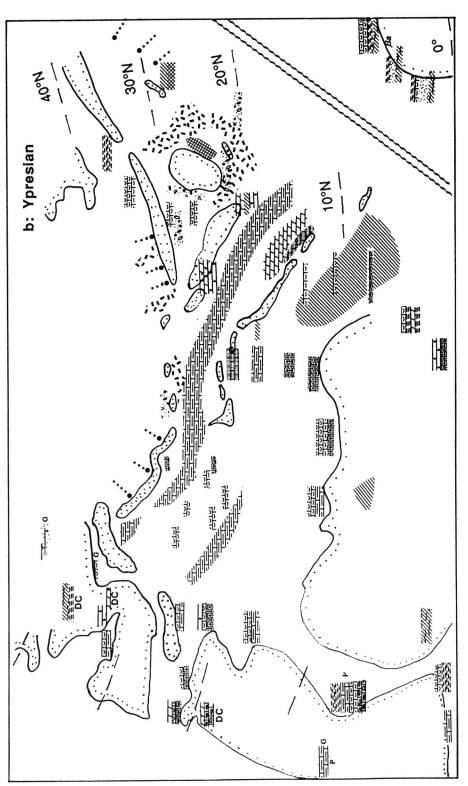

Fig. 5b

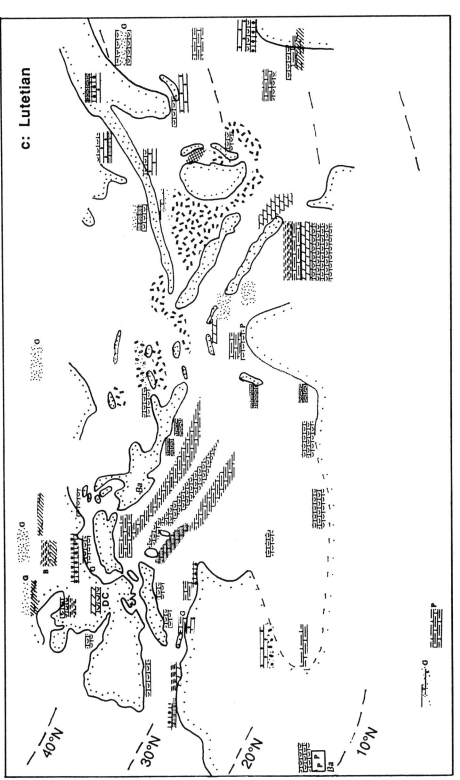

Fig. 5c

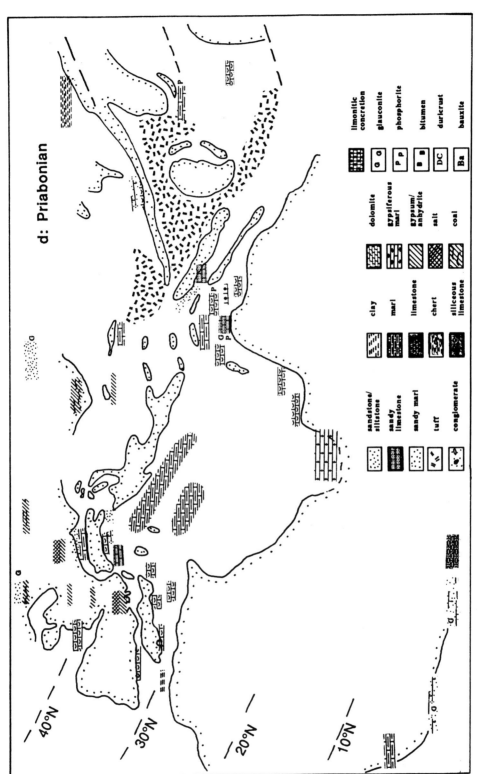

Fig. 5d

The climatic evolution in Europe during the Eocene was diverging. Cyclic changes documented by alternating formation of duricrusts and lateritic pedogenesis signify that the climate was warm and switched periodically from semiarid to humid conditions. *Cavelier and Pomerol* [1979] and *Thiry* [1981, 1989] reconstructed the climatic evolution from pedogenetic and sedimentologic analyses of the Paris Basin. According to these authors the climate was semiarid and warm to temperate during the Thanetian and lower Ypresian. By the late Ypresian the climate turned toward more humid conditions. After the marine ingression into the Paris Basin (Lutetian), the Priabonian continental facies again reflect increased aridity. However, during the marine ingression, salinity of marine waters of the Paris Basin increased at certain times, as indicated by benthic foraminiferal assemblages [*Murray and Wright*, 1974]. In all western European continental settings, similar changes are observed during the Lutetian and Priabonian. In southern France the existing large perennial lakes, located near the Massiv Central, illustrate that the yearly precipitation exceeded evaporation at least during the Lutetian and early Priabonian. However, small ephemeral lakes intercalated in the fluvial fans indicate seasonality in the evaporation/precipitation balance. The sediments from the Pyrenean-Provincal Basin reflect humid to semiarid climatic conditions during the Paleocene [*Freytet and Plaziat*, 1982]. But locally the presence of mangrove molluscs and related faunas [*Plaziat*, 1970] indicates that early Eocene climates of southern France were, at times, more humid than during the Paleocene and that seasonal rainfall existed [*Freytet and Plaziat*, 1982].

Several brown coal seams from the late Paleocene through the Eocene prove that the humidity in central and eastern Europe was generally high. Humid conditions extended at these latitudes far to the west. Lithofacies analyses from southern England documented semihumid to humid conditions for the Paleocene and the Eocene [*Plint*, 1983]. This conclusion is supported by studies of benthic foraminiferal assemblages which indicate increased freshwater runoff to the Hampshire Basin at least during the Eocene [*Murray and Wright*, 1974]. Different thermophile lithofacies as well as pollen and spores indicate, however, that temperatures were rather warm. *Valeton* [1983a] describes deep chemical weathering from many central European sites (e.g., kaolinite beds from Meissen, Saxonia; Mn crusts in the Bohemian Massif, Spessart). The northernmost occurrence of karst-bauxite is confined to Hungary and Yugoslavia which was located at approximately 40°–35°N by the late Paleocene. Iron pisolites, considered as equivalents of laterites, are observed further to the northwest. In Figure 4 their temporal distribution is shown in the Helvetic facies belt [*Herb*, 1988; *Kerckhoven*, 1980].

The temperature pattern, however, was definitely different when compared between the late Paleogene

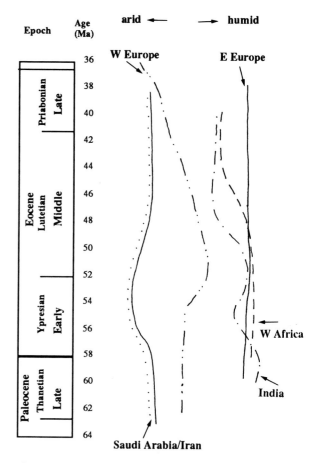

Fig. 6. Schematic climatic trends in the Tethyan realm during the early Paleogene. The humidity/aridity curve is based on spatial and temporal distribution of climatically indicative facies (bauxite, coal, evaporite, glauconites, Fe pisolites, silicified wood, and duricrusts; for discussion, see text). The western and eastern European aridity/humidity curves are based on sediment facies, mainly clay mineralogy and coal deposits, reported by *Freytet and Plaziat* [1982], *Thiry* [1989], and *Hohl* [1985].

and the present. Annual average temperatures were warmer at higher latitudes in the late Paleogene then at present. This is indicated by the intensive chemical weathering between 40° and 50°N during Ypresian and at least part of the Lutetian. Investigations of pollen and spore assemblages support this conclusion [*Krutzsch*, 1967; *Hochuli*, 1984]. Hochuli shows that even during the late Eocene, seasonality of average temperature was very low in western and central Europe. It had increased considerably only by the early Oligocene.

WHEN DID THE TETHYS BECOME A POTENTIAL BOTTOM WATER SOURCE AREA?

The spatial distribution of areas with arid, semiarid, and humid climatic conditions is shown in Figure 7. Comparing the spatial extent of areas where arid condi-

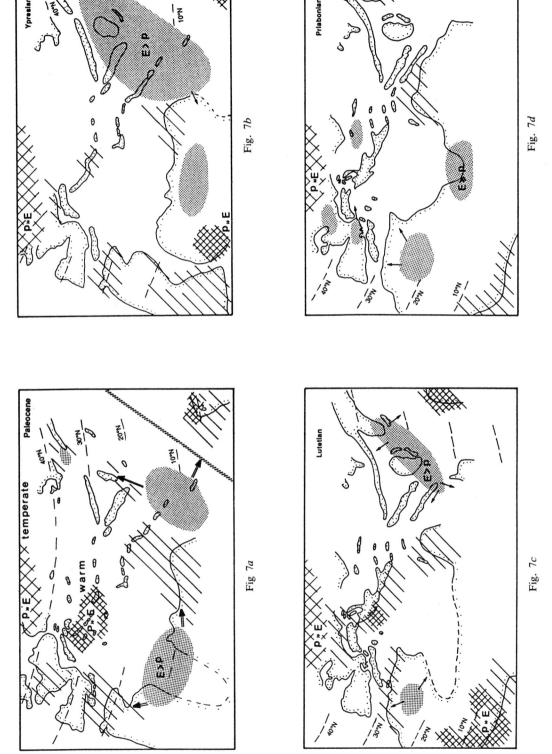

Fig. 7. Spatial distribution of Tethyan regions where humid and arid conditions prevailed during (a) Paleocene, (b) Ypresian, (c) Lutetian, and (d) Priabonian. Crosshatched areas reflect warm and temperate, humid conditions (narrow spacing: warm; wide spacing: temperature). Hatched signature represents the semiarid regions. Dotted shading indicates warm arid conditions. E, evaporation; P, precipitation.

tions prevailed, its broadest expansion is reached in the eastern Tethys during the Ypresian (Figure 7). Before and after this time, its spatial extension was more limited, although by the early/middle (?) Paleocene the arid zone may have expanded in northern Africa, as indicated by the arrows, for a short time interval. In the western Tethys, progressively increased aridity developed only during the late Eocene. Some small basins, for example, the Paris Basin and the Possagno Basin of the southern Alps, showed hypersaline conditions at certain times [*Murray and Wright*, 1974; *Oberhänsli et al.*, 1983]. This climatic change could be related to a regional alteration of the atmospheric circulation pattern. Today the atmospheric circulation is governed by the heat budget and the topography on continents. By the late Lutetian/early Priabonian the tectonic activities in the western Tethys may have increased orogeny. As a result of the major tectonic activities at the northern Tethyan margin, the Alpine chains, the Carpathian Mountains, and the Pontides were elevated to critical levels such as to cause fundamental changes in atmospheric circulation at the northern rim of the Tethys.

This information is critical to an understanding of the paleoceanography of the Tethys. We need answers to the following question: When and at what rate was warm saline bottom water formed during the early Tertiary in the Tethyan shallow water and oceanic realms? Bottom water formation in low-latitude areas resulted when evaporation exceeded precipitation and freshwater runoff from continents over large areas. Such bottom water formation was enhanced by the semiisolation of such seaways to the world ocean, preventing the immediate mixing of normal oceanic waters with saline surface water.

The deepwater connection between the western and eastern Tethys was interrupted at the beginning of the Tertiary. Major interoceanic gateways between the Tethys and the Atlantic and Indian oceans existed throughout the Paleocene and Eocene. A continuous deepwater exchange existed in the Indian Ocean whereas in the Atlantic it was interrupted by the late Eocene. As the continental collision progressed during the early Paleogene, the southern margin of Eurasia was reshaped. Previous major oceanic basins were fragmented and survived as small, occasionally isolated basins. Shelf areas were widespread at the northern and northeastern margin of the African/Arabian continent during the Paleocene and Ypresian. A size reduction of these shallow water areas started by the Lutetian age. The Tunisian/Libyan embayment was reshaped by the folding of the North African hinterland. In the southeastern Tethys a major change was linked to the slow emergence of the Saudi Arabian Peninsula which probably began by the end of the Cretaceous/early Paleocene (Oman Mountains). Fully continental conditions were established there by the Priabonian.

However, bottom water formation does not depend

solely on the conditions in the oceanic realm where dense water forms; the density structure of the subsurface water masses is also crucial. Despite its importance, the question is not further discussed in the context of this paper.

In general, the humidity/aridity pattern observed in the Tethys and its adjacent continents is similar to the present-day pattern. Thus we might expect that warm saline surface water formed in these regions where arid and semiarid conditions prevailed and sank to greater depth owing to higher density. During the early Tertiary the dense water probably left the Tethys in both directions (the Atlantic and Indian oceans). The mechanisms should basically be the same as those in the modern eastern Mediterranean Sea where dense water is produced and sinks and later is exported as intermediate water to the Atlantic. If subsurface density structure allowed the sinking of saline surface water, then bottom water formed in the eastern Tethys during most of the early Tertiary, although the formation rate may have been quite variable. Formation rates were probably the highest during the Ypresian. The newly formed saline bottom water mostly left the Tethyan Ocean at shallow bathyal depths in the direction of the Indian Ocean. In the western Tethys, formation of bottom water might have slightly increased only during the late Eocene, although water exchange at intermediate depth levels with the Atlantic Ocean probably also existed during much of the Paleocene and Eocene. I assume that these exchange rates were proportional to the present-day water exchange between the Atlantic Ocean and the Mediterranean Sea. Only during the Ypresian and probably during the Priabonian was the export of dense surface water to greater depth of the Indian Ocean and Atlantic Ocean, respectively, intensified. The "ocean general circulation model" studies by *Barron and Peterson* [1990] are in favor of this conclusion. They show for the Eocene a deep water flow which was leaving the Tethys in the direction of the Indian Ocean and reached 1500-m depth at 30°S.

Acknowledgments. I am very grateful to Jim Kennett, who sponsored my participation at the August 1991 meeting on Global Change in the Antarctic Climate, in Santa Barbara, and encouraged me to write this contribution. Reviews by A. M. Ziegler and J. T. Parrish strengthened the discussion. I am indebted to the Deutsche Forschungsgemeinschaft, which provides grants for my research.

REFERENCES

Adams, C. G., D. E. Lee, and R. B. Rosen, Conflicting isotopic and biotic evidence for tropical sea-surface temperatures during the Tertiary, *Palaeogeogr. Palaeoclimatol. Palaeoecol.*, *77*, 289–313, 1990.

Auboin, J., X. Le Pichon, and A. S. Monin (Eds.), Evolution of the Tethys, *Tectonophysics*, *123*, 1–315, 1986.

Bardossy, G., and G. J. J. Aleva, Lateritic bauxites, *Dev. Econ. Geol.*, *27*, 1–624, 1990.

Barron, E. J., and W. H. Peterson, The Cenozoic ocean circulation based on ocean general circulation model results, *Palaeogeogr. Palaeoclimatol. Palaeoecol.*, *83*, 1–28, 1990.

Baturin, G. N., Phosphorites on the sea floor, *Dev. Sedimentol.*, *33*, 1–343, 1982.

Benson, R. H., The origin of the psychrosphere as recorded in changes of deep-sea ostracode assemblages, *Lethaia*, *8*, 69–83, 1975.

Berggren, W. A., D. V. Kent, J. J. Flynn, and J. A. Van Couvering, Cenozoic geochronology, *Geol. Soc. Am. Bull.*, *96*, 1407–1418, 1985.

Brass, G. W., J. R. Southam, and W. H. Peterson, Warm saline bottom water in the ancient ocean, *Nature*, *296*, 620–623, 1982.

Brückner, W. D., Laterite and bauxite profiles of West Africa as an index of rhythmical climatic variations in the tropical belt, *Eclogae Geol. Helv.*, *50*, 239–254, 1957.

Cavelier, C., and C. Pomerol, Chronologie et interprétation des événements tectoniques Cénozoiques dans le Basin de Paris, *Bull. Soc. Geol. Fr.*, *VII*, *21*, 33–48, 1979.

Cerujavska, S., Fossil spores and pollen grains of the Priabonian coal basins in Bulgaria and their climatic interpretation, *Abh. Zentr. Geol. Inst.*, *10*, 177–180, 1967.

Chamberlin, T. C., On a possible reversal of deep-sea circulation and its influence on geologic climate, *J. Geol.*, *14*, 363–373, 1906.

Chamley, H., North Atlantic clay sedimentation and paleoenvironment since the Late Jurassic, in *Deep Sea Drilling Results in the Atlantic Ocean: Continental Margins and Paleoenvironment, Maurice Ewing Ser.*, vol. 3, edited by M. Talwani, W. Hay, and W. B. F. Ryan, pp. 342–361, AGU, Washington, D. C., 1979.

Corliss, B. H., Deep-sea benthonic foraminiferal turnover near the Eocene/Oligocene boundary, *Mar. Micropaleontol.*, *6*, 367–384, 1981.

Dercourt, J. (coord.), Géologie des pays européens; Espagne, Grèce, Italie, Portugal, Yougoslavie, *Int. Geol. Congr. 26th*, 1–393, 1980.

Dercourt, J., et al., Geological evolution of the Tethys belt from the Atlantic to the Pamir since the Lias, *Tectonophysics*, *123*, 241–315, 1986.

Freytet, P., and J.-C. Plaziat, Continental carbonate sedimentation and pedogenesis—Late Cretaceous and early Tertiary of southern France, *Contrib. Sedimentol.*, *12*, 1–213, 1982.

Gordon, W. A., Distribution by latitude of Phanerozoic evaporite deposits, *J. Geol.*, *83*, 671–684, 1975.

Grubic, A., Répartition paléogéographique des bauxites dans les Dinarides Yougoslaves, *Proc. Congr. Int. ICSOBA 3rd*, 145–150, 1972.

Hay, W. W., Paleoceanography: A review for the GSA centennial, *Geol. Soc. Am. Bull.*, *100*, 1934–1956, 1988.

Herb, R., Eocaene Paläogeographie und Paläotektonik des Helvetikums, *Eclogae Geol. Helv.*, *81*, 611–657, 1988.

Hochuli, P. A., Correlation of middle and late Tertiary sporomorph assemblages, *Paleobiol. Cont.*, *XIV*, 301–314, 1984.

Hohl, R. (Ed.), *Die Entwicklungsgeschichte der Erde*, pp. 1–703, Interdruck Leipzig, Verlag W. Dausien, Hanau, Germany, 1985.

Horrell, M. A., Energy balance constraints on 18-O based paleo-sea surface temperature estimates, *Paleoceanography*, *5*, 339–348, 1990.

Kazmin, V. G., I. M. Sbortshikov, L.-E. Ricou, L. P. Zonenshain, J. Boulin, and A. L. Knipper, Volcanic belts as markers of the Mesozoic-Cenozoic active margin of Eurasia, *Tectonophysics*, *123*, 123–152, 1986.

Kemp, E. M., Tertiary climatic evolution and vegetation history in the southeast Indian Ocean region, *Palaeogeogr. Palaeoclimatol. Palaeoecol.*, *24*, 169–204, 1978.

Kennett, J. P., and L. D. Stott, Proteus and Proto-Oceanus: Ancestral Paleogene oceans as revealed from Antarctic stable isotope results, ODP Leg 113, *Proc. Ocean Drill. Program Sci. Results*, *113*, 865–880, 1990.

Kerckhoven, C., C. Caron, J. Charollais, and J.-L. Pairis, Panorama des séries synorogéniques des Alpes occidentales, *Mem. BRGM*, *107*, 234–255, 1980.

Kozary, M. T., J. C. Dunlap, and W. E. Humphrey, Incidence of saline deposits in geologic time, *Spec. Pap. Geol. Soc. Am.*, *88*, 43–57, 1968.

Krutzsch, W., Der Florenwechsel im Alttertiär Mitteleuropas auf Grund von sporenpaläontologischen Untersuchungen, *Abh. Zentr. Geol. Inst.*, *10*, 17–37, 1967.

McGowran, B., The tertiary of Australia: Foraminiferal overview, *Mar. Micropaleontol.*, *4*, 235–264, 1979.

Mead, G. A., D. A. Hodell, and P. F. Ciesielski, Late Eocene to early Oligocene vertical oxygen isotopic gradients in the South Atlantic: Implications for warm saline deep water, in press, 1992.

Miller, K. G., R. G. Fairbanks, and G. S. Mountain, Tertiary oxygen isotope synthesis, sea level history and continental margin erosion, *Paleoceanography*, *2*, 1–19, 1987.

Mohr, B., Palynologischer Nachweis eines Mangrovenbiotops in der Südatlas-Randzone (Marokko) und seine paläoökologische Bedeutung, *Doc. Nat.*, *33*, 20–28, 1986.

Murray, J. W., and C. A. Wright, Paleogene foraminiferida and palaeoecology, Hampshire and Paris basins and the English Channel, *Spec. Pap. Palaeontol.*, *14*, 1–129, 1974.

Oberhänsli, H., and K. J. Hsü, Paleocene-Eocene paleoceanography, *Geodyn. Ser.*, *15*, 85–100, 1986.

Oberhänsli, H., A. Grünig, and R. Herb, Oxygen and carbon isotope study in the late Eocene sediments of Possagno (northern Italy), *Riv. Ital. Paleontol. Stratigr.*, *89*, 377–394, 1983.

Parrish, J. T., A. M. Ziegler, and C. R. Scotese, Rainfall patterns and the distribution of coals and evaporites in the Mesozoic and Cenozoic, *Palaeogeogr. Palaeoclimatol. Palaeoecol.*, *40*, 67–101, 1982.

Plaziat, J. C., Huîtres de mangrove et peuplements littoraux de l'Eocène inférieur des Corbières, les mangroves fossiles comme élément déterminant de paléoécologie littorale et de paléoclimatologie, *Geobios*, *3*, 7–27, 1970.

Plint, A. G., Facies, environments and sedimentary cycles in the middle Eocene, Bracklesham Formation of the Hampshire Basin: Evidence for global sea-level changes?, *Sedimentology*, *30*, 625–653, 1983.

Pomerol, C. (Ed.), *Stratotypes of Paleogene Stages*, pp. 1–301, International Union of Geological Sciences, Committee on Stratigraphy, International Subcommittee, Paleogene Stratigraphy, Paris, 1981.

Prasad, G., A review of the early Tertiary bauxite event in South America, Africa and India, in Geology for Development, Mineral Resources and Exploration Potential of Africa, *J. Afr. Earth Sci.*, *1(3-4)*, 305–313, 1983.

Robert, C., Climats et courants Cénozoiques dans l'Atlantique sud d'après l'étude des minéraux argileux (Legs 3, 39, et 40 DSDP), *Oceanol. Acta*, *3*, 369–376, 1980.

Ronov, A., V. Kain, and A. Balushovsky, *Atlas of Lithological-Paleogeographical Maps of the World: Mesozoic and Cenozoic of Continents and Oceans*, pp. 1–80, USSR Academy of Sciences, Leningrad, 1989.

Royden, L. H., and T. Bàldi, Early Cenozoic tectonics and paleogeography of the Pannonian and surrounding regions, *Mem. Am. Assoc. Pet. Geol.*, *45*, 1–16, 1988.

Savostin, L. A., J.-C. Sibuet, L. P. Zonenshain, X. Le Pichon, and M.-J. Roulet, Kinematic evolution of the Tethys belt from the Atlantic Ocean to the Pamirs since the Triassic, *Tectonophysics*, *123*, 1–36, 1986.

Shackleton, N. J., and A. Boersma, The climate of the Eocene ocean, *J. Geol. Soc. London*, *138*, 153–158, 1981.

Stöcklin, J., Salt deposits of the Middle East, *Spec. Pap. Geol. Soc. Am.*, *88*, 157–181, 1968.

Tapponier, P., Evolution tectonique du système alpin en Méditerranée, poinçonnement et ecrasement rigide-plastique, *Bull. Soc. Geol. Fr.*, *VII*, 19, 437–460, 1977.

Thiry, M., Sédimentation continentale et altérations associées: Calcitisations, ferruginisations et silicifications les argiles plastiques du Sparnacien du Bassin de Paris, *Sci. Géol. Mem.*, *64*, 1–173, 1981.

Thiry, M., Geochemical evolution and paleoenvironments of the Eocene continental deposits in the Paris Basin, *Palaeogeogr. Palaeoclimatol. Palaeoecol.*, *70*, 153–163, 1989.

Trappe, J., Microfacies zonation and spatial evolution of a carbonate ramp: Marginal Moroccan phosphate sea during the Paleogene, *Geol. Rundsch.*, *81*, 105–126, 1992.

Trümpy, R., *Geology of Switzerland*, part A, pp. 1–180, Wepf and Co., Basel, Switzerland, 1980.

Valeton, I., Klimaperioden lateritischer Verwitterung und ihr Abbild in den synchronen Sedimentationsräumen, *Z. Dtsch. Geol. Ges.*, *134*, 413–452, 1983a.

Valeton, I., Paleo-environment of lateritic bauxites with verti-cal and lateral differentiation, *Geol. Soc. Spec. Publ. London*, *11*, 77–91, 1983b.

Wildi, W., La chaîne tello-rifaine (Algérie, Maroc, Tunisie): Structure, stratigraphie et évolution du Trias au Miocène, *Rev. Geol. Dyn. Geogr. Phys.*, *24*, 201–297, 1983.

Wolfe, J. A., A paleobotanical interpretation of Tertiary climates in the northern hemisphere, *Am. Sci.*, *66*, 694–703, 1978.

Wüst, G., Gesetzmässige Wechselbeziehung zwischen Ozean und Atmosphäre in der zonalen Verteilung der Oberflächensalzgehalte, Verdunstung und Niederschläge, *Arch. Meteorol. Geophys. Bioklimatol.*, Ser. A, 7, 305–328, 1954.

Zaklinskaja, E. D., The early Paleogene flora of the northern hemisphere and paleofloristic provinces of this age, *Abh. Zentr. Geol. Inst.*, *10*, 183–187, 1967.

Ziegler, A. M., A. L. Raymond, T. C. Gierlowski, M. A. Harrell, D. B. Rowley, and A. L. Lottes, Coal, climate and terrestrial productivity: The Present and Early Cretaceous compared, *Geol. Soc. Spec. Publ. London*, *32*, 25–49, 1987.

(Received December 14, 1991;
accepted May 4, 1992.)

THE ANTARCTIC PALEOENVIRONMENT: A PERSPECTIVE ON GLOBAL CHANGE
ANTARCTIC RESEARCH SERIES, VOLUME 56, PAGES 185–202

LATE EOCENE–OLIGOCENE SEDIMENTATION IN THE ANTARCTIC OCEAN, ATLANTIC SECTOR (MAUD RISE, ODP LEG 113, SITE 689): DEVELOPMENT OF SURFACE AND BOTTOM WATER CIRCULATION

LISELOTTE DIESTER-HAASS[1]

Alfred Wegener Institut, 2850 Bremerhaven, Germany

Variations in opal and planktonic foraminifera accumulation rates in sediments from Ocean Drilling Program Site 689 on Maud Rise, Antarctic Ocean, indicate that at least 18 "productivity" cycles of 400–500 kyr are present in the Oligocene sediments and two cycles in the investigated part of the late Eocene. Accumulation rates of planktonic foraminifera are low, and their tests are easily dissolved during high-productivity periods, because of the high supply of organic matter. Clastic material is present only in high-productivity periods. Cyclic cooling-warming with a period of 400–500 kyr in the Antarctic area and related cyclic sea level changes are assumed to be responsible for these sedimentological changes. In four short periods of roughly 100- to 400-kyr duration (28–28.1, 29.1–29.2, 33.6–34, and 38–38.4 Ma), excellent carbonate preservation occurred in spite of high productivity in surface waters. This is tentatively attributed to the presence of warm saline deep water (WSDW). Intense carbonate dissolution during three periods (31–32, 34.4–35, and 37–38 Ma), in spite of low productivity, is attributed to the presence of Antarctic bottom water (AABW) in the area of Site 689. The study suggests that in the late Eocene–Oligocene, a four-layered ocean existed in the Weddell Sea area, with AABW, WSDW, Antarctic intermediate water, and a surface water layer and that these water masses shifted vertically and latitudinally during the Oligocene.

INTRODUCTION

The Oligocene was an interval of major paleoceanographic and paleoclimatic change. The early Oligocene is marked by a large decrease in temperature [*Shackleton and Kennett*, 1975; *Kennett and Stott*, 1990; *Mead et al.*, 1992], and ice cover on the Antarctic continent increased significantly during this period [*Kennett and Stott*, 1990; *Wei*, 1991; *Ehrmann and Mackensen*, 1992]. The first appearance of Antarctic Bottom Water (AABW) during this time had a significant impact on the deepwater circulation in all oceans (*Kennett and Stott* [1990] and references listed below).

These events left distinct physical and chemical signatures in ocean sediments. In particular, the Maud Rise, in the Atlantic sector of the Antarctic Ocean, contains a well-preserved sequence of Eocene-Oligocene sediments. The long-term trends of Eocene-Oligocene sedimentation and oceanography have been described previously by *Barker et al.* [1990], *Stott et al.* [1990], *Kennett and Stott* [1990], and *Diester-Haass* [1991*a*]. In this study I investigated the high-resolution information on bottom and surface water circulation contained in those sediments that were deposited during the intervals of major turnover. The question of the existence or nonexistence of warm saline deep water (WSDW) [*Brass et al.*, 1982; *Kennett and Stott*, 1990; *Thomas*, 1990; *Mead et al.*, 1992], the development of Antarctic Bottom Water, as well as the problem of cyclicity of paleoceanographic events will be examined here.

Location of Site 689, Maud Rise

Maud Rise is located at about 62°–66°S, 0°–5°E in the Atlantic sector of the Antarctic Ocean, 1000 km south of the present polar front, 100 km north of the Antarctic divergence, and 700 km north of the Antarctic continent. Maud Rise reaches depths as shallow as 1800 m with surrounding water depths greater than 5000 m. Sediments on Maud Rise have been drilled at two water depths: Ocean Drilling Program (ODP) Site 689 was drilled near the crest of Maud Rise in 2080-m water depth, and ODP Site 690, in 2914 m; for studies of the history of vertical water mass stratification, see *Barker et al.* [1988] (Figure 1).

---

[1]Now at Fachrichtung Geographie, Universität des Saarlandes, 6600 Saarbrücken, Germany.

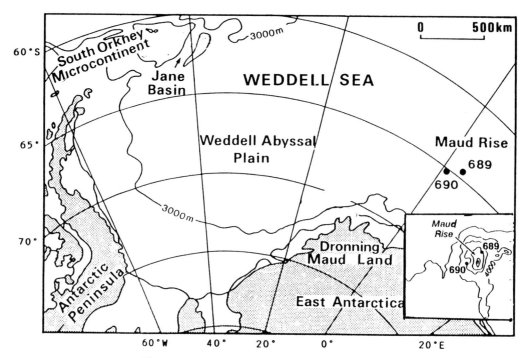

Fig. 1.   Location of ODP Leg 113 sites 689 and 690.

Samples and Procedures

Site 689 is the first and southernmost site in the Antarctic Ocean that contains a relatively complete sediment sequence with well-preserved siliceous and benthic and planktonic calcareous microfossils. The site is isolated from terrigenous supply from the nearby continent that would complicate the sedimentary record.

The Oligocene sediments at Maud Rise Site 689 consist of light colored diatom- and radiolarian-bearing nannofossil oozes [*Barker et al.*, 1988]. Core photographs and visual descriptions indicate that the sediments consist of alternating lighter and darker units. The results from this study show that darker sediments have greater opal abundances than lighter ones. Samples were collected at 20-cm intervals, and the coarse fractions were analyzed. The sediments (10 cm³) were freeze dried, weighed, and wet sieved on 40- and 63-μm sieves. The residue (<40 μm) was dried and weighed, and the >63-μm fraction was dry sieved into five subfractions (63–125, 125–250, 250–500, 500–1000, and >1000 μm). About 800 grains (if present) were counted from each fraction and assigned to one of about 25 biogenous, terrigenous, and authigenic categories. The composition of the 40- to 63-μm fraction and of the sand fraction was calculated, following the method of *Sarnthein* [1971], by multiplying the percentage of each component in each fraction with the weight percent of the fraction of the total sand fraction. Finally, these

products (percent component times percent weight of fraction) are summed and give the percentage of the total sand fraction component.

Parameters Used in Interpretation

The coarse fraction data acquired in this study provide information on the changes in carbonate dissolution, the variations in abundance of radiolaria and diatoms, and paleoceanographic significance of these changes.

Accumulation rates of the main components have been calculated according to the formula

$$\text{comp. MAR} = \text{LSR} \times \text{DBD} \times \% \text{ comp.}/100$$
$$\times 1000 \quad \text{mg/cm}^2 \text{ kyr}$$

where comp. MAR is the mass accumulation rate of the component in mg/cm² kyr; LSR is the linear sedimentation rate, calculated from the ages given by *Kennett and Stott* [1990, p. 879] and *Thomas et al.* [1990] and meter below seafloor data in centimeters per kiloyear (Figure 2); DBD is dry bulk density in grams per cubic meter from *Barker et al.* [1988] calculated from wet density and porosity data with the formula WBD − 1.026 × porosity/100 (Figure 3); and "comp." is the percentage of component of total sediment.

The amount of opal derived from diatoms and radiolaria in marine sediments is considered to be an indica-

Site 689B, linear sedimentation rate, LSR

Site 689, bulk mass accumulation rate, MAR

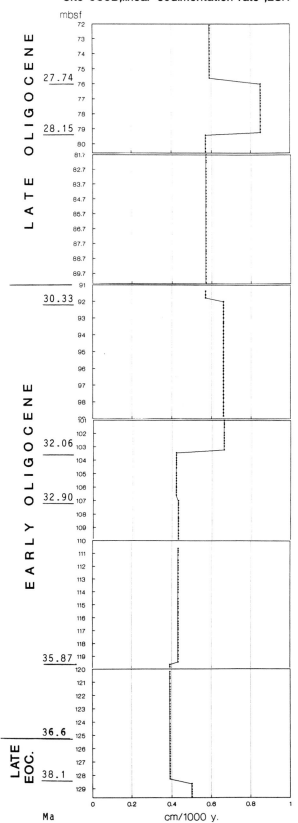

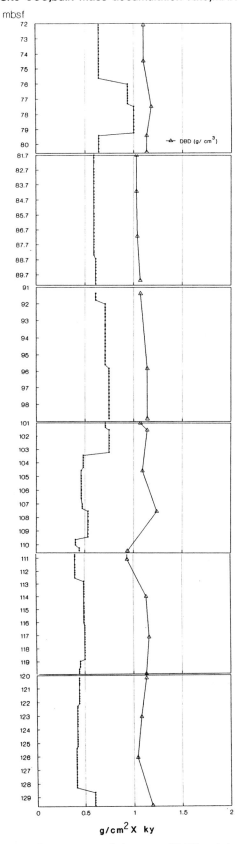

Fig. 2.   Linear sedimentation rates (LSR) with ages used for calculation of MAR from *Thomas et al.* [1990] and *Kennett and Stott* [1990].

Fig. 3.   Bulk mass accumulation rates (MAR) and dry bulk density (DBD) from *Barker et al.* [1988].

tor of surface water nutrient concentrations [*Goll and Bjørklund*, 1974; *Berger*, 1976; *Leinen et al.*, 1986; *Pokras*, 1986; *Baldauf and Barron*, 1990]. Higher nutrient levels in surface waters result in greater diatom (opal) productivity and less opal dissolution. The >40-μm-sized opal fraction is only a small part of the total opal input, but variations in >40-μm opal size fractions are proportional to those of the bulk sediment [*Leinen et al.*, 1986] and can thus be used as a reliable indicator of opal content.

Dissolution of carbonate was determined in these sediments by determining the extent of fragmentation of planktonic foraminifera and the benthic/planktonic foraminiferal ratio [*Berger*, 1973; *Thunell*, 1976]. Although both dissolution parameters yielded similar results, the benthic/planktonic foraminiferal ratio was more sensitive to dissolution. Fragmentation of planktonic foraminifera occurs from sample preparation and from diagenesis, for example, cementation with increasing depth in the sediments [*Diester-Haass*, 1988], thus affecting fragmentation indices.

Dissolution of carbonate can be produced by two different mechanisms: (1) by the chemical properties of bottom waters, leading to a decrease in carbonate ion content and thus an increase in carbonate dissolution with water depth, until the carbonate compensation depth (CCD) is reached, where all carbonate is dissolved [*Broecker*, 1981]; (2) by CO_2 generated from the decomposition of sedimentary organic matter, which dissolves the carbonate tests. The supply of organic matter to the seafloor generally decreases with increasing water depth [*Dymond and Lyle*, 1985; *Diester-Haass*, 1991*b*].

During the investigated period of the late Eocene to late Oligocene, the water depth of the site did not change enough to contribute markedly to variations in carbonate dissolution. The observed variations in carbonate dissolution can be attributed either to considerable changes in the provenance and composition of bottom water masses and/or to changes in the rates of supply of organic matter.

Only very few data on organic carbon content in the Eocene-Oligocene of Site 689 are available [*Egeberg and Abdullah*, 1990] and these are below the detection limit. Diagenetic overprint destroyed the original signal. On the basis of observations of Modern and Neogene sediments, it is also likely that during the Paleogene increasing productivity in surface waters, reflected for instance by increasing opal MAR, is accompanied by increasing organic matter supply to the seafloor and corresponding increase in the strength of carbonate dissolution [*Thiede et al.*, 1982; *Diester-Haass et al.*, 1986; *Oberhänsli et al.*, 1990; *Diester-Haass*, 1983, 1991*b*]. On the basis of this analysis, I infer variations in organic matter supply in the absence of organic carbon data. I also infer that changes in carbonate dissolution in one site, which remained in the same

water depth and in the same water mass, allow determination of changes in organic matter supply rates.

RESULTS

The following results have been obtained from sediments of Hole 689B, cores 9 through 14. The age of the sediments is latest Eocene in Core 14 (38.1 Ma) to latest Oligocene in Core 9 (27.5 Ma). Ages are from *Thomas et al.* [1990] and *Kennett and Stott* [1990] (Figure 2). The concentration of the 40- to 63-μm and >63-μm fractions varies between 1% and 11% (Figure 4). These values are very low but provide reliable data on varying radiolarian and diatom abundances and on the proportion and preservation of benthic and planktonic foraminifera.

The variation in sand fraction percentages indicates a correlation between carbonate preservation, accumulation rate of planktonic foraminifera, and percent sand fraction. In most of the units with maximum percent sand values the preservation of planktonic foraminifera is good and MAR is highest.

The concentration of siliceous and carbonate microfossils in the sand fraction (Figures 5, 6, and 7) is highly variable, for example, radiolarian percentages vary from <10% up to 80% (Figure 5). In intervals characterized by high radiolarian concentrations, diatoms are also present (Figure 5) (and silicoflagellates which are not shown here, because they occur mainly in <63-μm fractions). Planktonic foraminiferal percentages vary antithetically to opal abundances and range between 65% and <5% (Figure 6). Benthic organisms vary between <5% and 60% (Figure 7).

Accumulation rates of radiolaria and diatoms are shown in Figure 8. Opal maxima are characterized by the highest accumulation rates of radiolaria (16–28 mg/cm^2 kyr). Diatom accumulation rates in the maxima are 2–12 mg/cm^2 kyr.

Accumulation rates of benthic foraminifera (Figure 9) show variations between 0.4 and 8 mg/cm^2 kyr in the late Oligocene and 0.4 to 12 mg/cm^2 kyr in the early Oligocene–late Eocene. Planktonic foraminiferal MARs, however, range from <0.4 up to 32 mg/cm^2 kyr.

In general, the maxima in planktonic foraminiferal accumulation rates correspond with minima in radiolarian accumulation rates and vice versa (units marked with a vertical line in Figure 9). Units marked with a different signature in Figure 9 are those in which accumulation rates of both components vary in concert.

Benthic foraminiferal accumulation rate variations are similar to radiolarian accumulation rate changes with maxima in benthic foraminiferal accumulation rates occurring with maxima in opal MAR.

The extent of dissolution of calcareous tests was determined by means of the benthic/planktonic foraminiferal ratio and the fragmentation of planktonic foraminifera (Figure 10). Both curves exhibit similar temporal

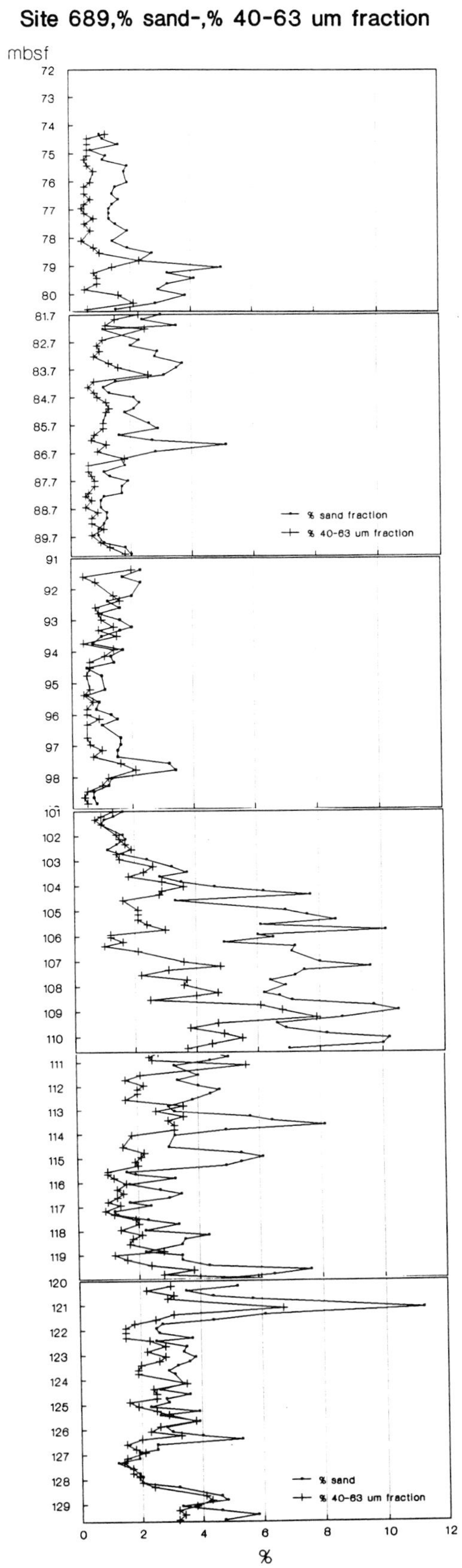

Site 689,% sand-,% 40-63 um fraction

mbsf

Site 689, % diatoms and % radiolaria

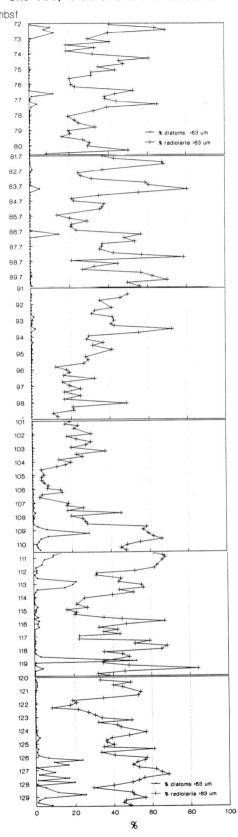

Fig. 4. Percent sand and percent 40- to 63-μm fraction in Oligocene sediments (cores 9–14).

Fig. 5. Composition of the sand fraction (percent diatoms and radiolaria).

Site 689, % plankton foraminifera

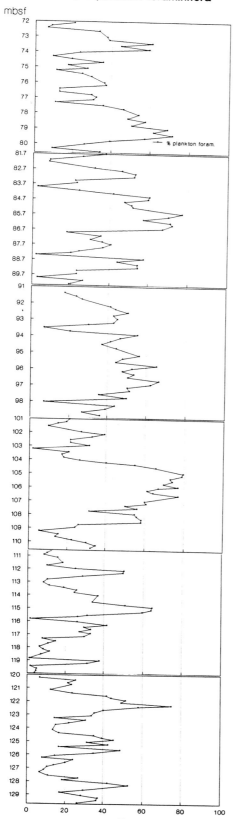

Fig. 6. Composition of the sand fraction (percent planktonic foraminifera).

Site 689, % benthos

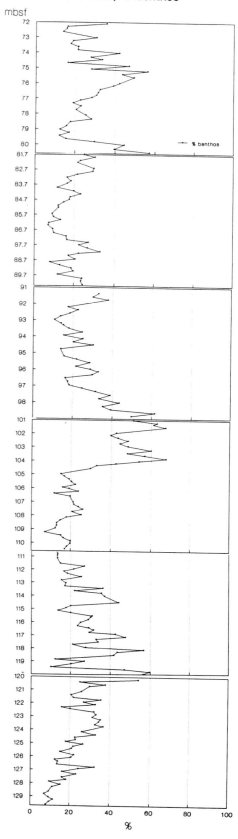

Fig. 7. Composition of the sand fraction (percent benthos: foraminifera plus sponges plus ostracods plus echinoids).

Site 689B,accumulation rates diat,rad.

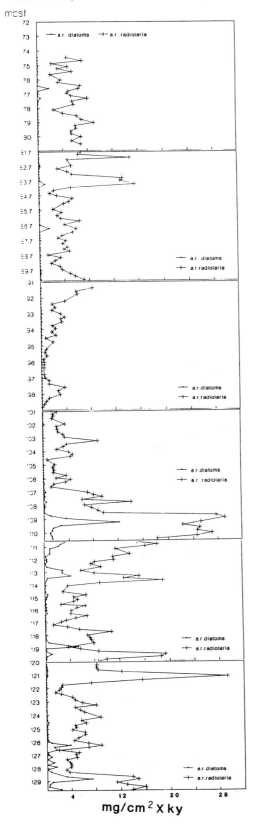

mg/cm^2 x ky

variations, but the variations in the benthic/planktonic foraminiferal ratios have a much larger amplitude.

The value of benthic/planktonic foraminiferal ratios in the Oligocene paleowater depth of about 1500 m [*Barker et al.*, 1990] would be about 5–10, if no dissolution had occurred [*Berger and Diester-Haass*, 1988]. The higher values found in these sediments indicate that very strong carbonate dissolution has occurred in those intervals in which opal contents are highest (Figure 5). Planktonic foraminifera are best preserved in the opal-poor units. Strong dissolution of planktonic foraminifera is always related to low plankton foraminiferal accumulation rates and vice versa (Figure 10).

Foraminifera are the dominant (>75%) constituent of the benthic fraction. Sponges, and sometimes ostracods and echinoids, comprise the remainder of the benthos (Figure 11). Accumulation rates of sponge spicules (Figure 12) covary with those of radiolaria and are probably dissolution controlled: the higher the radiolarian input, the better the opal preservation.

Accumulation rates of ostracods (Figure 13) parallel the carbonate dissolution curve and thus show that ostracod abundances are most likely dissolution controlled. Accumulation rates of echinoids (mainly spines) do not show an obvious relation to any of the other parameters (Figure 13). They are absent between 96 and 117 meters below seafloor (mbsf).

Bolboforma spp. occurs in cores 13 and 14 of Hole 689B in strongly varying percentages. The accumulation rates are up to 2.8 mg/cm^2 kyr, and maxima correlate to radiolarian accumulation rate maxima (Figure 14).

The accumulation rates of terrigenous matter (>63 μm, i.e., quartz, mica, rock fragments) (Figure 15) are very low (maxima of 0.16 mg/cm^2 kyr), and often no terrigenous matter >40 μm is present. These clastic particles always occur in opal-rich units (Figure 5), whereas opal-poor units do not contain clastic material. Volcanic glass (Figure 16) is rare (0–2% of the sand or 40- to 63-μm fraction). Pyrite is absent in the late Eocene–early Oligocene (Figure 17) and appears between 101 and 110 mbsf with values up to 18%, varying between 0 and 4% at 72–99 mbsf. Authigenic minerals (Figure 17) occur only between 120 and 121 mbsf.

The calculation of MAR of individual components suffers from the closed-sum problem; if one component decreases, the other must increase. This is especially important for the MARs of opal and planktonic foraminifera, which vary antithetically in most intervals. Since MARs change in intervals with constant LSR, these variations are due to changes in percent opal and planktonic foraminifera. Some general reflections help to solve the problem of differentiating between closed-sum problems and real production changes.

Fig. 8.   (Opposite) Accumulation rates (MAR) of diatoms and radiolaria (>63 μm).

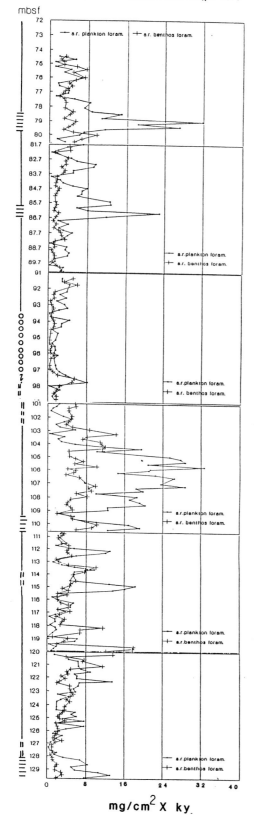

Site 689B,accumulation rates b.,pl. for.

mg/cm^2 X ky.

The following thoughts are the basis for the interpretation of the data:

The key to understanding is the carbonate dissolution, which increases without exception in all units with low percent and low MARs of planktonic foraminifera. I conclude that variations in planktonic foraminiferal percent values and MARs are driven by carbonate dissolution.

What was responsible for the numerous short-term changes in carbonate dissolution? As was previously mentioned, changes in water depth can be excluded. Changes in organic matter supply or changes in bottom water mass chemistry are possible mechanisms to explain the numerous changes in carbonate dissolution and planktonic foraminiferal MARs. Changes in bottom water mass chemistry most probably occurred (see the discussion sections). However, these changes were less frequent and less regular than those of the organic matter supply. Organic carbon contents are below detection limits because of diagenesis [*Egeberg and Abdullah*, 1990]. Therefore my inferences cannot be confirmed by data. The increase in MAR of radiolaria in the levels with strong carbonate dissolution and low MAR of planktonic foraminifera might be either a simple dilution phenomenon or a real productivity increase. The latter assumption is confirmed by the observation that diatoms and silicoflagellates occur only in the opal maxima units. Preservation of these solution-sensitive organisms is only possible in high-productivity areas/periods, where SiO_2 input is so high as to compensate for pore water undersaturation. If only varying dilution by planktonic foraminifera were responsible for the MAR variations of opal and productivity had remained constant, diatoms and silicoflagellates should be present in all samples. So two factors pointing to higher productivity and higher organic C input coincide in the opal-rich units: strong carbonate dissolution and occurrence of solution-sensitive opal skeletons.

These considerations led to the conclusion that the variations in MAR of opal and plankton foraminifera are not simple dilution phenomena in a closed-sum system, but that they reflect changes in productivity and thus in oceanography that are described below.

DISCUSSION

The variations in accumulation rates of carbonate and siliceous microfossils are interpreted to be the result of

Fig. 9. (Opposite) MAR of benthic and planktonic foraminifera. Vertical line, maxima in plankton foraminiferal MAR correlate with minima in radiolarian MAR. Horizontal lines, maxima in plankton foraminiferal MAR and good preservation correlate with maxima in radiolarian MAR (influence of WSDW?). Double vertical lines, small plankton foraminiferal MAR and strong carbonate dissolution correlate with low radiolarian MAR (influence of AABW?). Circles, low-productivity period, low MARs of radiolaria and foraminifera, good carbonate preservation.

Site 689 B,benth/pl.foram. ratio;fragm.

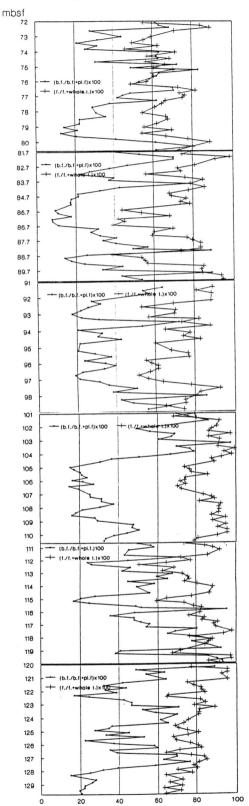

689B, composition of the benthos

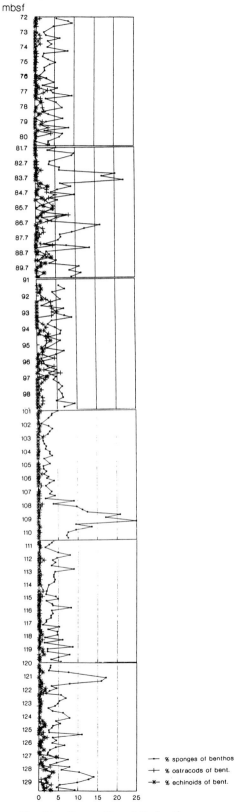

Fig. 10.   Dissolution of carbonates: benthos/plankton foraminiferal ratio, calculated as (percent benthos/percent benthos
+ percent plankton) × 100. Fragmentation of planktonic foraminifera, calculated as (percent fragments/percent fragments +
percent whole tests) × 100.

Fig. 11.   Composition of the benthos.

Site 689B,accumulation rates fish,sponge

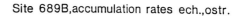

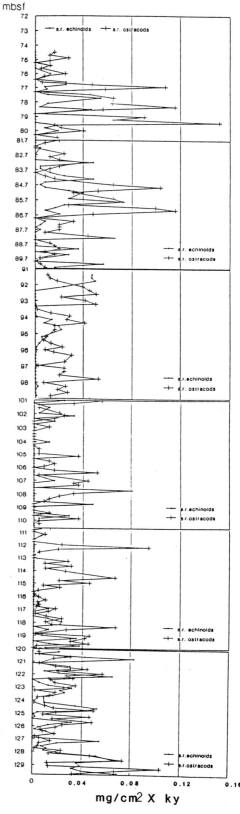

Fig. 12.   Accumulation rates of sponges and fish debris.

Fig. 13.   Accumulation rates of ostracods and echinoderms.

Site 689B, accum. rates Bolboforma

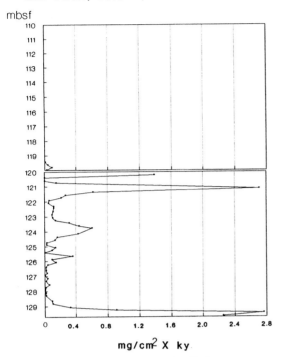

Fig. 14.   Accumulation rates of *Bolboforma* spp.

Site 689B, accumulation rate terrig.mat.

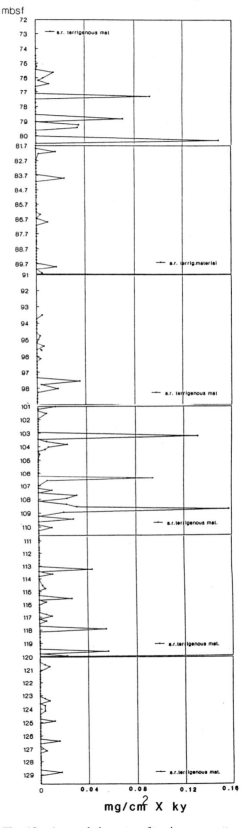

Fig. 15.   Accumulation rates of terrigenous matter.

productivity variations and dissolution which, in turn, are controlled by the chemistry of surface and bottom waters.

Productivity Cycles and Related Cycles in Carbonate Dissolution

The sediment composition shows cyclic variations during the late Eocene–Oligocene of Site 689 (Figures 5–7). Eighteen zones of maxima in diatom and radiolarian content are separated by opal minima with less than 20–30% opal. These changes occur 18 times in the 36.6 to 27.5 Ma interval, i.e., at a mean duration of about 500 kyr for each cycle. When calculating cycle length for the early and late Oligocene separately, durations of 530 and 362 kyr are found. I assume a duration of 400–500 kyr. High diatom accumulation rates are attributed to high surface water productivity. The increased supply of diatoms to the seafloor during periods of high productivity protects the siliceous microfossils against dissolution, by compensating the SiO_2 undersaturation of the pore waters, so that accumulation rates of all opal-producing organisms increase with increasing productivity of surface waters.

The strong variations during the Oligocene between high and low surface water productivity cannot be attributed to a polar front and its latitudinal migrations, as it is in the present-day and late Glacial ocean [*Corliss and Thunell*, 1983]. The similarity of floras from the Maud Rise and the Falkland Plateau in the Oligocene

Site 689 B, % volcanic glass

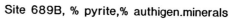
Site 689B, % pyrite,% authigen.minerals

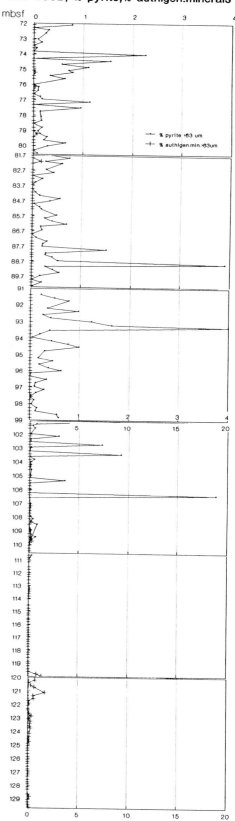

Fig. 16. Percent volcanic glass in the sand and 40- to 63-μm fraction.

Fig. 17. Percent pyrite and percent authigenic minerals in the sand and 40- to 63-μm fraction. Note the change in scale at 99 mbsf.

excludes a strong zonal isolating feature like the present polar front [*Kennett and Barker*, 1990]. A circum-Antarctic circulation began probably after the opening of the Drake Passage, which is considered to have occurred between 30 and 22 Ma [*Kennett*, 1977; *Wise et al.*, 1985; *Barker and Burrell*, 1977].

The productivity cycles in the Oligocene are attributed to cyclic variations in latitudinal temperature gradients. It is assumed that cooling in the Antarctic area leads to stronger temperature gradients, with more vigorous atmospheric and oceanic circulation, more turnover, and thus more nutrient supply from deeper water layers to the surface.

The occurrence of 18 productivity cycles in the Oligocene (27.3–36.6 Ma) can be interpreted as a consequence of at least 18 cooling/warming periods in the Antarctic area on a time scale of about 400–500 kyr. A similar cyclicity of about 0.5 m.y. of cool-warm oscillations is described by *Keller* [1983]. Shorter periodicities (e.g., 44 and 26 kyr) as described by *Mead et al.* [1986] for Oligocene sediments from the South Atlantic could not be found because the sampling interval was too large for the sedimentation rate at this site.

The radiolarian accumulation rate maxima correlate with plankton foraminiferal accumulation rate minima and vice versa in those intervals marked by a vertical line on the left side of Figure 9 (for ages, see Figure 19, "normal conditions"). These units also exhibit the strongest dissolution of planktonic foraminifera (Figure 10). Increased productivity does not only imply higher opal accumulation rates to the seafloor, but probably also higher organic matter input. This organic matter is decomposed by bacteria and CO_2 is generated, which dissolves the calcareous planktonic foraminiferal tests. The fact that high productivity enhances carbonate dissolution explains the direct contrast of the curves of opal and plankton foraminiferal accumulation rates and the correlation between opal content and carbonate dissolution.

These results allow me to conclude that plankton foraminiferal accumulation rates are dissolution controlled. The question of how much productivity variations affect the MAR record cannot be answered here, because dissolution destroys the information needed to determine this, except when low accumulation rates correlate with good preservation (see discussion below).

In the investigated Oligocene sequence, 27.3–36.6 Ma, 18 productivity cycles have been found, which I attribute to cyclic temperature variations and thus variations in the water masses, which affect chemistry and hence microfossil production and dissolution. The average duration is about 400–500 kyr. Each cycle comprises a high productivity interval, attributed to cooler temperatures in the Antarctic area with high opal accumulation rates, strong carbonate dissolution, and thus low accumulation rates of planktonic foraminifera and a warmer, low-productivity interval with good preservation and high accumulation rates of planktonic foraminifera. On the basis of these data, the productivity in the early Oligocene was nearly twice as high as in the late Oligocene (opal accumulation rate maxima up to 30 mg/cm^2 kyr compared with maxima of 15 mg/cm^2 kyr.

Several sharp, longer-term changes in productivity are seen in the MAR records, and these changes coincide with major $\delta^{18}O$ changes and with changes in the eustatic sea level curve of *Haq et al.* [1987] (Figure 18). The 36 Ma event [*Haq et al.*, 1987] or the 35.8 Ma event [*Miller et al.*, 1991; *Keller*, 1983] consists of a major regression and $\delta^{18}O$ shift toward heavier values (3°–4°C temperature drop) [*Kennett and Stott*, 1990; *Stott et al.*, 1990; *Mead et al.*, 1992] and increase in Antarctic ice volume [*Wei*, 1991] which coincides with the 122–121 mbsf major increase in opal accumulation rates. Within 45 cm, i.e., 117 kyr, a fundamental change in oceanography occurred, from a low to a very high productivity ocean. Accumulation rates of radiolaria increased by a factor of 4. Another high-productivity period as determined in this study at about 34–33.5 Ma, 112–109 mbsf, is seen only in very weak excursions to higher $\delta^{18}O$ and lower sea level.

At about 33 Ma (109–108 mbsf) at Site 689, a change toward a low-productivity ocean (decrease in radiolarian accumulation rate by a factor of 3) took place within 20 cm, i.e., about 50 kyr. This event might correspond to the warm interval of *Miller et al.* [1991] and *Stott et al.* [1990].

The major increase in productivity (radiolarian accumulation rate increase by a factor of 5) at 92–91 mbsf (i.e., about 30 Ma) within 100 cm (i.e., 150 kyr) probably is part of the Oi2 event of *Miller et al.* [1991], which was dated at 31.5 (32.5)–28 Ma and corresponds to *Keller*'s [1983] major faunal turnover and rapid cooling. It also coincides with the major sea level drop of *Haq et al.* [1987] and *Schlanger and Premoli-Silva* [1986] and the ice growth period of *Miller et al.* [1987] and *Miller et al.* [1991]. The productivity maximum at 28.5 Ma correlates to a sea level drop [*Haq et al.*, 1987] and the upper part of the Oi2 cooling event [*Miller et al.*, 1991].

The comparison of productivity changes in Site 689 and $\delta^{18}O$ and sea level variations from other areas reveals that the changes are simultaneous but that a small change of one parameter is accompanied by a large change of another parameter.

A correlation of the numerous small-scale productivity variations with sea level and oxygen isotope data cannot yet be established because high-resolution data are not available in these other studies and because age constraints are too sparse. Future investigations will have to test the assumption that cyclic temperature variations are responsible for the productivity cycles in the Antarctic Ocean and that these correlate with ice volume and sea level changes in the Oligocene.

In the data there are three exceptions to the general observation that high (low) planktonic foraminiferal MARs correlate with low (high) opal MARs.

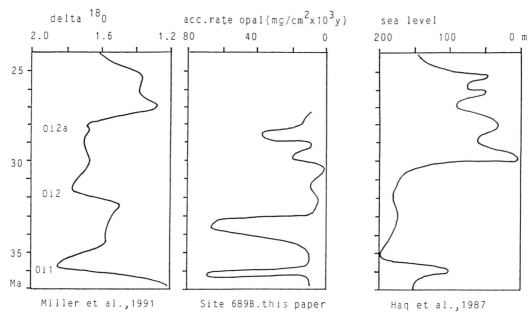

Fig. 18.   Synthesis showing paleoceanographic results for the Oligocene: (left) $\delta^{18}O$ variations of the benthic foraminifer *Cibicidoides* spp. in the Oligocene in the South Atlantic [*Miller et al.*, 1991]; (middle) productivity variations in the Maud Rise area (this paper) expressed by radiolarian accumulation rates in mg/cm^2 kyr; (right) sea level curve of *Haq et al.* [1987].

Excellent preservation of planktonic foraminifera in high-productivity periods: Influence of WSDW? (exception 1). There are four intervals of about 100- to 400-kyr duration, marked with horizontal lines in Figure 9 (and see Figure 19), where high accumulation rates of planktonic foraminifera and related good carbonate preservation (low benthic/planktonic foraminiferal ratios) coincide with high radiolarian accumulation rates. Why did the organic matter supplied to the sea floor together with the opal not dissolve the plankton foraminifera in these intervals? A different bottom water mass may have protected carbonate shells against dissolution. In the middle Eocene, warm saline deep water (WSDW) has been suggested to have filled the Antarctic Ocean basin [*Kennett and Stott*, 1990; *Mead et al.*, 1992] (for a different opinion, see *Thomas* [1990], *Miller et al.* [1991], and *Mackensen and Ehrmann* [1992]). The time when the WSDW disappeared from the Maud Rise area is still under discussion. *Kennett and Stott* [1990] suggest this event happened at 28 Ma, although it may have disappeared during some periods in the Oligocene. *Diester-Haass* [1991a] suggested that WSDW no longer influenced sedimentation on the flank of the Maud Rise (Site 689) after 37 Ma.

The following scenario is suggested to account for the preservation of foraminifera during high-productivity intervals (Figure 19). In the Eocene the WSDW filled the Weddell Sea basin but did not reach Site 689 near the crest of the Maud Rise [*Diester-Haass*, 1991a]. In

the latest Eocene, formation of AABW set in, and this water mass flowed below the WSDW, which was shifted up and down and latitudinally corresponded to the intensity of AABW formation (Figure 19, area 4). Oxygen isotope data of *Mead et al.* [1992] show that in the Oligocene Site 689 was in the center of a cooler water mass and that at depths of Site 690 (~1000 m below Site 689 at the flank of the Maud Rise) WSDW existed. This water mass might have been shifted upward during the four periods with good carbonate preservation and high surface water productivity (Figure 19, area 2) and protected planktonic foraminifera against dissolution, in spite of high organic C input, because WSDW is saturated with respect to $CaCO_3$.

There is no evidence from other sites that might argue for or against this assumption. Investigations with a high time resolution are needed from other sites in the Southern Ocean.

Strong dissolution of planktonic foraminifera in low-productivity periods: Is this the result of AABW? (exception 2). In Figure 9, three intervals are marked with double vertical lines, where low accumulation rates of planktonic foraminifera coincide with low radiolarian accumulation rates (Figure 8) and very strong carbonate dissolution (Figure 10). What mechanism is responsible for the dissolution of carbonate in a low-productivity environment? It can be assumed that when opal production is low, organic matter input also is low. In this case a different water mass may be responsible for the

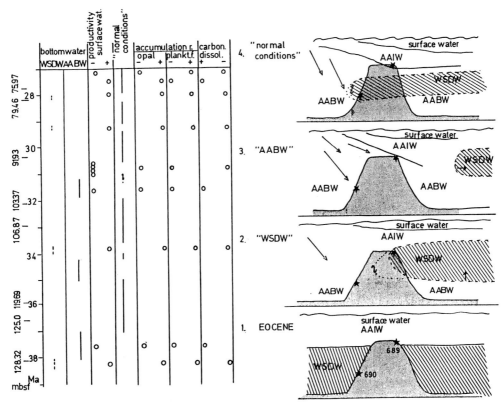

Fig. 19. Tentative hydrographic reconstruction for the Maud Rise area in the late Eocene–Oligocene, based on sedimentological data of Site 689. (Left) Ages in Ma and mbsf. "Bottom water": WSDW, hydrographic reconstruction 2 (right) is assumed for periods marked with dots; AABW, hydrographic reconstruction 3 (right) is assumed for periods marked with vertical line. Productivity in surface water of specific intervals: low or high. "Normal conditions": hydrographic reconstruction 1 (right) is assumed for periods marked with vertical line. Accumulation rate of opal and plankton foraminifera in the specific intervals (low-high). Carbonate dissolution in the specific intervals (strong-weak).

carbonate dissolution. The hypothetical hydrographic reconstruction (Figure 19) suggests that in these intervals formation of AABW was so intense that the layer of AABW thickened and reached the top of the Maud Rise. WSDW was pushed northward and upward. AABW strongly dissolved tests of planktonic foraminifera. This hypothesis needs to be tested by further detailed studies from other Southern Ocean sites.

Good preservation and low accumulation rates of planktonic foraminifera in a low-productivity period (exception 3). In Figure 9 an interval marked with circles contains sediments with low accumulation rates of benthos, radiolaria, and planktonic foraminifera; carbonate dissolution is relatively weak. It is assumed that these conditions reflect a low-productivity environment with a sluggish circulation and no vertical turnover, preventing a supply of nutrients from deeper water masses. Thus production of both opal and carbonate test producing organisms was low. Low organic C supply prevented major dissolution. The site has been in the area of AAIW (Figure 19, area 4).

This case with low accumulation rates and good preservation of planktonic foraminifera is the only one that allows a statement on production: low accumulation rates are a consequence not of dissolution but of low production.

Accumulation Rates of Benthic Organisms: Paleoceanographic Significance

The accumulation rates of benthic foraminifera vary between 0.4 and 8 mg/cm^2 kyr in the late Oligocene and 0.8 and 18 mg/cm^2 kyr in the early Oligocene. These variations are not controlled by carbonate dissolution (Figures 9 and 10), suggesting that benthic foraminifera are much less sensitive to dissolution than planktonic forms [*Arrhenius*, 1952].

The variations, however, are closely related to surface water productivity as expressed by radiolarian and diatom accumulation rates. The cyclic productivity changes as well as the longer-term trends are similar for the benthic foraminifera and radiolarian/diatom accu-

mulation rates. The only exception is the low accumulation in benthic foraminifera at 109 mbsf, where opal
MAR is at a maximum.

The early Oligocene benthic foraminifera accumulation rates of up to 18 mg/cm^2 kyr correspond to the
highest radiolarian accumulation rates, whereas in the
late Oligocene accumulation rates of both groups are
smaller. The low planktonic foraminifera and radiolarian accumulation rates interval (94–97 mbsf) is also
marked by a minimum in benthic foraminifera accumulation rates. These relationships reflect the dependency
of benthic productivity on surface water production.

Sponge spicule accumulation rates (Figure 12) follow
closely those of radiolaria and suggest a dissolution
dependency. As more opal is supplied to the sediments,
the better the preservation of all siliceous material.
Production differences are hidden by these dissolution
effects.

Ostracod accumulation rates are very low (0–0.16
mg/cm^2 kyr) and clearly follow the carbonate dissolution curve (Figure 10). Thus little can be said about
production variations.

Echinoderm accumulation rates (<0.08 mg/cm^2 kyr)
are partly influenced by carbonate dissolution. Their
rare occurrences are mostly in well-preserved units.

The Bolboforma Problem

Specimens of the genus *Bolboforma* spp. occur in the
investigated sequence of Core 14 and lowermost Core
13H of Hole 689B. The last appearance is in Core
13H-6, 133–135 cm (last occurrence described by *Kennett and Kennett* [1990]: Core 14H, 110 cm). The
abundance of *Bolboforma* spp. follows the radiolarian
accumulation rate variations: the higher the radiolarian
accumulation rate, the more *Bolboforma* exist, and vice
versa (Figures 8 and 14). In the interval from 126 to 119
mbsf, the maxima in *Bolboforma* and radiolarian accumulation rate coincide with dissolution maxima of
planktonic foraminifera and suggest that the dissolution
resistant *Bolboforma* [*Müller et al.*, 1985] are relatively
enriched. In the interval from 128 to 126 mbsf, however,
where AABW is assumed to be responsible for strong
carbonate dissolution although productivity (radiolarian
accumulation rates) is low, *Bolboforma* is rare. And in
the interval from 130 to 128 mbsf, where WSDW has
been suggested to inhibit carbonate dissolution in spite
of high productivity, *Bolboforma* is abundant, as are
radiolaria.

These findings suggest that *Bolboforma* abundances
are controlled more by surface water parameters than
by carbonate dissolution and thus bottom water effects.
The simultaneous increase and decrease of radiolaria
and *Bolboforma* points to temperature or nutrient effects. My interpretation points to cooler and more
nutrient-rich conditions during these periods. This is
opposite to the findings of *Kennett and Kennett* [1990],
who believe that *Bolboforma* is associated with more
temperate conditions and disappears with cooling.

Terrigenous Matter and Its Paleoceanographic Significance

The temporal distribution of terrigenous matter accumulation rates (Figure 15), although very low, points to
a relationship between climate, productivity, and clastic
supply. Terrigenous matter occurs in 19 of the opal
maxima units; only three exceptions exist, at 75.5, 79.5,
and 106 mbsf, in Hole 689B.

Opal-rich periods are those where a strong latitudinal
temperature gradient is assumed, which leads to high
turnover and mixing in the oceans. The increased temperature gradient is assumed to originate from intense
cooling in the Antarctic area. Perhaps ice growth was
intensified and led to a eustatic sea level lowering
[*Grobe et al.*, 1990]. A consequence could be an enhanced erosion of the continent, with more and coarser
clastic material reaching the Antarctic Ocean. Opal-
poor units were deposited in warmer intervals with high
sea level and weak temperature gradients and low
productivity.

The small amounts and small grain sizes of terrigenous matter make a supply by ice drift improbable. The
paleoceanographic reconstruction for the Oligocene by
Wei and Wise [1990] assumes a position of the Maud
Rise in a warmer surface return current of the Weddell
Sea Gyre and thus explains the absence of ice-rafted
detritus.

CONCLUSIONS

The high-resolution investigation of a late Eocene–
Oligocene profile from the Maud Rise, Site 689, revealed 18 high-low productivity cycles. High-productivity periods are characterized by high opal accumulation
rates and low planktonic foraminiferal accumulation
rates as a consequence of strong carbonate dissolution.
Terrigenous matter is present. Low-productivity periods have low accumulation rates of opal and high
planktonic foraminiferal accumulation rates as a consequence of excellent carbonate preservation.

The paleoceanographic-paleoclimatic explanation for
these findings is a strong latitudinal temperature gradient, strong cooling in Antarctica, high mixing rate in the
ocean, and nutrient supply to the surface waters in
high-productivity periods. Higher productivity increased opal production and preservation but led to
strong carbonate dissolution as a consequence of higher
organic matter supply. It is possible that during these
cooler periods a eustatic sea level lowering enhanced
the erosion of the Antarctic continent. The phenomenon
may explain the occurrence of clastic material of >40
μm in the sediments from high-productivity periods.

This general rule, explaining the coincidence of high-opal and low-plankton foraminiferal accumulation rates, has been interrupted in several units by changes in bottom water characteristics. It is assumed that in the late Eocene–Oligocene a three- or four-layer ocean existed in the Weddell Sea area (Figure 19), where AABW, WSDW, and AAIW were shifted latitudinally and vertically, thus exerting their different influences on the Maud Rise sediments.

Four periods (28.0–28.1, 29.1–29.2, 33.6–34.0, and 38.0–38.4 Ma) with excellent carbonate preservation in spite of high surface water productivity are attributed to the influence of WSDW (Figure 19). Three periods (31–32, 34.4–35.0, and 37–38.0 Ma) of strong carbonate dissolution in spite of low surface water productivity are tentatively explained by AABW that reached the top of the Maud Rise and caused dissolution of planktonic foraminifera.

A very low productivity period (30.4–31 Ma) with low accumulation rates of opal and carbonate fossils and good carbonate preservation is attributed to a period with sluggish surface circulation and weak turnover.

Acknowledgments. I thank my colleagues from the Alfred Wegener Institut and H. Oberhänsli, Mainz, for discussions and D. Fütterer for providing working facilities. Thanks are due to W. Ehrmann, D. Fütterer, T. Janecek, A. Mackensen, and an anonymous reviewer, who helped with comments and corrections of the manuscript. The samples were provided through the assistance of the Ocean Drilling Program. Financial support of the Deutsche Forschungsgemeinschaft is gratefully acknowledged (grant Fu119/161).

REFERENCES

Arrhenius, G., Sediment cores from the East Pacific, *Rep. Swed. Deep Sea Exped. 1947–1948*, *5*, 1–228, 1952.

Baldauf, J. G., and J. A. Barron, Evolution of biosiliceous sedimentation patterns—Eocene through Quaternary: Paleoceanographic response to polar cooling, in *Geological History of the Polar Oceans: Arctic Versus Antarctic*, edited by U. Bleil and J. Thiede, pp. 575–607, Kluwer Academic Press, Dordrecht, Netherlands, 1990.

Barker, P. F., and J. Burrell, The opening of the Drake Passage, *Mar. Geol.*, *25*, 15–34, 1977.

Barker, P. F., et al., Leg 113, *Proc. Ocean Drill. Program Initial Rep.*, *113*, 785 pp., 1988.

Barker, P. F., et al., Leg 113, *Proc. Ocean Drill. Program Sci. Results*, *113*, 1033 pp., 1990.

Berger, W. H., Deep-sea carbonates: Pleistocene dissolution cycles, *J. Foraminiferal Res.*, *3*, 187–195, 1973.

Berger, W. H., Biogenous deep-sea sediment preservation and interpretation, in *Chemical Oceanography*, vol. 5, edited by J. P. Riley and R. Chester, pp. 265–388, Academic, San Diego, Calif., 1976.

Berger, W. H., and L. Diester-Haass, Paleoproductivity: The benthic planktonic ratio in foraminifera as a productivity index, *Mar. Geol.*, *81*, 15–25, 1988.

Brass, G. W., J. R. Southam, and W. H. Peterson, Warm saline bottom water in the ancient ocean, *Nature*, *296*, 620–623, 1982.

Broecker, W. S., Glacial to interglacial changes in ocean and atmosphere chemistry, in *Climatic Variations and Variability: Facts and Theories*, edited by A. Berger, pp. 111–112, D. Reidel, Hingham, Mass., 1981.

Corliss, B. H., and R. C. Thunell, Carbonate sedimentation beneath the Antarctic circumpolar current during the late Quaternary, *Mar. Geol.*, *51*, 293–326, 1983.

Diester-Haass, L., Differentiation of high oceanic fertility in marine sediments caused by coastal upwelling and/or river discharge off NW Africa during the late Quaternary, in *Coastal Upwelling: Its Sedimentary Record*, NATO Conf. Ser., vol. 10B, edited by J. Thiede and E. Suess, pp. 399–419, Plenum, New York, 1983.

Diester-Haass, L., Sea-level changes, carbonate dissolution and history of the Benguela Current in the Oligocene-Miocene off southwest Africa (DSDP Site 362, Leg 40), *Mar. Geol.*, *79*, 213–242, 1988.

Diester-Haass, L., Eocene/Oligocene paleoceanography in the Antarctic Ocean, Atlantic sector (Maud Rise, ODP Leg 113, sites 689B and 690B), *Mar. Geol.*, *100*, 249–276, 1991*a*.

Diester-Haass, L., Rhythmic carbonate content variations in Neogene sediments above the oceanic lysocline, in *Cycles and Events in Stratigraphy*, edited by G. Einsele, W. Ricken, and A. Seilacher, pp. 94–109, Springer, New York, 1991*b*.

Diester-Haass, L., A. Meyers, and P. Rothe, Light-dark cycles in opal-rich sediments near the Plio-Pleistocene boundary, DSDP Site 532, Walvis Ridge Continental Terrace, *Mar. Geol.*, *73*, 1–23, 1986.

Dymond, J., and M. Lyle, Flux comparisons between sediments and sediment traps in the eastern tropical Pacific: Implications for atmospheric CO_2 variations during the Pleistocene, *Limnol. Oceanogr.*, *30*, 699–712, 1985.

Egeberg, P. K., and M. I. Abdullah, The diagenetic factors controlling the dissolved organic carbon (DOC) in pore water from deep-sea sediments (ODP Leg 113, Weddell Sea), *Proc. Ocean Drill. Program Sci. Results*, *113*, 169–177, 1990.

Ehrmann, W. U., and A. Mackensen, Sedimentological evidence for the formation of an East Antarctic ice sheet in Eocene/Oligocene time, *Palaeogeogr. Palaeoclimatol. Palaeoecol.*, *93*, 85–112, 1992.

Goll, R. M., and K. R. Bjørklund, Radiolaria in surface sediments of the South Atlantic, *Micropaleontology*, *20*, 38–75, 1974.

Grobe, H., D. K. Fütterer, and V. Spiess, Oligocene to Quaternary sedimentation processes on the Antarctic continental margin, *Proc. Ocean Drill. Program Sci. Results*, *113*, 121–131, 1990.

Haq, B. U., J. Hardenbol, and P. R. Vail, Chronology of fluctuating sea-levels since the Triassic (250 million years ago to present), *Science*, *235*, 1156–1167, 1987.

Keller, G., Paleoclimatic analyses of middle Eocene through Oligocene planktic foraminiferal faunas, *Palaeogeogr. Palaeoclimatol. Palaeoecol.*, *43*, 73–94, 1983.

Kennett, D. M., and J. P. Kennett, *Bolboforma* Daniels and Spiegler, from Eocene and lower Oligocene sediments, Maud Rise, Antarctica, *Proc. Ocean Drill. Program Sci. Results*, *113*, 667–673, 1990.

Kennett, J. P., Cenozoic evolution of Antarctic glaciation, the circum-Antarctic ocean, and their impact on global paleoceanography, *J. Geophys. Res.*, *82*, 3843–3860, 1977.

Kennett, J. P., and P. F. Barker, Latest Cretaceous to Cenozoic climate and oceanographic developments in the Weddell Sea, Antarctica: An ocean-drilling perspective, *Proc. Ocean Drill. Program Sci. Results*, *113*, 937–960, 1990.

Kennett, J. P., and L. D. Stott, Proteus and Proto-Oceanus: Ancestral Paleogene oceans as revealed from Antarctic stable isotopic results, ODP Leg 113, *Proc. Ocean Drill. Program Sci. Results*, *113*, 865–880, 1990.

Leinen, M., D. Cwienk, R. G. Heath, P. Biscaye, V. Kolla, J. Thiede, and J. P. Dauphin, Distribution of biogenic silica and quartz in recent deep-sea sediments, *Geology*, *14*, 199–203, 1986.

Mackensen, A., and W. Ehrmann, Middle Eocene through Early Oligocene climate history and paleoceanography in the Southern Ocean: Stable oxygen and carbon isotopes from ODP sites on Maud Rise and Kerguelen Plateau, *Mar. Geol.*, in press, 1992.

Mead, G. A., L. Tauxe, and J. J. LaBreque, Oligocene paleoceanography of the South Atlantic: Paleoclimatic implications of sediment accumulation rates and magnetic susceptibility measurements, *Paleoceanography*, *1*, 273–284, 1986.

Mead, G. A., D. A. Hodell, and P. F. Ciesielski, Late Eocene to early Oligocene vertical oxygen isotopic gradients in the South Atlantic: Implications for warm saline deep water, *Palaeogeogr. Palaeoclimatol. Palaeoecol.*, in press, 1992.

Miller, K. G., R. G. Fairbanks, and G. S. Mountain, Tertiary oxygen isotope synthesis, sea-level history, and continental margin erosion, *Paleoceanography*, *2*, 1–19, 1987.

Miller, K. G., J. D. Wright, and R. G. Fairbanks, Unlocking the ice house: Oligocene-Miocene oxygen isotopes, eustasy, and margin erosion, *J. Geophys. Res.*, *96*, 6829–6848, 1991.

Müller, C., D. Spiegler, and L. Pastouret, The genus *Bolboforma* Daniels and Spiegler in the Oligocene and Miocene sediments of the North Atlantic and Northern Europe, *Initial Rep. Deep Sea Drill. Proj.*, *80*, 669–675, 1985.

Oberhänsli, H., P. Heinze, L. Diester-Haass, and G. Wefer, Upwelling off Peru during the last 430.000 yr and its relationship to the bottom-water environment, as deduced from coarse grain-size distributions and analyses of benthic foraminifers at holes 679D, 680B, and 681B, Leg 112, *Proc. Ocean Drill. Program Sci. Results*, *112*, 369–390, 1990.

Pokras, E. M., Preservation of fossil diatoms in Atlantic sediment cores: Controlled by supply rate, *Deep Sea Res.*, *33*, 893–902, 1986.

Sarnthein, M., Oberflächensedimente im Persischen Golf und Golf von Oman, II, Quantitative Komponentenanalyse der Grobfraktion, *Meteor Forschungsergeb. Reihe C*, *1*, 1–113, 1971.

Schlanger, S. O., and I. Premoli-Silva, Oligocene sea-level falls recorded in mid-Pacific atoll and arcipelagic apron settings, *Geology*, *14*, 392–395, 1986.

Shackleton, N. J., and J. P. Kennett, Paleotemperature history of the Cenozoic and the initiation of Antarctic glaciation: oxygen and carbon isotope analyses in DSDP sites 277, 279 and 281, *Initial Rep. Deep Sea Drill. Proj.*, *29*, 713–719, 1975.

Stott, L. D., J. P. Kennett, N. J. Shackleton, and R. M. Corfield, The evolution of Antarctic surface waters during the Paleogene: Inferences from the stable isotopic composition of planktonic foraminifers, ODP Leg 113, *Proc. Ocean Drill. Program Sci. Results*, *113*, 849–863, 1990.

Thiede, J., E. Suess, and P. J. Müller, Late Quaternary fluxes of major sediment components to the sea-floor at the northwest African continental slope, in *Geology of the Northwest African Continental Margin*, edited by U. von Rad, K. Hinz, M. Santhein, and E. Seibold, pp. 605–631, Springer-Verlag, New York, 1982.

Thomas, E., Late Cretaceous through Neogene deep-sea benthic foraminifers (Maud Rise, Weddell Sea, Antarctica), *Proc. Ocean Drill. Program Sci. Results*, *113*, 571–594, 1990.

Thomas, E., et al., Upper Cretaceous–Paleogene stratigraphy of sites 689 and 690, Maud Rise (Antarctica), *Proc. Ocean Drill. Program Sci. Results*, *113*, 901–914, 1990.

Thunell, R. C., Optimum indices of calcium carbonate dissolution in deep-sea sediments, *Geology*, *4*, 525–528, 1976.

Wei, W., Evidence for an earliest Oligocene abrupt cooling in the surface waters of the Southern Ocean, *Geology*, *19*, 780–783, 1991.

Wei, W., and S. W. Wise, Middle Eocene to Pleistocene calcareous nannofossils recovered by Ocean Drilling Program Leg 113 in the Weddell Sea, *Proc. Ocean Drill. Program Sci. Results*, *113*, 639–673, 1990.

Wise, S. W., A. M. Gombos, and J. P. Muza, Cenozoic evolution of polar water masses, southwest Atlantic Ocean, in *South Atlantic Paleoceanography*, edited by K. J. Hsü and K. J. Weissert, pp. 283–324, Cambridge University Press, New York, 1985.

(Received December 6, 1991;
accepted May 27, 1992.)

GEOTECHNICAL STRATIGRAPHY OF NEOGENE SEDIMENTS: MAUD RISE AND KERGUELEN PLATEAU

FRANK R. RACK

Department of Oceanography, Texas A&M University, College Station, Texas 77843
Ocean Drilling Program, College Station, Texas 77845

ALAN PITTENGER

Department of Oceanography, Texas A&M University, College Station, Texas 77843

The dominant feature of pelagic sediment accumulation patterns at intermediate water depths in the Southern Ocean is the sequential northward migration and replacement of a calcareous ooze sedimentary facies by a siliceous ooze facies, following the late Oligocene opening of Drake Passage and the subsequent initiation of the Antarctic Circumpolar Current. Discrete measurements of physical properties (density, porosity, and water content) and continuous whole-core measurements of bulk density and acoustic compressional velocity are evaluated for six sites drilled in the Southern Ocean during Ocean Drilling Program (ODP) Leg 113 (sites 689 and 690), Leg 119 (Site 744), and Leg 120 (sites 747, 748, and 751). These sites are all located on submarine rises in open ocean (pelagic) depositional environments south of the Polar Front, a major oceanographic and sedimentologic boundary. High-resolution profiles derived from core measurements of bulk density and compressional velocity are used to correlate between adjacent drill holes on Maud Rise, where multiple overlapping sedimentary sequences were recovered using the advanced hydraulic piston core barrel. Stratigraphic age models are applied to the bulk density data to provide a temporal framework for a discussion of paleoceanographic events and to constrain future acoustic research studies. The early to middle Miocene intervals at sites 689 and 690 are marked by distinct sequences of meter-scale, alternating carbonate and siliceous units separated by several proposed stratigraphic hiatuses. At least six middle to late Miocene dissolution or productivity events are identified in an accumulation of nannofossil ooze on Maud Rise; others are noted at the ODP sites in the Raggatt Basin and on the southwestern flank of the Kerguelen Plateau. An understanding of the possible interactions between the Antarctic Circumpolar Current system and seafloor topography, as modified by progressive plate tectonic motions and glacio-eustatic fluctuations, is used to develop a hypothesis for explaining the observed sediment accumulation patterns at intermediate water depths of the Southern Ocean during the Neogene.

INTRODUCTION

Research about the mass physical and geotechnical properties of marine sediments (index properties, consolidation characteristics, permeability, and microfabric), and their associations with acoustic properties, is essential for seafloor mapping activities. This information also has important applications in the science of paleoceanography. Expeditions to the Southern Ocean by the Ocean Drilling Program (ODP) have provided new materials for attempting these types of investigations.

Most researchers agree that the central problem of paleoceanography is the history of the circulation of the ocean. The wind-driven and thermohaline circulation are linked by definition to the temperature and salinity distribution in the ocean. The circulation of the ocean is linked to both atmospheric circulation and to biological

communities. The distribution and relative abundance of organisms in surface waters are determined by the oxygenation and fertility of the oceans, which are controlled by ocean circulation [*Berger*, 1981].

This paper explores the potential contribution to paleoceanographic studies that can be made by integrating (1) discrete measurements of the physical properties of sediment core samples and (2) nondestructive, closely spaced whole-core measurements of bulk density and acoustic compressional wave velocity.

Bulk density data collected with the gamma ray attenuation porosity evaluator (GRAPE) are shown to be extremely useful as a paleoceanographic tool since they provide a temporal resolution far surpassing that of most shipboard or shore-based studies. Velocity data are used in combination with the bulk density data to make correlations between lithologies at adjacent ODP

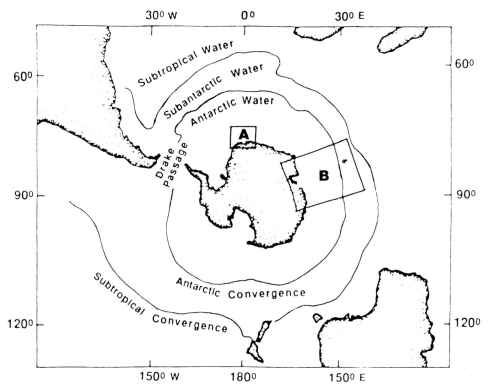

Fig. 1. General location map showing study areas (A, Maud Rise; B, Kerguelen Plateau) [after *Kennett*, 1978]. Detailed bathymetry and site locations for Maud Rise and Kerguelen Plateau are provided in Figures 3 and 12, respectively. Major water masses (Subtropical, Subantarctic, and Antarctic) and oceanographic fronts (Subtropical Convergence and Antarctic Convergence or Polar Front) are also indicated. The two fronts form the generalized northern and southern boundaries for the Antarctic Circumpolar Current.

holes and sites. These data are also used to detect diagenetic zones within the sediments and to identify possible stratigraphic hiatuses in sediment cores of late Oligocene to Quaternary age. Possible stratigraphic hiatuses are suggested by the reversals, inflections, and discontinuities in the downhole bulk density trends.

The emphasis of this study is on cores recovered with the advanced hydraulic piston corer (APC) by the Ocean Drilling Program during legs 113, 119, and 120 (Figure 1 and Table 1). Acoustic compressional velocity data measured on whole-round sediment cores using the P wave logger (PWL) are used in combination with the

GRAPE bulk density data to (1) make correlations between adjacent holes at a given ODP site, (2) detect changes in lithology and diagenetic zones within the sediments, and (3) identify potential stratigraphic hiatuses. The ODP drill sites will be discussed in two groups, divided according to their relative geographic location (Maud Rise and Kerguelen Plateau).

Seafloor sediments are basically a two-phase composite material, consisting of granular solids and pore fluids, which undergo consolidation as more materials accumulate above them. The deterioration of environmental conditions in the Southern Ocean since the late Oligocene caused the progressive replacement of marine organisms which construct skeletons out of calcium carbonate by organisms which construct their skeletons out of opal or biogenic silica. The primary assumption of this paper is that the changes in bulk density trends reflect changing oceanographic and depositional conditions on local or regional scales. The sedimentary responses to prevailing environmental conditions are preserved as inflections, discontinuities, and reversals in physical properties trends and changes in the state of consolidation measured from sediment cores.

TABLE 1. Site Locations

Leg	Site	Latitude	Longitude	Depth, mbsl
113	689	64°31.01′S	03°06.00′E	2080
113	690	65°09.62′S	01°12.29′E	2914
119	744	61°34.66′S	80°35.46′E	2307
120	747	54°48.68′S	76°47.64′E	1695
120	748	58°26.45′S	78°58.89′E	1290
120	751	57°43.56′S	79°48.89′E	1634

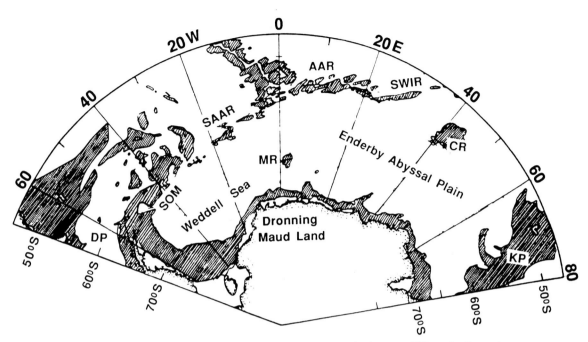

Fig. 2. Bathymetric map of the Atlantic-Indian Basin from Drake Passage (DP) to the Kerguelen Plateau (KP) [after *Gordon et al.*, 1981]. Other bathymetric features include the South American–Antarctic Ridge (SAAR), African-Antarctic Ridge (AAR), Southwest Indian Ridge (SWIR), Conrad Rise (CR), Maud Rise (MR), and South Orkney Margin (SOM). Hachured regions are above the 4000-m bathymetric contour.

Using the concept of sedimentary facies, one can evaluate vertical geologic sequences by associating particular sediment layers with inferences about past prevailing environmental and/or depositional conditions. Deep-sea carbonates are known to exhibit numerous fine-scale acoustic reflectors [*Mayer*, 1979*a*, *b*, 1980]. In the Equatorial Pacific, these types of acoustic horizons have been interpreted to represent paleoceanographic events [*Mayer et al.*, 1986; *Theyer et al.*, 1989]. This study presents a first attempt at applying these techniques to sediments in the Southern Ocean south of the Polar Front. The objective of this research is to gain an improved understanding of the depositional history of biogenous, pelagic materials at intermediate water depths of the Southern Ocean during the past 30 million years.

Studies of the physical and geotechnical properties of marine sediments at a number of geographic locations in the Atlantic-Indian Basin (Figure 2) may provide an improved understanding of localized patterns of sedimentation and latitudinal variations in sediment facies, as ocean circulation patterns adjusted to changes in tectonic and climatic boundary conditions.

METHODS

Index Properties

Measurements of index properties (bulk density, dry density, grain density, porosity, and water content) and calcium carbonate data were obtained from Ocean Drilling Program shipboard analyses detailed in the work of *Barker et al.* [1988], *Barron et al.* [1989], and *Schlich et al.* [1989].

The sediment index properties at each ODP site are displayed graphically to provide a general understanding of the broad geotechnical trends that are observed in the plots of shipboard measurements. These data have been supplemented by additional shore-based tests, measurements, and observations to create a framework for regional comparisons among drill sites and to develop plausible geological constraints for paleoceanographic hypotheses and models. Laboratory corrections to these data, which compensate for the effects of compaction and rebound in the sediments upon core recovery, are relatively minor (1–5%) for the depths of burial encountered at these sites [see *Hamilton*, 1976].

Calcium carbonate analyses were performed both on board the JOIDES *Resolution* and onshore using a Coulometrics CO_2 coulometer with an accuracy of 0.15%. Laboratory calcium carbonate measurements for ODP sites 689 and 690 are from *O'Connell* [1990], where the reader can also find a more thorough explanation of the laboratory procedures involved in this method. Age versus sub-bottom-depth models of sediment accumulation for each of the ODP sites are taken from the integrated geomagnetic-biostratigraphic syntheses of *Gersonde et al.* [1990], *Barron et al.* [1991], and *Harwood et al.* [1992].

Whole-Core Measurements

The gamma ray attenuation porosity evaluator is a whole-core measuring system which can provide rapid, closely spaced measurements of the attenuation of a gamma ray beam (at 1-cm intervals) as it passes through the diameter of a core surrounded by a plastic core liner. The density of the core material is calculated from the gamma ray attenuation. The values of saturated bulk density provided by the GRAPE are used as the primary source of high-resolution data for this study. Care was taken to ensure the quality of data by editing out core and section breaks and by comparing the density data with core photographs to identify zones of sediment disturbance. The reader is referred to *Boyce* [1976] for a full description of the GRAPE apparatus.

Bulk density variations in the pelagic sediments from Maud Rise and Kerguelen Plateau primarily arise from downhole fluctuations in the relative abundance and preservation state of microfossil skeletons. The microfossil skeletons found in these sediments are generally composed of either biogenic silica (opal) or calcium carbonate. The observed downhole bulk density variations and gradients are largely attributed to differences between the grain densities of amorphous silica (1.8–2.1 g/cm^3) and calcium carbonate (2.72 g/cm^3), respectively, and by changes in the relative composition and texture of the bulk sediment.

Relative changes in bulk density are used to correlate between adjacent boreholes at a given site on Maud Rise and to suggest improved depth-to-density and density-to-age correlations between ODP sites 689 and 690. Comparisons are also made between the bulk density profiles from the two Maud Rise sites and the bulk density data collected at 2-cm intervals from ODP sites 744, 747, 748, and 751 on the central and southern Kerguelen Plateau.

A second continuous whole-core measurement system is the *P* wave Logger. This system measures acoustic compressional wave velocity through the plastic core liner at an interval of about 2 cm using the travel time method [*Schultheiss and McPhail*, 1989]. These data are used in combination with the GRAPE bulk density data to (1) make correlations between boreholes, (2) detect changes in lithology and diagenetic zones within the sediments, and (3) identify the position of possible stratigraphic hiatuses.

RESULTS: GENERAL COMMENTS

The physical attributes of pelagic sediments are determined jointly by (1) the number of sources for the input of sedimentary material, (2) the rate of input from each source, (3) the relative preservation of these ma-

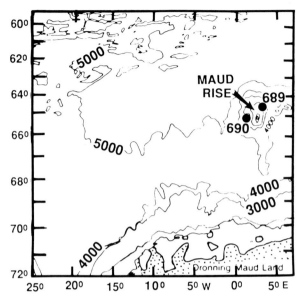

Fig. 3. Bathymetric chart of the eastern Weddell Sea showing the location of ODP sites 689 and 690 on Maud Rise [after *Barker et al.*, 1988].

terials as they pass through the water column and are deposited on the seafloor, and (4) postdepositional processes.

In an open ocean setting, such as the site locations on Maud Rise and Kerguelen Plateau, the rate of supply of sedimentary materials and hydrodynamic processes jointly determine where sediment will accumulate on the seafloor. Rates of primary production in these open ocean settings depend on environmental conditions which regulate the water temperature, nutrient levels, light intensity in the upper water column, and the distribution and extent of sea ice. These conditions vary seasonally, annually, and on longer time scales.

Relationships among density, porosity, and carbonate content for marine sediments are well known [e.g., *Bryant et al.*, 1974, 1981; *Hamilton*, 1976; *Mayer*, 1979a, b, 1980, 1991; *Lee*, 1982]. This knowledge of established index property relationships can be used in association with additional laboratory techniques, such as step-loaded consolidation tests, to map the sedimentary response to environmental perturbations in a particular region of the ocean. In this study, GRAPE bulk density data are used to develop a high-resolution analysis of the first continuously cored pelagic sequences recovered with the APC from the Southern Ocean, south of the Polar Front (Antarctic Convergence).

MAUD RISE: ODP SITES 689 AND 690

Sites 689 and 690 are located in water depths of 2000 and 2900 m, respectively (see Table 1 and Figure 3). The sediments recovered from Maud Rise, in the northeast-

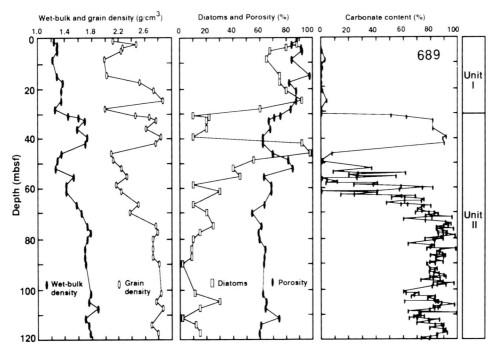

Fig. 4. Summary plots of index properties, percent diatoms, and carbonate content versus subbottom depth at ODP Site 689 above 120 mbsf. The transition from a nannofossil ooze to a diatom ooze takes place in a number of steps above 80 mbsf in the late Oligocene and early Neogene. Carbonate data are from *O'Connell* [1990]. Lithologic units are shown to the right of the profiles.

ern Weddell Sea, are primarily calcareous and siliceous pelagic oozes with very small amounts of ice-rafted debris in the upper part of the lithologic sequence. The ODP sites drilled on Maud Rise are separated from the continent of Antarctica by a deep bathymetric trough (Figure 3).

Lithostratigraphic Unit I (0–31 meters below seafloor (mbsf)), at Site 689 near the crest of Maud Rise, is composed of biogenic siliceous ooze ranging in age from late Miocene to Quaternary (Figure 4). Lithologic Unit II (31–149 mbsf) is dominated by a mixture of biogenic siliceous and calcareous oozes of late Eocene to late Miocene age [*Barker et al.*, 1988]. Sediments recovered below 149 mbsf consist mainly of nannofossil ooze and chalk, with varying amounts of foraminifers comprising from 15% to 55% of the sediment in some stratigraphic intervals.

At Site 690 a thin Pleistocene to Quaternary calcareous layer (0–2.4 mbsf) overlies a 22-m-thick late Miocene to Pleistocene diatom ooze (Unit I) (Figure 5). Underlying Lithostratigraphic Unit I, Unit II (24.4–92.9 mbsf) is characterized by a variety of pure and mixed biogenic siliceous and calcareous biogenous oozes. Unit II is divided into two subunits at ~54 mbsf; the relative numbers of diatom skeletons increase above this level, while fluctuations among the different microfossil species increase in the upper subunit. Lithostratigraphic

Unit III (92.9–137.8 mbsf) and Unit IV (137.8–281.1 mbsf) consist of Paleocene to Eocene calcareous biogenic sediment, with increased amounts of terrigenous sediment in Unit IV (quartz, clay, and mica increase in the interval from 137.8 to 177.3 mbsf) [*Barker et al.*, 1988].

Split-Core Measurements

Temporal fluctuations in the accumulation of low-density (high-porosity) biosiliceous microfossil skeletons and fragments, relative to calcareous skeletons, are suggested as the dominant feature of the observed fluctuations in downhole physical properties at the ODP sites on Maud Rise. High measured porosity values are directly correlated with increases in diatoms and radiolarians relative to nannofossils and foraminifers in these sediments (Figures 4 and 5). The number of silicoflagellates in the sediment is only important in discrete zones within the upper lithologic unit at each site [*Barker et al.*, 1988].

Comparisons among index property profiles show that the sediment recovered above 90 mbsf at Site 689 and above 70 mbsf at Site 690 exhibits similar trends (Figures 4 and 5). The differences in the relative magnitude of laboratory carbonate values between these two sites are attributed to the effects of water depth–

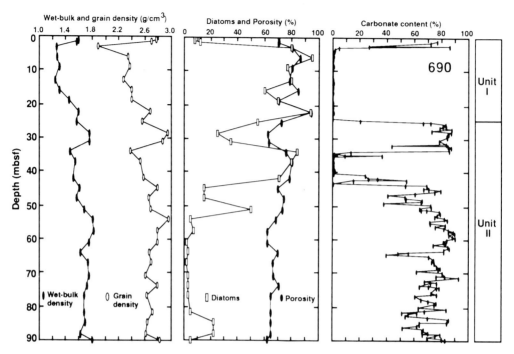

Fig. 5. Summary plots of index properties, percent diatoms, and carbonate content versus subbottom depth at ODP Site 690 above 90 mbsf. The sediments within this depth interval are late Oligocene and younger in age. The transition from a nannofossil ooze to a diatom ooze takes place in a number of steps above 55 mbsf. Carbonate data are from *O'Connell* [1990]. Lithologic units are shown to the right of the profiles.

dependent microfossil preservation. Lithologic changes are clearly observed as changes in index property trends in these profiles; however, the spacing of measurements is inadequate for conducting high-resolution (centimeter scale) paleoceanographic studies. Therefore GRAPE bulk density data collected at 1-cm intervals are used to provide a detailed proxy record of local and perhaps regional geochemical and environmental changes at intermediate water depths of the Southern Ocean.

Whole-Core Measurements

Profiles of GRAPE bulk density data are shown for the upper 100 m of sediment at Site 689 (Hole 689B and Hole 689D; Figures 6 and 7) and at Site 690 (Hole 690B and Hole 690C; Figures 8 and 9). These profiles clearly show the high-amplitude (0.2–0.4 g/cm^3) changes of bulk density which occur within narrow, vertically stratified depth intervals (from ~0.5 to 1.5 m thick). The observed changes in bulk density trends arise from fluctuations in the relative accumulation and preservation of biogenic, siliceous and calcareous microfossil skeletons. Compressional wave velocities measured by the PWL are also shown in summary figures for each hole.

Discrete depth intervals along the density profiles have been correlated to the paleomagnetic age models

of *Speiss* [1990], which also forms the basis for the integrated model of *Gersonde et al.* [1990]. The magnetic interpretations of *Speiss* [1990] are modified here, to allow for the assumed synchroneity of the downhole bulk density peaks and troughs at similar stratigraphic intervals from ODP sites 689 and 690 (Table 2). The geomagnetic age models and density profiles from each of the "B holes" at the two sites have been intercompared to identify potential discrepancies arising from a reliance on single data point interpretations of geomagnetic polarity reversals [*Speiss*, 1990]. These revised ages are then applied to the previously unpublished GRAPE bulk density data collected from cores at holes 689D and 690C.

The dominant features of the GRAPE density curves from holes 689B and 689D are identified from 100 mbsf to the seafloor (Figures 6 and 7). Similar density profiles are observed at holes 690B and 690C; however, because of the decreased accumulation rates at Site 690, they occur at a shallower subbottom depth and appear closer together (Figures 8 and 9).

There is a prominent density "low" in the late Oligocene, observed as a decreasing density trend from 87 mbsf to ~82 mbsf in each of the two profiles from Site 689. This feature is better developed at Hole 689D than at Hole 689B because a core break at Hole 689B

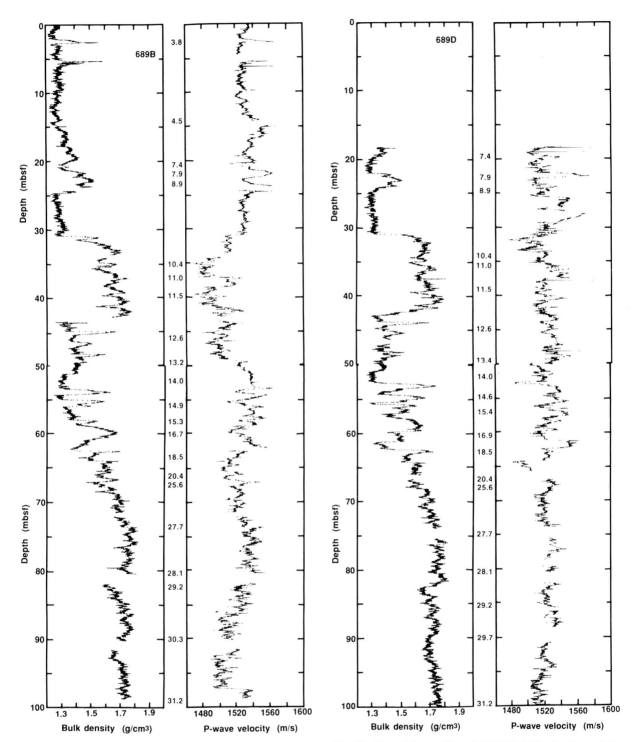

Fig. 6. Downhole profiles of GRAPE bulk density and *P* wave velocity versus subbottom depth at ODP Hole 689B from 100 mbsf to the seafloor. Approximate stratigraphic age assignments in millions of years ago (Ma) are given to the right of the density curve.

Fig. 7. Downhole profiles of GRAPE bulk density and *P* wave velocity versus subbottom depth at ODP Hole 689D from 100 mbsf to the seafloor. Approximate stratigraphic age assignments in millions of years ago are given to the right of the density curve.

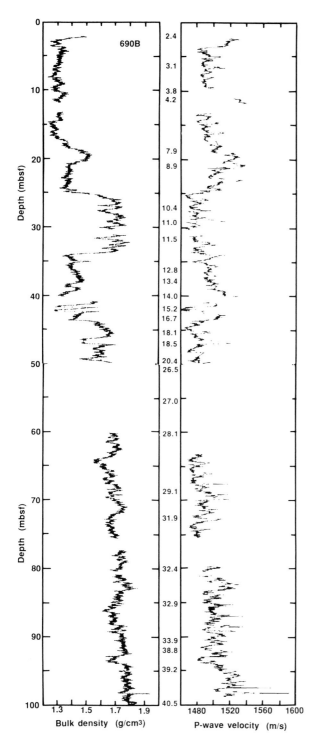

Fig. 8. Downhole profiles of GRAPE bulk density and *P* wave velocity versus subbottom depth at ODP Hole 690B from 100 mbsf to the seafloor. Approximate stratigraphic age assignments in millions of years ago are given to the right of the density curve.

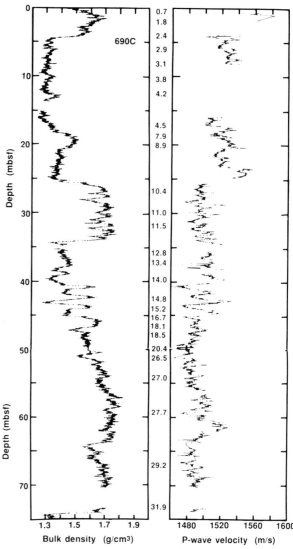

Fig. 9. Downhole profiles of GRAPE bulk density and *P* wave velocity versus subbottom depth at ODP Hole 690C from 75 mbsf to the seafloor. Approximate stratigraphic age assignments in millions of years ago are given to the right of the density curve.

disrupts that profile. The density low is associated with an increase in diatoms from ~10% to 25% over the time interval from ~30 Ma to 28 Ma (million years ago) based on shipboard smear slide descriptions. There is a corresponding increase in acoustic wave velocities across this interval (from 1490 m/s to 1530 m/s), best observed at Hole 689B.

From ~75 mbsf to ~52 mbsf, there is an overall trend of decreasing bulk density values and increasing opal content in the sediments at Site 689. An extended stratigraphic hiatus from 25.6 Ma to 20.4 Ma is placed at ~67 mbsf in Hole 689B and at ~50 mbsf in Hole 690B; this hiatus is similar in age to the proposed opening of

Drake Passage at 23.5 ± 2 Ma [*Barker and Burrell*, 1977]. The generally decreasing density trend is interrupted by a series of high-amplitude density peaks (trend reversals) in the early to middle Miocene at both Site 689 and Site 690. The increased values of bulk density correspond to discrete high-carbonate beds which show up as lighter-colored sediment in core photographs; these changes are best observed at Hole 690C (Figure 10). The relatively low velocity, carbonate rich intervals alternate with higher-velocity, carbonate poor intervals, which are identified by decreased bulk density values.

A broad density peak is centered at 60 mbsf at Hole 689B (~17 Ma to 18.5 Ma) (Figure 6). This peak is observed at Hole 689D as a flattened and stepped feature in the density profile (Figure 7). The flattened aspect of this peak is also observed at holes 690B and 690C near 46 mbsf (Figures 8 and 9); however, the upper part of this peak is truncated by a core break at Hole 690C. *Speiss* [1990] and *Gersonde et al.* [1990] have identified two stratigraphic hiatuses within this interval at Site 689. One hiatus is placed near the crest of the peak, and the second hiatus is placed at the upper boundary, just below a series of narrow density peaks and troughs.

There are a series of three narrow density peaks in the interval from ~56 mbsf to ~52 mbsf at Hole 689D. Only the two deeper peaks are observed at Hole 689B owing to a core break at 53 mbsf where about 0.5 m of sediment is missing. The lowermost peak at Hole 689D is truncated by a core break between Core 689D-4H and Core 689D-5H (~56.7 mbsf) where approximately 0.5 m of sediment is missing. These three density peaks are also truncated at Hole 690B in the first section of a core (Figure 8) but are clearly observed at Hole 690C (Figure 9). The low-carbonate intervals separating these density peaks (high carbonate) are believed to be associated with dissolutional hiatuses in the middle Miocene, some of which have been proposed by *Ablemann* [1990]. Similar hiatuses were proposed by *Keller and Barron* [1987] and may be related to eustatic sea level fluctuations arising from changes in ice volume on Antarctica [*Haq et al.*, 1987].

There is a small density "step" at ~51 mbsf at holes 689B and 689D. A similar feature is located near 39 mbsf at holes 690B and 690C. The approximate age of this feature at each site is ~13.6 Ma. There is a second narrow density low above this feature at Site 690 which is not observed at Site 689.

Three narrow density peaks are observed at Site 689 from 49 mbsf to 43 mbsf. Only one of these peaks is observed at Site 690 in the corresponding interval from 38 mbsf to 34 mbsf. Their absence at Site 690 suggests carbonate dissolution on the slope of Maud Rise below the backtracked water depth of Site 689 during this time period.

The dominant features of the density profiles in the upper 45 mbsf at Site 689 are a broad density high from 43 mbsf to ~30.5 mbsf associated with a largely monospecific (*Reticulofenestra perplexa*) nannofossil ooze and a smaller density peak centered at 23 mbsf (Figures 6 and 7). The corresponding intervals at Site 690 are at 34 mbsf to ~25 mbsf and 19.5 mbsf (Figures 8 and 9).

There are a number of lower-density intervals within the high-carbonate *R. perplexa* ooze sequence which can be correlated between adjacent APC holes and also between Site 689 and Site 690. *Wei and Wise* [1990] attribute fluctuations in the relative abundance and preservation of *R. perplexa* in these intervals to changes in surface water mass temperature over Maud Rise during the late Miocene. A regression equation, comparing the GRAPE density data and laboratory determinations of calcium carbonate from *O'Connell* [1990], was used to estimate the percentage of calcium carbonate (at 1-cm resolution) for the interval from 23 mbsf to 35 mbsf at holes 690B and 690C (Figure 11). The calculated values are within 10–15% of the laboratory determinations. The fluctuations may reflect changes in primary productivity between biosiliceous and calcareous species and/or enhanced dissolution of calcium carbonate.

The paleomagnetic age model of *Speiss* [1990] was not consistent across the *R. perplexa* ooze interval at sites 689 and 690, owing to low magnetic intensities and problems associated with the identification of geomagnetic polarity reversals from discrete measurements of sediment samples using a spinner magnetometer [*Speiss*, 1990]. This paper suggests that these interpretation problems can be resolved by making a few modifications to the geomagnetic polarity identifications at Site 689 and by then applying them consistently to the data at Site 690 (Table 2). The similarities among the density profiles at these two sites are used to constrain the age-depth correlation between age models.

There are large decreases in bulk density across the lithologic transition from nannofossil ooze to diatom ooze; these are observed at ~32 mbsf at Site 689 and at ~25 mbsf at Site 690. At these two sites the measured acoustic velocities increase from Lithologic Unit II to Unit I; the velocities are generally higher in the few meters above this transition. The age of this transition is suggested to be approximately 9.8 Ma, assuming constant sediment accumulation.

A multiple density peak centered at 23 mbsf at Hole 689B is identified by a velocity inversion between two velocity peaks (Figure 6). There is a thin carbonate-enriched bed, bounded by stratigraphic hiatuses above and below, that is associated with this velocity inversion. There appears to be no increase in carbonate associated with this density feature at Site 690 (~20 mbsf), according to laboratory determinations of calcium carbonate [*O'Connell*, 1990].

TABLE 2. Paleomagnetic Age Models for ODP Sites 689 and 690

Age, Ma	Magnetochron	Depth, mbsf*	
		Hole 689 B	Hole 690 B
3.88	base of C2AR-3	3.38	10.23
3.97	base of C3 N-1	4.40	10.48
4.10	base of C3 R-1	8.79	11.21
4.24	base of C3 N-2	9.63	11.78
4.40	base of C3 R-2	11.45	
4.47	base of C3 N-3	11.72	
4.57	base of C3 R-3	15.17	18.32
4.77	base of C3 N-4	16.92	
	lower C3 R-4		
4.85–6.17	hiatus	17.65?	18.83† (~4.6–>7.6 Ma)?
6.37	base of C3AR-2	18.09	hiatus
6.50	base of C3BN	18.67	
6.70	base of C3BR	18.92	
7.41	base of C4 N-1/2/3	20.17	21.07† (<8.4–>8.9 Ma)
7.90	base of C4 R-3	22.15	hiatus
8.50	base of C4AN-1/2	23.65	
8.71	base of C4AR-2	23.81	
8.80	base of C4AN-3	24.17	
8.92	base of C4AR-3	24.67	
10.42	base of C5 N-1/2	34.55 (35.0)	28.03
11.03	base of C5 R-2	37.55	(29.53)† (10.7–11.38 Ma)
11.09	base of C5 N-3	37.93	(29.78) hiatus
11.55	base of C5 R-3	45.93 (39.7)	31.72‡ (11.0–11.4 Ma)
11.73	base of C5AN-1	46.41 (40.3)	32.45
11.86	base of C5AR-1	46.68 (41.9)	33.47
12.12	base of C5AN-2	47.18	33.95
12.83	base of C5AR-2/3/4	48.68	36.35
13.01	base of C5AN-5	49.18	36.95
13.20	base of C5AR-5	49.68	37.20
13.46	base of C5AN-6	50.91	37.97‡ (12.8–13.8 Ma)
13.69	base of C5AR-6	51.93	38.97
14.08	base of C5AN-7	52.41	39.95
14.20	base of C5AR-7	52.85	40.20
	lower C5AN-8	55.15	
14.66–14.93	hiatus	55.28	
	upper C5BR-1	55.40	
15.13	base of C5BR-1	56.53	
15.27	base of C5BN-2	57.40	
	lower C5BR-2	57.15	
15.49–>16.22	hiatus	58.75	
	upper C5CN	58.85	
	lower C5CN	59.40	
<16.98–18.02	hiatus	59.65	···‡ (15.1–17.6 Ma)
18.12	base of C5DR-1	60.03	45.41 hiatus
18.14	base of C5DN-2?	60.40	45.68
18.56	base of C5DR-2	61.76	46.68
	lower C5E-C6	66.75	
<20.45–>25.60	hiatus	66.86	51.28† (20.5–26.56 Ma)
	upper C7 R-1	66.97	hiatus
25.67	base of C7 R-1	67.11	
25.97	base of C7 N-2	67.88	
26.38	base of C7 R-2	68.61	
26.56	base of C7AN	68.88	
26.86	base of C7AR	69.38	53.25
26.93	base of C8 N-1	70.11	54.75
27.01	base of C8 R-1	70.38	55.00
27.74	base of C8 N-2	75.98	60.11
28.15	base of C8 R-2	79.46	60.99
29.21	base of C9 N-1/2	85.05	68.48
29.73	base of C9 R-2	89.82	
30.33	base of C10N-2	91.93	
31.23	base of C10R-2	100.23	
32.06	base of C11N-2	103.38	

TABLE 2.   (continued)

Age, Ma	Magnetochron	Depth, mbsf*	
		Hole 689 B	Hole 690 B
32.46	base of C11R-2	104.38	80.76
32.90	base of C12N	106.88	84.01
35.29	base of C12R	116.71	
35.87	base of C13N-1/2	119.70	
37.24	base of C13R-2	128.33	
37.68	base of C15N-1/2	134.02	
38.10	base of C15R-2	135.77	
38.34	base of C16N-1	139.45	
38.50	base of C16R-1	139.55	
39.24	base of C16N-2/3	144.77	95.70
39.53	base of C16R-3	145.02	96.59
40.70	base of C17N-1/2	151.73	

*Values in parentheses are adjusted depth from bulk density correlation.
†Magnetostratigraphic hiatus.
‡Biostratigraphic hiatus.

KERGUELEN PLATEAU: ODP SITES 747, 748, 751, AND 744

The Kerguelen Plateau is divided into two distinct domains just north of 57°S [Houtz et al., 1977; Schlich et al., 1989]. The Kerguelen-Heard plateau extends northward to about 54°S and includes the sea surface expression of the plateau at Kerguelen, Heard, and McDonald islands. To the south, the southern Kerguelen Plateau extends from 57°S to 64°S where it merges into the Princess Elizabeth Trough to the north of Prydz Bay, Antarctica.

Four sites (Sites 747, 748, 751 and 744), drilled on the Kerguelen Plateau during ODP Legs 119 and 120, are discussed in this study (Table 1; Figure 12). The sediments recovered from these sites generally consist of siliceous, calcareous and mixed biogenic oozes. The lithologies encountered at these sites are comparable to those found on Maud Rise, far to the west of Kerguelen Plateau, although at a more southerly geographic location. GRAPE bulk density and porosity data are used to compare the results from ODP sites on Kerguelen Plateau.

ODP Site 747

Description. Site 747 lies on a northwest trending basement high bordering the northern extension of the 77°E Graben in the transition zone between the northern and southern parts of the Kerguelen Plateau (Figure 12). It is located approximately 400 km north of Site 751 and 500 km south of the position of the present-day Antarctic Convergence or Polar Front (Figure 1). The sediment cover above basement thickens to the west and northwest of the site, but it thins to the southeast and northeast [Schlich et al., 1989].

A relatively continuous calcareous ooze and chalk sequence below ~34.5 mbsf (Lithologic Unit II) con-tains abundant, well-preserved benthic and planktonic foraminifers and calcareous nannoplankton of Oligocene to late Miocene age (Figure 13). This lower unit is overlain by an increasingly diatomaceous, nannofossil ooze with foraminifers [Schlich et al., 1989]. Abundant siliceous microfossils are restricted to the late Miocene–Pleistocene sediments at Site 747, while ice-rafted debris and dropstones are only abundant above ~20 mbsf. This is in marked contrast with Site 751, drilled to the south in the Raggatt Basin, where highly variable density profiles are observed and where siliceous microfossils are frequently abundant throughout the Miocene [Schlich et al., 1989] (Figure 14).

Index Properties. Figure 13 shows the pattern of sharp breaks in density, porosity, and carbonate content trends near 34.5 mbsf. These trend reversals occur across the interval separating Lithologic Unit I and Unit II. Several stratigraphic hiatuses are proposed between ~36 mbsf and ~20 mbsf by the integrated paleomagnetic-biostratigraphic age model of Harwood et al. [1992]. Four stratigraphic hiatuses have been proposed within the interval from ~36 mbsf to ~20 mbsf. These hiatuses are identified as 8.0–5.8 Ma (~36 mbsf), 5.3–4.4 Ma (~30 mbsf), 4.1–3.4 Ma (~28 mbsf), and 3.1–2.2 Ma (28–24 mbsf). The broad density peak centered at ~35 mbsf, bracketed by lower density intervals above and below, is associated with two late Miocene hiatuses. The sawtooth pattern of density shifts in the upper 30 mbsf may reflect some of the younger hiatuses. Other hiatuses are proposed at ~67 mbsf (13–12.2 Ma) and possibly one near 85 mbsf (~15.8 Ma ?). These latter two hiatuses are better observed in the GRAPE bulk density profile shown in Figure 15. The late Oligocene/ Miocene boundary at Hole 747A is located near 127 mbsf, where a 10% increase in porosity is positioned between two porosity lows (Figure 13).

The GRAPE bulk density data from 150 mbsf to the

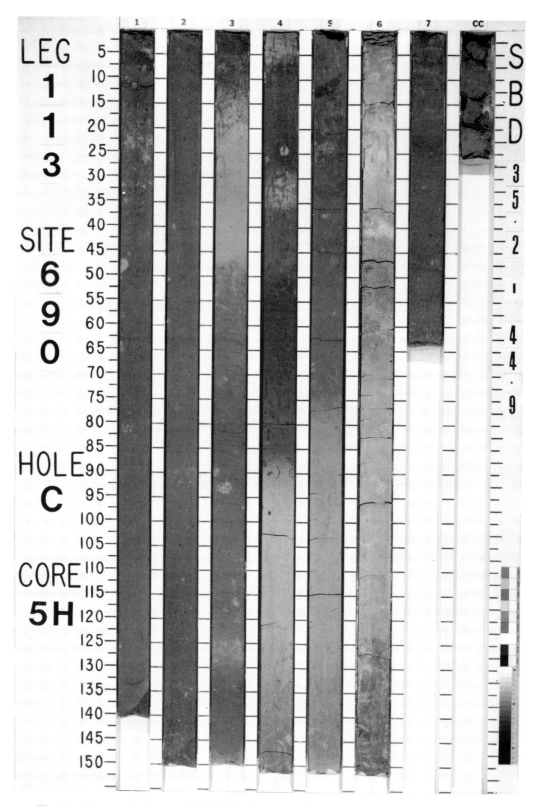

Fig. 10.   Photograph of Core 113-690C-5H, illustrating lighter-colored, carbonate-enriched beds corresponding to a series of bulk density peaks in the middle Miocene below 40 mbsf. Darker-colored intervals correspond to lower-carbonate beds and decreased values of bulk density. (Refer to Figure 9 for downhole density profile versus subbottom depth.)

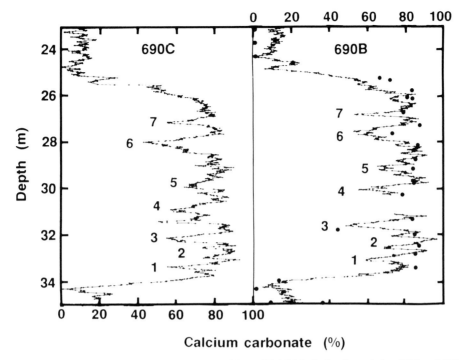

Fig. 11. Profiles of calcium carbonate calculated from GRAPE bulk density at holes 690B and 690C from 23 mbsf to 35 mbsf. Solid dots plotted next to the curve for Hole 690B represent laboratory measurements of calcium carbonate from *O'Connell* [1990]. Seven lower-carbonate zones are identified within this high-carbonate sequence. The paleomagnetic age control within this interval is poor; therefore the carbonate fluctuations are used to correlate between holes and sites on Maud Rise.

seafloor at Hole 747A are presented in Figure 15. The change from a predominantly calcareous lithology below 34.5 mbsf to an overlying mixed diatom-nannofossil ooze from 34.5 mbsf to the seafloor is observed by changes in the gradient of bulk density data. The GRAPE bulk density decreases from 1.75 g/cm^3 at 34.5 mbsf to 1.25 g/cm^3 near the seafloor.

Relatively constant values of density (and carbonate content) are observed from 34 mbsf to 80 mbsf, with the exception of two discrete low-density intervals located above and just below ~40 mbsf. Below 150 mbsf the sediment changes to a nannofossil chalk with increased amounts of foraminifers. Age assignments from *Harwood et al.* [1992] are included in Figure 15 to provide a means of comparison between sites on Kerguelen Plateau and Maud Rise. The relationships between bulk density and carbonate content for these sites are similar to those presented earlier in the profiles from ODP sites 689 and 690 (F. R. Rack, unpublished data).

ODP Site 748

Description. Site 748 is located on the southern Kerguelen Plateau in the western part of the Raggatt Basin (in the lee of Banzare Bank) (Figure 12). To the northwest, the Raggatt Basin abruptly terminates along nor-

mal faults at the 77°E Graben. Farther south, the basement outcrops at Banzare Bank where the Neogene sediments are truncated [*Colwell et al.*, 1988; *Schlich et al.*, 1989; *Coffin et al.*, 1990]. The sedimentary cover thickens to the east, toward the central part of the Raggatt Basin (near Site 751). Site 748 has an extended hiatus in the middle Miocene [*Harwood et al.*, 1992], possibly indicating erosion by current flow around the boundary of the Raggatt Basin.

Hiatuses at Site 748 lie near ~60 mbsf (22.6–22 Ma and 21.3–19.5 Ma), ~38 mbsf (18–11.3 Ma), ~22 mbsf (10.2–9.3 Ma), ~10.5 mbsf (8.0–4.0 Ma), and ~5 mbsf (2.8–2.0 Ma) [*Harwood et al.*, 1992]. Other hiatuses may also exist near ~90 mbsf, ~100 mbsf, and ~110 mbsf (early Oligocene) and below ~130 mbsf (late Eocene), based on sharp changes in porosity observed at these subbottom depths.

Index Properties. The GRAPE bulk density data from 0 to 150 mbsf at Hole 748B are presented in Figure 16. The transition to a diatom ooze occurs just below 10 mbsf at Site 748. A pair of density peaks separated by a density low are observed between ~15 mbsf and 10 mbsf. The positions of the proposed hiatuses discussed above are illustrated in this profile as sharp changes in density trends. The sharp change in density at ~115 mbsf marks an abrupt increase in cool water nannofossil

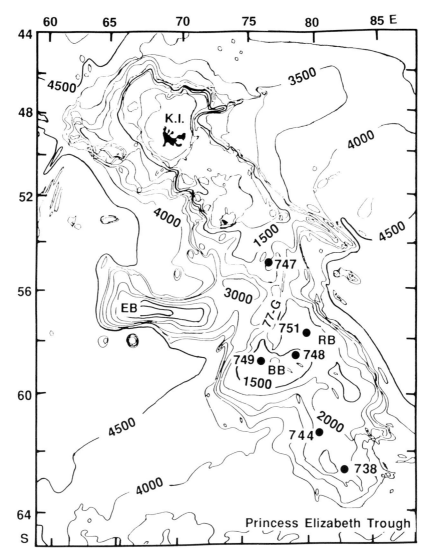

Fig. 12. Bathymetry of Kerguelen Plateau modified from *Schlich et al.* [1987]. ODP sites 747, 748, 751, and 744 are indicated by solid dots. Also shown are Kerguelen Island (K.I.), Elan Bank (EB), the 77°E Graben (77-G), Banzare Bank (BB), Raggatt Basin (RB), and Princess Elizabeth Trough.

taxa at the top of paleomagnetic Subchron C13R in the earliest Oligocene [*Wei*, 1991]. Other fluctuations in these profiles may be correlated between widely separated drill sites as additional supporting biostratigraphic data become available.

ODP Site 751

Description. Site 751 is located in the central part of the Raggatt Basin in approximately 1634 m of water. The geophysical profiles across this region show continuous, medium-amplitude parallel reflectors of presumed Neogene age that are restricted to the highest topographic portion of the Raggatt Basin. The Neogene sediments are interpreted to be truncated in all direc-

tions by erosional processes or nondeposition [*Colwell et al.*, 1988; *Coffin et al.*, 1990]. A 166.2-m section of lower Miocene to upper Pleistocene mixed biosiliceous and calcareous ooze was recovered at this site [*Schlich et al.*, 1989]. Diatoms dominate the sequence above 40 mbsf, where ice-rafted debris is also common; diatoms are abundant in restricted intervals below 40 mbsf.

Several Neogene hiatuses are identified at Site 751 by *Harwood et al.* [1992]. These are generally located near ~132 mbsf (16.3–15.6 Ma), ~104 mbsf (12.7–12.2 Ma), ~40–45 mbsf (8.5–6.0 Ma), and ~36 mbsf (5.7–4.4 Ma).

Index Properties. A detailed analysis of the geotechnical properties and sedimentary microfabric of the sediments at Site 751 is presented by *Rack and Palmer-*

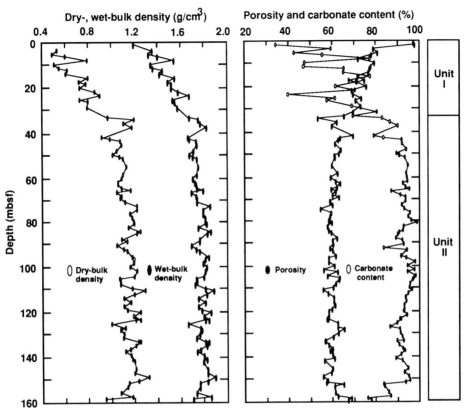

Fig. 13.   Density, porosity, and laboratory values of percent calcium carbonate versus subbottom depth at Hole 747A on Kerguelen Plateau. Lithologic units are shown to the right of the index properties profiles.

Julson [1992]. This paper expands on their analysis by discussing how the GRAPE density profile at Site 751 compares with the ones presented for sites 747, 748, and 744. Discrete index property measurements at Site 751 were obtained at a sample spacing of about 0.5 m (Figure 14). A sample spacing of ~1.5 m or more, for routine shipboard index properties measurements, was used at the other sites in this study (Figures 4, 5, and 13). The density (and calcium carbonate) data suggest some strong correlations among the sedimentary sequences drilled on Maud Rise (sites 689 and 690) and at Site 751 on Kerguelen Plateau, particularly within the early to middle Miocene, where high-amplitude changes in density are observed.

Discrete sediment samples were examined using a scanning electron microscope (SEM) to illustrate how these fluctuations in sediment properties are related to changes in sedimentary microfabric. The changes in fabric are caused by shifts in the accumulation of biosiliceous and calcareous microfossil skeletons and by dissolution of microfossil tests (for examples of SEM images, see *Rack and Palmer-Julson* [1992] and *Rack et al.* [1992]).

The GRAPE density data, from 150 mbsf to the seafloor at Site 751, are presented in Figure 17. The density profile at Site 751 illustrates the high variability

that is found in the lithologic composition of these sediments during the Neogene. Below ~40 mbsf, the sediments are primarily composed of nannofossils; however, the relative abundance of diatoms fluctuates greatly. These density fluctuations are most pronounced within several narrow zones between 130 mbsf and 100 mbsf, where the carbonate content of the sediment is observed to vary by as much as 70%. The dissolution-related decreases in GRAPE bulk density between 130 mbsf and 100 mbsf are clearly observed in this profile [*Rack and Palmer-Julson*, 1992].

Decreasing trends in bulk density are observed from 110 mbsf to ~100 mbsf and from ~95 mbsf to ~60 mbsf. These intervals have increased porosities and corresponding relative increases in the percentage of diatom skeletons in the sediment. The density signature noted at sites 747 and 748 (from 8.2 to 8.9 Ma) may also be observed at Site 751; however, biostratigraphic evidence suggests a hiatus across portions of this interval just above 45 mbsf [*Harwood et al.*, 1992]. The major lithologic change to a diatom ooze is a sharp decrease in bulk density above 40.1 mbsf. The sediments in the upper lithologic unit are commonly disturbed and thus provide limited useful information from the bulk density profile.

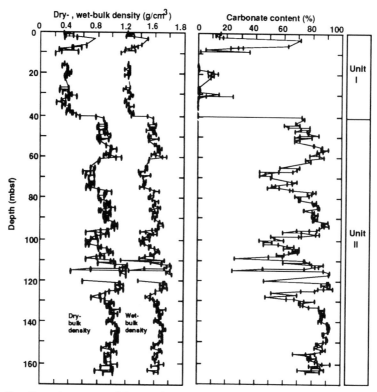

Fig. 14. Density and percent calcium carbonate versus subbottom depth at Hole 751A, in the Raggatt Basin on Kerguelen Plateau. Lithologic units are shown to the right of the index properties profiles.

ODP Site 744

Description. ODP Site 744 is located in 2307 m of water on the western boundary of the southern Kerguelen Plateau. Overlapping sequences of Neogene sediments were recovered at holes 744A and 744B above 100 mbsf. Hole 744A is characterized by poor sediment recovery above 80 mbsf, while Hole 744B exhibits good recovery of the sediment to 75 mbsf [*Barron et al.*, 1989]. Lithologic Unit I, a mixed carbonate and diatom ooze, is found above 23.2 mbsf and 21.5 mbsf at Hole 744A and Hole 744B, respectively. Lithologic Unit II is described as a nannofossil ooze with varying, but generally small, amounts of foraminifers and diatoms, which extends to ~176 mbsf at Hole 744A. Foraminifers are more abundant from 125 mbsf to about 160 mbsf, where they comprise 20–30% of the total composition of the sediment.

Age assignments at Site 744 are taken from *Barron et al.* [1991]. Hiatuses are proposed at ~22 mbsf (5.6–4.2 Ma), ~24 mbsf (8.9–6.0 Ma), 53 mbsf (12.2–11.2 Ma), ~55 mbsf (13.4–12.5 Ma), ~97 mbsf (24–21.3 Ma), ~99 mbsf (26.3–24.5 Ma), and ~118 mbsf (32–28.5 Ma). Many of these hiatuses may be observed in the bulk density profiles described below, although several are located at core breaks or in poorly recovered intervals at this site.

The density profiles at Site 744 exhibit density fluctuations similar to those seen at Site 751, although the amplitude of these fluctuations is smaller at Site 744.

Index Properties. The GRAPE bulk density profiles for Hole 744A and Hole 744B are presented in Figures 18 and 19, respectively. The bulk density data at Hole 744A can be tenuously correlated with the density profile from Hole 744B in several intervals; however, the best correlation is between the density low at ~62 mbsf at Hole 744A and a corresponding bulk density low near ~65 mbsf at Hole 744B.

The fluctuating density trends from 75 mbsf to 50 mbsf may represent the sequence of dissolution events identified in the early to middle Miocene at Site 751. Further examination of the Miocene intervals at these sites may provide additional information about potential depth-dependent dissolution patterns at the boundary between intermediate water and deepwater masses across the southern Kerguelen Plateau.

DISCUSSION

Introduction

In pelagic settings such as the oceanic regions above Maud Rise and Kerguelen Plateau, the rate of supply of

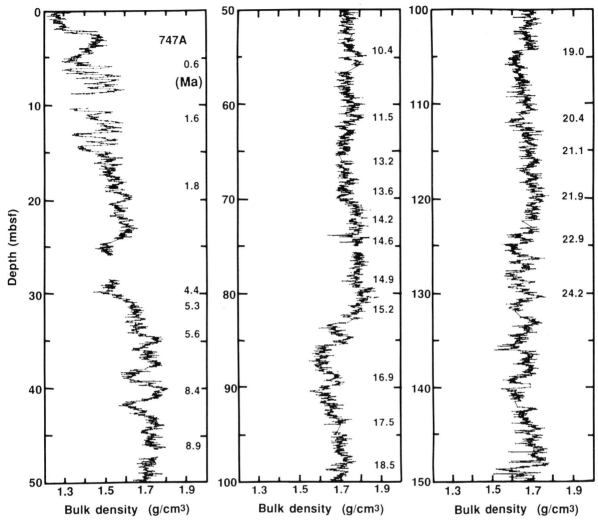

Fig. 15. GRAPE bulk density versus subbottom depth curve for ODP Hole 747A from 150 mbsf to the seafloor. Approximate stratigraphic age assignments in millions of years ago are given to the right of the density curve for each 50-m interval.

sedimentary materials and the hydrodynamic and geochemical conditions at the seafloor determine where sediment will accumulate. The rates of primary phytoplankton productivity and grazing by zooplankton are largely determined by ocean circulation patterns and by environmental conditions (for example, water temperature, nutrient levels, sea ice extent, and light intensity). These conditions vary seasonally, annually, and on longer time scales. Additional constraints on the geologic record include the vertical and lateral movements of tectonic barriers, climatic fluctuations, and changes in southern hemisphere ice volume and the extent of sea ice on longer time scales (tens of thousands to millions of years).

The deterioration of environmental conditions in the Southern Ocean since the late Oligocene has resulted in the sequential replacement of a sedimentary facies composed of calcareous nannofossils and foraminifers by a sedimentary facies dominated by diatoms and other siliceous skeletons and fragments. The geologic record of these fluctuations in sedimentary facies is observed as one moves vertically through the accumulated sequences of microfossil skeletons and fragments or laterally (equatorward) away from Antarctica. Several distinct intervals of sediment dissolution and erosion are observed at the ODP sites located south of ~56°S, both on Kerguelen Plateau and Maud Rise, possibly indicating an expansion of southern source water masses or an increase in the strength of eastward-flowing waters within the Antarctic Circumpolar Current (ACC) system and their recirculation to the southwest in the Weddell Gyre.

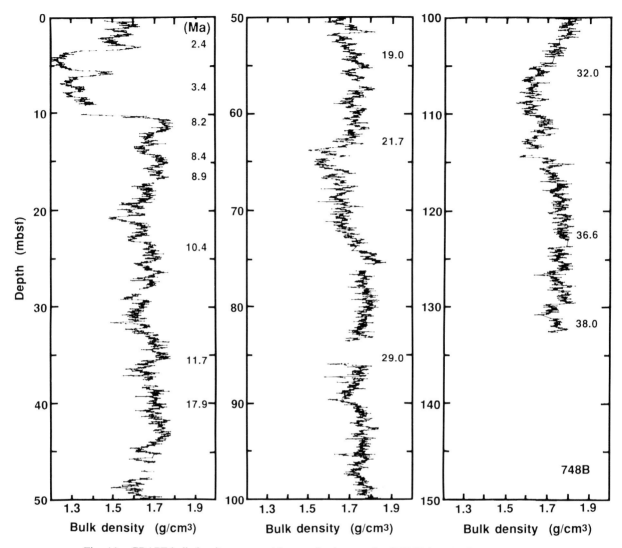

Fig. 16. GRAPE bulk density versus subbottom depth curve for ODP Hole 748B from 150 mbsf to the seafloor. Approximate stratigraphic age assignments in millions of years ago are given to the right of the density curve for each 50-m interval.

In the following discussion, comparisons will be made between the previously described results from the ODP sites on Maud Rise and those from Kerguelen Plateau. These comparisons are used to develop some working hypotheses concerning the evolution of ocean circulation in the Atlantic and Indian sectors of the Southern Ocean.

GRAPE Bulk Density: Maud Rise

The accumulation of low-density (high-porosity) biosiliceous microfossil skeletons and fragments, relative to the accumulation of calcareous skeletons, is suggested as the predominant feature of observed fluctuations in downhole physical properties in the sediments on Maud Rise and also on Kerguelen Plateau. High water content and porosity values (low bulk density) are directly linked to increases in diatom and radiolaria skeletons relative to nannofossils and foraminifers. The percentage of silicoflagellates is important only in discrete zones of the upper lithologic unit at each site on Maud Rise [*Barker et al.*, 1988]. The zones of increased biosiliceous sediment are also associated with increases in compressional wave velocities and darker-colored sediment in the upper 75 m at these sites.

Diester-Haass [1991] has provided a thorough analysis of coarse fraction sedimentologic indicators of paleoceanographic effects at sites 689 and 690 for the Eocene to late Oligocene. In her study, the carbonate

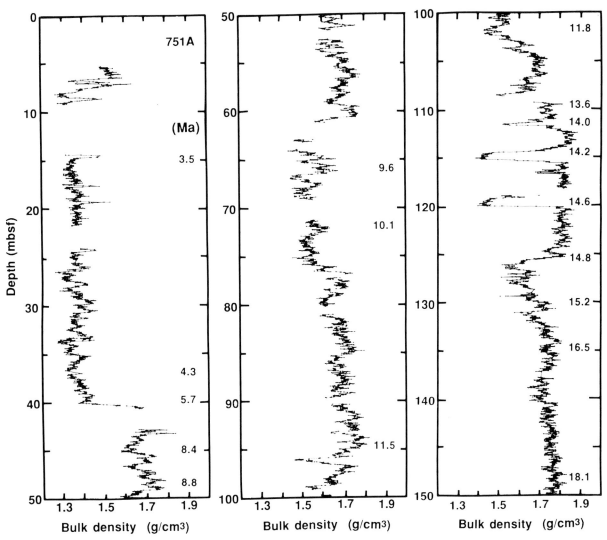

Fig. 17. GRAPE bulk density versus subbottom depth curve for ODP Hole 751A from 150 mbsf to the seafloor. Approximate stratigraphic age assignments in millions of years ago are given to the right of the density curve for each 50-m interval.

dissolution history at these sites is analyzed in terms of surface water productivity and bottom water characteristics. She suggests that more vigorous upwelling conditions and increased atmospheric circulation in the early Oligocene caused higher opal production and higher organic matter supply to the seafloor, thus causing more carbonate dissolution. Current winnowing is also suggested as an important process affecting sedimentation during the Oligocene, at Site 690 [Diester-Haass, 1991].

In the Neogene, changes in sediment color, composition, and bulk density profiles seem to indicate an alternation between southern source and northern source effects on biogenous accumulation patterns. The expansion of southern source waters is associated with

increased diatom accumulation, dark-colored sediment, and low density. The southward incursion of northern source waters is associated with the preservation of calcareous microfossils, lighter-colored sediment, and higher bulk density values.

The observed changes in downhole density trends arise from fluctuations in the relative accumulation and preservation of biogenous, siliceous, and calcareous microfossil skeletons in the sediment. Discrete shipboard bulk density measurements compare favorably with the fine-scale bulk density measurements obtained using GRAPE. The similarities between the Neogene calcium carbonate curves from ODP sites 689 and 699 [O'Connell, 1990] and the shape of the corresponding density profiles are clear. It is observed that the abso-

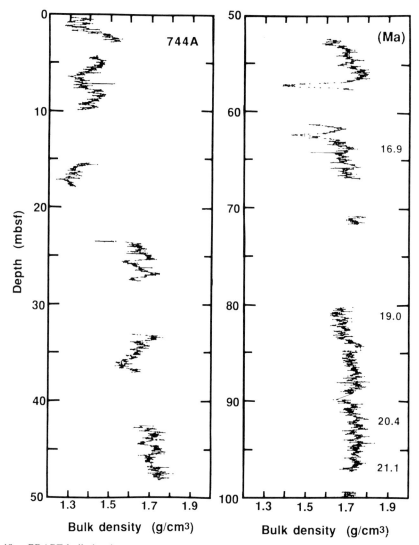

Fig. 18. GRAPE bulk density versus subbottom depth curve for ODP Hole 744A from 100 mbsf to the seafloor. Approximate stratigraphic age assignments in millions of years ago are given to the right of the density curve for each 50-m interval.

lute carbonate content is slightly less at the deeper site (Site 690) than at the shallower site (Site 689) on Maud Rise. The decreases in carbonate content in Neogene age sediments are associated with a decreasing diversity of nannofossil species and a transition to an assemblage dominated by one or two polar end-member species [Wise, 1988].

Figures 20 and 21 show the GRAPE bulk density profiles from holes 689B, 689D, 690B, and 690C versus age in millions of years ago (Ma) for most of the Neogene. The similarities in the density data from these holes are readily apparent. The age models from the B holes have been applied as consistently as possible to the additional holes from each site to illustrate the common features of the density profiles. There are still some correlation problems and inconsistencies left to be resolved, especially in the early Miocene portion of these profiles, but the overall fit among the individual sequences is extremely good.

The upper boundary of the late Oligocene to early Miocene hiatus, associated with the postulated deepwater opening of Drake Passage, is associated with a density peak centered at ~20 Ma. The correlation of this density peak in Figures 20 and 21 is in need of some revision, as it is observed at a slightly younger position at Hole 690B than at Hole 689B. This peak is difficult to identify at holes 689D and 690C. The age-density inconsistencies are due to the assumption of constant sediment accumulation in the age models, as discussed earlier in this paper (Table 2).

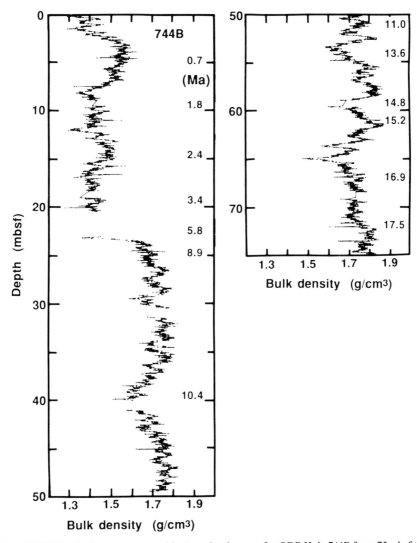

Fig. 19. GRAPE bulk density versus subbottom depth curve for ODP Hole 744B from 75 mbsf to the seafloor. Approximate stratigraphic age assignments in millions of years ago are given to the right of the density curve.

There is a low-density interval centered at ~18.5 Ma that is observed at both sites following a sharp decrease in density values, which may reflect a migration of water masses or an erosional pulse that has not been recognized previously. The overlying interval to ~15.5 Ma appears somewhat differently at each site. There is a hiatus at Site 689 from ~17 Ma to ~15.5 Ma, although the interval from ~18 Ma is missing at Hole 689D owing to a missing 0.5 m of sediment between two cores. At Site 690, two separate hiatuses have been identified. There is a period of sediment accumulation centered at ~17 Ma between these two hiatuses at this site.

There is a series of three density peaks (four at Hole 689D?) that are observed from ~15.5 Ma to ~14.2 Ma at both sites. Inconsistencies in core recovery at holes 689B and 690B resulted in the interpretation of problematic age assignment over these intervals in the age models for these sites, by *Gersonde et al.* [1990]. The uppermost density peak at Hole 689B is missing at a core break, while this sequence at Hole 690B is truncated and condensed in the first section of a core. The more complete intervals are found at holes 689D and 690C at each site.

There are also differences in the pattern of density fluctuations between Site 689 and Site 690 in the time interval from 13.5 Ma to ~12 Ma that seem to indicate that there was dissolution of carbonate with increasing water depth on Maud Rise. Three density peaks that are observed at Site 689 are missing at Site 690 within this interval. In the overlying sequence, relative fluctuations

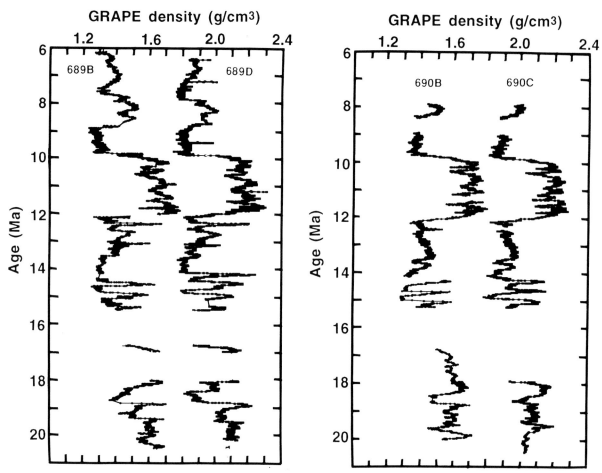

Fig. 20. GRAPE bulk density versus age in millions of years ago from 21 Ma to 6 Ma for holes 689B and 689D. The profile from Hole 689D has been shifted to the right (by 0.5 g/cm^3) to allow for comparisons between profiles.

Fig. 21. GRAPE bulk density versus age in millions of years ago from 21 Ma to 6 Ma for holes 690B and 690C. The profile from Hole 690C has been shifted to the right (by 0.5 g/cm^3) to allow for comparisons between profiles.

in bulk density are observed throughout the overall high-density zone from ~12 Ma to 9.8 Ma. These relative density fluctuations may reflect shifts in the position of water mass boundaries over Maud Rise, causing changes in the accumulation of siliceous sediments or the episodic dissolution of calcareous microfossils within this sequence.

The late Miocene sequences at sites 689 and 690 are quite different. At Site 689, there is a multiple density peak associated with a thin carbonate-enriched bed centered at ~8.4 Ma; this carbonate layer is not observed at Site 690, although a single density peak is also observed at this site on the slope of Maud Rise. The increases in density that are not related to the carbonate layer are likely caused by the increased accumulation of ice-rafted debris. This density feature may therefore be interpreted as representing an increase in sea ice extent in the region of Maud Rise, which may correspond to the expansion of the West Antarctic Ice Shelf. The thin carbonate bed at Site 689 may represent a change in the

position of the calcium compensation depth in this region or an increase in calcareous productivity and flux to the seafloor during this time interval.

GRAPE Bulk Density: Kerguelen Plateau

The analysis of physical properties data from ODP sites indicates a period of environmental change in the Southern Ocean during the late Oligocene to late Miocene. Sites 747 and 751 on Kerguelen Plateau are located in 1695 m and 1634 m of water, respectively. Site 751 is positioned approximately 400 km to the south of Site 747, in the center of the Raggatt Basin. These two sites are separated by a bathymetric depression in the central region of Kerguelen Plateau known as the 56° Saddle. Dramatic differences in lithology and sediment accumulation have been observed between these two sites in the Miocene. These differences are observed in the density profiles shown in Figure 22.

At Site 751, several possible dissolution events have

GRAPE density (g/cm³)

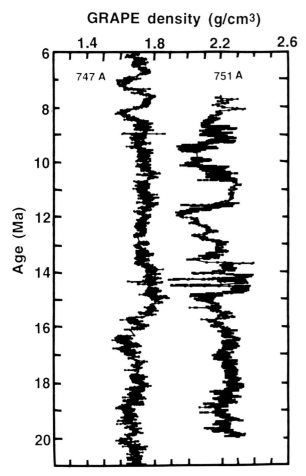

Fig. 22. GRAPE bulk density versus age in millions of years ago (Ma) from 21 Ma to 6 Ma for holes 747A and 751A. The profile from Hole 751A has been shifted to the right (by 0.5 g/cm³) to allow for comparisons between profiles.

been identified in the middle Miocene (~16 Ma to 12 Ma) where there are large fluctuations in the relative accumulation of calcareous and siliceous microfossil skeletons [*Schlich et al.*, 1989]. This suggests that there were certain oceanographic or sedimentologic processes which were restricted to latitudes south of about 55°S in the Southern Ocean, during the early to middle Miocene. Possible processes are changes in the extent of sea ice, changes in the intensity or location of currents, and changes in biological productivity, among others.

To the north, at ODP Site 747, high-carbonate sediments were preserved throughout the early Neogene without significant evidence for dissolution of microfossils until the late Miocene. In the late Miocene (during the interval of ~8 Ma to 6 Ma), there are sharp bulk density and porosity fluctuations (near 32 mbsf) which are accompanied by a series of hiatuses. There is a subsequent increase in the diatom content of the overlying late Miocene and younger sediment. These

changes are reflected at the other sites located to the south, by a change from dominantly calcareous to dominantly biosiliceous sediments. The differences in sediment lithology and GRAPE bulk density between ODP sites 747 and 751 are suggestive of the presence of a paleofront or water mass boundary across the central Kerguelen Plateau during the Miocene. Shifts in the latitudinal position and longevity of this postulated paleofront or boundary may explain the differences in the density records between these two sites.

A particularly interesting feature of the density profile at Site 747 is the high-amplitude density oscillations from ~15 mbsf to ~5 mbsf (Figure 15). It is suggested that these fluctuations may be indicative of current winnowing during the Pleistocene. The generalized geographic location of Site 747 may have also been subject to intermittent, high-velocity current flow since the late Miocene. These inferences are consistent with theories and experiments pertaining to the interaction of ocean currents with submarine topographic relief as presented in a number of papers [*Boyer*, 1975; *McCartney*, 1976; *Boyer and Zhang*, 1989; *Pratt*, 1989; *Wolff and Olbers*, 1989; *Treguier and McWilliams*, 1990; D. L. Boyer et al., "Laboratory simulation of bathymetric effects on the Antarctic Circumpolar Current," submitted to *Journal of Geophysical Research*, 1992]. Geotechnical comparisons between sites 747, 748, 751, and 744 may provide insight into the interactions of the ACC with the topography of the central Kerguelen Plateau during the Miocene.

Woodruff and Chambers [1991] have observed strong dissolutional effects between 60 mbsf and 52 mbsf at Hole 744B (Figure 19). They identify several stratigraphic hiatuses within this lower density interval. *Woodruff and Savin* [1991] have recently demonstrated the presence of dissolution events and brief hiatuses in the middle Miocene at Site 744 (from 75 mbsf to 50 mbsf) using an analysis of stable isotope records. They postulate that there were latitudinal exchanges and mixing of southern component and northern component water masses in the South Atlantic and Indian oceans during the middle to late Miocene.

It is likely that the dissolutional events observed at the ODP sites in the Raggatt Basin and the southern Kerguelen Plateau were caused by the intersection of sharp water property gradients with the seafloor. These high-gradient regions are typically associated with the position of oceanographic fronts or boundaries. The similarity of the pattern of facies changes, and hiatuses between the Raggatt Basin and Maud Rise suggest a linkage between the two regions. It is suggested that the potential causative mechanism for at least some of these events, both on Maud Rise and on Kerguelen Plateau, was the simultaneous acceleration of the ACC and Weddell Gyre. These accelerations may have been associated with changes in the ACC system following the two proposed ridge-arc collisions in the western

Weddell Sea, at ~16 Ma and 7–8 Ma [*Barker et al.*, 1984]. There may be temporal associations between these collisions and climatic events such as the expansion of the East Antarctic Ice Sheet in the middle Miocene and the establishment of the West Antarctic Ice Shelf in the late Miocene and early Pliocene, although these interpretations are highly speculative, based on the data presented here.

The lithologic change from calcareous to dominantly biosiliceous sediment accumulation occurs about 1.5 m.y. earlier on Maud Rise (at ~9.8 Ma), relative to the Raggatt Basin on Kerguelen Plateau (<8.0 Ma). The differences in the age of this transition may be explained by the more southerly position of the Maud Rise and by the production of a southern source intermediate water mass in the Weddell Sea [*Kennett and Barker*, 1990]. A discrete 0.5-m calcareous ooze layer within the biosiliceous ooze at Site 689 is bounded by two hiatuses at 8.2 Ma and ~7.9 Ma. This thin carbonate bed may reflect a southward penetration of warmer surface waters, prior to the northward advance of the West Antarctic Ice Shelf. Further examination of the Miocene intervals at these sites may provide additional information about water depth–dependent dissolution patterns at the boundary between intermediate water and deepwater masses, across the southern Kerguelen Plateau.

Ocean Currents, Seismic Facies, and Sediment Body Geometry

The Weddell Gyre provides a recirculation path for waters at the southern boundary of the ACC. After turning south and westward near 30°E longitude, these recirculated waters traverse the southern Weddell-Enderby Basin as they flow toward Maud Rise in the northeastern Weddell Sea. *Jacobs and Georgi* [1977] and *Gordon et al.* [1978, 1981] have suggested that the Weddell Gyre may have extended further eastward in the past, if oceanographic conditions were somewhat different. The geometry of the Weddell-Enderby Abyssal Plain suggests that deep, eastward flowing currents south of Conrad Rise would intersect the western slope of the central Kerguelen Plateau. Below ~2000-m water depth, the flow would be deflected to the north, south, or back to the west.

The early formation of the Weddell Gyre along the southern boundary of the ACC may have resulted from a "blocking" of eastward flowing waters by the Kerguelen Plateau within the lower water column. This type of "blocking" is proposed by *Rhines* [1989], for the generalized case of an eastward flowing current encountering a meridional submarine plateau. The northern boundary for the transition between eastward flowing and recirculated waters may have been located just west of the central Kerguelen Plateau, near Elan Bank, during the early and middle Miocene. This oceanographic transition zone may have migrated northward in

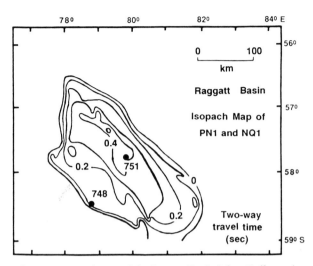

Fig. 23.   Isopach map of the Raggatt Basin [after *Coffin et al.*, 1990]. Contours are in seconds of two-way travel time. The seismic facies are thickest in the center of the Raggatt Basin near Site 751. The facies thin toward the boundaries of the basin, as in the direction of Site 748.

the middle to late Miocene as Drake Passage deepened, the lateral ocean ridge boundaries migrated northward, and the volume and extent of ice around Antarctica fluctuated.

Neogene seismic facies in the Raggatt Basin are truncated in all directions as a result of erosional processes or the nondeposition of sediment [*Colwell et al.*, 1988; *Coffin et al.*, 1990]. The youngest identified seismic facies (NQ1) is largely confined to the highest topographic portion of the Raggatt Basin; sequence PN1 is more widespread but thins toward the boundaries of the basin [*Coffin et al.*, 1990]. The geometry and relative thickness of Neogene sediments in the Raggatt Basin have been discussed by *Coffin et al.* [1990]; they present isopach maps, derived from multichannel seismic data across the Raggatt Basin, that can be used to constrain geologic and oceanographic interpretations of Neogene environmental conditions in this region of the Kerguelen Plateau.

The isopach map shown in Figure 23 illustrates the acoustic thickness of seismic sequences PN1 and NQ1 in seconds of two-way travel time. This map is used to show the relative thickness of Neogene sedimentary units at ODP sites 748 and 751 in the Raggatt Basin. There is a maximum thickness of Neogene sediment in the center of the Raggatt Basin, near Site 751. These units thin in the direction of Site 748, toward Banzare Bank on the western boundary of the basin. The interpretation of the geometry of these deposits is that sediment was directed to the center of the Raggatt Basin as a current circled around the periphery. This oceanic flow regime may have resulted in the nondeposition or erosion of sediment at Site 748 during certain time intervals during the middle and late Miocene. The evidence for erosion on the lee side of Banzare Bank, near the location of Site 748, is consistent with models of current flow over topography [*Boyer and Zhang*, 1989, 1990; D. L. Boyer et al., submitted, 1992].

Generalized boundary conditions for Neogene flow regimes across the Kerguelen Plateau are provided by the absence of Neogene sediment at ODP Site 749 on Banzare Bank (long Neogene hiatus) and the apparent erosion along the boundary of the Raggatt Basin at ODP Site 748 during extended periods of the Miocene. Prior to the late Oligocene, seismic horizons extended across the entire Raggatt Basin without interruption. Since the late Oligocene, truncated sedimentary horizons have been shaped into a northwest-southeast oriented elongate sediment body [*Coffin et al.*, 1990]. This suggests that the paleoflow conditions changed after the late Oligocene, possibly supporting the presence of a standing eddy in the Raggatt Basin downstream of Banzare Bank.

Boyer and Zhang [1989, 1990] present the results of laboratory fluid dynamics experiments which describe the interactions of ocean currents and seafloor features. They note a depth-dependent interaction of currents with isolated topographic features, such as seamounts. Their study focuses on the effects of the time-varying oscillations in a free-stream current and the resulting characteristic patterns of flow. It is suggested here that the use of these kinds of laboratory models is important for unravelling the sedimentary history of the Raggatt Basin and Banzare Bank on Kerguelen Plateau.

A discussion of Neogene seismic reflection profiles, along the continental margin of Antarctica in the Ross Sea, has recently been presented by *Bartek et al.* [1991]. These authors suggest strong relationships between the seismic sequence boundaries identified in the Ross Sea and the timing of the advances and retreats of the East Antarctic Ice Sheet during the Neogene. Confirmation of the timing and nature of these glacially derived sediment deposits on the Antarctic continental margin awaits further scientific drilling. The relative ages that have been assigned to the seismic sequence boundaries in the Ross Sea are similar to some of the dissolutional events which are observed in the sediments from sites 689, 690, and 751. Future research should compare downhole profiles of index properties presented in this study with the published eustatic sea level curves of *Haq et al.* [1987] to investigate the relationships between property trends and inferred changes in sea level.

Interactions of Currents With Bottom Topography

Two different models for the interaction of eastward flowing jets with bottom topography which have application to the ACC system have recently been published in the literature. *Pratt* [1989] suggests that there is a critical control of eastward zonal jets by bottom topog-

raphy. Variations in flow are observed as transitions from subcritical to supercritical flow regimes, variations in the width of individual fronts, and changes in the separation distance between high-velocity current cores. He advances a mechanism for the "upstream influence" of critically controlled flow regimes over large horizontal scales by the propagation of long potential vorticity waves (Rossby waves). In this theory, large-scale topography exercises an influence over the general circulation as a whole, as opposed to exerting a purely local influence. *Pratt* [1989] suggests that this sort of process may be observed in the ACC system in the regions of Kerguelen Plateau, Macquarie Ridge, and possibly Drake Passage.

A different kind of approach to the interactions of currents with topography is presented by *Rhines* [1989]. In this model, transfers of topographic potential vorticity are suggested as a major control over broad baroclinic flows, with diffusional effects being secondary. A particularly interesting aspect of Rhines' model is that "blocking" of the oncoming flow by topography can induce an elongated gyre circulation. This process could effectively send an eastward zonal flow back to the west at higher latitude. The blocking effect also promotes a closed "gyre" circulation over the topographic obstacle and advects water from the south and north into this topographic gyre.

The similarities between the sedimentary records from Maud Rise and from Raggatt Basin on Kerguelen Plateau have been interpreted with respect to theoretical models of biological productivity and current interactions with seafloor topography. The data are used to support a hybrid model combining aspects of the two models by *Rhines* [1989] and *Pratt* [1989], where deeper levels of the eastward circulation are effectively blocked by the Kerguelen Plateau and become recirculated waters in the Weddell Gyre, while the shallower levels are intensified through the central region of Kerguelen Plateau by bathymetric interactions. These effects are modified through time as atmospheric circulation evolves, the velocity of the eastward oceanic flow fluctuates, and sea ice expands northward away from Antarctica.

CONCLUSIONS

An understanding of oceanographic processes in the Southern Ocean can be used to help in the interpretation of sedimentologic and geotechnical data obtained from deep-sea cores. The physical and geotechnical properties of sediments from ODP sites in the Southern Ocean have been used in this study to speculate on the possible existence of a paleoceanic front and/or a sedimentologic boundary across the central Kerguelen Plateau in the early to middle Miocene.

The sediment sequences recovered from the six Ocean Drilling Program (ODP) sites discussed in this paper represent the first-ever long-term geologic data from intermediate water depths within the Antarctic zone (i.e., south of the Polar Front and north of the Antarctic Continental Shelf). As such, they provide the best available geologic record of the early Neogene evolution of the circumpolar ocean circulation across a wide range of latitudes within the Atlantic and Indian sectors of the Southern Ocean.

An analysis of physical property trends can be used (1) to create a framework for regional comparisons among ODP sites, (2) to map lateral and vertical stratigraphic changes in sedimentary facies, (3) to make inferences about the geologic history of a particular region, (4) to examine the effects of natural processes on marine depositional environments, and (5) to develop geological constraints on paleoceanographic and paleoclimatic models.

Among other causes, the bulk density variations observed in the sediments from Maud Rise are likely explained by a combination of processes, including (1) fluctuations in ice volume on East Antarctica, (2) changes in ocean circulation and upwelling associated with tectonic events (such as the opening of the Drake Passage), and (3) changes in climatic gradients associated with fluctuations in atmospheric circulation and precipitation patterns, interhemispheric linkages, and orbital forcing. These potential paleoenvironmental records from the Southern Ocean may provide a framework for comparisons with similar age sequences in the equatorial Pacific [Mayer et al., 1986; Berger and Meyer, 1987; Theyer et al., 1989; Mayer, 1991].

An understanding of the physical properties of the major biogenous sediment facies in the Southern Ocean provides the basic information for investigating (1) the development of local and basinal patterns of sediment accumulation and (2) latitudinal variations in the deposition of particular sediment facies in the geologic record. A detailed analysis of sediment facies associations may also provide a framework for future research aimed at combining this geologic information with high-resolution seismic data to consider larger paleoceanographic questions.

The opening of Drake Passage and the subsidence of various oceanic edifices and continental fragments within the Southern Ocean have fundamentally altered the large-scale circulation of this region. The intent of this research was to contribute to an understanding of the timing and effects of these tectonic and oceanographic changes in the southern hemisphere, from the late Oligocene to the late Miocene, a relatively poorly known interval in southern high-latitude paleoceanography and Antarctic glacial history [Wise et al., 1991]. It is suggested here that the latitudinal movements of paleoceanographic fronts and water mass boundaries may have been constrained by the location and shape of topographic features and the effects of their interactions with ocean currents. A mechanism is proposed which

combines a relative strengthening of the Antarctic Circumpolar Current system and increases in the vigor of the Weddell Gyre to explain the distribution of hiatuses and the changes in sediment facies associations that are observed in the geologic record at these ODP sites.

There is some evidence to support the hypothesis that high-frequency fluctuations in the size of the Antarctic Ice Sheet may have occurred during the Neogene [Robin, 1988; Barron et al., 1989; Harwood et al., 1992; Wise et al., 1991; Hambrey et al., 1991; Anderson et al., 1991]. However, the response of the ice sheet to climate warming (cooling) may be significantly complicated by a number of nonlinear feedback mechanisms, thereby making it difficult to identify the important causative factors for these fluctuations [Berger and Mayer, 1987]. The postulated establishment of a temperate glacial regime in the Oligocene, and its transition to a polar glacial regime in the middle Miocene [Anderson et al., 1991; Bartek et al., 1991], may provide constraints on the evolution of Antarctic water masses and the effects of enhanced thermohaline circulation near the Antarctic continent. Further to the north, fluctuations in the intensity of the Antarctic Circumpolar Current system may also provide a logical mechanism for explaining sedimentary events and hiatuses.

The analysis of physical properties data from these ODP sites indicates that the late Oligocene to late Miocene period was a time of environmental change in the Southern Ocean. The observed differences in sediment lithology at ODP sites 747 and 751 especially are suggestive of the presence of a paleofront or water mass boundary across the central Kerguelen Plateau during the Miocene. This front may have moved northward in the late Miocene to a position in the Indian Ocean which is closer to the observed modern position of the Polar Front, as the climatic situation deteriorated.

The conclusions of this study include the following:

1. Accumulations of biogenous sediments have distinct physical properties which are determined by their microfabric. Siliceous oozes create a relatively open fabric of interlocking sand- and silt-sized diatom skeletons and siliceous fragments; these oozes exhibit high measured porosity values (75–90%) and low values of bulk density (1.2–1.4 g/cm^3). Nannofossil oozes are made up of silt- and clay-sized calcareous particles with lower porosity values (<75%) and higher values of bulk density (1.6–1.8 g/cm^3). The biogenous sediments that accumulated on the Kerguelen Plateau exhibit physical and mechanical properties that are similar to the sediments from Maud Rise.

2. Temporal fluctuations in the size, shape, and composition of microfossil skeletons are observed as reversals, inflections, and discontinuities in the trends of index properties (bulk density, porosity); as changes in the consolidation characteristics and permeability of the sediment; and as qualitative differences in the microfabric.

3. The assumption of a two-component pelagic sedimentary input, composed of biogenous siliceous and calcareous microfossils, was generally valid at these sites, thus permitting bulk density variations to reflect changes in first-order lithologic composition. High-amplitude sharp changes in bulk density are often associated with the dissolution of carbonate skeletons and the creation of a lag deposit of robust siliceous tests and fragments. Gradual changes in density are marked by shifts in the relative abundances of siliceous and calcareous microfossils.

4. Nondestructive measurements of bulk density and compressional velocity can be effectively used to correlate between adjacent ODP boreholes or sites in a particular geographic region. These correlations may then be used to evaluate the presence/absence of stratigraphic hiatuses and to map their extent, when used in conjunction with high-resolution seismic profiles. Comparisons of discrete measurements of porosity with empirical curves for various types of sediment [Hamilton, 1976] are also valuable for identifying the potential downhole position of stratigraphic hiatuses in unconsolidated sediments. There is often a sharp decrease in porosity followed by a porosity peak at the level of a stratigraphic discontinuity.

5. GRAPE bulk density data can be used to provide empirical determinations of percent calcium carbonate in high-latitude pelagic environments. The correspondence that is observed between empirically derived estimates of calcium carbonate and laboratory determinations is within 15–20%, for most sediment intervals in the upper 100 mbsf at sites 689 and 690. A systematic error in the calculation of bulk density by the GRAPE software has recently been identified; this error may account for much of the variance in carbonate estimates. The empirical technique breaks down somewhat if the two-component assumption is violated (for example, by the addition of a significant clay or ice-rafted component).

Acknowledgments. We deeply appreciate the reviews of Suzanne O'Connell and Kate Moran which greatly improved this manuscript. This project received USSAC funding under grant 20241.

REFERENCES

Ablemann, A., Oligocene to middle Miocene radiolarian stratigraphy of southern high latitudes from Leg 113, sites 689 and 690, Maud Rise, *Proc. Ocean Drill. Program Sci. Results, 113*, 675–708, 1990.

Anderson, J. B., L. R. Bartek, and M. A. Thomas, Seismic and sedimentological record of glacial events on the Antarctic Peninsula shelf, in *Geological Evolution of Antarctica*, edited by M. R. A. Thomson, J. A. Crame, and J. W. Thomson, pp. 687–692, Cambridge University Press, New York, 1991.

Barker, P. F., and J. Burrell, The opening of Drake Passage, *Mar. Geol., 25*, 15–34, 1977.

Barker, P. F., P. L. Barber, and E. C. King, An early Miocene ridge crest-trench collision on the South Scotia Ridge near 36°W, *Tectonophysics, 102*, 315–332, 1984.

Barker, P. F., et al., Leg 113, *Proc. Ocean Drill. Program Initial Rep., 113*, 785 pp., 1988.

Barron, J., et al., Leg 119, *Proc. Ocean Drill. Program Initial Rep., 119*, 942 pp., 1989.

Barron, J., J. G. Baldauf, E. Barrera, J.-P. Caulet, B. Huber, B. H. Keating, D. Lazarus, H. Sakai, H. R. Theirstein, and W. Wei, Biochronologic and magnetochronologic synthesis of Leg 119 sediments from the Kerguelen Plateau and Prydz Bay, Antarctica, *Proc. Ocean Drill. Program Sci. Results, 119*, 813–848, 1991.

Bartek, L. R., P. R. Vail, J. B. Anderson, P. A. Emmet, and S. Wu, Effect of Cenozoic ice sheet fluctuations in Antarctica on the stratigraphic signature of the Neogene, *J. Geophys. Res., 96*, 6753–6778, 1991.

Berger, W. H., Paleoceanography: The deep-sea record, in *The Sea*, vol. 7, *The Oceanic Lithosphere*, edited by C. Emiliani, pp. 1437–1519, John Wiley, New York, 1981.

Berger, W. H., and L. A. Mayer, Cenozoic paleoceanography 1986: An introduction, *Paleoceanography, 2*, 613–623, 1987.

Boyce, R. E., Definitions and laboratory techniques of compressional sound velocity parameters and wet-water content, wet-bulk density, and porosity parameters by gravimetric and gamma ray attenuation techniques, *Initial Rep. Deep Sea Drill. Proj., 33*, 931–958, 1976.

Boyer, D. L., Numerical analysis of laboratory experiments on topographically controlled flow, in *Numerical Models of Ocean Circulation*, pp. 327–339, National Academy Press, Washington, D. C., 1975.

Boyer, D. L., and X. Zhang, Time-dependent rotating stratified flow past isolated topography, in *Mesoscale/Synoptic Coherent Structures in Geophysical Turbulence*, edited by J. C. J. Nihoul and B. M. Jamart, pp. 655–670, Elsevier, New York, 1989.

Boyer, D. L., and X. Zhang, Motion of oscillatory currents past isolated topography, *J. Phys. Oceanogr., 20*, 1425–1448, 1990.

Bryant, W. R., A. P. Deflache, and P. K. Trabant, Consolidation of marine clays and carbonates, in *Deep Sea Sediments: Physical and Mechanical Properties*, edited by A. L. Inderbitzen, pp. 209–244, Plenum, New York, 1974.

Bryant, W. R., R. H. Bennett, and C. E. Katherman, Shear strength, porosity, and permeability of oceanic sediments, in *The Sea*, vol. 7, *The Oceanic Lithosphere*, edited by C. Emiliani, pp. 1555–1616, John Wiley, New York, 1981.

Coffin, M. F., M. Munschy, J. B. Colwell, R. Schlich, H. L. Davies, and Z.-G. Li, Seismic stratigraphy of the Raggatt Basin, southern Kerguelen Plateau: Tectonic and paleoceanographic implications, *Geol. Soc. Am. Bull., 102*, 563–579, 1990.

Colwell, J. B., M. F. Coffin, C. J. Pigram, H. L. Davies, H. M. J. Stagg, and P. J. Hill, Seismic stratigraphy and evolution of the Raggatt Basin, southern Kerguelen Plateau, *Mar. Pet. Geol., 5*, 75–81, 1988.

Diester-Haass, L., Eocene/Oligocene paleoceanography in the Antarctic Ocean, Atlantic sector (Maud Rise, ODP Leg 113, sites 689B and 690B), *Mar. Geol., 100*, 249–276, 1991.

Gersonde, R., A. Abelmann, L. H. Burckle, N. Hamilton, D. Lazarus, K. McCartney, P. O'Brien, V. Speiss, and S. W. Wise, Jr., Biostratigraphic synthesis of Neogene siliceous microfossils from the Antarctic Ocean, ODP Leg 113 (Weddell Sea), *Proc. Ocean Drill. Program Sci. Results, 113*, 915–936, 1990.

Hambrey, M. J., W. U. Ehrmann, and B. Larsen, Cenozoic glacial record of the Prydz Bay Continental Shelf, East Antarctica, *Proc. Ocean Drill. Program Sci. Results, 119*, 77–132, 1991.

Hamilton, E. L., Variations of density and porosity with depth in deep-sea sediments, *J. Sediment. Petrol.*, *46*, 280–300, 1976.

Haq, B. U., J. Hardenbol, and P. R. Vail, Chronology of fluctuating sea levels since the Triassic, *Science*, *235*, 1156–1159, 1987.

Harwood, D. M., et al., Neogene integrated magnetobiostratigraphy of the central Kerguelen Plateau, ODP Leg 120, *Proc. Ocean Drill. Program Sci. Results*, *120*, 1031–1052, 1992.

Houtz, R. E., D. E. Hayes, and R. G. Markl, Kerguelen Plateau bathymetry, sediment distribution, and crustal structure, *Mar. Geol.*, *25*, 95–130, 1977.

Keller, G., and J. A. Barron, Paleodepth distribution of Neogene deep-sea hiatuses, *Paleoceanography*, *2*, 697–713, 1987.

Kennett, J. P., The development of planktonic biogeography in the Southern Ocean during the cenozoic, *Mar. Micropaleontol.*, *3*, 301–345, 1978.

Kennett, J. P., and P. F. Barker, Latest Cretaceous to Cenozoic climate and oceanographic developments in the Weddell Sea, Antarctica: An ocean-drilling perspective, *Proc. Ocean Drill. Program Sci. Results*, *113*, 937–960, 1990.

Lee, H. J., Bulk density and shear strength of several deep-sea calcareous sediments, in *Geotechnical Properties, Behavior, and Performance of Calcareous Soils*, edited by K. R. Demars and R. C. Chaney, pp. 54–96, American Society for Testing and Materials, Philadelphia, Pa., 1982.

Mayer, L. A., Deep-sea carbonates: Acoustic, physical, and stratigraphic properties, *J. Sediment. Petrol.*, *49*, 819–836, 1979*a*.

Mayer, L. A., The origin of fine-scale acoustic stratigraphy in deep-sea carbonates, *J. Geophys. Res.*, *84*, 6177–6184, 1979*b*.

Mayer, L. A., Deep-sea carbonates: Physical property relationships and the origin of high-frequency acoustic reflectors, *Mar. Geol.*, *38*, 165–183, 1980.

Mayer, L. A., Extraction of high-resolution carbonate data for paleoclimate reconstruction, *Nature*, *352*, 148–150, 1991.

Mayer, L. A., T. H. Shipley, and E. L. Winterer, Equatorial Pacific seismic reflectors as indicators of global oceanographic events, *Science*, *233*, 761–764, 1986.

McCartney, M. S., The interaction of zonal currents with topography with applications to the Southern Ocean, *Deep Sea Res.*, *23*, 413–427, 1976.

O'Connell, S. B., Variations in Upper Cretaceous and Cenozoic calcium carbonate percentages, Maud Rise, Weddell Sea, Antarctica, *Proc. Ocean Drill. Program Sci. Results*, *113*, 971–984, 1990.

Pratt, L. J., Critical control of zonal jets by bottom topography, *J. Mar. Res.*, *47*, 111–130, 1989.

Rack, F. R., and A. Palmer-Julson, Sediment microfabric and physical properties record of late Neogene Polar Front migration, Site 751, *Proc. Ocean Drill. Program Sci. Results*, *120*, 179–205, 1992.

Rack, F. R., W. R. Bryant, and A. Palmer-Julson, Microfabric and physical properties of deep-sea high latitude carbonate oozes, in *Carbonate Microfabrics*, edited by R. Rezak and D. Lavoie, Springer-Verlag, New York, in press, 1992.

Rhines, P. B., Deep planetary circulation and topography: Simple models of midocean flows, *J. Phys. Oceanogr.*, *19*, 1449–1470, 1989.

Robin, G. deQ., The Antarctic ice sheet, its history and response to sea level and climatic changes over the past 100 million years, *Palaeogeogr. Palaeoclimatol. Palaeoecol.*, *67*, 31–50, 1988.

Schlich, R., M. Coffin, M. Munschy, M. H. J. Stagg, Z. G. Li, and K. Revell, Bathymetric chart of the Kerguelen Plateau, scale 1:5,000,000, Bur. of Miner. Resour., Geol. and Geophys., Canberra, 1987.

Schlich, R., et al., Leg 120, *Proc. Ocean Drill. Program Initial Rep.*, *120*, 648 pp., 1989.

Schultheiss, P. J., and S. D. McPhail, An automated *P*-wave logger for recording fine-scale compressional wave velocity structures in sediments, *Proc. Ocean Drill. Program Sci. Results*, *108*, 407–413, 1989.

Speiss, V., Cenozoic magnetostratigraphy of Leg 113 drill sites, Maud Rise, Weddell Sea, Antarctica, *Proc. Ocean Drill. Program Sci. Results*, *113*, 261–315, 1990.

Theyer, F., E. Vincent, and L. A. Mayer, Sedimentation and paleoceanography of the central equatorial Pacific, in *The Geology of North America*, vol. N, *The Eastern Pacific Ocean and Hawaii*, edited by E. L. Winterer, D. M. Hussong, and R. W. Decker, pp. 347–372, Geological Society of America, Boulder, Colo., 1989.

Treguier, A. M., and J. C. McWilliams, Topographic influences on wind-driven flow in a B-plane channel: An idealized model for the Antarctic Circumpolar Current, *J. Phys. Oceanogr.*, *20*, 321–343, 1990.

Wei, W., Evidence for an earliest Oligocene abrupt cooling in the surface waters of the Southern Ocean, *Geology*, *19*, 780–783, 1991.

Wei, W., and S. W. Wise, Jr., Middle Eocene to Pleistocene calcareous nannofossils recovered by Ocean Drilling Program Leg 113 in the Weddell Sea, *Proc. Ocean Drill. Program Sci. Results*, *113*, 639–666, 1990.

Wise, S. W., Mesozoic-Cenozoic history of calcareous nannofossils in the region of the Southern Ocean, *Palaeogeogr. Palaeoclimatol. Palaeoecol.*, *67*, 157–179, 1988.

Wise, S. W., Jr., J. R. Breza, D. M. Harwood, and W. Wei, Paleogene glacial history of Antarctica, in *Controversies in Modern Geology: Evolution of Geological Theories in Sedimentology, Earth History and Tectonics*, edited by D. W. Muller, J. A. McKensie, and H. Weissert, pp. 133–171, Academic, San Diego, Calif., 1991.

Wolff, J.-O., and D. J. Olbers, The dynamical balance of the Antarctic Circumpolar Current studied with an eddy resolving quasigeostraphic model, in *Mesoscale/Synoptic Coherent Structures in Geophysical Turbulence*, edited by J. C. J. Nihoul and B. M. Jamart, pp. 435–458, Elsevier, New York, 1989.

Woodruff, F., and S. R. Chambers, Middle Miocene benthic foraminiferal oxygen and carbon isotopes and stratigraphy: Southern Ocean Site 744, *Proc. Ocean Drill. Program Sci. Results*, *119*, 935–938, 1991.

Woodruff, F., and S. M. Savin, Mid-Miocene isotope stratigraphy in the deep sea: High-resolution correlations, paleoclimatic cycles, and sediment preservation, *Paleoceanography*, *6*, 755–806, 1991.

(Received December 16, 1991;
accepted June 3, 1992.)

CENOZOIC GLACIAL HISTORY OF THE ROSS SEA REVEALED BY INTERMEDIATE RESOLUTION SEISMIC REFLECTION DATA COMBINED WITH DRILL SITE INFORMATION

JOHN B. ANDERSON

Department of Geology and Geophysics, Rice University, Houston, Texas 77251

LOUIS R. BARTEK

Department of Geology, University of Alabama, Tuscaloosa, Alabama 35487

Intermediate resolution seismic reflection profiles, in conjunction with existing Deep Sea Drilling Project Leg 28 drill sites, provide a record of ice sheet grounding episodes on the Ross Sea continental shelf. By late Oligocene–early Miocene time, the shelf was deeply scoured by grounded ice sheets resulting in a foredeepened topography. Other evidence of ice sheets having grounded on the continental shelf by this time includes glacial troughs and till tongues. The most dramatic glacial erosion and deposition began in the early Miocene, when troughs as large as those of the modern shelf were carved, indicating the establishment of a full-bodied ice sheet with ice streams similar in size to those of the present. Waxing and waning of the ice sheet culminated in the mid-Miocene with a major erosion episode and the creation of the Mid-Miocene Unconformity. The middle-late Miocene section that is situated above this unconformity in the eastern Ross Sea shows features that, for the most part, indicate glacial-marine sedimentation and infrequent grounding events. Middle-late Miocene strata are thin to absent in the western Ross Sea. Several widespread unconformities separate middle and late Miocene strata from Pliocene-Pleistocene strata. During the Pliocene-Pleistocene the frequency of ice sheet grounding events increased; at least seven grounding episodes are recognized in the eastern Ross Sea. In the western Ross Sea the increased frequency of grounding resulted in the amalgamation of erosional surfaces such that the Pliocene-Pleistocene section is incomplete, except perhaps on the outer shelf. The increased frequency of grounding on the shelf is believed to have been in response to higher-frequency eustatic changes caused by expansion and contraction of northern hemisphere ice sheets and possibly the East Antarctic ice sheet. All of the seismic units recognized in this study can be sampled by shallow drilling when coupled with the seismic data.

INTRODUCTION

Seismic studies have revealed that three sedimentary basins occupy the Ross Sea continental shelf: the Eastern Basin, the Central Trough, and the Victoria Land Basin (*Davey* [1985]; Figure 1). These basins include thick Neogene sequences that record the history of glaciation in the region. During Deep Sea Drilling Project (DSDP) Leg 28, four sites were drilled on the continental shelf to examine this sedimentary record [*Hayes and Frakes*, 1975]. Three of these sites (sites 270, 271, and 272) were drilled on the western flank of the Eastern Basin, and one site (Site 273) was drilled on the northern edge of the Central Trough (Figure 1). Three other sites were drilled in McMurdo Sound (MSSTS 1 and CIROS 1 and 2, Figure 1) as part of a joint New Zealand and U.S. program. To date, these are the only drill sites on the West Antarctic continental shelf.

Unfortunately, recovery at the DSDP Leg 28 shelf sites was poor, and the interpretation of the recovered strata has remained problematic with regard to the onset of continental glaciation in the region [*Hayes and Frakes*, 1975; *Balshaw*, 1981; *Savage and Ciesielski*, 1983; *Leckie and Webb*, 1983; *Hambrey and Barrett*, 1992]. For example, massive diamictites, dating back to the late Oligocene–early Miocene, were recovered at these sites, but the subglacial versus glacial-marine origin of these diamictites remains problematic [*Barrett*, 1975; *Balshaw*, 1981; *Hambrey and Barrett*, 1992]. This is an important controversy because a subglacial (beneath the ice sheet) origin for these diamictites implies expansion of a marine ice sheet onto the continental shelf; a glacial-marine origin for these deposits may simply imply the presence of much smaller tidewater glaciers in the region. The CIROS 1 site penetrated tills of early Oligocene age [*Barrett*, 1989]; however, the

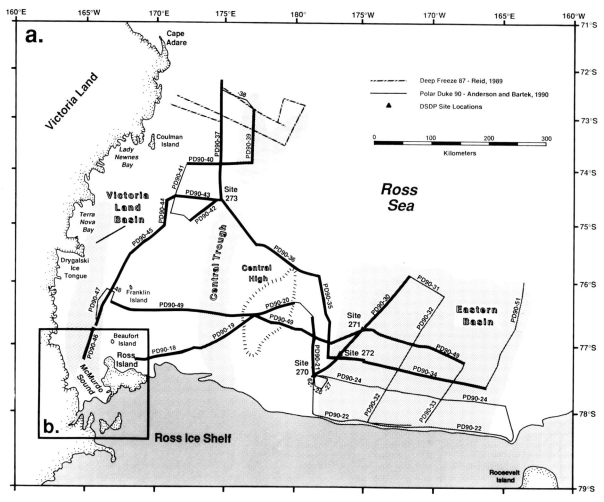

Fig. 1*a*. Location of *Polar Duke* 90 (PD 90) seismic tracklines and DSDP Leg 28 drill sites. Also shown are the locations of the Eastern Basin, Central Trough, Victoria Land Basin, and Central High.

question remains as to whether these tills are the product of local or continental glaciation.

There also are important questions concerning the stability of the Antarctic Ice Sheet following its formation. Recent discoveries of recycled marine diatoms in Pliocene tills (Sirius Group), believed to be derived from East Antarctic subglacial basins, suggest that the East Antarctic Ice Sheet retreated from these basins at some point during the late Pliocene to early Pleistocene [*Harwood*, 1985, 1986]. Further, the occurrence of *Nothofagus* plant fossils in these deposits has led to the suggestion that the Antarctic climate was warm enough to support rooted vegetation in some regions throughout the Pliocene [*Webb and Harwood*, 1991]. These ideas have not gone uncontested [*Clapperton and Sugden*, 1990; *Burckle and Pokras*, 1991].

While the recovery at sites 270 through 273 was too sparse to allow confident paleoclimatic interpretations from lithologic units alone, these sites provide the sole

biostratigraphic framework for interpreting seismic data of the West Antarctic continental shelf. If these data are of suitable quality and resolution, they can provide answers to some of the questions stated above. For example, the first ice sheets to ground on the continental shelf should have left a record, in the form of glacial erosion surfaces and subglacial deposits. Subsequent grounding events should likewise be recorded. If the Antarctic climate experienced extreme variations during the Pliocene-Pleistocene, we might expect to see evidence of ice sheet waxing and waning during that period. If the climate was warm enough to support *Nothofagus* forests, there might be incised fluvial valleys and fluvial deltas in this portion of the stratigraphic column. This type of analysis calls for seismic data with sufficient resolution to recognize these features, yet with adequate penetration to image the late Cenozoic stratigraphic column.

Conventional seismic data do not provide adequate

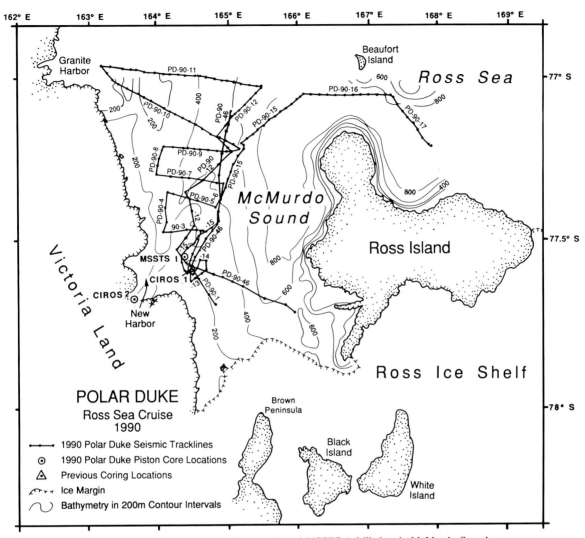

Fig. 1b.   Seismic lines, CIROS 1 and 2, and MSSTS 1 drill sites in McMurdo Sound.

resolution for imaging glacial and glacial-marine facies, incised fluvial valleys, or small fluvial deltas, especially in the relatively thin Pliocene-Pleistocene sequences. The Pliocene-Pleistocene section at Site 271 is only 250 m thick; this interval consists of five reflectors on multichannel seismic (MCS) profiles through this site [*Hinz and Block*, 1983; *Cooper et al.*, 1991]. Some attempts have been made to interpret Antarctica's glacial record from the overall stacking patterns of seismic sequences [*Larter and Barker*, 1989; *Cooper et al.*, 1991], but this approach needs to be verified with higher-resolution data and drilling.

Prior to the U.S. Antarctic Program's 1990 season (USAP 1990), an investigation was made of existing high-resolution seismic data from the Ross Sea [*Bartek*, 1989]. This data set consists of discontinuous lines that are often of poor quality and limited to a few hundred milliseconds penetration; the data proved to be unsuit-

able for accomplishing the objectives stated above [*Bartek*, 1989].

The USAP 1990 seismic survey was designed specifically to provide the best possible stratigraphic resolution of the Oligocene and younger section. Some of the questions addressed during this cruise were as follows: (1) does the Neogene section contain glacial erosion surfaces and/or subglacial seismic facies, and if so, when did these events take place; (2) is there seismic evidence of fluvial valley incision and fluvial delta development that might have resulted from supposed major interglacial warming episodes during the Pliocene-Pleistocene [*Webb and Harwood*, 1991]; (3) does the observed increase in shelf edge progradation that occurred in early Miocene time, as seen in MCS records and interpreted by *Cooper et al.* [1991] to signal the first advance of ice sheets to the continental shelf edge, correspond to the first glacial erosional and depositional features on the shelf observed in intermediate resolution records; and (4) are the Oligocene diamictite-bearing intervals at CIROS 1 associated with regional glacial erosion surfaces? This paper addresses all but the last of these questions. The McMurdo Sound data are still under investigation.

Another goal of USAP 1990 was to identify areas where shallow drill sites might best be located so as to sample the most complete and representative Neogene sequences, particularly where expanded Pliocene-Pleistocene sequences can be recovered.

Our correlations between DSDP sites and seismic records come directly from the published biostratigraphic records of these sites [*Hayes and Frakes*, 1975; *Savage and Ciesielski*, 1983; *Leckie and Webb*, 1983]. It is hoped that the results reported in this paper may provide the stimulus for micropaleontologists to reexamine these sites using more current biostratigraphic zonations and better constrain the timing of events outlined in this paper.

METHODS

During the 1990 austral summer, nearly 6000 km of high-resolution seismic data were acquired on the continental shelf and in McMurdo Sound using the R/V *Polar Duke* (Figure 1). The planned cruise track relied upon existing data sets to avoid the structural features that disrupt the stratigraphic sequences on the shelf, especially in the western Ross Sea. Stratigraphic correlations are facilitated where tracklines cross existing drill sites on the shelf and in McMurdo Sound (Figure 1).

Data were acquired using a 150-in.[3] bubble-free generator/injector (GI) air gun. This new source provided an average stratigraphic resolution of 8.0 m, compared with an average of 70 m for conventional air gun records from the same area [*Hinz and Block*, 1983; *Cooper et al.*, 1987]. A 100-in.[3] water gun was used during a brief period of service to the air gun. These two sources provided records that are very similar in terms of stratigraphic resolution; however, the GI gun provided greater subbottom penetration (between 1.0 and 1.5 s two-way travel time for the GI gun compared with 1.0 s for the water gun). The data were acquired using a single-channel streamer at an 8-s firing interval and prefiltered to include data between 20 and 600 Hz. Table 1 presents the steps taken in processing of data.

SEISMIC EXPRESSION OF ICE SHEET GROUNDING EVENTS

This type of study requires acquisition of seismic data with sufficient stratigraphic resolution to identify features and seismic facies that can be interpreted as being either subglacial or glacial-marine in origin. Thus the instrumentation used to acquire the data is dictated by our knowledge of glacial and glacial-marine sedimentary features and facies. Our analysis was aided by recent

TABLE 1. 1990 Ross Sea Data Processing

Step	Description
a	Removal of dc bias.
b	Compensation for loss of amplitude due to geometric spreading, using spherical divergence corrections.
c	Application of a 50-to 140-Hz Butterworth filter selected following extensive filter tests.
d	Vertical stacking of four traces.
e	Display of excessively noisy traces as dead traces, and the removal of sections repeated owing to boat circles during down time.
f	Final sections displayed as variable area wiggle trace plots with 40 stacked traces per inch and at a vertical scale of 3.5 in./s. Longer lines plotted in panels each consisting of no more than 3200 shot points.

The 1990 Ross Sea cruise data were processed during the summer of 1990 at Rice University using CogniSeis Development's DISCO data processing software.

studies that have concentrated on characterizing seismic facies in northern high-latitude regions [*King and Fader*, 1986; *King et al.*, 1991; *Stoker*, 1990; *Vorren et al.*, 1990; *Solheim et al.*, 1990; *Belknap and Shipp*, 1991] and by a much improved understanding of glacial and glacial-marine sedimentary processes, especially those that occur on the Antarctic continental shelf [*Chriss and Frakes*, 1972; *Anderson et al.*, 1980, 1984, 1991; *Elverhoi*, 1984; *Domack*, 1988; *Hambrey et al.*, 1991]. We also relied on many descriptions of glacial-marine outcrops in evaluating the scale of features that might be imaged in the subsurface. See *Anderson* [1983] for a detailed discussion and bibliography on ancient glacial-marine deposits.

The following is a brief description of some of the key seismic features and facies that we used to identify ice sheet grounding events on the Ross Sea continental shelf. We have acquired seismic data from the Antarctic Peninsula continental shelf and the northwestern Weddell Sea that show these same features.

Glacial Erosion Surfaces

In our opinion, the most reliable criterion for recognizing glacial grounding events is evidence of foredeepening and overdeepening of the shelf (Figure 2). Additional evidence exists as widespread, irregular and deep erosional surfaces (glacial troughs) which indicate maximum scouring in a direction both transverse and parallel to the shelf break (Figure 3). Glacial troughs typically are many tens of kilometers wide and indicate erosion depths of many tens to hundreds of meters, much larger and deeper than incised fluvial valleys [e.g., *Berryhill et al.*, 1987; *Thomas and Anderson*, 1988]. Such large-scale erosion surfaces and foredeepened topography are typical of modern high-latitude continental shelves and are attributed to erosion by ice sheets that once grounded on these shelves [*Shepard*, 1931; *Holtedahl*, 1970; *Johnson et al.*, 1982; *Anderson*, 1991]. Large troughs dominate the modern topography of the Ross Sea; they were carved by large ice streams that drain the West Antarctic Ice Sheet [*Anderson et al.*, 1992].

Till Tongues

Large (kilometers to tens of kilometers in length and width), thick (tens of meters), acoustically massive sedimentary bodies are common in the late Cenozoic sequences of the Ross Sea (Figure 3). These bodies are similar to features observed on the Canadian and Norwegian continental shelves ("till tongues" of *King and Fader* [1986] and *King et al.* [1991]). *King and Fader* [1986] describe a mechanism for formation of till tongues which involves buoyancy line migration of grounded ice sheets. In the Antarctic, these bodies usually rest directly on glacial erosional surfaces (Figure 3), further supporting a subglacial origin.

Subglacial Deltas and Shelf Margin Delta-Fan Systems

Studies of Ice Stream B, one of the major ice streams flowing into the Ross Ice Shelf, indicate that erosion rates at the base of ice streams are high and require relatively rapid deposition at or near the grounding line [*Alley et al.*, 1989]. Highly unconsolidated, water-saturated sediment is transported at the base of the ice stream and is deposited at the grounding line as a "subglacial delta" with topset beds comprised of till and foreset and bottomset beds comprised of sediment gravity flow deposits and glacial-marine sediments [*Alley et al.*, 1989]. The deltas prograde through a "conveyor belt recycling" mechanism which, in turn, allows the grounding line to advance across its surface.

Hayes and Davey [1975] described a large (75 km long) geomorphological feature with seaward dipping clinoforms on the continental shelf of the western Ross Sea that they interpreted as a delta. Similar features were observed during USAP 1990. In dip section the features that we interpret as subglacial deltas show prograding clinoforms that downlap onto glacial erosion surfaces (Figure 4). They typically lack stratification in their updip (proximal) portions and grade downdip into acoustically laminated deposits in the bottomset (prodelta) portions of the delta (Figure 4).

Shelf margin delta-fan systems differ from subglacial deltas in their shelf margin position, but also in their more prominent acoustic lamination (Figure 5). This geometry implies a slightly different mechanism from that of subglacial delta development, one that involves greater glacial-marine sedimentation and sediment grav-ity flow transport near the grounding line. The updip portions of late Pleistocene shelf margin deltas are foredeepened surfaces of erosion (Figure 5).

PREVIOUS STUDIES OF ROSS SEA GLACIAL HISTORY

In this section we briefly review the results of previous studies of the Ross Sea's glacial history. We confine our discussion to the Ross Sea area; the proxy (deep sea) record of glacial history and the continental record are largely ignored. To attempt to summarize these records would be an enormous undertaking and beyond the scope of this paper. *Hambrey and Barrett* [1992] provide a summary of the deep-sea and continental records of Antarctic glaciation.

Early seismic surveys of the Ross Sea led to the discovery of a widespread erosion surface which was termed the Ross Sea Unconformity (RSU) [*Houtz and Meijer*, 1970; *Houtz and Davey*, 1973]. All four of the DSDP Leg 28 shelf sites penetrated an unconformity, assumed to be the RSU, and marked by Pliocene-Pleistocene diamictites resting on older (Miocene and Pliocene) strata (Figure 6); the RSU was interpreted by DSDP shipboard scientists as the first major advance of grounded ice on the continental shelf, approximately 5.0 Ma [*Hayes and Frakes*, 1975]. *Savage and Ciesielski* [1983] reexamined DSDP sites 272 (eastern Ross Sea) and 273 (western Ross Sea) and interpreted the RSU as having been caused by the first northward expansion of grounded ice onto the continental shelf sometime during the time interval between 13.8 and 4.0 Ma. They also dated a major increase in ice-rafted debris at Site 274, a deep-sea site situated northwest of the Ross Sea, between ~8.8 and 10.3 Ma. These combined data were taken as evidence that initial development of the West Antarctic Ice Sheet occurred sometime after 10.0 Ma. Multichannel seismic surveys in the Ross Sea by *Hinz and Block* [1983] and *Sato et al.* [1984] also led to the conclusion that ice sheets did not ground on the continental shelf prior to the Pliocene.

The oldest glacial-marine deposits at Site 270 rest unconformably on late Oligocene green (glauconitic) sand deposited in a littoral environment with no apparent glacial ice in the vicinity. The age of the glauconitic sand is constrained by K-Ar dates from 26 to 28 Ma old [*McDougall*, 1975]. Foraminifera from the glauconitic sand and the glacial sediments that overly it indicate a major increase in water depth, a minimum of 250 m, between the time these two units were deposited [*Leckie and Webb*, 1983]. *Leckie and Webb* [1983] argue that this deepening is due to subsidence and that the late Oligocene–early Miocene strata at Site 270 were deposited by glacial-marine processes. *Bartek et al.* [1991]

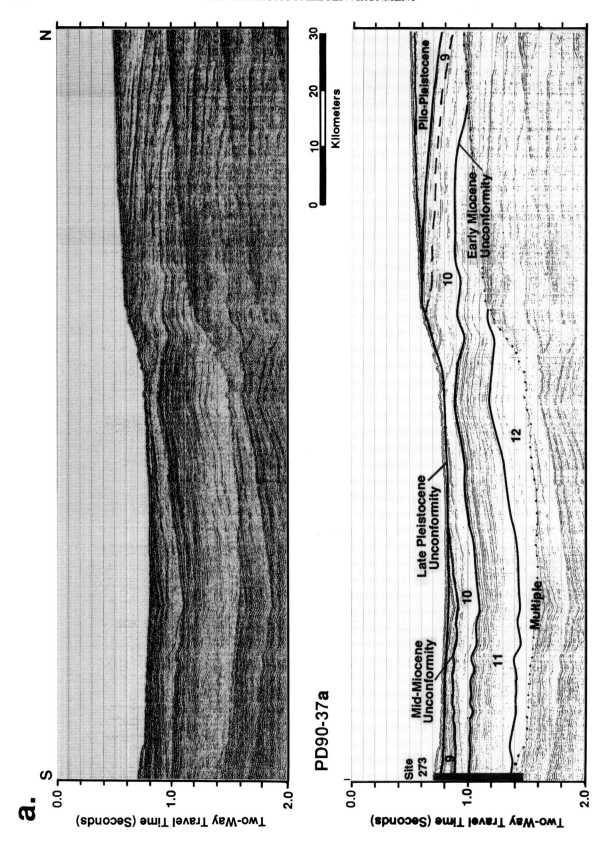

PD90-37a

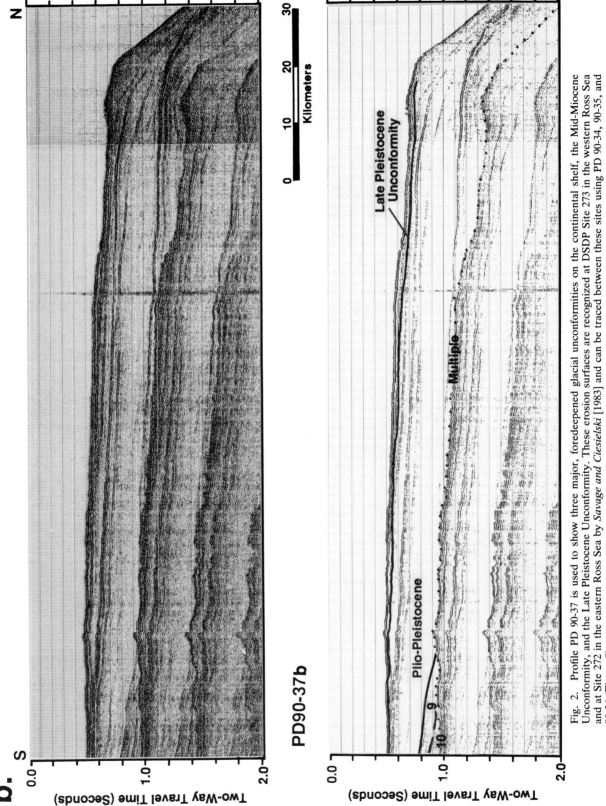

Fig. 2. Profile PD 90-37 is used to show three major, foredeepened glacial unconformities on the continental shelf, the Mid-Miocene Unconformity, and the Late Pleistocene Unconformity. These erosion surfaces are recognized at DSDP Site 273 in the western Ross Sea and at Site 272 in the eastern Ross Sea by *Savage and Ciesielski* [1983] and can be traced between these sites using PD 90-34, 90-35, and 90-36 (Figure 8).

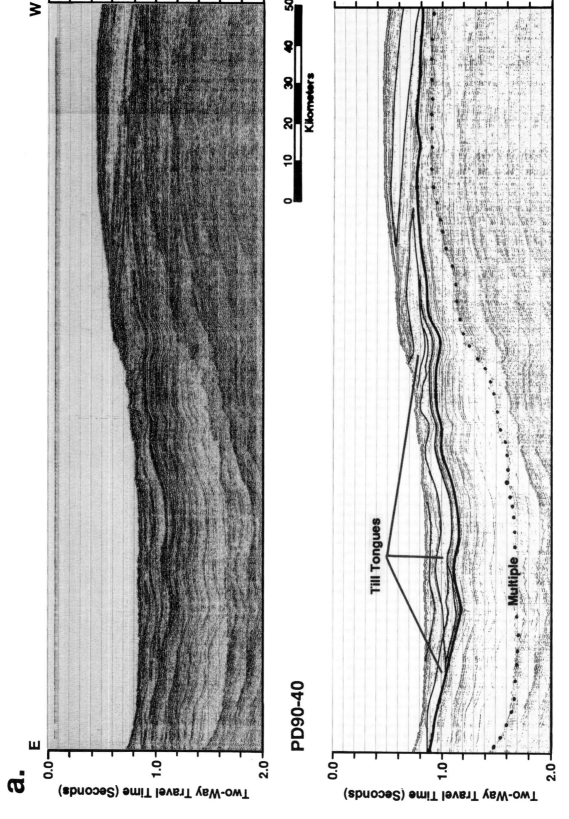

Fig. 3a. Profile PD 90-40, a strike line that crosses profile PD 90-37 (Figure 1), is used to illustrate glacial troughs and the till tongues associated with these troughs.

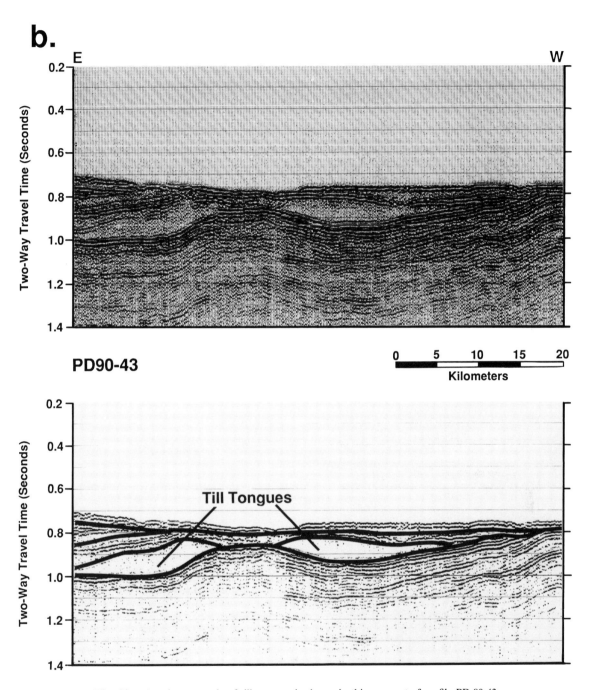

Fig. 3*b*.   Another example of till tongues is shown in this segment of profile PD 90-43.

argued that glacial erosion perhaps caused deepening of the shelf at Site 270, based on a significant change in sedimentary facies across this boundary.

The oldest (late Oligocene–early Miocene) glacial diamictite in Site 270 occurs at a subbottom depth of 260 to 270 m. It was interpreted as a glacial-marine unit by *Barrett* [1975] and as a possible subglacial deposit by *Balshaw* [1981]. The deposits at Site 270 have recently been reinterpreted by *Hambrey and Barrett* [1992], and they suggest that lodgment till may occur as deep as 350 m at this site.

The Miocene and younger deposits sampled at the DSDP sites consist mainly of pebbly mudstones with variable concentrations of ice-rafted material and diatoms. Petrographic studies of these deposits indicate a West Antarctic origin [*Barrett*, 1975]. *Balshaw* [1981]

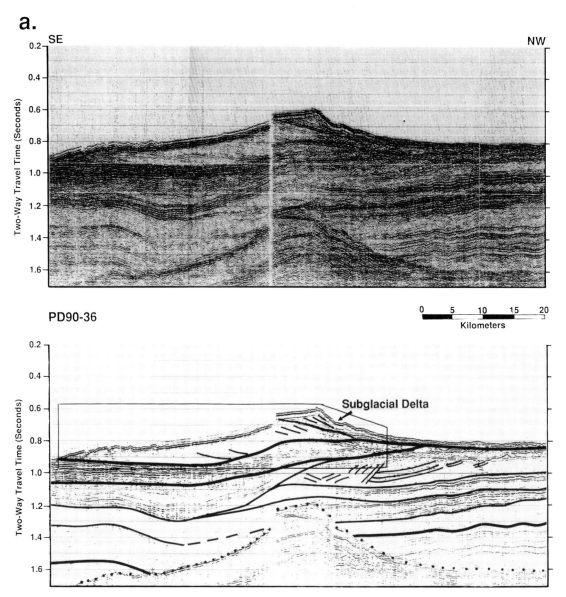

Fig. 4a. This segment of profile PD 90-36 shows a prominent geomorphic feature interpreted to be a subglacial delta. The clinoforms within this feature indicate progradation in a seaward direction. Features with similar geometry and thicknesses are seen in seismic Unit 10 in an area east of the Central High (on profiles PD 90-21 and 90-35).

recognized several massive diamictites within the late Oligocene through Miocene sequence (Figure 6).

In McMurdo Sound, MSSTS 1 penetrated glacial deposits ranging back to late Oligocene age [*Barrett and McKelvey*, 1986]. CIROS 1 penetrated early Oligocene diamictites with glacially striated pebbles [*Robinson et al.*, 1987; *Barrett et al.*, 1989]. A notable change in the character of sedimentation occurs around the mid-Oligocene and indicates an increased glacial influence after this time [*Barrett et al.*, 1989]. *Bartek et al.* [1991] suggest that the Mid-Oligocene Unconformity at CIROS

1 corresponds to the U6 unconformity on the central and eastern shelf, indicating that the Mid-Oligocene Unconformity marks a continental glaciation.

In a recent paper by *Cooper et al.* [1991], a change from aggradational to progradational acoustic geometry of paleoshelf edge reflections that occurs in the early Miocene is interpreted as marking the onset of major advances of the grounded ice to the continental shelf edge. *Bartek et al.* [1991] take a somewhat different view, arguing that the overall acoustic geometry of the Ross Sea is similar to other passive continental margins

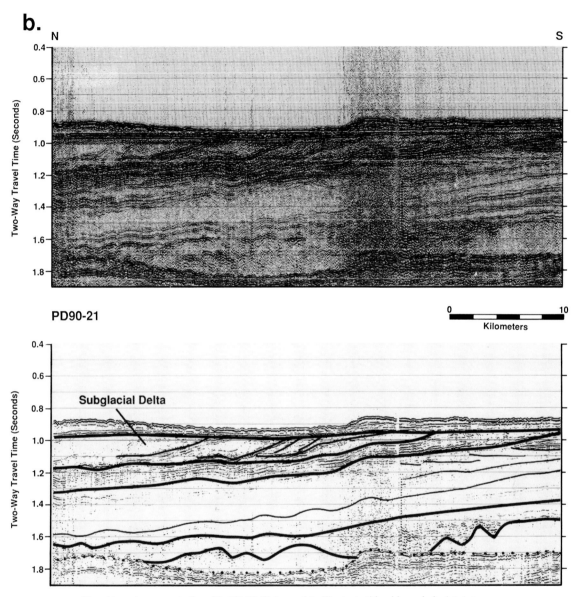

Fig. 4*b*.   A segment of profile PD 90-21 is used to illustrate this older subglacial delta.

of the world and that this implies a eustatic influence on these sequences.

RESULTS

Eastern Basin

Our seismic data from the Eastern Basin show 13 seismic units that we can map across the basin. Figure 6 provides a correlation between these units and drill sites, as well as a brief description of each seismic unit. Seismic lines *Polar Duke* (PD) 90-30 (a dip line through the DSDP sites) and PD 90-49 (a strike line) are used to illustrate these seismic units (Figure 7).

The oldest seismic unit (Unit 13) strongly onlaps acoustic basement and lies below a regional unconformity (U6 unconformity of *Hinz and Block* [1983]) (Figure 6). This unit is believed to correlate to the late Oligocene interval at Site 270 (DSDP units 2J, 3, and 4). The foraminifera in these units indicate deposition at water depths of less than 100 m [*Leckie and Webb*, 1983]. Only along the flanks of the basin does Unit 13 occur above the bottom multiple; it is not possible to trace it for more than a few tens of kilometers. Unit 13 shows no conspicuous features of glacial erosion or sedimentation.

Unit 12 is the oldest seismic unit with good evidence of subglacial erosion and deposition. It is characterized by thick, acoustically massive, wedge-shaped bodies

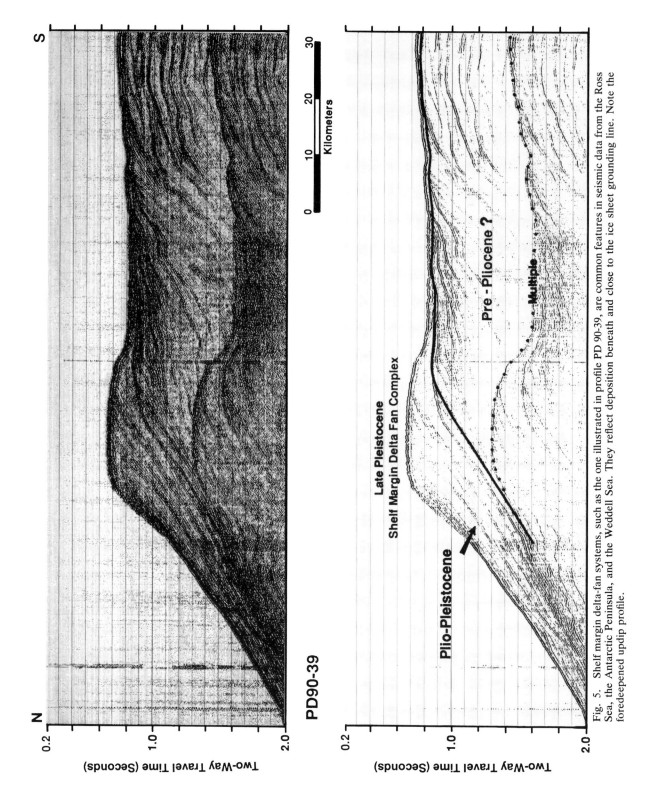

Fig. 5.   Shelf margin delta-fan systems, such as the one illustrated in profile PD 90-39, are common features in seismic data from the Ross Sea, the Antarctic Peninsula, and the Weddell Sea. They reflect deposition beneath and close to the ice sheet grounding line. Note the foredeepened updip profile.

(till tongues), tens to hundreds of meters thick, and tens of kilometers in extent (Figure 7). Its contact with Unit 13 is an erosional surface (U6 of *Hinz and Block* [1983]) that shows evidence of foredeepening (Figure 7, line PD 90-30a).

Site 270 sampled Unit 12 (DSDP units 2I through 2A), and *Leckie and Webb* [1983] interpreted the foraminifera in the lower half of this seismic unit to be late Oligocene–early Miocene in age and the upper half of the unit to be of early Miocene age. Foraminifera from the lower part of Unit 12 indicate paleowater depths of between 150 m and 300 m. The foraminiferal assemblage from the top of this unit indicates a water depth of up to 500 m. The cores recovered from this interval penetrated mainly pebbly mudstone in which diatoms are rare to absent and the ice-rafted component varies from 10% to greater than 20% sand-sized debris and from <5% to 20% pebbles [*Barrett*, 1975]. Two massive diamictites occur within this interval [*Balshaw*, 1981] (Figure 6).

Units 12 and 11 are separated by an unconformity that is basinal in extent, characterized by strong toplap along the margins of the basin, and shows evidence of foredeepened topography (Figure 7, line PD 90-30a). This surface corresponds to *Hinz and Block*'s [1983] U5 unconformity. Unit 11 is acoustically laminated and thickens dramatically into the basin, where the contact with Unit 12 is a downlap surface. This unit is situated between the sampled intervals of Site 270 and Site 272, which indicates that it is early Miocene in age (Figure 6). On the basis of its seismic character, Unit 11 is interpreted as consisting predominantly of highstand glacial-marine deposits; there is little evidence of subglacial erosion and deposition within this unit, however, and the unconformity on which it rests probably is of glacial origin.

An unconformity of basinal extent separates Unit 10 and Unit 11; the top of Unit 11 has been deeply eroded and is onlapped by Unit 10 (Figure 7). The foredeepened topography of this surface is shown in Figure 7 on line PD 90-30a. This surface is correlated to the U4A unconformity of *Cooper et al.* [1991].

The base of Unit 10 was sampled at Site 272 and was dated by *Savage and Ciesielski* [1983] as 19.2 Ma. The upper portion of this unit also contains early Miocene microfossils [*Savage and Ciesielski*, 1983] (Figure 6). Given these age constraints, we refer to the unconformity between Unit 10 and Unit 11 as the Early Miocene Unconformity and suggest that it may correspond to the 22.5 Ma major sequence boundary of *Vail et al.* [1977] (Figure 6).

Unit 10 contains the most spectacular subglacial features of all the units on the shelf. It includes broad (tens of kilometers), deep (many tens of meters) erosional surfaces, huge till tongues, and a large subglacial delta (Figure 4b). Along the western margin of the Eastern Basin, Unit 10 displays striking progradation

from south to north and erosion surfaces with relief of up to 400 ms infilled with mainly acoustically massive deposits (Figure 8b). Locally, clinoform sets, interpreted as subglacial deltas, infill these erosion surfaces (Figure 4b).

The dominant lithology of cores recovered from Unit 10 at Site 272 is diatomaceous pebbly mudstone with an average of 10% to 20% sand-sized, ice-rafted debris and <5% pebbles. This site did not penetrate any of the till tongues seen in the seismic records.

Units 10 and 9 are separated by a major unconformity, which we call the Mid-Miocene Unconformity (Figures 7c, 8b, and 6). Using profiles 34, 35, and 36, this unconformity was traced from the Eastern Basin to the Central Trough (Figure 8). This unconformity was recognized at both Site 272 and Site 273 by *Savage and Ciesielski* [1983], who called it the Miocene Disconformity. Their biostratigraphic work at these two sites shows that this unconformity spans the time interval 18.2 to 14.1 Ma.

Unit 9 is, for the most part, an acoustically laminated unit (Figures 7a, 7c, 8a, and 8b). This unit was sampled at Site 272 and interpreted to be middle Miocene in age [*Savage and Ciesielski*, 1983]. The sediments are predominantly fine grained with an ice-rafted debris content that varies between 10% and 20% for sand-sized debris and <5% pebbles. Diatoms are relatively abundant. Two diamictites occur within this interval [*Balshaw*, 1981]. Otherwise, the strong acoustic lamination that characterizes most of this unit implies glacial-marine sedimentation.

Gentle downlap characterizes the contact between Unit 8 and Unit 9. Unit 8 contains small till tongues (Figure 7a) that extend seaward into strong but discontinuous reflectors. Unit 8 was not sampled by the DSDP site, but a mid-Miocene to early Pliocene age is indicated by its stratigraphic position relative to sites 272 and 271 (Figure 6).

Pliocene-Pleistocene strata are mostly confined to the middle and outer shelf of the eastern Ross Sea. DSDP sites 270 and 272 sampled only a thin (<30 m) Pliocene-Pleistocene section, which the seismic records show to be an amalgamation of several units bounded by erosional surfaces on the inner shelf (Figure 7a). Site 271 sampled the upper part of the Pliocene-Pleistocene section (Figures 7a and 7c). Our calculations indicate that Unit 7 was sampled at the base of Site 271, which indicates a Pliocene age for this unit.

A glacial unconformity separates seismic units 7 and 8. This unconformity separates Pliocene strata from mid-Miocene to early Pliocene strata. We suggest that this unconformity corresponds to a major sequence boundary at 5.5 Ma [*Wornardt and Vail*, 1991] and refer to it as the Late Miocene/Early Pliocene Unconformity (Figure 6).

Units 7 through 5 are mostly progradational, whereas the four younger Pliocene-Pleistocene units (units 1

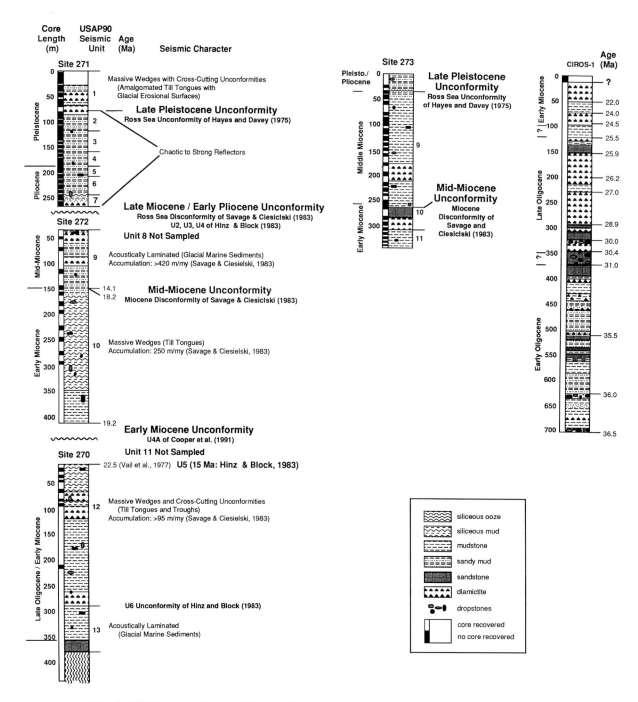

Fig. 6.   This figure provides a highly generalized summary of lithologic and biostratigraphic data from DSDP Leg 28 drill sites in the Ross Sea. The summary is based on the descriptions of the cores from these sites by the original shipboard scientists [*Hayes and Frakes*, 1975] and on published results from later investigations of these cores [*Balshaw*, 1981; *Savage and Ciesielski*, 1983; *Leckie and Webb*, 1983]. Also shown are the associated seismic units, as defined by this study, and major unconformities recognized in previous seismic investigations [*Houtz and Davey*, 1973; *Hayes and Davey*, 1975; *Hinz and Block*, 1983; *Sato et al.*, 1984; *Cooper et al.*, 1991].

through 4) are mainly aggradational (Figure 7*b*). Smooth unconformities and semiregional, continuous high-amplitude reflectors generally define the boundaries between these units [*Alonso et al.*, 1992]. In most cases, these bounding surfaces are planar to subhorizontal when viewed in a dip section (Figure 7*b*); however, strike sections show broad erosional surfaces (tens of kilometers wide), similar in scale to the modern troughs of the region (Figures 7c and 7d). Units 2 through 6 display very similar reflection patterns which range from chaotic to strongly stratified. Unit 5 has U-shaped channels that are between 1.0 and 6.0 km in width and have relief of tens of meters (Figure 9). The channel geometry and size fall within the range of subglacial valleys [Armentrout, 1983; Ashley et al., 1985] and incised stream valleys of the Texas-Louisiana shelf [Berryhill et al., 1987; Thomas and Anderson, 1988].

Sample recovery at Site 271 was very poor (7%). On the basis of material recovered, the Pliocene section consists mainly of pebbly, sandy mudstone with generally less than 5% pebbles and from 20% to 40% sand-sized, ice-rafted debris. Diatoms are rare in all but the basal portion of the section.

The youngest seismic unit in the Eastern Basin, Unit 1, is separated from underlying units by a widespread unconformity (Figure 7). Our data show that on the inner shelf of the eastern Ross Sea and over most of the western Ross Sea, this unconformity is an amalgamation of several unconformities (U2 through U4 of Hinz and Block [1983] and Ross Sea Unconformity of Houtz and Meijer [1970], Hayes and Davey [1975], and Savage and Ciesielski [1983]) (Figure 7). We recommend that this most recent Pleistocene unconformity be named the Late Pleistocene Unconformity (Figure 6). The deposits beneath the Late Pleistocene Unconformity range from Pleistocene at the shelf edge to early Miocene at DSDP Site 270 (Figure 7).

The DSDP drill sites penetrated diamictites within Unit 1. This unit contains a mixed microfossil assemblage, but detailed analyses of this assemblage yielded a Pleistocene age [Savage and Ciesielski, 1983; Leckie and Webb, 1983]. Unit 1 bears all of the features of subglacial erosion and deposition: troughs, till tongues, hummocky surfaces, hyperbolic reflector patterns, moundlike features with internal chaotic reflector patterns, and massive facies that grade seaward into well-stratified facies (see also Karl et al. [1987] and Alonso et al. [1992]).

Piston cores have penetrated the top of Unit 1 and recovered massive diamictons interpreted as tills [Kellogg et al., 1979; Anderson et al., 1980] overlain by glacial-marine sediments containing abundant siliceous biogenic material. Radiocarbon dates on the upper glacial-marine unit yielded ages of less than 17,000 years [Anderson et al., 1992].

Western Ross Sea

The USAP 1990 seismic coverage of the western Ross Sea is more restricted than the coverage in the Eastern Basin (Figure 1). Only one drill site (DSDP Site 273) exists on the open shelf. Recovery at this site was poor (37%) and consists of a monotonous sequence of pebbly mudstone. The oldest sediments recovered at Site 273

are of early Miocene age [Savage and Ciesielski, 1983] (Figure 6).

Direct correlation of the late Oligocene–early Miocene units of the Victoria Land Basin with those of the Eastern Basin is difficult because of stratigraphic pinchouts of the older section against the Central High [Hinz and Block, 1983; Cooper et al., 1987] and because the Miocene section has been deeply eroded in the western Ross Sea (Figure 10).

Lines PD 90-35 and 90-36 (Figure 8) were acquired to the north of the Central High (Figure 1) and were used to correlate the DSDP sites of the Eastern Basin with Site 273. Units 10 and 9 were mapped along the entire length of lines PD 90-35 and 90-36 (Figure 8); this tie is constrained by the Miocene Disconformity that was identified at DSDP sites 272 and 273 by Savage and Ciesielski [1983], which we call the Mid-Miocene Unconformity (Figure 6).

Lines PD 90-43 through 90-46 were acquired in an effort to tie the seismic units at Site 273 with the MSSTS and CIROS sites in McMurdo Sound (Figure 1). Much of the seafloor traversed by these lines is underlain by late Oligocene and early Miocene strata that are situated at or near (within 100 ms) the seafloor (Figure 11). A widespread unit was mapped throughout the western Ross Sea by Cooper et al. [1987], their Unit V1, that was assigned an early Miocene to Recent age. The base of Unit V1 was correlated to a prominent unconformity (TM unconformity of Bartek [1989]). This unconformity was mapped as far north as Site 273 [Bartek, 1989], where it corresponds to the Miocene Disconformity of Savage and Ciesielski [1983]. Thus the TM unconformity apparently corresponds to the Mid-Miocene Unconformity of the eastern Ross Sea. The problem lies in attempting to trace this unconformity south of approximately 77°, where basement highs and faults disrupt the Cenozoic strata (Figure 11). An ongoing investigation is aimed at tying the seismic stratigraphy of McMurdo Sound, which is constrained by MSSTS 1 and CIROS 1 and 2, with the shelf to the north. Figure 11 provides preliminary results of this work.

One significant difference between the western and eastern Ross Sea is the limited distribution of Pliocene-Pleistocene strata on the western shelf. Site 273 recovered only a thin veneer of Pliocene-Pleistocene deposits (Figure 6), and the seismic profiles show that the Pliocene-Pleistocene section is confined to the outer shelf and upper slope (Figures 2 and 5). Lines 37 (Figure 2) and 39 (Figure 5) show a Pliocene-Pleistocene sequence up to 400 ms thick near the shelf edge, where a prominent shelf margin delta-fan complex exists.

DISCUSSION

Strong evidence of subglacial erosion and deposition exists in the late Oligocene–early Miocene (Unit 12) strata of the Eastern Basin in the form of large till

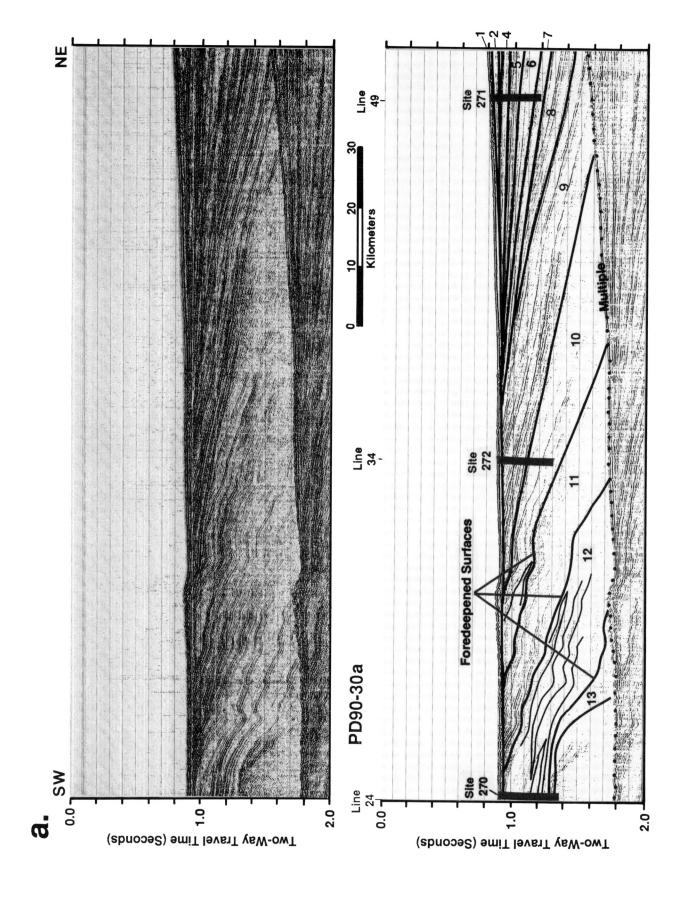

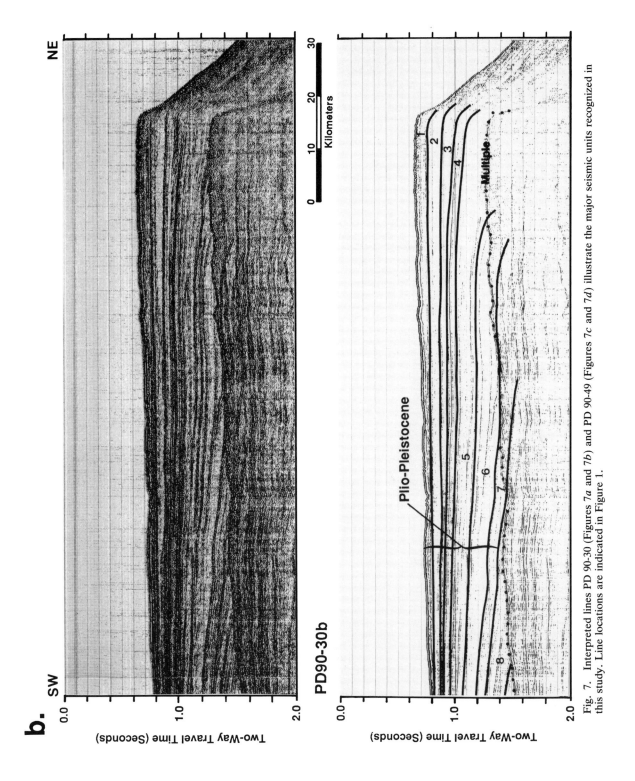

Fig. 7. Interpreted lines PD 90-30 (Figures 7a and 7b) and PD 90-49 (Figures 7c and 7d) illustrate the major seismic units recognized in this study. Line locations are indicated in Figure 1.

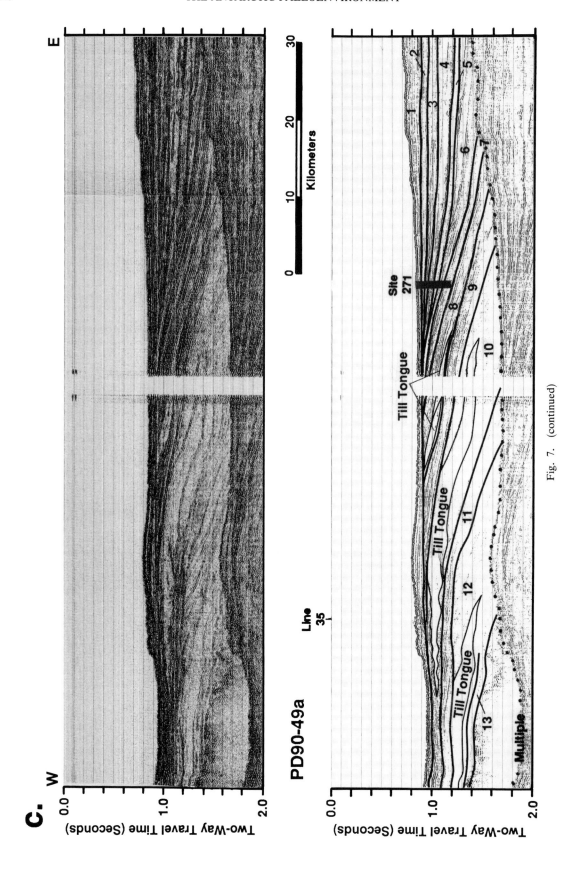

Fig. 7.   (continued)

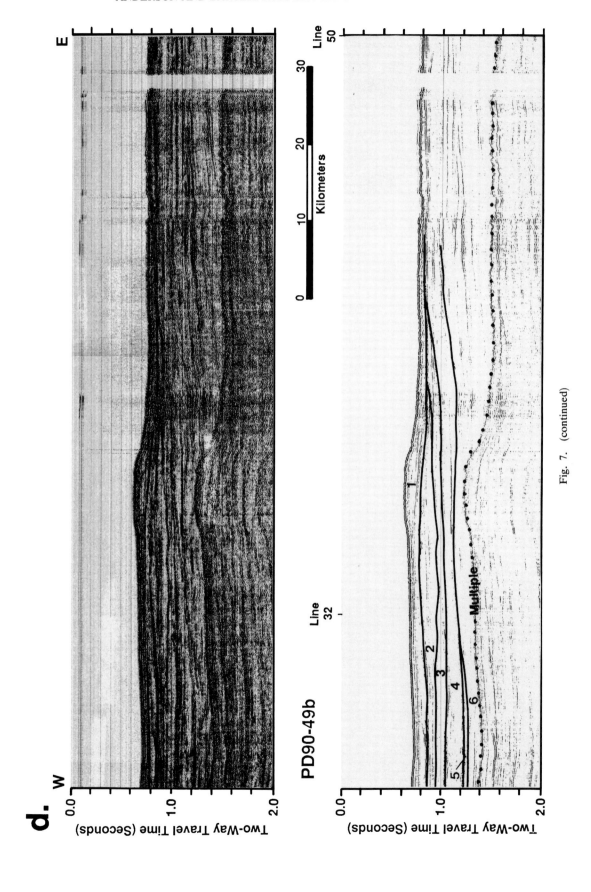

Fig. 7. (continued)

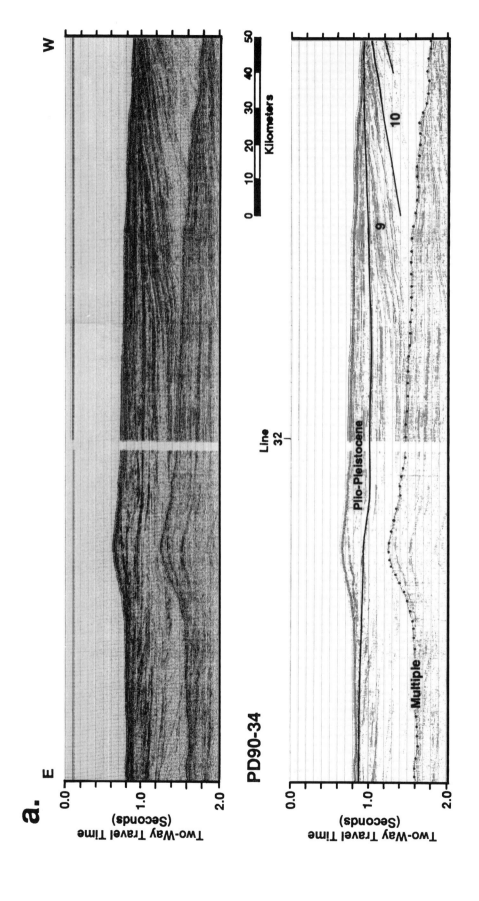

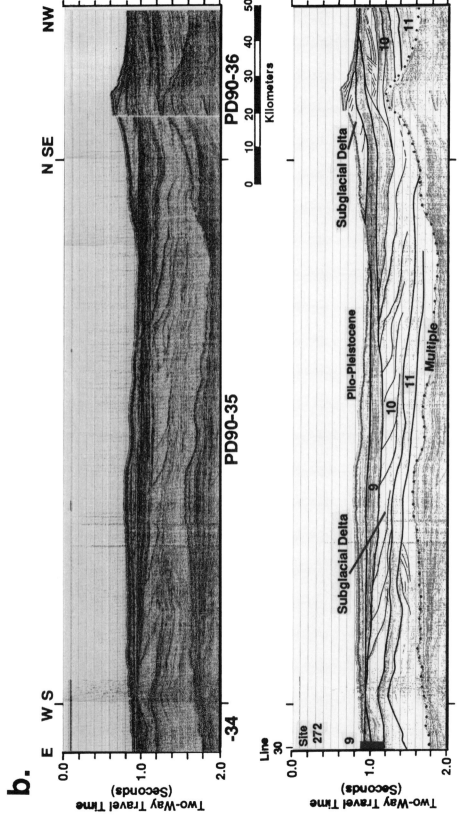

Fig. 8. Seismic profiles PD 90-34, 90-35, and 90-36 are used to trace the Mid-Miocene Unconformity from the Eastern Basin to the Central Trough. This unconformity also is recognized at DSDP sites 272 and 273 as the Miocene Disconformity by *Savage and Ciesielski* [1983].

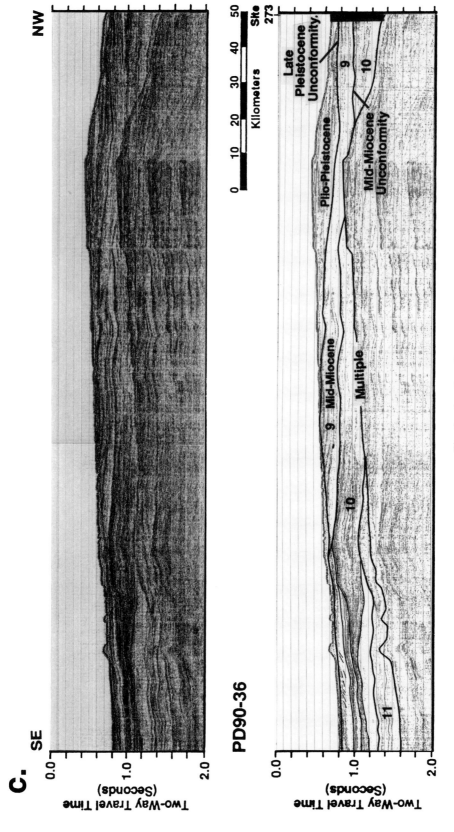

Fig. 8. (continued)

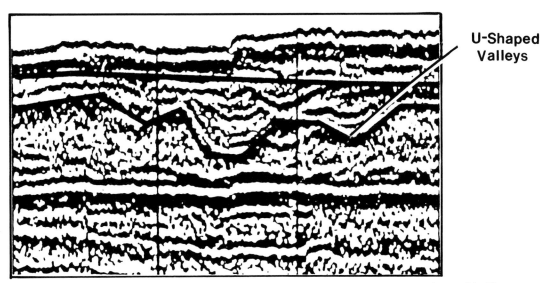

Fig. 9.   U-shaped channels in seismic Unit 5 have depth to width ratios that are consistent with either subglacial tunnel valleys or incised fluvial valleys.

tongues and glacial erosional surfaces. There is evidence that the inner shelf was foredeepened during this time interval, and foraminifera at Site 270 indicate deepening from just above sea level to 300 to 500 m below sea level [*Leckie and Webb*, 1983]. Leckie and Webb attribute this dramatic change in water depth to subsidence, but our seismic data show glacial erosion surfaces rather than onlapping sequences; the latter would result from gradual shelf submergence. Large till tongues are observed at least as far north as DSDP Site 271, where Unit 12 dips below the level of penetration in our data. We estimate the age of Unit 12 to be between 24 Ma and 22.5 Ma (Figure 6). This corresponds to the Mi 1 $\delta^{18}O$ event of *Miller et al.* [1991], which they interpret as the product of a "major increase in ice volume" that occurred between approximately 24.5 Ma and 23.5 Ma.

Unit 12 was sampled at Site 270, and the cores from this unit are characterized by an abundance of fine-grained terrigenous material and a paucity of diatoms; diamictites occur near the middle and top of the sampled interval (Figure 6).

Currently, diatomaceous compound glacial-marine sediments blanket the Ross Sea continental shelf [*Anderson et al.*, 1984]. *Anderson and Ashley* [1991] argue that late Quaternary diatomaceous glacial marine sediments are unique to the polar regions of Antarctica where dilution by meltwater-derived silts is limited. We interpret the paucity of diatoms in Unit 12 as evidence for a more temperate glacial setting during the late Oligocene–early Miocene. This interpretation is not inconsistent with ice sheets having grounded on the continental shelf during this time interval; grounding events do not necessarily imply a polar climate. For

example, during the late Wisconsin glacial maximum, ice sheets were grounded on the west coast of North America as far south as the Puget Lowlands [*Domack*, 1988] and on the east coast as far south as Maine [*Belknap and Shipp*, 1991], yet these areas are not thought to have experienced a polar climate.

The most dramatic evidence of ice sheet grounding in the eastern Ross Sea is associated with Unit 10, which is an early Miocene unit. Unit 10 represents a period of repeated ice sheet advances onto the continental shelf that resulted in a foredeepened and overdeepened shelf topography with relief equal to that of the present seafloor. The width to depth ratios of glacial troughs in Unit 10 are roughly equal to those of the present Ross Sea. This implies that ice streams as large as those of the present existed by early Miocene time.

The base of Unit 10 is a widespread and major unconformity (Early Miocene Unconformity) that occurs within the early Miocene section of sites 270 and 272 (Figure 6). We believe that this unconformity corresponds to the 22.5 Ma sequence boundary of *Vail et al.* [1977]. We also correlate this surface to *Cooper et al.*'s [1991] U4A unconformity which represents a change from aggradational to progradational stacking of shelf edge sequences. Cooper and his colleagues interpret this change as signaling large fluctuations in the amount of grounded ice and entrained sediment reaching the outer shelf since early Miocene time. They do not discount the possibility of earlier grounding episodes on the inner shelf. Our results are consistent with those of Cooper and his colleagues; however, in the case of the Ross Sea the association of increased progradation and ice sheet grounding could be fortuitous. *Bartek et al.* [1991] point out that a similar change

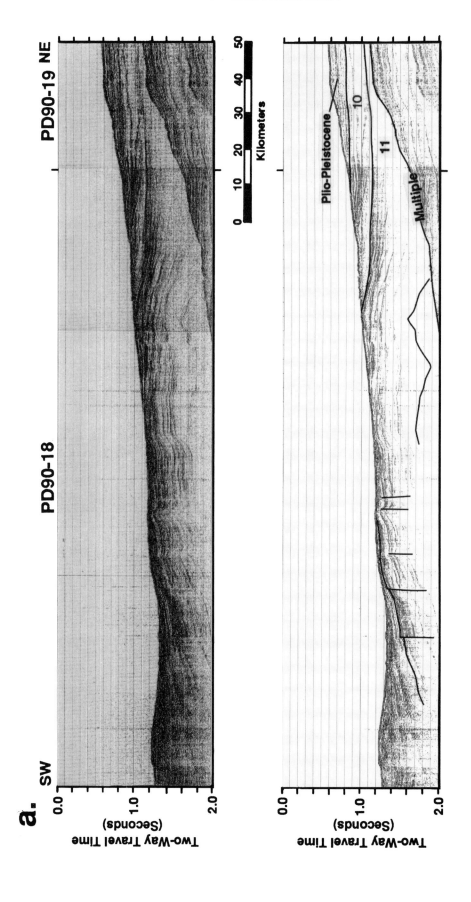

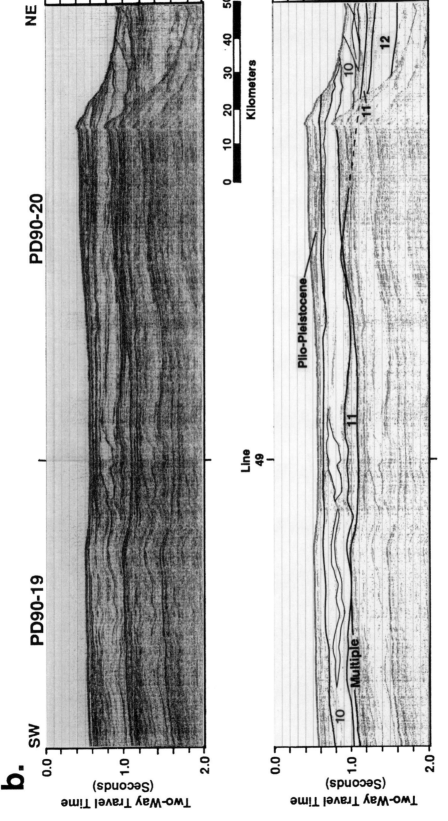

Fig. 10. Seismic profiles PD 90-18, 90-19, and 90-20 (Figures 10a and 10b) and PD 90-49 (Figures 10c and 10d) provide an east-west transect across the continental shelf. Note that the Miocene section has been eroded deeply in the western Ross Sea.

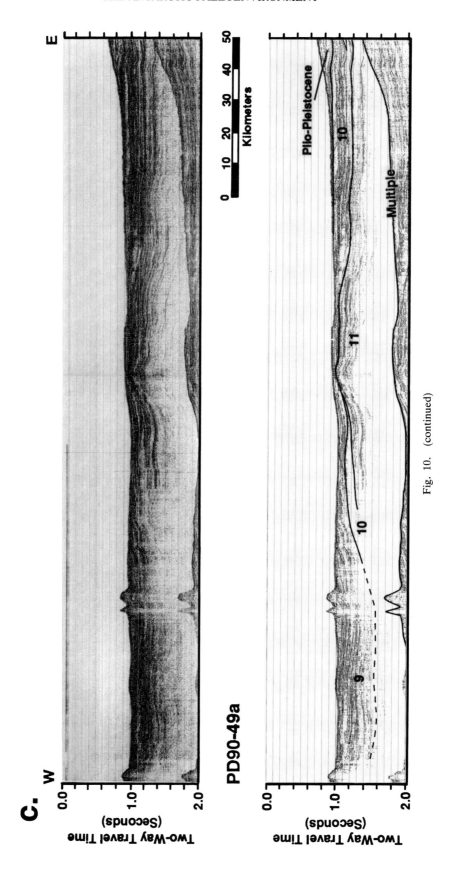

Fig. 10. (continued)

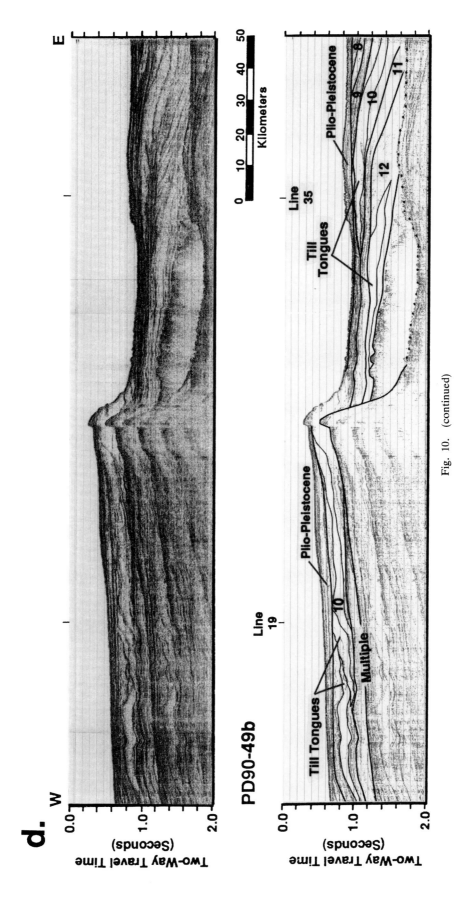

Fig. 10. (continued)

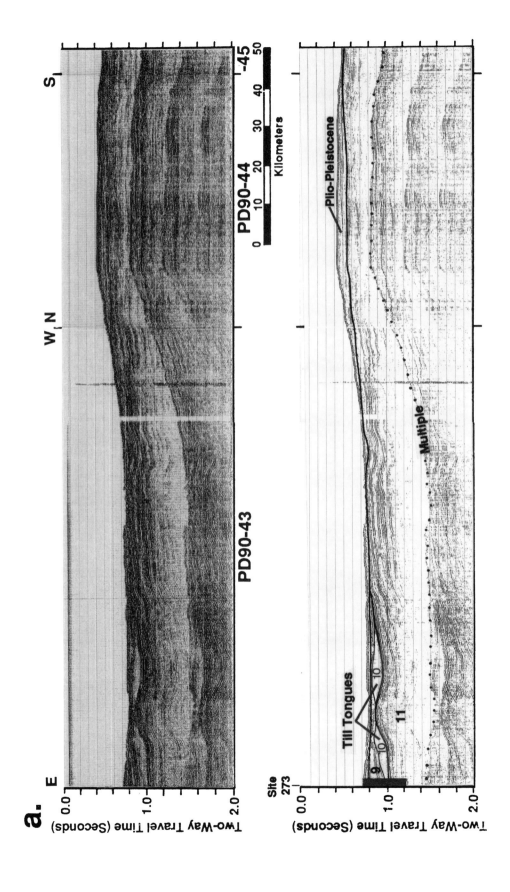

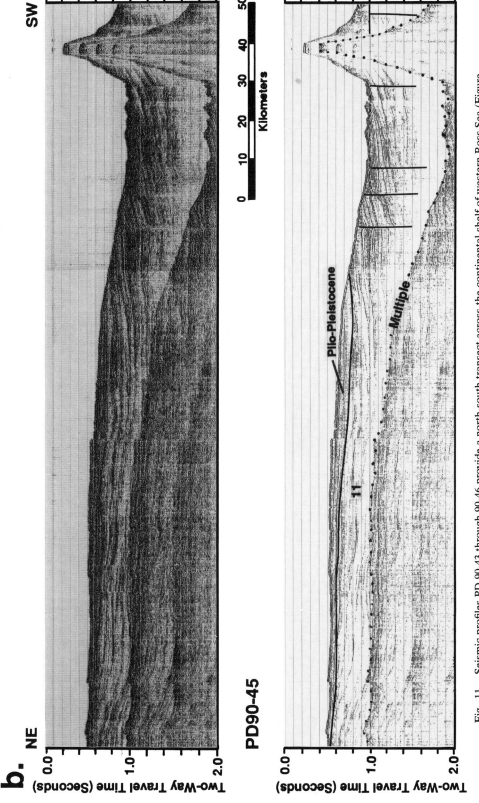

Fig. 11. Seismic profiles PD 90-43 through 90-46 provide a north-south transect across the continental shelf of western Ross Sea (Figure 1). The seafloor in this region is characterized by thin mid-Miocene and younger strata south of approximately the latitude of Site 273 and by a seaward thickening wedge of Pliocene-Pleistocene strata north of this site.

c.

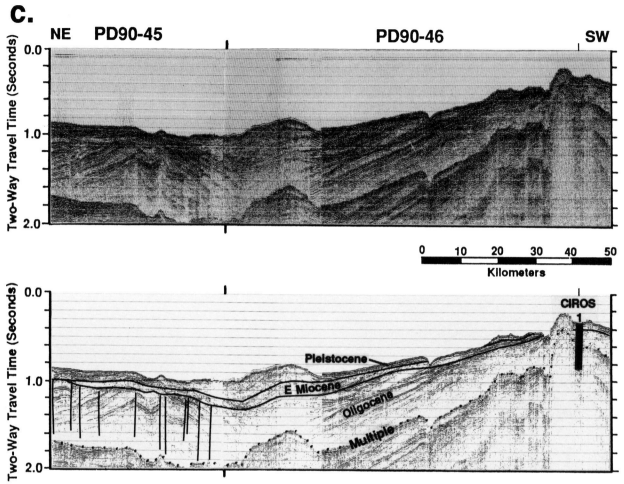

Fig. 11   (continued)

in sequence stratigraphy at this time occurs in other, low-latitude passive margin settings. For example, the Cenozoic sequence stratigraphy of the Ross Sea is virtually identical to that of the Western Flores Sea of Indonesia (Figure 7 of *Bartek et al.* [1991]), where shelf margin sequences shift from aggradational to progradational at approximately 21.0 Ma.

Unit 10 is believed to represent a long episode (22.5 to 18.2 Ma) of ice sheet advance and retreat across the continental shelf. The $\delta^{18}O$ record for this time interval indicates fluctuations in ice volume [*Miller et al.*, 1991] of a magnitude and frequency that could be due to changes in the volume of the West Antarctic Ice Sheet.

The abundance of diatoms in Unit 10, relative to Unit 12, may indicate that the flux of fine-grained terrigenous material from the continent was diminished. This, in turn, implies that the interglacial climate was much colder by early Miocene time. The interval of glacial erosion and deposition represented by Unit 10 culminated with the development of the Mid-Miocene Unconformity, which appears to be shelf-wide [*Savage and*

Ciesielski, 1983; *Cooper et al.*, 1987; *Bartek et al.*, 1991; this study]. At approximately 14 Ma, deposition at Site 272 was renewed [*Savage and Ciesielski*, 1983]. The Mid-Miocene Unconformity coincides with an overall fall in global sea level [*Haq et al.*, 1987] and shift in the global oxygen isotope record that is indicative of significant ice buildup on Antarctica [*Shackleton and Kennett*, 1975; *Miller et al.*, 1991].

The late Miocene section was not sampled at any of the Ross Sea DSDP sites. Seismic Unit 8 occurs within the unsampled interval between these sites and represents this time interval. Based on the seismic character of Unit 8, this was a time of predominantly glacial-marine sedimentation on the shelf.

The oxygen isotope record [*Shackleton and Kennett*, 1975; *Miller et al.*, 1991] suggests that after the major buildup of ice on Antarctica in the mid-Miocene, the ice sheet maintained its full-bodied state. Most of this ice must have been situated on East Antarctica. Our data show that the West Antarctic Ice Sheet retreated from the eastern Ross Sea shelf after the mid-Miocene grounding episode. In fact, there were apparently few grounding events during the late Miocene. Perhaps the overdeepened shelf had to be raised by glacial-marine sedimentation before the next grounding event could occur. Late Miocene strata are virtually lacking on the western shelf; so the record of glacial activity in this area during late Miocene time is unknown.

The Miocene-Pliocene boundary marks a change in the style of deposition on the eastern Ross Sea continental shelf, one in which individual unit thicknesses decrease dramatically and the overall stacking geometry of sequences shifts from strongly progradational to strongly aggradational (Figure 7b). Seven seismic units are associated with the Pliocene-Pleistocene interval sampled at DSDP Site 271; each is separated by a widespread glacial erosion surface [Alonso et al., 1992]. The increased frequency of grounding events on the continental shelf during the Pliocene-Pleistocene is attributed to a higher frequency of eustatic rise and fall [Haq et al., 1987]. Thus by Pliocene time the marine ice sheet was strongly influenced by sea level, probably more so than climate, and its advances across the continental shelf were, therefore, probably in harmony with northern hemisphere glaciations.

The sparsely sampled Pliocene glacial-marine deposits of Site 271 contain only minor amounts of diatomaceous material [Hayes and Frakes, 1975]. This implies warmer interglacial conditions relative to the present [Anderson and Ashley, 1991]. U-shaped valleys were imaged in one Pliocene unit (Unit 5); these are interpreted as subglacial tunnel valleys, but a fluvial origin cannot be completely ruled out.

Houtz and Davey [1970] were the first to recognize a shelf-wide unconformity, the Ross Sea Unconformity, which they believed to mark the first major episode of glacial erosion on the continental shelf. Hayes and Davey [1975] estimated that as much as 800 m of strata had been removed from the inner shelf during this erosional episode. Our data show that the Ross Sea Unconformity is an amalgamation of several erosion surfaces that converge updip and along strike (Figures 2 and 7); we recommend that this most recent unconformity be named the Late Pleistocene Unconformity.

The Ross Sea shelf averages a depth of 500 m; so the thin nature of Pliocene-Pleistocene strata in the western Ross Sea probably is not due to a lack of space. Rather, the seismic data suggest that glacial erosion has removed most of the younger strata from the western shelf. That is to say, the Pliocene-Pleistocene was a period of net erosion on the western Ross Sea continental shelf. Future drilling programs aimed at acquiring a good Pliocene-Pleistocene section should focus on the Eastern Basin outer shelf and not the western Ross Sea.

It is noteworthy that the proxy record of glaciation in the Ross Sea is very different from the one we have described. The nearest deep-sea drill site is Site 274, which is located off the Wilkes Land continental margin. *Savage*

and Ciesielski [1983] examined the cores from this site and found that the first significant influx of ice-rafted debris at this site occurred at 10.3 to 8.8 Ma, which they concluded was the first grounding event in the Ross Sea.

CONCLUSIONS

1. Ice sheets first grounded on the eastern continental shelf of the Ross Sea in the late Oligocene–early Miocene, resulting in overdeepened and foredeepened shelf topography. The dominance of terrigenous fine-grained material in glacial-marine sediments recovered from this stratigraphic interval at DSDP Site 270 implies interglacial conditions warmer than those of the present.

2. By early Miocene time an ice sheet, with ice streams equal in size to those of the present, grounded on the continental shelf. This grounding event resulted in shelf topography similar to that of the present. Glacial marine sediments from this stratigraphic interval contain significant concentrations of diatomaceous material, and this implies a more polar interglacial climate.

3. A mid-Miocene unconformity is recognized in both our seismic data and at DSDP sites 272 and 273 [Savage and Ciesielski, 1983]. This unconformity marks an episode of ice sheet expansion across the entire Ross Sea continental shelf. This grounding episode is believed to correspond to a major buildup of ice on East Antarctica that is recorded in the oxygen isotope record [Shackelton and Kennett, 1975; Miller et al., 1991].

4. The middle-late Miocene seismic facies of the eastern Ross Sea indicate that this was a period of mainly glacial-marine sedimentation and infrequent grounding events. Glacial erosion has removed much of the mid-Miocene and younger section from the western Ross Sea shelf; so it is not known whether the East Antarctic Ice Sheet was grounded there during this time interval.

5. Beginning in the Pliocene, the frequency of ice sheet grounding events on the shelf increased. The Pliocene-Pleistocene section includes seven units and records at least that number of ice sheet grounding events. This probably reflects the strong impact of higher-frequency eustatic fluctuations, caused by the onset of northern hemisphere glaciation, on the stability of the marine ice sheet. Improved biostratigraphic zonation of the Pliocene section at DSDP Site 271 could provide a better record of when these grounding events occurred.

6. Possible fluvial valleys occur within one unit (Unit 5) of the Pliocene sequence; although the favored interpretation is that these valleys are subglacial tunnel valleys. The Pliocene cores from Site 271 recovered glacial-marine sediments with only minor diatomaceous material, and this may imply warmer interglacial conditions than those that exist today.

7. The late Pleistocene section in both the eastern and western Ross Sea consists of an amalgamation of tillites and glacial erosional surfaces.

8. The USAP 1990 seismic data show that ice sheet

grounding has resulted in widespread erosional surfaces. In some areas these surfaces are overlain by subglacial deposits, but commonly they are infilled by glacial marine deposits. Thus drilling alone does not provide an accurate record of ice sheet grounding episodes. Indeed, the DSDP drill sites on the Ross Sea shelf provided a very poor record of grounding events.

9. Glacial erosion during the late Pleistocene has cut deep into the early Miocene and younger strata of the Ross Sea; so older strata can be sampled by shallow drilling. Our USAP 1990 seismic data could provide an important regional data base for planning such a drilling program, but more detailed site surveys using high-resolution seismic sources will be needed to better constrain the stratigraphy of individual sites.

10. The proxy record of ice rafting at Site 274, the closest deep-sea site to the Ross Sea, indicates that ice sheets did not advance onto the continental shelf until Pliocene time [*Savage and Ciesielski*, 1983]. This clearly is not the case and raises serous doubts about the validity of ice-rafting records from deep-sea sites as indicators of Antarctica's glacial history.

Acknowledgments. Alan Cooper helped us plan our cruise by providing access to MCS records. He also provided us with a thorough review of the manuscript which proved to be very helpful. Adrien Pascouet and Jim Hedger of Seismic Systems, Inc., provided the GI gun and trained us to use and maintain it. Rick Shiftman of Teledyne, Inc., provided last minute assistance in streamer configuration and repair. Lastly, the highly professional crew of the *Polar Duke* deserves a special word of thanks for making this a productive and pleasant cruise. This project was funded by the National Science Foundation, Division of Polar Programs, under grant DPP 8818523.

REFERENCES

Alley, R. B., D. D. Blankenship, S. T. Rooney, and C. R. Bentley, Sedimentation beneath ice shelves—The view from Ice Stream B, *Mar. Geol.*, *85*, 101–120, 1989.

Alonso, B., J. B. Anderson, J. T. Diaz, and L. R. Bartek, Pliocene-Pleistocene seismic stratigraphy of the Ross Sea: evidence for multiple ice sheet grounding episodes, in *Contributions to Antarctic Research III, Antarct. Res. Ser.*, edited by D. H. Elliot, AGU, Washington, D. C., in press, 1992.

Anderson, J. B., Ancient glacial marine deposits: Their spatial and temporal distribution, in *Glacial-Marine Sedimentation*, edited by B. F. Molnia, pp. 3–94, Plenum, New York, 1983.

Anderson, J. B., The Antarctic continental shelf: Results from marine geological investigations, in *The Geology of Antarctica*, edited by R. Tingey, Oxford University Press, New York, 1991.

Anderson, J. B., and G. M. Ashley, Glacial marine sedimentation, paleoclimatic significance: A discussion, in Glacial Marine Sedimentation: Paleoclimatic Significance, *Spec. Pap. Geol. Soc. Am.*, *261*, 223–226, 1991.

Anderson, J. B., D. D. Kurtz, E. W. Domack, and K. M. Balshaw, Glacial and glacial marine sediments of the Antarctic continental shelf, *J. Geol.*, *88*, 399–414, 1980.

Anderson, J. B., C. F. Brake, and N. C. Myers, Sedimentation in the Ross Sea, Antarctica, *Mar. Geol.*, *57*, 295–333, 1984.

Anderson, J. B., D. S. Kennedy, M. J. Smith, and E. W. Domack, Sedimentary facies associated with Antarctica's floating ice masses, in Glacial Marine Sedimentation: Paleoclimatic Significance, *Spec. Pap. Geol. Soc. Am.*, *261*, 1–25, 1991.

Anderson, J. B., S. S. Shipp, L. R. Bartek, and D. E. Reid, Evidence for a grounded ice sheet on the Ross Sea continental shelf during the late Pleistocene and preliminary paleodrainage reconstruction, in *Contributions to Antarctic Research III, Antarct. Res. Ser.*, edited by D. H. Elliot, AGU, Washington, D. C., in press, 1992.

Armentrout, J. M., Glacial lithofacies of the Neogene Yakataga Formation, Robinson Mountains, southern Alaska Coast Range, in *Glacial Marine Sedimentation*, edited by B. F. Molina, pp. 629–665, Plenum, New York, 1983.

Ashley, G. M., J. Shaw, and N. D. Smith, Glacial sedimentary environments, *SEPM Short Course*, *16*, 245 pp., 1985.

Balshaw, K. M., Antarctic glacial chronology reflected in the Oligocene through Pliocene sedimentary section in the Ross Sea, Ph.D. dissertation, 140 pp., Rice Univ., Houston, Tex., 1981.

Barrett, P. J., Textural characteristics of Cenozoic preglacial and glacial sediments at Site 270, Ross Sea, Antarctica, *Initial Rep. Deep Sea Drill. Proj.*, *28*, 757–767, 1975.

Barrett, P. J. (Ed.), Antarctic Cenozoic history from the CIROS-1 drillhole, McMurdo Sound, *DSIR Bull. N. Z.*, *245*, 254 pp., 1989.

Barrett, P. J., and B. C. McKelvey, Stratigraphy, in Antarctic Cenozoic History From the MSSTS-1 Drillhole, McMurdo Sound, *DSIR Bull. N. Z.*, *237*, 9–51, 1986.

Barrett, P. J., M. J. Hambrey, D. M. Harwood, A. R. Pyne, and P.-N. Webb, Synthesis, in Antarctic Cenozoic History From the CIROS-1 Drillhole, McMurdo Sound, *DSIR Bull. N. Z.*, *245*, 241–251, 1989.

Bartek, L. R., Sedimentology and stratigraphy of McMurdo Sound and the Ross Sea, Antarctica: Implications for glacial history and analysis of high latitude marginal basins, Ph.D. dissertation, 416 pp., Rice Univ., Houston, Tex., 1989.

Bartek, L. R., P. R. Vail, J. B. Anderson, P. A. Emmet, and S. Wu, Effect of Cenozoic ice sheet fluctuations in Antarctica on the stratigraphic signature of the Neogene, *J. Geophys. Res.*, *96*, 6753–6778, 1991

Belknap, D. F., and R. C. Shipp, Seismic stratigraphy of glacial marine units, Maine inner shelf, in Glacial Marine Sedimentation: Paleoclimatic Significance, *Spec. Pap. Geol. Soc. Am.*, *261*, 137–158, 1991.

Berryhill, H. L., J. R. Suter, and N. S. Hardin, Late Quaternary facies and structure of the northern Gulf of Mexico: Interpretation from seismic data, *AAPG Stud. Geol.*, *23*, 289 pp., 1987.

Burckle, L. H., and E. M. Pokras, Implications of a Pliocene stand of *Nothofagus* (southern beech) within 500 km of the south pole, *Antarct. Sci.*, *3*, 389–403, 1991.

Chriss, T., and L. A. Frakes, Glacial marine sedimentation in the Ross Sea, in *Antarctic Geology and Geophysics*, edited by R. J. Adie, pp. 747–762, Universitetsforlaget, Oslo, 1972.

Clapperton, C. M., and D. E. Sugden, Late Cenozoic glacial history of the Ross Embayment, Antarctica, *Quat. Sci. Rev.*, *9*, 253–272, 1990.

Cooper, A. K., F. J. Davey, and J. C. Behrendt, Seismic stratigraphy and structure of the Victoria Land Basin, western Ross Sea, Antarctica, in *The Antarctic Continental Margin: Geology and Geophysics of Western Ross Sea, Earth Sci. Ser.*, vol. 5B, edited by A. K. Cooper and F. J. Davey, pp. 27–76, Circum-Pacific Council for Energy and Mineral Resources, Houston, Tex., 1987.

Cooper, A. K., P. J. Barrett, K. Hinz, V. Traube, G. Leitchenkov, and H. M. J. Stagg, Cenozoic prograding sequences of the Antarctic continental margin: A record of glacioeustatic and tectonic events, *Mar. Geol.*, *102*, 175–213, 1991.

Davey, F. J., The Antarctic margin and its possible hydrocarbon potential, *Tectonophysics*, *114*, 443–470, 1985.

Domack, E. W., Biogenic facies in the Antarctic glacimarine environment: Basis for a polar glacimarine summary, *Palaeogeogr. Palaeoclimatol. Palaeoecol.*, *63*, 357–372, 1988.

Elverhoi, A., Glaciogenic and associated marine sediments in the Weddell Sea, fjords of Spitsbergen and the Barents Sea: A review, *Mar. Geol.*, *57*, 53–88, 1984.

Hambrey, M. J., and P. J. Barrett, The Cenozoic sedimentary and climatic record from the Ross Sea region of Antarctica, in *The Antarctic Paleoenvironment: A Perspective on Global Change, Part 2, Antarct. Res. Ser.*, edited by J. P. Kennett and D. A. Warnke, AGU, Washington, D. C., in press, 1992.

Hambrey, M. J., W. U. Ehrmann, and B. Larsen, The Cenozoic glacial record from the Prydz Bay continental shelf, East Antarctica, *Proc. Ocean Drill. Program Sci. Results*, *119*, 77–132, 1991.

Haq, B. L., J. Hardenbol, and P. R. Vail, Chronology of fluctuating sea levels since the Triassic, *Science*, *235*, 1156–1167, 1987.

Harwood, D. M., Late Neogene climatic fluctuations in the high southern latitudes: Implications of a warm Gauss and deglaciated Antarctic continent, *S. Afr. J. Sci.*, *81*, 239–241, 1985.

Harwood, D. M., Recycled siliceous microfossils from the Sirius Formation, *Antarct. J. U. S.*, *21*, 101–103, 1986.

Hayes, D. E., and F. J. Davey, A geophysical study of the Ross Sea, Antarctica, *Initial Rep. Deep Sea Drill. Proj.*, *28*, 887–907, 1975.

Hayes, D. E., and L. A. Frakes, General synthesis: Deep Sea Drilling Project 28, *Initial Rep. Deep Sea Drill. Proj.*, *28*, 919–942, 1975.

Hinz, K., and M. Block, Results of geophysical investigations in the Weddell Sea and in the Ross Sea, Antarctica, *Proc. World Pet. Congr. 11th, PD2*, 79–91, 1983.

Holtedahl, O., On the geomorphology of the West Greenland shelf, with general remarks on the marginal channels problem, *Mar. Geol.*, *8*, 155–172, 1970.

Houtz, R. E., and F. J. Davey, Seismic profiler and sonobuoy measurements in the Ross Sea, Antarctica, *J. Geophys. Res.*, *78*, 3448–3468, 1973.

Houtz, R. E., and R. Meijer, Structure of the Ross Sea shelf from profiler data, *J. Geophys. Res.*, *75*, 6592–6597, 1970.

Johnson, G. L., J. R. Vanney, and P. Hayes, the Antarctic continental shelf—A review, in *Antarctic Geoscience*, edited by C. Craddock, pp. 22–27, University of Wisconsin Press, Madison, 1982.

Karl, H. A., E. Reimnitz, and B. D. Edwards, Extent and nature of Ross Sea unconformity in the western Ross Sea, Antarctica, in *The Antarctic Continental Margin: Geology and Geophysics of Western Ross Sea, Earth Sci. Ser.*, vol. 5B, edited by A. K. Cooper and F. J. Davey, pp. 77–92, Circum-Pacific Council for Energy and Mineral Resources, Houston, Tex., 1987.

Kellogg, T. B., R. S. Truesdale, and L. E. Osterman, Late Quaternary extent of the West Antarctic ice sheet: New evidence from Ross Sea cores, *Geology*, *7*, 249–253, 1979.

King, L. H., and G. Fader, Wisconsinan glaciation of the continental shelf, southeast Atlantic Canada, *Bull. Geol. Surv. Can.*, *363*, 72 pp., 1986.

King, L. H., K. Rokoengen, G. Fader, and T. Gunleiksrud, Till-tongue stratigraphy, *Geol. Soc. Am. Bull.*, *103*, 637–659, 1991.

Larter, R. D., and P. F. Barker, Seismic stratigraphy of the Antarctic Peninsula Pacific margin: A record of Pliocene-Pleistocene ice volume and paleoclimate, *Geology*, *17*, 731–734, 1989.

Leckie, R. M., and P.-N. Webb, Late Oligocene–early Miocene glacial record of the Ross Sea, Antarctica: Evidence from DSDP Site 270, *Geology*, *11*, 578–582, 1983.

McDougall, I., Potassium-argon dating of glauconite from a greensand drilled at Site 270 in the Ross Sea, DSDP Leg 28, *Initial Rep. Deep Sea Drill. Proj.*, *28*, 1071–1072, 1975.

Miller, K. G., J. D. Wright, and R. G. Fairbanks, Unlocking the ice house: Oligocene-Miocene oxygen isotopes, eustasy, and margin erosion, *J. Geophys. Res.*, *96*, 6829–6848, 1991.

Reid, D. E., Late Cenozoic glacial-marine, carbonate and turbidite sedimentation in the northwestern Ross Sea, Antarctica, M.A. thesis, 179 pp., Rice Univ., Houston, Tex., 1989.

Robinson, P. H., A. R. Pyne, M. J. Hambrey, K. J. Hall, and P. J. Barrett, Core log, photographs and grain size analyses from the CIROS-1 drillhole western McMurdo Sound, Antarctica, *Antarct. Data Ser. 14*, 241 pp., Victoria Univ. of Wellington, Wellington, 1987.

Sato, S., N. Asakura, T. Saki, N. Oikawa, and Y. Kaneda, Preliminary results of geological and geophysical survey in the Ross Sea and in the Dumont D'urville Sea, off Antarctica, *Mem. Natl. Inst. Polar Res. Spec. Issue Jpn.*, *33*, 66–92, 1984.

Savage, M. L., and P. F. Ciesielski, A revised history of glacial sediments in the Ross Sea region, in *Antarctic Earth Science*, edited by R. L. Oliver, P. R. James, and J. B. Jago, pp. 555–559, Cambridge University Press, New York, 1983.

Shackleton, N. J., and J. P. Kennett, Paleotemperature history of the Cenozoic and the initiation of Antarctic glaciation: Oxygen and carbon isotope analyses in DSDP sites 277, 279, and 281, *Initial Rep. Deep Sea Drill. Proj.*, *29*, 743–755, 1975.

Shepard, F. P., Glacial troughs of the continental shelf, *J. Geol.*, *39*, 345–360, 1931.

Solheim, A., L. Russwurm, A. Elverhoi, and B. Nyland, Glacial geomorphic features in the northern Barents Sea: Direct evidence for grounded ice and implications for the pattern of deglaciation and late glacial sedimentation, in Glacimarine Environments: Processes and Sediments, *Geol. Soc. Spec. Publ. London*, *53*, 253–268, 1990.

Stoker, M. S., Glacially-influenced sedimentation on the Hebridean slope, northwestern United Kingdom continental margin, in Glacimarine Environments: Processes and Sediments, *Geol. Soc. Spec. Publ. London*, *53*, 349–362, 1990.

Thomas, M. A., and J. B. Anderson, The effect and mechanism of episodic sea level events: The record preserved within late Wisconsinan–Holocene incised valley-fill sequences, *Trans. Gulf Coast Assoc. Geol. Soc.*, *38*, 399–406, 1988.

Vail, P. R., R. M. Mitchum, Jr., R. G. Todd, J. M. Widmier, S. Thompson, J. B. Sangree, J. N. Bubb, and W. G. Hatelid, Seismic stratigraphy and global changes of sea level, in Seismic Stratigraphy—Applications to Hydrocarbon Exploration, *Mem. Am. Assoc. Pet. Geol.*, *26*, 49–212, 1977.

Vorren, T. O., E. Lebesbye, and K. B. Larsen, Geometry and genesis of the glacigenic sediments in the southern Barents Sea, in Glacimarine Environments: Processes and Sediments, *Geol. Soc. Spec. Publ. London*, *53*, 269–288, 1990.

Webb, P.-N., and D. M. Harwood, Late Cenozoic glacial history of the Ross embayment, Antarctica, *Quat. Sci. Rev.*, *10*, 215–223, 1991.

Wornardt, W. W., and P. R. Vail, Revision of the Plio-Pleistocene cycles and their application to sequence stratigraphy of shelf and slope sediments in the Gulf of Mexico, in *Sequence Stratigraphy as an Exploration Tool, Concepts and Practices in the Gulf Coast*, Program and Extended Abstracts of *Eleventh Annual Research Conference*, pp. 391–397, Gulf Coast Section of the Society of Economic Paleontologists and Mineralogists, Earth Enterprises, Austin, Tex., 1990.

(Received January 2, 1992;
accepted June 5, 1992.)

TOWARD A HIGH-RESOLUTION STABLE ISOTOPIC RECORD OF THE SOUTHERN OCEAN DURING THE PLIOCENE-PLEISTOCENE (4.8 TO 0.8 MA)

DAVID A. HODELL AND KATHRYN VENZ

Department of Geology, University of Florida, Gainesville, Florida 32611

We report a near-continuous, stable isotopic record for the Pliocene-Pleistocene (4.8 to 0.8 Ma) from Ocean Drilling Program Site 704 in the sub-Antarctic South Atlantic (47°S, 7°E). During the early to middle Pliocene (4.8 to 3.2 Ma), variation in $\delta^{18}O$ was less than ~0.5‰, and absolute values were generally less than those of the Holocene. These results indicate some warming and minor deglaciation of Antarctica during intervals of the Pliocene but are inconsistent with scenarios calling for major warming and deglaciation of the Antarctic ice sheet. The climate system operated within relatively narrow limits prior to ~3.2 Ma, and the Antarctic cryosphere probably did not fluctuate on a large scale until the late Pliocene. Benthic oxygen isotopic values exceeded 3‰ for the first time at 3.16 Ma. The amplitude and mean of the $\delta^{18}O$ signal increased at 2.7 Ma, suggesting a shift in climate mode during the latest Gauss. The greatest $\delta^{18}O$ values of the Gauss and Gilbert chrons occurred at ~2.6 Ma, just below a hiatus that removed the interval from ~2.6 to 2.3 Ma in Site 704. These results agree with those from Subantarctic Site 514, which suggest that the latest Gauss (2.68 to 2.47 Ma) was the time of greatest change in Neogene climate in the northern Antarctic and Subantarctic regions. During this period, surface water cooled as the Polar Front Zone (PFZ) migrated north and perennial sea ice cover expanded into the Subantarctic region. Antarctic ice volume increased and the ventilation rate of Southern Ocean deep water decreased during glacial events after 2.7 Ma. We suggest that these changes in the Southern Ocean were related to a gradual lowering of sea level and a reduction in the flux of North Atlantic Deep Water (NADW) with the initiation of ice growth in the northern hemisphere. The early Matuyama Chron (~2.3 to 1.7 Ma) was marked by relatively warm climates in the Southern Ocean except for strong glacial events associated with isotopic stages 82 (2.027 Ma), 78 (1.941 Ma), and 70 (1.782 Ma). At 1.67 Ma (stage 65/64 transition), surface waters cooled as the PFZ migrated equatorward and oscillated about a far northerly position for a prolonged interval between 1.67 and 1.5 Ma (stages 65 to 57). Beginning at ~1.42 Ma (stage 52), all parameters ($\delta^{18}O$, $\delta^{13}C$, %opal, %$CaCO_3$) in Hole 704 become highly correlated with each other and display a very strong 41-kyr cyclicity. This increase in the importance of the 41-kyr cycle is attributed to an increase in the amplitude of the Earth's obliquity cycle that was likely reinforced by increased glacial suppression of NADW, which may explain the tightly coupled response that developed between the Southern Ocean and the North Atlantic beginning at ~1.42 Ma (stage 52).

INTRODUCTION

Although Antarctica with the adjacent Southern Ocean is one of the most important components of the climate system, knowledge of its history pales in comparison with the high-latitude northern hemisphere and North Atlantic Ocean. This lopsided, northern hemisphere view of late Neogene climate evolution requires balance from high-resolution study of sequences from the Southern Ocean. The main stumbling block has been the paucity of good quality sequences available from the Southern Ocean that are suitable for high-resolution paleoclimatic research. Problems associated with incomplete core recovery, the presence of hiatuses, and carbonate preservation at the high latitudes of the southern hemisphere have hampered efforts at obtaining a continuous paleoclimatic record. Expeditions by the

Ocean Drilling Program (ODP) in the South Atlantic and Indian sectors of the Southern Ocean during 1987–1988 (legs 113, 114, 119, and 120) were designed to recover sequences that would be suitable for high-resolution studies of Neogene paleoclimatology. Of the 32 sites drilled on these four legs, only one site, ODP Site 704 (Leg 114), has sufficient stratigraphic continuity and carbonate content during the Pliocene-Pleistocene to be suitable for high-resolution stable isotopic work. The scarcity of long continuous, carbonate-bearing sequences from the high latitudes of the southern hemisphere underscores the need for future drilling in the Subantarctic sector.

Here we present new isotopic and carbonate data from Hole 704A and Hole 704B that augment results from previous studies [*Hodell and Ciesielski*, 1991; *Froelich et al.*, 1991]. *Hodell and Ciesielski* [1991]

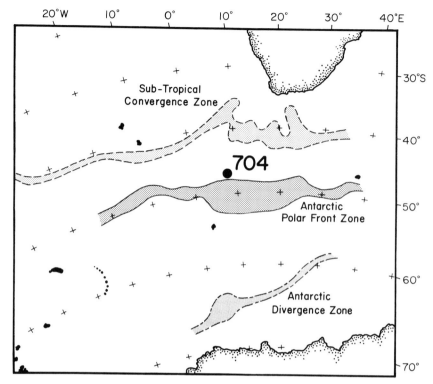

Fig. 1.   Position of ODP Site 704 in the eastern Subantarctic South Atlantic (46°52.8′S, 7°25.3′E) relative to major hydrographic fronts (modified after *Lutjeharms* [1985]).

presented all isotopic data versus subbottom depth in Hole 704A and did not convert to composite depth or an age scale because of incomplete core recovery and lack of an adequate chronology. Here we have spliced Hole 704A and Hole 704B together to produce a composite depth section that is reasonably complete for the interval from 4.8 to 0.8 Ma. The older, late Miocene record of Site 704 has been discussed previously by *Müller et al.* [1991], and isotopic results for the late Pleistocene (i.e., Brunhes) are forthcoming [*Hodell*, 1992]. When completed, the isotopic record from Site 704 will provide a high-resolution southern hemisphere climate signal spanning the last 8.6 million years.

In this study, we compare $\delta^{18}O$ results of planktic and benthic foraminifers from the Pliocene-Pleistocene (4.8 to 1.0 Ma) with isotopic measurements of the same species from the top (i.e., Holocene) of Hole 704A. This comparison permits the placement of relative constraints on temperature and ice volume fluctuations during the older part of the record and provides several new insights into the climatic evolution of the Southern Ocean during the Pliocene-Pleistocene.

SITE 704: LOCATION AND OCEANOGRAPHY

Site 704 is positioned just north of an area of strong hydrographic gradients associated with the Polar Front Zone (PFZ) (Figure 1). The PFZ is bounded by the

Subantarctic Front to the north and by the Antarctic Polar Front to the south. The average width of the PFZ in the South Atlantic is 670 km, and it is centered generally at 45°S with a latitudinal span of roughly ±2.5° [*Lutjeharms*, 1985]. The PFZ separates cold, nutrient-rich Antarctic surface water to the south from warmer, less nutrient-rich Subantarctic Surface Water to the north. Siliceous productivity (mainly diatoms) is relatively high in the nutrient-rich surface waters south of the Subantarctic Front, whereas calcareous production dominates to the north. The PFZ represents a transition zone, therefore, from pure diatom ooze to the south near the Antarctic Polar Front to a mixed siliceous-calcareous ooze to the north near the Subantarctic Front [*Cooke and Hays*, 1982; *Burckle and Cirilli*, 1987]. South of the Antarctic Polar Front, sediments consist of silty diatomaceous clay with the terrigenous component becoming progressively important to the south. This lithologic boundary appears to coincide with the spring sea ice limit [*Burckle et al.*, 1982; *Burckle and Cirilli*, 1987]. Because of the close proximity of Site 704 to the PFZ, the accumulation of opal and carbonate in the sediments should be sensitive to past changes in its position. Other factors, such as dissolution, can also affect the proportion of carbonate and opal in the sediments.

Because of the temperature dependence of the $\delta^{18}O$

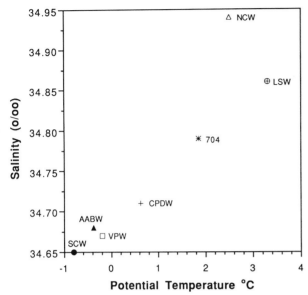

Fig. 2. Potential temperature versus salinity diagram for major water masses in the world's oceans. NCW, Northern Component Water; LSW, Labrador Sea Water; SCW, Southern Component Water; AABW, Antarctic Bottom Water; CPDW, Circumpolar Deep Water; VPW, Vema Passage Water; 704, temperature (1.852°C) and salinity (34.79‰) at ~2500 m at GEOSECS station 92 (46°11′S, 14°36′E) near the position of ODP Site 704. Data for water mass properties are those summarized by *Oppo and Fairbanks* [1987, Table 1].

of precipitated calcite, the PFZ is also a region of steep gradients in the surface equilibrium $\delta^{18}O$ values of calcite. Between approximately 40° and 60°S latitude, the $\delta^{18}O$ of equilibrium calcite increases by 3‰, which represents nearly half of the predicted equator-to-pole gradient in $\delta^{18}O$ calcite [*Charles and Fairbanks*, 1990]. Even subtle changes in the position of the PFZ should be recorded by the $\delta^{18}O$ of planktic foraminifers.

The relatively shallow water depth of Site 704 (2532 m) places it above the carbonate lysocline and calcite compensation depth and within a mixing zone between North Atlantic Deep Water (NADW) and Circumpolar Deep Water (CPDW). A temperature-salinity plot reveals that Site 704 is located on a mixing line that connects Southern Component Water (SCW), CPDW, and deep waters formed in the high-latitude North Atlantic (Labrador Sea Water and Northern Component Water; Figure 2). A salinity profile of the Agulhas Basin clearly shows the tongue of high-salinity NADW as it penetrates into the Southern Ocean, inserts itself into CPDW, and upwells at the Antarctic Divergence (Figure 3). NADW is a major source of heat and salt to the Southern Ocean today, and it continually renews CPDW in the South Atlantic sector [*Jacobs et al.*, 1985; *Gordon*, 1981; *Oppo and Fairbanks*, 1987]. Because of the difference in $\delta^{13}C$ values between NADW (~1‰ at its core) and CPDW (0.4‰ at its core [*Kroopnick*,

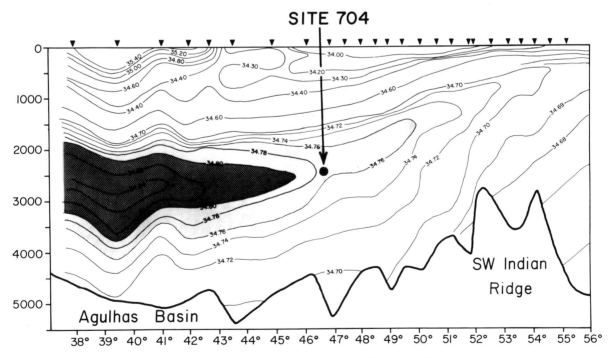

Fig. 3. Vertical salinity profile (contour interval of 0.1‰) of the eastern Subantarctic sector of the Southern Ocean (modified after *Gordon and Molinelli* [1986]). Note the tongue of high-salinity NADW (shaded) as it penetrates into the circumpolar region, inserts itself into the Antarctic Circumpolar Current, and upwells at the Antarctic Divergence. Site 704 is located in a mixing zone between NADW and CPDW.

1985]), the carbon isotopic ratios of benthic foraminifers at Site 704 should be highly sensitive to past changes in the mixing ratio of these water masses.

METHODS

Stable Isotopes

Isotopic methods were the same as those reported by *Hodell and Ciesielski* [1991] except that new measurements were made by reacting samples at 90°C instead of 70°C. We tested for offsets between new and old data by analyzing duplicate samples from several intervals. Curiously, we found new and old data for *Neogloboquadrina pachyderma* to be consistently offset by 0.3‰, yet found no such offset between new and old data for *Cibicides*. The offset in *N. pachyderma* remains inexplicable, but we have corrected the *N. pachyderma* data reported by *Hodell and Ciesielski* [1991] by adding 0.3‰ to $\delta^{18}O$ values. No correction was applied to benthic oxygen isotopic data or carbon isotopic results. The complete isotopic data set is reported in the appendix Tables A1–A3. Analytical precision for the new data, estimated by the 1-sigma error of 139 analyses of an internal carbonate working standard (CM-UF), was ±0.07 for $\delta^{18}O$ and ±0.03 for $\delta^{13}C$.

Planktic isotopic analyses consisted of measurements of two species, *N. pachyderma* and *Globigerina bulloides*. *N. pachyderma* (sinistrally coiled morphotype) has been shown to be a reliable recorder of $\delta^{18}O$ and $\delta^{13}C$ values of surface waters in the modern South Atlantic sector of the Southern Ocean [*Charles and Fairbanks*, 1990]. Core top values of $\delta^{18}O$ and $\delta^{13}C$ of *N. pachyderma* follow those expected for calcite precipitated at a constant offset from isotopic equilibrium with surface waters [*Labeyrie et al.*, 1986; *Charles and Fairbanks*, 1990]. The offset in $\delta^{18}O$ values between *N. pachyderma* and equilibrium calcite is estimated to be −0.4‰ [*Labeyrie et al.*, 1986; *Charles and Fairbanks*, 1990]. The calculated equilibrium $\delta^{18}O_{calcite}$ at 47°S is ~2.4‰ [*Charles and Fairbanks*, 1990], which gives a predicted Holocene value of *N. pachyderma* of ~2.0‰ at Site 704. The measured $\delta^{18}O$ of *N. pachyderma* during the Holocene (stage 1) in Hole 704A is 2.21‰ [*Hodell*, 1992], which is slightly greater than that predicted. For $\delta^{13}C$, a constant offset of −1‰ exists between the $\delta^{13}C$ of *N. pachyderma* and ΣCO_2 [*Charles and Fairbanks*, 1990]. The measured Holocene $\delta^{13}C$ of *N. pachyderma* at Site 704 is 0.89‰.

Because of the scarcity of *N. pachyderma* in sediments older than ~3.4 Ma, we measured specimens of *G. bulloides* that ranged in size from 212 to 295 μm. The measured Holocene value for *G. bulloides* from Site 704 is 2.30‰, which agrees well with measured values of 2.39‰ for *G. bulloides* from the top of Core E49-23, located at the same latitude (47°S) as Site 704 but in the southern Indian Ocean [*Howard and Prell*, 1992]. The measured Holocene $\delta^{13}C$ of *G. bulloides* at Site 704 is

0.37‰. *G. bulloides* precipitates calcite out of carbon isotopic equilibrium by −1.53‰ relative to the $\delta^{13}C$ of the ΣCO_2 [*Williams et al.*, 1977]. In intervals where the records of *N. pachyderma* and *G. bulloides* overlap, we found no significant difference in $\delta^{18}O$ values (<0.1‰) between the two species, but $\delta^{13}C$ values of *G. bulloides* were consistently less than *N. pachyderma* by ~0.5‰.

All benthic isotopic results were measured on specimens of the epibenthic foraminifers *Planulina wuellerstorfi*, *Cibicidoides kullenbergi*, or *Cibicides* sp. from the >150-μm size fraction. For simplicity, hereinafter, we refer to this mixture of related genera as *Cibicides*. *P. wuellerstorfi* and *C. kullenbergi* have been shown to be isotopically similar and can be used interchangeably within analytical error [*Duplessy et al.*, 1970; *Graham et al.*, 1981; *Woodruff et al.*, 1981]. *Hodell and Warnke* [1991] calculated the equilibrium $\delta^{18}O_{calcite}$ value for benthic foraminifers at Site 704 to be between 3.58 and 3.65‰. Assuming an oxygen isotopic disequilibrium for *P. wuellerstorfi* of −0.64‰ [*Shackleton and Opdyke*, 1973], the predicted Holocene value should be between 2.94 and 3.01‰. Indeed, the measured $\delta^{18}O$ of *Cibicides* from the top of Hole 704 is 3.0‰ [*Hodell*, 1992]. Carbon isotopes of *P. wuellerstorfi* and *C. kullenbergi* reflect the $\delta^{13}C$ of dissolved bicarbonate with no offset [*Graham et al.*, 1981]. The measured Holocene $\delta^{13}C$ value for *Cibicides* at the top of Hole 704A is 0.64‰.

Stratigraphy

Although core recovery was good at Site 704, missing section at core breaks, double-cored intervals, and compressed or expanded sections are still common with the advanced piston corer (APC) and extended core barrel (XCB) systems even under ideal conditions [*Ruddiman et al.*, 1986]. The soft, water-saturated sediments of holes 704A and 704B were easily disturbed during coring and handling in relatively high seas. To obtain as complete a section as possible, we spliced the records between holes 704A and 704B together by correlating the percent carbonate records from the two holes. The "road map" and splices for holes 704A and 704B used to produce the composite depth section for Site 704 are given in Figure 4. The composite section often switches back and forth between the holes in order to fill gaps and avoid disturbed intervals (e.g., soupy sediments, double-cored intervals, etc.). For each section, we attempted to use the most continuous, least disturbed interval from either of the two holes. The only major change made to the stratigraphy of the core was the removal of a 14-m section from Hole 704A between 121 and 135 meters below seafloor (mbsf). We suspect this section of core was disturbed by either slumping or coring because $\delta^{18}O$ values are constant over this 14-m section and do not correlate with standard isotopic zonations of Site 607 or Site 677. Furthermore, the

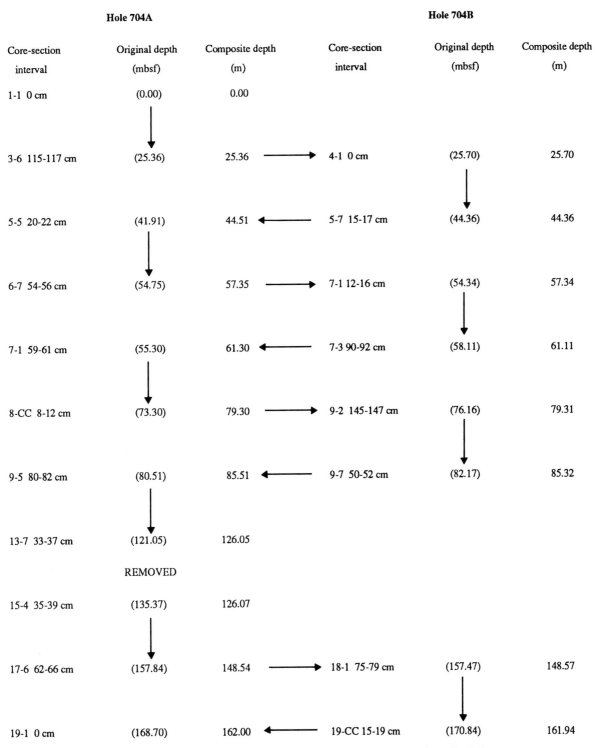

Fig. 4. Roadmap used to construct composite depth section for ODP Site 704.

sediments are normally magnetized throughout this interval, which would result in an anomalously long Olduvai Subchron (37 m) if this section of core were retained.

The age model for Site 704 was derived mainly by a combination of magnetostratigraphy and oxygen isotope stratigraphy. All age-depth pairs used to construct the chronology are given in Table 1 and plotted in Figure 5.

TABLE 1.   Age-Depth Control Points Used to Establish
Chronology in Site 704

Stage/Datum	Original Depth, mbsf	Composite Depth, m	Age, Ma
25	35.00 (B)	35.00	0.859
26	36.20 (B)	36.20	0.881
27	38.40 (B)	38.40	0.906
28	39.10 (B)	39.10	0.922
29	40.20 (B)	40.20	0.946
30	43.20 (B)	43.20	0.965
31	44.40 (B)	44.40	0.985
32	42.42 (A)	45.02	1.007
33	43.20 (A)	45.80	1.022
34	44.40 (A)	47.00	1.047
35	47.00 (A)	49.60	1.068
36	50.00 (A)	52.60	1.088
37	51.70 (A)	54.30	1.110
38	52.30 (A)	54.90	1.130
39	54.00 (A)	56.60	1.153
40	54.60 (B)	57.60	1.173
41	57.40 (B)	60.40	1.192
42	56.50 (A)	62.50	1.212
43	57.70 (A)	63.70	1.229
44	59.10 (A)	65.10	1.254
45	60.60 (A)	66.60	1.273
46	62.70 (A)	68.70	1.294
47	65.00 (A)	71.00	1.313
48	66.00 (A)	72.00	1.335
49	67.50 (A)	73.50	1.359
50	68.40 (A)	74.40	1.371
51	69.30 (A)	75.30	1.394
52	70.10 (A)	76.10	1.416
53	72.50 (A)	78.50	1.434
54	73.00 (A)	79.00	1.455
55	77.86 (B)	80.71	1.478
56	80.06 (B)	83.21	1.499
57	81.76 (B)	84.91	1.514
58	82.45 (A)	87.45	1.537
58	85.25 (A)	90.25	1.550
59	85.75 (A)	90.75	1.560
60	89.35 (A)	94.35	1.576
61	91.55 (A)	96.55	1.596
62	92.35 (A)	97.35	1.616
63	94.55 (A)	99.55	1.636
64	97.35 (A)	102.35	1.662
65	101.05 (A)	106.05	1.682
66	104.35 (A)	109.35	1.700
67	106.05 (A)	111.05	1.718
68	107.15 (A)	112.15	1.741
69	108.65 (A)	113.65	1.759
70	113.25 (A)	118.25	1.782
71	116.25 (A)	121.25	1.801
72	116.85 (A)	121.25	1.829
73	117.43 (A)	122.43	1.853
74	118.64 (A)	123.64	1.862
75	135.57 (A)	126.27	1.883
76	137.07 (A)	127.77	1.908
77	138.65 (A)	129.35	1.922
78	143.25 (A)	133.95	1.941
79	144.31 (A)	135.01	1.985
80	146.15 (A)	136.85	2.001
81	147.31 (A)	138.01	2.013
82	149.75 (A)	143.00	2.027
83	152.30 (A)	144.06	2.049
84	153.36 (A)	146.18	2.069
85	155.48 (A)	147.69	2.091
86	156.99 (A)	148.57	2.110
87	157.47 (A)	149.60	2.138
88	160.65 (B)	151.75	2.152
89	161.45 (B)	152.25	2.177
90	162.05 (B)	153.15	2.198
91	162.35 (B)	153.45	2.221
92	163.05 (B)	154.15	2.238
93	163.85 (B)	154.95	2.260
94	165.15 (B)	156.25	2.275
95	165.65 (B)	156.75	2.301
top of core 704B-19X	166.15 (B)	157.25	~2.600
top Kaena	176.10 (A)	169.40	2.920
base Kaena	177.80 (A)	171.10	2.990
top Mammoth	179.10 (A)	172.40	3.080
base Mammoth	181.50 (A)	174.80	3.180
Gauss/Gilbert	186.64 (A)	179.94	3.400
top Cochiti	195.55 (A)	188.85	3.880
base Cochiti	201.94 (A)	195.24	3.970
top Nunivak	204.19 (A)	197.49	4.100

TABLE 1.   (continued)

Stage/Datum	Original Depth, mbsf	Composite Depth, m	Age, Ma
base Nunivak	210.35 (A)	203.65	4.240
top Sidufjall	211.99 (A)	205.29	4.400
base Sidufjall	213.24 (A)	206.54	4.470
top Thvera	215.40 (A)	208.70	4.570
base Thvera	219.47 (A)	212.77	4.770
Gilbert/Chron 5	224.76 (A)	218.06	5.350

(A) and (B) designate Hole 704A and Hole 704B.

During the Gauss and Gilbert chrons, sedimentation rates average 23 m/m.y., whereas sedimentation rates average 100 m/m.y. during the Matuyama. This increase in sedimentation rates that occurs across the Gauss/Matuyama transition has been discussed previously and related to the proximity of the PFZ [*Hodell and Ciesielski*, 1990, 1991; *Froelich et al.*, 1991]. Such high sedimentation rates during the Matuyama are not uncommon for the Southern Ocean because the PFZ is a region of high productivity and flux of calcareous and siliceous microfossils to the sediments. Because of variations in sedimentation rate, the sampling frequency of the isotopic record varies from approximately 10 kyr during the Gilbert, to 5 kyr during the Gauss, to between 3 and 4 kyr during the Matuyama Chron.

The magnetostratigraphy of Site 704 is excellent below 140 mbsf where sediments permitted the use of the pass-through cryogenic magnetometer [*Hailwood and Clement*, 1991]. For the Gilbert and Gauss chrons, we used the polarity reversal boundaries determined by *Hailwood and Clement* [1991] with the exception of the Gauss/Matuyama boundary (Table 1). In Hole 704A the Gauss/Matuyama boundary was identified at 168.45 mbsf between cores 114-704A-18X and 114-704A-19X. Because of the occurrence of a hiatus at the Gauss/Matuyama boundary [*Fenner*, 1991], we estimated the age of the top of Core 704B-19X (166.15 mbsf) by extrapolating the sedimentation rate for the Gauss and derived an age of ~2.6 Ma (Table 1). This age agrees fairly well with the last abundant appearance datum of *Nitzchia weaveri* at 169.6 mbsf dated at 2.63 Ma [*Fenner*, 1991].

Above 140 mbsf the quality of the magnetic data is poor because of very weak magnetic intensities and the soft, water-saturated nature of the sediments [*Hailwood and Clement*, 1991]. For the Matuyama Chron, the age model for Site 704 was derived mainly by correlation of oxygen isotopic records between Site 704 and Site 607 [*Ruddiman et al.*, 1989; *Raymo et al.*, 1989]. The depth of the midpoint of each isotopic stage was designated in Hole 704A, and the age of each stage was assigned according to the "TP607" chronology (Table 1) [*Ruddiman et al.*, 1989; *Raymo et al.*, 1989]. An alternative

Age (Ma)

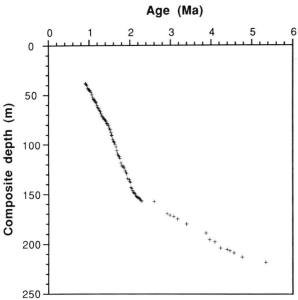

Fig. 5.   Age versus depth plot of isotopic and paleomagnetic control points listed in Table 1. Note the hiatus between ~2.6 and 2.3 Ma and increase in sedimentation rates across the unconformity.

chronology for the lower Pleistocene has been proposed by *Shackleton et al.* [1991] based upon orbital tuning of isotopic variations in ODP Site 677. This alternative time scale implies that radiometric dates of polarity reversal boundaries above the Gauss are too young by 5 to 7%. To convert the time scale used here to the alternative time scale of *Shackleton et al.* [1991], all ages should be increased by ~6%.

A normal polarity event was identified between 153.5 and 152 mbsf in Hole 704A and between 151.5 and 149 mbsf in Hole 704B [*Hailwood and Clement*, 1991]. This event was interpreted by *Hodell and Ciesielski* [1991] to represent the Réunion Event with age boundaries of 2.04 and 2.01 Ma. In Site 609 the Réunion Event occurs between stages 79 and 81 [*Raymo et al.*, 1989]. Our interpretation of stages in the lower Matuyama of Site 704 is consistent with the inferred correlation between isotopic stages and the Réunion Event in Site 609.

The identification of the top and bottom of the Olduvai Subchron is problematic in Site 704. The bottom of the Olduvai Subchron occurs at the base of stage 71 in sites 607 and 677 [*Raymo et al.*, 1989; *Shackleton et al.*, 1991]. In Hole 704A, stage 71 has been identified at 116.0 mbsf (Table 1). A polarity reversal does occur at 116 mbsf, but the reversed polarity below this level is based upon one discrete sample measurement only [*Hailwood and Clement*, 1991]. In addition, we have removed a normally magnetized interval below this level between ~121 and 135 because we suspect it has been disturbed. We suggest that the base of the Olduvai

Subchron should be placed at the reversal between 116 and 118 mbsf in Hole 704A.

On the basis of discrete sample measurements made aboard ship, the top of the Olduvai (1.66 Ma) was constrained to occur between 92.86 and 94.54 mbsf [*Ciesielski et al.*, 1988, p. 640]. Subsequent resampling and analysis of Hole 704A for paleomagnetics elevated the position of the top of the Olduvai Subchron to between 85.3 and 89.5 mbsf [*Hailwood and Clement*, 1991]. The polarity pattern in Hole 704A is normal between 89 and 93 mbsf, reversed between ~93 and 94 mbsf, and normal below ~94 mbsf [*Hailwood and Clement*, 1991, p. 377]. The top of the Olduvai has been correlated to the base of stage 63 in sites 607 and 677 [*Raymo et al.*, 1989; *Shackleton et al.*, 1991]. In Hole 704A the base of stage 63 occurs from ~94.84 to 99.05 mbsf. This agrees with the shipboard position of the top of the Olduvai between 92.86 and 94.54 mbsf but disagrees with the shore-based placement of the boundary between 85.3 and 89.5 mbsf. We suggest, therefore, that the top of the Olduvai Subchron should be placed at the polarity reversal between 93 and 94 mbsf in Hole 704A.

A normal polarity event, represented by one point only, was identified at 71 mbsf in Hole 704A [*Hailwood and Clement*, 1991]. This normal event falls in stage 52 according to the isotopic zonation. The Gilsa Event has been tied to stage 52 by correlation of carbonate variations between Site 609 and Site 607 [*Ruddiman et al.*, 1989]. We suggest, therefore, that the brief normal event at 71 mbsf in Hole 704A may represent the Gilsa Event.

The top of the record is constrained by the Jaramillo Subchron that was identified between 38.41 and 44.27 mbsf in Hole 704B [*Hailwood and Clement*, 1991]. The top of the Jaramillo occurs near the stage 26/27 transition, whereas the bottom occurs in stage 31 [*Shackleton et al.*, 1991]. The interpretation of isotope stages in Site 704 is consistent with this correlation between oxygen isotope stages and the boundaries of the Jaramillo Subchron.

For the most part, the identification of isotope stages in Site 704 is consistent with existing paleomagnetic data. We acknowledge, however, that the age model is not completely satisfactory because of the poor quality of the paleomagnetic data above 140 mbsf. Many of the problems with the existing chronology, however, can only be overcome by reoccupying Site 704 under better sea state conditions and recovering less disturbed sequences that are amenable to whole-core magnetic measurements. Even with its shortcomings, Site 704 remains one of the most complete sequences available from the Southern Ocean with which to study the climatic evolution of the Antarctic/Subantarctic region during the late Neogene.

RESULTS

Oxygen Isotopes

The planktic and benthic oxygen isotopic records span the interval from 4.8 to 0.8 Ma (Figure 6). Variations in oxygen isotopes are compared with measured Holocene $\delta^{18}O$ values (bold vertical lines in Figure 6) from the top of Hole 704A. From 4.8 to 3.2 Ma, benthic $\delta^{18}O$ values are always less than 3‰ (the Holocene value) and exhibit low variability with a range of about 0.5‰ (Figure 6). Similarly, $\delta^{18}O$ values of G. bulloides are also generally less than those of the Holocene (Figure 6), except for the interval from ~4.1 to 3.9 Ma when $\delta^{18}O$ values exceed 2.5‰. Minimum planktic $\delta^{18}O$ values are also 0.5 to 0.6‰ less than those of the Holocene.

Benthic $\delta^{18}O$ values first exceed those of the Holocene during the Mammoth Subchron at 3.16 Ma (Figure 6). After this event, glacial stages exceeding 3‰ are common, and the $\delta^{18}O$ signal is marked by higher mean values and greater amplitude fluctuations than the preceding period (Figure 6). Distinct glacial/interglacial cycles are evident during the late Gauss with an amplitude of ~0.6‰ (Figure 6). Mean benthic $\delta^{18}O$ values increase again at ~2.75 Ma, and values commonly exceed 3.5‰ thereafter. The greatest benthic $\delta^{18}O$ values of the Gilbert and Gauss chrons occur at 2.6 Ma, just below a hiatus that removed the latest Gauss and earliest Matuyama in Site 704 (Figure 6).

During the late Gauss the $\delta^{18}O$ record of G. bulloides shows a pattern similar to the Cibicides signal (Figure 6). Planktic $\delta^{18}O$ values progressively increase between 2.8 and 2.6 Ma and commonly exceed 3.0‰ after 2.7 Ma. Across the Gauss/Matuyama transition, mean $\delta^{18}O$ values for both planktic and benthic foraminifers are permanently offset from those of the Gilbert and early Gauss chrons. The amplitudes of the signals also increase across the Gauss/Matuyama transition.

An expanded $\delta^{18}O$ record for the Matuyama Chron is presented in Figure 7 with stage identifications according to the isotopic taxonomy of Site 607 [Ruddiman et al., 1989; Raymo et al., 1989]. From 2.3 to 1.8 Ma, the benthic $\delta^{18}O$ record is generally marked by a low-amplitude signal with weak glacial stages. Exceptions include glacial stages 88, 86, 82, and 78, which represent strong glaciations in the Southern Ocean and elsewhere [Raymo et al., 1989; Shackleton et al., 1991]. The true amplitude of the $\delta^{18}O$ signal is probably not reflected in stages 82 and 78 (Figure 7), because these glaciations are marked by very low carbonate content and preservation of Cibicides is sporadic. The carbonate and opal data suggest that these glacial stages are much stronger than indicated by the $\delta^{18}O$ signal (Figure 7). Stage 70 is a strong, double-peaked glacial event in the Southern Ocean. Following this event, $\delta^{18}O$ values decrease abruptly during stage 69, and the magnitude of glacial benthic $\delta^{18}O$ values progressively increase beginning with low values in stage 68 and culminating in maximum

values (>4‰) in stage 58. Stage 58 (1.537 Ma) is a very intense glacial episode in the Southern Ocean and is followed by an abrupt decrease in benthic $\delta^{18}O$ at the stage 58/57 transition. Glacial stage 56 is very subdued at Site 704, as it is in other records [Ruddiman et al., 1989; Shackleton et al., 1991]. Beginning with stage 52 (1.416 Ma), the oxygen isotopic record shows an exceptionally pure 41-kyr signal, and the $\delta^{18}O$ signal becomes highly correlated with other parameters (e.g., $\delta^{13}C$, %opal, %CaCO_3, etc.).

The planktic $\delta^{18}O$ record of N. pachyderma generally parallels the benthic signal except for the interval between 2.3 and 2.0 Ma (Figure 7), which is marked by high-frequency variations in planktic $\delta^{18}O$ that do not correspond to the characteristic 41-kyr cycle of the benthic record. Above stage 78, most all isotopic stages are recognized in Hole 704 and correlated with the oxygen isotopic stratigraphy of Site 607 in the North Atlantic (Figure 7). The amplitude of the planktic $\delta^{18}O$ signal varies considerably during the Matuyama. Exceptionally strong glacial $\delta^{18}O$ events occur during stages 62, 58, 54, 52, 50, 48, 46, 38, and 36. Again, we suggest that the magnitude of the planktic $\delta^{18}O$ maxima is underestimated for stages 82 and 78, because of sporadic preservation of N. pachyderma during these exceptionally low carbonate intervals. From ~1.45 to 0.95 Ma, the planktic $\delta^{18}O$ signal is marked by a particularly pure 41-kyr cycle. After stage 30 the signal amplitude is reduced during the Jaramillo Subchron.

Carbon Isotopes

During the Gilbert and Gauss chrons, benthic $\delta^{13}C$ values average ~0.5‰ and range from 0 to +1‰ (Figure 8). During the late Gauss the minima in $\delta^{13}C$ values become progressively more extreme such that the lowest $\delta^{13}C$ values are recorded just prior to the hiatus spanning the Gauss/Matuyama boundary. Across the boundary, mean benthic $\delta^{13}C$ values decrease by ~0.5‰, and the signal exhibits higher-frequency variation in the lower Matuyama than in the preceding Gauss or Gilbert chrons (Figure 8). At ~1.6 Ma, mean $\delta^{13}C$ values decrease again by another ~0.5‰.

In order to compare the planktic $\delta^{13}C$ records, carbon isotopic values were corrected to the $\delta^{13}C$ of ΣCO_2 by adding 1‰ to the $\delta^{13}C$ of N. pachyderma and 1.53‰ to the $\delta^{13}C$ of G. bulloides. The $\delta^{13}C$ signal during the Gilbert Chron is marked by large, long-term variations in $\delta^{13}C$. For example, low carbon isotopic values occur from 4.8 to 4.6 Ma, from 4.3 to 4.15 Ma, and from 3.9 to 3.4 Ma, whereas relatively high $\delta^{13}C$ values occur from 4.6 to 4.35 Ma and from 4.15 to 3.9 Ma. Unlike the benthic $\delta^{13}C$ record, there is no decrease in planktic $\delta^{13}C$ values across the Gauss/Matuyama transition.

The detailed carbon isotopic record for the Matuyama Chron is shown in Figure 9. The benthic $\delta^{13}C$ signal shows a clear pattern of $\delta^{13}C$ minima associated with

ODP Site 704

$\delta^{18}O$ (o/oo, pdb)

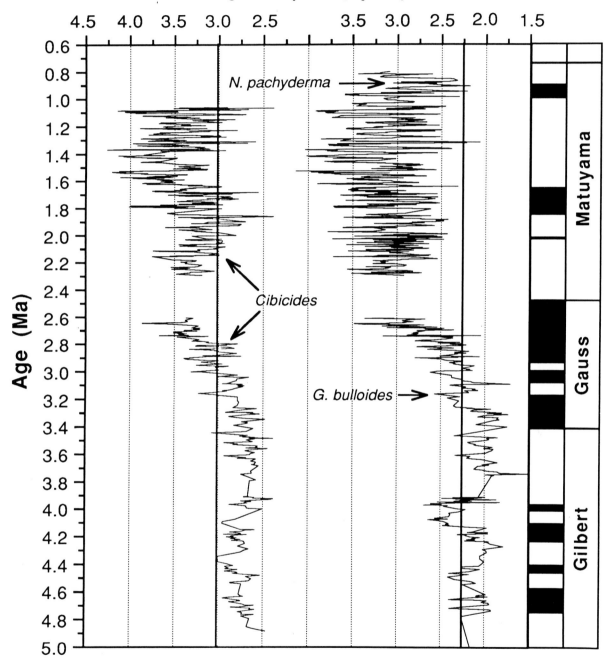

Fig. 6. Oxygen isotopic results of the benthic (*Cibicides*) and planktic (*G. bulloides* and *N. pachyderma*) foraminifers for the Gilbert through Matuyama chrons in Site 704. Oxygen isotopic results are relative to Pee Dee Belemnite (PDB) and are not corrected for species departure from equilibrium. The bold vertical lines represent the measured Holocene $\delta^{18}O$ value for the respective taxa at the top of Hole 704A. Magnetostratigraphy is after *Hailwood and Clement* [1991].

ODP Site 704

$\delta^{18}O$ (o/oo, pdb)

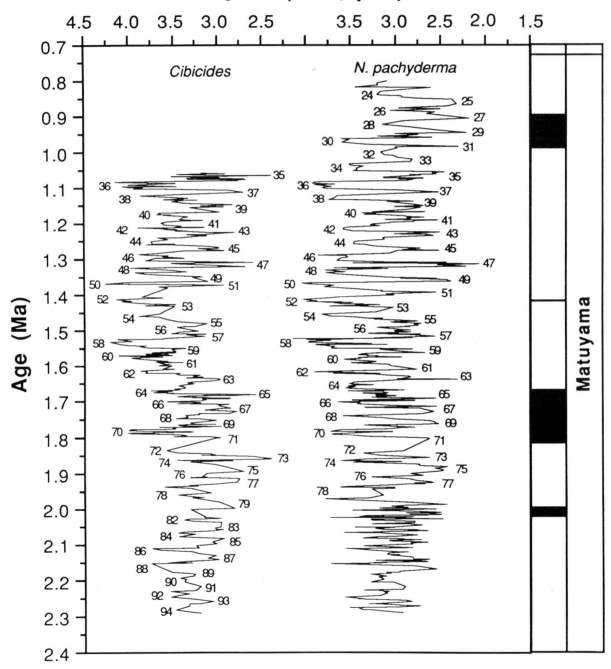

Fig. 7.   Expanded oxygen isotopic record for *Cibicides* and *N. pachyderma* during the Matuyama Chron. Isotopic stage numbering is according to the taxonomy of *Raymo et al.* [1989] and *Ruddiman et al.* [1989].

$\delta^{18}O$ maxima (glacial stages) throughout the entire length of the record. The oxygen isotopic stage designations are provided for comparison (Figure 9). The character of the benthic $\delta^{13}C$ record clearly changes beginning with stage 52 (1.416 Ma) when benthic $\delta^{13}C$ values are commonly less than $-1‰$ during glacial stages after 1.42 Ma and exhibit a strong 41-kyr cyclicity (Figure 9).

ODP Site 704
$\delta^{13}C$ (o/oo, pdb)

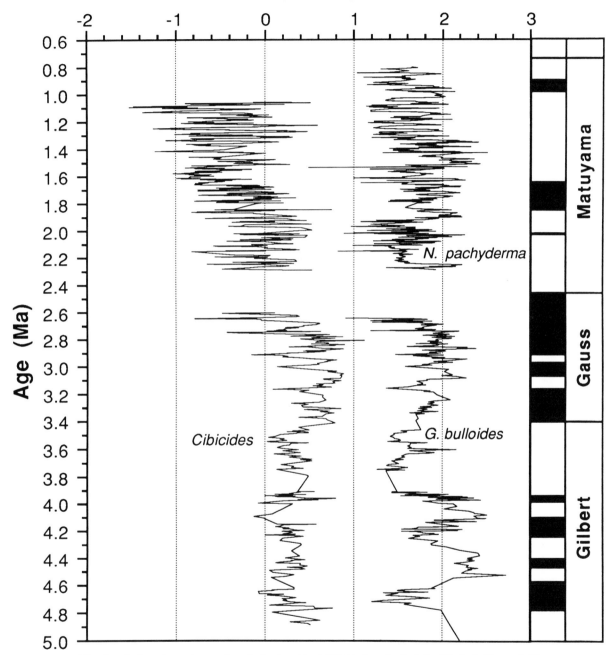

Fig. 8. Carbon isotopic results of the benthic (*Cibicides*) and planktic (*G. bulloides* and *N. pachyderma*) foraminifers for the Gilbert through Matuyama chrons in Site 704. Planktic $\delta^{13}C$ values have been corrected to the $\delta^{13}C$ of ΣCO_2 by adding 1.0‰ to *N. pachyderma* data and 1.53‰ to *G. bulloides* data. No correction was applied to *Cibicides* data. Magnetostratigraphy is after *Hailwood and Clement* [1991].

ODP Site 704
$\delta^{13}C$ (o/oo, pdb)

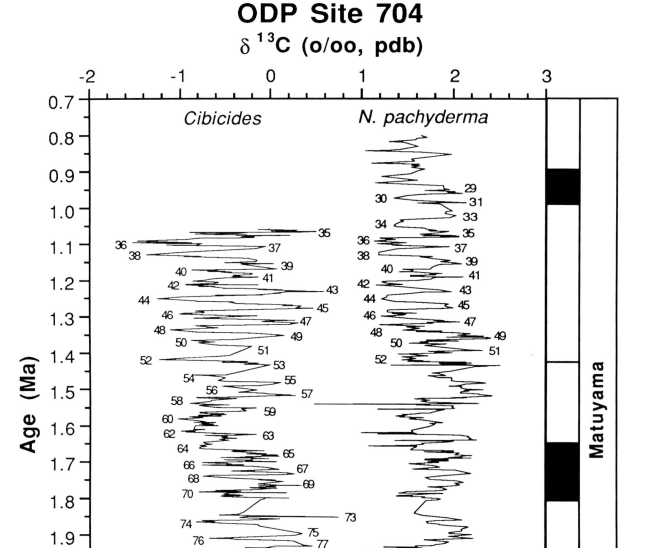

Fig. 9. Expanded carbon isotopic record for *Cibicides* and *N. pachyderma* during the Matuyama Chron. Carbon results of *N. pachyderma* have been corrected to the $\delta^{13}C$ of $\sum CO_2$ by adding 1.0‰. Stage numbers are those defined by oxygen isotopic record in Figure 7.

The planktic $\delta^{13}C$ record during the Matuyama does not always correlate with the benthic $\delta^{13}C$ signal (Figure 9). This is especially true for the early Matuyama when planktic $\delta^{13}C$ variation bears little resemblance to the benthic signal between ~2.3 and 1.5 Ma. Beginning with stage 52 (1.416 Ma), however, the planktic $\delta^{13}C$ signal becomes highly correlated with the benthic one, and distinct glacial-interglacial cycles are evident with a periodicity of 41 kyr (Figure 9).

Calcium Carbonate

The early Gilbert was marked by very low carbonate values (<20%) that represent the tail end of a carbonate dissolution event that lasted from 5.5 to 4.8 Ma (Figure 10) [*Froelich et al.*, 1991; *Müller et al.*, 1991]. During the Gilbert and Gauss chrons, carbonate content was fairly high, averaging 72.5%. During this period, intervals of relatively low carbonate values include 4.87–4.7, 4.5–4.45, 4.1–3.9, 3.13, 2.9, 2.65, and 2.6 Ma (Figure 10). These carbonate lows correspond with maxima in the percent opal record [*Froelich et al.*, 1991]. Times of notably high carbonate values include 4.7–4.1, 3.7–3.25, 3.3–3.27, and 2.85–2.83 Ma (Figure 10).

Across the Gauss/Matuyama transition, a dramatic increase occurs in the bulk-accumulation rates of all sedimentary components including carbonate, opal, and ice-rafted debris [*Froelich et al.*, 1991; *Warnke and Allen*, 1991; *Allen and Warnke*, 1991]. In Figure 11 the carbonate signal for the Matuyama (2.3 to 0.8 Ma) was filtered using a five-point running mean to smooth the high-density carbonate signal. The first strong carbonate minimum (opal maximum) coincides with glacial stage 82 (Figure 11). Glacial stages 78 and 70 are also marked by particularly low carbonate values. At 1.67 Ma (stage 65/64 transition), an abrupt decrease occurs in carbonate content from ~70 to 15%. This decrease coincides with an increase in opal [*Froelich et al.*, 1991] and an increase in organic carbon content of the sediment [*Ciesielski et al.*, 1988, p. 651]. These lithologic changes are reflected in the physical properties of the sediments by a decrease in grain and wet bulk density, an increase in water content, and increased porosity [*Meinert and Nobes*, 1991]. These changes in physical properties are also apparent in the geophysical logging data at ~100 mbsf [*Mwenifumbo and Blangy*, 1991]. Carbonate content remains exceptionally low from stage 64 to stage 58 and then increases (Figure 11). Beginning with stage 50 (1.371 Ma), the carbonate signal begins to show very strong 41-kyr cyclicity (Figure 11).

DISCUSSION

Because of the long length of the isotopic and carbonate records, the discussion is divided into three parts: (1) the Gilbert and early Gauss chrons (4.8 to 3.2 Ma),

(2) the late Gauss and Gauss/Matuyama transition (3.2 to 2.3 Ma), and (3) the Matuyama Chron (2.3 to 0.8 Ma).

Gilbert and Early Gauss Chrons (4.8 to 3.2 Ma)

During the Pliocene prior to ~3.0 Ma, global climate is generally considered to have been warmer than today, and various mechanisms have been proposed to explain these "warmer-than-present" climates [*Crowley*, 1991]. Glaciers and sea ice were greatly reduced or absent in the northern hemisphere, and ice volume fluctuations were largely restricted to Antarctica. *Hodell and Warnke* [1991] reviewed the evidence for early Pliocene warmth in the Southern Ocean, which included the invasion of low- and mid-latitude species into the Southern Ocean between 4.8 and 4.0 Ma [*Ciesielski and Weaver*, 1974; *Keany*, 1978; *Abelmann et al.*, 1990]. Although it is generally agreed that the Southern Ocean was warmer than today during parts of the Pliocene, widely diverging opinions exist concerning the dynamics of the Antarctic ice sheet during this period. At one extreme, a minimum ice configuration has been proposed with much of East Antarctica deglaciated based upon the occurrence of marine diatoms and *Nothofagus* floras in the Sirius till of the Transantarctic Mountains [*Webb and Harwood*, 1991, and references therein]. An alternate view suggests that the Antarctic ice sheet has remained relatively stable in its present form since the late Miocene [*Clapperton and Sugden*, 1990, and references therein] or may have increased in size under warmer climates. *Prentice and Matthews* [1991] proposed that warmer temperatures should increase moisture supply and precipitation over the continent, leading to ice growth via the "snow-gun" effect.

The oxygen isotopic data presented here provide some constraints on the magnitudes of temperature and ice volume change during the Pliocene. Warming and deglaciation produce unidirectional trends in oxygen isotopes and both act to decrease $\delta^{18}O$. At most, the planktic and benthic $\delta^{18}O$ values during the Pliocene were 0.5 to 0.6‰ less than the Holocene (Figure 6). This can be interpreted as either that temperatures were up to 2.5°C warmer than today or that ice volume was less than today. It is most likely that the $\delta^{18}O$ signal reflects a combination of warmer temperature and decreased ice volume. *Webb and Harwood* [1991] proposed than Antarctic ice volume was reduced to 1/3 the present size of the ice sheet during the early Pliocene and that sea surface temperatures warmed considerably (>5°C) at high latitudes. The magnitude of these combined ice volume and temperature effects on $\delta^{18}O$ would be more than 1.5‰. Because planktic and benthic $\delta^{18}O$ values are at most only 0.6‰ less than the Holocene during the Pliocene, the oxygen isotopic data from Site 704 do not support such extensive warming and deglaciation of Antarctica as proposed by *Webb and Harwood* [1991].

It is difficult to evaluate the ice sheet growth hypoth-

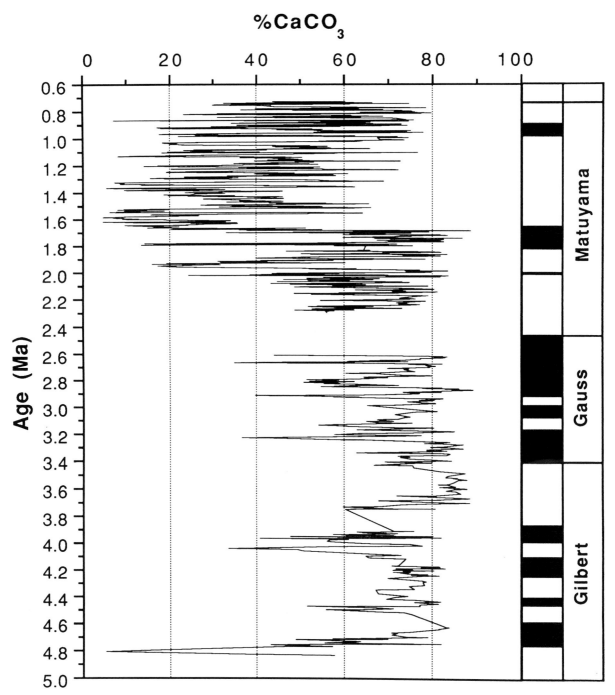

Fig. 10.   Variations in the percent $CaCO_3$ of the sediments of Site 704 during the Gilbert through Matuyama chrons. Carbonate data are after *Froelich et al.* [1991], and magnetostratigraphy is after *Hailwood and Clement* [1991].

esis using oxygen isotopic data alone because the "snow gun hypothesis" results in opposing effects on the $\delta^{18}O$ signal [*Prentice and Matthews*, 1991]. Increased sea surface temperatures would decrease $\delta^{18}O$, whereas increased ice volume would increase $\delta^{18}O$. This would result in a damped $\delta^{18}O$ signal during times

of warming and ice growth. Estimates from faunal analyses suggest that sea surface temperatures were 5° to 10°C warmer during the early Pliocene than today [*Ciesielski and Weaver*, 1974; *Abelmann et al.*, 1990]. The lowest planktic $\delta^{18}O$ values were only 0.5–0.6‰ less than today, which translates into ~2.0°–2.5°C of

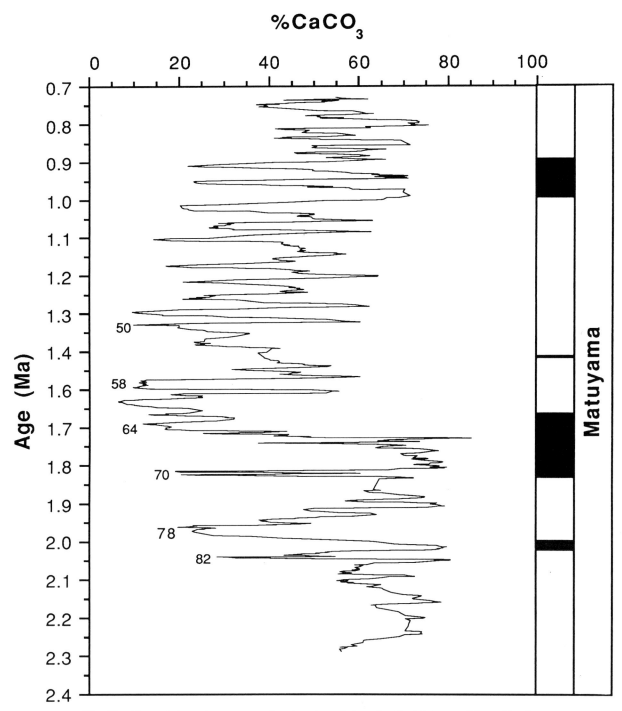

Fig. 11.   Five-point running average of carbonate variations for the Matuyama Chron. Numbers represent selected oxygen isotope stages.

warming. This implies that either the faunal data over-estimate temperature or the isotopic data underestimate the degree of warming. This discrepancy might be explained by the snow gun effect if the decrease in $\delta^{18}O$ values expected from 5°–10°C warming was offset by an

increase in the $\delta^{18}O$ of seawater due to increased ice volume during the early Pliocene.

On the basis of the oxygen isotopic record of Site 704, we suggest that neither scenario (i.e., massive deglaciation nor ice sheet buildup larger than today) is likely to have occurred during the Pliocene Epoch prior to 3.2 Ma. The low mean $\delta^{18}O$ values (i.e., less than the Holocene) and dampened amplitude of the benthic $\delta^{18}O$ signal suggest that temperatures were warmer than today in the Southern Ocean during the Pliocene, but the climate system operated within relatively narrow limits (±0.5‰). It is our view that the Antarctic cryosphere system did not fluctuate widely during the Pliocene prior to 3.2 Ma. The oxygen isotopic signal can accommodate some warming and minor deglaciation of Antarctica, but it is inconsistent with major warming (>5°C warmer than today) and massive deglaciation (2/3 less ice volume than today) of the Antarctic ice sheet. Likewise, it is also unlikely than Antarctic ice volume was much greater than today during the Pliocene prior to 3.2 Ma.

This view is also supported by *Kennett and Barker* [1990], who visualized more or less stable ice shelves in West Antarctica during the Pliocene after 4.8 Ma based upon the cessation of turbidity-current activity in the Weddell Sea. Other authors have also proposed that cold, polar desert conditions permanently set in during the late Miocene and the Antarctic glacier system did not fluctuate on a large scale until northern hemisphere glaciation lowered sea level during the late Pliocene [*Denton et al.*, 1983, 1984; *Clapperton and Sugden*, 1990, and references therein].

Although planktic and benthic $\delta^{18}O$ values are generally less than those of the Holocene from 4.8 to 3.2 Ma, the signals are marked by distinct maxima and minima implying low-amplitude, glacial-interglacial cycles. Two periods of particular interest are the "higher-than-average" planktic $\delta^{18}O$ values between 4.1 and 3.9 Ma and "lower-than-average" $\delta^{18}O$ values between ~3.7 and 3.3 Ma. Between 4.1 and 3.9 Ma, planktic $\delta^{18}O$ values were clearly greater than the Holocene, and carbonate content was relatively low. *Froelich et al.* [1991] suggested that the PFZ was likely at a more northerly position during this interval. This period roughly correlates with a sea level lowstand (between 4.0 and 3.7 Ma [*Haq et al.*, 1987]) and with the widespread occurrence of fragments of the giant diatom *Ethmodiscus* in the Weddell Sea [*Abelmann et al.*, 1990].

An interval of consistently low $\delta^{18}O$ values of planktic and benthic foraminifers is recorded between ~3.7 and 3.3 Ma during the middle Pliocene (Figure 6). This interval is marked by high carbonate and low opal accumulation, suggesting that the PFZ was stationary at a far southerly position [*Froelich et al.*, 1991]. A similar conclusion was reached by *Westall and Fenner* [1991] based upon studies of Site 699. The middle Pliocene has been described elsewhere as a period of global warmth when sea level was generally higher than today and sea surface temperatures were warmer than today in the North Atlantic [*Cronin and Dowsett*, 1991, and references therein].

Late Gauss and the Gauss/Matuyama Transition (3.2 to 2.3 Ma)

The interval from the middle to late Gauss (~3.2 to 2.4 Ma) represents a transitional period linking the generally warmer conditions before 3.2 Ma with the colder, high-amplitude climatic variations after 2.4 Ma. Prior to ~2.6 Ma, the cryosphere is believed to have been largely unipolar and restricted to Antarctica [*Hodell and Warnke*, 1991]. The lack of significant ice-rafted debris in the high-latitude North Atlantic prior to 2.57 Ma argues against large continental ice sheets on the northern hemisphere, although glaciers had reached the ocean and produced some ice rafting before that time [*Jansen and Sjøholm*, 1991]. It was not until the late Pliocene (~2.4 Ma) that large ice sheets developed on the northern hemisphere and the cryosphere became truly bipolar. Study of the transitional late Gauss period (3.2 to 2.5 Ma) is important, therefore, for understanding the events that led up to the onset of major northern hemisphere glaciation at ~2.4 Ma [*Shackleton et al.*, 1984a].

Records of ice-rafted debris and oxygen isotopes from the North Atlantic suggest than northern hemisphere glaciation developed gradually over several hundred thousand years during the late Pliocene [*Keigwin*, 1986; *Raymo et al.*, 1989]. *Prell* [1985] showed that the Pliocene $\delta^{18}O$ record of Hole 572 was marked by two modes of variability with a boundary at ~2.9 Ma. *Keigwin* [1986] noted a sequential enrichment in $\delta^{18}O$ values at ~3.1, 2.7, and 2.6 Ma, culminating in a 30% increase in signal amplitude at 2.5–2.4 Ma near the Gauss/Matuyama boundary. It is this date of 2.4 Ma that has been generally accepted as the onset of northern hemisphere glaciation [*Shackleton et al.*, 1984a], but this event is better viewed as an increase in the variability of ice volume that permitted delivery of ice-rafted debris (IRD) to the open North Atlantic [*Raymo et al.*, 1989]. The 2.4-Ma event has been shown to consist of a package of three strong glacial events at 2.404, 2.362, and 2.317 Ma with a characteristic period of 41 kyr [*Raymo et al.*, 1989; *Sikes et al.*, 1991].

Whereas the late Pliocene climatic transition is well known from the northern hemisphere (*Shackleton et al.* [1984a], *Keigwin* [1986], *Raymo et al.* [1989], and *Jansen and Sjøholm* [1991], among others), the nature of this event is poorly understood from the high latitudes of the southern hemisphere. The isotopic record from Site 704 provides an opportunity to compare the interhemispheric relationships of climatic and oceanographic changes during the late Pliocene transition.

In Site 704, benthic $\delta^{18}O$ values first exceed the Holocene at 3.2 Ma in the Mammoth Subchron of the Gauss (Figure 6). This event has been recognized previously in several oxygen isotopic records and interpreted as representing bottom water cooling coupled with minor glaciation [*Keigwin*, 1986; *Prell*, 1984; *Hodell et al.*, 1985]. A further increase in mean planktic and benthic $\delta^{18}O$ values occurs at ~2.75 Ma (Figure 6), suggesting a shift in climate mode during the latest Gauss. A similar shift in the pattern of climate variability has been reported at ~2.73 Ma from Site 806B from the Ontong Java Plateau [*Jansen et al.*, 1992]. This $\delta^{18}O$ increase at ~2.75 Ma is also evident in the benthic oxygen isotopic record of Site 607 [*Raymo et al.*, 1989, 1992]. Following this increase in Site 704, the greatest planktic and benthic $\delta^{18}O$ values of the Gauss and Gilbert chrons are recorded between 2.6 and 2.7 Ma, just below a hiatus that removed the latest Gauss and earliest Matuyama (Figure 6).

Hodell and Ciesielski [1991, p. 415] cautioned about the likelihood of a hiatus and/or missing section in Site 704 at the Gauss/Matuyama boundary, but the duration of the missing section was unknown. We estimate that the hiatus spans the interval from ~2.6 to 2.3 Ma and is approximately 300 kyr in duration. Missing in the hiatus are the three prominent oxygen isotopic events (stages 100, 98, and 96) that mark the "onset" of northern hemisphere glaciation.

Recognition of this hiatus in Site 704 provides several new insights into the timing of paleoceanographic changes in the Southern Ocean. *Hodell and Ciesielski* [1990, 1991] reported that the time of greatest paleoceanographic change in Site 704 was coincident with the Gauss/Matuyama boundary at 2.47 Ma. It is now apparent that the timing of many of these paleoceanographic changes began earlier during the latest Gauss at ~2.75 Ma and continued across the Gauss/Matuyama transition. Among the changes observed during this interval were the following: (1) an order of magnitude increase in accumulation rates of biogenic sedimentary components [*Froelich et al.*, 1991], (2) a several-fold increase in the accumulation of IRD [*Warnke and Allen*, 1991; *Allen and Warnke*, 1991], (3) a 0.5‰ increase in mean $\delta^{18}O$ values of planktic and benthic foraminifers and an increase in signal amplitude (Figure 6), and (4) a 0.5‰ decrease in average benthic $\delta^{13}C$ values (Figure 8). These changes have been interpreted previously as reflecting a northward advance of the PFZ, an increase in carbonate and siliceous productivity, an increase in ice volume on Antarctica, and a decrease in the ventilation rate of deep water in the Southern Ocean [*Hodell and Ciesielski*, 1990, 1991].

These data from Site 704 are consistent with faunal interpretations from Site 514 (western Subantarctic South Atlantic), suggesting that the latest Gauss (2.68 to 2.47 Ma) was the time of greatest change in Neogene climate of the northern Antarctic and Subantarctic sectors of the Southern Ocean [*Ciesielski and Grinstead*, 1986]. At Site 514, radiolarian assemblages indicate that cooling began at ~2.83 Ma and culminated with the northward advance of Antarctic surface waters over Site 514 by 2.67 Ma [*Ciesielski and Grinstead*, 1986]. This migration of the PFZ at 2.67 Ma coincided with maximum accumulation rates of IRD (200 mg/cm²/10³ yr) in Hole 514 [*Bornhold*, 1983]. The highest Antarctic factor loadings indicate that maximum cooling oc-

curred between 2.58 and 2.47 Ma [*Ciesielski and Grinstead*, 1986]. Accompanying this cooling during the latest Gauss was an intensification of deepwater circulation that resulted in increased erosion throughout much of the Southern Ocean [*Ledbetter and Ciesielski*, 1986]. This intensification of circulation apparently resulted in erosion or nondeposition at Site 704 between 2.6 and 2.3 Ma.

Results from sites 704 and 514 suggest that major paleoceanographic changes in Southern Ocean paleoceanography began at ~2.7 Ma in the late Gauss. *Ciesielski and Grinstead* [1986] suggested these climatic changes in the Southern Ocean preceded the onset of northern hemisphere glaciation (NHG) by several 100 kyr (assuming that NHG began at 2.4 Ma). However, new data from the Norwegian Sea indicate that IRD accumulation increased by 2 to 3 orders of magnitude at 2.57 Ma [*Jansen and Sjøholm*, 1991], which coincides with the beginning of the interval of maximum cooling in the Subantarctic Hole 514 (the highest Antarctic factor loadings are estimated at 2.58, 2.56, 2.52, and 2.47 Ma).

To a first approximation (i.e., 10^5 years), it appears that climatic and oceanographic changes were synchronous between the high-latitude North Atlantic and Southern Ocean. *Hodell and Ciesielski* [1990] reviewed various mechanisms that might explain the interhemispheric coupling of the polar oceans during the climate transition of the late Pliocene. Two important coupling mechanisms include sea level variation and changes in the flux of NADW production. *Ciesielski and Grinstead* [1986] suggested that Antarctic ice sheets reached their maximum configuration during the late Gauss and were unable to expand further because they had reached sea level. In this model, northern and southern hemisphere ice sheets are interlocked via sea level variation. The initiation of ice growth in the northern hemisphere at 2.57 Ma would have produced a gradual lowering of sea level that would have triggered grounding ice advance and thickening of the marine-based parts of the Antarctic ice sheet [*Denton and Hughes*, 1983]. Support for this mechanism comes from the synchronous increase in $\delta^{18}O$ and ice-rafted detritus accumulation in Site 704, which suggests that the climatic transition during the late Pliocene involved ice volume accumulation on both Antarctica and northern hemisphere continents [*Hodell and Ciesielski*, 1991]. Results from ODP Leg 113 (Weddell Sea) also support major ice growth on Antarctica and the expansion of sea ice during the latest Gauss [*Kennett and Barker*, 1990; *Burckle et al.*, 1990].

Changes in production rate of NADW have also been proposed as an interhemispheric coupling mechanism because NADW provides an important source of heat and salt to the Southern Ocean. On the basis of carbon isotopic gradients between the North Atlantic, Southern Ocean, and Pacific, *Hodell and Ciesielski* [1990, 1991] inferred that the production of NADW decreased with the onset of northern hemisphere glaciation. *Raymo et al.* [1992] have examined these carbon isotopic signals in detail and have concluded that NADW production was strong between 3.0 and 2.7 Ma in the late Pliocene but was considerably reduced during glacial stages after 2.7 Ma. The weakening of NADW during glaciations after 2.7 Ma may have provided strong positive feedback in the climate system that fueled the transition from a unipolar to a bipolar cryosphere.

Matuyama Chron (2.3 to 0.8 Ma)

The climatic response of the high latitudes of the northern hemisphere during the Matuyama Chron has been shown to be dominated by the 41-kyr cycle of orbital obliquity [Ruddiman et al., 1986, 1989]. We have identified most oxygen isotopic stages during the Matuyama according to the isotope taxonomy of Site 607 [Ruddiman et al., 1989; Raymo et al., 1989; Figure 7]. During the early Matuyama from 2.3 to 2.1 Ma, the planktic $\delta^{18}O$ record displays high-frequency variations but does not show the characteristic 41-kyr signal of the benthic record (Figure 7). This interval coincides with low-amplitude ice volume changes and the weak glacial stages 94, 92, and 90 in Site 607 [Raymo et al., 1989]. Froelich et al. [1991] suggested that the PFZ was migrating far southward during this period, based upon low opal and high carbonate accumulation rates. It is possible that the planktic record was more affected by the high-frequency precessional cycle during this relatively warm period during the early Matuyama.

Stages 82 (2.027 Ma) and 78 (1.941 Ma) were very strong glacial events in the Southern Ocean and represent the two strongest glacial events of the early Matuyama Chron in Site 607 [Raymo et al., 1989]. The magnitudes of these $\delta^{18}O$ maxima are probably subdued in Site 704 because of the lack of foraminifers in the extremely low-carbonate intervals. Changes in faunal assemblages, high opal content, and low carbonate accumulation suggest a northerly position of the PFZ during these strong glacial stages [Froelich et al., 1991]. With the exception of stage 70 (1.782 Ma), glacial stages were relatively weak between 1.9 and 1.7 Ma. Stage 70 was a strong, double-peaked glacial event in the Southern Ocean (Figure 11).

At 1.67 Ma (stage 65/64 transition), a major change occurred in lithologic and faunal parameters in Site 704, indicating a far northerly position of the PFZ (Figure 11). Between 1.67 and 1.5 Ma (stages 65 to 57), carbonate percentages are low and the sediments are dominated by biogenic silica, but the mass accumulation rate of biogenic sedimentary components remains low during this period [Froelich et al., 1991]. The organic carbon content of the sediments increased at 1.67 Ma [Ciesielski et al., 1988, p. 651], and floral and faunal assemblages indicate an increased abundance of upwelling diatoms (e.g., Thalassionema nitzschioides) [Ciesielski, 1991]. At 1.67 Ma the planktic foraminiferal assemblage becomes dominated by the polar form N. pachyderma (sinistral)

[Brunner, 1991]. The delivery of IRD to Site 704 was greatly reduced beginning at 1.67 Ma [Warnke and Allen, 1991], indicating that either the zone of most rapid melting of icebergs was located north of Site 704 or the efficiency of Antarctic glaciers as agents of erosion was reduced [Warnke et al., this volume]. The combined evidence suggests a prolonged northerly position of the PFZ between 1.67 and 1.5 Ma, such that Site 704 may have been positioned in the region south of the PFZ [Froelich et al., 1991; Warnke et al., this volume].

Beginning at ~1.42 Ma (stage 52), all parameters measured in Site 704 begin to covary and display a very strong 41-kyr signal. Benthic and planktic $\delta^{18}O$ records show high covariance during this interval (Figure 7), and there is a clear pattern of carbonate minima (opal maxima) associated with $\delta^{18}O$ maxima, which is typical of an "Atlantic-type" carbonate stratigraphy (Figure 11). Planktic and benthic $\delta^{13}C$ signals also begin to covary at 1.4 Ma (Figure 9) such that minima in $\delta^{13}C$ correspond with maxima in $\delta^{18}O$. This pattern is typical of that observed during the late Pleistocene of the Southern Ocean [Charles and Fairbanks, 1990].

Other studies have also reported a similar increase in the importance of the 41-kyr cycle beginning at ~1.5 to 1.4 Ma [Pisias and Moore, 1981; Ruddiman et al., 1989]. Based upon time series analysis of isotopic data from V28-239, Pisias and Moore [1981] divided the climate signal of the last 2 million years into three intervals: late Pleistocene (0.9 Ma to present), middle Pleistocene (1.45 to 0.9 Ma), and late Pliocene–early Pleistocene (2.0 to 1.45 Ma). The middle Pleistocene was dominated by a concentration of variance between 60 and 20 kyr, presumably at 41 kyr [Pisias and Moore, 1981]. In Site 607, Ruddiman et al. [1989] noted that the interval between 1.48 and 1.07 Ma displayed an exceptionally pure 41-kyr signal, similar to the oxygen isotopic records of Site 704 (Figure 7).

The simplest explanation for an increase in 41-kyr power between 1.4 and 1.0 Ma would be a linear response to intensified insolation forcing by the Earth's tilt cycle. There is a progressive increase in the amplitude of the 41-kyr signal during this interval [Berger and Loutre, 1988], such that the amplitude was relatively low during latest Pliocene (2.2 to 1.8 Ma) and gradually increased during the early Pleistocene (Figure 12). For example, minimal tilt values were less than 22.5° during 8 out of 12 cycles between 1.5 and 1.0 Ma, whereas tilt was less than 22.5° only once out of 16 cycles between 2.2 and 1.5 Ma. The alternative time scale of Shackleton et al. [1991] would give a different correlation between $\delta^{18}O$ and the tilt cycle than the "TP607" time scale used here. Future work will focus on directly testing the correlation between the $\delta^{18}O$ signal and the amplitude modulation of the obliquity cycle using both time scales.

To reconstruct patterns of deepwater circulation during the Matuyama, we compare the benthic carbon isotopic record of Site 704 (Southern Ocean) with similar records from Site 607 (North Atlantic) and Site 677

Obliquity

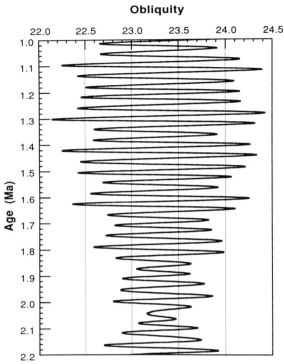

Fig. 12.   Variations in the Earth's obliquity cycle during the Matuyama Chron after *Berger and Loutre* [1988].

(Pacific) (Figure 13). Carbon isotopic gradients between the North Atlantic, Southern Ocean, and Pacific provide a means of monitoring the relative flux of NADW to the Southern Ocean [*Oppo and Fairbanks*, 1987]. Today, the distribution of $\delta^{13}C$ in deep waters of the Atlantic is controlled mainly by mixing processes where high $\delta^{13}C$ (>1‰) of NADW mixes with low $\delta^{13}C$ (0.4–0.5‰) of waters formed in the Southern Ocean. The deep waters of the Southern Ocean have $\delta^{13}C$ values that are intermediate between those of the North Atlantic and the Pacific, reflecting a mixture of waters derived from both North Atlantic and Indo-Pacific sources. For example, the modern carbon isotopic values at these sites are 0.9‰ at Site 607, 0.64‰ at Site 704, and −0.05‰ at Site 677. Today, the North Atlantic–Pacific gradient is ~1‰, the North Atlantic–Southern Ocean gradient is ~0.3‰, and the Southern Ocean–Pacific gradient is ~0.7‰.

In Figure 13 the carbon isotopic signals of *Cibicides* in Site 607 (North Atlantic, 3427 m) and Site 704 (Southern Ocean, 2532 m) are compared with the $\delta^{13}C$ record of Site 677 (eastern equatorial Pacific, 3461 m). The isotopic data from Site 677 are based primarily on analysis of *Uvigerina* [*Shackleton and Hall*, 1989], and carbon isotopic values were corrected to equilibrium by adding 0.9‰ [*Shackleton et al.*, 1984b]. Throughout the Matuyama, $\delta^{13}C$ values of North Atlantic Site 607 are always greater than either Site 704 or Site 677 (Figure 13), indicating that NADW was being produced

throughout this interval even during glacial periods [*Raymo et al.*, 1990]. From 2.3 to 1.8 Ma, $\delta^{13}C$ values of Site 704 fall about midway between North Atlantic Site 607 and Pacific Site 677. From 1.8 to 1.45 Ma, $\delta^{13}C$ values from Southern Ocean Site 704 overlap those of Pacific Site 677. Stage 52 (1.416 Ma) marks a change in the character of all three benthic $\delta^{13}C$ records (Figure 13). At this time, a strong depletion occurs in Site 607, and benthic $\delta^{13}C$ values in Site 704 fall below −1‰. From 1.42 to 1.0 Ma, $\delta^{13}C$ values in the Southern Ocean are generally less than those of the Pacific during glacial stages and are intermediate between the North Atlantic and the Pacific during interglacial stages (Figure 13).

The changes in carbon isotopic gradients are interpreted as reflecting stronger reductions of NADW beginning with stage 52 (1.416 Ma). This finding agrees with *Raymo et al.* [1990], who interpreted a significant decrease in the production of glacial NADW after ~1.5 Ma based upon carbon isotopic gradients between the North Atlantic and the Pacific. Changes in the production rate of NADW would have altered heat and salt transport to the Southern Ocean. Today, upwelling of warm deep water derived from the North Atlantic is important for the rapid retreat of sea ice during the spring and for basal melting of ice shelves. A reduction of NADW during glaciations, therefore, would decrease heat and salt input to the Southern Ocean and increase the northward extent of sea ice. *Crowley and Parkinson* [1988] modeled the effect of varying production rates of NADW on vertical heat flux of the Southern Ocean and changes in sea ice extent around Antarctica. They concluded that NADW variations can explain about 20–30% of the overall variance in Antarctic sea ice extent between 18 kyr and present. Although their model suggests that the overall influence of NADW may be relatively minor, the linkage was probably more significant at a period of 41 kyr [*Crowley and Parkinson*, 1988]. In addition, various models have predicted that reductions in NADW production may play a role in controlling atmospheric CO_2 levels [*Boyle*, 1988; *Broecker and Peng*, 1989]. We propose that intensified variations in NADW production after 1.5 Ma served as an important interhemispheric link between the polar oceans by directly altering heat and salt flux to the Southern Ocean and perhaps indirectly by influencing atmospheric CO_2. This mechanism might explain the tightly coupled response that developed between the Southern Ocean and the North Atlantic beginning at ~1.42 Ma.

After glacial stage 52, the $\delta^{13}C$ values of Site 704 (Southern Ocean) were often less than those of Site 677 (Pacific) (Figure 13). During some glaciations, Southern Ocean deep water was up to 0.6‰ lighter than Pacific deep water. Taken at face value, this would suggest that during glacial stages, the "oldest" deep water (i.e., furthest from its source) was located in the Southern Ocean and not in the deep Pacific as today. This interpretation is potentially complicated by the fact that

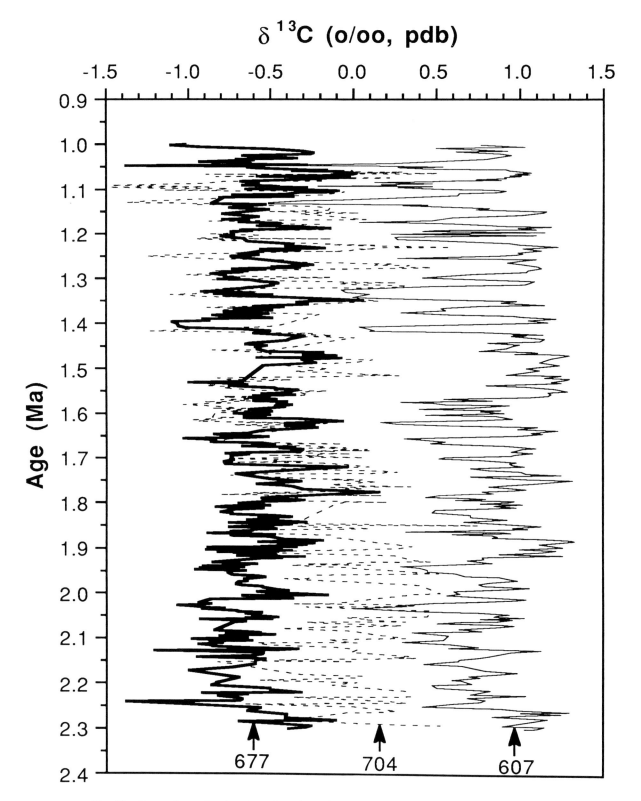

Fig. 13.   Comparison of carbon isotopic values of benthic foraminifers between Site 607 (North Atlantic, 3427 m), Site 704 (Southern Ocean, 2532 m), and Site 677 (eastern equatorial Pacific, 3461 m). Carbon isotopic values from Site 677 were measured on *Uvigerina* and have been corrected to *P. wuellerstorfi/Cibicides* by adding 0.9‰.

the Site 704A record is based on analyses of *Cibicides*, whereas the site 677 record is based largely on specimens of *Uvigerina*. If the relative carbon isotopic disequilibrium between these two species did not remain constant through time, then the negative $\delta^{13}C$ gradient between Holes 704A and 677 could be spurious. *Uvigerina* has an infaunal habitat, and its $\delta^{13}C$ may partly reflect pore water CO_2 that can be influenced by the flux of organic carbon to the sediments [*Zahn et al.*, 1986]. Although this possibility cannot be ruled out, reversals in the Southern Ocean–Pacific $\delta^{13}C$ gradient have also been reported for certain glacial intervals of the late Pleistocene based upon comparison of the same benthic foraminiferal species [*Curry et al.*, 1988; *Oppo et al.*, 1990]. If potential species disequilibrium effects can be ignored, then the Southern Ocean had the most depleted $\delta^{13}C$ values and presumably the highest nutrient and ΣCO_2 levels in the glacial ocean during the early Pleistocene. The negative ^{13}C gradients between the Southern Ocean and the Pacific could be explained by a source area of bottom water formation in the North Pacific as proposed by *Curry et al.* [1988] for the last glaciation. Alternatively, *Boyle* [1990] has suggested that the negative ^{13}C gradients may be related to increased productivity of unusually ^{13}C-depleted Antarctic plankton during glacial intervals. The $\delta^{13}C$ of organic matter in the Antarctic has been shown to be ~8‰ lower than plankton elsewhere in the ocean [*Sackett et al.*, 1965], presumably because of higher concentrations of dissolved aqueous CO_2 in cold Antarctic surface water [*Rau et al.*, 1989]. The high fractionation of Antarctic plankton would increase the efficiency of the biological pumping of ^{12}C from surface to deep water and could drive the $\delta^{13}C$ of Southern Ocean deep water more negative by several tenths of a per mil [*Boyle*, 1990]. No completely satisfactory explanation exists for the reversal in the $\delta^{13}C$ gradient between the Southern Ocean and the Pacific during glacial stages of the late Pleistocene, and investigation into this problem will likely be an important priority for future research.

CONCLUSIONS

New isotopic data and improved chronology of ODP Site 704 have allowed us to refine previous interpretations of the climatic evolution of the Southern Ocean during the Pliocene-Pleistocene [*Hodell and Warnke*, 1991; *Hodell and Ciesielski*, 1990, 1991]. Global climate is generally considered to have been warmer than today during the Pliocene (prior to 3.2 Ma), and the cryosphere is believed to have been unipolar and restricted to Antarctica. At Site 704, benthic $\delta^{18}O$ values were always equal to or less than the Holocene value (3.0‰) prior to 3.2 Ma, suggesting warmer conditions and/or less continental ice volume than today. Planktic $\delta^{18}O$ values were also generally less than the Holocene (2.3‰) during this period (except for the interval from

4.1 to 3.9 Ma). The amplitude of the planktic and benthic $\delta^{18}O$ signals was low (~0.5‰), and minimum values were never more than 0.5–0.6‰ less than those of the Holocene. While the oxygen isotopic record of Site 704 can accommodate some warming and minor deglaciation of Antarctica during the Pliocene, it is inconsistent with scenarios calling for major warming (e.g., >5°C warmer than today) and massive deglaciation (e.g., 2/3 less ice volume) on the Antarctic continent. We suggest that the Antarctic glacier system did not fluctuate on a large scale during the Pliocene (prior to 3.2 Ma) and the climate system operated within relatively narrow limits (i.e., 0.5‰ variation in $\delta^{18}O$).

It was not until the late Pliocene that large ice sheets developed on the northern hemisphere and the cryosphere became truly bipolar. During the late Gauss (~2.7 to 2.47 Ma), the Southern Ocean underwent a major climatic transition in the northern Antarctic and Subantarctic regions. Beginning at ~2.7 Ma, faunal assemblages indicate a northward advance of the PFZ [*Ciesielski and Grinstead*, 1986], accumulation rates of biogenic sedimentary components and ice-rafted debris increase markedly in the Subantarctic sector, oxygen isotopic values increase by ~0.5‰ (Figure 6), and benthic carbon isotopic values decrease by 0.5‰ (Figure 8). Maximum cooling is indicated between 2.58 and 2.47 Ma based upon radiolarian assemblages in Site 514 [*Ciesielski and Grinstead*, 1986]. Intensification of Antarctic circulation associated with this cooling apparently resulted in erosion or nondeposition at Site 704 between 2.6 and 2.3 Ma. Recognition of this hiatus implies that the numerous paleoceanographic changes reported by *Hodell and Ciesielski* [1990, 1991] as occurring at the Gauss/Matuyama boundary (2.47 Ma) may have occurred earlier during the latest Gauss (between 2.7 and 2.47 Ma).

The first significant increase in ice-rafted debris to the high-latitude North Atlantic occurred at 2.57 Ma when IRD accumulation increased by several orders of magnitude [*Jansen and Sjöholm*, 1991]. This increase in IRD accumulation in the Norwegian Sea at 2.57 Ma occurred during the time represented by the hiatus at Site 704 but coincided with the beginning of the highest Antarctic factor loadings in Hole 514 [*Ciesielski and Grinstead*, 1986]. We suggest that strong positive feedback mechanisms between the polar oceans may have contributed to changing the boundary conditions of the Earth's climate system during the late Pliocene. Beginning at ~2.6 Ma, the lowering of sea level by increased ice volume in the northern hemisphere stimulated ice advance along the Antarctic margin and interior with attendant delivery of IRD to the Southern Ocean. In addition, increased glacial suppression of NADW after 2.7 Ma may have decreased the heat and salt flux to the Southern Ocean [*Raymo et al.*, 1992].

The early Matuyama (~2.3 to 1.7 Ma) was generally warm in the Southern Ocean except during stages 82 (2.027 Ma), 78 (1.941 Ma), and 70 (1.782 Ma), which were strong glacial events marked by northerly advances of the PFZ. At 1.67 Ma (stage 65/64 transition), the PFZ migrated far equatorward and oscillated about a northerly position for a prolonged interval between 1.67 and 1.5 Ma (stages 65 to 57). Most parameters in Site 704 show a change in amplitude and frequency at ~1.42 Ma (stage 52), such that the interval between 1.4 and 0.9 Ma exhibits an exceptionally pure 41-kyr signal. This increase in the importance of the 41-kyr cycle between ~1.5 and 0.9 Ma has been noted previously

[*Pisias and Moore*, 1981; *Ruddiman et al.*, 1989] and has been attributed to obliquity forcing. Carbon isotopic gradients between the North Atlantic (Site 607), the Southern Ocean (Site 704), and the Pacific (Site 677) suggest that suppression of NADW intensified greatly during glacial periods after stage 52 (1.416 Ma). This may have served as an important interhemispheric link between the polar oceans and may explain the tightly coupled response that developed between the Southern Ocean and the North Atlantic beginning at 1.42 Ma.

TABLE A1.   Oxygen and Carbon Isotopic Results of *Neogloboquadrina pachyderma* From Site 704

Hole	Core	Section	Interval, cm	Depth, mbsf	Composite depth, m	Age, Ma	$\delta^{18}O$, ‰, pdb	$\delta^{13}C$, ‰, pdb
704B	4H	4	70-72	30.91	30.91	0.801	3.09	0.65
704B	4H	4	105-107	31.26	31.26	0.806	3.19	0.72
704B	4H	4	120-122	31.41	31.41	0.808	3.20	0.60
704B	4H	4	145-147	31.66	31.66	0.812	3.19	0.59
704B	4H	5	30-32	32.01	32.01	0.818	3.44	0.30
704B	4H	5	38-42	32.10	32.10	0.819	2.61	0.46
704B	4H	5	70-72	32.41	32.41	0.831	3.16	0.62
704B	4H	5	105-107	32.76	32.76	0.839	3.20	0.42
704B	4H	5	120-122	32.91	32.91	0.841	3.15	0.31
704B	4H	5	145-147	33.16	33.16	0.843	2.93	0.04
704B	4H	6	30-32	33.51	33.51	0.846	2.93	0.63
704B	4H	6	70-72	33.91	33.91	0.849	2.75	0.84
704B	4H	6	105-107	34.26	34.26	0.853	2.37	0.98
704B	5H	1	15-17	35.36	35.36	0.866	2.32	0.54
704B	5H	1	51-52	35.71	35.71	0.872	2.42	0.58
704B	5H	1	73-75	35.94	35.94	0.876	2.96	0.11
704B	5H	1	93-95	36.14	36.14	0.880	2.50	0.63
704B	5H	1	120-122	36.41	36.41	0.883	2.86	0.54
704B	5H	1	125-127	36.46	36.46	0.884	3.05	0.64
704B	5H	2	15-17	36.86	36.86	0.889	2.56	0.54
704B	5H	2	51-52	37.21	37.21	0.892	2.65	0.69
704B	5H	2	93-95	37.64	37.64	0.897	2.51	0.62
704B	5H	3	15-17	38.36	38.36	0.906	2.18	0.53
704B	5H	3	51-52	38.71	38.71	0.913	2.90	0.22
704B	5H	3	93-95	39.14	39.14	0.923	3.14	0.61
704B	5H	3	125-127	39.46	39.46	0.930	2.83	0.15
704B	5H	4	15-17	39.86	39.86	0.939	2.56	0.89
704B	5H	4	51-52	40.21	40.21	0.946	2.21	0.90
704B	5H	4	93-95	40.64	40.64	0.949	3.00	0.96
704B	5H	4	125-127	40.96	40.96	0.951	2.74	0.69
704B	5H	5	15-17	41.36	41.36	0.953	2.87	1.02
704B	5H	5	51-52	41.71	41.71	0.956	3.19	0.93
704B	5H	5	93-95	42.14	42.14	0.958	2.59	1.10
704B	5H	5	125-127	42.46	42.46	0.960	2.88	0.76
704B	5H	6	15-17	42.86	42.86	0.963	3.59	0.62
704B	5H	6	51-52	43.21	43.21	0.965	3.51	0.50
704B	5H	6	93-95	43.64	43.64	0.972	3.59	0.35
704B	5H	6	125-127	43.96	43.96	0.978	3.31	0.58
704B	5H	7	15-17	44.36	44.36	0.984	2.30	1.14
704B	5H	7	20-22	44.41	44.41	0.985	3.00	0.61
704A	5H	5	20-22	41.91	44.51	0.989	2.98	0.86
704A	5H	5	50-52	42.21	44.81	1.000	3.16	0.93
704A	5H	5	70-74	42.42	45.02	1.007	3.12	0.99
704A	5H	5	80-82	42.51	45.11	1.009	3.02	0.92
704A	5H	5	110-112	42.81	45.41	1.014	2.98	0.91
704A	5H	5	140-142	43.11	45.71	1.020	2.81	1.03
704A	5H	6	20-22	43.41	46.01	1.026	2.83	0.96
704A	5H	6	50-52	43.71	46.31	1.033	3.51	0.42
704A	5H	6	70-74	43.92	46.52	1.037	3.49	0.46
704A	5H	6	80-82	44.01	46.61	1.039	3.35	0.42
704A	5H	6	110-112	44.31	46.91	1.045	3.43	0.37
704A	5H	6	140-142	44.61	47.21	1.049	3.42	0.35
704A	5H	7	29-33	45.01	47.61	1.052	3.46	0.39
704A	6H	1	19-21	45.40	48.00	1.055	2.45	0.69

TABLE A1.   (continued)

Hole	Core	Section	Interval, cm	Depth, mbsf	Composite depth, m	Age, Ma	$\delta^{18}O$, ‰, pdb	$\delta^{13}C$, ‰, pdb
704A	6H	1	54-56	45.70	48.30	1.058	2.59	0.70
704A	6H	1	84-86	46.10	48.70	1.061	2.54	0.82
704A	6H	1	119-121	46.40	49.00	1.063	2.84	0.73
704A	6H	1	130-134	46.52	49.12	1.064	2.71	0.95
704A	6H	2	19-21	46.90	49.50	1.067	2.94	0.86
704A	6H	2	54-56	47.25	49.85	1.070	3.13	0.76
704A	6H	2	84-86	47.55	50.15	1.072	3.12	0.64
704A	6H	2	90-92	47.61	50.21	1.072	2.69	0.75
704A	6H	2	105-107	47.76	50.36	1.073	2.87	0.69
704A	6H	2	120-124	47.90	50.50	1.074	2.69	0.75
704A	6H	2	130-134	48.02	50.62	1.075	2.96	1.03
704A	6H	2	140-142	48.11	50.71	1.075	2.80	0.98
704A	6H	2	4-6	48.25	50.85	1.076	2.80	0.66
704A	6H	3	19-21	48.40	51.00	1.077	2.82	1.06
704A	6H	3	54-56	48.75	51.35	1.080	2.88	1.06
704A	6H	3	84-86	49.05	51.65	1.082	3.82	0.20
704A	6H	3	105-107	49.26	51.86	1.083	3.89	0.21
704A	6H	4	19-21	49.90	52.50	1.087	3.92	0.37
704A	6H	4	34-36	50.05	52.65	1.089	3.67	0.22
704A	6H	4	54-56	50.25	52.85	1.091	3.90	0.14
704A	6H	4	85-87	50.55	53.15	1.095	3.70	0.49
704A	6H	4	105-107	50.76	53.36	1.098	3.81	0.17
704A	6H	4	130-134	51.02	53.62	1.101	3.78	0.44
704A	6H	4	140-142	51.11	53.71	1.102	3.59	0.33
704A	6H	5	19-21	51.40	54.00	1.106	2.98	0.96
704A	6H	5	54-56	51.75	54.35	1.112	2.52	0.39
704A	6H	5	84-86	52.05	54.65	1.122	3.64	0.19
704A	6H	5	105-107	52.26	54.86	1.129	3.68	0.19
704A	6H	5	130-134	52.52	55.12	1.133	3.74	0.78
704A	6H	5	140-142	52.61	55.21	1.134	3.37	0.64
704A	6H	6	15-17	52.86	55.46	1.138	2.80	0.77
704A	6H	6	19-21	52.90	55.50	1.138	3.09	0.76
704A	6H	6	54-56	53.25	55.85	1.143	2.93	0.97
704A	6H	6	84-86	53.55	56.15	1.147	2.79	0.86
704A	6H	6	105-107	53.76	56.36	1.150	2.69	1.03
704A	6H	6	130-134	54.02	56.62	1.153	3.01	1.09
704A	6H	6	140-142	54.11	56.71	1.155	2.71	0.85
704A	6H	7	19-21	54.40	57.00	1.161	2.85	0.78
704A	6H	7	54-56	54.75	57.35	1.168	2.67	0.75
704B	7H	1	12-16	54.34	57.34	1.168	3.25	0.45
704B	7H	1	30-32	54.51	57.51	1.171	3.36	0.56
704B	7H	1	50-53	54.71	57.71	1.174	3.21	0.55
704B	7H	1	70-72	54.91	57.91	1.175	3.34	0.41
704B	7H	1	90-93	55.11	58.11	1.176	3.15	0.63
704B	7H	1	100-102	55.21	58.21	1.177	3.09	0.65
704B	7H	1	110-112	55.31	58.31	1.178	2.95	0.76
704B	7H	1	130-133	55.51	58.51	1.179	2.88	0.88
704B	7H	2	8-10	55.79	58.79	1.181	2.88	0.75
704B	7H	2	30-32	56.01	59.01	1.183	2.68	0.81
704B	7H	2	50-52	56.21	59.21	1.184	2.78	0.81
704B	7H	2	69-71	56.40	59.40	1.185	2.87	0.53
704B	7H	2	89-91	56.60	59.60	1.187	2.83	0.57
704B	7H	2	108-110	56.79	59.79	1.188	2.92	0.53
704B	7H	2	130-132	57.01	60.01	1.189	3.11	1.11
704B	7H	2	145-147	57.16	60.16	1.190	2.96	0.91
704B	7H	3	9-11	57.30	60.30	1.191	2.53	0.81
704B	7H	3	31-33	57.52	60.52	1.193	2.78	0.78
704B	7H	3	50-52	57.71	60.71	1.195	2.80	0.76
704B	7H	3	71-73	57.92	60.92	1.197	3.04	0.67
704B	7H	3	90-92	58.11	61.11	1.199	3.07	0.61
704A	7H	1	59-61	55.30	61.30	1.201	3.24	0.23
704A	7H	1	90-92	55.61	61.61	1.204	3.27	0.23
704A	7H	1	119-121	55.90	61.90	1.206	3.17	0.29
704A	7H	2	10-12	56.31	62.31	1.210	3.49	0.45
704A	7H	2	29-31	56.50	62.50	1.212	3.56	0.16
704A	7H	2	59-61	56.80	62.80	1.216	3.58	0.42
704A	7H	2	90-92	57.11	63.11	1.221	3.47	0.66
704A	7H	2	119-121	57.41	63.41	1.225	2.51	0.85
704A	7H	3	2-6	57.74	63.74	1.230	3.01	0.98
704A	7H	3	10-12	57.81	63.81	1.231	2.66	0.67
704A	7H	3	29-31	58.00	64.00	1.234	2.58	0.55

TABLE A1. (continued)

Hole	Core	Section	Interval, cm	Depth, mbsf	Composite depth, m	Age, Ma	$\delta^{18}O$, ‰, pdb	$\delta^{13}C$, ‰, pdb
704A	7H	3	59-61	58.30	64.30	1.240	3.14	0.28
704A	7H	3	90-92	58.61	64.61	1.245	3.12	0.28
704A	7H	3	119-121	58.90	64.90	1.250	3.45	0.22
704A	7H	4	10-12	59.31	65.31	1.257	3.46	0.37
704A	7H	4	29-31	59.50	65.50	1.259	3.33	0.61
704A	7H	4	59-61	59.80	65.80	1.263	3.24	0.95
704A	7H	4	90-92	60.11	66.11	1.267	3.00	0.91
704A	7H	4	119-121	60.40	66.40	1.270	2.72	0.91
704A	7H	5	2-6	60.74	66.74	1.274	2.92	1.00
704A	7H	5	10-12	60.81	66.81	1.275	2.52	0.89
704A	7H	5	29-31	61.00	67.00	1.277	2.98	0.71
704A	7H	5	59-61	61.30	67.30	1.280	3.02	0.29
704A	7H	5	90-92	61.61	67.61	1.283	3.06	0.28
704A	7H	5	119-121	61.90	67.90	1.286	3.40	0.29
704A	7H	6	2-6	62.24	68.24	1.289	3.86	0.60
704A	7H	6	10-12	62.31	68.31	1.290	3.54	0.33
704A	7H	6	29-31	62.50	68.50	1.292	3.54	0.48
704A	7H	6	50-52	62.71	68.71	1.294	3.52	0.46
704A	7H	6	90-92	63.11	69.11	1.297	3.60	0.22
704A	7H	7	2-6	63.74	69.74	1.303	3.65	0.55
704A	8H	1	19-23	64.41	70.41	1.308	3.07	0.90
704A	8H	1	29-31	64.50	70.50	1.309	2.43	0.79
704A	8H	1	47-51	64.69	70.69	1.310	3.47	0.92
704A	8H	1	67-69	64.88	70.88	1.312	2.35	1.07
704A	8H	1	80-82	65.01	71.01	1.313	2.07	0.72
704A	8H	1	92-96	65.14	71.14	1.316	2.53	0.79
704A	8H	1	110-112	65.31	71.31	1.320	2.22	0.20
704A	8H	1	123-125	65.44	71.44	1.323	3.68	0.61
704A	8H	1	126-130	65.48	71.48	1.324	3.67	0.69
704A	8H	1	139-141	65.60	71.60	1.326	3.08	0.49
704A	8H	2	5-9	65.77	71.77	1.330	3.78	0.64
704A	8H	2	29-31	66.00	72.00	1.335	3.55	0.45
704A	8H	2	40-44	66.12	72.12	1.337	3.77	0.88
704A	8H	2	50-52	66.21	72.21	1.338	3.72	0.60
704A	8H	2	64-68	66.36	72.36	1.341	3.63	1.00
704A	8H	2	84-88	66.56	72.56	1.344	3.47	1.19
704A	8H	2	110-112	66.81	72.81	1.348	3.18	1.05
704A	8H	2	124-126	66.95	72.95	1.350	3.09	1.33
704A	8H	2	139-141	67.10	73.10	1.353	2.54	0.93
704A	8H	2	144-148	67.16	73.16	1.354	2.94	1.37
704A	8H	3	4-8	67.26	73.26	1.355	2.64	1.41
704A	8H	3	15-19	67.37	73.37	1.357	2.43	1.39
704A	8H	3	30-32	67.51	73.51	1.359	2.41	0.95
704A	8H	3	47-49	67.68	73.68	1.361	2.38	0.71
704A	8H	3	65-69	67.87	73.87	1.364	3.35	1.06
704A	8H	3	80-83	68.01	74.01	1.366	4.04	0.64
704A	8H	3	92-96	68.14	74.14	1.368	3.97	0.73
704A	8H	3	110-113	68.31	74.31	1.370	3.84	0.52
704A	8H	3	124-126	68.45	74.45	1.372	3.68	0.93
704A	8H	3	139-141	68.60	74.60	1.376	3.79	0.69
704A	8H	4	30-32	69.01	75.01	1.387	3.14	1.17
704A	8H	4	40-44	69.12	75.12	1.389	3.16	1.32
704A	8H	4	50-52	69.21	75.21	1.392	2.55	0.73
704A	8H	4	64-68	69.36	75.36	1.396	3.10	0.99
704A	8H	4	80-82	69.51	75.51	1.400	3.01	0.40
704A	8H	4	83-85	69.54	75.54	1.401	3.15	0.53
704A	8H	4	100-102	69.71	75.71	1.405	3.63	0.73
704A	8H	4	110-112	69.81	75.81	1.408	3.73	0.44
704A	8H	4	124-126	69.95	75.95	1.412	4.02	0.60
704A	8H	4	139-141	70.10	76.10	1.416	3.87	0.44
704A	8H	5	5-9	70.27	76.27	1.417	3.92	0.67
704A	8H	5	16-19	70.37	76.37	1.418	3.99	0.59
704A	8H	5	30-32	70.51	76.51	1.419	3.59	0.65
704A	8H	5	42-44	70.63	76.63	1.420	3.81	0.75
704A	8H	5	50-52	70.71	76.71	1.421	3.47	0.50
704A	8H	5	64-68	70.86	76.86	1.422	3.68	1.14
704A	8H	5	80-82	71.01	77.01	1.423	3.61	0.93
704A	8H	5	90-92	71.11	77.11	1.424	3.61	1.20
704A	8H	5	110-112	71.31	77.31	1.425	3.35	0.85
704A	8H	5	121-125	71.43	77.43	1.426	3.37	1.01
704A	8H	5	126-130	71.48	77.48	1.426	3.44	1.14

TABLE A1.   (continued)

Hole	Core	Section	Interval, cm	Depth, mbsf	Composite depth, m	Age, Ma	$\delta^{18}O$, ‰, pdb	$\delta^{13}C$, ‰, pdb
704A	8H	5	139-141	71.60	77.60	1.427	3.17	0.87
704A	8H	6	14-18	71.86	77.86	1.429	3.60	1.22
704A	8H	6	1-3	71.91	77.91	1.430	3.73	0.32
704A	8H	6	1-3	71.91	77.91	1.430	3.35	0.41
704A	8H	6	30-32	72.01	78.01	1.430	3.31	0.92
704A	8H	6	39-43	72.11	78.11	1.431	3.44	1.51
704A	8H	6	50-52	72.21	78.21	1.432	3.27	1.35
704A	8H	6	64-68	72.36	78.36	1.433	3.22	1.36
704A	8H	6	80-82	72.51	78.51	1.434	3.02	1.24
704A	8H	6	110-112	72.81	78.81	1.447	3.21	0.97
704A	8H	6	120-124	72.92	78.92	1.452	3.76	0.77
704A	8H	6	124-128	72.96	78.96	1.453	3.82	0.84
704A	8H	cc	8-12	73.30	79.30	1.459	3.70	0.92
704B	9H	2	145-147	76.16	79.31	1.459	3.57	0.73
704B	9H	3	10-12	76.31	79.46	1.461	3.44	0.78
704B	9H	3	32-34	76.53	79.68	1.464	3.29	0.84
704B	9H	3	72-74	76.53	79.68	1.464	3.14	0.98
704B	9H	3	50-52	76.71	79.86	1.467	3.23	0.99
704B	9H	3	89-91	77.10	80.25	1.472	2.75	0.98
704B	9H	3	100-102	77.20	80.35	1.473	2.80	0.99
704B	9H	3	112-114	77.33	80.48	1.475	3.05	1.04
704B	9H	3	129-131	77.50	80.65	1.477	3.00	1.28
704B	9H	3	146-148	77.67	80.82	1.479	2.75	1.34
704B	9H	4	12-14	77.83	80.98	1.480	2.71	1.36
704B	9H	4	31-33	78.02	81.17	1.482	2.79	1.14
704B	9H	4	71-73	78.42	81.57	1.485	2.74	1.10
704B	9H	4	89-91	78.60	81.75	1.487	2.89	1.03
704B	9H	4	100-102	78.71	81.86	1.488	2.84	0.96
704B	9H	4	111-113	78.82	81.97	1.489	2.95	1.10
704B	9H	4	131-133	79.02	82.17	1.490	3.20	1.08
704B	9H	5	9-11	79.30	82.45	1.493	3.06	1.09
704B	9H	5	28-30	79.49	82.64	1.494	2.99	1.11
704B	9H	5	50-52	79.71	82.86	1.496	2.96	1.15
704B	9H	5	68-70	79.89	83.04	1.498	2.83	1.33
704B	9H	5	90-92	80.11	83.26	1.499	3.01	1.15
704B	9H	5	100-102	80.21	83.36	1.500	3.01	1.12
704B	9H	5	110-112	80.31	83.46	1.501	2.97	1.18
704B	9H	5	129-131	80.50	83.65	1.503	2.83	1.09
704B	9H	6	10-12	80.81	83.96	1.506	3.29	1.08
704B	9H	6	28-30	80.99	84.14	1.507	3.22	1.16
704B	9H	6	48-50	81.19	84.34	1.509	2.72	1.28
704B	9H	6	50-52	81.21	84.36	1.509	3.08	1.28
704B	9H	6	68-70	81.39	84.54	1.511	2.82	1.29
704B	9H	6	90-92	81.61	84.76	1.513	2.65	1.43
704B	9H	6	112-114	81.83	84.98	1.515	2.56	1.41
704B	9H	7	5-7	82.26	85.41	1.519	3.23	0.98
704B	9H	7	10-12	82.31	85.46	1.519	2.81	1.15
704B	9H	7	31-33	82.52	85.67	1.521	3.28	0.97
704B	9H	7	50-52	82.71	85.86	1.521	3.47	0.80
704A	9H	5	98-101	80.70	85.70	1.521	4.14	0.67
704A	9H	5	98-101	80.70	85.70	1.523	3.93	0.58
704A	9H	5	128-132	81.00	86.00	1.524	3.99	0.66
704A	9H	1	13-15	81.34	86.34	1.527	3.72	0.75
704A	9H	6	34-38	81.56	86.56	1.529	3.95	0.65
704A	9H	6	63-67	81.85	86.85	1.532	3.97	0.72
704A	9H	6	98-100	82.19	87.19	1.535	3.26	1.28
704A	9H	6	123-127	82.45	87.45	1.537	3.88	0.63
704A	9H	6	140-142	82.61	87.61	1.538	3.09	-0.51
704A	9H	7	3-7	82.75	87.75	1.538	3.89	0.63
704A	9H	7	33-37	83.05	88.05	1.540	3.86	0.60
704A	10H	1	3-7	83.25	88.25	1.541	3.74	1.01
704A	10H	1	32-36	83.54	88.54	1.542	3.76	0.98
704A	10H	1	62-66	83.84	88.84	1.543	3.67	1.00
704A	10H	1	93-97	84.14	89.14	1.545	3.71	0.99
704A	10H	1	132-136	84.54	89.54	1.547	3.70	1.01
704A	10H	2	3-7	84.75	89.75	1.548	3.71	1.02
704A	10H	2	33-37	85.05	90.05	1.549	3.15	0.98
704A	10H	2	109-111	85.30	90.30	1.551	2.81	0.18
704A	10H	2	63-67	85.35	90.35	1.552	3.17	0.91
704A	10H	2	92-96	85.64	90.64	1.558	2.90	0.80
704A	10H	2	123-127	85.95	90.95	1.561	2.67	0.75

TABLE A1.   (continued)

Hole	Core	Section	Interval, cm	Depth, mbsf	Composite depth, m	Age, Ma	$\delta^{18}O$, ‰, pdb	$\delta^{13}C$, ‰, pdb
704A	10H	2	142-146	86.14	91.14	1.562	3.21	0.73
704A	10H	3	3-7	86.25	91.25	1.562	3.16	0.65
704A	10H	3	32-36	86.54	91.54	1.564	3.37	0.58
704A	10H	3	63-67	86.85	91.85	1.565	3.37	0.45
704A	10H	3	95-99	87.17	92.17	1.566	3.41	0.50
704A	10H	3	124-128	87.46	92.46	1.568	3.30	0.45
704A	10H	4	3-7	87.75	92.75	1.569	3.31	0.38
704A	10H	4	33-37	88.05	93.05	1.570	3.19	0.47
704A	10H	4	62-66	88.34	93.34	1.572	3.21	0.42
704A	10H	4	92-96	88.64	93.64	1.573	3.08	0.56
704A	10H	4	123-127	88.95	93.95	1.574	2.93	0.68
704A	10H	4	142-146	89.14	94.14	1.575	3.41	0.56
704A	10H	5	3-7	89.25	94.25	1.576	3.38	0.54
704A	10H	5	28-32	89.50	94.50	1.577	3.40	0.52
704A	10H	5	58-62	89.80	94.80	1.580	3.57	0.66
704A	10H	5	123-127	90.45	95.45	1.586	3.21	0.85
704A	10H	5	140-144	90.62	95.62	1.588	3.25	0.89
704A	10H	6	3-7	90.75	95.75	1.589	3.49	0.75
704A	10H	6	28-32	91.00	96.00	1.591	3.36	0.81
704A	10H	6	63-67	91.35	96.35	1.594	3.07	0.79
704A	10H	6	93-97	91.65	96.65	1.598	3.05	0.74
704A	10H	6	133-137	92.05	97.05	1.609	2.76	0.55
704A	10H	7	3-7	92.25	97.25	1.614	3.62	0.57
704A	10H	7	19-21	92.40	97.40	1.616	3.90	0.00
704A	10H	7	24-28	92.46	97.46	1.617	3.70	0.60
704A	10H	7	45-47	92.66	97.66	1.619	3.27	0.18
704A	11H	1	5-9	92.77	97.77	1.620	3.17	0.83
704A	10H	7	62-66	92.84	97.84	1.620	3.69	0.74
704A	11H	1	29-33	93.01	98.01	1.622	3.23	1.08
704A	11H	1	63-67	93.35	98.35	1.625	3.08	1.18
704A	11H	1	98-100	93.69	98.69	1.628	2.85	1.16
704A	11H	1	129-131	94.00	99.00	1.631	2.83	1.20
704A	11H	2	3-7	94.25	99.25	1.633	2.89	1.18
704A	11H	2	20-22	94.41	99.41	1.635	2.93	1.26
704A	11H	2	35-39	94.57	99.57	1.636	2.98	0.96
704A	11H	2	50-52	94.71	99.71	1.637	2.31	0.37
704A	11H	2	62-66	94.84	99.84	1.639	2.97	1.01
704A	11H	2	88-92	95.10	100.10	1.641	3.15	0.94
704A	11H	2	124-128	95.46	100.46	1.644	3.46	0.66
704A	11H	3	3-7	95.75	100.75	1.647	3.47	0.68
704A	11H	3	34-36	96.05	101.05	1.650	3.51	0.58
704A	11H	3	110-112	96.31	101.31	1.652	3.24	0.08
704A	11H	3	63-67	96.35	101.35	1.653	3.44	0.61
704A	11H	3	94-98	96.66	101.66	1.656	3.52	0.59
704A	11H	3	124-128	96.96	101.96	1.658	3.44	0.57
704A	11H	4	3-7	97.25	102.25	1.661	3.55	0.54
704A	11H	4	33-37	97.55	102.55	1.663	3.50	0.54
704A	11H	4	64-68	97.86	102.86	1.665	3.28	1.01
704A	11H	4	98-102	98.20	103.20	1.667	2.90	1.00
704A	11H	4	123-125	98.44	103.44	1.668	3.08	0.69
704A	11H	5	3-7	98.75	103.75	1.670	3.52	0.89
704A	11H	5	33-37	99.05	104.05	1.671	3.13	1.07
704A	11H	5	67-69	99.38	104.38	1.673	3.13	1.05
704A	11H	5	97-101	99.69	104.69	1.675	3.08	1.03
704A	11H	5	124-128	99.96	104.96	1.676	3.36	1.02
704A	11H	6	3-7	100.25	105.25	1.678	3.21	1.21
704A	11H	6	33-37	100.55	105.55	1.679	3.10	1.13
704A	11H	6	63-67	100.85	105.85	1.681	3.22	1.07
704A	11H	6	85-89	101.07	106.07	1.682	3.03	1.01
704A	11H	6	127-129	101.48	106.48	1.684	2.99	0.95
704A	11H	6	142-146	101.64	106.64	1.685	3.21	1.20
704A	11H	7	3-7	101.75	106.75	1.686	3.10	0.82
704A	11H	7	36-40	102.08	107.08	1.688	3.26	0.70
704A	12H	1	3-7	102.25	107.25	1.689	2.95	0.95
704A	12H	1	27-31	102.49	107.49	1.690	2.55	0.69
704A	12H	1	63-67	102.85	107.85	1.692	2.86	0.96
704A	12H	1	94-98	103.16	108.16	1.694	3.18	0.83
704A	12H	1	124-128	103.46	108.46	1.695	3.48	0.73
704A	12H	2	3-7	103.75	108.75	1.697	2.78	0.85
704A	12H	2	63-67	104.35	109.35	1.700	3.64	0.52
704A	12H	2	94-98	104.66	109.66	1.703	3.39	0.66

TABLE A1.   (continued)

Hole	Core	Section	Interval, cm	Depth, mbsf	Composite depth, m	Age, Ma	$\delta^{18}O$, ‰, pdb	$\delta^{13}C$, ‰, pdb
704A	12H	2	124-128	104.96	109.96	1.706	3.42	0.80
704A	12H	3	3-7	105.25	110.25	1.710	3.03	0.86
704A	12H	3	34-38	105.56	110.56	1.713	2.54	0.69
704A	12H	3	59-63	105.81	110.81	1.715	3.34	0.56
704A	12H	3	85-89	106.07	111.07	1.718	2.72	1.03
704A	12H	3	125-127	106.46	111.46	1.727	2.57	1.20
704A	12H	4	3-7	106.75	111.75	1.733	2.70	0.90
704A	12H	4	29-33	107.01	112.01	1.738	3.58	0.70
704A	12H	4	57-61	107.29	112.29	1.743	3.13	0.76
704A	12H	4	93-97	107.65	112.65	1.747	3.01	1.04
704A	12H	4	128-132	108.00	113.00	1.751	2.63	1.12
704A	12H	5	3-7	108.25	113.25	1.754	2.57	1.02
704A	12H	5	34-38	108.56	113.56	1.758	2.55	1.08
704A	12H	5	67-71	108.89	113.89	1.760	2.52	1.03
704A	12H	5	94-98	109.16	114.16	1.762	2.71	0.94
704A	12H	6	3-7	109.75	114.75	1.764	3.34	0.73
704A	12H	6	43-47	110.15	115.15	1.766	3.11	0.89
704A	12H	6	93-97	110.65	115.65	1.769	2.78	0.92
704A	12H	6	124-128	110.96	115.96	1.771	2.93	0.94
704A	13H	1	3-7	111.75	116.75	1.774	3.06	0.85
704A	13H	1	33-37	112.05	117.05	1.776	3.37	0.63
704A	13H	1	63-67	112.35	117.35	1.777	3.60	0.50
704A	13H	1	92-94	112.63	117.63	1.779	3.68	0.46
704A	13H	1	127-129	112.98	117.98	1.781	3.72	0.41
704A	13H	2	3-7	113.25	118.25	1.782	3.03	0.89
704A	13H	2	34-38	113.56	118.56	1.784	3.30	0.72
704A	13H	2	64-68	113.86	118.86	1.786	3.54	0.55
704A	13H	2	92-94	114.13	119.13	1.788	3.71	0.39
704A	13H	2	127-131	114.49	119.49	1.790	3.69	0.39
704A	13H	3	3-7	114.75	119.75	1.791	3.52	0.70
704A	13H	3	33-37	115.05	120.05	1.793	3.39	0.66
704A	13H	3	63-67	115.35	120.35	1.795	3.31	0.83
704A	13H	3	92-96	115.64	120.64	1.797	3.01	0.87
704A	13H	3	126-130	115.98	120.98	1.799	2.65	0.84
704A	13H	4	3-7	116.25	121.25	1.801	2.62	0.71
704A	13H	4	33-37	116.55	121.55	1.835	2.91	0.58
704A	13H	4	63-67	116.85	121.85	1.841	3.35	0.67
704A	13H	4	93-97	117.15	122.15	1.847	3.17	0.94
704A	13H	4	121-125	117.43	122.43	1.853	2.62	1.10
704A	13H	5	3-7	117.75	122.75	1.855	2.83	0.82
704A	13H	5	33-37	118.05	123.05	1.858	3.33	0.71
704A	13H	5	63-67	118.35	123.35	1.860	3.43	0.72
704A	13H	5	93-96	118.64	123.64	1.862	3.60	0.73
704A	13H	5	127-129	118.98	123.98	1.865	2.93	0.85
704A	13H	6	3-7	119.25	124.25	1.867	3.47	0.86
704A	13H	6	33-37	119.55	124.55	1.869	3.08	0.98
704A	13H	6	63-67	119.85	124.85	1.872	2.73	1.17
704A	13H	6	92-96	120.14	125.14	1.874	2.74	1.16
704A	13H	6	121-125	120.43	125.43	1.876	2.72	1.12
704A	13H	7	3-7	120.75	125.75	1.879	2.42	1.09
704A	13H	7	33-37	121.05	126.05	1.881	2.48	1.10
704A	15H	4	35-39	135.57	126.27	1.883	2.56	1.10
704A	15H	4	63-67	135.85	126.55	1.888	2.47	0.96
704A	15H	4	63-67	135.98	126.68	1.890	2.47	0.96
704A	15H	4	92-96	136.14	126.84	1.893	2.58	1.21
704A	15H	4	124-128	136.46	127.16	1.898	2.83	0.95
704A	15H	5	3-7	136.75	127.45	1.903	2.69	1.22
704A	15H	5	35-39	137.07	127.77	1.908	3.26	0.69
704A	15H	5	92-96	137.63	128.33	1.913	2.85	0.96
704A	15H	6	3-7	138.25	128.95	1.918	2.75	0.90
704A	15H	6	35-37	138.56	129.26	1.921	2.62	0.81
704A	15H	6	63-67	138.85	129.55	1.923	2.58	0.79
704A	15H	6	93-97	139.15	129.85	1.924	2.73	0.64
704A	15H	6	123-127	139.45	130.15	1.925	2.79	0.59
704A	15H	7	2-6	139.74	130.44	1.927	2.67	0.61
704A	15H	7	32-36	140.04	130.74	1.928	2.88	0.25
704A	16H	1	22-26	140.44	131.14	1.929	3.00	0.67
704A	16H	2	19-23	141.91	132.61	1.935	3.57	0.08
704A	16H	2	28-42	142.05	132.75	1.936	3.61	0.22
704A	16H	5	3-7	142.25	132.95	1.937	3.01	0.71
704A	16H	2	117-121	142.89	133.59	1.940	3.01	0.66

TABLE A1.   (continued)

Hole	Core	Section	Interval, cm	Depth, mbsf	Composite depth, m	Age, Ma	$\delta^{18}O$, ‰, pdb	$\delta^{13}C$, ‰, pdb
704A	16H	3	3-7	143.25	133.95	1.941	3.26	0.78
704A	16H	3	43-47	143.65	134.35	1.958	3.13	0.19
704A	16H	3	64-67	143.85	134.55	1.966	3.28	0.57
704A	16H	3	70-74	143.92	134.62	1.969	3.78	0.28
704A	16H	3	80-82	144.01	134.71	1.973	3.44	-0.03
704A	16H	3	95-99	144.17	134.87	1.979	2.84	0.69
704A	16H	3	110-112	144.31	135.01	1.985	2.43	0.59
704A	16H	3	130-134	144.52	135.22	1.987	2.88	0.69
704A	16H	3	144-146	144.65	135.35	1.988	2.91	0.36
704A	16H	4	20-22	144.91	135.61	1.990	3.04	0.57
704A	16H	4	50-52	145.21	135.91	1.993	2.86	0.53
704A	16H	4	70-74	145.42	136.12	1.995	3.18	0.70
704A	16H	4	80-82	145.51	136.21	1.995	2.90	0.58
704A	16H	4	94-98	145.66	136.36	1.997	2.86	0.88
704A	16H	4	110-112	145.81	136.51	1.998	2.55	0.36
704A	16H	4	144-146	146.15	136.85	2.001	3.48	0.15
704A	16H	5	20-22	146.41	137.11	2.004	2.83	0.76
704A	16H	5	50-52	146.71	137.41	2.007	2.49	0.69
704A	16H	5	70-74	146.92	137.62	2.009	2.88	0.92
704A	16H	5	80-82	147.01	137.71	2.010	2.62	0.42
704A	16H	5	110-112	147.31	138.01	2.013	2.49	0.60
704A	16H	5	144-146	147.65	138.35	2.014	2.84	0.73
704A	16H	6	20-22	147.91	138.61	2.015	2.64	0.31
704A	16H	6	50-52	148.21	138.91	2.016	3.29	0.28
704A	16H	6	70-74	148.42	139.12	2.016	3.32	0.00
704A	17X	1	3-7	149.75	140.45	2.020	3.40	0.19
704A	17X	1	20-22	149.91	140.61	2.020	2.80	0.47
704A	17X	1	36-40	150.08	140.78	2.021	3.02	0.66
704A	17X	1	50-52	150.21	140.91	2.021	3.07	0.65
704A	17X	1	63-67	150.35	141.05	2.022	3.57	0.03
704A	17X	1	70-74	150.42	141.12	2.022	3.73	0.13
704A	17X	1	80-82	150.51	141.21	2.022	3.55	-0.11
704A	17X	1	93-97	150.65	141.35	2.022	2.93	0.65
704A	17X	1	125-129	150.97	141.67	2.023	2.93	0.63
704A	17X	1	135-137	151.07	141.77	2.024	2.83	0.63
704A	17X	2	3-7	151.25	141.95	2.024	3.00	0.70
704A	17X	2	20-22	151.41	142.11	2.025	2.79	0.57
704A	17X	2	33-37	151.55	142.25	2.025	2.88	0.77
704A	17X	2	50-52	151.71	142.41	2.025	2.73	0.84
704A	17X	2	63-67	151.85	142.55	2.026	2.47	0.71
704A	17X	2	70-74	151.92	142.62	2.026	2.89	0.63
704A	17X	2	80-82	152.01	142.71	2.026	2.88	0.56
704A	17X	2	95-99	152.17	142.87	2.027	3.14	0.52
704A	17X	2	110-112	152.31	143.01	2.027	3.27	0.23
704A	17X	2	119-123	152.41	143.11	2.029	3.26	0.41
704A	17X	2	140-142	152.61	143.31	2.033	2.78	0.92
704A	17X	3	3-7	152.75	143.45	2.036	2.79	1.25
704A	17X	3	20-22	152.91	143.61	2.040	3.26	0.72
704A	17X	3	20-22	152.91	143.61	2.040	2.85	0.59
704A	17X	3	37-41	153.09	143.79	2.043	2.70	0.99
704A	17X	3	64-68	153.36	144.06	2.049	3.05	1.11
704A	17X	3	70-74	153.42	144.12	2.050	3.45	0.95
704A	17X	3	80-82	153.51	144.21	2.050	3.01	0.82
704A	17X	3	110-112	153.81	144.51	2.053	3.22	0.50
704A	17X	3	126-130	153.98	144.68	2.055	3.13	0.94
704A	17X	3	140-142	154.11	144.81	2.056	2.76	0.76
704A	17X	4	3-7	154.25	144.95	2.057	2.86	0.93
704A	17X	4	20-22	154.41	145.11	2.059	2.64	0.69
704A	17X	4	50-52	154.71	145.41	2.062	3.12	0.84
704A	17x	4	63-67	154.85	145.55	2.063	3.28	0.74
704A	17X	4	70-74	154.92	145.62	2.064	3.57	0.77
704A	17X	4	80-82	155.01	145.71	2.065	3.17	0.53
704A	17X	4	110-112	155.31	146.01	2.067	3.10	0.43
704A	17X	4	126-130	155.48	146.18	2.069	2.87	0.72
704A	17X	4	140-142	155.61	146.31	2.071	2.76	0.62
704A	17X	5	20-22	155.91	146.61	2.075	2.98	0.31
704A	17X	5	33-37	156.05	146.75	2.077	3.22	0.56
704A	17X	5	50-52	156.21	146.91	2.080	3.36	0.22
704A	17X	5	63-67	156.35	147.05	2.082	2.91	0.95
704A	17X	5	70-74	156.42	147.12	2.083	2.96	0.98
704A	17X	5	80-82	156.51	147.21	2.084	2.71	0.77

TABLE A1.   (continued)

Hole	Core	Section	Interval, cm	Depth, mbsf	Composite depth, m	Age, Ma	$\delta^{18}O$, ‰, pdb	$\delta^{13}C$, ‰, pdb
704A	17X	5	96-100	156.68	147.38	2.086	3.06	0.82
704A	17X	5	110-112	156.81	147.51	2.088	3.07	0.48
704A	17X	5	127-131	156.99	147.69	2.091	2.64	0.91
704A	17X	5	140-142	157.11	147.81	2.094	2.83	0.88
704A	17X	6	3-7	157.25	147.95	2.097	2.81	0.72
704A	17X	6	20-22	157.41	148.11	2.100	3.18	0.23
704A	17X	6	27-31	157.49	148.19	2.102	3.35	0.55
704A	17X	6	50-52	157.71	148.41	2.107	2.81	0.57
704A	17X	6	62-66	157.84	148.54	2.109	3.17	0.77
704B	18X	1	75-79	157.47	148.57	2.110	3.51	-0.01
704B	18X	1	75-79	157.47	148.57	2.110	3.55	0.12
704B	18X	1	75-79	157.47	148.57	2.110	3.50	0.32
704B	18X	1	90-94	157.62	148.72	2.114	3.27	0.22
704B	18X	1	105-109	157.77	148.87	2.118	2.98	0.42
704B	18X	1	105-109	157.77	148.87	2.118	2.86	0.38
704B	18X	1	120-124	157.92	149.02	2.122	3.08	0.41
704B	18X	1	135-139	158.07	149.17	2.126	3.09	0.39
704B	18X	1	135-139	158.07	149.17	2.126	2.97	0.41
704B	18X	1	135-139	158.07	149.17	2.126	2.95	0.46
704B	18X	1	140-144	158.12	149.22	2.128	3.05	0.42
704B	18X	2	0-4	158.22	149.32	2.130	2.95	0.40
704B	18X	2	15-19	158.37	149.47	2.134	2.81	0.44
704B	18X	2	30-34	158.52	149.62	2.138	2.80	0.51
704B	18X	2	45-49	158.67	149.77	2.139	2.62	0.57
704B	18X	2	60-64	158.82	149.92	2.140	2.66	0.60
704B	18X	2	75-79	158.97	150.07	2.141	2.73	0.73
704B	18X	2	90-94	159.12	150.22	2.142	3.22	0.58
704B	18X	2	105-109	159.27	150.37	2.143	3.12	0.52
704B	18X	2	105-109	159.27	150.37	2.143	3.00	0.52
704B	18X	2	120-124	159.42	150.52	2.144	2.87	0.56
704B	18X	2	135-139	159.57	150.67	2.145	2.68	0.60
704B	18X	2	140-144	159.62	150.72	2.145	2.68	0.53
704B	18X	3	0-4	159.72	150.82	2.146	2.64	0.52
704B	18X	3	15-19	159.87	150.97	2.147	3.07	0.51
704B	18X	3	30-34	160.02	151.12	2.148	3.24	0.52
704B	18X	3	45-49	160.17	151.27	2.149	3.10	0.58
704B	18X	3	60-64	160.32	151.42	2.150	3.25	0.57
704B	18X	3	75-79	160.47	151.57	2.151	3.36	0.11
704B	18X	3	90-94	160.62	151.72	2.152	3.72	-0.18
704B	18X	3	105-109	160.77	151.87	2.158	2.78	0.60
704B	18X	3	120-124	160.92	152.02	2.166	2.54	0.53
704B	18X	3	135-137	161.06	152.16	2.173	2.94	0.44
704B	18X	3	140-144	161.12	152.22	2.176	2.97	0.66
704B	18X	3	140-144	161.12	152.22	2.176	3.10	0.54
704B	18X	4	0-4	161.22	152.32	2.179	3.05	0.64
704B	18X	4	0-4	161.22	152.32	2.179	3.10	0.56
704B	18X	4	15-19	161.37	152.47	2.182	3.22	0.55
704B	18X	4	15-19	161.37	152.47	2.182	3.27	0.51
704B	18X	4	30-34	161.52	152.62	2.186	3.10	0.52
704B	18X	4	45-49	161.67	152.77	2.189	3.19	0.54
704B	18X	4	60-64	161.82	152.92	2.193	3.15	0.58
704B	18X	4	75-79	161.97	153.07	2.196	3.13	0.58
704B	18X	4	90-94	162.12	153.22	2.203	3.27	0.52
704B	18X	4	90-94	162.12	153.22	2.203	3.03	0.48
704B	18X	4	105-109	162.27	153.37	2.215	2.89	0.56
704B	18X	4	120-124	162.42	153.52	2.223	2.92	0.46
704B	18X	4	135-139	162.57	153.67	2.226	3.35	0.50
704B	18X	4	135-139	162.57	153.67	2.226	3.20	0.52
704B	18X	4	140-144	162.62	153.72	2.228	3.17	0.61
704B	18X	4	140-144	162.62	153.72	2.228	3.17	0.51
704B	18X	5	0-4	162.72	153.82	2.230	3.07	0.57
704B	18X	5	0-4	162.72	153.82	2.230	3.14	0.50
704B	18X	5	15-19	162.87	153.97	2.234	3.09	0.59
704B	18X	5	30-34	163.02	154.12	2.237	3.12	0.63
704B	18X	5	45-49	163.17	154.27	2.241	3.17	0.57
704B	18X	5	60-64	163.32	154.42	2.245	3.52	0.61
704B	18X	5	60-64	163.32	154.42	2.245	3.56	0.67
704B	18X	5	75-79	163.47	154.57	2.250	3.34	0.81
704B	18X	5	75-79	163.47	154.57	2.250	3.37	0.91
704B	18X	5	75-79	163.47	154.57	2.250	3.42	0.92
704B	18X	5	90-94	163.62	154.72	2.254	2.88	1.22

TABLE A1.   (continued)

Hole	Core	Section	Interval, cm	Depth, mbsf	Composite depth, m	Age, Ma	$\delta^{18}O$, ‰, pdb	$\delta^{13}C$, ‰, pdb
704B	18X	5	105-109	163.77	154.87	2.258	2.82	1.12
704B	18X	5	120-124	163.92	155.02	2.261	3.02	1.10
704B	18X	5	135-139	164.07	155.17	2.263	3.22	0.95
704B	18X	5	140-144	164.12	155.22	2.263	3.29	0.91
704B	18X	6	0-4	164.22	155.32	2.264	3.24	0.85
704B	18X	6	0-4	164.22	155.32	2.264	3.29	0.99
704B	18X	6	15-19	164.37	155.47	2.266	3.04	1.15
704B	18X	6	30-34	164.52	155.62	2.268	2.94	1.15
704B	18X	6	45-49	164.67	155.77	2.269	2.72	1.11
704B	18X	6	60-64	164.82	155.92	2.271	2.76	0.97
704B	18X	6	75-79	164.97	156.07	2.273	2.98	0.73
704B	18X	6	90-94	165.12	156.22	2.275	3.51	0.36
704B	18X	6	90-94	165.12	156.22	2.275	3.42	0.38
704B	18X	6	105-109	165.27	156.37	2.281	3.36	0.70
704B	18X	6	105-109	165.27	156.37	2.281	3.26	0.71
704B	18X	6	120-124	165.42	156.52	2.289	2.92	0.92

TABLE A2.   Oxygen and Carbon Isotopic Results of *Globigerina bulloides* From Site 704

Hole	Core	Section	Interval, cm	Depth, mbsf	Composite depth, m	Age, Ma	$\delta^{18}O$, ‰, pdb	$\delta^{13}C$, ‰, pdb
704B	19X	2	0-4	167.72	158.82	2.641	3.22	-0.62
704B	19X	2	30-34	168.02	159.12	2.649	3.11	0.18
704B	19X	2	45-49	168.17	159.27	2.652	2.99	0.30
704B	19X	2	60-64	168.32	159.42	2.657	3.08	-0.14
704B	19X	2	75-79	168.47	159.57	2.661	3.34	-0.34
704B	19X	2	90-94	168.62	159.72	2.663	2.96	0.13
704B	19X	2	105-109	168.77	159.87	2.668	2.77	0.35
704B	19X	2	120-124	168.92	160.02	2.673	3.22	-0.32
704B	19X	2	135-137	169.06	160.16	2.677	2.65	0.25
704B	19X	2	140-142	169.11	160.21	2.678	2.67	0.21
704B	19X	3	0-4	169.22	160.32	2.681	2.72	0.25
704B	19X	3	15-19	169.37	160.47	2.685	2.73	0.34
704B	19X	3	30-34	169.52	160.62	2.689	2.77	0.48
704B	19X	3	45-49	169.67	160.77	2.693	2.83	0.23
704B	19X	3	60-64	169.82	160.92	2.697	2.85	0.40
704B	19X	3	75-79	169.97	161.07	2.701	2.78	0.34
704B	19X	3	90-94	170.12	161.22	2.705	2.77	0.29
704B	19X	3	105-109	170.27	161.37	2.709	2.83	0.25
704B	19X	3	120-124	170.42	161.52	2.712	2.80	0.24
704B	19X	3	135-139	170.57	161.67	2.716	2.85	0.26
704B	19X	CC	5-9	170.77	161.87	2.721	2.67	0.16
704B	19X	CC	15-19	170.87	161.97	2.724	2.97	0.23
704A	19X	1	6-10	168.78	162.08	2.727	2.88	0.39
704A	19X	1	20-22	168.91	162.21	2.731	2.05	0.36
704A	19X	1	30-34	169.02	162.32	2.734	3.16	-0.34
704A	19X	1	33-37	169.05	162.35	2.734	3.05	-0.16
704A	19X	1	50-52	169.21	162.51	2.739	2.36	0.66
704A	19X	1	60-62	169.31	162.61	2.741	2.49	0.61
704A	19X	1	75-77	169.46	162.76	2.745	2.47	0.54
704A	19X	1	87-91	169.59	162.89	2.749	2.36	0.46
704A	19X	1	110-112	169.81	163.11	2.754	2.44	0.36
704A	19X	1	124-128	169.96	163.26	2.758	2.50	0.44
704A	19X	1	140-142	170.11	163.41	2.762	2.58	0.55
704A	19X	1	143-147	170.15	163.45	2.763	2.70	0.45
704A	19X	2	1-5	170.23	163.53	2.765	2.72	0.52
704A	19X	2	20-22	170.41	163.71	2.770	2.79	0.25
704A	19X	2	30-34	170.52	163.82	2.773	2.57	0.51
704A	19X	2	33-37	170.55	163.85	2.774	2.55	0.53
704A	19X	2	46-50	170.68	163.98	2.777	2.42	0.61
704A	19X	2	50-52	170.71	164.01	2.778	2.29	0.48
704A	19X	2	60-62	170.81	164.11	2.781	2.28	0.42
704A	19X	2	76-80	170.98	164.28	2.785	2.53	0.35
704A	19X	2	93-97	171.15	164.45	2.790	2.25	0.58
704A	19X	2	110-112	171.31	164.61	2.794	2.39	0.56
704A	19X	2	125-129	171.47	164.77	2.798	2.21	0.41

TABLE A2.   (continued)

Hole	Core	Section	Interval, cm	Depth, mbsf	Composite depth, m	Age, Ma	$\delta^{18}O$, ‰, pdb	$\delta^{13}C$, ‰, pdb
704A	19X	2	140-142	171.61	164.91	2.802	2.27	0.39
704A	19X	3	27-29	171.98	165.28	2.811	2.71	0.43
704A	19X	3	30-34	172.02	165.32	2.813	2.77	0.30
704A	19X	3	60-62	172.31	165.61	2.820	2.35	0.46
704A	19X	3	62-66	172.34	165.64	2.821	2.51	0.33
704A	19X	3	82-86	172.54	165.84	2.826	2.49	0.41
704A	19X	3	94-98	172.66	165.96	2.829	2.65	0.17
704A	19X	3	110-112	172.81	166.11	2.833	2.50	0.53
704A	19X	3	112-116	172.84	166.14	2.834	2.50	0.46
704A	19X	3	128-132	173.00	166.30	2.838	2.34	0.40
704A	19X	3	140-143	173.12	166.42	2.842	2.25	0.42
704A	19X	4	1-5	173.23	166.53	2.844	2.18	0.40
704A	19X	4	20-22	173.41	166.71	2.849	2.34	0.07
704A	19X	4	30-34	173.52	166.82	2.852	2.27	0.15
704A	19X	4	34-38	173.56	166.86	2.853	2.33	0.17
704A	19X	4	50-52	173.71	167.01	2.857	2.21	0.75
704A	19X	4	60-62	173.81	167.11	2.860	2.34	0.66
704A	19X	4	66-70	173.88	167.18	2.862	2.30	0.64
704A	19X	4	86-90	174.08	167.38	2.867	2.34	0.85
704A	19X	4	110-112	174.31	167.61	2.873	2.44	0.48
704A	19X	4	124-128	174.46	167.76	2.877	2.51	0.24
704A	19X	4	140-142	174.61	167.91	2.881	2.06	0.48
704A	19X	5	1-5	174.73	168.03	2.884	2.37	0.51
704A	19X	5	20-22	174.91	168.21	2.889	2.30	0.44
704A	19X	5	30-34	175.02	168.32	2.892	2.79	0.34
704A	19X	5	42-44	175.13	168.43	2.894	2.63	0.32
704A	19X	5	50-52	175.21	168.51	2.897	2.64	0.32
704A	19X	5	60-65	175.33	168.63	2.900	2.49	0.45
704A	19X	5	98-100	175.69	168.99	2.909	2.44	-0.06
704A	19X	5	123-127	175.95	169.25	2.916	2.27	0.48
704A	19X	5	140-142	176.11	169.41	2.920	2.27	0.14
704A	19X	6	1-5	176.23	169.53	2.925	2.40	0.29
704A	19X	6	20-22	176.41	169.71	2.933	2.33	0.50
704A	19X	6	30-34	176.52	169.82	2.937	2.35	0.35
704A	19X	6	34-38	176.56	169.86	2.939	2.30	0.37
704A	19X	6	50-52	176.71	170.01	2.945	2.27	0.75
704A	19X	6	60-62	176.81	170.11	2.949	2.12	0.67
704A	19X	6	74-78	176.96	170.26	2.955	2.16	0.56
704A	19X	6	93-97	177.15	170.45	2.963	2.20	0.37
704A	19X	6	110-112	177.31	170.61	2.970	2.26	0.59
704A	19X	6	123-127	177.45	170.75	2.976	2.17	0.33
704A	19X	6	140-142	177.61	170.91	2.982	2.21	0.31
704A	19X	6	143-147	177.65	170.95	2.984	2.14	0.24
704A	19X	7	3-7	177.75	171.05	2.988	2.53	0.35
704A	19X	7	18-22	177.90	171.20	2.997	2.62	0.05
704A	20X	1	2-6	178.24	171.54	3.020	2.37	0.54
704A	20X	1	27-31	178.49	171.79	3.038	2.37	0.49
704A	20X	1	46-48	178.67	171.97	3.050	2.26	0.54
704A	20X	1	62-66	178.84	172.14	3.062	2.24	0.59
704A	20X	1	68-72	178.90	172.20	3.066	2.16	0.46
704A	20X	1	80-82	179.01	172.31	3.074	2.14	0.63
704A	20X	1	94-98	179.16	172.46	3.083	1.71	0.74
704A	20X	1	110-112	179.31	172.61	3.089	2.31	0.56
704A	20X	1	123-127	179.45	172.75	3.095	2.39	0.29
704A	20X	1	140-142	179.61	172.91	3.101	2.28	0.30
704A	20X	1	143-147	179.65	172.95	3.103	2.23	0.29
704A	20X	2	3-7	179.75	173.05	3.107	2.21	0.18
704A	20X	2	20-22	179.91	173.21	3.114	2.26	0.21
704A	20X	2	33-37	180.05	173.35	3.120	2.11	0.25
704A	20X	2	46-48	180.17	173.47	3.125	2.14	0.25
704A	20X	2	63-67	180.35	173.65	3.132	2.08	0.19
704A	20X	2	68-72	180.40	173.70	3.134	2.23	0.28
704A	20X	2	80-82	180.51	173.81	3.139	2.20	0.16
704A	20X	2	98-102	180.70	174.00	3.147	2.18	0.08
704A	20X	2	123-127	180.95	174.25	3.157	2.57	-0.17
704A	20X	3	20-22	181.41	174.71	3.176	2.28	0.33
704A	20X	3	33-37	181.55	174.85	3.182	2.35	0.26
704A	20X	3	46-48	181.65	174.95	3.186	2.42	0.37
704A	20X	3	94-98	182.16	175.46	3.208	2.29	0.41
704A	20X	4	28-32	183.00	176.30	3.244	2.35	0.56
704A	20X	4	46-48	183.17	176.47	3.251	2.32	0.36

TABLE A2. (continued)

Hole	Core	Section	Interval, cm	Depth, mbsf	Composite depth, m	Age, Ma	$\delta^{18}O$, ‰, pdb	$\delta^{13}C$, ‰, pdb
704A	20X	4	62-66	183.34	176.64	3.259	2.13	0.31
704A	20X	4	68-72	183.40	176.70	3.261	2.20	0.38
704A	20X	4	80-82	183.51	176.81	3.266	1.93	0.37
704A	20X	4	95-99	183.67	176.97	3.273	2.13	0.30
704A	20X	4	113-117	183.85	177.15	3.281	2.16	0.42
704A	20X	4	140-142	184.11	177.41	3.292	1.81	0.22
704A	20X	4	142-146	184.14	177.44	3.293	1.80	0.32
704A	20X	4	3-7	184.25	177.55	3.298	1.97	0.35
704A	20X	4	20-22	184.41	177.71	3.305	1.73	0.20
704A	20X	4	36-40	184.58	177.88	3.312	2.00	0.30
704A	20X	4	62-66	184.84	178.14	3.323	2.08	0.19
704A	20X	4	68-72	184.90	178.20	3.326	1.84	0.42
704A	20X	4	95-99	185.17	178.47	3.337	1.86	0.10
704A	20X	4	110-112	185.31	178.61	3.343	1.83	0.27
704A	20X	4	124-128	185.46	178.76	3.349	1.90	0.17
704A	20X	4	42-44	186.14	179.44	3.379	2.10	0.20
704A	20X	4	80-82	186.51	179.81	3.394	1.74	0.17
704A	21X	1	4-8	187.76	181.06	3.460	2.34	0.23
704A	21X	1	15-18	187.86	181.16	3.466	2.11	0.00
704A	21X	1	25-29	187.97	181.27	3.472	1.85	0.07
704A	21X	1	36-40	188.08	181.38	3.478	1.98	0.06
704A	21X	1	55-57	188.26	181.56	3.487	1.95	-0.08
704A	21X	1	64-68	188.36	181.66	3.493	1.99	-0.07
704A	21X	1	90-94	188.62	181.92	3.507	2.19	-0.12
704A	21X	1	103-105	188.74	182.04	3.513	2.17	-0.08
704A	21X	1	119-122	188.90	182.20	3.522	2.10	-0.07
704A	21X	1	129-132	189.00	182.30	3.527	2.12	-0.02
704A	21X	1	144-147	189.16	182.46	3.536	2.00	0.00
704A	21X	2	6-9	189.27	182.57	3.542	1.91	-0.15
704A	21X	2	15-18	189.36	182.66	3.547	1.95	0.04
704A	21X	2	24-27	189.46	182.76	3.552	2.01	0.30
704A	21X	2	35-38	189.56	182.86	3.557	1.90	0.11
704A	21X	2	55-58	189.76	183.06	3.568	1.92	0.11
704A	21X	2	65-68	189.87	183.17	3.574	1.95	0.08
704A	21X	2	90-93	190.12	183.42	3.587	1.99	0.12
704A	21X	2	102-104	190.23	183.53	3.593	2.14	0.08
704A	21X	2	119-121	190.40	183.70	3.603	2.33	0.63
704A	21X	2	130-132	190.51	183.81	3.608	2.16	0.10
704A	21X	3	6-9	190.77	184.07	3.622	2.20	0.38
704A	21X	3	15-18	190.86	184.16	3.627	2.14	0.32
704A	21X	3	24-27	190.95	184.25	3.632	2.03	0.07
704A	21X	3	51-53	191.22	184.52	3.647	1.91	0.01
704A	21X	3	65-68	191.36	184.66	3.654	1.85	0.05
704A	21X	3	92-95	191.64	184.94	3.669	2.03	-0.06
704A	21X	3	98-100	191.69	184.99	3.672	1.95	0.03
704A	21X	3	116-118	191.87	185.17	3.682	2.08	-0.10
704A	21X	4	5-8	192.26	185.56	3.703	2.12	-0.02
704A	21X	4	15-18	192.36	185.66	3.708	2.03	-0.13
704A	21X	4	25-28	192.47	185.77	3.714	1.97	-0.16
704A	21X	4	37-40	192.58	185.88	3.720	1.85	-0.07
704A	21X	4	53-57	192.75	186.05	3.729	1.83	0.08
704A	21X	4	65-69	192.87	186.17	3.736	1.50	-0.02
704A	21X	4	78-80	192.99	186.29	3.742	1.99	-0.27
704A	21X	CC	5-9	193.08	186.38	3.747	1.89	0.05
704A	21X	CC	15-19	193.20	186.50	3.753	1.92	-0.17
704A	22X	1	10-14	197.32	190.62	3.905	2.07	-0.04
704A	22X	1	30-34	197.52	190.82	3.908	1.99	-0.10
704A	22X	1	43-45	197.64	190.94	3.909	2.35	0.30
704A	22X	1	52-56	197.74	191.04	3.911	2.03	0.01
704A	22X	1	62-66	197.84	191.14	3.912	2.10	-0.10
704A	22X	1	75-79	197.97	191.27	3.914	2.48	0.17
704A	22X	1	100-104	198.22	191.52	3.918	2.27	0.16
704A	22X	1	118-120	198.39	191.69	3.920	2.23	0.10
704A	22X	1	148-151	198.70	192.00	3.924	1.86	0.54
704A	22X	2	10-14	198.82	192.12	3.926	2.00	0.05
704A	22X	2	25-28	198.96	192.26	3.928	2.32	0.51
704A	22X	2	42-44	199.13	192.43	3.930	2.20	0.34
704A	22X	2	52-56	199.24	192.54	3.932	2.23	0.22
704A	22X	2	61-65	199.33	192.63	3.933	2.18	0.38
704A	22X	2	74-77	199.45	192.75	3.935	2.14	0.00
704A	22X	2	83-87	199.55	192.85	3.936	2.14	0.17

TABLE A2.   (continued)

Hole	Core	Section	Interval, cm	Depth, mbsf	Composite depth, m	Age, Ma	$\delta^{18}O$, ‰, pdb	$\delta^{13}C$, ‰, pdb
704A	22X	2	90-94	199.62	192.92	3.937	2.34	0.15
704A	22X	2	117-120	199.88	193.18	3.941	2.19	0.70
704A	22X	2	132-135	200.04	193.34	3.943	2.53	0.61
704A	22X	2	146-150	200.18	193.48	3.945	1.82	0.76
704A	22X	3	9-12	200.30	193.60	3.947	2.16	0.31
704A	22X	3	30-34	200.52	193.82	3.950	2.33	0.65
704A	22X	3	44-46	200.65	193.95	3.952	2.42	0.83
704A	22X	3	53-56	200.75	194.05	3.953	2.59	0.69
704A	22X	3	62-65	200.84	194.14	3.955	2.61	0.67
704A	22X	3	73-76	200.95	194.25	3.956	2.48	0.44
704A	22X	3	99-101	201.20	194.50	3.960	2.55	0.40
704A	22X	3	133-135	201.54	194.84	3.964	2.69	0.80
704A	22X	4	10-14	201.82	195.12	3.968	2.41	0.91
704A	22X	4	32-34	202.03	195.33	3.975	2.46	0.25
704A	22X	4	62-64	202.33	195.63	3.993	2.31	0.44
704A	22X	4	72-74	202.43	195.73	3.998	2.21	0.60
704A	22X	4	100-103	202.72	196.02	4.015	2.29	0.64
704A	22X	4	119-121	202.90	196.20	4.025	2.40	0.56
704A	22X	4	133-135	203.04	196.34	4.034	2.38	0.49
704A	22X	5	10-12	203.31	196.61	4.049	2.40	0.77
704A	22X	5	31-32	203.52	196.82	4.061	2.56	0.83
704A	22X	5	43-45	203.64	196.94	4.068	2.41	0.84
704A	22X	5	53-55	203.74	197.04	4.074	2.64	0.92
704A	22X	5	62-64	203.83	197.13	4.079	2.52	0.98
704A	22X	5	73-75	203.94	197.24	4.086	2.46	0.83
704A	22X	5	106-108	204.27	197.57	4.102	2.40	0.88
704A	22X	5	119-121	204.40	197.70	4.105	2.44	0.97
704A	22X	6	6-8	204.77	198.07	4.113	2.39	0.49
704A	22X	6	17-19	204.88	198.18	4.116	2.47	0.38
704A	22X	CC	5-7	205.33	198.63	4.126	2.15	0.86
704A	22X	CC	15-17	205.53	198.83	4.130	1.97	0.28
704A	23X	1	12-14	206.83	200.13	4.160	2.13	0.66
704A	23X	1	28-30	206.99	200.29	4.164	2.19	0.43
704A	23X	1	38-40	207.09	200.39	4.166	1.96	0.39
704A	23X	1	53-55	207.24	200.54	4.169	2.09	0.20
704A	23X	1	73-75	207.44	200.74	4.174	1.96	0.18
704A	23X	1	92-94	207.63	200.93	4.178	2.29	0.13
704A	23X	1	102-104	207.73	201.03	4.180	2.30	0.01
704A	23X	1	119-121	207.90	201.20	4.184	2.20	0.70
704A	23X	1	137-139	208.08	201.38	4.188	2.21	0.61
704A	23X	2	5-7	208.26	201.56	4.192	2.18	0.65
704A	23X	2	25-27	208.46	201.76	4.197	2.16	0.01
704A	23X	2	35-38	208.57	201.87	4.200	2.08	0.22
704A	23X	2	46-48	208.67	201.97	4.202	2.05	0.19
704A	23X	2	65-67	208.86	202.16	4.206	2.14	0.25
704A	23X	2	85-87	209.06	202.36	4.211	2.10	0.25
704A	23X	2	95-97	209.16	202.46	4.213	2.14	0.38
704A	23X	2	115-117	209.36	202.66	4.218	2.20	0.33
704A	23X	2	125-127	209.46	202.76	4.220	2.11	0.43
704A	23X	3	10-14	209.82	203.12	4.228	2.22	0.14
704A	23X	3	28-30	209.99	203.29	4.232	1.95	0.18
704A	23X	3	43-45	210.14	203.44	4.235	1.99	0.22
704A	23X	3	53-55	210.24	203.54	4.237	1.96	0.18
704A	23X	3	73-75	210.44	203.74	4.249	1.93	0.09
704A	23X	3	92-94	210.63	203.93	4.267	1.78	0.42
704A	23X	3	113-115	210.84	204.14	4.288	2.02	0.34
704A	23X	3	122-124	210.93	204.23	4.297	1.98	0.35
704A	23X	3	136-138	211.07	204.37	4.310	1.97	0.52
704A	23X	4	10-12	211.31	204.61	4.334	2.05	0.67
704A	23X	4	33-35	211.54	204.84	4.356	2.05	0.88
704A	23X	4	55-57	211.74	205.04	4.376	2.00	0.90
704A	23X	4	73-75	211.94	205.24	4.395	2.03	0.76
704A	23X	4	80-82	212.01	205.31	4.401	2.04	0.82
704A	23X	4	91-93	212.12	205.42	4.407	2.06	0.75
704A	23X	4	102-104	212.23	205.53	4.413	2.26	0.84
704A	23X	4	123-125	212.44	205.74	4.425	2.24	0.73
704A	23X	4	132-134	212.53	205.83	4.430	2.05	0.61
704A	23X	5	10-12	212.83	206.13	4.447	2.17	0.77
704A	23X	5	33-35	213.04	206.34	4.459	2.40	0.74
704A	23X	5	43-45	213.14	206.44	4.464	2.30	0.78
704A	23X	5	53-55	213.24	206.54	4.470	2.17	0.88

TABLE A2.   (continued)

Hole	Core	Section	Interval, cm	Depth, mbsf	Composite depth, m	Age, Ma	$\delta^{18}O$, ‰, pdb	$\delta^{13}C$, ‰, pdb
704A	23X	5	58-62	213.30	206.60	4.473	2.26	0.76
704A	23X	5	73-75	213.44	206.74	4.479	2.29	0.81
704A	23X	5	90-92	213.61	206.91	4.487	2.41	0.74
704A	23X	5	132-134	214.03	207.33	4.507	2.33	0.69
704A	23X	6	11-13	214.32	207.62	4.520	2.15	1.20
704A	23X	CC	4-6	214.57	207.87	4.532	2.09	0.82
704A	23X	CC	14-16	214.67	207.97	4.536	2.18	0.69
704A	23X	CC	29-31	214.82	208.12	4.543	2.24	0.60
704A	24X	1	10-12	216.31	209.61	4.615	1.98	0.35
704A	24X	1	20-22	216.41	209.71	4.620	2.08	0.39
704A	24X	1	30-32	216.51	209.81	4.625	2.09	0.08
704A	24X	1	40-44	216.62	209.92	4.630	1.96	-0.02
704A	24X	1	52-54	216.73	210.03	4.635	2.11	-0.04
704A	24X	1	62-65	216.84	210.14	4.641	2.26	0.13
704A	24X	1	68-72	216.90	210.20	4.644	2.29	0.31
704A	24X	1	83-85	217.04	210.34	4.651	2.40	-0.20
704A	24X	1	88-90	217.09	210.39	4.653	2.12	-0.03
704A	24X	1	98-100	217.19	210.49	4.658	1.97	-0.06
704A	24X	1	121-123	217.42	210.72	4.669	2.03	-0.17
704A	24X	1	133-135	217.54	210.84	4.675	1.95	0.09
704A	24X	2	10-12	217.81	211.11	4.688	1.94	0.33
704A	24X	2	41-43	218.12	211.42	4.704	2.27	-0.20
704A	24X	2	62-64	218.33	211.63	4.714	2.39	-0.33
704A	24X	2	80-82	218.51	211.81	4.723	1.93	0.12
704A	24X	2	100-102	218.71	212.01	4.733	1.92	0.04
704A	24X	3	30-32	219.51	212.81	4.774	2.26	0.46
704A	24X	4	132-134	222.03	215.33	5.051	2.14	0.72
704A	24X	5	10-12	222.31	215.61	5.081	2.21	0.68
704A	24X	5	20-22	222.41	215.71	5.092	2.11	0.50
704A	24X	5	30-32	222.51	215.81	5.103	2.38	0.45
704A	24X	5	41-43	222.62	215.92	5.115	2.38	0.57

TABLE A3.   Oxygen and Carbon Isotopic Results of *Cibicides* From Site 704

Hole	Core	Section	Interval, cm	Depth, mbsf	Composite depth, m	Age, Ma	$\delta^{18}O$, ‰, pdb	$\delta^{13}C$, ‰, pdb
704A	6H	1	54-56	45.75	48.35	1.058	3.17	-0.01
704A	6H	1	54-56	45.75	48.35	1.060	3.17	0.00
704A	6H	1	84-86	46.05	48.65	1.060	2.91	-0.13
704A	6H	1	84-86	46.05	48.65	1.062	3.04	0.30
704A	6H	1	105-107	46.26	48.86	1.064	3.45	0.01
704A	6H	1	130-134	46.52	49.12	1.065	3.00	0.51
704A	6H	1	140-142	46.61	49.21	1.066	2.39	0.35
704A	6H	2	4-6	46.75	49.35	1.067	2.88	-0.89
704A	6H	2	19-21	46.90	49.50	1.067	3.51	-0.25
704A	6H	2	19-21	46.90	49.50	1.070	3.42	-0.35
704A	6H	2	54-56	47.25	49.85	1.072	2.97	-0.83
704A	6H	2	80-82	47.61	50.21	1.073	3.34	-0.56
704A	6H	2	105-107	47.76	50.36	1.074	2.76	-0.26
704A	6H	2	120-124	47.90	50.50	1.075	2.69	-0.38
704A	6H	2	130-134	48.02	50.62	1.075	3.32	0.22
704A	6H	2	140-142	48.11	50.71	1.076	3.33	0.02
704A	6H	3	4-6	48.25	50.85	1.077	2.67	-0.23
704A	6H	3	19-21	48.40	51.00	1.080	2.78	-0.32
704A	6H	3	54-56	48.75	51.35	1.080	2.76	-0.17
704A	6H	3	54-56	48.75	51.35	1.082	3.03	-0.19
704A	6H	3	84-86	49.05	51.65	1.083	4.04	-0.78
704A	6H	3	105-107	49.26	51.86	1.087	4.14	-0.92
704A	6H	4	19-21	49.90	52.50	1.091	3.47	-1.47
704A	6H	4	54-56	50.25	52.85	1.095	4.00	-0.87
704A	6H	4	84-86	50.55	53.15	1.095	3.46	-1.52
704A	6H	4	84-86	50.55	53.15	1.098	4.06	-1.08
704A	6H	4	105-107	50.76	53.36	1.098	3.97	-0.96
704A	6H	4	105-107	50.76	53.36	1.098	3.83	-0.75
704A	6H	4	130-134	51.02	53.62	1.101	3.92	-0.81
704A	6H	4	140-142	51.11	53.71	1.102	3.94	-0.79

TABLE A3.   (continued)

Hole	Core	Section	Interval, cm	Depth, mbsf	Composite depth, m	Age, Ma	$\delta^{18}O$, ‰, pdb	$\delta^{13}C$, ‰, pdb
704A	6H	4	140-142	51.11	53.71	1.102	3.63	-1.15
704A	6H	5	19-21	51.40	54.00	1.106	2.87	-0.05
704A	6H	5	54-56	51.75	54.35	1.112	2.70	-0.16
704A	6H	5	84-86	52.05	54.65	1.122	3.86	-0.95
704A	6H	5	105-107	52.26	54.86	1.129	3.32	-1.37
704A	6H	5	130-134	52.52	55.12	1.133	3.60	-0.77
704A	6H	5	140-142	52.61	55.21	1.134	3.34	-0.76
704A	6H	5	140-142	52.61	55.21	1.134	3.22	-0.95
704A	6H	5	140-142	52.61	55.21	1.134	3.36	-0.96
704A	6H	6	19-21	52.90	55.50	1.138	3.51	-0.25
704A	6H	6	54-56	53.25	55.85	1.143	3.34	-0.14
704A	6H	6	84-86	53.55	56.15	1.147	2.82	-0.15
704A	6H	6	105-107	53.78	56.38	1.150	3.18	-0.50
704A	6H	6	130-134	54.02	56.62	1.153	2.92	0.04
704A	6H	6	140-142	54.11	56.71	1.155	3.30	-0.32
704A	6H	7	54-56	54.75	57.35	1.168	2.97	0.08
704B	7H	1	30-32	54.51	57.51	1.171	3.66	-0.87
704B	7H	1	50-53	54.71	57.71	1.174	3.63	-0.49
704B	7H	1	70-72	54.91	57.91	1.175	3.68	-0.69
704B	7H	1	90-93	55.11	58.11	1.176	3.52	-0.42
704B	7H	1	100-102	55.21	58.21	1.177	3.60	-0.54
704B	7H	1	110-112	55.31	58.31	1.178	3.47	-0.44
704B	7H	1	130-133	55.51	58.51	1.179	3.39	-0.20
704B	7H	2	8-10	55.79	58.79	1.181	3.32	-0.19
704B	7H	2	30-32	56.01	59.01	1.183	3.38	-0.22
704B	7H	2	50-52	56.21	59.21	1.184	3.39	-0.22
704B	7H	2	69-71	56.40	59.40	1.185	3.40	-0.41
704B	7H	2	89-91	56.60	59.60	1.187	3.40	-0.33
704B	7H	2	108-110	56.79	59.79	1.188	3.43	-0.43
704B	7H	2	130-132	57.01	60.01	1.189	3.38	-0.36
704B	7H	2	145-147	57.16	60.16	1.190	3.32	-0.13
704B	7H	3	9-11	57.30	60.30	1.191	3.15	-0.17
704B	7H	3	31-33	57.52	60.52	1.193	3.41	-0.51
704B	7H	3	50-52	57.71	60.71	1.195	3.57	-0.64
704B	7H	3	71-73	57.92	60.92	1.197	3.60	-0.68
704B	7H	3	90-92	58.11	61.11	1.199	3.63	-0.72
704A	7H	1	59-61	55.30	61.30	1.201	3.60	-0.86
704A	7H	1	90-92	55.61	61.61	1.204	3.60	-0.57
704A	7H	1	119-121	55.90	61.90	1.206	3.40	-0.57
704A	7H	2	10-12	56.31	62.31	1.210	3.48	-0.92
704A	7H	2	10-12	56.31	62.31	1.210	3.14	-0.94
704A	7H	2	29-31	56.50	62.50	1.212	3.18	-0.13
704A	7H	2	29-31	56.50	62.50	1.212	3.41	-0.40
704A	7H	2	29-31	56.50	62.50	1.212	3.72	-0.90
704A	7H	2	29-31	56.50	62.50	1.212	3.89	-0.94
704A	7H	2	59-61	56.80	62.80	1.216	3.46	-0.61
704A	7H	2	59-61	56.80	62.80	1.216	3.50	-0.79
704A	7H	2	59-61	56.80	62.80	1.216	3.60	-0.67
704A	7H	2	90-92	57.11	63.11	1.221	3.33	-0.48
704A	7H	2	90-92	57.11	63.11	1.221	3.47	-0.52
704A	7H	2	90-92	57.11	63.11	1.221	3.38	-0.34
704A	7H	2	119-121	57.40	63.40	1.225	3.09	0.13
704A	7H	2	119-121	57.40	63.40	1.225	2.80	-0.13
704A	7H	2	119-121	57.74	63.74	1.230	3.12	0.59
704A	7H	3	10-12	57.81	63.81	1.231	3.12	0.29
704A	7H	3	10-12	57.81	63.81	1.231	3.34	0.18
704A	7H	3	29-31	58.00	64.00	1.234	3.11	0.04
704A	7H	3	29-31	58.00	64.00	1.234	3.12	0.11
704A	7H	3	29-31	58.00	64.00	1.234	3.17	0.12
704A	7H	3	59-61	58.30	64.30	1.240	3.41	-0.16
704A	7H	3	59-61	58.30	64.30	1.240	3.64	-0.08
704A	7H	3	59-61	58.30	64.30	1.240	3.57	-0.60
704A	7H	3	59-61	58.30	64.30	1.240	3.67	-0.16
704A	7H	3	90-92	58.61	64.61	1.245	3.55	-0.97
704A	7H	3	119-121	58.90	64.90	1.250	3.74	-1.25
704A	7H	4	2-6	59.24	65.24	1.256	3.73	-0.83
704A	7H	4	10-12	59.31	65.31	1.257	3.48	-0.81
704A	7H	4	30-32	59.50	65.50	1.259	3.79	-0.40
704A	7H	4	30-32	59.50	65.50	1.259	3.54	-0.46
704A	7H	4	59-61	59.80	65.80	1.263	3.11	-0.37
704A	7H	4	59-61	59.80	65.80	1.263	3.11	-0.38

TABLE A3. (continued)

Hole	Core	Section	Interval, cm	Depth, mbsf	Composite depth, m	Age, Ma	$\delta^{18}O$, ‰, pdb	$\delta^{13}C$, ‰, pdb
704A	7H	4	90-92	60.11	66.11	1.267	2.98	0.25
704A	7H	4	119-121	60.40	66.40	1.270	2.98	0.13
704A	7H	4	119-121	60.40	66.40	1.270	3.21	0.34
704A	7H	5	10-12	60.81	66.81	1.275	2.92	0.28
704A	7H	5	10-12	60.81	66.81	1.275	3.07	0.48
704A	7H	5	29-31	61.00	67.00	1.277	3.39	0.34
704A	7H	5	90-92	61.61	67.61	1.283	3.68	-0.83
704A	7H	5	119-121	61.90	67.90	1.286	3.88	-0.73
704A	7H	6	2-6	62.24	68.24	1.289	3.75	-0.73
704A	7H	6	10-12	62.31	68.31	1.290	3.55	-0.95
704A	7H	6	29-31	62.50	68.50	1.292	3.62	-1.01
704A	7H	6	50-52	62.71	68.71	1.294	3.51	-0.78
704A	7H	6	90-92	63.11	69.11	1.297	3.37	-0.31
704A	7H	6	90-92	63.11	69.11	1.297	3.69	-0.14
704A	7H	7	2-6	63.74	69.74	1.303	3.80	-0.85
704A	8H	1	29-31	64.50	70.50	1.309	2.59	0.27
704A	8H	1	47-51	64.69	70.69	1.310	3.52	-0.04
704A	8H	1	67-69	64.88	70.88	1.312	3.44	-0.09
704A	8H	1	92-96	65.14	71.14	1.316	3.08	0.31
704A	8H	1	110-112	65.31	71.31	1.320	2.68	0.21
704A	8H	1	123-125	65.44	71.44	1.323	3.97	-0.83
704A	8H	2	5-9	65.77	71.77	1.330	3.53	-0.57
704A	8H	2	29-31	66.00	72.00	1.335	3.34	-1.11
704A	8H	2	40-44	66.12	72.12	1.337	3.92	-1.03
704A	8H	2	64-68	66.36	72.36	1.341	3.80	-0.82
704A	8H	2	84-88	66.56	72.56	1.344	3.60	-0.19
704A	8H	2	110-112	66.81	72.81	1.348	3.16	0.09
704A	8H	2	124-126	66.95	72.95	1.350	3.27	0.16
704A	8H	3	47-49	67.68	73.68	1.361	3.10	-0.41
704A	8H	3	65-69	67.87	73.87	1.364	4.02	-0.88
704A	8H	3	92-96	68.14	74.14	1.368	4.26	-0.61
704A	8H	3	110-112	68.31	74.31	1.370	3.38	-0.80
704A	8H	3	124-126	68.45	74.45	1.372	2.93	-0.76
704A	8H	3	139-141	68.60	74.60	1.376	3.66	-0.60
704A	8H	4	5-9	68.77	74.77	1.380	3.56	-0.20
704A	8H	4	100-102	69.71	75.71	1.405	3.85	-0.45
704A	8H	4	110-112	69.81	75.81	1.408	3.61	-0.60
704A	8H	4	124-126	69.95	75.95	1.412	4.13	-0.93
704A	8H	5	5-9	70.27	76.27	1.417	3.96	-1.23
704A	8H	5	16-19	70.37	76.37	1.418	4.07	-0.98
704A	8H	5	64-68	70.86	76.86	1.422	3.93	-0.24
704A	8H	5	90-92	71.11	77.11	1.424	3.77	-0.53
704A	8H	5	121-125	71.43	77.43	1.426	3.47	-0.12
704A	8H	6	14-18	71.86	77.86	1.429	3.81	-0.29
704A	8H	6	39-43	72.11	78.11	1.431	3.48	0.01
704A	8H	6	120-124	72.92	78.92	1.452	3.65	-0.37
704A	8H	CC	8-12	73.30	79.30	1.459	3.88	-0.49
704B	9H	2	145-147	76.16	79.31	1.459	3.83	-0.85
704B	9H	3	10-12	76.31	79.46	1.461	3.66	-0.68
704B	9H	3	32-34	76.53	79.68	1.464	3.61	-0.54
704B	9H	3	50-52	76.71	79.86	1.467	3.58	-0.49
704B	9H	3	100-112	77.26	80.41	1.474	3.44	-0.58
704B	9H	3	129-131	77.50	80.65	1.477	3.29	-0.49
704B	9H	3	146-148	77.67	80.82	1.479	3.19	0.01
704B	9H	4	12-14	77.83	80.98	1.480	3.11	0.12
704B	9H	4	71-73	78.42	81.57	1.485	3.20	0.02
704B	9H	4	89-91	78.60	81.75	1.487	3.24	-0.08
704B	9H	4	100-102	78.71	81.86	1.488	3.38	-0.11
704B	9H	4	131-133	79.02	82.17	1.490	3.46	-0.53
704B	9H	5	90-102	80.16	83.31	1.500	3.21	-0.14
704B	9H	5	125-145	80.57	83.72	1.503	3.43	-0.23
704B	9H	6	10-12	80.81	83.96	1.506	3.48	-0.35
704B	9H	6	28-30	80.99	84.14	1.507	3.52	-0.22
704B	9H	6	48-50	81.19	84.34	1.509	3.13	-0.06
704B	9H	6	68-70	81.39	84.54	1.511	3.35	-0.01
704B	9H	6	90-92	81.61	84.76	1.513	3.24	0.24
704B	9H	6	102-104	81.73	84.88	1.514	3.24	0.07
704B	9H	6	112-114	81.83	84.98	1.515	3.14	0.28
704B	9H	6	130-132	82.01	85.16	1.516	3.20	0.01
704B	9H	7	5-7	82.26	85.41	1.519	3.49	-0.36
704B	9H	7	31-33	82.52	85.67	1.521	3.61	-0.82

TABLE A3.   (continued)

Hole	Core	Section	Interval, cm	Depth, mbsf	Composite depth, m	Age, Ma	$\delta^{18}O$, ‰, pdb	$\delta^{13}C$, ‰, pdb
704B	9H	7	50-52	82.71	85.86	1.523	3.61	-0.78
704A	9H	5	128-132	81.00	86.00	1.524	4.12	-0.37
704A	9H	6	34-38	81.56	86.56	1.529	3.96	-0.65
704A	9H	6	63-67	81.85	86.85	1.532	4.20	-0.65
704A	9H	7	3-7	82.75	87.75	1.538	4.09	-0.89
704A	10H	1	3-7	83.25	88.25	1.541	4.00	-0.45
704A	10H	1	32-36	83.54	88.54	1.542	3.80	-0.75
704A	10H	1	62-66	83.84	88.84	1.543	3.88	-0.79
704A	10H	1	132-136	84.54	89.54	1.547	3.81	-0.63
704A	10H	2	3-7	84.75	89.75	1.548	3.64	-0.59
704A	10H	2	33-37	85.05	90.05	1.549	3.36	-0.15
704A	10H	2	63-67	85.35	90.35	1.552	3.52	-0.32
704A	10H	2	92-96	85.64	90.64	1.558	3.47	-0.26
704A	10H	2	142-145	86.14	91.14	1.562	3.74	-0.78
704A	10H	3	3-7	86.25	91.25	1.562	3.47	-0.72
704A	10H	3	32-36	86.54	91.54	1.564	3.69	-0.78
704A	10H	3	63-67	86.85	91.85	1.565	3.81	-0.51
704A	10H	3	95-97	87.17	92.17	1.566	3.55	-0.66
704A	10H	3	124-128	87.46	92.46	1.568	3.64	-0.59
704A	10H	3	143-147	87.65	92.65	1.568	3.80	-0.48
704A	10H	4	3-7	87.75	92.75	1.569	3.51	-0.80
704A	10H	4	33-37	88.05	93.05	1.570	4.11	-0.80
704A	10H	4	62-66	88.34	93.34	1.572	3.85	-0.54
704A	10H	4	92-96	88.64	93.64	1.573	3.65	-0.71
704A	10H	4	123-127	88.95	93.95	1.574	3.62	-0.90
704A	10H	4	142-146	89.14	94.14	1.575	3.81	-0.83
704A	10H	5	3-7	89.25	94.25	1.576	3.95	-0.74
704A	10H	5	28-32	89.50	94.50	1.577	3.97	-0.81
704A	10H	5	58-62	89.80	94.80	1.580	3.86	-1.03
704A	10H	5	123-127	90.45	95.45	1.586	3.56	-0.74
704A	10H	5	140-144	90.62	95.62	1.588	3.70	-0.71
704A	10H	6	3-7	90.75	95.75	1.589	3.39	-0.85
704A	10H	6	3-7	90.75	95.75	1.589	3.67	-0.60
704A	10H	6	28-32	91.00	96.00	1.591	3.64	-0.76
704A	10H	6	63-67	91.35	96.35	1.594	3.59	-0.63
704A	10H	6	63-67	91.35	96.35	1.594	3.53	-0.79
704A	10H	6	93-97	91.65	96.65	1.598	3.66	-0.57
704A	10H	6	93-97	91.65	96.65	1.598	3.51	-0.83
704A	10H	6	133-137	92.05	97.05	1.609	3.62	-0.85
704A	10H	6	133-137	92.05	97.05	1.609	3.49	-0.72
704A	10H	7	3-7	92.25	97.25	1.614	3.86	-0.99
704A	10H	7	3-7	92.25	97.25	1.614	3.83	-0.83
704A	10H	7	24-28	92.46	97.46	1.617	3.79	-0.94
704A	11H	1	5-9	92.77	97.77	1.620	3.81	-0.44
704A	11H	1	5-9	92.77	97.77	1.620	3.42	-0.64
704A	11H	1	29-33	93.01	98.01	1.622	3.47	-0.15
704A	11H	1	63-67	93.35	98.35	1.625	3.42	-0.42
704A	11H	1	63-67	93.35	98.35	1.625	3.19	-0.72
704A	11H	1	98-100	93.69	98.69	1.628	3.27	-0.47
704A	11H	1	129-131	94.00	99.00	1.631	3.22	-0.54
704A	11H	1	129-131	94.00	99.00	1.631	3.16	-0.53
704A	11H	2	3-7	94.25	99.25	1.633	3.35	-0.37
704A	11H	2	3-7	94.25	99.25	1.633	3.24	-0.48
704A	11H	2	35-37	94.57	99.57	1.636	2.98	-0.56
704A	11H	2	62-66	94.84	99.84	1.639	2.97	-0.45
704A	11H	2	88-92	95.10	100.10	1.641	3.03	-0.57
704A	11H	2	124-128	95.46	100.46	1.644	3.27	-0.74
704A	11H	3	3-7	95.75	100.75	1.647	3.34	-0.73
704A	11H	3	34-38	96.06	101.06	1.650	3.26	-0.64
704A	11H	3	63-67	96.35	101.35	1.653	3.38	-0.78
704A	11H	3	94-98	96.66	101.66	1.656	3.44	-0.70
704A	11H	3	124-128	96.96	101.96	1.658	3.39	-0.77
704A	11H	4	3-7	97.25	102.25	1.661	3.36	-0.79
704A	11H	4	33-37	97.55	102.55	1.663	3.37	-0.72
704A	11H	4	64-68	97.86	102.86	1.665	3.49	-0.44
704A	11H	4	98-102	98.20	103.20	1.667	3.42	-0.06
704A	11H	4	123-125	98.44	103.44	1.668	3.51	-0.41
704A	11H	5	3-7	98.75	103.75	1.670	3.75	-0.43
704A	11H	5	33-37	99.05	104.05	1.671	3.51	-0.33
704A	11H	5	97-101	99.69	104.69	1.675	3.72	-0.04
704A	11H	5	124-128	99.96	104.96	1.676	3.57	-0.26

TABLE A3. (continued)

Hole	Core	Section	Interval, cm	Depth, mbsf	Composite depth, m	Age, Ma	$\delta^{18}O$, ‰, pdb	$\delta^{13}C$, ‰, pdb
704A	11H	5	130-134	100.02	105.02	1.676	3.50	-0.23
704A	11H	5	142-146	100.14	105.14	1.677	3.57	-0.15
704A	11H	6	3-7	100.25	105.25	1.678	3.51	-0.07
704A	11H	6	33-37	100.55	105.55	1.679	3.13	0.09
704A	11H	6	50-52	100.71	105.71	1.680	2.56	-0.09
704A	11H	6	80-82	101.01	106.01	1.682	2.77	-0.09
704A	11H	6	85-89	101.07	106.07	1.682	3.15	0.10
704A	11H	6	110-112	101.31	106.31	1.683	2.77	-0.27
704A	11H	6	127-129	101.48	106.48	1.684	3.30	-0.10
704A	11H	7	3-7	101.75	106.75	1.686	3.46	-0.55
704A	11H	7	36-40	102.08	107.08	1.688	3.29	-0.20
704A	12H	1	3-7	102.25	107.25	1.689	3.07	-0.01
704A	12H	1	3-7	102.25	107.25	1.689	3.28	-0.16
704A	12H	1	27-31	102.49	107.49	1.690	2.85	-0.20
704A	12H	1	63-67	102.85	107.85	1.692	2.88	-0.21
704A	12H	1	90-94	103.12	108.12	1.693	3.17	-0.51
704A	12H	1	94-98	103.16	108.16	1.694	3.13	-0.19
704A	12H	1	124-128	103.46	108.46	1.695	3.13	-0.27
704A	12H	2	3-7	103.75	108.75	1.697	3.17	0.07
704A	12H	2	33-37	104.05	109.05	1.698	3.17	-0.36
704A	12H	2	63-67	104.35	109.35	1.700	3.57	-0.75
704A	12H	2	80-82	104.51	109.51	1.702	3.19	-0.58
704A	12H	2	90-94	104.62	109.62	1.703	3.27	-0.43
704A	12H	2	124-128	104.96	109.96	1.706	3.18	-0.48
704A	12H	2	124-128	104.96	109.96	1.706	3.45	-0.29
704A	12H	2	140-142	105.11	110.11	1.708	2.87	-0.75
704A	12H	3	3-7	105.25	110.25	1.710	2.99	-0.19
704A	12H	3	34-38	105.56	110.56	1.713	2.97	-0.04
704A	12H	3	85-89	106.07	111.07	1.718	2.85	0.10
704A	12H	3	90-94	106.12	111.12	1.719	2.96	-0.01
704A	12H	3	110-112	106.31	111.31	1.723	3.14	-0.43
704A	12H	3	125-127	106.46	111.46	1.727	2.78	0.18
704A	12H	3	140-142	106.61	111.61	1.730	2.84	0.27
704A	12H	4	3-7	106.75	111.75	1.733	2.88	0.10
704A	12H	4	3-7	106.75	111.75	1.733	3.21	0.03
704A	12H	4	20-22	106.91	111.91	1.736	3.20	-0.65
704A	12H	4	29-33	107.01	112.01	1.738	3.36	-0.74
704A	12H	4	57-61	107.29	112.29	1.743	3.33	-0.37
704A	12H	4	90-94	107.62	112.62	1.747	3.47	0.06
704A	12H	4	93-97	107.65	112.65	1.747	3.21	-0.13
704A	12H	4	128-132	108.00	113.00	1.751	3.01	0.15
704A	12H	5	34-38	108.56	113.56	1.758	3.30	-0.13
704A	12H	5	90-94	109.12	114.12	1.761	3.29	0.07
704A	12H	5	94-98	109.16	114.16	1.762	3.24	0.35
704A	12H	5	110-112	109.31	114.31	1.762	3.02	0.20
704A	12H	6	3-7	109.75	114.75	1.764	3.39	-0.13
704A	12H	6	20-22	109.91	114.91	1.765	3.45	0.03
704A	12H	6	43-47	110.15	115.15	1.766	3.35	-0.04
704A	12H	6	43-47	110.15	115.15	1.766	3.43	0.06
704A	12H	6	50-52	110.21	115.21	1.767	3.38	0.07
704A	12H	6	80-82	110.51	115.51	1.768	2.96	-0.23
704A	12H	6	90-94	110.62	115.62	1.769	3.32	-0.18
704A	12H	6	93-97	110.65	115.65	1.769	3.18	-0.05
704A	12H	6	124-128	110.96	115.96	1.771	3.25	-0.10
704A	12H	6	124-128	110.96	115.96	1.771	3.66	0.08
704A	12H	6	140-142	111.11	116.11	1.771	3.45	0.03
704A	13H	1	3-7	111.75	116.75	1.774	3.51	-0.15
704A	13H	1	33-37	112.05	117.05	1.776	3.48	-0.52
704A	13H	1	63-67	112.35	117.35	1.777	3.97	-0.60
704A	13H	1	80-82	112.51	117.51	1.778	4.00	-0.32
704A	13H	1	92-94	112.63	117.63	1.779	3.99	-0.63
704A	13H	1	127-129	112.98	117.98	1.781	3.84	-0.79
704A	13H	1	140-142	113.11	118.11	1.781	3.64	-0.77
704A	13H	2	3-7	113.25	118.25	1.782	3.57	0.18
704A	13H	2	20-22	113.41	118.41	1.783	3.27	-0.45
704A	13H	2	34-38	113.56	118.56	1.784	3.31	-0.45
704A	13H	2	64-68	113.86	118.86	1.786	4.01	-0.52
704A	13H	2	92-94	114.13	119.13	1.788	3.71	-0.37
704A	13H	2	95-99	114.17	119.17	1.788	3.98	-0.33
704A	13H	2	127-131	114.49	119.49	1.790	3.59	-0.54
704A	13H	3	3-7	114.75	119.75	1.791	3.43	-0.30

Hole	Core	Section	Interval, cm	Depth, mbsf	Composite depth, m	Age, Ma	$\delta^{18}O$, ‰, pdb	$\delta^{13}C$, ‰, pdb
704A	13H	3	33-37	115.05	120.05	1.793	3.35	-0.55
704A	13H	3	63-67	115.35	120.35	1.795	3.39	-0.35
704A	13H	3	92-96	115.64	120.64	1.797	3.50	0.21
704A	13H	3	95-99	115.67	120.67	1.797	3.07	-0.05
704A	13H	3	126-130	115.98	120.98	1.799	2.97	-0.06
704A	13H	4	3-7	116.25	121.25	1.801	3.03	-0.10
704A	13H	4	33-37	116.55	121.55	1.835	3.58	-0.29
704A	13H	4	63-67	116.85	121.85	1.841	3.46	-0.35
704A	13H	4	80-82	117.01	122.01	1.844	3.08	-0.56
704A	13H	4	93-97	117.15	122.15	1.847	3.02	0.75
704A	13H	4	95-99	117.17	122.17	1.848	3.18	-0.27
704A	13H	4	110-112	117.31	122.31	1.851	2.60	-0.34
704A	13H	4	140-142	117.61	122.61	1.854	2.53	0.11
704A	13H	5	20-22	117.91	122.91	1.857	2.39	-0.24
704A	13H	5	50-52	118.21	123.21	1.859	2.42	-0.76
704A	13H	5	95-99	118.67	123.67	1.862	3.46	-0.61
704A	13H	5	110-112	118.81	123.81	1.863	2.83	-0.82
704A	13H	6	3-7	119.25	124.25	1.867	3.31	-0.47
704A	13H	6	33-37	119.55	124.55	1.869	3.18	-0.13
704A	13H	6	63-67	119.85	124.85	1.872	3.08	-0.13
704A	13H	6	95-99	120.17	125.17	1.874	3.02	-0.11
704A	15H	4	92-96	136.14	126.84	1.893	2.70	0.36
704A	15H	4	124-128	136.46	127.16	1.898	3.12	0.27
704A	15H	5	35-39	137.07	127.77	1.908	3.17	-0.67
704A	15H	5	35-39	137.07	127.77	1.908	3.12	-0.47
704A	15H	5	63-67	137.35	128.05	1.910	3.31	-0.36
704A	15H	5	63-67	137.35	128.05	1.910	3.05	-0.24
704A	15H	5	92-96	137.64	128.34	1.913	2.75	0.25
704A	15H	6	63-67	138.85	129.55	1.923	2.78	0.36
704A	15H	6	93-97	139.15	129.85	1.924	2.87	0.37
704A	15H	7	2-6	139.74	130.44	1.927	2.96	0.47
704A	15H	7	32-36	140.05	130.75	1.928	3.01	0.05
704A	16H	1	83-87	141.05	131.75	1.932	3.32	-0.05
704A	16H	1	90-94	141.15	131.85	1.932	3.23	-0.28
704A	16H	1	123-127	141.50	132.20	1.934	3.22	0.11
704A	16H	2	19-23	141.91	132.61	1.935	3.48	-0.32
704A	16H	2	38-42	142.10	132.80	1.936	3.60	-0.37
704A	16H	2	138-142	143.10	133.80	1.940	3.29	-0.05
704A	16H	3	3-7	143.25	133.95	1.941	3.29	-0.02
704A	16H	3	33-37	143.55	134.25	1.953	3.07	0.52
704A	16H	3	43-47	143.65	134.35	1.958	3.45	-0.22
704A	16H	3	64-67	143.85	134.55	1.966	3.15	-0.27
704A	16H	3	64-67	143.85	134.55	1.966	3.28	-0.41
704A	16H	3	95-99	144.17	134.87	1.979	2.98	0.41
704A	16H	4	94-98	145.66	136.36	1.997	2.81	0.53
704A	16H	4	128-132	146.00	136.70	2.000	3.30	0.36
704A	17X	1	36-40	150.08	140.78	2.021	3.14	-0.10
704A	17X	1	63-67	150.35	141.05	2.022	3.15	-0.35
704A	17X	1	93-97	150.65	141.35	2.022	3.12	0.24
704A	17X	2	33-37	151.55	142.25	2.025	2.97	0.39
704A	17X	2	80-82	152.01	142.71	2.026	3.05	0.18
704A	17X	2	95-99	152.17	142.87	2.027	3.32	0.26
704A	17X	2	119-123	152.41	143.11	2.029	3.38	-0.28
704A	17X	2	139-143	152.61	143.31	2.033	3.27	0.06
704A	17X	3	3-7	152.75	143.45	2.036	2.96	0.48
704A	17X	3	64-68	153.36	144.06	2.049	2.97	0.45
704A	17X	3	97-101	153.69	144.39	2.052	3.03	-0.04
704A	17X	3	126-130	153.98	144.68	2.055	3.02	0.09
704A	17X	4	154.25	154.25	144.95	2.057	2.94	0.08
704A	17X	4	34-38	154.56	145.26	2.060	3.05	0.32
704A	17X	4	63-67	154.85	145.55	2.063	3.38	-0.28
704A	17X	4	95-99	155.17	145.87	2.066	3.45	-0.11
704A	17X	4	126-130	155.48	146.18	2.069	3.25	0.30
704A	17X	5	33-37	156.05	146.75	2.077	3.44	-0.43
704A	17X	5	63-67	156.35	147.05	2.082	2.93	0.00
704A	17X	5	127-131	156.99	147.69	2.091	3.06	0.17
704A	17X	6	3-7	157.25	147.95	2.097	3.01	0.32
704A	17X	6	27-31	157.49	148.19	2.102	3.24	-0.15
704A	17X	6	62-66	157.84	148.54	2.109	3.21	0.03
704B	18X	1	75-79	157.47	148.57	2.110	3.74	-0.58
704B	18X	1	90-94	157.62	148.72	2.114	3.66	-0.52

TABLE A3. (continued)

Hole	Core	Section	Interval, cm	Depth, mbsf	Composite depth, m	Age, Ma	$\delta^{18}O$, ‰, pdb	$\delta^{13}C$, ‰, pdb
704B	18X	1	105-109	157.77	148.87	2.118	3.29	-0.11
704B	18X	1	120-139	158.00	149.10	2.124	3.21	0.03
704B	18X	2	0-4	158.22	149.32	2.130	3.02	-0.01
704B	18X	2	15-34	158.45	149.55	2.137	3.11	0.35
704B	18X	2	45-64	158.75	149.85	2.140	3.00	0.25
704B	18X	2	75-79	158.97	150.07	2.141	2.99	0.39
704B	18X	3	15-19	159.87	150.97	2.147	3.47	-0.36
704B	18X	3	30-34	160.02	151.12	2.148	3.48	-0.44
704B	18X	3	45-49	160.17	151.27	2.149	3.55	-0.30
704B	18X	3	60-64	160.32	151.42	2.150	3.29	0.01
704B	18X	3	75-79	160.47	151.57	2.151	3.47	-0.65
704B	18X	3	90-94	160.62	151.72	2.152	3.75	-0.82
704B	18X	3	140-144	161.12	152.22	2.176	3.52	-0.38
704B	18X	4	0-4	161.22	152.32	2.179	3.34	-0.31
704B	18X	4	15-19	161.37	152.47	2.182	3.25	0.06
704B	18X	4	30-34	161.52	152.62	2.186	3.27	0.03
704B	18X	4	60-64	161.82	152.92	2.193	3.43	-0.56
704B	18X	4	75-79	161.97	153.07	2.196	3.36	0.03
704B	18X	4	90-94	162.12	153.22	2.203	3.38	-0.12
704B	18X	4	105-109	162.27	153.37	2.215	3.19	0.34
704B	18X	4	135-139	162.57	153.67	2.226	3.24	0.36
704B	18X	4	140-144	162.62	153.72	2.228	3.28	0.06
704B	18X	5	0-4	162.72	153.82	2.230	3.54	-0.36
704B	18X	5	30-34	163.02	154.12	2.237	3.32	0.23
704B	18X	5	45-49	163.17	154.27	2.241	3.45	0.33
704B	18X	5	60-64	163.32	154.42	2.245	3.53	-0.65
704B	18X	5	75-79	163.47	154.57	2.250	3.28	-0.02
704B	18X	5	105-109	163.77	154.87	2.258	3.05	0.27
704B	18X	5	120-124	163.92	155.02	2.261	3.13	0.20
704B	18X	5	135-139	164.07	155.17	2.263	3.20	-0.03
704B	18X	6	0-4	164.22	155.32	2.264	3.32	-0.05
704B	18X	6	60-64	164.82	155.92	2.271	3.32	-0.17
704B	18X	6	105-109	165.27	156.37	2.281	3.48	-0.48
704B	18X	6	120-124	165.42	156.52	2.289	3.24	0.29
704B	18X	CC	5-9	165.77	156.87	2.290	3.19	0.53
704B	19X	1	5-9	166.27	157.37	2.603	3.37	0.11
704B	19X	1	15-19	166.37	157.47	2.606	3.29	-0.48
704B	19X	1	75-79	166.97	158.07	2.622	3.38	0.36
704B	19X	1	105-109	167.27	158.37	2.629	3.35	0.39
704B	19X	1	135-139	167.57	158.67	2.637	3.40	0.01
704B	19X	2	0-4	167.72	158.82	2.641	3.75	-0.53
704B	19X	2	15-19	167.87	158.97	2.645	3.86	-0.79
704B	19X	2	30-34	168.02	159.12	2.649	3.56	-0.05
704B	19X	2	60-64	168.32	159.42	2.657	3.34	-0.04
704B	19X	2	75-79	168.47	159.57	2.661	3.23	0.14
704B	19X	2	135-137	169.06	160.16	2.677	3.31	0.61
704B	19X	2	140-142	169.11	160.21	2.678	3.21	0.59
704B	19X	3	0-4	169.22	160.32	2.681	3.24	0.62
704B	19X	3	15-19	169.37	160.47	2.685	3.29	0.60
704B	19X	3	30-34	169.52	160.62	2.689	3.21	0.58
704B	19X	3	45-49	169.67	160.77	2.693	3.38	0.49
704B	19X	3	60-64	169.82	160.92	2.697	3.39	0.43
704B	19X	3	75-79	169.97	161.07	2.701	3.24	0.44
704B	19X	3	90-94	170.12	161.22	2.705	3.51	0.39
704B	19X	3	105-109	170.27	161.37	2.709	3.45	0.23
704B	19X	3	120-124	170.42	161.52	2.712	3.41	0.38
704B	19X	3	135-139	170.57	161.67	2.716	3.51	0.33
704B	19X	CC	5-9	170.70	161.80	2.720	3.36	0.22
704B	19X	CC	15-19	170.87	161.97	2.724	3.46	0.23
704A	19X	1	6-10	168.78	162.08	2.727	3.39	0.13
704A	19X	1	20-22	168.91	162.21	2.731	3.34	0.25
704A	19X	1	30-34	169.02	162.32	2.734	3.67	0.04
704A	19X	1	33-37	169.05	162.35	2.734	3.39	0.02
704A	19X	1	50-52	169.21	162.51	2.739	3.12	0.63
704A	19X	1	75-77	169.46	162.76	2.745	3.47	-0.43
704A	19X	1	87-91	169.59	162.89	2.749	3.41	-0.31
704A	19X	1	110-142	169.96	163.26	2.758	3.27	0.25
704A	19X	1	143-147	170.15	163.45	2.763	3.22	0.73
704A	19X	2	1-5	170.52	163.82	2.770	3.19	0.29
704A	19X	2	20-22	170.41	163.71	2.773	3.29	0.44
704A	19X	2	30-34	170.52	163.82	2.773	3.30	0.56

TABLE A3.   (continued)

Hole	Core	Section	Interval, cm	Depth, mbsf	Composite depth, m	Age, Ma	$\delta^{18}O$, ‰, pdb	$\delta^{13}C$, ‰, pdb
704A	19X	2	46-50	170.68	163.98	2.777	3.11	0.88
704A	19X	2	50-52	170.71	164.01	2.778	3.12	0.64
704A	19X	2	76-80	170.98	164.28	2.785	3.14	0.52
704A	19X	2	93-97	171.15	164.45	2.790	3.04	0.79
704A	19X	2	110-112	171.31	164.61	2.794	2.79	0.40
704A	19X	2	125-142	171.54	164.84	2.800	2.97	0.54
704A	19X	2	140-142	171.61	164.91	2.802	2.91	0.66
704A	19X	2	140-142	171.61	164.91	2.802	2.99	0.62
704A	19X	3	20-22	171.78	165.08	2.806	3.24	1.12
704A	19X	3	27-29	171.98	165.28	2.811	3.26	0.77
704A	19X	3	27-29	171.98	165.28	2.811	3.21	0.60
704A	19X	3	30-34	172.02	165.32	2.813	3.14	0.59
704A	19X	3	50-52	172.21	165.51	2.818	2.86	0.57
704A	19X	3	62-66	172.34	165.64	2.821	2.87	0.71
704A	19X	3	82-86	172.54	165.84	2.826	2.82	0.72
704A	19X	3	94-98	172.66	165.96	2.829	3.12	0.69
704A	19X	3	94-98	172.66	165.96	2.829	3.07	0.52
704A	19X	3	110-112	172.81	166.11	2.833	3.33	0.75
704A	19X	3	112-116	172.84	166.14	2.834	3.21	0.90
704A	19X	3	112-116	172.84	166.14	2.834	3.16	0.73
704A	19X	3	128-132	173.00	166.30	2.838	3.17	0.90
704A	19X	3	128-132	173.00	166.30	2.838	3.12	0.73
704A	19X	3	140-142	173.11	166.41	2.841	3.01	0.80
704A	19X	4	1-5	173.23	166.53	2.844	3.05	0.76
704A	19X	4	1-5	173.23	166.53	2.844	3.00	0.59
704A	19X	4	30-34	173.52	166.82	2.852	3.00	0.27
704A	19X	4	34-38	173.56	166.86	2.853	2.98	0.49
704A	19X	4	66-70	173.88	167.18	2.862	2.80	0.64
704A	19X	4	86-90	174.07	167.37	2.867	2.87	0.89
704A	19X	4	124-128	174.46	167.76	2.877	3.08	0.62
704A	19X	4	124-128	174.46	167.76	2.877	3.18	0.46
704A	19X	4	140-142	174.61	167.91	2.881	2.75	0.57
704A	19X	4	140-142	174.61	167.91	2.881	2.76	0.43
704A	19X	5	1-5	174.73	168.03	2.884	2.95	0.51
704A	19X	5	1-5	174.73	168.03	2.884	3.13	0.69
704A	19X	5	20-22	174.91	168.21	2.889	3.20	0.60
704A	19X	5	30-34	175.02	168.32	2.892	3.25	0.51
704A	19X	5	42-44	175.13	168.43	2.894	3.33	0.45
704A	19X	5	50-52	175.21	168.51	2.897	3.29	0.25
704A	19X	5	60-62	175.31	168.61	2.899	3.39	0.28
704A	19X	5	76-80	175.48	168.78	2.904	3.36	0.01
704A	19X	5	98-102	175.70	169.00	2.909	3.23	-0.15
704A	19X	5	110-112	175.81	169.11	2.912	3.17	0.10
704A	19X	5	123-127	175.95	169.25	2.916	3.00	0.20
704A	19X	5	140-147	176.13	169.43	2.921	3.08	0.26
704A	19X	6	1-5	176.23	169.53	2.925	3.20	0.20
704A	19X	6	20-22	176.41	169.71	2.933	3.07	0.41
704A	19X	6	34-38	176.56	169.86	2.939	2.96	0.52
704A	19X	6	60-62	176.81	170.11	2.949	3.25	0.81
704A	19X	6	74-78	176.96	170.26	2.955	2.98	0.70
704A	19X	6	74-78	176.96	170.26	2.955	3.01	0.72
704A	19X	6	74-78	176.96	170.26	2.955	2.89	0.74
704A	19X	6	93-97	177.15	170.45	2.963	2.96	0.67
704A	19X	6	93-97	177.15	170.45	2.963	3.04	0.70
704A	19X	6	123-127	177.45	170.75	2.976	3.01	0.59
704A	19X	6	140-142	177.61	170.91	2.982	3.02	0.61
704A	19X	6	143-147	177.65	170.95	2.984	3.04	0.52
704A	19X	6	143-147	177.65	170.95	2.984	2.99	0.58
704A	19X	7	3-7	177.75	171.05	2.988	3.11	0.41
704A	19X	7	18-22	177.90	171.20	2.997	3.31	0.15
704A	19X	7	18-22	177.90	171.20	2.997	3.20	0.14
704A	20X	1	2-6	178.24	171.54	3.020	2.79	0.56
704A	20X	1	20-22	178.41	171.71	3.032	3.01	0.81
704A	20X	1	27-31	178.49	171.79	3.038	2.81	0.78
704A	20X	1	27-31	178.49	171.79	3.038	2.66	0.71
704A	20X	1	46-48	178.67	171.97	3.050	2.80	0.89
704A	20X	1	62-66	178.84	172.14	3.062	2.80	0.84
704A	20X	1	68-72	178.90	172.20	3.066	2.82	0.87
704A	20X	1	94-98	179.16	172.46	3.083	2.91	0.85
704A	20X	1	94-112	179.24	172.54	3.086	2.96	0.80
704A	20X	1	110-112	179.31	172.61	3.089	2.73	0.66

TABLE A3.   (continued)

Hole	Core	Section	Interval, cm	Depth, mbsf	Composite depth, m	Age, Ma	$\delta^{18}O$, ‰, pdb	$\delta^{13}C$, ‰, pdb
704A	20X	1	123-127	179.45	172.75	3.095	2.80	0.67
704A	20X	1	140-142	179.61	172.91	3.101	2.71	0.85
704A	20X	1	143-147	179.65	172.95	3.103	2.79	0.86
704A	20X	2	33-37	179.75	173.05	3.107	2.73	0.76
704A	20X	2	33-37	180.05	173.35	3.120	2.82	0.79
704A	20X	2	46-48	180.17	173.47	3.125	2.84	0.58
704A	20X	2	16-48	180.17	173.47	3.125	2.84	0.66
704A	20X	2	63-67	180.35	173.65	3.132	2.89	0.61
704A	20X	2	63-67	180.35	173.65	3.132	2.92	0.66
704A	20X	2	68-72	180.40	173.70	3.134	2.94	0.65
704A	20X	2	68-72	180.40	173.70	3.134	2.96	0.64
704A	20X	2	80-82	180.51	173.81	3.139	2.91	0.54
704A	20X	2	98-102	180.70	174.00	3.147	3.02	0.75
704A	20X	2	123-127	180.95	174.25	3.157	3.23	0.37
704A	20X	2	123-127	180.95	174.25	3.157	3.21	0.30
704A	20X	2	140-142	181.11	174.41	3.164	3.09	0.12
704A	20X	2	140-142	181.11	174.41	3.164	3.08	0.09
704A	20X	2	143-147	181.15	174.45	3.165	3.06	0.23
704A	20X	3	3-7	181.25	174.55	3.170	2.99	0.45
704A	20X	3	3-7	181.25	174.55	3.170	3.00	0.50
704A	20X	3	33-37	181.55	174.85	3.182	2.79	0.39
704A	20X	3	94-98	182.16	175.46	3.208	2.79	0.66
704A	20X	3	121-125	182.43	175.73	3.220	2.77	0.65
704A	20X	4	28-32	183.00	176.30	3.244	2.79	0.69
704A	20X	4	62-66	183.34	176.64	3.259	2.84	0.61
704A	20X	4	80-82	183.51	176.81	3.266	2.57	0.30
704A	20X	4	95-99	183.67	176.97	3.273	2.70	0.49
704A	20X	4	113-117	183.85	177.15	3.281	2.81	0.41
704A	20X	4	140-142	184.11	177.41	3.292	2.92	0.43
704A	20X	4	142-146	184.14	177.44	3.293	2.61	0.52
704A	20X	5	3-7	184.25	177.55	3.298	2.68	0.67
704A	20X	5	20-22	184.41	177.71	3.305	2.54	0.86
704A	20X	5	36-40	184.58	177.88	3.312	2.72	0.44
704A	20X	5	62-66	184.84	178.14	3.323	2.89	0.73
704A	20X	5	68-72	184.90	178.20	3.326	2.84	0.62
704A	20X	5	95-99	185.17	178.47	3.337	2.71	0.78
704A	20X	5	110-112	185.31	178.61	3.343	2.48	0.69
704A	20X	6	3-7	185.75	179.05	3.362	2.74	0.65
704A	20X	6	20-22	185.91	179.21	3.369	2.67	0.38
704A	20X	6	22-26	185.94	179.24	3.370	2.67	0.49
704A	20X	6	42-44	186.14	179.44	3.379	2.73	0.69
704A	20X	6	62-66	186.34	179.64	3.387	2.83	0.70
704A	20X	6	110-112	186.81	180.11	3.409	2.61	0.79
704A	20X	7	3-7	187.25	180.55	3.433	2.69	0.48
704A	20X	7	20-22	187.41	180.71	3.441	3.08	0.38
704A	20X	7	28-32	188.00	181.30	3.460	2.73	0.25
704A	20X	7	44-46	188.16	181.46	3.466	2.39	0.49
704A	21X	1	4-8	187.76	181.06	3.472	2.67	0.50
704A	21X	1	15-18	187.87	181.17	3.473	2.76	0.31
704A	21X	1	25-29	187.97	181.27	3.473	2.63	0.46
704A	21X	1	28-32	188.00	181.30	3.478	2.59	0.42
704A	21X	1	36-40	188.08	181.38	3.482	2.65	0.49
704A	21X	1	45-49	188.17	181.47	3.482	2.66	0.37
704A	21X	1	55-57	188.26	181.56	3.487	2.59	0.35
704A	21X	1	64-68	188.36	181.66	3.493	2.69	0.18
704A	21X	1	90-94	188.62	181.92	3.507	2.61	0.12
704A	21X	1	103-105	188.74	182.04	3.513	2.94	0.04
704A	21X	1	119-122	188.90	182.20	3.522	2.78	0.22
704A	21X	1	129-132	189.00	182.30	3.527	2.84	0.17
704A	21X	1	144-147	189.16	182.46	3.536	2.56	0.25
704A	21X	2	6-9	189.28	182.58	3.542	2.58	0.02
704A	21X	2	15-18	189.36	182.66	3.547	2.56	0.08
704A	21X	2	24-27	189.46	182.76	3.552	2.69	0.49
704A	21X	2	35-38	189.56	182.86	3.557	2.45	0.30
704A	21X	2	45-47	189.66	182.96	3.563	2.60	0.28
704A	21X	2	55-58	189.76	183.06	3.568	2.66	0.28
704A	21X	2	65-68	189.86	183.16	3.573	2.70	0.21
704A	21X	2	90-93	190.11	183.41	3.587	2.59	0.14
704A	21X	2	102-104	190.23	183.53	3.593	2.64	0.20
704A	21X	2	119-121	190.40	183.70	3.603	2.65	0.38
704A	21X	2	130-132	190.51	183.81	3.608	2.94	0.20

TABLE A3. (continued)

Hole	Core	Section	Interval, cm	Depth, mbsf	Composite depth, m	Age, Ma	$\delta^{18}O$, ‰, pdb	$\delta^{13}C$, ‰, pdb
704A	21X	3	6-9	190.77	184.07	3.622	2.69	0.40
704A	21X	3	15-18	190.86	184.16	3.627	2.79	0.27
704A	21X	3	24-27	190.95	184.25	3.632	2.58	0.10
704A	21X	3	51-53	191.22	184.52	3.647	2.64	0.41
704A	21X	3	65-68	191.36	184.66	3.654	2.58	0.27
704A	21X	3	92-95	191.63	184.93	3.669	2.60	0.53
704A	21X	3	98-100	191.69	184.99	3.672	2.63	0.40
704A	21X	3	116-118	191.87	185.17	3.682	2.55	0.53
704A	21X	4	15-18	192.37	185.67	3.709	2.61	0.18
704A	21X	4	37-40	192.58	185.88	3.720	2.64	0.33
704A	21X	4	53-57	192.75	186.05	3.729	2.63	0.30
704A	21X	4	65-69	192.87	186.17	3.736	2.63	0.43
704A	21X	4	78-80	192.99	186.29	3.742	2.75	0.13
704A	21X	CC	5-9	193.77	187.07	3.784	2.58	0.47
704A	21X	CC	15-19	193.87	187.17	3.789	2.65	0.49
704A	22X	1	10-14	197.32	190.62	3.905	2.67	0.37
704A	22X	1	20-22	197.41	190.71	3.906	2.66	0.56
704A	22X	1	30-34	197.52	190.82	3.908	2.71	0.36
704A	22X	1	43-45	197.64	190.94	3.909	2.72	0.40
704A	22X	1	52-56	197.74	191.04	3.911	2.60	0.37
704A	22X	1	100-104	198.22	191.52	3.918	2.61	0.29
704A	22X	1	118-120	198.39	191.69	3.920	2.39	0.27
704A	22X	1	132-135	198.54	191.84	3.922	2.39	0.32
704A	22X	1	139-141	198.60	191.90	3.923	2.46	0.27
704A	22X	1	148-151	198.70	192.00	3.924	2.52	0.11
704A	22X	2	10-14	198.82	192.12	3.926	2.43	0.46
704A	22X	2	25-28	198.96	192.26	3.928	2.41	0.29
704A	22X	2	52-56	199.24	192.54	3.932	2.55	0.16
704A	22X	2	61-65	199.33	192.63	3.933	2.63	0.31
704A	22X	2	74-77	199.45	192.75	3.935	2.68	0.00
704A	22X	2	102-104	199.73	193.03	3.939	2.75	0.34
704A	22X	2	132-135	200.03	193.33	3.943	2.63	0.12
704A	22X	2	146-150	200.18	193.48	3.945	2.69	0.31
704A	22X	3	44-46	200.65	193.95	3.952	2.71	0.45
704A	22X	3	52-56	200.74	194.04	3.953	2.67	0.44
704A	22X	3	62-65	200.84	194.14	3.955	2.70	0.80
704A	22X	3	90-94	201.13	194.43	3.959	2.77	0.69
704A	22X	3	120-122	201.41	194.71	3.963	2.88	0.37
704A	22X	4	10-13	201.81	195.11	3.968	2.80	0.60
704A	22X	4	32-34	202.03	195.33	3.975	2.69	0.22
704A	22X	4	62-64	202.33	195.63	3.993	2.68	-0.08
704A	22X	4	72-74	202.43	195.73	3.998	2.51	0.31
704A	22X	5	53-55	203.74	197.04	4.074	2.81	0.08
704A	22X	5	81-83	204.02	197.32	4.090	2.95	-0.13
704A	22X	5	106-108	204.27	197.57	4.102	2.96	-0.02
704A	22X	6	6-8	206.27	199.57	4.147	2.62	0.58
704A	22X	6	5-7	206.26	199.56	4.147	2.75	0.14
704A	22X	6	15-17	206.36	199.66	4.149	2.59	0.32
704A	22X	6	25-27	206.46	199.76	4.152	2.70	0.21
704A	23X	1	12-14	206.83	200.13	4.160	2.74	0.17
704A	23X	1	28-30	206.99	200.29	4.164	2.87	0.13
704A	23X	1	38-40	207.09	200.39	4.166	2.85	0.12
704A	23X	1	73-75	207.44	200.74	4.174	2.75	0.27
704A	23X	1	92-94	207.63	200.93	4.178	2.74	0.20
704A	23X	1	102-104	207.73	201.03	4.180	2.94	0.21
704A	23X	1	119-121	207.90	201.20	4.184	2.68	0.32
704A	23X	1	137-139	208.08	201.38	4.188	2.80	0.32
704A	23X	2	5-7	208.26	201.56	4.192	2.86	0.44
704A	23X	2	25-27	208.46	201.76	4.197	2.82	0.22
704A	23X	2	35-38	208.56	201.86	4.199	2.90	0.19
704A	23X	2	46-48	208.67	201.97	4.202	2.91	0.28
704A	23X	2	65-67	208.86	202.16	4.206	2.79	0.31
704A	23X	2	95-96	209.16	202.46	4.213	2.87	0.25
704A	23X	2	115-117	209.36	202.66	4.218	2.77	0.20
704A	23X	2	125-127	209.46	202.76	4.220	2.75	0.34
704A	23X	3	12-14	209.83	203.13	4.228	2.85	0.00
704A	23X	3	28-30	209.99	203.29	4.232	2.72	0.36
704A	23X	3	43-45	210.14	203.44	4.235	2.71	0.43
704A	23X	3	73-75	210.44	203.74	4.249	2.70	0.20
704A	23X	3	92-94	210.63	203.93	4.267	2.78	0.17
704A	23X	3	113-115	210.84	204.14	4.288	2.66	0.40

TABLE A3. (continued)

Hole	Core	Section	Interval, cm	Depth, mbsf	Composite depth, m	Age, Ma	$\delta^{18}O$, ‰, pdb	$\delta^{13}C$, ‰, pdb
704A	23X	3	122-124	210.93	204.23	4.297	2.71	0.41
704A	23X	3	136-138	211.07	204.37	4.310	2.75	0.38
704A	23X	4	20-22	211.41	204.71	4.343	3.01	0.30
704A	23X	4	58-62	211.80	205.10	4.381	3.00	0.39
704A	23X	4	73-75	211.94	205.24	4.395	2.91	0.23
704A	23X	4	91-93	212.12	205.42	4.407	2.83	0.45
704A	23X	4	102-104	212.23	205.53	4.413	2.93	0.29
704A	23X	4	110-112	212.31	205.61	4.418	2.92	0.31
704A	23X	4	123-125	212.44	205.74	4.425	2.88	0.18
704A	23X	4	132-134	212.53	205.83	4.430	2.93	0.12
704A	23X	4	140-142	212.61	205.91	4.435	2.78	0.40
704A	23X	5	12-14	212.83	206.13	4.447	2.93	0.39
704A	23X	5	20-22	212.91	206.21	4.452	2.79	0.48
704A	23X	5	33-35	213.04	206.34	4.459	2.78	0.42
704A	23X	5	43-45	213.14	206.44	4.464	2.84	0.25
704A	23X	5	53-55	213.24	206.54	4.470	2.75	0.41
704A	23X	5	73-75	213.44	206.74	4.479	2.53	0.45
704A	23X	5	80-82	213.51	206.81	4.483	2.61	0.19
704A	23X	5	90-92	213.61	206.91	4.487	2.69	0.05
704A	23X	5	111-142	213.96	207.26	4.503	2.63	0.09
704A	23X	6	11-13	214.32	207.62	4.520	2.66	0.33
704A	23X	6	4-6	214.50	207.80	4.528	2.86	0.26
704A	23X	6	14-16	214.60	207.90	4.533	2.87	0.23
704A	23X	6	29-31	214.75	208.05	4.540	2.74	0.07
704A	24X	1	10-12	216.31	209.61	4.615	2.77	0.33
704A	24X	1	30-32	216.51	209.81	4.625	2.75	0.33
704A	24X	1	40-44	216.61	209.91	4.629	2.72	0.08
704A	24X	1	52-54	216.73	210.03	4.635	2.86	0.10
704A	24X	1	63-65	216.84	210.14	4.641	2.95	-0.08
704A	24X	1	98-100	217.19	210.49	4.658	2.76	-0.07
704A	24X	1	110-112	217.31	210.61	4.664	2.75	0.21
704A	24X	1	121-123	217.42	210.72	4.669	2.88	0.02
704A	24X	1	133-135	217.54	210.84	4.675	2.81	0.20
704A	24X	2	10-12	217.81	211.11	4.688	2.72	0.35
704A	24X	2	25-27	217.96	211.26	4.696	2.70	0.26
704A	24X	2	25-27	217.96	211.26	4.696	2.60	0.19
704A	24X	2	41-43	218.12	211.42	4.704	2.74	0.25
704A	24X	2	52-54	218.23	211.53	4.709	2.86	0.20
704A	24X	2	68-72	218.40	211.70	4.717	2.64	0.28
704A	24X	2	80-82	218.51	211.81	4.723	2.65	0.46
704A	24X	2	80-82	218.51	211.81	4.723	2.59	0.40
704A	24X	2	88-90	218.59	211.89	4.727	2.60	0.36
704A	24X	2	121-123	218.92	212.22	4.743	2.78	0.05
704A	24X	2	132-134	219.03	212.33	4.748	2.73	0.55
704A	24X	3	10-12	219.31	212.61	4.762	2.74	0.56
704A	24X	3	10-12	219.31	212.61	4.762	2.60	0.76
704A	24X	3	31-32	219.51	212.81	4.774	2.74	0.48
704A	24X	3	46-48	219.67	212.97	4.792	2.65	0.08
704A	24X	3	99-101	220.20	213.50	4.850	2.68	0.62
704A	24X	3	110-113	220.31	213.61	4.862	2.62	0.28
704A	24X	3	120-122	220.41	213.71	4.873	2.54	0.48
704A	24X	3	128-130	220.49	213.79	4.882	2.47	0.51

Acknowledgments. Reviews by C. Charles and an anonymous referee were helpful for improving the quality of the manuscript. Samples for this study were provided by the Ocean Drilling Program with funding from the National Science Foundation (NSF). This research was supported by NSF grants DPP-8717854 and OCE-8858012.

REFERENCES

Abelmann, A., R. Gersonde, and V. Speiss, Pliocene-Pleistocene paleoceanography of the Weddell Sea—Siliceous microfossil evidence, in *Geological History of the Polar Oceans: Arctic Versus Antarctic*, edited by U. Bleil and J. Thiede, pp. 729–759, Kluwer, Amsterdam, 1990.

Allen, C. P., and D. A. Warnke, History of ice rafting at ODP Leg 114 sites, Subantarctic South Atlantic, *Proc. Ocean Drill. Program Sci. Results*, *114*, 599–608, 1991.

Berger, A., and M. F. Loutre, New insolation values for the climate of the last 10 million years, *Sci. Rep. 1988/13*, Inst. d'Astron. et de Geophys. Georges Lemaire, Univ. Catholique de Louvain-la-Neuve, Louvain-la-Neuve, Belgium, 1988.

Bornhold, B. D., Ice-rafted debris in sediments from Leg 71, southwest Atlantic Ocean, *Initial Rep. Deep Sea Drill. Proj.*, *71*, 307–316, 1983.

Boyle, E. A., Vertical oceanic nutrient fractionation and glacial/interglacial CO_2 cycles, *Nature*, *331*, 55–56, 1988.

Boyle, E. A., Effect of depleted planktonic $^{13}C/^{12}C$ on bottom water during periods of enhanced relative Antarctic productivity (abstract), *Eos Trans. AGU*, *71*, 1357–1358, 1990.

Broecker, W. S., and T.-H. Peng, The cause of the glacial to interglacial atmospheric CO_2 change: A polar alkalinity hypothesis, *Global Biogeochem. Cycles*, *3*, 215–239, 1989.

Brunner, C., Latest Miocene to Quaternary biostratigraphy and paleoceanography of Site 704, subantarctic South Atlantic Ocean, *Proc. Ocean Drill. Program Sci. Results*, *114*, 201–216, 1991.

Burckle, L. H., and J. Cirilli, Origin of diatom ooze belt in the Southern Ocean: Implications for late Quaternary paleoceanography, *Micropaleontology*, *33*, 82–86, 1987.

Burckle, L. H., D. Robinson, and D. W. Cooke, Reappraisal of sea-ice distribution in Atlantic and Pacific sectors of the Southern Ocean at 18,000 yrs BP, *Nature*, *299*, 435–437, 1982.

Burckle, L. H., R. Gersonde, and N. Abrams, Late Pliocene-Pleistocene paleoclimate in the Jane Basin region: ODP Site 697, *Proc. Ocean Drill. Program Sci. Results*, *113*, 803–809, 1990.

Charles, C. D., and R. G. Fairbanks, Glacial to interglacial changes in the isotopic gradients of Southern Ocean surface water, in *Geological History of the Polar Oceans: Arctic Versus Antarctic*, edited by U. Bleil and J. Thiede, pp. 519–538, Kluwer, Amsterdam, 1990.

Ciesielski, P. F., Relative abundances and ranges of select diatoms and silicoflagellates from sites 699 and 704, Subantarctic South Atlantic, *Proc. Ocean Drill. Program Sci. Results*, *114*, 753–778, 1991.

Ciesielski, P. F., and G. P. Grinstead, Pliocene variation in the position of the Antarctic Convergence in the southwest Atlantic, *Paleoceanography*, *1*, 197–232, 1986.

Ciesielski, P. F., and F. M. Weaver, Early Pliocene temperature changes in the Antarctic seas, *Geology*, *2*, 511–515, 1974.

Ciesielski, P. F., et al., Leg 114, *Proc. Ocean Drill. Program Initial Rep.*, *114*, 815 pp., 1988.

Clapperton, C. M., and D. E. Sugden, Late Cenozoic glacial history of the Ross Embayment, Antarctica, *Quat. Sci. Rev.*, *9*, 253–272, 1990.

Cooke, D. W., and J. D. Hays, Estimates of Antarctic Ocean seasonal sea-ice cover during glacial intervals, in *Antarctic Geoscience*, edited by C. Craddock, pp. 1017–1025, University of Wisconsin Press, Madison, 1982.

Cronin, T. M., and H. J. Dowsett (Eds.), Pliocene Climates, *Quat. Sci. Rev.*, *10*(2/3), 296 pp., 1991.

Crowley, T. J., Modeling Pliocene warmth, *Quat. Sci. Rev.*, *10*,

Crowley, T. J., and C. L. Parkinson, Late Pleistocene variations in Antarctic sea ice II: Effect of interhemispheric deep ocean heat exchange, *Clim. Dyn.*, *3*, 93–105, 1988.

Curry, W. B., J. C. Duplessy, L. D. Labeyrie, and N. J. Shackleton, Changes in the distribution of $\delta^{13}C$ of deep water TCO_2 between the last glaciation and the Holocene, *Paleoceanography*, *3*, 317–341, 1988.

Denton, G. H., and T. J. Hughes, Milankovitch theory of ice ages: Hypothesis of ice-sheet linkage between regional insolation and global climate, *Quat. Res.*, *20*, 125–144, 1983.

Denton, G. H., D. E. Kellogg, T. B. Kellogg, and M. L. Prentice, Ice sheet overriding of the Transantarctic Mountains, *Antarct. J. U. S.*, *18*, 93–95, 1983.

Denton, G. H., M. L. Prentice, D. E. Kellogg, and T. B. Kellogg, Late Tertiary history of the Antarctic Ice Sheet: Evidence from the dry valleys, *Geology*, *12*, 263–267, 1984.

Duplessy, J. C., C. Lalou, and A. C. Vinot, Differential isotopic fractionation in benthic foraminifera and paleotemperatures reassessed, *Science*, *168*, 250–251, 1970.

Fenner, J. M., Late Pliocene–Quaternary quantitative diatom

stratigraphy in the Atlantic sector of the Southern Ocean, *Proc. Ocean Drill. Program Sci. Results*, *114*, 97–122, 1991.

Froelich, P. N., et al., Biogenic opal and carbonate accumulation rates in the subantarctic South Atlantic: The late Neogene of Meteor Rise Site 704, *Proc. Ocean Drill. Program Sci. Results*, *114*, 515–550, 1991.

Gordon, A., Seasonality of Southern Ocean sea ice, *J. Geophys. Res.*, *86*, 4193–4197, 1981.

Gordon, A. L., and E. J. Molinelli, Thermohaline and chemical distributions and the atlas data set, in *Southern Ocean Atlas*, edited by A. L. Gordon and E. J. Molinelli, pp. 1–11, Columbia University Press, New York, 1982.

Graham, D. W., B. H. Corliss, M. L. Bender, and L. D. Keigwin, Jr., Carbon and oxygen isotopic disequilibria of recent deep-sea benthic foraminifera, *Mar. Micropaleontol.*, *6*, 483–497, 1981.

Hailwood, E. A., and B. M. Clement, Magnetostratigraphy of sites 703 and 704, Meteor Rise, southeastern South Atlantic, *Proc. Ocean Drill. Program Sci. Results*, *114*, 367–386, 1991.

Haq, B. U., J. Hardenbol, and P. R. Vail, Chronology of fluctuating sea levels since the Triassic, *Science*, *235*, 1156–1167, 1987.

Hodell, D. A., Late Pleistocene Paleoceanography of the South Atlantic sector of the Southern Ocean: Linkage to NADW production, *Paleoceanography*, in press, 1992.

Hodell, D. A., and P. F. Ciesielski, Southern Ocean response to the intensification of northern hemisphere glaciation at 2.4 Ma, in *Geological History of the Polar Oceans: Arctic Versus Antarctic*, edited by U. Bleil and J. Thiede, pp. 707–728, Kluwer, Amsterdam, 1990.

Hodell, D. A., and P. F. Ciesielski, Stable isotopic and carbonate stratigraphy of the late Pliocene and Pleistocene of Hole 704A: Eastern subantarctic South Atlantic, *Proc. Ocean Drill. Program Sci. Results*, *114*, 409–435, 1991.

Hodell, D. A., and D. A. Warnke, Climatic evolution of the Southern Ocean during the Pliocene Epoch from 4.8 to 2.6 million years ago, *Quat. Sci. Rev.*, *10*, 205–214, 1991.

Hodell, D. A., D. F. Williams, and J. P. Kennett, Late Pliocene reorganization of deep vertical water-mass structure in the western South Atlantic: Faunal and isotopic evidence, *Geol. Soc. Am. Bull.*, *96*, 495–503, 1985.

Howard, W. R., and W. L. Prell. Late Quaternary surface circulation of the southern Indian Ocean and its relationship to orbital variations, *Paleoceanography*, *7*, 79–118, 1992.

Jacobs, S. S., R. G. Fairbanks, and Y. Horibe, Origin and evolution of water masses near the Antarctic continental margin: evidence from $H_2^{18}O/H_2^{16}O$ ratios in seawater, in *Oceanology of the Antarctic Continental Shelf, Antarc. Res. Ser.*, vol. 43, edited by S. S. Jacobs, pp. 59–85, AGU, Washington, D. C., 1985.

Jansen, E., and J. Sjöholm, Reconstruction of glaciation over the past 6 Myr from ice-borne deposits in the Norwegian Sea, *Nature*, *349*, 600–603, 1991.

Jansen, E., L. A. Mayer, J. Backman, R. M. Leckie, and T. Takayama, Evolution of climate cyclicity at site 806B—Oxygen isotope record, *Proc. Ocean Drill. Program Sci. Results*, *130*, in press, 1992.

Keany, J., Paleoclimatic trends in early and middle Pliocene deep-sea sediments in the Antarctic, *Mar. Micropaleontol.*, *3*, 35–49, 1978.

Keigwin, L. D., Jr., Pliocene stable isotope record of Deep Sea Drilling Project Site 606: Sequential events of ^{18}O enrichment beginning at 3.1 Ma, *Initial Rep. Deep Sea Drill. Proj.*, *94*, 911–920, 1986.

Kennett, J. P., and P. F. Barker, Latest Cretaceous to Cenozoic climate and oceanographic developments in the Weddell Sea, Antarctica: An ocean-drilling perspective, *Proc. Ocean Drill. Program Sci. Results*, *113*, 937–960, 1990.

Kroopnick, P. M., The distribution of ^{13}C in the world oceans, *Deep Sea Res.*, *32*, 57–84, 1985.

Labeyrie, L., J.-J. Pichon, M. Labracherie, P. Ippolito, J. Duprat, and J.-C. Duplessy, Melting history of Antarctica during the past 60,000 years, *Nature*, *322*, 701–706, 1986.

Ledbetter, M. T., and P. F. Ciesielski, Post-Miocene disconformities and paleoceanography in the Atlantic sector of the southern ocean, *Palaeogeogr. Palaeoclimatol. Palaeoecol.*, *52*, 185–214, 1986.

Lutjeharms, J. R. E., Location of frontal systems between Africa and Antarctica: Some preliminary results, *Deep Sea Res.*, *32*, 1499–1509, 1985.

Meinert, J., and D. Nobes, Physical properties of sediments beneath polar front upwelling regions in the subantarctic South Atlantic (Hole 704A), *Proc. Ocean Drill. Program Sci. Results*, *114*, 671–683, 1991.

Müller, D. W., D. A. Hodell, and P. F. Ciesielski, Late Miocene to earliest Pliocene (9.8–4.5 Ma) paleoceanography of the subantarctic southeast Atlantic: Stable isotopic, sedimentologic and microfossil evidence, *Proc. Ocean Drill. Program Sci. Results*, *114*, 671–683, 1991.

Mwenifumbo, C. J., and J. P. Blangy, Short-term spectral analysis of downhole logging measurements from Site 704, *Proc. Ocean Drill. Program Sci. Results*, *114*, 577–588, 1991.

Oppo, D. W., and R. G. Fairbanks, Variability in deep and intermediate water circulation of the Atlantic Ocean: Northern hemisphere modulation of the southern ocean, *Earth Planet. Sci. Lett.*, *86*, 1–15, 1987.

Oppo, D. W., R. G. Fairbanks, A. L. Gordon, and N. J. Shackleton, Late Pleistocene Southern Ocean δ^{13}C variability, *Paleoceanography*, *5*, 43–54, 1990.

Pisias, N. G., and T. C. Moore, Jr., The evolution of Pleistocene climate: A time series approach, *Earth Planet. Sci. Lett.*, *52*, 450–458, 1981.

Prell, W. L., Covariance patterns of foraminiferal δ^{18}O: An evaluation of Pliocene ice volume changes near 3.2 million years ago, *Nature*, *226*, 692–694, 1984.

Prell, W. L., Pliocene stable isotope and carbonate stratigraphy (holes 572C and 573A): Paleoceanographic data bearing on the question of Pliocene glaciation, *Initial Rep. Deep Sea Drill. Proj.*, *85*, 723–734, 1985.

Prentice, M. L., and R. K. Matthews, Tertiary ice sheet dynamics: The snow-gun hypothesis, *J. Geophys. Res.*, *96*, 6811–6827, 1991.

Rau, G. H., T. Takahashi, and D. J. Des Marais, Latitudinal variations in plankton δ^{13}C: Implications for CO_2 and productivity in past oceans, *Nature*, *341*, 516–518, 1989.

Raymo, M. E., W. F. Ruddiman, J. Backman, B. M. Clement, and D. G. Martinson, Late Pliocene variation in northern hemisphere ice sheets and North Atlantic Deep Water circulation, *Paleoceanography*, *4*, 413–446, 1989.

Raymo, M. E., W. F. Ruddiman, N. J. Shackleton, and D. W. Oppo, Evolution of Atlantic-Pacific δ^{13}C gradients over the last 2.5 m.y., *Earth Planet. Sci. Lett.*, *97*, 353–368, 1990.

Raymo, M. E., D. Hodell, and E. Jansen, Response of deep ocean circulation to initiation of northern hemisphere glaciation (3–2 M.Y.), *Paleoceanography*, in press, 1992.

Ruddiman, W. F., D. Cameron, and B. M. Clement, Sediment disturbance and correlation of offset holes drilled with the hydraulic piston corer, *Initial Rep. Deep Sea Drill. Proj.*, *94*, 615–634, 1986.

Ruddiman, W. F., M. E. Raymo, D. G. Martinson, B. M. Clement, and J. Backman, Pleistocene evolution of northern hemisphere climate, *Paleoceanography*, *4*, 353–412, 1989.

Sackett, W. M., W. R. Eckelmann, M. L. Bender, and A. W. H. Bé, Temperature dependence of carbon isotope composition in marine plankton and sediments, *Science*, *148*, 235–237, 1965.

Shackleton, N. J., and M. A. Hall, Stable isotope history of the Pleistocene at ODP Site 677, *Proc. Ocean Drill. Program Sci. Results*, *111*, 295–316, 1989.

Shackleton, N. J., and N. D. Opdyke, Oxygen isotope and palaeomagnetic stratigraphy of equatorial Pacific Core V28-238: Oxygen isotope temperatures and ice volumes on a 10^5 and 10^6 year scale, *Quat. Res.*, *3*, 39–55, 1973.

Shackleton, N. J., J. Backman, H. Zimmerman, D. V. Kent, M. A. Hall, D. G. Roberts, D. Schnitker, and J. Baldauf, Oxygen isotope calibration of the onset of ice-rafting and history of glaciation in the North Atlantic region, *Nature*, *307*, 620–623, 1984*a*.

Shackleton, N. J., M. A. Hall, and A. Boersma, Oxygen and carbon isotope data from Leg 74 foraminifers, *Initial Rep. Deep Sea Drill. Proj.*, *74*, 599–612, 1984*b*.

Shackleton, N. J., A. Berger, and W. R. Peltier, An alternative astronomical calibration of the lower Pleistocene timescale based on ODP Site 677, *Trans. R. Soc. Edinburgh*, *81*, 251–261, 1991.

Sikes, E. L., L. D. Keigwin, and W. B. Curry, Pliocene paleoceanography: Circulation and oceanographic changes associated with the 2.4-Ma glacial event, *Paleoceanography*, *6*, 245–257, 1991.

Warnke, D. A., and C. P. Allen, Ice rafting, glacial-marine sediments, and siliceous oozes: South Atlantic/subantarctic Ocean, *Proc. Ocean Drill. Program Sci. Results*, *114*, 589–598, 1991.

Warnke, D. A., C. P. Allen, D. W. Muller, D. A. Hodell, and C. A. Brunner, Miocene-Pliocene Antarctic glacial evolution as reflected in the sedimentary record: A synthesis of IRD, stable isotope, and planktonic foraminiferal indicators, this volume.

Webb, P. W., and D. M. Harwood, Late Cenozoic glacial history of the Ross Embayment, Antarctica, *Quat. Sci. Rev.*, *10*, 215–224, 1991.

Westall, F., and J. Fenner, The Pliocene-Recent Polar Front Zone in the South Atlantic: Changes in its position and sediment accumulation rates from OPD Leg 114, holes 699A, 701C, and 704B, *Proc. Ocean Drill. Program Sci. Results*, *114*, 609–648, 1991.

Williams, D. F., M. A. Sommer, and M. L. Bender, Carbon isotopic compositions of recent planktonic forminifera of the Indian Ocean, *Earth Planet. Sci. Lett.*, *36*, 391–403, 1977.

Woodruff, F., S. M. Savin, and R. G. Douglas, Miocene stable isotope record: A detailed deep Pacific Ocean study and its paleoclimatic implications, *Science*, *212*, 665–668, 1981.

Zahn, R., K. Winn, and M. Sarnthein, Benthic foraminiferal d13C and accumulation rates of organic carbon: *Uvigerina peregrina* group and *Cibicidoides wuellerstorfi*, *Paleoceanography*, *1*, 27–42, 1986.

(Received December 17, 1991;
accepted April 13, 1992.)

MIOCENE-PLIOCENE ANTARCTIC GLACIAL EVOLUTION: A SYNTHESIS OF ICE-RAFTED DEBRIS, STABLE ISOTOPE, AND PLANKTONIC FORAMINIFERAL INDICATORS, ODP LEG 114

Detlef A. Warnke,[1] Carl P. Allen,[1,2] Daniel W. Muller,[3] David A. Hodell,[4]
and Charlotte A. Brunner[5]

We have combined sedimentary, planktonic foraminiferal, and stable isotope data from Ocean Drilling Program Leg 114 drill sites (Subantarctic South Atlantic) to outline the Neogene evolution of Antarctic glaciation as seen from these sites. No ice-rafted debris (IRD) was observed in the Oligocene sequence. Ice-rafted debris first reached Site 699 (at latitude 51°32.531'S) at about 23.5 Ma, but only in very small amounts, indicating that at least some icebergs had reached a size and/or the Southern Ocean had cooled to the degree necessary to deliver debris to this site, located just south of the present-day Polar Front Zone (PFZ). Small amounts of IRD reached the site intermittently through the Miocene. At other sites, IRD first arrived at ~8.8 Ma. The latest Miocene (Chron 6) carbon shift is well documented at Site 704, located north of the present-day PFZ. The shift occurred during a time of relative climatic warmth, although fluctuations are clearly discernible. During this time, circulation changes occurred, causing the transfer of large volumes of organic carbon to the ocean. Strong tectonic pulses in the proto–Straits of Gibraltar and in the circum-Pacific are also implicated. The latest Miocene ice-rafting record of Site 699 in combination with the stable isotope record of Site 704 shows that ice-rafting episodes commenced after the carbon shift and may be positively correlated with positive oxygen isotope excursions, and negatively correlated with positive carbon isotope excursions. These correlations, if confirmed, suggest significant volume changes in Antarctic ice, which caused sea level changes and perhaps calcium carbonate dissolution in deep-sea sediments. The early Pliocene was a time of general global warmth (although the amount of warming that occurred in the Antarctic is in dispute), yet widespread ice-rafting episodes did occur. One interpretation of these relationships is that the "snow gun" was active, i.e., that warmer temperatures led to higher precipitation rates on Antarctica. These conditions caused the formation of wet-based, erosive outlet glaciers that delivered variable amounts of debris to parts of the Southern Ocean. Between 2.6 and 2.3 Ma, a great change in the measured parameters is noted: oxygen isotopic ratios shifted to higher values, ice rafting increased, and foraminiferal populations (dominated by globorotalids previously) began to exhibit an alternation between populations dominated by globorotalids and populations dominated by *Neogloboquadrina pachyderma* with low frequencies of *Globigerina* spp. Another change occurred at ~1.67 Ma, leading to still colder conditions as indicated by changes in foraminiferal assemblages and higher oxygen isotopic ratios but also to reduced ice rafting, perhaps caused by a change in the boundary conditions on Antarctica and a change in the nature of ablation. A climatic amelioration occurred at about 1.4 Ma, as evidenced by planktonic foraminiferal indicators. The other indicators, however, suggest only a minor amelioration, perhaps to some state intermediate between those prevailing just before and after 1.67 Ma. Later, Pleistocene variations are discernible and have been described by Westall and Fenner (1991).

INTRODUCTION

Ocean Drilling Program (ODP) Leg 114 occupied seven drill sites in the Subantarctic South Atlantic (Figure 1), in the general vicinity of the present-day Polar Front Zone (PFZ), approximated by the 50° latitude in this area. It was one of the aims of this leg to establish a history of ice-rafting episodes, at distances far enough from the source areas to yield a "clean," purely ice-rafted signal. It was hoped that the evolving history of ice rafting, together with the results of the other Southern Ocean legs, would lead to a better definition of the evolution of Antarctic glacial development.

Sedimentary analyses of samples from three drill sites (114-699, 114-701, and 114-704) resulted in a picture of the ice-rafting history at those sites; the results are

[1]Department of Geological Sciences, California State University, Hayward, Hayward, California 94542.

[2]Now with Engineering-Science, Inc., Berkeley, California 94710.

[3]Geologisches Institut, Zurich, Switzerland.

[4]Department of Geology, University of Florida, Gainesville, Florida 32611.

[5]Center for Marine Science, University of Southern Mississippi, Stennis Space Center, Mississippi 39529.

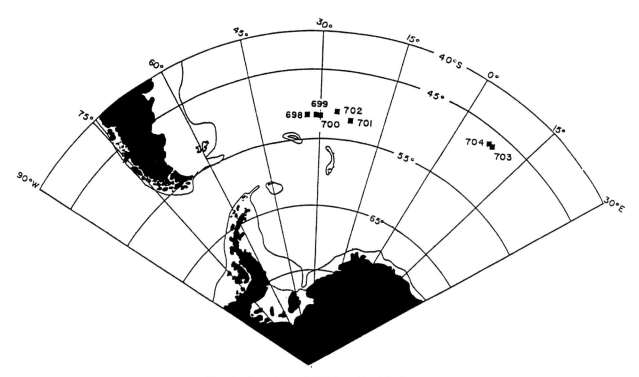

Fig. 1.   Locations of ODP Leg 114 drill sites

presented by *Allen and Warnke* [1991] and *Warnke and Allen* [1991]. In the present paper we attempt to integrate sedimentary data with stable isotope data and with planktonic foraminiferal analyses to furnish a more complete history of Neogene Antarctic glacial evolution. Because of the wide sample spacing for ice-rafted debris (IRD) analyses (e.g., ~7–30 ka for Site 701 and 30 ka at best for Hole 704A), this synthesis must be considered preliminary. Nevertheless, a general picture of Neogene Antarctic cryospheric evolution has emerged.

THE OLIGOCENE ENIGMA

Antarctica has experienced continental glaciations over tens of millions of years."Continental" refers to glaciations on a continental scale, to the degree that they are discernible in the stable isotope record. The antiquity of such glaciations was once mainly indicated by oxygen isotopes [see *Miller et al.*, 1987; *Wise et al.*, 1985; *Mathews and Poore*, 1980] and by scattered land-based evidence [e.g., *LeMasurier and Rex*, 1982; *Birkenmajer*, 1986] but was not generally accepted. Similarly, marine "proxy" indicators for glacial conditions are provided by ice-rafted debris were the subject of much debate because the mere presence of IRD contains no clues as to the size of the glaciers (although distance of the core locations from the source might). For instance, the occurrence of IRD in Eocene seg-

ments of a South Pacific core [*Geitzenauer et al.*, 1968] has been the subject of discussion [e.g., *Warnke and Hansen*, 1977] but did not find general credence as a first indicator of Antarctic glaciation of some extent, mainly because the ice-rafted origin of the terrigenous fraction was suspect. An analysis by *Gram* [1974] of several piston cores from the Pacific sector of the Southern Ocean reaffirmed the early age for Antarctic glaciation advocated by *Geitzenauer et al.* [1968] but likewise did not find general acceptance. Recent work by *Wei* [1992] has confirmed the original age assignments. The presence or absence of any ice-rafted debris in the Oligocene of Leg 114 is significant within this context.

Drilling by ODP in Prydz Bay and on the Kerguelen Plateau corroborated an Oligocene age for Antarctic glaciation [*Kennett and Barker*, 1991; *Wise et al.*, 1991; *Barron et al.*, 1991; *Breza and Wise*, 1991], confirming the results of earlier drilling efforts in the Ross Sea [*Hayes and Frakes*, 1975; *Leckie and Webb*, 1983]. Sediments beneath the glacial sediments of confirmed Oligocene age have also been interpreted to be glacigenic and to suggest Eocene glaciations [*Barron et al.*, 1991].

Some of the best "marine" evidence for Oligocene Antarctic glaciation is provided by the CIROS 1 drill hole, McMurdo Sound [*Barrett*, 1989]. The drill hole records two periods of glaciation, from 36 to 34.5 Ma and from 30.5 to about 22 Ma [*Barrett et al.*, 1989].

Glaciers were calving at sea level for most of the time represented by the core, as indicated by sediments interpreted by the authors to be glacial-marine. "The ice was temperate, not polar···" [*Barrett et al.*, 1989]. The authors also state that cool-temperate *Nothofagus* forests grew in the foothills of the Transantarctic Mountains throughout the time of deposition of these deposits. The authors correlated these glacial-marine deposits with those on Prydz Bay and concluded that a "wet-based" ice sheet of continental dimensions covered most of Antarctica, in seeming contradiction to the conclusion of *Mildenhall* [1989], who stated that glacial ice could not have been very extensive because of the continuous existence of forest taxa, at least along the coast. Mildenhall's assertion is supported by ODP Leg 113 drilling which yielded no evidence for continent-wide ice sheet formation prior to the Neogene [*Kennett and Barker*, 1990], at least none of any great temporal extent. A brief, intensive glaciation (indicated by one ice-rafting episode recorded from the Kerguelen Plateau [*Breza and Wise*, 1992; *Zachos et al.*, 1992]) is the exception. This episode is also recorded at ODP sites 738 and 744 [*Barron et al.*, 1991]. The conflicting results and interpretations concerning Oligocene Antarctic glacial evolution provided by ODP Leg 113 (Weddell Sea: no large-scale, pre-middle Miocene glaciation [see *Kennett and Barker*, 1990]) and Leg 119 (Prydz Bay: glaciation at least by Oligocene time) are presently unreconciled. Ice-rafted debris was searched for in the Oligocene samples available from Leg 114 sites, since even at these Subantarctic latitudes, the presence or absence of IRD helps constrain models of cryospheric development during the Paleogene.

Neogene Antarctic glacial evolution is recorded in drill sites of all Southern Ocean legs, but the sedimentary record seems to be best reflected in drill sites in the vicinity of the present-day PFZ, where much of the iceberg melting occurs (the so-called distal glacial-marine facies [*Molnia*, 1983]). Turbidity currents and other sediment transport mechanisms are excluded, and linear sedimentation rates can be calculated for use in the computation of mass accumulation rates.

METHODS

Data reports on which the following discussion is based are presented in Ocean Drilling Program volume 114 B [*Ciesielski et al.*, 1991] and in later publications, with the exception of foraminiferal frequency data (Table 1). The stratigraphic framework for this report is based on paleomagnetic, floral, and faunal analyses [*Ciesielski et al.*, 1988, 1991]. These results were combined into a unified stratigraphic framework for sites 699 and 701 by P. F. Ciesielski (written communication to D. A. Warnke, 1990) as reported by *Allen and Warnke* [1991, Tables 4 and 5] and reproduced in Tables 2 and 3. Details of the stratigraphy for Site 704 are presented in

the work of *Müller et al.* [1991], *Hodell and Ciesielski* [1991], and *Brunner* [1991], among others. The ages chosen for the IRD computations for Site 704 are shown in the work of *Allen and Warnke* [1991, Table 6] and are reproduced in Table 4 (note however that there is uncertainty in the placement of the top of the Olduvai Subchron [see *Hodell and Venz*, this volume]). These ages were used in the computation of linear sedimentation rates and subsequently of apparent mass accumulation rates of coarse-grained (0.25 to 2 mm) IRD as follows: dry bulk densities of samples were determined in the laboratory. These densities, multiplied by the sedimentation rate, yield mass accumulation rates. The coarse-grained fraction was isolated by sieving, and the IRD percentages were established by counting. These counts, combined with the mass accumulation rates, yield apparent mass accumulation rates of IRD ("apparent," because density differences were not taken into account) expressed in mg/cm^2/1000 yr. See *Allen and Warnke* [1991] for details. For isotope analyses of the Miocene record, sinistrally coiled specimens of *Neogloboquadrina pachyderma* were picked from the 150- to 250-μm size fraction. Where possible, monospecific or monogeneric assemblages were picked for the benthic record, mainly *Cibicidoides kullenbergi*. Prior to analysis, all samples were vacuum roasted for 1 hour at 400°C. Samples were reacted in a common acid bath of orthophosphatic acid at 70°C. Isotopic ratios of purified CO$_2$ were measured on line by a triple-collector VG isogas precision isotope ratio mass spectrometer (PRISM). All results are reported in standard delta notation. Analytical details, correction factors, etc., are reported by *Müller et al.* [1991]. Analyses on Pliocene-Pleistocene samples were carried out on the same instrument. For planktonic foraminifers, specimens of sinistrally coiled *N. pachyderma* were used; analyses of benthic foraminifers were performed on *Planulina wuellerstorfi* or on mixed species of *P. wuellerstorfi*, *C. kullenbergi*, and *Cibicidoides* spp. Analytical details, correction factors, etc., are provided by *Hodell and Ciesielski* [1991].

The foraminiferal frequency data (Table 1) are based on counts of the >250-μm fraction. Counting groups are globorotalids which are dominated by *Globorotalia puncticulata*, globigerinids which are dominated by *Globigerina bulloides*, and sinistral *Neogloboquadrina pachyderma*. A minimum of 250 specimens per sample were counted.

RESULTS

The Oligocene

None of the Oligocene portions of the cores raised during Leg 114 show any signs of IRD. However, the drill sites (Figure 1) are in the vicinity of the present-day Polar Front Zone (PFZ), perhaps too far to be reached

TABLE 1.   Relative Frequency of Planktonic Foraminiferal Taxa, Hole 704A

Depth, mbsf	Number of Tests Counted	Sinistral *Neogloboquadrina pachyderma*, %	*Globigerina bulloides*, %	*Globorotalia* spp., %
0.72	269	49	38	13
2.22	355	4	69	27
3.72	255	1	74	25
6.72	389	0	56	44
7.86	407	0	45	54
9.36	202	7	93	0
12.36	334	5	88	7
13.00	289	72	22	6
15.36	230	43	37	20
16.86	278	98	1	1
17.68	338	14	1	86
19.18	357	0	7	93
20.68	304	2	8	90
22.18	320	32	1	67
23.68	269	49	1	50
25.18	368	35	11	54
29.90	399	66	11	23
34.10	303	15	1	83
42.01	289	29	13	57
45.90	530	0	11	89
47.40	398	8	9	83
48.90	402	45	0	55
50.40	302	81	0	19
51.90	289	60	1	39
53.40	330	2	1	97
54.50	550	4	0	95
56.92	311	1	1	97
58.40	321	1	4	96
59.93	359	41	20	39
61.42	313	88	0	12
62.95	317	25	15	59
64.91	354	0	13	87
67.91	363	9	1	90
69.41	293	5	1	94
70.91	296	89	0	11
72.41	317	9	1	91
74.41	298	91	0	9
75.91	303	52	1	47
77.41	310	67	0	32
78.91	320	31	2	67
80.41	466	72	0	28
81.91	297	95	0	5
82.81	351	94	0	6
83.90	321	98	0	2
86.90	363	48	0	52
88.40	287	43	0	57
91.40	344	45	1	54
92.90	407	89	0	10
93.42	297	1	0	99
94.92	287	38	0	62
96.42	300	75	0	25
97.92	321	32	0	68
99.35	333	28	4	68
100.92	376	7	3	89
101.82	322	5	13	81
102.92	321	3	24	73
104.42	395	88	0	12
105.92	339	7	4	89
107.42	332	3	27	69
108.92	390	0	10	90
110.42	296	0	7	93

TABLE 1. (continued)

Depth, mbsf	Number of Tests Counted	Sinistral Neogloboquadrina pachyderma, %	Globigerina bulloides, %	Globorotalia spp., %
112.42	399	77	0	23
113.92	330	85	0	14
115.44	356	17	0	83
116.92	369	8	2	89
118.42	435	1	7	92
119.92	496	0	13	87
120.82	477	1	13	86
121.92	354	32	12	56
123.42	324	18	7	76
124.92	335	10	5	85
126.42	335	36	13	51
127.92	387	14	8	78
129.42	294	2	13	85
131.42	392	63	24	14
132.92	414	21	31	48
134.42	290	11	50	39
135.92	273	4	43	53
137.42	357	39	52	9
138.92	318	0	52	48
139.82	336	0	42	58
140.92	345	0	13	87
142.42	438	7	14	79
143.92	407	27	15	58
145.42	500	8	36	55
146.92	448	3	12	86
148.42	249	0	10	88
150.42	373	0	7	93
151.92	334	1	6	93
153.42	405	48	19	33
154.92	443	59	8	33
156.42	349	33	21	46
157.92	377	8	29	62
159.92	445	0	3	97
161.42	413	29	12	59
162.92	470	8	18	75
164.42	580	6	47	47
165.92	493	10	40	50
167.42	276	0	11	89
169.42	363	0	6	94
170.92	363	0	3	97
172.42	363	0	5	95
173.92	281	0	3	97
176.92	379	0	6	94
179.12	316	0	2	98
180.62	323	0	2	98
182.12	311	0	0	100
183.62	373	0	4	96

by icebergs calving from Antarctic glaciers during the Oligocene. An enigmatic gravelly layer of apparently Oligocene age occurs at Site 699 [see *Ciesielski et al.*, 1988]. However, this material is washed and graded, indicating that this particular type of "bedding" was introduced during the coring process. Because coring operations were carried out in heavy seas, this gravelly layer must be the result of massive downhole contamination and cannot be used as evidence for Oligocene ice rafting. Sediments above and below this gravelly layer do not contain any IRD.

The Miocene to 6 Ma

The first unquestionable evidence for ice rafting comes from higher levels at Site 699 (Figure 2). The oldest ice-rafted material was deposited at about 23.5 Ma (earliest Miocene), based on the revised stratigraphic framework of P. F. Ciesielski (personal communication, 1990). However, the amount of this material is small and accumulation rates were extremely low. Similarly, very small amounts of ice-rated material occur at levels in the core with the assigned ages of 14.25, 14.13,

TABLE 2. Hole 114-699A Age Model

Depth, mbsf	Age, Ma	Comment
0.12	0.195	LAAD *Hemidiscus karstenii*
4.90	0.420	FAAD *Hemidiscus karstenii*
8.16	0.620	LAD *Actinocyclus ingens*
10.59	0.730	Brunches/Matuyama
11.17	0.770	estimate based on sedimentation rate above hiatus
11.17	1.140	estimate based on sedimentation rate below hiatus*
11.75	1.200	top *Actinocyclus ingens* acme 1 (>10%)
19.59	1.660	top Olduvai
21.19	1.880	base Olduvai
29.45	2.470	Matuyama/Gauss
31.74	2.630	top *Nitzschia weaveri* acme 2 (>10%)
36.50	2.990	base *Cosmiodiscus insignis* acme (>10%)
39.03	3.180	first consistent *Coscinodiscus vulnificus*
40.69	3.400	Gauss/Gilbert
42.82	3.830	first consistent *Nitzschia interfrigidaria*
44.00	3.880	top Cochiti
45.27	4.190	FAD *Nitzschia angulata*
45.27	4.690	LAD *Dictyocha pygmaea*
53.61	5.350	Gilbert/C3AN
58.00	5.530	C3AN.33
61.00	5.680	C3AN.61
62.40	5.890	C3AN/C3AR
64.64	6.000	estimate†
64.64	10.000	estimate, within Chron C5N, lower to middle portion based on abundant *D. dimorpha*
66.59	10.420	
66.59	14.080	very crude; only one sample in this interval‡
68.36	14.300	
68.36	23.500	estimate, only slightly younger than 23.55 Ma
70.29	23.550	C6CN.3?
94.74	24.210	C6CN/C6CR

Double entries in depth column signify hiatuses. FAD, first appearance datum; FAAD, first abundant appearance datum; LAD, last appearance datum; LAAD, last abundant appearance datum.

*Between 1.20 and 1.66 Ma.

†In Chron C3AR, but younger than 6.00 Ma based on correlation to Hole 704B.

‡Sample is middle Miocene, probably lower *N. denticuloides* Zone.

TABLE 3. Site 114-701 Age Model

Depth, mbsf	Age, Ma	Comment
17.02	0.73	Brunches/Matuyama
31.50	0.91	top Jaramillo (as recognized in Hole 701C)
32.50	1.20	top *Actinocyclus ingens* acme (as recognized in Hole 701C)
37.00	1.45	base *Actinocyclus ingens* acme (as recognized in Hole 701C)
44.93	1.66	top Olduvai
50.95	1.88	base Olduvai
68.73	2.47	Matuyama/Gauss
71.40	2.63	LAAD (>10%) *Nitzschia weaveri* (top acme 2)
71.40	3.11	last consistent *Nitzschia interfrigidaria* (s.str.)
74.48	3.40	Gauss/Gilbert
99.05	4.10	top Nunivak
102.97	4.24	base Nunivak (based on Hole 701B)
106.83	4.40	top Sidufjell (based on Hole 701B)
110.59	4.47	base Sidufjell (based on Hole 701B)
113.99	4.57	top Thvera (based on Hole 701B)
116.10	4.77	base Thvera
140.27	5.68	C3AN.61
144.20	5.89	C3AN/C3AR
165.54	6.36	FAD *Cosmiodiscus insignis v. triangulata*
174.27	6.55	age above hiatus based on sedimentation rate*
174.27	7.60	age of hiatus based on sedimentation rate†
179.07	7.90	C4R/C4AN
183.98	8.21	C4AN.52
188.24	8.41	C4AN.50
194.76	8.50	C4AN/C4AR
204.02	8.92	C4AR/C5N
228.40	10.06	C5N/C5R
234.18	11.36	LAD *Cyrtocapsella tetrapera*

To a depth of 74.48 mbsf, Hole 701A values are given, except as noted. Hole 701A has not been studied in sufficient detail to determine depths of unconformities, and therefore they are inferred using data from Hole 701C. There may be some difference as the B/M boundary is shallower in Hole 701A (17.02 versus 23.00 mbsf) and the M/G boundary is deeper (68.73 versus 66.70 mbsf). These differences may alter IRD data near hiatuses. Below 74.48 mbsf, Hole 701C values are given except as noted. Hiatus depth at 71.40 mbsf inferred as midpoint between bracketing datums between 71.63 mbsf and 71.17 mbsf in Hole 701C. Depth of hiatus at 174.27 mbsf inferred as midpoint between bracketing samples examined (178.30 and 170.23 mbsf) (P. Ciesielski, written communication 1990).

*Sedimentation rate between 144.20 and 165.54 mbsf extrapolated to hiatus depth.

†Sedimentation rates between 8.21 and 7.90 Ma extrapolated.

and 10.40 Ma. All of these sediments are "sandwiched" between hiatuses (refer to Figure 2 for the position of hiatuses in the core). Because of these hiatuses, it is not known whether or not records of significant ice-rafting episodes were removed by later erosional episodes. The beginning of ice rafting in this part of the Southern Ocean has been estimated to be at about 8.7 Ma (Deep Sea Drilling Program (DSDP) Leg 71 [see *Ciesielski and Weaver*, 1983]). This age corresponds very well with an initial, minor ice-rafting episode reported by *Breza* [1992] for ODP Site 120-751 (~58°S) on the Kerguelen

Plateau. Unfortunately, that critical period of time is not represented at Site 699 because of a hiatus that extends from 6 to 10 Ma (see Figures 2 and 3). At Site 701, however, some ice rafting is recorded in this time interval, albeit by very low "apparent mass accumulation rates" (AMARs) (see Figure 4) just as on the

Kerguelen Plateau. It is unknown whether or not the minor ice-rafting episodes recorded at Site 701 and on the Kerguelen Plateau are time equivalent. In general, the interval from 9.8 to 6.4 Ma was marked by stable paleoceanographic conditions and relative warmth as indicated by planktonic oxygen isotopic values from Hole 704B [see *Müller et al.*, 1991]. The PFZ, if it existed at all, must have been far to the south of the drill sites. Very little IRD reached Site 699 and Site 701, and none reached Site 704.

Latest Miocene

Site 699 reveals the beginning of ice rafting above the 10–6 Ma hiatus. In Figure 5*a*, this record is plotted opposite the benthic oxygen isotope record from Hole 704B [*Müller et al.*, 1991]. In Figure 5*b* the planktonic carbon isotope record from Hole 704B has been plotted against the IRD record from Site 699. The time interval from about 6.4 to 6 Ma is the well-known Chron 6 "carbon shift" (see *Müller et al.* [1991] for a discussion). This interval contains the *Neobrunia* ooze at about 6.15 Ma as recorded at Site 701, interpreted to represent deposition of laminated, organic-carbon-rich sediment in a suboxic to anoxic environment. These conditions were perhaps caused by a massive meltwater lid and cessation of Antarctic Bottom Water (AABW) production [*Müller et al.*, 1991]. Neither Site 701 nor Site 704 shows significant ice rafting.

The carbon isotopic shift has been interpreted in

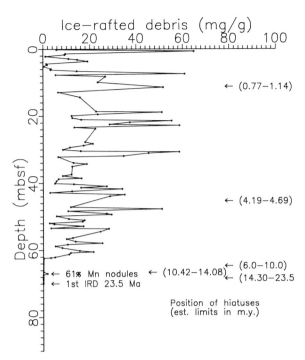

Fig. 2. Ice-rafted debris (IRD: quartz, feldspar, heavy minerals, lithic fragments) at Site 699, in milligrams of coarse-grained IRD per gram of sediment.

different ways: one of these interpretations is the transfer of organic matter to the ocean related to circulation changes and glacio-eustatic sea level drop that caused the termination of widespread "Monterey-type" sediment basins on the Pacific margin. However, the shift occurred during a time of relative warmth, although fluctuations are clearly discernible. Because of this relative warmth, strong tectonic pulses (i.e., in the circum-Pacific belt and the proto–Straits of Gibraltar) may also have played a major role [*Müller et al.*, 1991]. Other mechanisms are possible. Details are given by *Müller et al.* [1991].

The variability shown in the oxygen isotopic record during Chron 6 [*Hodell et al.*, 1991] suggests that cryospheric instability caused the production of large volumes of Antarctic Bottom Water, which in turn may have produced erosional episodes that removed the sedimentary record older than 5.9 Ma at Site 699. Ice rafting commenced at Site 699 at about 5.9 Ma, and it began at Site 701 at ~6 Ma (Figures 3 and 4), after the carbon isotopic shift. This relationship is an apparent reversal of the "Monterey hypothesis" [*Vincent and Berger*, 1985], i.e., it contradicts the hypothesis that decreased organic carbon burial leads to higher pCO_2 and warmer temperatures. Coincidentally, this time was also the beginning of deposition of the glacial-marine Yakataga Formation in Alaska [e.g., *Armentrout*, 1983], the deposition of glacial sediments on Iceland [*Vilmundardottir*, 1972], and the (unfortunately ill defined)

TABLE 4.   Hole 114-704A Age Model

Depth, mbsf	Age, Ma	Comment
1.14	0.195	LAAD *Hemidiscus karstenii*
3.52	0.258	FAD *Emiliana huxleyi*, mid Stage 8
7.64	0.423	FAAD *H. karstenii*
8.65	0.460	last consistent *Pseudoemiliana lacunosa*
13.25	0.620	LAD *Actinocyclus ingens*
34.51	0.730	Brunches/Matuyama
38.40	0.910	top Jaramillo
44.27	0.980	base Jaramillo
89.09	1.660	top Olduvai
168.45	2.470	Matuyama/Gauss
176.10	2.920	top Kaena
177.80	2.990	base Kaena
179.10	3.080	top Mammoth
181.50	3.180	base Mammoth
186.65	3.400	Gauss/Gilbert
198.55	3.880	top Cochiti
201.95	3.970	base Cochiti
204.20	4.100	top Nunivak
210.35	4.240	base Nunivak
212.00	4.400	top Sidufjell
213.25	4.470	base Sidufjell
215.40	4.570	top Thvera
219.45	4.770	base Thvera
224.76	5.350	Gilbert/C3AN

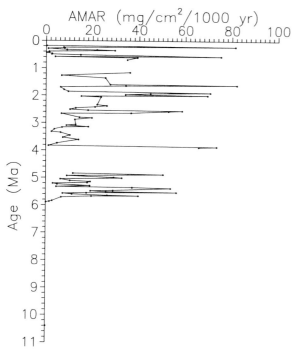

Fig. 3. Apparent mass accumulation rates (AMAR) of coarse-grained IRD at Site 699.

beginning of ice rafting at DSDP Site 344, near Spitsbergen [*Warnke*, 1982]. The latest Miocene was also a time of major South American glaciations [*Mercer and Sutter*, 1982].

Ice-rafting episodes at Site 699 (Figure 5*a*) between 5.9 and 4.8 Ma coincide with positive benthic oxygen isotope excursions. This is followed by a younger cluster of IRD peaks that exhibit no association with large, positive oxygen isotopic excursions. This younger interval is associated with strong calcium carbonate dissolution events [*Froelich et al.*, 1991; *Müller et al.*, 1991] and therefore is not well represented in the stable isotopic record. *Froelich et al.* [1991] described individual dissolution events as having occurred at the following times: 5.37, 5.29, 5.02, 4.93, and 4.81 Ma. Four of these episodes coincide with IRD peaks, a correlation that is unlikely to be fortuitous.

The dissolution event at 4.81 Ma at Site 704 occurs within a coring gap at Site 699. There is an additional IRD peak at 5.11 to 5.18 Ma that may coincide with a sequence boundary at 5.1 Ma [*Haq et al.*, 1987]. Calcium carbonate dissolution in the ocean is caused by several different processes. It is possible that these dissolution events resulted either directly or indirectly from a change in ocean alkalinity. These paleoceanographic changes were likely induced through sea level changes that disrupted the Mediterranean-Atlantic connection and led to desiccation and evaporite deposition

in the Mediterranean Sea: the so-called Messinian salinity crisis [*Müller et al.*, 1991].

There is a possible, although weak, negative correlation between IRD peaks and carbon isotopic excursions (see Figures 5*b* and 6). As a working hypothesis, we suggest that glacio-eustatic sea level changes may be responsible for carbon isotopic fluctuations and evaporite deposition between 5.9 and 4.8 Ma: lowered sea level episodes caused erosion of continental shelves and the transfer of isotopically light carbon previously sequestered on the shelves. Site 699 contains the most detailed IRD record for this particular interval but, because of sampling restrictions, has not been sampled in sufficient detail to reveal the necessary details of this transitional phase of Antarctic glaciation.

The Pliocene to the Gauss/Matuyama Boundary (2.47 Ma)

A hiatus above the discussed interval (i.e., above 4.8 Ma) as well as a lack of samples prevents delineation of further glacial development, at least at Site 699 (Figure 3). Site 701, on the other hand, shows variable ice-rafting activity above this level, with a noticeable peak (based on three data points) at ~4.4 Ma (Figure 4). The time interval between about 4.6 and 4.1 Ma is one of increased IRD fluxes as recorded at ODP Sit 751 on the Kerguelen Plateau [*Breza*, 1992] and ARA *Islas Orcadas* cores 0775-5 and 1176-66 from the South Atlantic [*Anderson*, 1985; *Breza*, 1992]. In the Weddell Sea the cessation of important turbidite deposition at Site 694 at

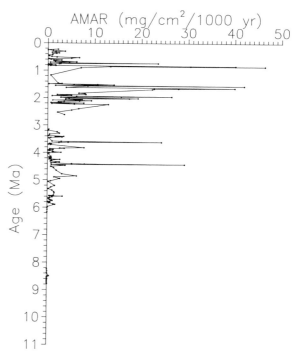

Fig. 4. Apparent mass accumulation rates of IRD, Site 701.

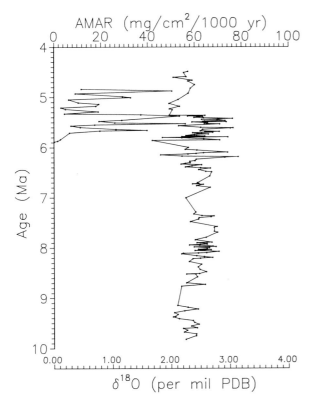

Fig. 5*a*. Apparent mass accumulation rates of IRD, Site 699, and the benthic oxygen isotope record from Hole 704B.

Fig. 5*b*. Apparent mass accumulation rates of IRD, Site 699, and the planktic carbon isotope record from Hole 704B.

about 4.8 Ma is interpreted to reflect the beginning of a permanent, relatively stable West Antarctic Ice Sheet (WAIS) [*Kennett and Barker*, 1990]. Prominent IRD peaks between about 4.8 and 3.6 Ma (Figure 4) reflect continuing ice rafting during the early Pliocene to Site 701. Delivery of this IRD may have been by large, tabular bergs from the WAIS, as originally postulated by *Ciesielski and Weaver* [1983]. Some IRD reached Site 704 beginning at 4.67 Ma.

There is, however, some difficulty with the idea of delivery of IRD by large tabular icebergs because the interval between about 4.6 and 4.2 Ma was relatively warm in the Subantarctic, as evidenced by low mean $\delta^{18}O$ values of planktonic and benthic foraminifera [*Hodell and Warnke*, 1991] and especially by high carbonate values [*Froelich et al.*, 1991]. *Abelmann et al.* [1990] even suggested a warming by as much as 5° to 10°C, compared with the present, on the basis of floral analyses. Presently available planktonic $\delta^{18}O$ values do not indicate such a significant warming but do suggest some mild warming. There is also evidence for a significant sea level rise [*Haq et al.*, 1987] during this interval, perhaps caused by disintegration of much of the Antarctic Ice Sheet, as advocated by *Webb et al.* [1984] and *Harwood* [1985] and supported by *Dowsett and Cronin* [1990]. The problem with this scenario is that the entire

negative excursion in the benthic foraminiferal isotope record would have to be attributed to ice volume changes rather than temperature, a hypothesis not accepted by *Hodell and Venz* [this volume].

As was stated above, the early Pliocene is a time of significant ice rafting in many parts of the Southern Ocean. For instance, *Anderson*'s [1985] analysis of piston core I07-5 from the South Atlantic (latitude 49°S) reveals five major closely spaced IRD episodes between 4.55 and 4.35 Ma and a minor episode at ~4.2 Ma, which is also the time of the first clear IRD signal at Site 704 (Figure 7). Figure 8 shows the core I07-5 episodes for comparison. These IRD episodes have been interpreted as representing either large volumetric expansions of the Antarctic Ice Sheet [*Anderson*, 1985] or large melting episodes [*Breza*, 1992]. It is likely that some warming occurred in Southern Ocean surface waters. This may have eliminated or reduced the size of the WAIS. At the same time, the higher surface water temperatures led to increased evaporation and snowfall which, in turn, caused a volumetric expansion of Antarctic ice. This is the so-called snow gun hypothesis of *Prentice and Matthews* [1991]. If this indeed occurred, the production of icebergs would have switched from ice shelves to rapidly moving outlet glaciers which have the potential of carrying much more debris [see *Drewry*

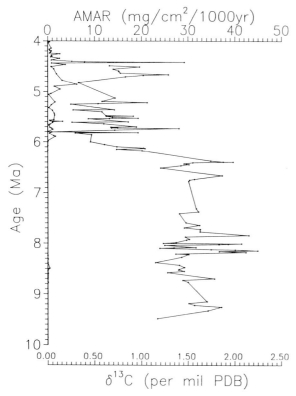

Fig. 6.   Apparent mass accumulation rates of IRD, Site 701, and the planktic carbon isotope record from Hole 704B.

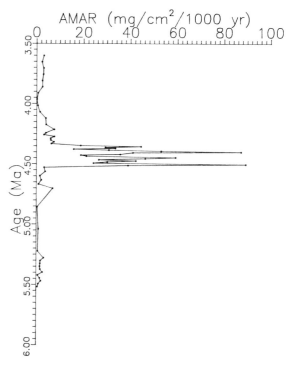

Fig. 8.   Early Pliocene ice-rafting episodes at piston core location I07-5 (48°51.2'S, 36°33.3'W). Data from *Anderson* [1985].

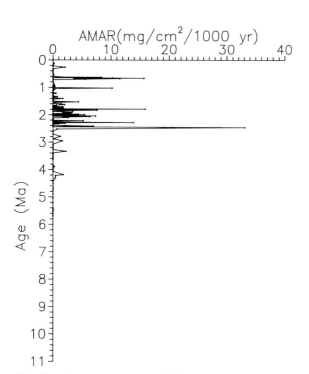

Fig. 7.   Apparent mass accumulation rates at Hole 704A.

and Cooper, 1981]. This change perhaps accounts not only for the widespread presence of IRD in the Southern Ocean, but also for the significant increase in IRD flux rates during the early Pliocene [*Anderson*, 1985], despite the postulated higher sea surface temperatures [*Hodell and Warnke*, 1991]. It is possible that delivery rates were faster because there was much less sea ice to impede iceberg dispersal (L. Krissek, personal communication, 1992).

We have not investigated the physical nature of the IRD itself to determine the type of all transport processes involved. A study by *Gram* [1974] of IRD of piston cores from the Pacific sector of the Southern Ocean revealed the presence of large numbers of quartz grains with surface features indicative of subaqueous transport. *Gram* [1974] interpreted this fact in terms of intermittent ice shelf recession during the Gauss and Gilbert chrons. In sum, the important question of the stability of the WAIS during the early Pliocene has as yet no definite answer because of conflicting lines of evidence. The problem is further reviewed by *Hodell and Warnke* [1991] and *Hodell and Venz* [this volume].

Between 4.1 and 3.7 Ma, temperatures decreased and the PFZ moved northward as shown by an increase of biogenic silica at Site 704 [*Froelich et al.*, 1991]. Still, the predominantly calcareous character of the sediment, as well as the dominance of globorotalids in the fora-

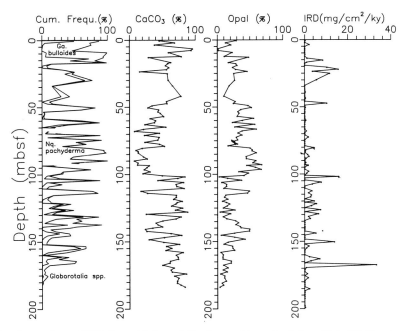

Fig. 9. Comparison of IRD data with foraminiferal assemblages (both from Hole 704A) and opal and calcium carbonate data (Hole 704A and Hole 704B spliced together). Data from *Froelich et al.* [1991], *Warnke et al.* [1989], and this paper. The increase in ice rafting at 168 mbsf and the change in foraminiferal assemblages are clearly indicated. Note further changes at 100 mbsf (~1.67 Ma). See text for discussion.

miniferal populations [*Brunner*, 1991], indicates that the PFZ lay far to the south of Site 704, with only occasional penetration of icebergs through the entire PFZ.

Benthic $\delta^{18}O$ values at Site 704 increased at 3.2 Ma [*Hodell and Ciesielski*, 1991], following an increase in biogenic silica and distinct ice-rafting episodes at ~3.3 Ma [*Froelich et al.*, 1991; *Allen and Warnke*, 1991]. The Gauss Chron (3.4–2.47 Ma) is characterized by increased planktonic and benthic $\delta^{18}O$ values and intermittent ice rafting at Site 704 (see Figure 7). This interval is missing at Site 701, but at Site 699 there was a general increase in IRD with superimposed maxima and minima (Figure 3). However, the reduced amount of fine sediments in this interval suggests that concentration of IRD may have resulted from winnowing by currents. Therefore IRD signals must be interpreted with caution.

The Early Matuyama Chron, to ~1.67 Ma

All measured parameters change significantly between 2.6 and 2.3 Ma, at a hiatus that spans the Gauss/Matuyama (G/M) boundary (2.47 Ma). The record is most complete at Site 704 (Figure 7), where the ice-rafting record is directly compared with other parameters measured at that site (Figure 9). In Hole 704A the G/M boundary was placed at 168.45 meters below seafloor (mbsf) [*Hailwood and Clement*, 1991] and has been correlated with the oxygen isotope record by

Hodell and Ciesielski [1991]. However, a hiatus occurs at this boundary [*Fenner*, 1991], eliminating ~300 kyr from the uppermost Gauss Chron [*Hodell and Venz*, this volume]. As a result, the change to more abundant IRD at this boundary is rather abrupt. Nevertheless, oxygen isotope values become higher during the Gauss Chron, indicating a cooling trend which climaxed in a rapid climatic deterioration near the G/M boundary. The changes were dramatic, seem to have been global in extent, and signal the beginning of the "time without a modern analog" (a time of significant cooling and glacial expansion in the northern hemisphere, yet not to the extent realized in the late Pleistocene; see *Raymo et al.* [1987]).

These changes involve a significant increase in IRD apparent mass accumulation rates between ~168 mbsf (~2.6 Ma) and 100 mbsf (~1.67 Ma) and a change in the planktonic-foraminiferal assemblages from one dominated by globorotalids to alternations between two assemblages, one dominated by globorotalids and one dominated by sinistrally coiled *Neogloboquadrina pachyderma* with accessory *Globigerina* spp. In addition, biogenic opal increased and carbonate decreased (Figure 9) [*Warnke et al.*, 1989; *Froelich et al.*, 1991; *Brunner*, 1991]; the amplitude of the benthic $\delta^{18}O$ variations increased [*Hodell and Ciesielski*, 1991], and a significant increase occurred in planktonic $\delta^{18}O$ values [*Hodell and Ciesielski*, 1991; *Warnke et al.*, 1989]. The alternations in foraminiferal assemblages indicate significant shifts in the position of the PFZ with respect to Site 704. The high melting rates indicated by the IRD apparent mass accumulation rates, combined with the characteristics of the foraminiferal assemblages, suggest that the PFZ had reached the drill site, with the Subantarctic Front (SAF) migrating over the drill site on several occasions, thereby bringing it squarely into the zone of major iceberg melting. In general, IRD apparent mass accumulation rates decrease markedly from Site 699 in the west to Site 704 in the east (compare Figures 3, 4, and 7; see also Figure 5 of *Allen and Warnke* [1991]). These relationships indicate that the icebergs traveled from west to east in the Antarctic Circumpolar Current after their escape from the Weddell Sea Gyre [see *Westall and Fenner*, 1991].

The interval between 168 and 100 mbsf (~2.6 and 1.67 Ma) is also characterized by cycles formed by rapidly fluctuating percentages of calcium carbonate and biogenic opal (Figure 9) and by the highest carbonate and silica accumulation rates encountered at this site [*Froelich et al.*, 1991]. Differences in cyclicity are evident between the IRD and foraminiferal data and the geochemical data sets which can only be resolved by additional sampling and analyses. The high accumulation rates of biogenic particles indicate alternating optimal growth conditions for calcareous and siliceous plankton: upwelling near the northern boundary of the PFZ, yet with temperatures still favorable to calcareous plankton growth, alternating with cooler temperatures favoring growth of siliceous plankton (for a review of diatom growth conditions, see *Warnke and Allen* [1991]). Another change to still higher $\delta^{18}O$ values occurs at 100 mbsf (about 1.67 Ma), coincident with a marked decrease in IRD apparent mass accumulation rates and the change to planktonic foraminiferal assemblages dominated by *N. pachyderma* (Figure 9). A shift at the same level is particularly evident in the carbonate and silica record.

The interval between about 2.6 Ma and 1.67 Ma represents an intermediate stage in the development of the Antarctic cryosphere and is the southern hemisphere equivalent of the "time without modern analog" described by *Raymo et al.* [1987]. In the North Atlantic, this interval is marked by widespread ice-rafting episodes (although ice rafting started earlier in regions closer to the source areas [see *Jansen et al.*, 1988; *Krissek*, 1989; *Henrich et al.*, 1989; *Warnke*, 1982]). At about 2.5 Ma, mean opal flux rates increased dramatically at equatorial Atlantic ODP sites 662, 663, and 664 [*Ruddiman and Janecek*, 1989] and were interpreted to have resulted from changes in equatorial circulation patterns which in turn were caused by changes in the Atlantic sector of the Southern Ocean. The results presented here support this suggestion.

The "1.67 Ma Event"

Above 100 mbsf at Site 704 (1.67 Ma), further changes in foraminiferal assemblages, oxygen isotope values, and biogenic sediments suggest a further decrease in surface water temperatures. This decrease was related to further expansion of the polar ice sheets, accompanied by a reduction in bulk mass accumulation rates at Site 704 [*Froelich et al.*, 1991] and a northward shift in the average position of the PFZ. The dominance of *Neogloboquadrina pachyderma* from 100 to 73 mbsf (~1.4 Ma) and the remarkable increase in opal percentages [*Froelich et al.*, 1991] suggest location of the site south of the PFZ, at least temporarily. The dominance of diatoms and decreased rates of opal accumulation [*Froelich et al.*, 1991] are strong indications for the position of the site in the low-productivity zone south of the PFZ [*Warnke and Allen*, 1991]. Delivery of IRD to the site was greatly reduced, perhaps because of a change in the boundary conditions on Antarctica [*Warnke*, 1970] and the increased importance of floating ice shelves to ablation, because modern ice shelves are virtually devoid of debris beyond the grounding line [*Drewry and Cooper*, 1981]. This interpretation contradicts *Ciesielski et al.* [1983], who proposed that only such large, tabular bergs could transport debris so far north. Whether or not the role of such tabular bergs as agents of erosion and transport has changed over time remains a topic for further research.

From 1.4 to 1.2 Ma (73 to 59 mbsf) the foraminiferal assemblages suggest surface water conditions relatively similar to those between ~2.5 and 1.67 Ma (Figure 9). However, calcium carbonate and opal contents suggest a much lesser degree of amelioration, perhaps to a stage intermediate between conditions just below and above 100 mbsf (1.67 Ma). In any event, floral analyses presented by *Westall and Fenner* [1991] indicate a gradual shift of the PFZ to the south during much of the Matuyama Chron. A detailed description of PFZ migrations during the Brunhes Chron is provided by *Westall and Fenner* [1991].

SUMMARY

1. IRD is absent in all Oligocene material examined for Leg 114. IRD first reached one of the drill sites, Site 699, at about 23.5 Ma, but only in very small amounts. Similarly small amounts reached the site intermittently through much of the Miocene. IRD first reached Site 701 around 8.8 Ma, similar to the first reported occurrence of IRD on the Kerguelen Plateau [*Breza*, 1992]. The oxygen isotope record of Hole 704B indicates that much of the Miocene, up to 6.4 Ma, was a time of relative stability, but with some fluctuations. The Polar Front Zone, if it existed, was located south of the drill sites.

2. The Chron 6 carbon isotopic shift occurred between 6.4 and 6.0 Ma, and signaled a profound change in all measured parameters. It was a time of relative

warmth (particularly at 6.15 Ma), although with major fluctuations. A large amount of isotopically light carbon returned to the ocean, caused by changes in oceanic circulation and tectonic events in the proto–Straits of Gibraltar and on the Pacific margin [see *Müller et al.*, 1991]. No significant ice rafting is recorded at sites 701 and 704.

3. Immediately after the carbon shift, ice rafting started at sites 699 and 701. The record of ice rafting is best developed at Site 699 for the latest Miocene and suggests that ice-rafting episodes are positively correlated with oxygen isotope excursions and perhaps negatively correlated with carbon isotope excursions. Likely explanations for these postulated correlations are significant Antarctic glacial expansions that lowered sea level, led to erosion of shallow continental shelves, and played a role in the isolation of the Mediterranean Sea. This isolation caused the deposition of the upper evaporites and may have caused a change in the alkalinity of the ocean which in turn may have led to several dissolution events. Four of the recorded dissolution events [*Froelich et al.*, 1991] can be correlated with IRD peaks at Site 699.

4. A time of climatic amelioration followed, accompanied by minor ice rafting. Beginning at about 4.4 Ma, IRD supply increased at sites 699 and 701, but only minor ice rafting occurred at Site 704. The entire interval from about 4.8 to 4.2 Ma was characterized by relatively warm conditions, although the degree of warming is still much debated. At that time, the West Antarctic Ice Sheet may have disappeared, but the warmer water temperatures and ensuing higher rates of evaporation may have led to a volumetric expansion of the East Antarctic Ice Sheet and increased iceberg calving from rapidly moving outlet glaciers. These particular types of icebergs may be responsible for the widespread occurrence of IRD recognized in several drill sites and piston core locations. Lack of sea ice may have aided dispersal mechanisms. The oxygen isotope record shows cooling/ice growth during the remainder of the Gauss Chron.

5. All measured parameters changed significantly between 2.6 and 2.3 Ma during a hiatus at Site 704 that spans the Gauss/Matuyama boundary. Increase in ice rafting is particularly prominent at Site 704, where it coincides with an oxygen isotope shift to higher values and a change in the planktonic foraminiferal assemblages from one dominated by globorotalids below to alternations between two assemblages, one dominated by globorotalids and one dominated by sinistrally coiled *N. pachyderma* with accessory *Globigerina* spp. In addition, opal percentages [*Froelich et al.*, 1991] as well as bulk, carbonate, and opal accumulation rates increased dramatically. These factors indicate oscillations of the PFZ around a mean position that shifted to the north. As a consequence, the Subantarctic Front swept across the drill site, and the drill site during these

northward excursions was within the PFZ, the zone of major iceberg melting. Icebergs traveled from west to east as indicated by decreasing IRD mass accumulation rates from Site 699 to Site 704. At the same time, changes occurred in the North Atlantic, beginning the "time without a modern analog" [*Raymo et al.*, 1987].

6. Another change occurred at about 1.67 Ma (100 mbsf) as documented at Site 704, marked by decreased ice rafting, higher oxygen isotope ratios (and greater amplitudes of oscillations), foraminiferal populations dominated by *N. pachyderma*, and increased opal percentages (yet decreased opal mass accumulation rates). These changes indicate that the "average" geographic position of the PFZ shifted to the north of Site 704, bringing the site into the cold, low-productivity zone south of the PFZ. Additional factors in the reduction of delivery of IRD may have been changing glacial boundary conditions on Antarctica and the increased importance of floating ice shelves in ice sheet ablation, because modern ice shelves generally do not carry any debris beyond the grounding line.

7. Some climatic amelioration occurred at about 1.4 Ma, as indicated by the planktonic foraminiferal indicators, but calcium carbonate and opal percentages do not suggest a return to conditions similar to those that prevailed during the Gauss Chron. Instead, these indicators suggest conditions intermediate between those just before and after 1.67 Ma. IRD flux remained low or absent, perhaps indicating that floating ice shelves had become at least semipermanent fixtures of the Antarctic ice margin. Later, Pleistocene climatic fluctuations are described by *Westall and Fenner* [1991].

Acknowledgments. We thank Larry Krissek, Jim Kennett, and Woody Wise for detailed and constructive critiques, which greatly improved the manuscript.

REFERENCES

Abelmann, A., R. Gersonde, and V. Spiess, Pliocene-Pleistocene paleoceanography in the Weddell Sea—Siliceous microfossil evidence, in Bleil, U. and J. Thiede, *Geological History of the Polar Oceans: Arctic Versus Antarctic*, edited by U. Bleil and J. Thiede, pp. 729–759, Kluwer Academic Publishers, Amsterdam, 1990.

Allen, C. P., and D. A. Warnke, History of ice-rafting at ODP Leg 114 sites, Subantarctic South Atlantic, *Proc. Ocean Drill. Program Sci. Results*, *114*, 599–607, 1991.

Anderson, D. M., Pliocene paleoceanography of the Southern Ocean and the development of the West Antarctic Ice Sheet, M.S. thesis, 42 pp., San Jose State Univ., San Jose, Calif., 1985.

Armentrout, J. M., Glacial lithofacies of the Neogene Yakataga Formation, Robinson Mountains, southern Alaska Coast Range, Alaska, in *Glacial-Marine Sedimentation*, edited by B. F. Molnia, pp. 629–665, Plenum, New York, 1983.

Barrett, P. J. (Ed.), Antarctic Cenozoic history from the CIROS-1 drillhole, McMurdo Sound, *DSIR Bull. N. Z.*, *245*, 254 pp., 1989.

Barrett, P. J., M. J. Hambrey, D. M. Harwood, A. R. Pyne, and P. N. Webb, Synthesis, in Antarctic Cenozoic History

From the CIROS-1 Drillhole, McMurdo Sound, *DSIR Bull. N. Z.*, *245*, 241–251, 1989.

Barron, J. A., B. Larsen, and J. G. Baldauf, Evidence for late Eocene to early Oligocene Antarctic glaciation and observations on late Neogene glacial history of Antarctica: Results from Leg 119, *Proc. Ocean Drill. Program Sci. Results*, *119*, 869–891, 1991.

Birkenmajer, K., Onset and course of Tertiary glaciation (abstract), in *Second International Congress on Paleoceanography, Abstracts*, 1986.

Breza, J. R., High resolution study of ice-rafted debris, ODP Leg 120, Site 751 southern Kerguelen Plateau, *Proc. Ocean Drill. Program Sci. Results*, *120*, 207–221, 1992.

Breza, J. R., and S. W. Wise, Jr., Lower Oligocene ice-rafted debris on the Kerguelen Plateau: Evidence for East Antarctic continental glaciation, *Proc. Ocean Drill. Program Sci. Results*, *120*, 161–178, 1992.

Brunner, C. A., Latest Miocene to Quaternary biostratigraphy and paleoceanography of Site 704 in the Subantarctic South Atlantic, *Proc. Ocean Drill. Program Sci. Results*, *114*, 201–215, 1991.

Ciesielski, P. F., and F. M. Weaver, Neogene and Quaternary paleoenvironmental history of Deep Sea Drilling Project Leg 71 sediments, southwest Atlantic Ocean, *Initial Rep. Deep Sea Drill. Proj.*, *71*, 461–477, 1983.

Ciesielski, P. F., et al., Leg 114, *Proc. Ocean Drill. Program Initial Rep.*, *114*, 815 pp., 1988.

Ciesielski, P. F., et al., Leg 114, *Proc. Ocean Drill. Program Sci. Results*, *114*, 826 pp., 1991.

Dowsett, H. J., and T. M. Cronin, High eustatic sea level during the middle Pliocene: Evidence from the southeastern U.S. Atlantic coastal plain, *Geology*, *18*, 435–438, 1990.

Drewry, D. J., and A. P. R. Cooper, Processes and models of Antarctic glaciomarine sedimentation, *Ann. Glaciol.*, *2*, 117–122, 1981.

Fenner, J. M., Late Pliocene–Quaternary quantitative diatom stratigraphy in the Atlantic sector of the Southern Ocean, *Proc. Ocean Drill. Program Sci. Results*, *114*, 97–121, 1991.

Froelich, P. N., et al., Biogenic opal and carbonate accumulation rates in the Subantarctic South Atlantic: The late Neogene of Meteor Rise Site 704, *Proc. Ocean Drill. Program Sci. Results*, *114*, 515–550, 1991.

Geitzenauer, K. R., S. V. Margolis, and D. S. Edwards, Evidence consistent with Eocene glaciation in a South Pacific deep-sea sedimentary core, *Earth Planet. Sci. Lett.*, *4*, 173–177, 1968.

Gram, R., Mineralogical changes in Antarctic deep-sea sediments and their paleo-climatic significance, Ph.D. dissertation, 288 pp., Fla. State Univ., Tallahassee, 1974.

Hailwood, E. A., and B. M. Clement, Magnetostratigraphy of ODP sites 703 and 704, Meteor Rise, southeastern South Atlantic, *Proc. Ocean Drill. Program Sci. Results*, *114*, 367–385, 1991.

Haq, B. U., J. Hardenbol, and P. R. Vail, Chronology of fluctuating sea levels since the Triassic, *Science*, *235*, 1156–1167, 1987.

Harwood, D. M., Late Neogene climatic fluctuations in the southern high latitudes: Implications of a warm Pliocene and deglaciated Antarctic continent, *S. Afr. J. Sci.*, *81*, 239–241, 1985.

Hayes, D. E., and L. A. Frakes, General synthesis, Deep Sea Drilling Project, Leg 28, *Initial Rep. Deep Sea Drill. Proj.*, *28*, 919–924, 1975.

Henrich, R., Glacial/interglacial cycles in the Norwegian Sea: Sedimentology, paleoceanography, and evolution of late Pliocene to Quaternary northern hemisphere climate, *Proc. Ocean Drill. Program Sci. Results*, *104*, 189–232, 1989.

Hodell, D. A., and P. F. Ciesielski, Stable isotopic and carbonate stratigraphy of the late Pliocene and Pleistocene of Hole 704 A: Eastern Subantarctic South Atlantic, *Proc. Ocean Drill. Program Sci. Results*, *114*, 409–435, 1991.

Hodell, D. A., and K. Venz, Toward a high-resolution stable isotopic record of the Southern Ocean during the Pliocene-Pleistocene (4.8 to 0.8 Ma), this volume.

Hodell, D. A., and D. A. Warnke, Climatic evolution of the Southern Ocean during the Pliocene Epoch from 4.8 to 2.6 million years ago, *Quat. Sci. Rev.*, *10*, 205–214, 1991.

Hodell, D. A., D. W. Muller, P. F. Ciesielski, and G. A. Mead, Synthesis of oxygen and carbon isotopic results from ODP Site 704: Implications for major climatic-geochemical transitions during the late Neogene, *Proc. Ocean Drill. Program Sci. Results*, *114*, 475–480, 1991.

Jansen, E., U. Bleil, R. Henrich, L. Kringstad, and B. Slettemark, Paleoenvironmental changes in the Norwegian Sea and the northeast Atlantic during the last 2.8 m.y.: Deep Sea Drilling Project/Ocean Drilling Program sites 610, 642, 643 and 644, *Paleoceanography*, *3*, 563–581, 1988.

Kennett, J. P., and P. F. Barker, Climatic and oceanographic developments in the Weddell Sea, Antarctica, since the latest Cretaceous: An ocean-drilling perspective, *Proc. Ocean Drill. Program Sci. Results*, *113*, 937–960, 1990.

Krissek, L. A., Late Cenozoic records of ice-rafting at ODP sites 642, 643 and 644, Norwegian Sea: Onset, chronology, and characteristics of glacial/interglacial fluctuations, *Proc. Ocean Drill. Program Sci. Results*, *104*, 61–74, 1989.

Leckie, R. M., and P. N. Webb, Late Oligocene–early Miocene glacial record of the Ross Sea, Antarctica: Evidence from DSDP Site 270, *Geology*, *11*, 578–582, 1983.

LeMasurier, W. E., and D. C. Rex, Volcanic record of Cenozoic glacial history in Marie Byrd Land and western Ellsworth Land: Revised chronology and evaluation of tectonic factors, in *Antarctic Geoscience*, edited by C. Craddock, pp. 725–734, University of Wisconsin Press, Madison, 1982.

Matthews, R. K., and R. Z. Poore, Tertiary $\delta^{18}O$ record and glacio-eustatic sea level fluctuations, *Geology*, *8*, 501–504, 1980.

Mercer, J. H., and J. F. Sutter, Late Miocene–earliest Pliocene glaciation in southern Argentina: Implications for global ice-sheet history, *Palaeogeogr. Palaeoclimatol. Palaeoecol.*, *38*, 185–206, 1982.

Mildenhall, D. C., Terrestrial palynology, in Antarctic Cenozoic History From the CIROS-1 Drillhole, McMurdo Sound, *DSIR Bull. N. Z.*, *245*, 119–127, 1989.

Miller, K. G., R. G. Fairbanks, and G. S. Mountain, Tertiary oxygen isotope synthesis, sea level history, and continental margin erosion, *Paleoceanography*, *2*, 1–19, 1987.

Molnia, B., Distal glacial-marine sedimentation: Abundance, composition, and distribution of North-Atlantic Ocean Pleistocene ice-rafted sediment, in *Glacial-Marine Sedimentation*, edited by B. Molnia, pp. 593–626, Plenum, New York, 1983.

Müller, D. W., D. A. Hodell, and P. F. Ciesielski, Late Miocene to earliest Pliocene (9.8–4.5 Ma) paleoceanography of the Subantarctic southeast Atlantic: Stable isotopic, sedimentologic and microfossil evidence, *Proc. Ocean Drill. Program Sci. Results*, *114*, 459–474, 1991.

Prentice, M. L., and R. K. Matthews, Tertiary ice sheet dynamics: The snow gun hypothesis, *J. Geophys. Res.*, *96*, 6811–6827, 1991.

Raymo, M. E., W. F. Ruddiman, and B. M. Clement, Pliocene-Pleistocene paleoceanography of the North Atlantic at Deep Sea Drilling Project Site 609, *Initial Rep. Deep Sea Drill. Proj.*, *94*, 895–901, 1987.

Ruddiman, W. F., and T. R. Janecek, Pliocene-Pleistocene biogenic and terrigenous fluxes at equatorial Atlantic sites 662, 663 and 664, *Proc. Ocean Drill. Program Sci. Results*, *108*, 211–240, 1989.

Vilmundardottir, E. G., Report on geological investigations of Jokulsa in Fljotsdal Valley, summer 1970 (in Icelandic), 26 pp., Natl. Energy Authority, Orkustofnun, Reykjavik, 1972.

Vincent, E., and W. H. Berger, Carbon dioxide and polar cooling in the Miocene: The Monterey Hypothesis, in *The Carbon Cycle and Atmospheric CO₂: Natural Variations Archean to Present, Geophys. Monogr. Ser.*, vol. 32, edited by E. T. Sundquist and W. S. Broecker, pp. 455–468, AGU, Washington, D. C., 1985.

Warnke, D. A., Glacial erosion, ice rafting, and glacial-marine sediments: Antarctica and the Southern Ocean, *Am. J. Sci.*, *269*, 276–294, 1970.

Warnke, D. A., Pre–middle Pliocene sediments of glacial and periglacial origin in the Norwegian-Greenland seas: Results of D.S.D.P. Leg 38, *Earth Evol. Sci.*, *2*, 69–78, 1982.

Warnke, D. A., and C. P. Allen, Ice rafting, glacial-marine sediments, and siliceous oozes: South Atlantic/Subantarctic, *Proc. Ocean Drill. Program Sci. Results*, *114*, 589–607, 1991.

Warnke, D. A., and M. E. Hansen, Sediments of glacial origin in the area of operations of D.S.D.P. Leg 38 (Norwegian-Greenland seas): Preliminary results from sites 336 and 344, *Naturforsch. Ges. Freiburg Br. Pfannenstiel Gedenkband*, *67*, 371–392, 1977.

Warnke, D. A., C. P. Allen, D. A. Hodell, and C. A. Brunner, Possible recognition of the Sirius till equivalent, marine (STEM) in ODP Leg 114 cores (abstract), *Geol. Soc. Am. Abstr. Programs*, *21*, A330, 1989.

Webb, P.-N., B. C. McKelvey, J. H. Mercer, and L. D. Stott,

Cenozoic marine sedimentation and ice-volume variation on the East Antarctic craton, *Geology*, *12*, 287–291, 1984.

Wei, W., Calcareous nannofossil stratigraphy and reassessment of the Eocene glacial record in Subantarctic piston cores of the southeast Pacific, *Proc. Ocean Drill. Program Sci. Results*, *120*, 1093–1104, 1992.

Westall, F., and J. Fenner, Pliocene-Holocene Polar Front Zone in the South Atlantic: Changes in its position and sediment-accumulation rates from holes 699A, 701C, and 704B, *Proc. Ocean Drill. Program Sci. Results*, *114*, 609–646, 1991.

Wise, S. W., A. M. Gombos, and J. P. Muza, Cenozoic evolution of polar water masses, southwest Atlantic Ocean, in *South Atlantic Paleoceanography*, edited by K. J. Hsu and H. J. Weissert, pp. 283–324, Cambridge University Press, New York, 1985.

Wise, S. W., J. R. Breza, D. M. Harwood, and W. Wei, Paleogene glacial history of Antarctica, in *Controversies in Modern Geology*, edited by D. W. Müller, J. A. McKenzie, and H. Weissert, pp. 134–171, Academic, San Diego, Calif., 1991.

Zachos, J. C., J. R. Breza, and S. W. Wise, Early Oligocene ice sheet expansion on Antarctica: Stable isotope and sedimentological evidence from Kerguelen Plateau, southern Indian Ocean, *Geology*, *20*, 569–573, 1991.

(Received December 13, 1991;
accepted June 16, 1992.)

THE ANTARCTIC PALEOENVIRONMENT: A PERSPECTIVE ON GLOBAL CHANGE
ANTARCTIC RESEARCH SERIES, VOLUME 56, PAGES 327–347

A LATE NEOGENE ANTARCTIC GLACIO-EUSTATIC RECORD, VICTORIA LAND BASIN MARGIN, ANTARCTICA

Scott E. Ishman

U.S. Geological Survey, Reston, Virginia 22092

Hugh J. Rieck

U.S. Geological Survey, Flagstaff, Arizona 86001

Glaciomarine and marine sediments recovered at Dry Valley Drilling Project (DVDP) sites 10 and 11 represent the most complete Miocene and Pliocene (~6 Ma to ~2.8 Ma) sequence to date from the Antarctic continent. These cores document changing glacial and sea level conditions in the Victoria Land Basin throughout much of the late Neogene. Widely fluctuating bathymetries and environmental conditions indicated from benthic foraminifer and diatom distributions provide a detailed and well-constrained late Neogene sea level and glacial history of the Antarctic region. These records are the first to contain a late Neogene sea level history from the Antarctic region. DVDP sediments record high-frequency sea level changes, with amplitudes equal to third-order eustacy cycles. The timing of these cycles is constrained through the combination of micropaleontologic (foraminifer and diatom) and magnetostratigraphic analyses. The resolution attained using these data is of the order of 10^5 years. Major hiatuses (H1, H2, and H3) correlate with significant events on the Haq et al. (1987) curve (cycles 3.4, 3.6, and 3.7). The timing of these eustatic fluctuations, and the fluctuations in glacial/nonglacial conditions in the Taylor Valley region of the Victoria Land Basin throughout the late Neogene, significantly affect the interpretations of ice volume history of Antarctica. The paleoenvironmental record from DVDP 10 and 11 provides evidence for an unstable late Neogene East Antarctic Ice Sheet and its influence on global sea level.

INTRODUCTION

Seismic stratigraphy that has been applied to sediment sequences on continental margins has allowed development of global eustacy curves [*Vail et al.*, 1977]. Since their introduction, these curves have proved to be a valuable chronostratigraphic tool and are increasingly used to correlate major paleoceanographic events. Two processes most widely accepted as causes of sea level fluctuations are changes in volume of the ocean basins due to variations in rates of ocean floor spreading and major fluctuations of high-latitude continental ice sheets. *Pitman* [1978], among others, is a major proponent for sea level fluctuation due to changes in the volume of oceanic crust and rates of spreading at the mid-ocean ridges. The longevity (~10^6 to 10^7 years) and amplitude of the first- and second-order *Haq et al.* [1987] sea level fluctuations conform to the rates and volume changes associated with ocean basin evolution. Other researchers, beginning at least as early as 1888 [*Thomson*, 1888], have considered high-latitude ice volume variation, in particular, the waxing and waning of the large continental ice sheets of Antarctica, as a major

factor driving eustatic sea level fluctuation. The timing and amplitude of sea level fluctuations suggested in the third-order cycles of *Haq et al.* [1987] conform to the rates and ice volume changes associated with glacial-interglacial episodes. These two processes can be distinguished from one another in geologic and paleoceanographic records.

Proxy data reflecting paleoceanographic events range from stable isotope studies [*Kennett*, 1985; *Matthews and Poore*, 1980; *Miller et al.*, 1985, 1987] to deep-sea hiatuses [*Keller et al.*, 1987; *Keller and Barron*, 1983, 1986] and biofacies analyses [*Olsson et al.*, 1980, 1987]. These and similar studies, generally from low- to mid-latitude sites, have been used to test the seismic stratigraphic framework on which global eustatic inferences are founded and to interpret late Cenozoic Antarctic glacial and climatic events. In particular, the relatively short-term (10^5 years) third-order cycles are likely to have been affected by fluctuations in the volume of the Antarctic ice sheets. However, little direct evidence of the timing and extent of major Cenozoic glaciations in Antarctica has been available to evaluate the glacial component of eustatic fluctuations.

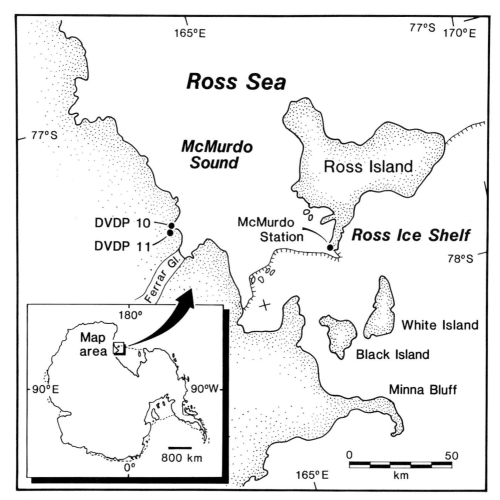

Fig. 1. Map showing the locations of DVDP sites 10 and 11 in Taylor Valley, McMurdo Sound, Antarctica.

OBJECTIVES

This study focuses on the most complete late Miocene and Pliocene nearshore sedimentary record obtained to date, from the Victoria Land Basin margin, Antarctica. Benthic foraminifer, diatom, sedimentologic, paleomagnetic, and apatite fission track data are employed to achieve the following objectives: (1) define a chronostratigraphy (resolution of <1 m.y.) for the late Neogene record of Taylor Valley, Antarctica; (2) interpret paleobathymetric fluctuations for the Taylor paleofjord; (3) evaluate eustatic and tectonic controls on the depositional facies identified; and (4) develop a eustatic model for the Victoria Land Basin region that can be compared with global eustatic curves for a part of the late Neogene (~6.0 to ~2.8 Ma). Results are used to evaluate the glacial aspect of the global eustacy curve from a locality proximal to and directly affected by Antarctic continental ice sheet fluctuations, as opposed to proxy studies

based on deep-sea records from low- to mid-latitude sites.

LOCATION AND GEOLOGIC SETTING

This study is based on drill core material collected during the Dry Valley Drilling Project [*McGinnis*, 1981]. Dry Valley Drilling Project (DVDP) sites 10 and 11 are located along the axis of lower (eastern) Taylor Valley, which extends for a distance of 100 km from the polar plateau and East Antarctic Ice Sheet (EAIS), through the Transantarctic Mountains (TAMs), and into McMurdo Sound at the margin of the Victoria Land Basin (Figure 1).

Lower Taylor Valley is cut into the crystalline basement complex of Precambrian and Cambrian metamorphic rocks of the Skelton Group and the Precambrian Granite Harbor intrusive complex [*Gunn and Warren*,

1962]. Farther inland, this crystalline basement is nonconformably overlain by near-horizontal Paleozoic and lower Mesozoic sediments of the Beacon Supergroup, which are intruded by the Jurassic Ferrar Dolerite [*Gunn and Warren*, 1962]. Late Tertiary mafic volcanic rocks are widespread in McMurdo Sound to the south and east.

Previous study of DVDP sites 10 and 11 shows that a thick sequence of upper Miocene to Holocene marine, glaciomarine, glacial, fluvio-glacial, and lacustrine sediments fills lower Taylor Valley [*McKelvey*, 1981]. These late Neogene strata were interpreted to have been deposited in a deep paleofjord [*Webb and Wrenn*, 1982; *Ishman and Webb*, 1988]. Because of continued late Cenozoic uplift of the Transantarctic Mountains [*Fitzgerald et al.*, 1986], the complete stratigraphic sequence records a marine to subaerial transition. This uplift tended to protect lower Taylor Valley from possible erosion during expansions of the EAIS and Taylor Glacier. Perhaps more importantly, the uplift and concurrent growth of Ross Island prevented erosion during late Pliocene(?) and Quaternary advances and groundings of the Ross Ice Shelf that destroyed late Neogene records across most of the Ross Sea continental shelf. Core DVDP site 10 was drilled about 50 m inland from the present coastline, from an elevation of 2.5 m above sea level to a depth of 188 m below surface (mbs) (185.5 m below sea level). Core recovery was 83.4%. DVDP site 11, located 3 km inland, was drilled from an elevation of 80 m above sea level to a depth of 328 mbs (248 m below sea level (mbsl)). Core recovery was 94.1%. Neither core reached crystalline basement rocks.

Sedimentologic, paleontologic, paleomagnetic, and gravity data have been employed in deciphering the chronology, depositional environments, and paleobathymetry of these upper Neogene strata. The interbedded marine, glaciomarine, and glacial sediments of the sequence provide detailed and direct evidence of interactions between Taylor Glacier, fed from the East Antarctic Ice Sheet, and the marine record of the Ross Ice Shelf along the Victoria Land Basin margin. These sediments are the focus of this study.

GENERAL SEDIMENTOLOGY AND STRATIGRAPHY

This study focuses on the Miocene and Pliocene stratigraphic intervals from the bottoms of the two cores up to about 194 mbs in DVDP site 11 and up to about 135 mbs in DVDP site 10. However, some comment will be made concerning the upper sections of each core. The lower intervals have been divided into two distinct parts on the basis of lithologic descriptions and sedimentological analyses conducted by *McKelvey* [1981], *Powell* [1981], and *Porter and Beget* [1981]. The lower part of the study interval in DVDP site 11 (328 mbs up to 255.00

mbs), unit 8 of *McKelvey* [1981], is dominated (81%) by massive diamictites, with minor interbeds of sandy mudstones, well-sorted sands, and pebbly gravels (Figure 2). The upper part of the study interval in DVDP site 11 (255.00 mbs up to 194.00 mbs), unit 7 of *McKelvey* [1981], is composed of varied lithologies ranging from mudstone to breccia and contains only about 38% diamictite (Figure 2). Strata from about 194.00 mbs to the ground surface also are varied but are progressively sandier upward. A western Taylor Valley provenance was inferred for sediments greater than 200.00 mbs, composed mainly of sediments derived from the Skelton Group and Granite Harbor Complex [*Porter and Beget*, 1981]. Less than 200.00 mbs a dominant Ross Sea–McMurdo Sound provenance was inferred from the presence of sediments derived from McMurdo Group volcanics [*Porter and Beget*, 1981].

The lower part of DVDP site 10 (186.00–167.20 mbs) lacks the thick sequence of diamictites present in the lower part of DVDP site 11. In contrast, the lithologies of the lower part of DVDP site 10 (greater than 167.20 mbs), units 5.33 to 5.12 of *McKelvey* [1981], are dominated by sandy mudstones, some of which contain dispersed pebbles; these represent a more distal facies less strongly influenced by ice-proximal and ice-contact depositional processes. Less lithologic variation occurs in the lower interval of DVDP site 10, suggesting the paleoenvironment was less sensitive to fluctuations of Taylor Glacier. Lithologies in the upper part of the lower interval of DVDP site 10 (167.20–135.66 mbs), units 5.11 to 4.7 of *McKelvey* [1981], are varied and resemble the lithologies of the upper part of the DVDP site 11 interval, i.e., sandy mudstones, sands, pebbly gravels, breccias, and some diamictites (Figure 2). Diamictites in DVDP site 10 are for the most part restricted to portions of the core above 125.00 mbs (younger than ~2.5 Ma). Provenance of the sediments is similar to DVDP site 11, a western Taylor Valley source for sediments greater than 154 mbs and a dominant Ross Sea–McMurdo Sound component less than 154 mbs [*Porter and Beget*, 1981; *Purucker et al.*, 1981].

On the basis of grain size distribution and sorting, *Powell* [1981] characterized six major sediment types from DVDP sites 10 and 11 (types I through VI) and inferred probable environments of deposition for each. Sediment types VI (poorly sorted gravely muddy sands) and V (poorly sorted gravely sandy muds) are interpreted by *Powell* [1981] as representing ice marginal marine environments, i.e., floating ice tongue and iceberg zone deposition. Other evidence suggesting glacial and glaciomarine deposition of the massive diamictons (type VI) includes thickness and sedimentary structures such as striated clasts, striated bedding plane surfaces, overcompacted sediments, and fine horizontal bedding. The massive diamictons are considered to represent fluctuations of Taylor Glacier into and out of McMurdo Sound, as ice-contact deposits (basal tills), deposited

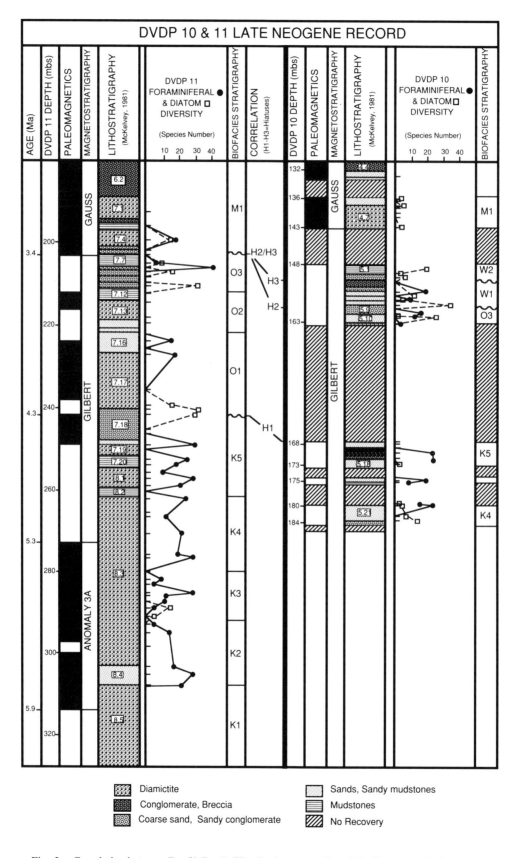

Fig. 2. Correlation between Dry Valley Drilling Project cores 10 and 11. Shown are the interpreted ages of the core (this paper), magnetostratigraphy (this paper), lithostratigraphy [*McKelvey*, 1981], foraminifer (solid circles) and diatom (open squares) diversity data (this paper), and biofacies stratigraphy (this paper). The hachured pattern represents no core recovery.

beneath several hundred meters of ice filling the deep Taylor paleofjord, to water-laid tills from a floating Taylor Glacier ice tongue of unknown thickness. Other sediment types do not necessarily reflect the presence of significant amounts of ice and are regarded as nonglacial sediments. In the lower portions of the cores (greater than 200.00 mbs and 157.00 mbs in DVDP site 11 and site 10, respectively), the bases of some of the major diamictite units are noted as sharp [*McKelvey*, 1981] and in at least one place as unconformable, while the overlying nonglacial strata rest with gradational contact upon the glacial units.

BENTHIC FORAMINIFER BIOFACIES ANALYSIS

Four benthic foraminifer assemblages were defined on the basis of their dominant taxa (revised from *Ishman and Webb* [1988]) (Figure 3). The assemblages have been used to define biofacies zones based on their stratigraphic longevity and paleoecologic significance. Further subdivision of these zonations (K1 through M1) represents environmental facies changes associated with fluctuating glacial conditions within the Taylor Fjord (Figure 2). The letters designated to these units are taken from the New Zealand Neogene ages Kapitean (K), Opoitian (O), Waipipian (W), and Mangapanian (M). Each event (designated by a letter and a number) represents an environmental cycle interpreted from faunal and floral changes.

Ammoelphidiella uniforamina Zone

The lowermost unit of DVDP site 10 and site 11 (less than 167.20 and 242.00 mbs, respectively) is marked by the presence of the *Ammoelphidiella uniforamina* Assemblage. This assemblage is dominated by the species *Epistominella exigua* Parker and *Globocassidulina subglobosa* (Brady). Additional major components of this assemblage are *Ehrenbergina glabra* (Heron-Allen and Earland), *E. pupa* (d'Orbigny), *Nonionella iridea* Heron-Allen and Earland, *Astrononion* aff. *A. antarcticus* Parr, and *Ammoelphidiella uniforamina* (D'Agostino) (Figure 3). Unique to this assemblage is a high percentage of miliolids (up to 45% of the fauna).

The *Ammoelphidiella uniforamina* Assemblage has close affinities to Antarctic margin deepwater assemblages [*Uchio*, 1960; *Pflum*, 1966; *Fillon*, 1974; *Anderson*, 1975; *Osterman and Kellogg*, 1979; *Ward*, 1984]. Estimates on depth limitations of the faunas range from a minimum of 300 mbsl to a maximum of 900 mbsl (Figure 4). These depth estimates are not unreasonable considering modern fjord depths in the Antarctic Peninsula, southern Chile, and New Zealand range up to 1050 m [*Freeland et al.*, 1980]. However, it should be noted that with increasing data on faunal-environmental associations (based on living foraminifer data), bathymetric relationships become much more ambiguous.

Although the depth estimates presented here are done so with confidence based on faunal compositions, additional paleoenvironmental interpretation will be included to support these claims.

The occurrence of this assemblage is punctuated by barren zones throughout the stratigraphic interval it defines, representing cyclic environmental events. Fluctuations in species abundance within the assemblage is indicative of changing environmental conditions. Five biofacies events, K1–K5 (Figure 2), are present in the *Ammoelphidiella uniforamina* Zone. These are all present in DVDP site 11 with only the latter two (K4 and K5) occurring in DVDP site 10. These biofacies units and the faunal and floral trends within them represent paleoenvironmental fluctuations induced by fluctuating glacial conditions in Taylor Fjord. However, not all of the trends that would be considered a complete cycle exist within each of the DVDP units. The units are characterized by two distinct foraminifer assemblages: a high-diversity (>15 species) fauna and a low-diversity (<15 species) fauna. The high-diversity fauna is typically dominated by *E. exigua*, with *G. subglobosa* and *N. iridea* of secondary importance. The diversity of this assemblage is enhanced by the frequency of nodosariids, common to the Ross Sea region. However, the low-diversity fauna is dominated by *G. subglobosa*, *E. pupa* and *E. glabra*, *Cassidulinoides porrectus*, *C. parkerianus*, *Quinqueloculina stalkeri*, and *Astrononion* aff. *A. antarcticus*, with *E. exigua* remaining abundant.

Facies unit K3 demonstrates the most complex sequence of biofacies evolution within the units. This unit is initiated by the occurrence of diatoms (from *Brady* [1979] and *Harwood* [1986]) replaced by the occurrence of the low-diversity benthic foraminifer fauna (Figure 2). This pattern is common in the units where diatoms occur, with their presence preceding the occurrence of the benthic foraminifer faunas. Within facies unit K3 the low-diversity foraminifer assemblage is progressively replaced by the high-diversity, *E. exigua* dominated assemblage. The cycle is concluded with the regression back to the low-diversity, *G. subglobosa*, *Ehrenbergina* spp., dominated assemblage. Although this complete biotic sequence is not fully represented in biofacies units K1, K2, K4, and K5, the high- and/or low-diversity components can be easily recognized.

Previous study of Pleistocene benthic foraminifera from the Ross Sea [*Osterman and Kellogg*, 1979] suggests an association between the occurrence of *Ehrenbergina glabra* and ice conditions. They conclude, however, that the occurrence of calcareous foraminifers is coincident with open water conditions and relate this with high surface productivity. The association of *E. glabra* with sponge mat, boulder, and open water biotopes [*Bernhard*, 1987], all found throughout the McMurdo Sound–Ross Sea, indicates its distribution is related to environmental stability and nutrient availability. These similarities also exist for the species *Globo-*

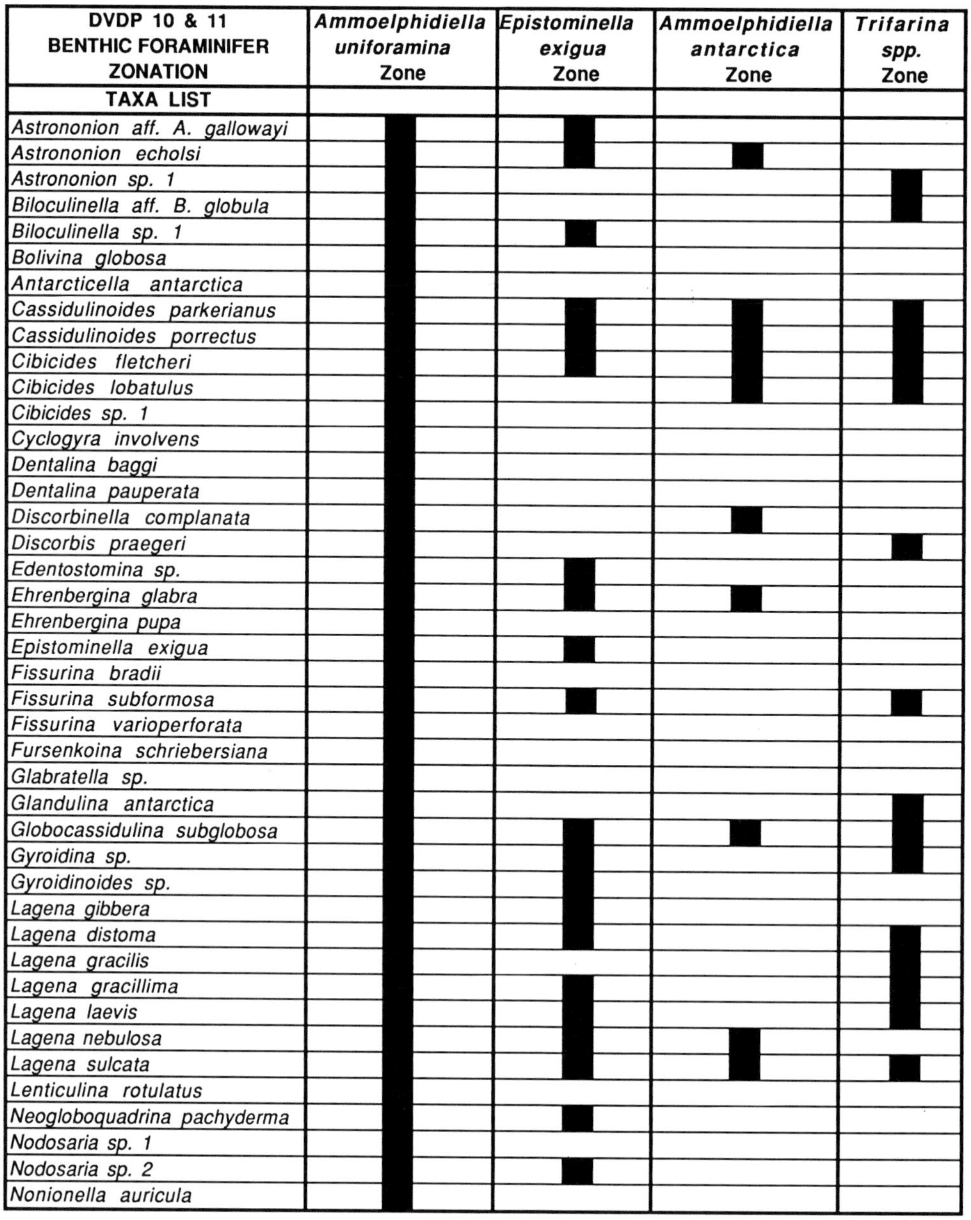

DVDP 10 & 11 BENTHIC FORAMINIFER ZONATION TAXA LIST	Ammoelphidiella uniforamina Zone	Epistominella exigua Zone	Ammoelphidiella antarctica Zone	Trifarina spp. Zone
Astrononion aff. A. gallowayi	■	■		
Astrononion echolsi	■		■	
Astrononion sp. 1	■			■
Biloculinella aff. B. globula	■			
Biloculinella sp. 1	■	■		
Bolivina globosa	■			
Antarcticella antarctica	■			
Cassidulinoides parkerianus	■	■	■	
Cassidulinoides porrectus	■			
Cibicides fletcheri	■	■	■	
Cibicides lobatulus	■		■	
Cibicides sp. 1	■			
Cyclogyra involvens	■			
Dentalina baggi	■			
Dentalina pauperata	■			
Discorbinella complanata	■		■	
Discorbis praegeri	■			■
Edentostomina sp.	■	■		
Ehrenbergina glabra	■	■	■	
Ehrenbergina pupa	■			
Epistominella exigua	■	■		
Fissurina bradii	■			
Fissurina subformosa	■	■		■
Fissurina varioperforata	■			
Fursenkoina schriebersiana	■			
Glabratella sp.	■			
Glandulina antarctica	■			
Globocassidulina subglobosa	■	■	■	■
Gyroidina sp.	■	■		■
Gyroidinoides sp.	■	■		■
Lagena gibbera	■	■		■
Lagena distoma	■	■		■
Lagena gracilis	■	■		■
Lagena gracillima	■	■		■
Lagena laevis	■	■		■
Lagena nebulosa	■	■	■	■
Lagena sulcata	■	■		■
Lenticulina rotulatus	■			
Neogloboquadrina pachyderma	■	■		
Nodosaria sp. 1	■			
Nodosaria sp. 2	■	■		
Nonionella auricula	■			

Fig. 3. The distribution of foraminifers in DVDP cores 10 and 11 with respect to the biofacies zonations. The solid bars represent occurrence; no bar represents absence.

cassidulina subglobosa, whose distribution in the Mc-Murdo Sound–Ross Sea region [*Osterman and Kellogg*, 1979] is closely associated with areas containing high organic carbon in the surface sediments [*Dunbar et al.*, 1985]. *Ehrenbergina* spp. and *Globocassidulina* spp. are common components of Holocene and supra–ice shelf sediments in the McMurdo Sound area. Sponge spicules are a major biogenic component of these sediments. *Bernhard* [1987] defined benthic foraminifer biotopes for Explorers Cove, McMurdo Sound. *Ehrenbergina glabra* is a prominent taxa within the sponge mat biotope.

Bernhard [1987] suggests that this association is based on the stability and high nutrient availability of the microhabitat. The abundance of robust miliolids, however, is suggestive of a deeper water habitat.

Epistominella exigua Zone

Above 167.20 and 242.00 mbs in DVDP site 10 and site 11, respectively, a significant faunal (Figure 3) as well as floral change occurs [*Harwood*, 1986]. The new assemblage ranges in DVDP site 10 from 163.00 to

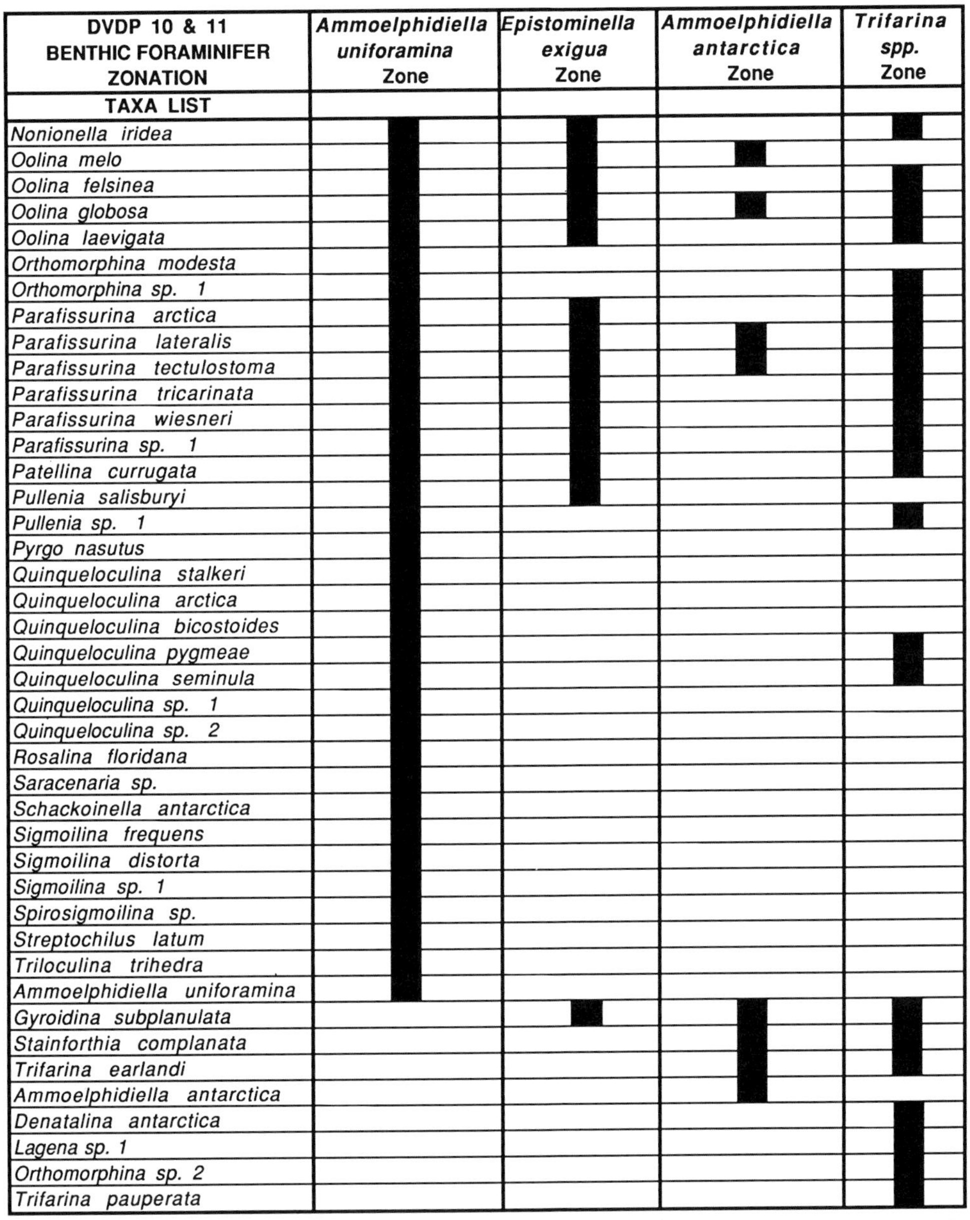

DVDP 10 & 11 BENTHIC FORAMINIFER ZONATION TAXA LIST	*Ammoelphidiella uniforamina* Zone	*Epistominella exigua* Zone	*Ammoelphidiella antarctica* Zone	*Trifarina spp.* Zone
Nonionella iridea	■	■		■
Oolina melo	■		■	
Oolina felsinea	■			
Oolina globosa	■		■	
Oolina laevigata	■			
Orthomorphina modesta	■			
Orthomorphina sp. 1	■			■
Parafissurina arctica	■	■		■
Parafissurina lateralis	■	■	■	■
Parafissurina tectulostoma	■	■		■
Parafissurina tricarinata	■	■		■
Parafissurina wiesneri	■	■		■
Parafissurina sp. 1	■	■		■
Patellina currugata	■	■		■
Pullenia salisburyi	■	■		■
Pullenia sp. 1	■			■
Pyrgo nasutus	■			■
Quinqueloculina stalkeri	■			■
Quinqueloculina arctica	■			■
Quinqueloculina bicostoides	■			■
Quinqueloculina pygmeae	■			■
Quinqueloculina seminula	■			■
Quinqueloculina sp. 1	■			■
Quinqueloculina sp. 2	■			■
Rosalina floridana	■			■
Saracenaria sp.	■			■
Schackoinella antarctica	■			■
Sigmoilina frequens	■			■
Sigmoilina distorta	■			■
Sigmoilina sp. 1	■			■
Spirosigmoilina sp.	■			■
Streptochilus latum	■			■
Triloculina trihedra	■			■
Ammoelphidiella uniforamina	■			■
Gyroidina subplanulata		■	■	■
Stainforthia complanata			■	■
Trifarina earlandi			■	■
Ammoelphidiella antarctica			■	■
Denatalina antarctica				■
Lagena sp. 1				■
Orthomorphina sp. 2				■
Trifarina pauperata				■

Fig. 3. (continued)

156.00 mbs and in DVDP site 11 from 242.00 to 202.00 mbs and is referred to as the *Epistominella exigua* Assemblage (Figure 2). As with the *Ammoelphidiella uniforamina* Assemblage, this assemblage is dominated by *Epistominella exigua* and *Globocassidulina subglobosa*. The assemblage is marked by a high species diversity (Figures 2, 3) that may suggest unfavorable/unstable environmental conditions [*Buzas*, 1972]. Significant to this zone is the absence of the miliolids, so common in the *Ammoelphidiella uniforamina* Zone, and the species *Ammoelphidiella uniforamina* (Figure 3).

The *Epistominella exigua* Zone is also punctuated by several interval zones throughout DVDP sites 10 and 11. Three biofacies units are defined for this zone, O1, O2, and O3 (Figure 2). As in the previous zone, high-diversity benthic foraminifer assemblages prevail with intermittent barren intervals, suggesting fluctuating environmental conditions. Diatom occurrence followed by the introduction of the high-diversity benthic foraminifer assemblages describes the biofacies cycles O1 and O3 within Taylor Fjord. Although biostratigraphically this zone is younger than the *A. uniforamina* Zone, the major benthic foraminifer components of the two faunas are analogous to each other (Figure 3).

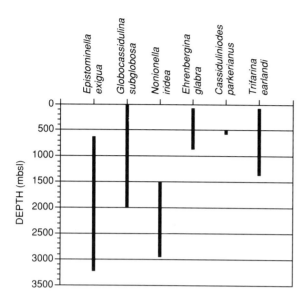

Fig. 4. Modern depth ranges for Southern Ocean benthic foraminifer species associations modified from *Murray* [1991] using data from *Uchio* [1960], *McKnight* [1962], *Pflum* [1966], *Kennett* [1968], *Anderson* [1975], *Osterman and Kellogg* [1979], *Quilty* [1985], and *Ward et al.* [1987].

Ammoelphidiella antarctica Zone

The third assemblage occurs only in a short interval in DVDP site 10 (~156.36–154.20 mbs) and represents the *Ammoelphidiella antarctica* Zone. This assemblage is distinguished by the occurrence of middle late Pliocene marker species *Ammoelphidiella antarctica* Conato and Segre (Figure 3). Other significant species are the common *Cassidulinoides* spp., *Globocassidulina subglobosa* and *Ehrenbergina glabra* (Figure 3). Also present in this assemblage is the species *Cibicides lobatulus* Walker and Jacob (Figure 3). This species often occurs in an attached life mode, indicating higher-energy/less stable conditions at the sediment water interface. This assemblage is represented by low species diversity. The zone includes two biofacies units, W1 and W2. Facies unit W1 includes a low-diversity benthic foraminifer fauna and diatom flora. Benthic foraminifer occurrence is preceded by diatom occurrence (as observed in the previous units containing both biogenic components). However, the diatoms become more consistent (Figure 2) suggesting prolonged open water conditions and higher surface productivity. Again, the low-diversity foraminifer assemblage composition is analogous to those already described, with the exception of the presence of *Ammoelphidiella antarctica* and *Cibicides lobatulus*, and is followed by a high-diversity fauna. Biofacies unit W2 is distinguished by its absence of foraminifers and presence of diatoms (Figure 2).

Trifarina spp. Zone

The last assemblage recognized in the DVDP sediments is the *Trifarina* spp. Assemblage. This assemblage occurs in the uppermost sediments studied from DVDP site 11 (202.00–193.00 mbs) and represents the *Trifarina* spp. Zone. The assemblage is dominated by the calcareous benthic species *Trifarina earlandi* (Parr), *T. pauperata* (Heron-Allen and Earland), *Stainforthia concava* (Höglund), *Globocassidulina subglobosa*, and *Epistominella exigua* (Figure 3). Faunal associations with modern Ross Sea assemblages (Figure 4) indicate that this assemblage represents shallower conditions (>300 mbsl) than represented by the *A. uniforamina* Assemblage. The occurrence of the benthic taxa *Cibicides lobatulus* suggests high bottom water current activity. The occurrence of the above species in Holocene sediments from the Ross Sea [*Ward et al.*, 1987; *Osterman and Kellogg*, 1979; *Kennett*, 1968] with high organic carbon contents suggests associations with high-productivity and/or low-oxygen conditions. This zone includes facies unit M1, marked by its high-diversity benthic foraminifer assemblage and the coeval occurrence of diatoms and benthic foraminifers (in DVDP site 11), not common in the preceding units. This facies unit is represented in DVDP site 10 by the occurrence of diatoms only.

AGE DETERMINATION

Paleontologic age estimates of the DVDP sites 10 and 11 sediments are best constrained by diatom data [*Brady*, 1979; *Harwood*, 1986]. However, some foraminifer data have provided useful biostratigraphic information. These data have been utilized to constrain and resolve the paleomagnetic record for cores 10 and 11. The magnetostratigraphy for DVDP site 10 and DVDP site 11 is based on the reinterpretation and revised correlation (this paper) of the original paleomagnetic data of *Elston and Bressler* [1981] and *Purucker et al.* [1981]. On the basis of our data we propose the presence of three hiatuses in the DVDP sites 10 and 11 records, H1, H2, and H3. Their ages and significance are discussed in the following sections.

The lowest two zones of normal polarity (314.00–273.00 mbs; Figure 2) in the *Ammoelphidiella uniforamina* Zone we interpret as Anomaly 3A (5.89–5.35 Ma; *Berggren et al.* [1985]). Two planktonic species, *Neogloboquadrina pachyderma* (Ehrenberg) and *Streptochilus* aff. *S. latum* Brönnimann and Resig, co-occur in the upper part (249.28–249.18 mbs in DVDP site 11) of the lower stratigraphic section (*Ammoelphidiella uniforamina* Zone) and indicate a late Miocene to early Pliocene age (~5 Ma) [*Kennett and Srinivasan*, 1983]. This is supported by the co-occurrence of diatom species *Thalassiosira torokina* and *Actinocyclus ingens* (late Miocene) (Figure 5) in the absence of Pliocene marker species *Nitzschia praeinterfrigidaria*, *N. barro-*

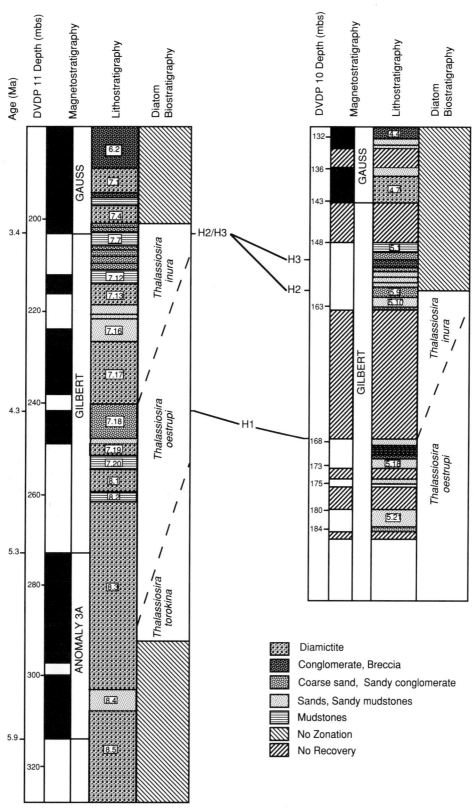

Fig. 5. Pliocene diatom biostratigraphic zonations for the Southern Ocean, Ross Sea region as applied to DVDP sites 10 and 11 (modified from *Harwood* [1986] using the diatom biostratigraphic zonations for the Southern Ocean established by *Baldauf and Barron* [1991]).

nii, Thalassiosira inura, and other lower Pliocene diatoms [*Brady*, 1979; D. M. Harwood, personal communication, 1992], which places the *A. uniforamina* Zone in the *T. torokina* diatom Zone of *Baldauf and Barron* [1991] (Figure 5). The 6.1- to ~5.0-Ma interval mentioned above lies just below our identification of the base of the Thvera Normal Polarity Subchronozone of the Gilbert Reversed Polarity Chronozone and thus is inferred to have been deposited during part of Chron 3, shortly before about 4.77 Ma [*Berggren et al.*, 1985]. Based on sequence and proportion of the polarity zonation in the absence of significant recognizable gaps below 249.00 mbs, the base of core DVDP site 11 appears to have been deposited at about 6.1 Ma (time scale of *Berggren et al.* [1985]) (Figure 2).

The base of the *Epistominella exigua* Zone is distinguished on the basis of significant faunal and floral changes at about 167.20 mbs in DVDP site 10 and about 242.00-m depth in DVDP site 11. Biostratigraphic analysis suggests a major hiatus at these horizons, unconformity H1 (Figures 2, 6, and 7). Across this horizon the benthic foraminifer species diversity decreases by 65% (Figure 8), and the diatom diversity decreases by 17%. The occurrence above this horizon of marine diatoms *Thalassiosira torokina*, *T. oestrupi*, and *Nitzschia praeinterfrigidaria* suggests an early Pliocene age [*Brady*, 1979; D. M. Harwood, personal communication, 1992], placing the *E. exigua* Zone (sediments up to 156.36 mbs in DVDP site 10 and up to 202.00 mbs in DVDP site 11) in the *T. oestrupi* diatom Zone of *Baldauf and Barron* [1991] (Figure 5). The disappearance of *Ammoelphidiella uniforamina* (a benthic Miocene indicator) across this boundary suggests a transition into the Pliocene. The major lithologic change in DVDP site 11 shown by *McKelvey* [1981] between the underlying interbedded and bioturbated conglomerate, sandstone, and breccia and the overlying massive diamictite (Figure 2) with a sharp, irregular base suggests a major break at this horizon. In DVDP site 10 the lithologic difference is less pronounced but is marked by a change from dominantly sandy mudstones to sandstones and pebble conglomerates. Also in DVDP site 11, a normal to reversed polarity zone boundary coincides with unconformity H1. This horizon terminates the first normal (Thvera) subchronozone of the Gilbert and thus constrains the younger age limit of these normal polarity, underlying nonglacial strata in DVDP site 11 to between 4.77 and 4.57 Ma [*Berggren et al.*, 1985]. Extrapolation from below suggests interruption beginning near the younger limit of this range, at about 4.57 Ma. The polarity zonations of the two cores for the *E. exigua* Zone (Figures 2, 6, 7) correlate with the latter half of the Gilbert Reversed Polarity Chron (late Chron C3), from about 4.24 to 3.40 Ma [*Berggren et al.*, 1985]. This suggests that H1 may represent a hiatus of ~330,000 years (4.57–4.24 Ma).

The third biostratigraphic zone, the *Ammoelphidiella*

antarctica Zone, is recognized only in DVDP site 10 between about 156.36 and 154.20 mbs. This interval is bounded by unconformities H2 below and H3 above (Figures 2, 6, 7) and is characterized by distinct faunal and floral assemblages. The presence of the calcareous benthic foraminifer *Ammoelphidiella antarctica* Conato and Segre (=*Trochoelphidiella onyxi* [*Webb*, 1972, 1974]), found in middle Pliocene sediments from Wright Valley [*Webb*, 1972, 1974], Deep Sea Drilling Project (DSDP) Site 273 [*D'Agostino*, 1980], and the Scallop Hill Formation of Brown Peninsula, Black Island, and White Island, McMurdo Sound [*Webb and Andreasen*, 1986; *Webb*, 1974], suggests a middle Pliocene age for this interval. The assemblage of the *A. antarctica* Zone of DVDP site 10 correlates closely with the microfauna of the "Pecten gravel" of *Webb* [1974] in Wright Valley. Diversities from this zone and the "Pecten gravel" samples are comparable, at 21 and 16 species, respectively. In addition, the presence of shallow water, attached forms (i.e., *Cibicides* spp. and *Rosalina* spp.) demonstrates the similarities of these faunas. Other beds in Wright Valley correlative with the "Pecten gravel" are thought to postdate volcanic cones dated between 4.2 and 3.7 Ma [*Webb*, 1974; *Armstrong*, 1978] and are overlain by alpine glacial deposits containing dated volcanic material ranging in age from 3.4 to 2.5 Ma [*Webb*, 1974]. The "Pecten gravel" (correlative to the Scallop Hill Formation) is, in places, underlain by volcanic rock dated at 3.8 Ma [*Webb and Andreasen*, 1986]. The reversed polarity beds of the *A. antarctica* Zone were most likely deposited very near the end of the Gilbert Chron (3.40 Ma [*Berggren et al.*, 1985]); thus the *A. antarctica* Zone in DVDP site 10 appears to range from about 3.7 or 3.8 Ma to 3.4 Ma. Diatom data from *Brady* [1979] and *Harwood* [1986] also support a middle Pliocene age for this zone.

Poor faunal and floral control and increasingly subaerial paleoenvironments of sediments above 135.66 mbs depth in DVDP site 10 and above 194.00 mbs depth in DVDP site 11 make biostratigraphic age determination of these upper strata more difficult. The benthic foraminifer assemblages that are present have strong affinities with Pleistocene and Holocene faunas from the Ross Sea. However, stratigraphic and limited paleontologic constraints on the polarity zonation of DVDP site 11 (Figure 7) allow confident correlation with the Matuyama Reversed Polarity Chron (2.48 Ma to 730 ka [*Mankinen and Dalrymple*, 1979]) and the Brunhes Normal Polarity Chron (<730 ka [*Mankinen and Dalrymple*, 1979]). Identification of the Olduvai Normal Polarity Subchronozone (1.87–1.67 Ma [*Mankinen and Dalrymple*, 1979]) within the Matuyama is supported by correlation with a corresponding interval of normal polarity in another core drilled about 1 km west of DVDP site 11 (unpublished data). Within and just below that interval, abundant in situ material of the barnacle *Bathylasma corolliforma* (Hoek), known in Antarctica

APPARENT ACCUMULATION RATES OF SEDIMENT AT DVDP 10

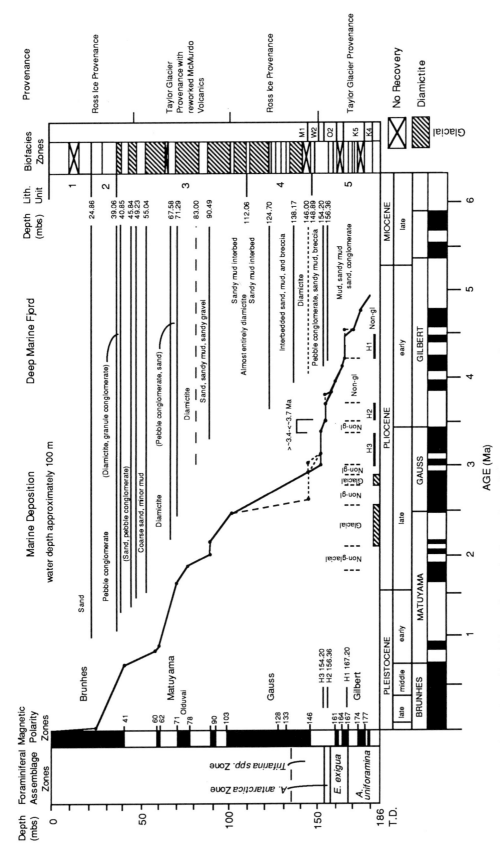

Fig. 6.   Depositional history diagram for Dry Valley Drilling Project Core 10. The figure incorporates published data and data presented in this paper to reflect the depositional history at DVDP site 10. The left vertical axis represents core depth (in meters), foraminifer zonations, and magnetostratigraphy. The right vertical axis represents lithostratigraphy, biofacies, and provenance. The horizontal axis is the geologic time scale with paleomagnetics [from *Berggren et al.*, 1985]. The curve represents the apparent accumulation rates of sediments with sediment types and interpretations of depositional environments.

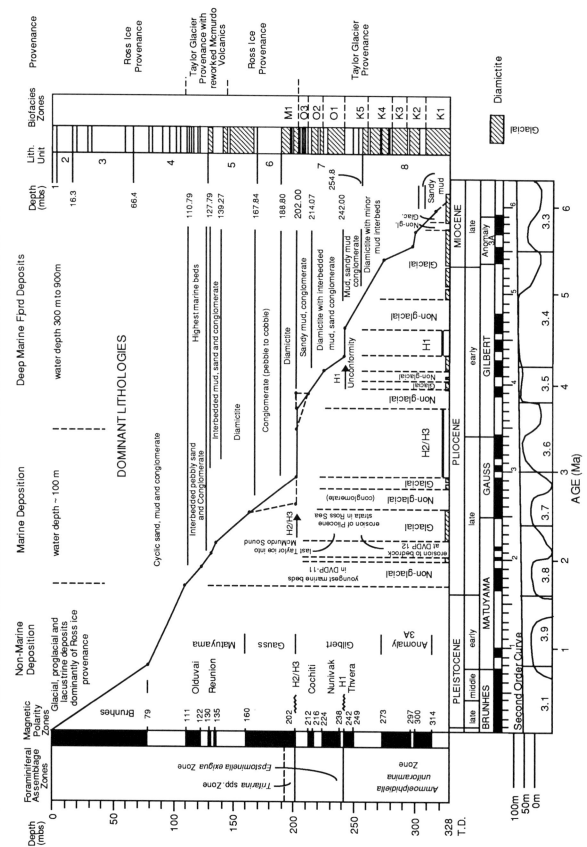

APPARENT ACCUMULATION RATES OF SEDIMENT AT DVDP 11

Fig. 7. (Opposite) Depositional history diagram for Dry Valley Drilling Project Core 11. The figure incorporates published data and data presented in this paper to reflect the depositional history at DVDP site 11. The left vertical axis represents core depth (in meters), foraminifer zonations, and magnetostratigraphy. The right vertical axis represents lithostratigraphy, biofacies, and provenance. The horizontal axis is the geologic time scale with paleomagnetics [from *Berggren et al.*, 1985] and the second- and third-order eustacy curves for the late Miocene to Pleistocene [*Haq et al.*, 1987]. The apparent accumulation rates of sediment curve include sediment types and interpretations of depositional environments.

only from lower Pleistocene and younger sediments [*Newman and Ross*, 1971], constrains the age of these correlative intervals of normal polarity in DVDP site 10 and DVDP site 11 to the Olduvai Subchron. A ^{14}C date of about 5.8 ka from shell material at about 25-m depth in DVDP site 10 was reported by *Stuiver et al.* [1976].

DISCUSSION

The DVDP site 10 and site 11 records provide a sensitive paleoenvironmental record of ice conditions on the continent and sea level conditions on the margin. The distinct benthic foraminifer zones of DVDP sites 10 and 11 are punctuated by barren intervals that suggest fluctuating paleoenvironmental conditions. Their stratigraphic distribution is not restricted to sediment type, organic carbon, or mutual occurrence with other marine biogenic material (sponge spicules, diatoms, etc.). The systematic occurrence of the foraminifer assemblages, distribution of diatoms, and various lithologies suggest a complex Neogene paleoceanographic-paleoclimatic history for the Victoria Land Basin. Paleontologic and associated radiometric age estimates now allow confident correlation of the polarity zonations of these cores with the polarity time scale, thereby refining age control to the degree necessary for correlation with proxy records.

Studies have compared direct evidence of paleoclimatic and eustatic changes from the Antarctic continent and margin with records derived from proxy (mid- to low-latitude) data. *Barrett et al.* [1987, 1989] and *Harwood et al.* [1989] correlate paleobathymetric trends, inferred from the micropaleontologic and sedimentologic data from the late Oligocene to early Miocene records of the MSSTS 1 and CIROS 1 cores, western McMurdo Sound, with the eustacy curve of *Haq et al.* [1987]. A strong correlation exists with the second-order cycles, and a possible correlation exists with third-order cycles, suggesting glacio-eustatic and/or tectonic fluctuations. The CIROS 1 drill hole contains a well-constrained paleoenvironmental record for the late Paleogene to early Neogene Antarctic margin [*Barrett et al.*, 1991]. *Bartek et al.* [1991] suggest that the complex sequence stratigraphy from the Ross Sea is directly influenced by rapidly fluctuating ice conditions during the Neogene and that the glacial events can be correlated with the Neogene global eustatic record. Dry Valley Drilling Project cores 10 and 11 provide a detailed late Neogene Antarctic glacial and glacio-eustatic record, within a fairly well constrained tectono-eustatic setting. This provides a record of Antarctic paleoenvironmental fluctuation to compare with other proxy records of southern hemisphere ice volume change.

Sea Level Record and Hiatuses

The biofacies units described in this paper (K1–M1) indicate highly fluctuating paleoenvironmental condi-

DVDP 11

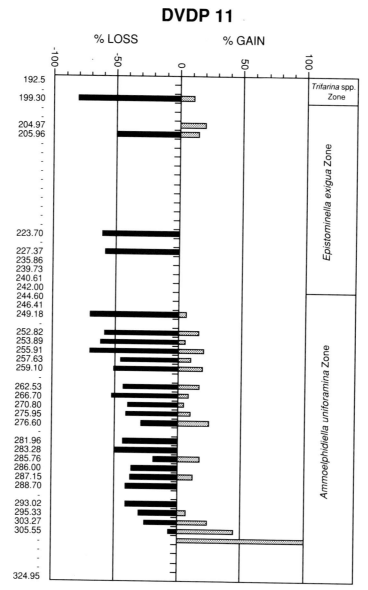

Fig. 8.   Diagram showing the percent loss of benthic foraminifer taxa and percent appearance of previously absent species with benthic foraminifer assemblage zones.

tions. The variability in diatom assemblages, benthic foraminifer faunas, and lithology suggests changes in open water, bathymetric, and glacial conditions within the Taylor paleofjord. This variability is of the order of 10^5 years, within the resolution of the *Haq et al.* [1987] third-order eustacy curve. On the basis of our data and stratigraphic data [*McKelvey*, 1981; *Powell*, 1981], we can correlate the DVDP record to specific late Neogene sea level events in the *Haq et al.* [1987] curve.

Greater than 308.00 mbs in DVDP site 11 is dominated by diamictite with intraformational deformation (slickenslides and shearing) [*McKelvey*, 1981], indicating loading and deposition as a lodgement till. This

interval is barren of biogenic material, further suggesting subglacial deposition. As was stated before, the base of the core is interpolated at ~6.1 Ma. Correlation of this glacial interval of DVDP site 11 with the *Haq et al.* [1987] Cenozoic sea level curve suggests it corresponds to the latter part of the lowstand of their third-order cycle 3.3 (6.3–5.8 Ma) (Figure 7). Deposition of this massive till may also coincide with a portion of the Ross Sea Unconformity of *Hayes et al.* [1975] and *Savage and Ciesielski* [1983], which *Karl et al.* [1987] believe may in fact represent a complex of erosional events whose timing remains unresolved. This unit is conformably overlain by a ~5-m-thick interval (307.45–302.79

mbs) of fossiliferous sandy mudstone with few clasts [McKelvey, 1981]. This suggests a transition into marine conditions, coincident with the early part of the sea level highstand of cycle 3.3 (5.5–5.8 Ma) [Haq et al., 1987] (Figure 7).

The interval from 302.79 to ~255.00 mbs in DVDP site 11 is marked by the almost continuous presence of benthic foraminifers. Variation in abundance and diversity indicates fluctuating environmental conditions (K2–K4) associated with glaciomarine deposition. The occurrence of marine diatoms represents a brief interval of more open marine conditions followed by a return to an ice-covered glaciomarine environment. This interval (302.79–255.00 mbs) ranges from 5.5 to 4.9 Ma and correlates with the Haq et al. [1987] lowstand between 5.5 and 5.0 Ma. These sediments grade into a fossiliferous (benthic foraminifers) sequence (255.00–242.00 mbs) of mudstones, sandy mudstones, and conglomerates [McKelvey, 1981] and represent a transition into marine deposition, coincident with cycle 3.4 (~5.0 to ~4.2 Ma) (Figure 7).

Unconformity H1 truncates biofacies K5 and is the boundary between the Ammoelphidiella uniforamina Zone (upper Miocene) and the overlying Epistominella exigua Zone (lower Pliocene) (242.00 mbs). Across this horizon the benthic foraminifer diversity decreases by 65% (Figure 8), and the diatom diversity decreases by 17%. A major lithologic change occurs from a bioturbated sandy conglomerate to a massive diamictite (Figure 2) with a sharp irregular base [McKelvey, 1981]. The unconformity is overlain by sediments estimated at between 4.24 and 4.57 Ma. Given these age constraints the time represented by H1 is approximately 330 kyr. This places H1 within the high sea level cycle 3.4 of Haq et al. [1987] (Figure 7). The presence of bioturbation at the top of K5 may suggest that H1 represents a prolonged period of nondeposition, but this seems unlikely given the proximity to a tectonically active margin (Transantarctic Mountains). However, the timing of this unconformity is coincident with Ross Sea unconformity U2 that Hinz and Block [1983] date at approximately 3.7–4.3 Ma. The age estimates on U2 have been questioned [Kellogg et al., 1979; Savage and Ciesielski, 1983]; therefore if H1 and U2 can be correlated, the dating of U2 may be resolved.

In DVDP site 11, unconformity H1 is overlain by a sequence of diamictites interbedded with mudstones, sandy mudstones, sands, and conglomerates (242.00–202.00 mbs). These sediments contain benthic foraminifers of the Epistominella exigua Zone, as well as diatoms. The diamictites represent increased glaciomarine deposition. The interbeds indicate weaker ice influence and possibly greater terrestrial input. The sediments within the interbeds show a gradational coarsening upward in the section to 202.00 mbs [McKelvey, 1981], where the sequence is truncated by an unconformity (H3). DVDP site 10 displays a similar

sedimentologic pattern throughout the E. exigua Zone (167–156 mbs) that is terminated by unconformity H2. This suggests that progressive shallowing occurred, leading up to an erosional event.

Unconformity H3 is present in both DVDP site 10 (154.20 mbs) and DVDP site 11 (202.00 mbs). A significant benthic foraminifer event occurs across this boundary with a 53% loss in species. H3 also occurs at a reversed to normal polarity change in Core 11. The hiatus, determined from DVDP site 10, represents approximately 400 kyr (~3.4 to ~3.0 Ma). Because of its presence in both the DVDP site 10 and site 11 records, the magnitude of this erosional event was much greater than H2. The timing and extent of H3 make it reasonable to correlate it with the major sea level drop during the early part of cycle 3.7 (3.0–2.7 Ma) of Haq et al. [1987]. Assuming that the chronology is correct, this unconformity may also extend into the Ross Sea record and represent U1 of Hinz and Block [1983]. Unconformity H3 is overlain by diatom-bearing mudstone, sand, and gravel indicating marine conditions. In DVDP site 11, H3 is overlain by a benthic foraminifer and diatom-bearing diamictite, suggesting an ice marginal marine setting.

Subsequent shallowing with periodic subaerial conditions is evident throughout the remainder of the DVDP records. Some records of intermittent marine incursions into Taylor Valley in the Pleistocene exist. The youngest identifiable marine sediments in DVDP site 11 (~128.00 to ~111.00 mbs) correlate with a corresponding interval in Core ETV 10 (about 1 km west of DVDP site 11; unpublished data) that contains abundant (up to 40%) in situ barnacle and bivalve shell material of Pliocene and Pleistocene age (~2.0 Ma). Sediments above these levels are not directly interpretable in terms of eustatic fluctuation.

Evidence exists for at least three unconformities, H1–H3, in the DVDP site 10 and site 11 records. Bartek et al. [1991] have constructed a seismostratigraphic and lithostratigraphic framework on which they interpret rapid glacial fluctuations of marine-based ice sheets in the Ross Sea. They have used their framework to correlate cores from DSDP Site 270 and CIROS 1. They concluded that the Ross Sea Neogene stratigraphic signature, defined by seismostratigraphic and lithostratigraphic records, can be correlated to the Haq et al. [1987] third-order eustacy cycles, inferring their glacio-eustatic nature. DVDP cores 10 and 11 record similar high-frequency glacio-eustatic fluctuations with significant erosional events. The stratigraphic model proposed by Bartek et al. [1991] may help explain the erosional mechanism for H1, H2, and H3. They suggest that because of the short time interval between glacial advances and retreats and slow subsidence, parts of the depositional sequence were eroded. This places much emphasis on the boundary conditions (uplift, subsidence, basal ice conditions, ice volume fluctuations) when interpreting eustatic records. Models prescribed for Antarctic glaciomarine deposition [*Anderson*, 1972; *Drewry and Cooper*, 1981; *Kellogg and Kellogg*, 1988; *Alley et al.*, 1989; *Anderson et al.*, 1991] demonstrate the importance of the basal ice conditions with respect to the preservation of sediments (deposition or erosion). Renewed sedimentologic studies of the DVDP sediments may help to place them into the stratigraphic framework described in one of the more recent glaciomarine depositional models.

Tectono-Eustacy

The Victoria Land Basin margin is bounded by the Transantarctic Mountains (TAMs) and has been tectonically active throughout late Mesozoic and Cenozoic time. The orogenic activity has produced a regional tectono-eustatic background upon which the record of shorter-term glacio-eustatic fluctuations is overlaid. The discrimination between eustatic and regional tectonic sea level fluctuation from the DVDP records can be used to help determine timing and rates of uplift in the southern TAMs. Paleoelevations are important in determining ice volume flow through the TAMs, as well as the potential for interior basins and seaways to develop during times of major interglacials. This has major implications for the validity of the proposed origin and emplacement of the Sirius Group [*Webb et al.*, 1984] in the TAMs.

The uplift history of the northern Transantarctic Mountain region is becoming better constrained. *Fitzgerald et al.* [1986] have calculated average uplift rates for this region of the order of 100 m/m.y., post–50 Ma. Other workers [*Webb*, 1972; *Drewry*, 1975; *Webb and Wrenn*, 1982; *Denton et al.*, 1984] have estimated Cenozoic uplift rates for the TAMs ranging from 40 to 200 m/m.y. *Behrendt and Cooper* [1991] suggest that the uplift rates along the TAMs are variable. They estimate a post–early Pliocene uplift rate of ~1 km/m.y. for the rift shoulder of the TAMs, adjacent to the Victoria Land Basin margin. *Mercer* [1968] suggested uplift rates greater than 1 km/m.y. for specific blocks within the TAMs. Paleobathymetric estimates based on benthic foraminifers from DVDP sites 10 and 11 yield an average uplift rate of 82 m/m.y. for the late Neogene (post–6 Ma) [*Ishman and Webb*, 1988]. However, this overall average rate has been subdivided into 42 m/m.y. for the late Miocene and an increased rate of 125 m/m.y. for the post-Miocene [*Ishman and Webb*, 1988]. An average minimum post-Pliocene uplift rate of about 90 m/m.y. is inferred from magnetostratigraphic correlation of Core ETV 10 (unpublished data). The uplift record accounts for most of the gradual and accelerated shallowing of the Taylor Valley sediments throughout the late Neogene. It does not, however, explain the rapid faunal and floral transitions in DVDP sites 10 and 11, nor the lithologic variations reflecting nonglacial/glacial alternations.

Rather, the regional tectonic uplift is responsible for gradual shallowing, which mirrors the second-order eustatic curve of the last 5 m.y. shown by *Haq et al.* [1987] (Figure 7). This second-order duration and magnitude of tectonic shallowing on the margin of the Victoria Land Basin may suggest similar eustatic mechanisms associated with intraplate stresses as described by *Cloetingh et al.* [1985] and *Cloetingh* [1988]. It may also be suggested that the preservation of the higher-order cycles (third order) are a result of a combination of uplift and increased sedimentation rates. *Fulthorpe* [1991] has suggested that preservation of high-resolution sea level records is greatly improved with increased sedimentation rates combined with reduced subsidence. The effects of increased or constant sedimentation rates with uplift, as is observed for the Victoria Land Basin margin, would create a similar high-resolution record of sea level fluctuation, as is preserved in the DVDP records.

The presence of the *Trifarina* spp. Zone boundary (154.20 and 202.00 mbs in DVDP site 10 and site 11, respectively) is of particular importance with respect to uplift of the TAMs. A major benthic foraminifer faunal transition occurs at the lower boundary of this zone, with a greater than 15% species gain and greater than 50% species loss (Figure 8), and represents the introduction of the Modern McMurdo Sound benthic foraminifer fauna. Provenance studies indicate a change in source area and thus a change in ice dynamics, from EAIS influence (i.e., Taylor Glacier) to West Antarctic Ice Sheet influence (Ross Ice Shelf). We interpret that this was a critical time in the evolution of the TAMs. The uplift rates estimated from the benthic foraminifer assemblages in the DVDP cores indicate increased uplift after ~3.4 Ma. An increase in uplift would produce a damming effect on the outflow of Taylor Glacier, thus reducing its influence in Taylor Fjord and allowing incursion of the Ross Ice Shelf into the paleofjord. The inferred timing of rapid uplift of the TAMs is consistent with renewed cooling or climatic deterioration from the Sirius Group [*Webb and Harwood*, 1991] and Site 704 [*Hodell and Warnke*, 1991] records. *Behrendt and Cooper* [1991] suggest that rapid uplift of the rift shoulder of the TAMs in the Pliocene resulted in the growth of the EAIS. This response was due to changes in ice flow and atmospheric conditions. Although the uplift rate indicated by the data from the DVDP records is an order of magnitude less than that proposed by *Behrendt and Cooper* [1991], the mechanism for EAIS volume increase is very similar.

Climatic History

Evidence for significant ice volume on the Antarctic continent since approximately 31 Ma exists from Antarctic records [*Barrett et al.*, 1989, 1991; *Hambrey et al.*, 1989] and is interpreted from Southern Ocean deep-sea records (for a summary, see *Kennett and Barker* [1990]). The evolution of the Antarctic cryosphere, through the Miocene, is well documented in the deep-sea records from the world's oceans [*Barker and Burrell*, 1977; *Kennett*, 1977, 1986; *Prentice and Matthews*, 1988; *Miller et al.*, 1987]. This includes the major expansion of the East Antarctic Ice Sheet in early middle Miocene (*Shackleton and Kennett* [1975], *Savin et al.* [1975], and others). Late Miocene development of the West Antarctic Ice Sheet [*Mercer*, 1976; *Ciesielski et al.*, 1982] is interpreted from sedimentologic changes recorded in Ocean Drilling Program (ODP) Site 696 in the Weddell Sea [*Anderson et al.*, 1986; *Kennett and Barker*, 1990]. The late Miocene development of the West Antarctic Ice Sheet may be recorded in the lowermost diamictite in DVDP site 11 (328.00–307.45 mbs). The nature of this diamictite represents a strong glacial influence at about 6.1 Ma that continues through the remainder of the upper Miocene section in DVDP site 11 (307.45–202.00 mbs). The early Pliocene (~5–4 Ma) records from southern high latitudes indicate climatic amelioration [*Bandy et al.*, 1971; *Ciesielski and Weaver*, 1974; *Kennett*, 1986; *Hodell and Kennett*, 1986; *Hodell and Warnke*, 1991]. Sea surface temperature estimates from faunal data for the earliest Pliocene suggest temperatures 5°–10°C warmer than present [*Ciesielski and Weaver*, 1974; *Abelmann et al.*, 1990]. The data from the DVDP cores are consistent with this interpretation by showing abundant benthic foraminifer occurrence throughout the 5–4 Ma interval. Isotopic records from the Southern Ocean [*Kennett*, 1986; *Hodell and Kennett*, 1986; *Hodell and Warnke*, 1991] show a distinct increase in benthic foraminifer $\delta^{18}O$ values at approximately 3.6 to 3.4 Ma. This shift ranges from +0.58‰ [*Kennett*, 1986] at DSDP Site 590 to approximately +0.60‰ [*Hodell and Warnke*, 1991] at ODP Site 704. These isotopic shifts are followed by progressive increases in $\delta^{18}O$ to about 2.4 Ma, when the next significant positive isotopic excursion occurs. These data suggest significantly warm conditions in Antarctica during the early to middle Pliocene. Data collected from the Antarctic continent and sediments collected from beneath West Antarctic ice streams [*Webb et al.*, 1984; *Harwood*, 1986; *Pickard et al.*, 1988; *Webb and Harwood*, 1991; *Scherer*, 1991] suggest dynamic East Antarctic and West Antarctic ice sheets that waxed and waned rapidly throughout the late Neogene. Much debate exists over the timing and magnitude of ice volume change throughout the middle and late Pliocene [*Denton et al.*, 1984, 1989; *Harwood*, 1985, 1986; *Webb et al.*, 1984]. Diatoms recovered from clasts within the Sirius Group indicate intervals of open marine conditions throughout the early to middle late Pliocene (~4.8 to ~3.1–2.5 Ma) in East Antarctic basins [see *Webb and Harwood*, 1991]. Ice sheet models of *Oerlemans* [1982] suggest that for mean annual sea level temperatures between +2° and +5°C the EAIS is maintained at a level

that allows for restricted marine incursions into the East Antarctic interior. This is the same time interval within the DVDP records representing less ice-influenced deposition. However, the DVDP records also indicate rapidly changing depositional environments (K5–O3), suggesting a much less stable glacial/climatic environment. These records also strongly correlate with the timing of the most significant sea level highstands in the Pliocene (cycles 3.6 and 3.7 of *Haq et al.* [1987]).

Isotopic records from both ODP Site 704 [*Hodell and Ciesielski*, 1990; *Hodell and Warnke*, 1991] and DSDP Site 590 [*Kennett*, 1986] show significant positive $\delta^{18}O$ shifts at approximately 2.6–2.4 Ma and are coincident with the initiation of northern hemisphere ice buildup. The DVDP site 10 and site 11 records indicate progressive climatic deterioration from ~3.4 Ma to present. This is consistent with *Webb and Harwood*'s [1991] interpretation of the Sirius data, which place the deposition of the Sirius Group between ~3.0 and 2.6 Ma followed by glacial conditions approaching modern.

CONCLUSIONS

Dry Valley Drilling Project cores 10 and 11 provide us with the most complete late Miocene and Pliocene record for the Victoria Land Basin margin, Antarctica. These cores reflect a complex and detailed glacial history, recorded in very nearshore glaciomarine sediments that were directly affected by fluctuations of the East Antarctic Ice Sheet and the West Antarctic Ross Ice Shelf. The glacio-eustatic record is overlaid on a background of reasonably well-constrained tectonic uplift. The combined use of benthic foraminifer and marine diatom biostratigraphy, radiometric age constraints on closely correlative deposits, and stratigraphic limitations and correlations between the two cores have allowed confident correlation of their revised paleomagnetic polarity zonations with the polarity time scale. Over most of the cored sections, age resolution is of the order of 10^5 years. Benthic foraminifer, diatom, and sedimentological data suggest the following:

1. Throughout the late Neogene the Transantarctic Mountains have continued to rise. Average uplift rates calculated from apatite fission track data and minimum estimates from paleobathymetric ranges of barnacles and bivalves give estimates of the order of 100 m/m.y. Benthic foraminifer data indicate differential uplift rates of 42 m/m.y. (pre–middle Pliocene) and 125 m/m.y. (post–middle Pliocene). This regional tectono-eustatic trend at DVDP sites 10 and 11 parallels the second-order trend of the *Haq et al.* [1987] curve and forms the background upon which a paleoenvironmental record of much higher frequency is recognized.

2. The paleoenvironmental record interpreted from DVDP sites 10 and 11 is in agreement with recent paleoceanographic data from the Southern Ocean for the Pliocene. These data suggest limited ice conditions

in the Pliocene by the presence of benthic foraminifer assemblages and marine diatom floras. The faunal, floral, and sedimentologic data from DVDP sites 10 and 11 indicate highly fluctuating glaciomarine/marine conditions in the Victoria Land Basin margin throughout the Pliocene. Major paleoenvironmental events are characterized by either (1) diamictites or (2) major unconformities. In general, glacial intervals are gradationally overlain by nonglacial strata. Transitions into nonglacial intervals grade from diamictites (basal till or glaciomarine diamictite) to interbedded pebbly sands, sandy mudstones, and mudstones, first with the appearance of open water marine diatoms followed by low-diversity then high-diversity benthic foraminifer assemblages. The unconformities represent major erosional/nondepositional events in the late Neogene. H1 and H2/H3 are separated, in DVDP site 11, by 20 m of sediment. A late early to early late Pliocene section is preserved between unconformities H2 and H3 in DVDP site 10. Unconformable contacts have been recognized in seismic records from the western Ross Sea basin. Unconformities H1–H3 may be coincident with unconformities recognized in the Ross Sea (U1 and U2).

3. The number and timing of paleoenvironmental oscillations recognized in DVDP cores 10 and 11 (~100,000 years) generally correlate well with the third-order cycles of the *Haq et al.* [1987] curve. Major erosional events and diamictite units correspond with major sea level lowstands. Nonglacial and glaciomarine biogenic sediments generally accumulated during sea level highstands.

4. Brief glacial fluctuations (represented by K2, K3, and K4), not recognized in the eustatic curve, particularly in cycles 3.8 and 3.5, can be recognized in the cores. We cannot determine the significance of these paleoenvironmental oscillations of frequency greater than the third-order cycles without additional records from around the Antarctic continent. Most of these higher-frequency episodes are recognized more easily in the more expanded and ice-proximal record of DVDP site 11 but are not so clearly reflected in DVDP site 10. Thus these higher-frequency events may indicate fluctuations of the East Antarctic Ice Sheet and consequent expansions of Taylor Glacier, or they may be attributed to local alpine glacial fluctuations and be of only local or perhaps regional significance. In the latter case, such events may correlate with Pliocene and Pleistocene events proposed for the dry valleys and Ross Sea [e.g., *Calkin et al.*, 1970; *Nichols*, 1971; *Webb*, 1972; *Bull and Webb*, 1973; *Denton et al.*, 1989]. Clearly, DVDP sites 10 and 11 demonstrate the value of nearshore, ice-marginal records for detailed interpretation of Antarctic glacial history.

Acknowledgments. This work was supported by the National Science Foundation through grants to Peter-Noel Webb (DPP 82-14174) and Donald P. Elston (81-20877). Additional support was provided through a Smithsonian Institution Graduate Fellowship Award to S.E.I., and by the U.S. Geological Survey to H.J.R. The authors would like to thank T. M. Cronin, D. P. Elston, C. P. Hart, D. M. Harwood, J. P. Kennett, L. E. Osterman, R. Z. Poore, and P.-N. Webb for their comments and critical review of the manuscript. L. H. Burckle also reviewed the manuscript. Thanks goes to J. Nagy for his graphics support and the Byrd Polar Research Center for their support of this research. Finally, congratulations are due to the drillers and participants of the Dry Valley Drilling Project and all the subsequent drilling projects in the region for providing us with invaluable physical records from the Antarctic continental margin despite the often adverse working conditions.

REFERENCES

Abelmann, A., R. Gersonde, and V. Spiess, Pliocene-Pleistocene paleoceanography in the Weddell Sea—Siliceous microfossil evidence, in *Geological History of the Polar Oceans: Arctic Versus Antarctic*, edited by U. Bleil and J. Thiede, pp. 729–759, Kluwer, Amsterdam, 1990.

Alley, R. B., D. D. Blankenship, S. T. Rooney, and C. R. Bentley, Sedimentation beneath ice shelves: The view from ice stream B, *Mar. Geol.*, 85, 101–120, 1989.

Anderson, J. B., Nearshore glacial-marine deposition of modern sediments from the Weddell Sea, *Nature*, 240, 189–192, 1972.

Anderson, J. B., Ecology and distribution of foraminifera in the Weddell Sea of Antarctica, *Micropaleontology*, 21, 69–96, 1975.

Anderson, J. B., R. Wright, and B. A. Andrews, Weddell Fan and associated abyssal plain, Antarctica: Morphology, sediment processes and factors influencing sediment supply, *Geo Mar. Lett.*, 6, 121–129, 1986.

Anderson, J. B., L. R. Bartek, and M. A. Thomas, Seismic and sedimentologic record of glacial events on the Antarctic Peninsula, in *Geological Evolution of Antarctica*, edited by M. R. A. Thomson, J. A. Crame, and J. W. Thomson, pp. 287–291, Cambridge University Press, New York, 1991.

Armstrong, R. L., K-Ar dating: Late Cenozoic McMurdo Volcanic Group and dry valley glacial history, Victoria Land, Antarctica: *N. Z. J. Geol. Geophys.*, 21, 685–698, 1978.

Baldauf, J. G., and J. A. Barron, Diatom biostratigraphy: Kerguelen Plateau and Pridz Bay regions of the Southern Ocean, *Proc. Ocean Drill. Program Sci. Results*, 119, 547–598, 1991.

Bandy, O. L., R. E. Casey, and R. C. Wright, Late Neogene planktonic zonation, magnetic reversals, and radiometric dates, Antarctic to the tropics, in *Antarctic Oceanography I*, *Antarct. Res. Ser.*, vol. 15, edited by J. L. Reid, pp. 1–26, AGU, Washington, D. C., 1971.

Barker, P. F., and J. Burrell, The opening of the Drake Passage, *Mar. Geol.*, 25, 15–34, 1977.

Barrett, P. J., D. P. Elston, D. M. Harwood, B. C. McKelvey, and P.-N. Webb, Cenozoic glaciation, tectonism, and sea-level change from MSSTS-1 on the margin of the Victoria Land Basin, Antarctica, *Geology*, 15, 634–637, 1987.

Barrett, P. J., M. J. Hambrey, D. M. Harwood, A. R. Pyne, and P.-N. Webb, Synthesis, in Antarctic Cenozoic History From the CIROS-1 Drillhole, McMurdo Sound, *DSIR Bull. N. Z.*, 245, 241–253, 1989.

Barrett, P. J., M. J. Hambrey, and P. R. Robinson, Cenozoic glacial and tectonic history from CIROS-1, McMurdo Sound, in *Geological Evolution of Antarctica*, edited by M. R. A. Thomson, J. A. Crame, and J. W. Thomson, pp. 651–656, Cambridge University Press, New York, 1991.

Bartek, L. R., P. R. Vail, J. B. Anderson, P. A. Emmet, and S. Wu, Effect of Cenozoic ice sheet fluctuations in Antarctica on the stratigraphic signature of the Neogene, *J. Geophys. Res.*, *96*, 6753–6778, 1991.

Behrendt, J. C., and A. Cooper, Evidence of rapid Cenozoic uplift of the shoulder escarpment of the Cenozoic West Antarctic rift system and a speculation on possible climatic forcing, *Geology*, *19*, 315–319, 1991.

Berggren, W. A., D. V. Kent, J. J. Flynn, and J. A. Van Couvering, Cenozoic geochronology, *Geol. Soc. Am. Bull.*, *96*, 1407–1418, 1985.

Bernhard, J. M., Foraminiferal biotopes in Explorers Cove, McMurdo Sound, Antarctica, *J. Foraminiferal Res.*, *17*, 286–297, 1987.

Brady, H. T., The dating and interpretation of diatom zones in Dry Valley Drilling Project holes 10 and 11 Taylor Valley, South Victoria Land, Antarctica, in Proceedings of the Seminar III on Dry Valley Drilling Project, *Mem. Natl. Inst. Polar Res. Spec. Issue Jpn.*, *13*, 150–164, 1979.

Bull, C., and P.-N. Webb, Some recent developments in the investigation of the glacial history and glaciology of Antarctica, in *Paleoecology of Africa, the Surrounding Islands and Antarctica*, VIII, edited by E. M. van Zinderen Bakker, Sr., pp. 55–84, Balkema, Cape Town, 1973.

Buzas, M. A., Patterns of species diversity and their explanations, *Taxon*, *21*, 275–286, 1972.

Calkin, P. E., R. H. Behling, and C. Bull, Glacial history of Wright Valley, southern Victoria Land, Antarctica, *Antarct. J. U. S.*, *1*, 22–27, 1970.

Ciesielski, P. F., and F. M. Weaver, Early Pliocene temperature changes in the Antarctic seas, *Geology*, *2*, 511–515, 1974.

Ciesielski, P. F., M. T. Ledbetter, and B. B. Elwood, The development of Antarctic glaciation and the Neogene paleoenvironment of the Maurice Ewing Bank, *Mar. Geol.*, *46*, 1–51, 1982.

Cloetingh, S., Intraplate stresses: A tectonic cause for third-order cycles in apparent sea level?, *Spec. Publ. Soc. Econ. Paleontol. Mineral.*, *42*, 21–29, 1988.

Cloetingh, S., H. McQueen, and K. Lambeck, On tectonic mechanism for regional sea level variations, *Earth Planet. Sci. Lett.*, *75*, 157–166, 1985.

D'Agostino, A., Foraminiferal biostratigraphy, paleoecology, and systematics of DSDP Site 273, Ross Sea, Antarctica, M.S. thesis, 124 pp., Northern Ill. Univ., DeKalb, Ill., 1980.

Denton, G. H., M. L. Prentice, D. E. Kellogg, and T. B. Kellogg, Late Tertiary history of the Antarctic Ice Sheet: Evidence from the dry valleys, *Geology*, *12*, 263–267, 1984.

Denton, G. H., J. G. Bockheim, S. C. Wilson, J. E. Leide, and B. G. Andersen, Late Quaternary ice-surface fluctuations of Beardmore Glacier, Transantarctic Mountains, *Quat. Res.*, *31*, 183–209, 1989.

Drewry, D. J., Initiation and growth of the East Antarctic Ice Sheet, *J. Geol. Soc. London*, *131*, 255–273, 1975.

Drewry, D. J., and A. P. Cooper, Processes and models of Antarctic glaciomarine sedimentation, *Ann. Glaciol.*, *2*, 117–122, 1981.

Dunbar, R. B., J. B. Anderson, E. W. Domack, and S. S. Jacobs, Oceanographic influences on sedimentation along the Antarctic continental shelf, in *Oceanology of the Antarctic Continental Shelf*, *Antarct. Res. Ser.*, vol. 43, edited by S. S. Jacob, pp. 291–312, AGU, Washington, D. C., 1985.

Elston, D. P., and S. L. Bressler, Magnetic stratigraphy of DVDP drill cores and late Cenozoic history of Taylor Valley, Transantarctic Mountains, Antarctica, in *Dry Valley Drilling Project*, *Antarct. Res. Ser.*, vol. 33, edited by L. D. McGinnis, pp. 413–426, AGU, Washington, D. C., 1981.

Fillon, R. H., Late Cenozoic foraminiferal paleoecology of the Ross Sea, Antarctica, *Micropaleontology*, *20*, 129–151, 1974.

Fitzgerald, P. G., M. Sandiford, P. J. Barrett, and A. J. W. Gleadow, Asymmetric extension associated with uplift and subsidence of the Transantarctic Mountains and Ross Embayment, *Earth Planet. Sci. Lett.*, *81*, 67–78, 1986.

Freeland, H. J., D. M. Farmer, and C. D. Levings, Fjord oceanography, *NATO Conf. Ser. 4*, *4*, 715 pp., 1980.

Fulthorpe, C. S., Geological controls on seismic sequence resolution, *Geology*, *19*, 61–65, 1991.

Gunn, B. M., and G. Warren, Geology of Victoria Land between the Mawson and Mulock glaciers, Antarctica, *N. Z. Geol. Surv. Bull.*, *71*, 157 pp., 1962.

Hambrey, M. J., B. Larsen, W. U. Ehrmann, and ODP Leg 119 Shipboard Scientific Party, Forty million years of Antarctic glacial history yielded by Leg 119 of the Ocean Drilling Program, *Polar Rec.*, *25*, 99–106, 1989.

Haq, B. U., J. Hardenbol, and P. R. Vail, The new chronostratigraphic basis of Cenozoic and Mesozoic sea level cycles, *Spec. Publ. Cushman Found. Foraminiferal Res.*, *24*, 7–13, 1987.

Harwood, D. M., Late Neogene climatic fluctuations in the high-southern latitudes: Implications of a warm Gauss and deglaciated Antarctic continent, *S. Afr. J. Sci.*, *81*, 239–241, 1985.

Harwood, D. M., Diatom biostratigraphy and paleoecology with a Cenozoic history of Antarctic ice sheets, Ph.D. dissertation, 592 pp., Ohio State Univ., Columbus, Ohio, 1986.

Harwood, D. M., P. J. Barrett, A. R. Edwards, H. J. Rieck, and P.-N. Webb, Biostratigraphy and chronology, in Antarctic Cenozoic History From the CIROS-1 Drillhole, McMurdo Sound, *DSIR Bull. N. Z.*, *245*, 231–241, 1989.

Hayes, D. E., and L. A. Frakes, et al., Leg 28, *Initial Rep. Deep Sea Drill. Proj.*, *28*, 1017 pp., 1975.

Hinz, K., and M. Block, Results of geophysical investigations in the Weddell Sea and in the Ross Sea, Antarctica, in *Proceedings of the 11th World Petroleum Congress, London*, PD 2, pp. 79–91, John Wiley, New York, 1983.

Hodell, D. A., and P. F. Ciesielski, Southern Ocean response to the intensification of northern hemisphere glaciation at 2.4 Ma, in *Geological History of the Polar Oceans: Arctic Versus Antarctic*, edited by U. Bleil and J. Thiede, pp. 707–728, Kluwer, Amsterdam, 1990.

Hodell, D. A., and J. P. Kennett, Late Miocene–early Pliocene stratigraphy and paleoceanography of the South Atlantic and southwest Pacific oceans: A synthesis, *Paleoceanography*, *1*, 285–311, 1986.

Hodell, D. A., and D. A. Warnke, Climatic evolution of the Southern Ocean during the Pliocene epoch from 4.8 to 2.6 million years ago, *Quat. Sci. Rev.*, *10*, 205–214, 1991.

Ishman, S. E., and P.-N. Webb, Late Neogene benthic foraminifera from the Victoria Land Basin margin, Antarctica: Application to glacio-eustatic and tectonic events, in Benthos '86, *Rev. Paleobiol. Spec. Vol.*, *2*, 523–551, 1988.

Karl, H. A., E. Reimnitz, and B. D. Edwards, Extent and nature of the Ross Sea Unconformity in the western Ross Sea, Antarctica, in A. P. Cooper and F. J. Davey (Editors), *The Antarctic Continental Margin: Geology and Geophysics of the Western Ross Sea*, *Earth Sci. Ser.*, vol. 5B, edited by A. P. Cooper and F. J. Davey, pp. 77–92, Circum-Pacific Council for Energy and Mineral Resources, Houston, Tex., 1987.

Keller, G., and J. A. Barron, Paleoceanographic implications of Miocene deep-sea hiatuses, *Geol. Soc. Am. Bull.*, *94*, 590–613, 1983.

Keller, G., and J. A. Barron, Paleoceanographic implications of middle Eocene to Pliocene deep-sea hiatuses, in *Second International Conference on Paleoceanography Abstracts With Program*, Woods Hole Oceanographic Institution, Woods Hole, Mass., 1986.

Keller, G., T. Herbert, R. Dorsey, S. D'Hondt, M. Johnsson, and W. R. Chi, Global distribution of late Paleogene hiatuses, *Geology*, *15*, 199–203, 1987.

Kellogg, T. B., and D. E. Kellogg, Antarctic cryogenic sediments: Biotic and inorganic facies of ice shelf and marine-based ice sheet environments, *Palaeogeogr. Palaeoclimatol. Palaeoecol.*, *67*, 51–47, 1988.

Kellogg, T. B., R. S. Truesdale, and L. E. Osterman, Late Quaternary extent of the West Antarctic Ice Sheet: New evidence from Ross Sea cores, *Geology*, *7*, 249–253, 1979.

Kennett, J. P., The fauna of the Ross Sea: Ecology and distribution of foraminifera, part 6, *DSIR Bull. N. Z.*, *186*, 48 pp., 1968.

Kennett, J. P., Cenozoic evolution of Antarctic glaciation, the circum-Antarctic Ocean, and their impact on global paleoceanography, *J. Geophys. Res.*, *82*, 3843–3860, 1977.

Kennett, J. P. (Ed.), The Miocene ocean: Paleoceanography and biogeography, *Mem. Geol. Soc. Am.*, *163*, 337 pp., 1985.

Kennett, J. P., Miocene to early Pliocene oxygen and carbon isotope stratigraphy of the southwest Pacific, DSDP Leg 90, *Initial Rep. Deep Sea Drill. Proj.*, *90*, 1383–1411, 1986.

Kennett, J. P., and P. F. Barker, Latest Cretaceous to Cenozoic climate and oceanographic developments in the Weddell Sea, Antarctica: An ocean-drilling perspective, *Proc. Ocean Drill. Program Sci. Results*, *113*, 937–960, 1990.

Kennett, J. P., and S. Srinivasan, *Neogene Planktonic Foraminifera*, 265 pp., Hutchinson Ross, Stroudsburg, Pa., 1983.

Mankinen, E. A., and B. G. Dalrymple, Revised geomagnetic polarity time scale for the interval 0–5 m.y. B.P., *J. Geophys. Res.*, *84*, 615–626, 1979.

Matthews, R. K., and R. Z. Poore, Tertiary ^{18}O record and glacio-eustatic sea-level fluctuation, *Geology*, *8*, 501–504, 1980.

McGinnis, L. D. (Ed.), *Dry Valley Drilling Project*, Antarct. Res. Ser., vol. 33, 465 pp., AGU, Washington, D. C., 1981.

McKelvey, B. C., The lithologic logs of DVDP cores 10 and 11, eastern Taylor Valley, in *Dry Valley Drilling Project*, Antarct. Res. Ser., vol. 33, edited by L. D. McGinnis, pp. 63–94, AGU, Washington, D. C., 1981.

McKnight, W. M., The distribution of foraminifera on parts of the Antarctic coast, *Bull. Am. Paleontol.*, *44*, 65–154, 1962.

Mercer, J. H., Glacial geology of the Reedy Glacier area, *Geol. Soc. Am. Bull.*, *79*, 471–486, 1968.

Mercer, J. H., Glacial history of southernmost South America, *Quat. Res.*, *6*, 125–166, 1976.

Miller, K. G., G. S. Mountain, and B. E. Tucholke, Oligocene glacio-eustacy and erosion on the margins of the North Atlantic, *Geology*, *13*, 10–13, 1985.

Miller, K. G., R. G. Fairbanks, and G. S. Mountain, Tertiary oxygen isotope synthesis, sea-level history, and continental margin erosion, *Paleoceanography*, *2*, 1–19, 1987.

Murray, J. W., *Ecology and Paleoecology of Benthic Foraminifera*, 397 pp., John Wiley, New York, 1991.

Newman, W. A., and A. Ross (Eds.), *Antarctic Cirripedia*, Antarct. Res. Ser., vol. 14, 257 pp., AGU, Washington, D. C., 1971.

Nichols, R. L., Glacial geology of the Wright Valley, McMurdo Sound, in *Research in the Antarctic*, edited by L. O. Quam, pp. 293–340, American Association for the Advancement of Science, Washington, D. C., 1971.

Oerlemans, J., Response of the Antarctic ice sheet to a climatic warming: A model study, *J. Climatol.*, *2*, 1–11, 1982.

Olsson, R. K., K. G. Miller, and T. E. Ungrady, Late Oligocene transgression of middle Atlantic coastal plain, *Geology*, *8*, 549–554, 1980.

Olsson, R. K., A. J. Melillo, and B. L. Schreiber, Miocene sea level events in the Maryland coastal plain and the offshore Baltimore Canyon Trough, *Spec. Publ. Cushman Found. Foraminiferal Res.*, *24*, 85–97, 1987.

Osterman, L. E., and T. B. Kellogg, Recent benthic foraminiferal distributions from the Ross Sea, Antarctica: Relations to ecologic and oceanographic conditions, *J. Foraminiferal Res.*, *9*, 250–269, 1979.

Pflum, C. E., The distribution of foraminifera in the eastern Ross Sea, Amundsen Sea, and Bellingshausen Sea, Antarctica, *Bull. Am. Paleontol.*, *50*, 151–209, 1966.

Pickard, J., D. A. Adamson, D. M. Harwood, G. H. Miller, P. G. Quilty, and R. K. Dell, Early Pliocene marine sediments, coastline, and climate of East Antarctica, *Geology*, *16*, 158–161, 1988.

Pitman, W. C., Relationship between eustacy and stratigraphic sequences of passive continental margins, *Geol. Soc. Am. Bull.*, *89*, 1389–1403, 1978.

Porter, S. C., and J. E. Beget, Provenance and depositional environments of late Cenozoic sediments in permafrost cores from lower Taylor Valley, Antarctica, in *Dry Valley Drilling Project*, Antarct. Res. Ser., vol. 33, edited by L. D. McGinnis, pp. 351–363, AGU, Washington, D. C., 1981.

Powell, R. D., Sedimentation conditions in Taylor Valley, Antarctica, inferred from textural analysis of DVDP cores, in *Dry Valley Drilling Project*, Antarct. Res. Ser., vol. 33, edited by L. D. McGinnis, pp. 331–349, AGU, Washington, D. C., 1981.

Prentice, M. L., and R. K. Matthews, Cenozoic ice volume history: Development of a composite oxygen isotope record, *Geology*, *16*, 963–966, 1988.

Purucker, M. E., D. P. Elston, and S. L. Bressler, Magnetic stratigraphy of late Cenozoic glaciogenic sediments from drill cores, Taylor Valley, Transantarctic Mountains, Antarctica, in *Dry Valley Drilling Project*, Antarct. Res. Ser., vol. 33, edited by L. D. McGinnis, pp. 109–129, AGU, Washington, D. C., 1981.

Quilty, P. G., Distribution of foraminiferids in sediments of Prydz Bay, Antarctica, *Spec. Publ. South Aust. Dep. Mines Energy*, *5*, 329–340, 1985.

Savage, M. L., and P. F. Ciesielski, A revised history of glacial sedimentation in the Ross Sea region, in *Antarctic Earth Science*, edited by R. L. Oliver, P. R. James, and J. P. Jago, pp. 555–559, Cambridge University Press, New York, 1983.

Savin, S. M., R. G. Douglas, and F. G. Stehli, Tertiary marine paleotemperatures, *Geol. Soc. Am. Bull.*, *86*, 1499–1510, 1975.

Scherer, R. P., Quaternary and Tertiary microfossils from beneath Ice Stream B: Evidence for a dynamic West Antarctic Ice Sheet history, *Global Planet. Change*, *4*, 395–412, 1991.

Shackleton, N. J., and J. P. Kennett, Paleotemperature history of the Cenozoic and the initiation of Antarctic glaciation: Oxygen and carbon isotope analyses in DSDP sites 277, 279 and 281, *Initial Rep. Deep Sea Drill. Proj.*, *29*, 743–756, 1975.

Stuiver, M., G. H. Denton, and H. W. Borns, Carbon-14 dates of Adamussium colbecki (Mollusca) in marine deposits at New Harbor, Taylor Valley, *Antarct. J. U. S.*, *11*, 86–88, 1976.

Thomson, W., Polar ice-caps and their influence on changing sea levels, *Trans. Geol. Soc. Glasgow*, *8*, 322–340, 1888.

Uchio, T., Benthonic foraminifera of the Antarctic Ocean, in Special Publications From the Seto Marine Biological Laboratory, *Biol. Results Jpn. Antarct. Res. Exped.*, *12*, 20 pp., 1960.

Vail, P. R., R. M. Mitchum, Jr., R. G. Todd, J. M. Widmier, S. Thompson III, J. B. Sangree, J. N. Bubb, and W. G. Hatelid, Seismic stratigraphy and global changes in sea level, in Seismic Stratigraphy—Applications to Hydrocarbon Exploration, *Mem. Am. Assoc. Pet. Geol.*, *26*, 49–205, 1977.

Ward, B. L., Distribution of modern benthic foraminifera of

McMurdo Sound, Antarctica, Ph.D. thesis, 212 pp., Victoria Univ., Wellington, 1984.

Ward, B. L., P. J. Barrett, and P. Vella, Distribution and ecology of foraminifera in McMurdo Sound, Antarctica, *Palaeogeogr. Palaeoclimatol. Palaeoecol.*, *58*, 139–153, 1987.

Webb, P.-N., Wright Fjord, Pliocene marine invasion of an Antarctic dry valley, *Antarct. J. U. S.*, *7*, 226–234, 1972.

Webb, P.-N., Micropaleontology, paleoecology and correlation of the Pecten Gravels, Wright Valley, Antarctica, and description of *Trochoelphidiella onyxi* n. gen., n. sp., *J. Foraminiferal Res.*, *4*, 184–199, 1974.

Webb, P.-N., and J. E. Andreason, K/Ar dating of volcanic material associated with the Pliocene Pecten Conglomerate (Cockburn Island) and Scallop Hill Formation (McMurdo Sound), *Antarct. J. U. S.*, *21*, 59–61, 1986.

Webb, P.-N., and D. M. Harwood, Late Cenozoic glacial history of the Ross Sea Embayment, Antarctica, *Quat. Sci. Rev.*, *10*, 215–223, 1991.

Webb, P.-N., and J. H. Wrenn, Upper Cenozoic micropaleontology and biostratigraphy of eastern Taylor Valley, Antarctica, in *Antarctic Geoscience*, edited by C. Craddock, pp. 1117–1122, University of Wisconsin Press, Madison, Wis., 1982.

Webb, P.-N., D. M. Harwood, B. C. McKelvey, J. H. Mercer, and L. D. Stott, Cenozoic marine sedimentation and ice volume variation on the East Antarctic craton, *Geology*, *12*, 287–291, 1984.

(Received January 17, 1992;
accepted May 1, 1992.)

LATE QUATERNARY CLIMATIC CYCLES AS RECORDED IN SEDIMENTS FROM THE ANTARCTIC CONTINENTAL MARGIN

HANNES GROBE AND ANDREAS MACKENSEN

Alfred Wegener Institute for Polar and Marine Research, D-2850 Bremerhaven, Germany

To reveal the late Quaternary paleoenvironmental changes at the Antarctic continental margin, we test a lithostratigraphy adjusted to a stable isotope record from the eastern Weddell Sea. The stratigraphy is used to produce a stacked sedimentological data set of 11 sediment cores. We derive a general model of glaciomarine sedimentation and paleoenvironmental changes at the East Antarctic continental margin during the last two climatic cycles (300 kyr). The sedimentary processes considered include biological productivity, ice rafting, current transport, and gravitational downslope transport. These processes are controlled by a complex interaction of sea level changes and paleoceanographic and paleoglacial conditions in response to changes of global climate and local insolation. Sedimentation rates are mainly controlled by ice rafting which reflects mass balance and behavior of the Antarctic ice sheet. The sedimentation rates decrease with distance from the continent and from interglacial to glacial. Highest rates occur at the very beginning of interglacials, i.e., of oxygen isotope events 7.5, 5.5, and 1.1, these being up to 5 times higher than those during glacials. The sediments can be classified into five distinct facies and correlated to different paleoenvironments: at glacial terminations (isotope events 8.0, 6.0, and 2.0), the Antarctic cryosphere adjusts to new climatic conditions. The sedimentary processes are controlled by the rise of sea level, the destruction of ice shelves, the retreat of sea ice, and the recommenced feeding of warm North Atlantic Deep Water (NADW) to the Circumpolar Deep Water (CDW). During peak warm interglacial periods (at isotope events 7.5, 7.3, 5.5, and 1.1), the CDW promotes warmer surface waters and thus the retreat of sea ice which in turn controls the availability of light in surface waters. At distinct climatic thresholds, local insolation might also influence sea ice distribution. Primary productivity and bioturbation increase, the calcite compensation depth rises, and carbonate dissolution occurs in slope sediments also in shallow depth. Ice shelves and coastal polynyas favor the formation of very cold and saline Ice Shelf Water which contributes to bottom water formation. During the transition from an interval of peak warmth to a glacial episode (isotope stages 7.2–7.0 and 5.4–5.0), the superimposition of both intense ice rafting and reduced bottom currents produces a typical facies which occurs with a distinct lag in the time of response of specific sedimentary processes to climatic change. With the onset of a glacial episode (at isotope events 7.0 and 5.0) the Antarctic ice sheet expands owing to the lowering of sea level with the extensive glaciations in the northern hemisphere. Gravitational sediment transport becomes the most active process, and sediment transfer to the deep sea is provided by turbidity currents through canyon systems. During Antarctic glacial maxima (isotope stages 6.0 and 4.0–2.0) the strongly reduced input of NADW into the Southern Ocean favors further advances of the ice shelves far beyond the shelf break and the continuous formation of sea ice. Below ice shelves and/or closed sea ice coverage contourites are deposited on the slope.

INTRODUCTION

Antarctica represents a unique sedimentary environment. The Antarctic Ocean is covered for most of the year by sea ice which, together with the presence of a huge continental ice sheet, controls nearly all environmental conditions, resulting in markedly different sedimentation processes. With its extensive ice shelves in the Weddell Sea and the Ross Sea, the Southern Ocean is one of the Earth's principal sources of oceanic bottom waters and plays a key role in governing global oceanic circulation and climate. Concerning the sedimentation processes at the Antarctic continental margin, the understanding of the behavior of ice shelves is crucial.

Research on the Cenozoic cryospheric development of Antarctica was strengthened during recent years by the Ocean Drilling Project (ODP). The long-term record since Cretaceous time was reconstructed on southern Indian Ocean ODP legs 119 and 120, and particularly on legs 113 and 114 into the subpolar and polar South Atlantic. Site 693, drilled during Leg 113 [*Barker et al.*, 1988, 1990], is located in our investigation area off Cape Norvegia in the eastern Weddell Sea (Figure 1). Investigations of Miocene/Pliocene sediments from the South Atlantic have focused on glacial evolution and paleoceanographic changes during the Neogene [*Abelmann et al.*, 1990; *Hodell and Ciesielski*, 1990]. During the last decade there was an extensive sampling program carried out by gravity and piston coring from R/V *Po-*

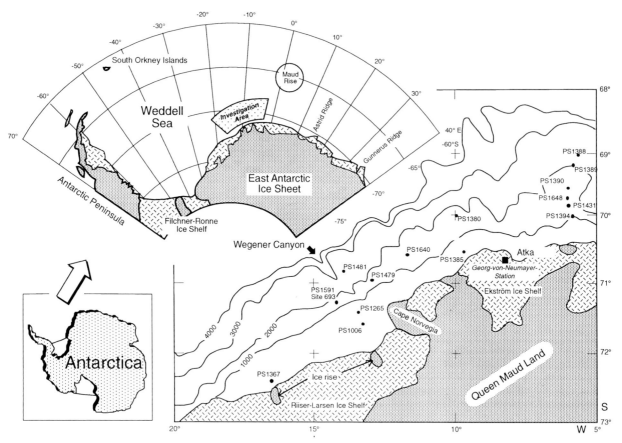

Fig. 1. Investigation area at the Antarctic continental margin in the eastern Weddell Sea. Four gravity cores from the shelf and 11 cores from the slope have been investigated. Core PS1591 is located at the same position as ODP Site 693, completing its undisturbed uppermost Quaternary sequence. Canyons, such as the Wegener Canyon in the investigation area, play an important role in transport processes from the shelf into the deep sea. Further sites, where similar sedimentary records occur, are indicated at the continental margin south of the South Orkney Islands, at Maud Rise, and around Astrid and Gunnerus ridges.

larstern mostly focusing on the Quaternary sediment sequence of the Atlantic part of the Southern Ocean.

The distribution of Quaternary sediments in the Southern Ocean results from a complex interaction of glaciological, oceanographic, and biological processes. Therefore the first marine geological work on Quaternary sediments in this area focused on sedimentation processes [*Anderson*, 1972; *Anderson et al.*, 1979, 1983b; *Elverhøi*, 1981; *Elverhøi and Roaldset*, 1983]. Several models were established regarding those processes to explain the cryospheric depositional processes in the marine environment and the behavior of the Antarctic ice sheet on the basis of the analysis of Pleistocene sediments [*Anderson et al.*, 1983a; *Domack*, 1982; *Drewry and Cooper*, 1981; *Kellogg and Kellogg*, 1988; *Orheim and Elverhøi*, 1981]. Only a few models were developed concerning paleoclimatic and paleoceanographic reconstructions [*Kellogg et al.*, 1979; *Weaver*, 1973]. More recently, sedimentation pat-

terns on the continental margin around the Weddell Sea were investigated in order to reveal environmental variations of the hydrosphere and cryosphere and to unravel the Quaternary climatic history of Antarctica [*Grobe*, 1986; *Haase*, 1986; *Fütterer et al.*, 1988; *Cordes*, 1990; *Fütterer and Melles*, 1990; *Grobe et al.*, 1990b; *Mackensen et al.*, 1990; *Melles*, 1991].

Our results, although based on the investigation of the late Quaternary deposits from the eastern Weddell Sea only, contribute to the understanding of the late Neogene glacial evolution of Antarctica as a whole. Records similar to those described in this paper were found in sediments of the Antarctic continental margin between South Orkney and Gunnerus Ridge, covering nearly a third of the Antarctic coast line. Stratigraphic problems were solved by the development of a lithostratigraphy which allows correlation of lithologic changes with a stable isotope record. The lithostratigraphy was applied toward the late Pleistocene chronology. We used the

$\delta^{18}O$ record of Core PS1388, measured on both plank-tonic and benthic foraminifera, which currently is the only record from the area substantially south of the Polar Front that can be correlated with the global isotope stratigraphy [*Mackensen et al.*, 1989] (Figure 2).

A high-resolution stratigraphy and an extended sedimentological data set of 11 cores with inferred high sedimentation rates and a dense sample spacing allow stacking of sedimentological parameters. The parameter stacks, in correlation with global climate and local insolation, were used to develop a detailed model of the sedimentary environment during a glacial and an interglacial stage. Special emphasis in the interpretations is given to the environmental changes due to sea ice cover and movements of the Antarctic ice edge in response to the two pronounced Quaternary climatic cycles covering the last 300 kyr.

MATERIAL AND METHODS

Sediments were recovered by gravity corer during *Polarstern* expeditions ANT IV/3 [*Fütterer*, 1987], ANT V/4 [*Miller and Oerter*, 1990], and ANT VI/3 [*Fütterer*, 1988] on profiles perpendicular to the Antarctic continental margin between 9° and 15°W. Eleven cores from the slope with a total length of 113 m were chosen for a detailed interpretation of the stratigraphic and sedimentological data sets because of good resolution and paleoenvironmental signals during the last 300 kyr (Figure 1, Table 1). The following interpretations will focus on the last two climatic cycles (indicated by bold lines in Figure 3) even if bottom ages of some of the cores can be much higher. Core sites were mainly situated in the middle of the continental slope terrace, in water depth between 2000 and 2800 m.

Cores PS1394, PS1431, PS1648, PS1390, PS1389, and PS1388 were located on a S-N profile on the slope off Atka Iceport with slightly increasing water depth between 2000 and 2500 m (Figure 1). In contrast, cores PS1591, PS1479, PS1640, and PS1380 were situated on a line mostly parallel to both the isobaths and the ice edge, resulting in an almost constant distance to the shelf. Core PS1481 was located further north off Cape Norvegia and has a basal age of up to 4 m.y. Owing to its proximity to the Wegener Canyon, several hiatuses are present, but the uppermost sequence of the last 300 kyr remained undisturbed. Gravity Core PS1591 was located at the same position and complemented Site 693 by recovering undisturbed the uppermost Quaternary sedimentary sequence. With the short cores PS1006, PS1265, PS1367, and PS1385 on the shelf (Figure 1, Table 1), only the Holocene sediments were recovered.

Cores were sampled with a mean sampling interval of 10 cm, providing stratigraphic resolution of between 1000 and 10,000 years, depending on sedimentation rates. In total, 1265 samples were taken (Table 1) and analyzed for carbon, grain size, clay mineralogy, and components of the coarse fraction. Sediment structure and ice-rafted debris were investigated on X radiographs of 1-cm-thick sediment slices. The gravel fraction >2 mm was counted in the radiographs and expressed as numbers per 10 cm^3 [*Grobe*, 1987]. Other sedimentological data were produced using standard methods as described by *Grobe et al.* [1990*b*]. The most important parameters (carbonate, radiolaria, illite, silt) of each core are plotted versus depth in Figure 3. Core PS1648 was chosen to show all parameters in detail (Figure 4). Stable oxygen and carbon isotope measurements were made on planktonic foraminiferal tests of sinistrally coiled *Neogloboquadrina pachyderma* with samples containing on average six specimens from the 125- to 250-μm size fraction. The foraminifera were measured using an automatic carbonate preparation device connected on line to a Finnigan MAT 251 mass spectrometer.

The peak warm times of interglacials correspond to the isotopic events 1.1, 5.5, 7.3/7.5, 9.3, and 11. In Antarctic glaciomarine sediments, only the facies deposited during these climatic events differ significantly from the rest of the Pleistocene deposits. Therefore when dealing with terms like "glacial" and "interglacial" a difference between a chronostratigraphically defined interglacial (e.g., stage 5) and the occurrence of a specific sedimentological facies (e.g., event 5.5), interpreted as being deposited during an interglacial, has to be kept in mind.

AREA OF INVESTIGATION

Physiography

The Antarctic continental margin in the eastern part of the Weddell Sea has been surveyed with swath sonar echo-sounding systems during several *Polarstern* expeditions [e.g., *Fütterer*, 1987, 1988; *Miller and Oerter*, 1990]. In particular, the bottom topography of the area off Queen Maud Land between 16°W and 2°W is well known. The slope can be divided into five major characteristic morphological units. The partly overdeepened continental shelf has a maximum water depth of between 300 and 400 m and a distinct shelf break. This is followed seaward by a very steep and narrow upper continental slope with inclinations of up to 16°. The transition to a gently inclined midslope bench occurs from west to east in decreasing water depths between 1700 m and 1200 m. The bench is 50 to 100 km wide, dipping seaward at 1.5° to about 3000-m water depth. The lower slope is characterized by the steep, clifflike Explora Escarpment [*Hinz and Krause*, 1982; *Henriet and Miller*, 1990], locally showing maximum slope inclinations of 30° [*Fütterer et al.*, 1990]. The continental rise ends in the Weddell Abyssal Plain at about 4400-m water depth.

The continental slope is deeply dissected by at least

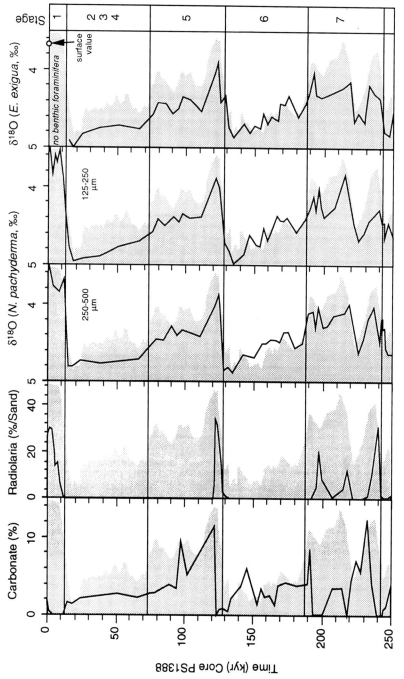

Fig. 2. Stable oxygen isotope record of Core PS1388 during the last two climatic cycles [from *Mackensen et al.*, 1989]. Measurements were made on two size fractions of the planktonic foraminifera *Neogloboquadrina pachyderma* and on the benthic foraminifera *Epistominella exigua*. The occurrence of radiolaria and carbonate, which is mainly produced by *N. pachyderma*, shows a good correlation with the climatic cycles. The chronostratigraphy of *Martinson et al.* [1987] is shaded in gray for comparison. The isotope stratigraphy of this core in correlation with the lithologic changes was used to establish a lithostratigraphy and correlate and date the other cores.

TABLE 1. Basic Data on the Sediment Cores Investigated

Core Site	Longitude, deg	Latitude, deg	Depth, m	Length of Core, cm	No. of Samples	Bottom Ages, kyr	Sedimentation Rate Interglacial, cm/kyr	Sedimentation Rate Glacial, cm/kyr
				Shelf				
PS1006	−13.272	−71.493	235	55	9	Holocene		
PS1265	−13.408	−71.352	230	31	5	Holocene		
PS1367	−16.517	−72.333	303	238	25	Holocene		
PS1385	−9.617	−70.483	328	95	12	Holocene		
				Slope				
PS1380	−9.983	−70.000	2060	945	100	450	5.8	0.7
PS1388	−5.883	−69.033	2517	1238	141	1000	5.2	1.1
PS1389	−5.967	−69.217	2259	909	93	410	4.2	0.5
PS1390	−6.383	−69.617	2798	990	108	520	4.7	0.6
PS1394	−6.667	−70.083	1945	910	100	123	21.3	3.4
PS1431	−6.570	−69.817	2458	935	99	244	7.9	1.2
PS1479	−13.404	−70.549	2322	1180	118	2100	1.8	0.6
PS1481	−13.963	−70.830	2505	1076	154	3990	2.1	0.3
PS1591	−14.555	−70.834	2361	1220	153	2290	2.9	0.3
PS1640	−11.657	−70.339	2101	1020	111	390	6.8	1.5
PS1648	−6.525	−69.740	2531	863	88	340	6.8	1.1

The shelf cores only recover the Holocene sediment sequence. The time included depends on sedimentation rates and the occurrence of hiatus. Sedimentation rates were calculated using a lithostratigraphy in correlation with an oxygen isotope stratigraphy.

one major submarine canyon system, i.e., the Wegener Canyon off Cape Norvegia (Figure 1) which was mapped by systematic bathymetric surveys [*Fütterer et al.*, 1990]. The tributary gullies of the canyon cut into the upper slope with the easternmost one continuing across the shelf and beneath the ice shelf. The formation of the canyon is structurally controlled by the sill of the Explora Escarpment and is considered to be of early late Miocene age [*Barker et al.*, 1988]. During the Quaternary the canyon was particularly active during glacial episodes [*Fütterer et al.*, 1990]. This indicates that the canyon played a major role in gravitational downslope transport processes during the Pleistocene climatic cycles.

Of paramount importance for the understanding of the sedimentary processes in this area is detailed information about the slope morphology that is largely responsible for distinct sedimentary patterns and the accumulation of undisturbed Quaternary sediment sequences close to the continent. The Quaternary sequence on most parts of the terrace in the intercanyon areas is less affected by turbidites and mass flow events than is usually the case in this area [*Wright and Anderson*, 1982; *Anderson et al.*, 1986]. At least during the last two climatic cycles, the flat part of the slope was not a source area for sediment gravity flows to the deep sea. Sediment supply to the abyssal plain was presumably restricted to canyon systems like the Wegener Canyon during glacials. The importance of sediment gravity flow has to be taken into account in dealing with more long-term geological records. High sediment supply to

form a fan, as described in the eastern Weddell Sea, apparently occurred before the development of the glacial shelf topography and during a more temperate glacial setting [*Anderson et al.*, 1986].

In this part of the Weddell Sea, a broad terrace in the middle of the continental slope favors relatively undisturbed sedimentation and thus provides excellent conditions for paleoenvironmental interpretations of the late Quaternary climatic cycles. In addition, the sediments deposited in water depth of less than 3500 m have a significant carbonate content because they are mostly deposited above the calcite compensation depth (CCD) and thus provide calcareous foraminifera for use in measuring stable isotopes [*Mackensen et al.*, 1989].

Modern Sediment Distribution Patterns

The most important recent environmental conditions and sedimentological parameters on shelf and slope of the study area were investigated in relation to the association of benthic foraminifera [*Mackensen et al.*, 1990]. The reconstruction of the distribution patterns of sediments and foraminifera were made on the basis of samples which were taken with a vented box corer, providing undisturbed surface material.

In general, the surface sediments at the eastern continental margin consist of silty and clayey muds with varying amounts of coarse sand and gravel. The highest values of sand and gravel were found on the shelf which is a different sedimentary environment compared with the conditions on the slope. On the shelf during settling,

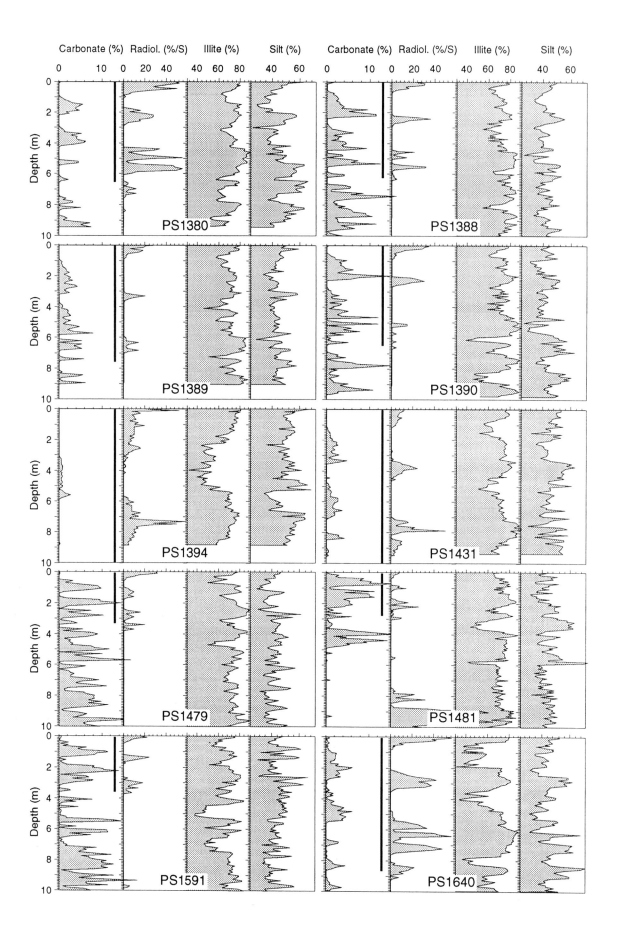

some of the clay and fine silt are kept in suspension and transported by the Antarctic Coastal Current (ACC), leading to an enrichment of the coarser grades in the bottom sediments [Elverhøi and Roaldset, 1983]. Biogenic particles constitute an integral part of the glacio-marine deposits. Sponge spicules are abundant in the shelf sediments where an extensive sponge fauna was observed; carbonate is mainly produced by bryozoans [Voss, 1988]. Sedimentation rates are relatively low (1–5 cm/kyr) [Elverhøi, 1981; Elverhøi and Roaldset, 1983].

Sediment echo soundings on the upper steep slope, well below the shelf edge, provide evidence that the slope is covered by a coarse, presumably residual sediment. The current-winnowed silts and clays are entrained by contour currents and transported along the continental slope [Anderson et al., 1979]. Down to a depth of about 1600 m, the sand-size fraction exceeds 40%. On the midslope bench and the lower slope, down to the abyssal plain, the mud content (<63 μm) exceeds 50% of the bulk sediment; gravel and sand are also present, but only in minor proportions.

Clay minerals are represented by chlorite, illite, and smectite. Because of the absence of any chemical weathering processes on the continent, kaolinite is only found in traces in all Quaternary sediments ($<5\%$, e.g., PS1648, Figure 4) and is of no paleoclimatic significance. The kaolinite found may have been produced by weathering during an earlier warm humid climate and locally stored in ancient sedimentary rocks.

The carbonate content of the surface sediments can be grouped in four depth zones. On the shelf, a carbonate content of up to 9.5% of the bulk dry sediment consists of bryozoans, bivalves, gastropods, brachiopods, solitary corals, and benthic foraminifera. At the shelf break and on the upper slope, down to about 2000 m, the carbonate content never exceeds 1%. This depth zone corresponds to the range of the core of the Weddell Deep Water (WDW), which is a branch of the Circumpolar Deep Water (CDW) [Foldvik et al., 1985; Anderson, 1975]. On the slope terrace, between 2000 and 3500 m, carbonate content is between 3 and 13%. This is almost exclusively produced by planktonic foraminifera, supplemented by some other marine organisms. No detrital carbonate was found. The recent depth of the CCD is ~4000 m. Below this depth, virtually no carbonate was found in the surface sediments [Anderson, 1975; Mackensen et al., 1990]. Siliceous particles, consisting of radiolaria, diatoms, sponge spicules, and

silicoflagellates were found in all surface samples in varying amounts.

STRATIGRAPHY

Owing to the paucity of biogenic carbonate, cores from south of the Antarctic Polar Front are not included in the construction of the Quaternary isotope chronostratigraphy [Imbrie et al., 1984; Prell et al., 1986; Martinson et al., 1987]. So far, all interpretations of southern high-latitude glacial and interglacial paleoenvironments [i.e., Ledbetter and Ciesielski, 1986; Burckle and Abrams, 1987; Fütterer et al., 1988; Pudsey et al., 1988] suffer from the lack of a detailed stratigraphy due to low levels of carbonate. Even a coring device with a large volume cannot provide sufficient foraminiferal tests to allow any accelerator mass spectrometer (AMS) dating. Since a few years ago, δ^{18}O and δ^{13}C isotopes can be measured on extremely small samples (20 μg) because of the development of mass spectrometers with an automatic preparation device, which only need three to six tests for one measurement.

The cores discussed in this paper include the first from the Antarctic continental margin in which sufficient foraminiferal tests were found and stable isotopes have been measured. Core PS1388 is from 69°S from the slope at a water depth of 2536 m. The stable isotope record was generated from benthic (Epistominella exigua) as well as planktonic species (Neogloboquadrina pachyderma) of two different size fractions (Figure 2). Although this record is somewhat affected by diagenesis, it is complete and is typical of the deep-sea record elsewhere. This is the only record of the Antarctic continental margin, spanning the last 300 kyr, which could be correlated in detail with the global isotope stratigraphy [Mackensen et al., 1989].

The interpretation of the isotopic data in the other cores in most cases remains difficult because of strong diagenetic alterations as shown in Core PS1648 (Figure 4). Low values of 3.6 to 3.7‰ for δ^{18}O were found in the surface sediments of all cores. In general, values lower than 4‰ are only found in the Holocene sequence. Samples with pre-Holocene ages are in the range of 4 to 5‰ and do not attain lower values even in those sediments obviously formed during the warmest intervals of isotopic stages 5 or 7. Diagenetic dissolution within the sedimentary column has altered the isotopic composition of the tests toward higher values by selectively removing the isotopically lighter carbonate [Savin and Douglas, 1973; Berger and Killingley, 1977; Wu et al., 1990]. Differential dissolution occurs probably as a result of a very low carbonate content of the sediment. Masking of the original isotope signal, as discussed in detail by Grobe et al. [1990b], was most intense during times of high productivity, when most of the carbonate was dissolved on the seafloor because of its deposition at or below the lysocline or even in the range of the CCD.

Fig. 3. (Opposite) Percentages of carbonate, radiolaria (percent of the sand fraction), illite, and silt in the individual cores plotted versus depth. Bold lines indicate the range of the last 300 kyr which was included in our investigations. Alternating occurrence of carbonate and silica can be observed in each of the cores. The distribution of illite nearly resembles the shape of an oxygen isotope curve. The percentage of silt shows highest values during peak warm times.

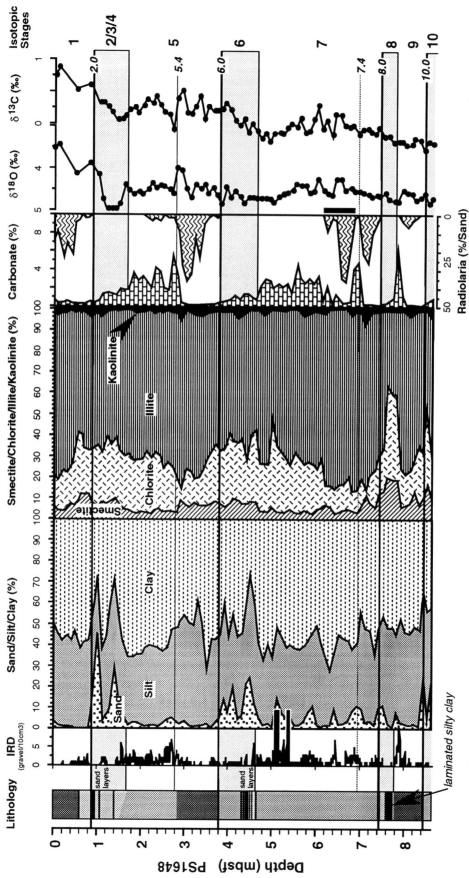

Fig. 4. Sedimentological parameters of Core PS1648 from the middle of the midslope bench off Atka Iceport. The core at its base is 340 kyr old, and mean sedimentation rates are 2.5 cm/kyr. It shows the typical record of the continental margin sediments during the last two climatic cycles. As in most cores of the Antarctic continental slope, oxygen stable isotope data are only of restricted value because of the strong diagenetic alteration of the foraminiferal tests. Stage boundaries have been defined by correlation of significant lithologic changes with the stable isotope record of Core PS1388 [Mackensen et al., 1989].

TABLE 2. Age Models for All Cores Including an Undisturbed Pleistocene Sediment Sequence of the Last 350 kyr

Age, kyr	PS1380, m	PS1388, m	PS1389, m	PS1390, m	PS1394, m	PS1431, m	PS1479, m	PS1481, m	PS1591, m	PS1640, m	PS1648, m
12.05	0.95	0.60	0.70	0.60	3.20	1.30	0.40	0.20	0.50	0.80	1.00
73.91	1.40	1.25	1.00	0.95	5.30	2.05	0.80	0.40	0.70	1.70	1.70
110.79	1.90	1.90	2.90	1.80	6.40	3.20	1.15	0.70	1.20	2.50	2.60
129.84	3.00	2.90	3.70	2.70		4.70	1.50	1.10	1.75	3.80	3.90
180.00	3.30	3.50	4.10	3.20		5.60	1.75	1.30	2.00	4.80	4.60
205.00	4.20	4.45	5.80	4.20		7.10	2.20	1.70	2.40	5.70	5.80
224.89	5.20	5.10	6.40	5.10		8.10	2.70	2.10	2.80	6.80	6.80
244.18	6.20	5.80	7.10	5.90		9.40	3.10	2.40	3.30	7.80	7.50
282.00	6.40	6.00	7.30	6.20			3.20	2.55	3.45	8.30	7.75
339.00	7.40	6.60	8.10	6.80			3.55	3.20	3.90	9.40	8.60
362.00	7.60	6.80	8.30	7.00			3.60		4.00	9.70	

Ages are from *Imbrie et al.* [1984] and *Martinson et al.* [1987].

Because of these severe problems with most of the isotopic data, we developed a lithostratigraphy as a tool for correlating the late Pleistocene sediments. The comparison of the isotopic record of Core PS1388 with the lithological parameters and biogeneous components (Figures 2 and 3) has shown that significant changes occur at distinct times of global climatic change. Those prominent lithologic changes were correlated with the isotopic events of the stable isotopic record of Core PS1388 (Figure 2). The changes were found to be similar in all late Pleistocene sediments around the Weddell Sea between the Antarctic Peninsula and 35°E [*Brehme*, 1991; *Melles*, 1991; *Fütterer*, 1991]. Variations of the biogenic constituents can be correlated between cores and are synchronous within their resolution. A further improvement of the lithostratigraphy was made feasible using the lithologic variations of the terrigenous detritus. Clay mineral composition and grain size parameters are influenced by local conditions and sources but show significant distribution patterns which could also be correlated in detail between cores (Figure 3).

Age Models

To compute age models for the last 300 kyr, stratigraphic fixed points were defined where specific lithologic changes can be correlated with distinct events of the isotope chronostratigraphy. The correlation of the sedimentological parameters between cores and the correlation of lithologic changes with the isotopic events of Core PS1388 (Figure 2) provide a sufficient data base to calculate sedimentation rates and, finally, to stack the most important parameters. The age models assume both zero age for core tops containing a well-defined Holocene section and constant sedimentation rates between isotopic events. Stratigraphic fix points (Table 2) up to event 8.0 were derived from the compiled chronostratigraphy of *Martinson et al.* [1987]; further stratigraphic calculations were based on the SPECMAP data

set [*Imbrie et al.*, 1984] with stage boundaries as defined by *Prell et al.* [1986]. Sedimentation rates across stages 2 to 4 and events 5.1 to 5.4 in all cores were integrated, because in this case stage boundaries could not be determined by lithologic changes.

Isotopic events at the glacial terminations (2.0, 6.0, 8.0, and 10.0) can easily be defined by significant lithologic changes and were placed at a depth where decreasing carbonate content correlates with the first occurrence of silica, indicating a drastic increase in productivity with the onset of the interglacial. The lithology of event 5.5 consists of an opal peak followed by relatively high carbonate values in the uppermost part of this horizon. Opal decreases to zero values toward event 5.4, which was placed at the top of the first carbonate occurrence within stage 5. Other lithologic changes, such as an increase in clay content and a significant minimum in the warm time illite peak, are in correspondence and support the definition of event 5.4 (e.g., PS1648, Figure 4).

To solve problems with sedimentation rates around isotopic event 7.1, insolation, as calculated by *Berger* [1978], was taken into account (Figure 5). Sedimentation rates during event 7.1 were too high when using the correspondence of lithologic changes, in particular, the decrease in carbonate content, with stage boundary 6/7. Instead, this lithologic change was correlated with the decline in insolation during the southern summer at 70°S around 180 kyr, shifting the age for the lithologic change around event 7.0 to about 10 kyr later (Figure 5a). Calculations of the influence of orbitally induced insolation changes on Antarctic sea ice cover suggest influence on Pleistocene sea ice distribution [*Crowley and Parkinson*, 1988]. Less sea ice might have affected sedimentation processes by extending interglacial conditions, especially surface water productivity, into the beginning of glacial stage 6.

One further stratigraphic fixed point, not available

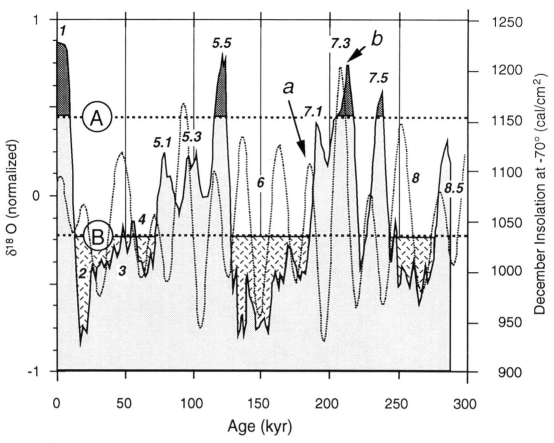

Fig. 5.   Summer insolation [*Berger*, 1978] in the investigation area plotted together with the oxygen isotope stratigraphy of *Martinson et al.* [1987]. Slight differences between global climate and the sedimentary record, as found in this study, can probably be explained by local summer insolation. During distinct sensitive times (*a*, *b*) of global climatic change, the local insolation influences sea ice conditions and thus primary productivity. Levels A and B indicate climatic thresholds at which sediment composition drastically changes.

from the standard chronostratigraphy, was defined at 205 kyr representing the "shoulder" between event 7.2 and event 7.3 (Figure 5*b*). This boundary was placed by interpolation at one of the significant changes in productivity, defined by a change from silica to carbonate microfossil preservation. In addition, this change correlates with a typical decrease in illite content within the drastic change of the clay mineral association between event 7.3 and event 7.2. Both lithologic changes were found to be significant in all cores (Figure 7; see arrows).

Event 7.4 was defined at a carbonate peak within stage 7 (Figures 2 and 3). This is the only occurrence of carbonate which cannot be correlated to a lithological change in clay mineralogy or grain size within the full sedimentary record of the last 300 kyr. It is obvious that the short duration of the cold event 7.4 has had an effect only on the sea-ice-controlled productivity, not on the

continental-ice-controlled processes of terrigenous sedimentation.

The section between 205 kyr and event 7.4, in the curve of *Martinson et al.* [1987], marked by the pronounced warm event 7.3, was found to be composed of two peaks in silica content in most of the cores with sufficient resolution (e.g., Figure 4, Core PS1648, bold line; Figure 3, Core PS1388, 4.4–5.0 m). The double peak is indicative of two events of high productivity. There might be two reasons for the discrepancy with the global chronostratigraphy: (1) the resolution of the stacked isotope record of *Martinson et al.* [1987] might not be sufficient in this range, or (2) this is a local effect, produced by the superimposition of local insolation and global climate. The peak value of insolation in the investigation area (70°S) is somewhat shifted by 7 kyr in relation to the chronostratigraphic peak value (Figure 5*b*), which is a further indication that productivity is

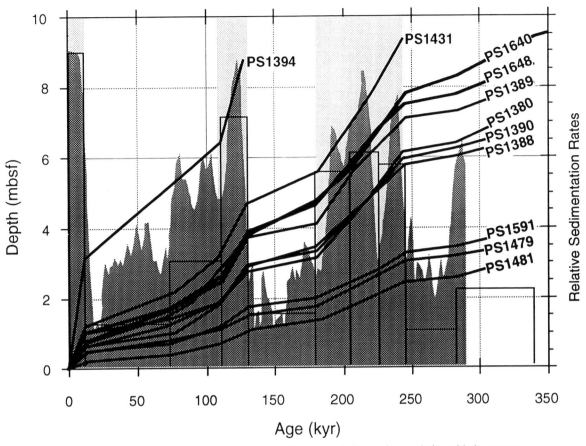

Fig. 6. Age versus depth plot showing sedimentation rates of all cores in correlation with the oxygen isotope stratigraphy of *Martinson et al.* [1987]. Highest sedimentation rates are found close to the continent in Core PS1394. Generally, sedimentation rates decrease with distance from the shelf and in the vicinity of the Wegener Canyon. The histogram plotted as an underlay shows relative sedimentation rates calculated from the stacked record of all cores, indicating 3 to 4 times higher values during interglacial compared with glacial periods.

controlled by an interaction of global climate and local insolation changes in the high-latitude Antarctic Ocean.

Sedimentation Rates

Integrated sedimentation rates (Table 1, Figure 6) were calculated for distinct glacial and interglacial periods. Sedimentation rates in general decrease with distance from the continent and within a climatic cycle from interglacial to glacial. Rates are highest very close to the continent owing to the higher input of terrigenous material (Core PS1394). Low sedimentation rates were found in the vicinity of the Wegener Canyon (cores PS1591, PS1479, and PS1481).

Sedimentation rates during the last 300 kyr, presented as a relative data stack (Figure 6, histogram), indicate that peak warm episodes, including isotopic stages/ events 1, 5.5, and 7, are 4 to 5 times higher in comparison with glacials integrated over all sites. Because rates

are calculated as mean values between stratigraphic fix points, it is difficult to determine detailed sedimentation rates within an individual climatic event, although the Holocene exhibits the highest values compared with other warm events. Within a climatic cycle the Holocene mainly reflects sedimentary conditions at the very beginning of an interglacial up to the climatic optimum. This indicates that sedimentation rates are highest during the terminations and the first few thousand years of an interglacial. The rates then decrease following the climatic optimum. This produces a sawtooth pattern of the sedimentation rate during peak warm intervals when a warm climatic event rapidly follows a glacial maximum.

The sedimentation rates during stage 9 are low in comparison with later stages, because rates are integrated for the full stage. In this case, event 8.5 was defined as belonging to interglacial stage 9. In addition, sequences deeper than about 5 m below the surface are increasingly compressed by the gravity coring procedure [*Melles*, 1991]; thus the apparent reduction of sedimentation rates is most probably artificial.

STACKED SEDIMENTOLOGICAL RECORD

A composite record was established for the interpretation of the continental margin sediments by stacking the specific sedimentological parameters. Detailed analysis of 11 cores (Figures 3 and 4), including the last two climatic cycles, permits the use of a method which is new in this context. Stratigraphic fixed points, as defined above, were used to compute an age model for each of the cores. After calculating the age of every individual sample, each parameter was calculated in equidistant time steps for the last 300 kyr to allow stacking of parameters between cores. Each data set was normalized to values between 0 and 1 by division by the maximum value. Individual data for each time step and each parameter on all cores were added and divided by the number of cores. Plots of four selected parameters (carbonate, radiolaria, illite, silt) versus time show a good correlation between all cores (Figure 7). Strongest variations between cores occur in the terrigenous parameters during glacials. Relative changes of clay mineral composition and grain size are mainly controlled by the glaciological processes on the shelf and are therefore dependent on the distance of the core location from the shelf break.

The stacked sedimentological records were plotted versus time together with the standard chronostratigraphic curve of *Martinson et al.* [1987]. The records show the relative changes of each parameter and represent mean values of slope sediment composition (Figure 8). Individual signals will be enhanced toward the shelf; others will be masked at the deeper slope. Thus interpretation of the stacked record will provide more general information about sedimentation processes at the slope. Some variations are representative at least for the margin of the eastern Weddell Sea and some for most of the Antarctic coastline of the Atlantic part of the Southern Ocean.

Opal

The percentage of radiolarians in the sand fraction was determined by coarse fraction analysis. Investigations of smear slides have shown that the occurrence of radiolarians can be correlated in detail with the siliceous microfossil content in the silt fraction such as diatoms and silicoflagellates [*Grobe and Kuhn*, 1987]. Thus the quantity of radiolaria could be used to predict the occurrence of silica in the sediments, because this parameter can easily be assessed and radiolaria are more resistant against solution.

The siliceous microfossils, as indicated by the radio-

larian peaks, are typically concentrated in distinct horizons which were found to occur in all cores (Figures 2 and 3). Siliceous microfossils occur at the glacial terminations and reach peak values close to the interglacial climate optimum. Opal decreases to zero again when the climatic curve drops below a distinct threshold (Figure 5a). In cores which included the full Pleistocene sedimentological record, silica was found only during the last five interglacial stages (Figure 3, e.g., cores PS1388, PS1479, and PS1591). The sole exception was found in Core PS1481, where high silica values below 7.50 m are due to the Pliocene age of the lower section of this core [*Grobe et al.*, 1990a].

Carbonate

Calcium carbonate content is mainly derived from the planktonic foraminifera *Neogloboquadrina pachyderma* (sinistral), some benthic foraminifera, and very few coccoliths and ostracods. *N. pachyderma* is a permanent component of Antarctic sea ice which may contain up to 1000 individuals per liter of melted ice [*Dieckmann et al.*, 1991]. Carbonate percent can generally mimic the oxygen isotope curve, which means that it reflects paleoclimatic history. However, the amount of carbonate preserved in the sediment can be influenced by dissolution. Carbonate values decrease to almost zero in sediments rich in silica (Figures 3 and 4). During peak warm times, both opal and carbonate are produced in surface waters, but because of high organic matter fluxes, calcareous shells are not preserved. We explain the lag of decreasing carbonate relative to the oxygen isotope curve at stage boundary 7/6 (Figure 8) with an insolation maximum which might have temporally extended interglacial sea ice conditions (Figure 5a).

Ice-Rafted Debris

Close to the continent, ice-rafted debris (IRD) is one of the most important parameters indicating the response of the Antarctic ice sheet to climatic change. The highest occurrence of IRD is found during interglacial stages. During warm episodes, the increasing IRD content shows a time lag relative to the terminations of up to several thousands of years (Figure 8). During stage 7, IRD increases at event 7.4, and in stage 5 IRD increases at event 5.4; these are the highest amounts of IRD found within the Holocene. Stage 7 shows higher values than stage 5 in almost all cores investigated. Coarse and medium sand follows the distribution of gravel.

Grain Size

The sediments are glaciomarine and thus consist of sandy mud with varying amount of gravel and rocks. Most information on transport mechanisms, such as current activity and ice rafting, can be derived from the silt/clay ratio and from the grain size distribution of the

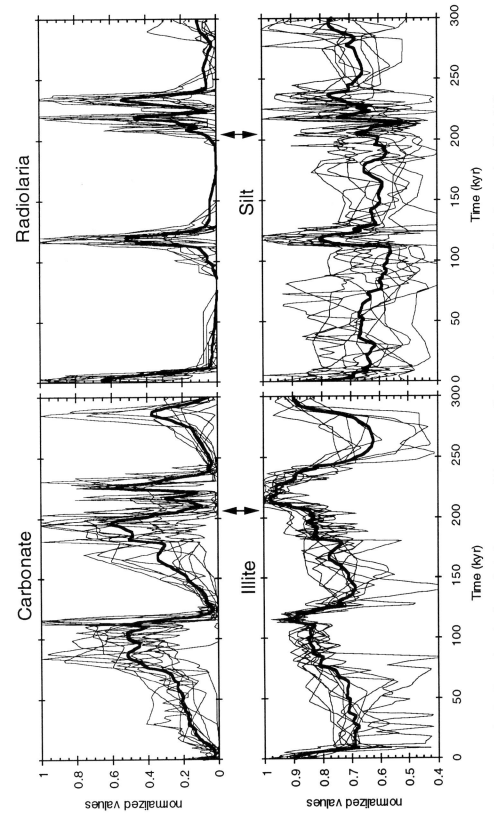

Fig. 7. Four selected sedimentological parameters of 11 cores normalized to values between 0 and 1 and plotted versus time. The bold line is the resulting stack of each parameter in all cores. Strongest variations between cores occur in the terrigenous parameters during glacials. Relative changes of clay mineral composition and grain size are mainly controlled by the glaciological processes on the shelf and are thus depending on the distance of the core location from the shelf edge.

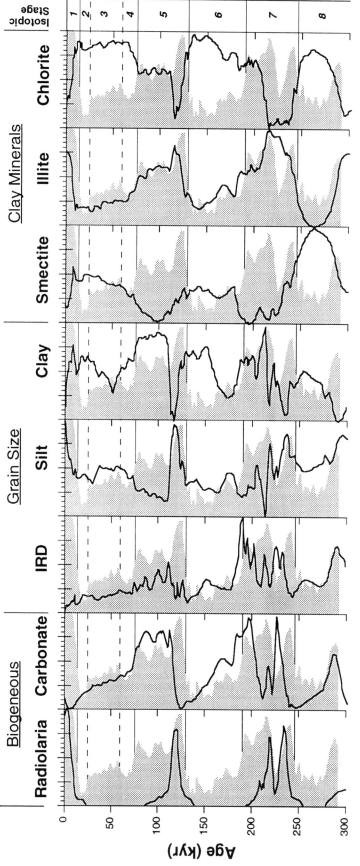

Fig. 8. Stacked record of important sedimentological parameters of all cores investigated. The oxygen isotope stratigraphy after *Martinson et al.* [1987] is shown by shading. Isotopic stage boundaries are indicated.

silt fraction (2–63 μm). Diatom frustules, which were found exclusively during peak warm events, are a minor part of the silt fraction and are not important to the grain size distribution. The amount of terrigenous silt has maxima during the warmest climatic events, although a time lag of about 5 kyr between glacial termination and changing grain size values can be observed (Figure 8). Within the transition from glacial to interglacial conditions, the mean of the silt fraction changes from 7.4 to 6.6 phi. The corresponding change in skewness of the bulk sediment from 0 to −0.2 is due to a decrease in the amount of clay. Both parameters indicate changes in the fine tails of the grain size distributions toward coarser sediments when the ratio of current-derived to ice-rafted material changes in favor of the latter.

Clay Minerals

Close to the continent, the clay mineral associations are sensitive indicators of ice rafting and of the extent of ice coverage. In particular, smectite, which is carried long distances by currents, can be considered to be the "marine sediment noise." A high content of smectite was found in the clay fraction deposited during times of maximal ice expansion (Figure 8). With the termination of the glacial, the smectite content does not show a sudden change due to a lagged and slowly increasing dilution by ice-rafted illite (Figures 3 and 8). Smectite reaches minimum values during the midpoint of the following interglacial. Chlorite increases with the end of an interglacial and reaches peak values during glacial maxima, indicating an enhanced input of shelf sediments.

The peak value of smectite during glacial stage 8 (Figures 4 and 8) is a local effect, produced by a setting when the adjacent ice shelf and/or closed sea ice conditions cover the slope area for a distinct time span. Current-transported clays and some downslope sediment movement produce a clay association, consisting of smectite and minor chlorite. Ice rafting and biological productivity are insignificant contributors to this sedimentary process.

CLASSIFICATION OF SEDIMENT FACIES

Correlation between the different sedimentological parameters allows the identification of five different sediment facies. The facies can be related to different sedimentary environments which are controlled by the slope morphology and by global and local climatic changes. The most important parameters used for the classification are grain size distribution, including the gravel fraction as an indicator of ice-rafting activity, the association of clay minerals (chlorite, illite, and the smectite group), siliceous microfossil and carbonate content, and finally the sediment textures as observed in the X-rayed core slices.

Holocene Shelf Facies

During the late Holocene, on the continental shelf a sediment facies is deposited that includes overconsolidated diamictons and residual glacial-marine sediments. The water depth at the present-day ice shelf edge varies between 200 and 400 m depending on the locations of ice rises and capes. The mean depth of the overdeepened shelf is around 300 m. From the investigation of long piston cores it is known that basal tills appear to be most widespread on the Antarctic continental shelf. They are overlain by glacial-marine sediments in most areas [Anderson et al., 1980, 1991]. Owing to the high content of gravel (up to 30%), it was not possible to recover the basal till in the investigation area, and thus in this paper only the uppermost sequence of the Holocene sedimentary record was analyzed. Despite this lack, the composition of the Holocene shelf sediments is crucial for the reconstruction of sediment transport processes at the continental margin.

The short cores of up to 0.6-m length show that the acoustically reflective sediment is a mostly unfossiliferous, apparently overconsolidated diamicton. In many places the surface of the shelf is modified by iceberg scour. The diamicton is deposited by melting processes beneath the ice shelf, most probably close to the grounding line, and is defined as a transitional glacial-marine sediment [Anderson et al., 1980]. Fine fractions of the grain size distributions are subsequently winnowed by currents, producing a residual glacial-marine sediment (Anderson et al. [1980]; Figure 9, Holocene shelf facies). The settling fines (<30 μm) are entrained by the Antarctic Coastal Current, which has velocities of up to 17 cm/s [Carmack and Foster, 1975]. Parts of the fine sediments are transported off the shelf, where they contribute to the grain size distribution of the slope sediments forming a compound glaciomarine sediment (e.g., carbonate facies, Figure 9). A portion may be transported further with the Weddell Gyre to the SE to the Filchner Trough, where thick laminated clays were found [Melles, 1991].

In addition, the mineralogical composition of the clay fraction remaining on the shelf is indicative of a residual glaciomarine sediment (Figure 10). The weight percentage of chlorite is about twice as high there as on the slope and the abyssal plain. All terrigenous material deposited on the shelf is directly delivered from Queen Maud Land by the continental ice. The geology of the coastal areas of the eastern Weddell Sea was reconstructed from dropstones, dredged on the shelf [Oskierski, 1988] and from pebbles and sands in tills [Andrews, 1984; Anderson et al., 1991]. Close to the investigated area the classification in different petrographical provinces shows mainly Mesozoic, in most cases Jurassic, basalts with different degrees of alteration [Rex, 1967; Juckes, 1972; Peters, 1989]. The source of chlorite may

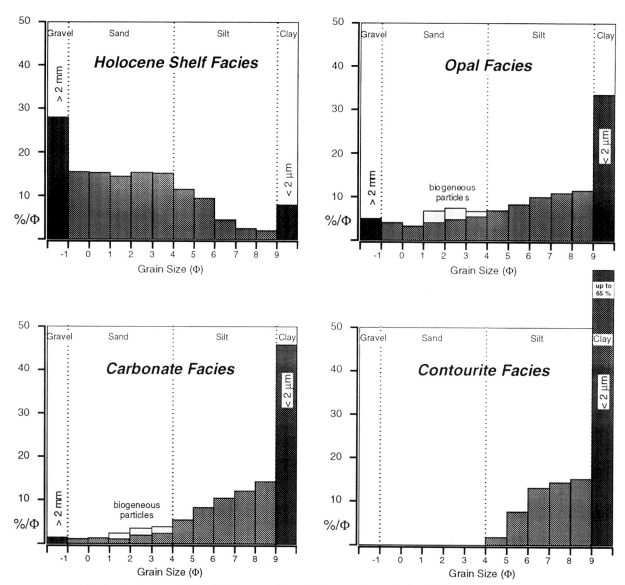

Fig. 9.   Typical grain size distributions of four different sedimentological facies. In the Holocene shelf facies the fine fractions are partly winnowed by the Antarctic Coastal Current. The opal facies is deposited on the slope during peak warm times. The carbonate facies can be found during moderate interglacial to glacial conditions, whereas the contourite facies only occurs during times of closed sea ice conditions, presumably during further advances of the ice shelves. In all slope facies, the fine fractions, missing on the shelf, are added to the ice-rafted component, depending on the contribution of the two transport processes, ice rafting and current transport.

thus be the basic to ultrabasic rocks, where weathered olivine and pyroxene have formed chlorite.

Determination of sedimentation rates of the shelf sediments is difficult and was done only at one site (Core 206, 420 m, Norwegian Antarctic Research Expedition 1978/1979 [*Elverhøi and Roaldset*, 1983]) where sufficient carbonate was found to use ^{14}C analysis. The analysis of a coral gave a rate of about 3 cm/kyr. However, it is evident that sedimentation rates will vary

greatly along the ice edge in correspondence with the morphological features of the shelf and the ice shelf and the intensity of the ACC. On the shelf, it remains problematic in getting long-term information about the behavior of the Antarctic ice sheet because it is difficult to penetrate the pebbly sediment with a gravity coring device and to produce any stratigraphy. Though piston cores up to 10 m in length have been taken on the Ross Sea and Weddell Sea shelf, they suffer from the lack of

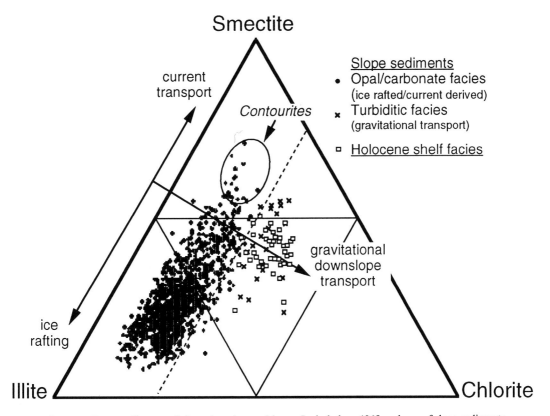

Fig. 10.   Ternary diagram of clay mineral assemblages. Included are 1265 analyses of slope sediments and 51 samples of the shelf deposits. The slope sediments plotted on the left-hand side of the dashed line consist mainly of illite and smectite. Their quantitative relation is controlled by the ratio of current transport to ice rafting. The contourite facies, deposited only by current transport, shows highest smectite values. On the contrary, sediments deposited during peak warm times and intense ice rafting are dragged more toward the illite corner. Typical of the shelf sediments is a higher content of chlorite. Such assemblages with a high chlorite content are found on the slope only in the turbiditic facies. The turbiditic facies represents small proximal turbidites, triggered and fed from the shelf by the advancing ice edge.

any stratigraphic information, and investigations are concentrating on a detailed description of the different sediment facies [*Anderson et al.*, 1980]. The only Cenozoic long-term sequences, recovered on the Antarctic shelf, were drilled in McMurdo Sound (MSSTS, CIROS, DVDP [see *Hambrey and Barrett*, 1992]), during Deep Sea Drilling Project Leg 28 on the Ross Sea shelf [*Hayes et al.*, 1975] and during ODP Leg 119 in Prydz Bay [*Hambrey et al.*, 1991].

Opal Facies (Peak Warm Interglacial Sediments)

The opal facies contains a significant number of siliceous skeletons such as diatoms, radiolarians, and some sponge spicules. Sediments of this facies can clearly be separated from other sediment types by its biogeneous content (Figures 8 and 9). The facies is almost barren of carbonate, with the exception of very few planktonic foraminifera. In places, these were just

sufficient for one single stable isotope analysis. The sediment includes occasional dropstones. Intense bioturbation masks all primary sediment structures. Arenaceous benthic foraminifera occur together with the opal facies but were found only in surface sediments [*Mackensen et al.*, 1990]. The tests are diagenetically destroyed at larger sediment depth.

In the investigated area the uppermost surface layer of the sediments between 300 and 10 cm, depending on the sedimentation rate, is rich in siliceous hard parts. This facies represents the Holocene deposits and is also typical for the modern sedimentary environment. Correlation of the opal facies between cores is supported by the $\delta^{18}O$ data, which are unaltered in the Holocene sequence, presenting a well-defined stage 1.

The opal facies was found only in five horizons which correspond to the peak warm events of the last interglacial stages 1, 5, 7, 9, and 11. This is probably because during the last 425 kyr, the glacial/interglacial climatic amplitudes [*Ruddiman et al.*, 1989] were higher than in earlier times and climatic optima provided totally ice-free conditions in the summer months and thus higher biogeneous silica fluxes. We assume that the variations in opal content are the result of varying siliceous primary productivity rather than a result of changes in deep-sea opal dissolution. This is in agreement with the findings of *Mortlock et al.* [1991], from sites further north close to the Polar Frontal Zone.

The clay mineral composition is characterized by peak values of illite, which can be attributed to transport mechanisms, e.g., ice rafting (Figure 10). Grain size distributions with a low clay content, a silt distribution with peak values in the range of coarse to medium silt, and increasing values of IRD (gravel and sand) toward the top are typical of the opal facies. The minimum clay and fine silt values are due to less deposition resulting from increased current activities (Figures 8 and 9).

A further important process is the impoverishment of the fine fraction by enhanced bioturbation. In most marine environments, bioturbation is sufficiently active to modify the deposited sediments as a result of various biological activities. One of the most important effects of the organisms is remixing of the beds so that particles are continuously resuspended and can be entrained by currents. Experiments with poorly sorted sediments, similar to glaciomarine deposits, have shown that grain size distribution, particularly in the fine tail range, is a reflection of the efficiency with which the bed was mixed [*Singer and Anderson*, 1984].

Carbonate Facies (Moderate Interglacial to Glacial Sediments)

During visual core description of most of the cores, cyclic lithologic changes were observed. The different sediment colors varied between light olive gray and brownish gray. Typical features of this facies are a light olive gray sediment color (5Y5/2, Rock Color Chart) and a significant amount of carbonate. Carbonate content is controlled by a single species of planktonic foraminifera (*Neogloboquadrina pachyderma*) which was up to 98% left coiling. The carbonate content may reach up to 25% and depends directly on the distance from the continent because of the extent of dilution by terrigenous material. Benthic calcareous foraminifera and very few coccoliths and ostracods may make up <1% of the carbonate. Their contribution to the carbonate content is unimportant, but their frequency can be correlated with the number of planktonic foraminifera.

Ice-rafted material, such as dropstones and sand, is common in the carbonate facies. Grain size distribution, as well as clay mineral composition, change by a small extent with climate. The amount of clay and the chlorite content both increase as the climate deteriorates. Bio-

turbation is of variable intensity and in most places masks the primary sedimentary structures.

Turbiditic Facies (Ice Advance Deposits)

The turbiditic facies consist of small proximal turbidites, which are organized in graded layers of gravel, sand, and silt, and only occur on the continental terrace close to the upper steep slope. The sediment is almost barren of biogeneous particles and has a characteristic mineralogical composition which is dominated by chlorite and smectite (Figure 10). The high chlorite/illite ratio, a low content of glauconite, and higher values of rock fragments and gravel show a relationship to the residual glaciomarine sediments of the shelf. Sediment facies colors, described from different core locations on the shelf close by (see, e.g., *Anderson et al.* [1981], *Kaharoeddin et al.* [1980], and *Elverhøi and Roaldset* [1983]) correspond well with the typical olive gray color of the turbiditic facies (N3-5Y3/2, Rock Color Chart).

The turbiditic facies is thought to be material from moraines, transported and released by the advancing ice sheet during sea level lowstand. The sediment was sorted during gravitational transport down the upper slope as observed by sediment textural analysis of gravity flow deposits from this area, showing size sorting and transformation of glacial sediments into sorted sand [*Wright and Anderson*, 1982]. Larger cobbles are deposited close to the shelf/slope transition as can be observed in the sediment echo soundings showing only a hard and very uneven surface reflector. Large dropstones were also abundant in biological dredge hauls in this area [*Voss*, 1988]. The upper slope is covered with gravel and sand, whereas most of the finer fractions are deposited on the landward side of the slope bench, where the slope gradient decreases. At a greater distance from the source area, on the seaward side of the bench, no distal part of the turbidites was observed. Those sediments may have been caught and transported to the deep sea in branches of canyons, cutting into the upper slope (e.g., Wegener Canyon [*Fütterer et al.*, 1990]).

Contourite Facies

The contourite facies consists of thin to medium beds containing horizontal laminae of mud and silty clay. No grain sizes coarser than medium silt occur (Figure 9). Descriptions of the grain size distributions of the contourites in this area were first published by *Anderson et al.* [1979]. The fine sediments originate from the winnowing processes at the upper slope and are, when reaching increasing water depth, transported and deposited by weak bottom currents. It is known from bottom photographs that the bottom currents in the Weddell Sea flow as contour currents parallel to the bathymetric contours of the continental margin, transporting sediments to the south, around the Weddell Sea, and finally along the Antarctic Peninsula into the South Atlantic [*Hollister and Elder*, 1969].

A facies originating from contour currents will mostly be deposited during times of enhanced sea ice conditions when other sedimentation processes, such as ice rafting and gravitational transport, become unimportant. Peak values of smectite are typical of the contourite facies. Smectite increases in all clay fractions when the amount of current-derived material increases (Figure 10). In addition, the missing IRD content suggests deposition under a closed sea ice cover. The laminated structure is the consequence of the lack of any bioturbation. The environment is hostile to burrowers because of a break in primary productivity in surface waters due to limited availability of light below the ice. Contourites were found in all cores taken from less than about 100 km from the shelf break and can be correlated between cores.

The ice edge reaches the continental shelf break during sea level lowstand. During further advances above the slope, a very dense, sometimes closed, cover of sea ice may have stabilized the ice shelf. Lower temperatures in the Antarctic Ocean waters hampered melting from below the ice shelf. Both processes encourage further seaward advances of the continental ice, finally in building up ice shelves more than 100 km broad and floating above the slope. It is obvious that the contourite facies is deposited during such strong ice conditions, but it is difficult to specify the contribution of sea ice and/or an ice shelf. During enhanced glacial conditions the sea ice cover is probably attached to the ice shelf margin for longer time intervals, and thus the sedimentological facies, deposited below, is the same.

PALEOCEANOGRAPHY, PALEOPRODUCTIVITY, AND PALEOGLACIOLOGY

The Southern Ocean today is a mixing reservoir for incoming North Atlantic Deep Water (NADW) and recirculated water from the Pacific and Indian oceans. The contribution of the NADW to the CDW has changed during the climatic cycles, particularly when production of bottom water in the North Atlantic nearly ceased owing to sea ice coverage [*Oppo and Fairbanks*, 1987]. NADW is largely composed of upper ocean waters and thus in the North Atlantic is the most nutrient-depleted deepwater mass formed in the oceans today. In the South Atlantic it still has high $\delta^{13}C$ values [*Kroopnick*, 1985]; therefore up to 50% of the glacial/interglacial $\delta^{13}C$ amplitude in the Southern Ocean is due to changes in the contribution of NADW [*Oppo et al.*, 1990]. Together with climate-controlled global changes between carbon reservoirs, changes in NADW input into the Southern Ocean are the most important reasons for the changes in $\delta^{13}C$ observed in high southern latitude cores (e.g., Core PS1648, Figure 4).

Variations in the relative flux of NADW to the Southern Ocean influence the properties of the CDW and surface water [*Corliss*, 1982; *Charles and Fairbanks*, 1990]. Because NADW is a heat source for the Antarctic Ocean waters, it consequently influences the formation of sea ice. The amount of sea ice coverage in turn is crucial to the availability of light in surface waters and hence is a main factor in controlling primary productivity. This is particularly important, because in Southern Ocean surface waters no nutrient limitation occurs between the Polar Front and the Antarctic Divergence [*Defelice and Wise*, 1981]. Variations in productivity during climatic cycles correspond to the extent of sea ice in a way similar to Recent seasonal processes, in which seasonal changes in sea ice cover control productivity and thus biogenic silica flux. This is inferred from sediment trap studies in the Antarctic Ocean [*Dunbar*, 1984; *Wefer et al.*, 1990].

The sequences of the opal and carbonate facies indicate that cyclicity of primary production varies in response to the glacial/interglacial changes. The events of higher productivity during peak warm times, indicated by maxima in silica content and intense bioturbation, and the variations of carbonate can be explained in terms of interaction of sea ice coverage, deepwater convection, and water masses. On the eastern Weddell Sea continental margin, the WDW appears to influence the depth range of lysocline and CCD [*Anderson*, 1975; *Mackensen et al.*, 1990]. However, the depth of the CCD depends on the properties of water masses and the productivity in surface waters. A high flux of organic matter increases the CO_2 content in the interstitial waters and thus the solution of carbonate. During a climatic cycle the CCD oscillates between 4000 m and 2000 m. It reaches the lowest water depth during the most intense productivity.

The first investigations in this area had already shown a significant amount of carbonate in the Weddell Sea margin sediments, consisting mainly of planktonic foraminifera [*Anderson*, 1972]. But the high carbonate content of up to 30% in Quaternary sediments of the continental margin off Queen Maud Land and carbonate values of up to 80% on Maud Rise [*Cordes*, 1990] were presumed to be related to ice-free waters in the Weddell Sea Polynya [*Grobe*, 1986], which had been observed for several years on satellite microwave images [*Zwally et al.*, 1983]. Furthermore, anomalous diatom abundance at Maud Rise was interpreted as resulting from the absence of sea ice in the polynya, permitting increased photosynthesis [*Defelice and Wise*, 1981]. The apparent cyclical variation in carbonate content was inferred to represent the presence and absence of the polynya through time [*Fütterer et al.*, 1988].

Now there is subsequent evidence from other cores which shows that carbonate is common in Pleistocene sediments along the Antarctic coastline of the Weddell Sea, at least between 50°W and 40°E. Pleistocene sediments, rich in foraminiferal carbonate, were found on the shelf of the South Orkney microcontinent [*Brehme*, 1991], on a seamount within the Weddell Sea [*Kuhn et al.*, 1992], and on Astrid and Gunnerus ridges [*Fütterer*, 1991] in water depths of <4000 m (Figure 1). This observation was also made on ODP Leg 113 drill sites [*Barker et al.*, 1990], and therefore the hypothesis of the influence of the Weddell Sea polynya on biogeneous sedimentation seemed to be weakened.

Variations of atmospheric CO_2 during Pleistocene climatic cycles were possibly controlled by Southern Ocean productivity, such that more effective "biological pump mechanisms" during glacial times cause the Southern Ocean to be a sink for atmospheric CO_2 [*Keir*, 1988]. Enhanced productivity during glacials should be caused by higher solar radiation and reduced circulation of surface waters [*Sundquist and Broecker*, 1985] or by increased atmospheric dust fallout which supplies iron to iron-limited Antarctic surface waters [*Martin*, 1990]. On the other hand, a reduced consumption of nutrients was postulated because of the extensive sea ice coverage during glaciations [*Mix and Fairbanks*, 1985]. This may result in an increasing amount of preformed nutrients in low latitudes which may enhance productivity there [*Sarnthein et al.*, 1988].

The glacial sediments analyzed in this study with little or no biogenic components, reduced bioturbation, low IRD content and consisting mainly of mud clearly indicate strongly reduced productivity and sedimentation rates during glaciations in high southern latitudes around the Antarctic continent. This evidence is consistent with other recent studies, indicating low productivity in the Southern Ocean during glacial times [*Labeyrie and Duplessy*, 1985; *Mackensen et al.*, 1989; *Grobe et al.*, 1990b; *Mortlock et al.*, 1991]. Estimates of sea ice distribution during the last glacial maximum show continuous ice coverage south of 60°S [*Hays*, 1978; *Cooke and Hays*, 1982]. During these times the availability of light in surface waters was strictly reduced by a snow-covered sea ice, which will have been the most significant factor limiting productivity in the Antarctic Ocean.

Glacial transport and supply of terrigenous sediments within the Antarctic Ocean are almost exclusively accomplished by rafting by icebergs. Transport by sea ice, an important sediment distribution process in the Arctic [*Wollenburg*, 1991], is missing owing to the absence of sediment supply by meltwater and rivers.

Two different types of icebergs in the Antarctic Ocean can be distinguished as a result of calving processes. Calving from glaciers, which is mainly a process of the Antarctic Peninsula, produces icebergs during most of a climatic cycle which transport a significant amount of sediment [*Drewry and Cooper*, 1981]. The sediment is incorporated at their base or by sedimentation on the glacier. However, in relation to the mass budget of the Antarctic ice sheet and the production of icebergs from ice shelves, icebergs calved from glaciers are subordi-

nate. Nevertheless, they have to be taken into account when calculating the sediment supply to the ocean, particularly in the vicinity of the Antarctic Peninsula.

Calving from ice shelves produces the typical Antarctic tabular iceberg which is mostly free of any sediment load [*Denton et al.*, 1971]. When the continental ice crosses the grounding line, after a short distance, ablation processes reach peak values. An ablation rate of up to 2 m/yr was calculated below ice shelves close to the grounding line [*Thomas*, 1973; *Kipfstuhl*, 1991]. The ice, lost at the base of the ice shelves, is partly replaced by the accumulation of snow on their surface. The basal debris load is lost rapidly on the way to the calving line or beyond. It is obvious that calving from broad ice shelves produces "clean" icebergs, almost free of any sediment load.

The NADW transports heat into the Antarctic Ocean. When the production of NADW is switched off during glacials, the lower water temperatures hamper basal melting of ice shelves. The coastal ice margins remain thicker, resulting in a seaward extension of the ice shelf and consequently in an increase of the grounded parts of the Antarctic ice sheet. The sediment load, incorporated at the base of the ice, will be transported further off the shelf and contributes to sedimentation on the slope.

Three transport processes by ice are possible around the East Antarctic Ice Sheet. The ice sheet is mainly surrounded by ice shelves, delivering a significant amount of sediment to the sea; two of them are controlled by sea level variations and are thus directly coupled to global climatic change. These are as follows:

1. In the landward part of an ice shelf, parts of the coastline exist where the grounding line coincides with the calving line. These "islands" act as stabilizing pivots which contribute in controlling size and extension of the ice shelf (e.g., Figure 1: Cape Norvegia and ice rises). Icebergs calving close to such areas incorporate some detrital debris.

2. Falls in sea level during glaciations reduce the size of the ice shelves shifting seaward from the grounding line. When sea level is at its lowest, the grounding line may reach the shelf edge in some places. Subglacial sediments will be bulldozed down the slope by episodical movements of the ice edge or may melt out close to the grounding line and deliver sediment directly to the slope.

3. The rapid rise in sea level during the transition from glacial to interglacial causes large parts of the marginal ice sheet to lift up and thus become an ice shelf. A resulting increase in calving processes produces a large number of icebergs, which transport some shelf sediments, still frozen at their bases, to the deep sea.

The ice-rafted fraction of sediments in the investigation area is delivered by two processes. First, the petrographical composition of dropstones on the slope is mainly granitic, indicating its origin from East Antarctica. Icebergs from calving areas further away are driven by the Circum-Antarctic Current and, when entrained by the Weddell Gyre, deliver sediments from somewhere on the East Antarctic coast. Since the full cryospheric development of the Antarctic ice sheet, those icebergs may have contributed to ice rafting at any time during the late Neogene, irrespective of glacial or interglacial conditions. This granitic IRD signal is found in all sediment facies in varying amounts, producing the typical glaciomarine composition of the Antarctic Ocean sediments.

In contrast, close to the continent, the amount and composition of sediments is controlled by the ice streams and source rocks mainly of the hinterland of the investigation area, which are Jurassic basalts. Material from the adjacent coastal areas contributes to the continental margin sediments at all water depths but is most important in shallower depths and close to the shelf. The sediments from the shelf are mainly responsible for the distinct and pronounced cycles in sediment composition on the upper slope. With increasing distance from the continent, the cyclicity becomes increasingly masked by pelagic sedimentation. Variations of the detrital components are strongest during glacials (Figure 7).

From the observations above, it is inferred that the composition of the detrital mineral suites of the Antarctic continental margin sediments mainly depends on the petrographic nature of the source rocks. The source of the terrigenous detritus changes with processes and distance from the continent. Because the East Antarctic craton largely consists of crystalline rocks, mica-illite and some chlorite associated with quartz, feldspar, and various mafic minerals are the most frequent species in the clay fraction. An influence of the specific basic to ultrabasic source rocks in Queen Maud Land can be recognized only very close to the continent. As a result of the exclusively physical weathering processes, kaolinite occurs only in traces, and the amount of the nonlayered minerals in the clay fraction is much higher there than in lower latitudes.

The clay mineral composition was found to be a sensitive indicator to changes in sediment transport. Each clay mineral is found in all facies but in varying amounts. The ternary diagram of clay mineral composition shows that 20% illite, 10% chlorite, and 10% smectite are minimum values in each facies (Figure 10). Clays are permanently delivered by current transport, as this is the only process which contributes to all facies. One mechanism responsible for changes in clay mineral composition appears to be dominated by differential settling and thus grain size sorting, but the final composition is controlled by some contribution of all of the three main sediment-delivering processes: ice rafting, current transport, and gravitational movement downslope. During the different stages of climatic cycles, each of these processes is dominant at some time.

SEDIMENTATION MODEL

Interpretation of the late Pleistocene sediment sequence from the Antarctic continental slope can be used to synthesize a general relationship between sediment facies and paleoclimatic, paleoceanographic and paleoglacial conditions. We propose a model for glaciomarine sedimentation during glacial and interglacial stages, based on a stacked sedimentological record from the eastern Weddell Sea continental margin covering the last 300 kyr. Very similar records were found in all sediments between South Orkney (45°W) and Gunnerus Ridge (35°E), an area which covers roughly a third of the Antarctic coastline (Figure 1). Therefore we suggest that our sedimentation model (Figures 11 and 12), including distinct paleoenvironmental changes and different sedimentary processes, applies to most of East Antarctica and the Weddell Sea area.

The model fits best for the last four climatic cycles (stages 1 to 11). Typical of this time period are a pronounced 100-kyr periodicity, abrupt terminations of glacials, and a trend of decreasing $\delta^{18}O$ values during interglacials when compared with Pleistocene data older than 420 kyr. However, our model may partly apply to early and middle Pleistocene deposits, although the cyclic changes of the global climate are generally less pronounced. This condition has a major influence on sea ice distribution during the southern summer.

Glacial Termination and Peak Warm Interglacial

The short transition from glacial to interglacial induces substantial changes in the sedimentary environment of the Antarctic Ocean. Several processes are triggered by two most important changes: the rise of sea level and the increasing contribution of NADW to the CDW.

The rising sea level causes the marginal parts of the Antarctic ice sheet to float, thereby producing broad, unstable ice shelves (Figure 11). The concurrent recession of the grounding line is sedimentologically identified on the shelf by the facies boundary between basal till and glacial-marine sediment [Anderson et al., 1980]. A reduced sea ice coverage close to the continent also contributes to destabilization, as does the warmer CDW, which provides heat for increased basal melting [Potter and Paren, 1985; Hodell and Ciesielski, 1990]. Within a very short geological time, the large ice shelves are increasingly reduced by intense calving processes. A large number of icebergs transport sediments, frozen at their base from the shelf contact, to the deep sea. This intense ice rafting is the main sedimentary process during the initial phase of an interglacial and is responsible for an increase in sedimentation rates by nearly an order of magnitude above those of glacial conditions. Sedimentation rates decrease slightly when ice shelves are reduced and glacial conditions and the mass budget of the ice sheet are more stable.

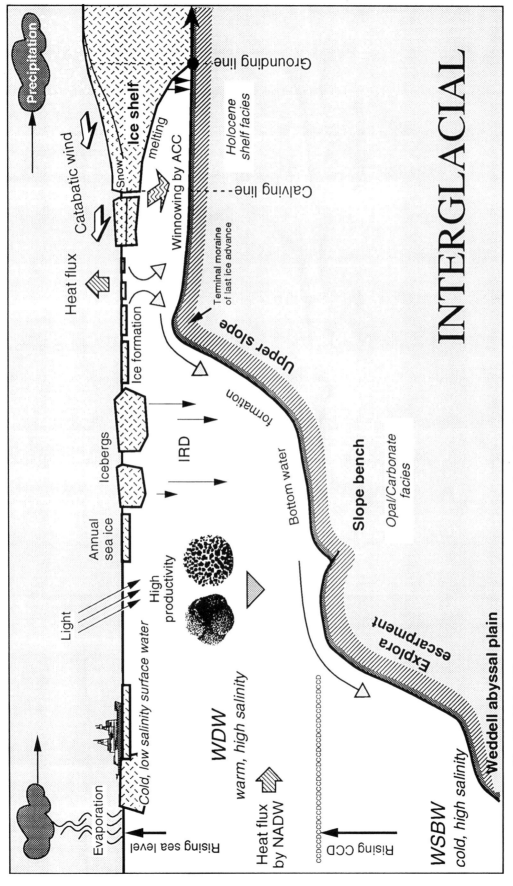

Fig. 11. Model of the environmental settings at the Antarctic continental margin during interglacial conditions. Included are the processes which are caused by the postglacial sea level rise, such as the retreat of the grounding line accompanied by intense calving and ice rafting. WDW, Weddell Deep Water (a branch of the Circumpolar Deep Water); NADW, North Atlantic Deep Water; WSBW, Weddell Sea Bottom Water; CCD, carbonate compensation depth; ACC, Antarctic Coastal Current; IRD, ice-rafted debris.

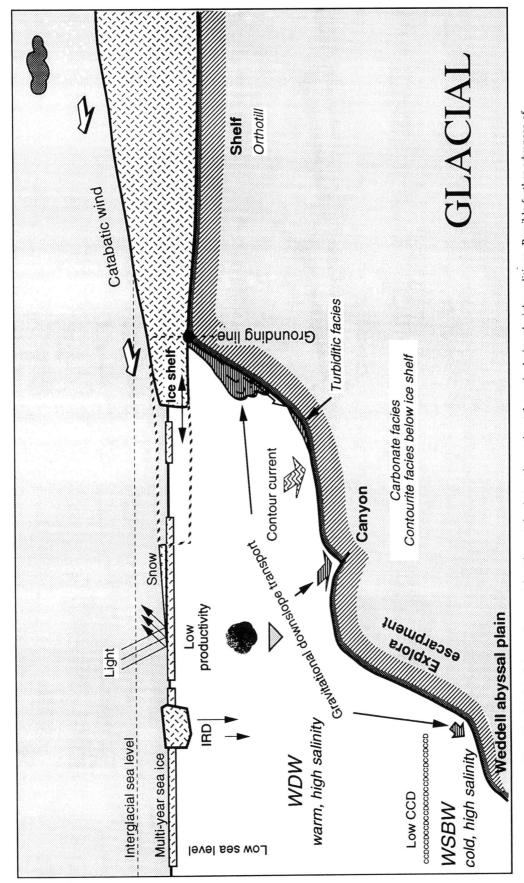

Fig. 12. Model of the environmental settings at the Antarctic continental margin during glacial conditions. Possible further advances of the ice margin during more severe glacial settings are indicated. Below those ice advances or below a closed sea ice cover, a contourite facies is deposited on the slope.

The contribution of NADW to Antarctic Ocean waters was much greater during interglacials than during glacials, at least over the past 550 kyr [e.g., *Duplessy et al.*, 1988; *Oppo et al.*, 1990]. With the glacial termination, the production of NADW in the North Atlantic recommences, feeding water to the Antarctic Ocean, with a short time delay of about 1 kyr. An increase in the flux of NADW, a major source of heat and salt, also has profound consequences for sedimentation processes. The upwelling of warmer CDW promotes the retreat of sea ice in spring. With increasing availability of light in the surface waters, primary production increases to peak values and so favors the burial of a significant amount of siliceous microfossils. When there is high productivity in surface waters, *N. pachyderma* contributes substantially to the plankton. But the high input of organic matter to the deep water increases the CO_2 content which, in turn, shoals the CCD. As a consequence, sediments of the peak warm intervals contain very few or no calcareous foraminifera (Figure 8). Additionally, higher export productivity favors secondary benthos activity, and consequently bioturbation is observed in sediment texture and grain size distribution (Figure 9).

The current regime at the continental margin became enhanced during peak warm times because the retreating continental ice edge and the reduction in sea ice favor wind-driven processes and oceanic circulation. Katabatic winds play an important role in the formation of coastal polynyas and hence sea ice formation above the shelf. This process, together with the increasing size of ice shelves, favors the formation of Ice Shelf Water (ISW), which flows down the slope thereby contributing to the formation of Weddell Sea Bottom Water (WSBW) [*Foldvik and Gammelsrød*, 1988]. These enhanced current activities are responsible for the deposition of relative coarse sediments, dominated by a coarse to medium silt. Very little clay contributes to the sedimentation, which there mainly consists of ice-rafted illite (Figure 10).

The response time of individual processes at the glacial termination differs. The lag of productivity might be only a few thousand years (3–5 kyr) and is dependent on the adjustment time of the oceanographic and sea ice conditions of the Antarctic Ocean. Close to the shelf break, the drastic changes in the terrigenous sedimentation stop when the postglacial ice shelves are reduced to a stable size. The retreat of the ice margin may last some thousand years.

The response time of those sedimentary processes which are directly linked to the mass budget and behavior of the Antarctic ice sheet might be longer (10–15 kyr). This can be observed by the general distribution of IRD in the slope sediments, which shows a significant delay in occurrence after glacial terminations and even after the following climatic optima (Figure 8). One explanation for the delayed response of specific sedi-

mentary processes might be the less sensitive behavior of the high southern latitude environment, where some glacial and hydrographic conditions may change more sluggishly and adjust to a changing climate with a time lag due to the presence of a huge ice sheet acting as a climatic buffer.

Moderate Interglacial to Glacial Transition

Toward the end of the interglacial warm period, the biogenic portion of the sediments changes from siliceous to calcareous microfossils. During a climatic cycle, large parts of the Antarctic Ocean are, for most of the time, continuously covered with ice, reducing the availability of light and hence primary production in the surface waters (Figure 12). The onset of calcareous sediment deposition in the Weddell Sea region is most probably linked to a decrease in the CO_2 content of deepwater masses. Lower export productivity due to reduced photosynthesis in surface waters lowers the CCD to about 4000 m and allows carbonate in sediments above this depth to be preserved. There will still be seasonal variations of the sea ice, but they are presumably restricted to the northern marginal areas of the Antarctic Ocean. In this context, continuous ice coverage means the presence of sea ice throughout the year but does not mean closed ice conditions. Short-term changes in the amount and distribution of sea ice are locally influenced by the intensity of katabatic winds, tidal currents, seasonal climatic variations, and presumably local insolation.

During the transition from interglacial to glacial, a delayed response of the sedimentary processes to climatic deterioration must be considered. A high amount of clay and fine silt is delivered by ice calving and rafting. These processes still contribute intensively to sedimentation because of continuing ice flow that is a result of the increased mass of the ice sheet during the former interglacial. In contrast, there is a much more rapid response to climatic cooling in the processes controlling bottom water formation. The production of cold ice shelf water and thus bottom current activity becomes reduced by increasing sea ice coverage and the lowering of sea level, which diminishes the extent of the ice shelves. The result of the superimposition of both intense ice rafting and lowered bottom current velocity is the deposition of much higher amounts of fine fractions in the post warm period sediments (Figures 8 and 9). In addition, peak values of illite in the clay fraction indicate that the terrigenous detritus is mainly delivered by ice rafting. A higher proportion of coarse IRD is found only at the beginning of this climatic transition. IRD decreases with the onset of the glacial maxima.

The composition of the facies deposited during the transition from moderate interglacial to early glacial conditions is quite similar. Only in this case, a decrease in carbonate content might indicate reduced productivity because of enhanced sea ice conditions. Very close to the continent, increasing gravitational transport down the slope, triggered by the advancing ice edge, is responsible for a higher amount of coarse detritus, changes in the composition of clay mineralogy, and an increase in sedimentation rates. In contrast, sedimentation rates decrease on the lower slope.

Glacial Maxima

During extensive glaciations on the northern hemisphere the Antarctic ice sheet also expands aerially. Thus during Pleistocene time the ice margins oscillate across the entire continental shelf, and the grounding lines repeatedly coincide with the shelf edge (Figure 12). Expansion and grounding of ice shelves occur in parts of East Antarctica as well as West Antarctica and take place predominantly in areas with broad continental shelves. The expansion is caused by sea level lowering as the result of the buildup of large ice sheets on Scandinavia, Siberia and North America. Thus the Antarctic ice sheet is more indirectly coupled to global cooling through processes occurring in the northern hemisphere.

From the sedimentological interpretations, no direct information can be revealed about the total mass budget of the Antarctic ice sheet. New results from a computer simulation of the behavior of the continental ice during the last climatic cycle show that in addition to the extending ice shelves, the total ice volume increases by about 10^{16} m^3 [*Huybrechts*, 1992]. This result is consistent with earlier findings [i.e., *Denton and Huges*, 1981]. The model also shows that fluctuations are primarily driven by changes in eustatic sea level. This causal relationship implies that changes of the ice sheet depend on the magnitude and duration of the sea level depression during the final stage of a glacial period [*Huybrechts*, 1992].

During glacials, most of the Antarctic Ocean is covered with ice, and temperatures are lower. This ice cover results in less evaporation there, and consequently there is less snow precipitation and accumulation on the continent. On the basis of ^{10}Be concentrations in the Vostok ice core, it is concluded that precipitation rates in the Antarctic were roughly halved during the last glaciation [*Yiou et al.*, 1985]. Even a thinning of the ice sheet has occurred in some areas [*Andersen*, 1990]. Furthermore, in the Transantarctic Mountains small local glaciers retreat during glacials, mainly because of changes in precipitation [*Denton et al.*, 1991].

The sea level lowering during glacial maxima makes shelf sediments available for ice erosion and subsequent redistribution and sedimentation on the upper slope. Gravitational sediment transport is the most active mechanism during sea level lowstands. Canyon systems such as the Wegener Canyon, branching up to the shelf

break, serve as preferred pathways for sediment transfer to the deep sea mainly by turbidity currents (Figure 12). Thus it was mainly during glacial episodes that deep-sea terrigenous sedimentation was strongly influenced by ice margin processes. As a result, depositional areas like the Weddell Fan [*Anderson et al.*, 1986] are fed with sediments.

Reduced NADW and thus a diminished heat transfer into the CDW have consequences on the amount of melting of both sea ice and ice shelves. During harsh climatic conditions the sea ice cover may be closed and attached to the ice shelf. Snow, mainly collected and delivered by katabatic winds from the inland ice, accumulates on the sea ice. This process adds to the overall thickness of the thick, multiyear ice. Thick and stable sea ice conditions help to increase the size of the ice shelf areas by stabilization of their margins and protect them against weakening by waves. Further expansion and stabilization of ice shelves is also favored by cooler shelf water, inhibiting basal melting. All processes may cause further advances of the ice shelf edge, up to 100 km from the shelf break, producing a broad ice shelf above the slope. During such peak glacial conditions, contourites are deposited on the slope below the ice (Figures 9 and 12).

Acknowledgments. We acknowledge the assistance and excellent cooperation of R/V *Polarstern*'s crew and master during the cruises. The authors are thankful to Hans Hubberten and Gerhard Kuhn for discussions and providing laboratory conditions. Laboratory work was undertaken mainly with the assistance of Margret Meyer zu Uptrup and Susanne Wiebe-Kawaletz. We thank Werner Ehrmann, Martin Melles, Mike Kendall, John Andrews, and an anonymous reviewer for improving the manuscript. Part of this work was supported by the Deutsche Forschungsgemeinschaft (grant Sp 296/1-1). This is publication 546 of the Alfred Wegener Institute for Polar and Marine Research and contribution 37 of the Sonderforschungsbereich 261 at Bremen University.

REFERENCES

Abelmann, A., R. Gersonde, and V. Spiess, Pliocene-Pleistocene paleoceanography in the Weddell Sea—Siliceous microfossil evidence, in *Geological History of Polar Oceans: Arctic Versus Antarctic, NATO/ASI Ser. C*, edited by U. Bleil and J. Thiede, pp. 729–759, Kluwer Academic Press, Dordrecht, Netherlands, 1990.

Andersen, B. G., Cenozoic glacier fluctuations in polar regions—Terrestrial records from Antarctica and the North Atlantic sector of the Arctic, in *Geological History of Polar Oceans: Arctic Versus Antarctic, NATO/ASI Ser. C*, edited by U. Bleil and J. Thiede, pp. 245–254, Kluwer Academic Press, Dordrecht, Netherlands, 1990.

Anderson, J. B., The marine geology of the Weddell Sea, Ph.D. thesis, *Contrib. 35*, 222 pp., Fla. State Univ., Tallahassee, 1972.

Anderson, J. B., Factors controlling $CaCO_3$ dissolution in the Weddell Sea from foraminiferal distribution patterns, *Mar. Geol., 19*, 315–332, 1975.

Anderson, J. B., D. D. Kurtz, and F. M. Weaver, Sedimentation on the Antarctic continental slope, in Geology of Continental Slopes, *Spec. Publ. Soc. Econ. Paleontol. Mineral.*, *27*, 265–283, 1979.

Anderson, J. B., D. D. Kurtz, E. W. Domack, and K. M. Balshaw, Glacial and glacial marine sediments of the Antarctic continental shelf, *J. Geol.*, *88*, 399–414, 1980.

Anderson, J. B., S. B. Davis, E. W. Domack, D. D. Kurtz, K. M. Balshaw, and R. Wright, Marine sediment core descriptions, IWSOE 68, 69, 70, Deep Freeze 79, 60 pp., Rice Univ., Houston, Tex., 1981.

Anderson, J. B., C. Brake, E. W. Domack, N. C. Myers, and J. Singer, Sedimentary dynamics of the Antarctic continental margin, in *Antarctic Earth Science*, edited by R. L. Oliver, P. R. James, and J. B. Jago, pp. 387–389, Australian Academy of Science, Canberra, 1983a.

Anderson, J. B., C. Brake, E. W. Domack, N. C. Myers, and R. Wright, Development of a polar glacial-marine sedimentation model from Antarctic Quarternary deposits and glaciological information, in *Glacial-Marine Sedimentation*, edited by B. F. Molina, pp. 233–264, Plenum, New York, 1983b.

Anderson, J. B., R. Wright, and B. Andrews, Weddell fan and associated abyssal plain, Antarctica: Morphology, sediment processes, and factors influencing sediment supply, *Geo Mar. Lett.*, *6*, 121–129, 1986.

Anderson, J. B., B. A. Andrews, L. R. Bartek, and E. M. Truswell, Petrology and palynology of Weddell Sea glacial sediments: Implications for subglacial geology, in *Geological Evolution of Antarctica*, edited by M. R. A. Thomson, J. A. Crame, and J. W. Thomson, pp. 231–235, Cambridge University Press, New York, 1991.

Andrews, B. A., Petrology of Weddell Sea glacial sediments: Implications for provenance and glacial history, *Antarct. J.*, *19*, 92–94, 1984.

Barker, P. F., et al., Leg 113, *Proc. Ocean Drill. Program Initial Rep.*, *113*, 785 pp., 1988.

Barker, P. F., et al., Leg 113, *Proc. Ocean Drill. Program Sci. Results*, 1033 pp., 1990.

Berger, A. L., Long-term variations of caloric insolation resulting from the Earth's orbital elements, *Quat. Res.*, *9*, 139–167, 1978.

Berger, W. H., and J. S. Killingley, Glacial-Holocene transition in deep sea carbonates: Selective dissolution and the stable isotope signal, *Science*, *197*, 563–566, 1977.

Brehme, I., Die Dokumentation der Bodenwasserströmung in den Sedimenten des nordwestlichen Weddell Meeres, *Rep. Polar Res.*, *110*, 92 pp., 1991.

Burckle, L. H., and N. Abrams, Regional Pliocene–early Pleistocene hiatuses in the Southern Ocean—Diatom evidence, *Mar. Geol.*, *77*, 207–218, 1987.

Carmack, E. C., and J. D. Foster, On the flow of water out of the Weddell Sea, *Deep Sea Res.*, *22*, 711–724, 1975.

Charles, C. D., and R. G. Fairbanks, Glacial to interglacial changes in isotopic gradients of Southern Ocean surface water, in *Geological History of Polar Oceans: Arctic versus Antarctic*, edited by U. Bleil and J. Thiede, pp. 729–760, NATO/ASI Series C: Dordrecht, the Netherlands (Kluwer Academic Press), 1990.

Cook, D. W., and J. D. Hays, Estimates of Antarctic Ocean seasonal sea-ice cover during glacial intervals, in *Antarctic Geoscience*, edited by C. Craddock, pp. 1017–1025, University of Wisconsin Press, Madison, 1982.

Cordes, D., Sedimentologie und Paläomagnetik an Sedimenten der Maudkuppe (Nordöstliches Weddellmeer), *Rep. Polar Res.*, *71*, 121 pp., 1990.

Corliss, B., Linkage of North Atlantic and Southern Ocean deep-water circulation during glacial intervals, *Nature*, *298*, 458–460, 1982.

Crowley, T. J., and C. L. Parkinson, Late Pleistocene varia-

tions in Antarctic sea ice, I, Effect of orbital insolation changes, *Clim. Dyn.*, *3*, 85–91, 1988.

Defelice, D. R., and S. W. Wise, Surface lithofacies, biofacies, and diatom diversity patterns as models for delineation of climatic change in the southeast Atlantic Ocean, *Mar. Micropaleontol.*, *6*, 29–70, 1981.

Denton, G. H., and T. J. Huges (Eds.), *The Last Great Ice Sheets*, 484 pp., John Wiley, New York, 1981.

Denton, G. H., R. L. Armstrong, and M. Stuiver, The late Cenozoic glacial history of Antarctica, in *The Late Cenozoic Glacial Ages*, edited by K. K. Turekian, pp. 267–306, Yale University Press, New Haven, Conn., 1971.

Denton, G. H., M. L. Prentice, and L. H. Burckle, Late Cenozoic history of the Antarctic ice sheet, in *The Geology of Antarctica*, edited by R. J. Tingley, pp. 365–433, Oxford University Press, New York, 1991.

Dieckmann, G. S., M. Spindler, M. A. Lange, S. F. Ackley, and H. Eicken, Antarctic sea ice: A habitat for the foraminifer Neogloboquadrina pachyderma, *J. Foraminiferal Res.*, *21*, 182–189, 1991.

Domack, D. J., Sedimentology of glacial and glacial-marine deposits on the George V–Adelie continental shelf, East Antarctica, *Boreas*, *1*, 79–97, 1982.

Domack, D. J., J. B. Anderson, and D. D. Kurtz, Clast shape as an indicator of transport and depositional mechanisms in glacial marine sediments: George V continental shelf, Antarctica, *J. Sediment. Petrol.*, *50*, 813–820, 1980.

Drewry, D. J., and A. P. Cooper, Processes and models of Antarctic glaciomarine sedimentation, *Ann. Glaciol.*, *2*, 117–122, 1981.

Dunbar, R. B., Sediment trap experiments on the Antarctic continental margin, *Antarct. J. U. S.*, *19*, 70–71, 1984.

Duplessy, J. C., N. J. Shackleton, R. G. Fairbanks, L. Labeyrie, D. Oppo, and N. Kalles, Deepwater source variations during the last climatic cycle and their impact on the global deepwater circulation, *Paleoceanography*, *3*, 343–360, 1988.

Elverhøi, A., Evidence for a late Wisconsin glaciation of the Weddell Sea, *Nature*, *293*, 641–642, 1981.

Elverhøi, A., and E. Roaldset, Glaciomarine sediments and suspended particulate matter, Weddell Sea shelf, Antarctica, *Polar Res.*, *1*, 1–21, 1983.

Foldvik, A., and T. Gammelsrød, Notes on Southern Ocean hydrography, sea-ice and bottom water formation, *Palaeogeogr. Palaeoclimatol. Palaeoecol.*, *67*, 3–17, 1988.

Foldvik, A., T. Gammelsrød, and T. Tørresen, Circulation and water masses on the southern Weddell Sea shelf, in *Oceanology of the Antarctic Continental Shelf, Antarct. Res. Ser.*, vol. 43, edited by S. S. Jacobs, pp. 5–20, AGU, Washington, D. C., 1985.

Fütterer, D. K., Die Expedition Antarktis IV mit FS *Polarstern* 1985/86, *Rep. Polar Res.*, *33*, 204 pp., 1987.

Fütterer, D. K., Die Expedition Antarktis VI mit FS *Polarstern* 1987/88, *Rep. Polar Res.*, *58*, 267 pp., 1988.

Fütterer, D. K., Die Expedition Antarktis VIII mit FS *Polarstern* 1989/1990, *Rep. Polar Res.*, *90*, 231 pp., 1991.

Fütterer, D. K., and M. Melles, Sediment patterns in the southern Weddell Sea: Filchner Shelf and Filchner Depression, in *Geological History of Polar Oceans: Arctic Versus Antarctic, NATO/ASI Ser. C*, edited by U. Bleil and J. Thiede, pp. 381–401, Kluwer Academic Press, Dordrecht, Netherlands, 1990.

Fütterer, D. K., H. Grobe, and S. Grünig, Quaternary sediment patterns in the Weddell Sea: Relations and environmental conditions, *Paleoceanography*, *3*, 551–561, 1988.

Fütterer, D. K., G. Kuhn, and H. W. Schenke, Wegener Canyon bathymetrie and results from rock dredging near ODP sites 691–693, eastern Weddell Sea, Antarctica, *Proc. Ocean Drill. Program Sci. Results*, *113*, 39–48, 1990.

Grobe, H., Spätpleistozäne Sedimentationsprozesse am antarktischen Kontinentalhang vor Kapp Norvegia, östliche Weddell See, *Rep. Polar Res.*, *27*, 120 pp., 1986.

Grobe, H., A simple method for the determination of ice-rafted debris in sediment cores, *Polarforschung*, *57*, 123–126, 1987.

Grobe, H., and G. Kuhn, Sedimentation processes at the Antarctic continental margin, in The Expedition Antarktis-IV of RV "POLARSTERN" 1985/86, *Rep. Polar Res.*, *33*, 80–84, 1987.

Grobe, H., D. K. Fütterer, and V. Spiess, Oligocene to Quaternary processes on the Antarctic continental margin, ODP Leg 113, Site 693, *Proc. Ocean Drill. Program Sci. Results*, *113*, 121–131, 1990*a*.

Grobe, H., A. Mackensen, H.-W. Hubberten, V. Spiess, and D. K. Fütterer, Stable isotope record and late Quaternary sedimentation rates at the Antarctic continental margin, in *Geological History of Polar Oceans: Arctic Versus Antarctic, NATO/ASI Ser. C*, edited by U. Bleil and J. Thiede, pp. 539–572, Kluwer Academic Press, Dordrecht, Netherlands, 1990*b*.

Haase, G. M., Glaciomarine sediments along the Filchner-Ronne Ice Shelf, southern Weddell Sea—First results of the 1983/1984 Antarktis-II/4 expedition, *Mar. Geol.*, *72*, 241–258, 1986.

Hambrey, M. J., and P. J. Barrett, The Cenozoic sedimentary and climatic record from the Ross Sea region of Antarctica, in *The Antarctic Paleoenvironment: A Perspective on Global Change, Part 2, Antarct. Res. Ser.*, edited by J. P. Kennett and D. A. Warnke, AGU, Washington, D. C., in press, 1992.

Hambrey, M. J., W. U. Ehrmann, and B. Larsen, Cenozoic glacial record of the Prydz Bay Continental Shelf, East Antarctica, *Proc. Ocean Drill. Program Sci. Results*, *119*, 77–132, 1991.

Hayes, D. E., et al., Leg 28, *Initial Rep. Deep Sea Drill. Proj.*, *28*, 1017 pp., 1975.

Hays, J. D., A review of the late Quaternary climatic history of Antarctic seas, in *Antarctic Glacial History and World Paleoenvironments*, edited by Z. Bakker, pp. 57–71, 1978.

Henriet, J. P., and H. Miller, Some speculations regarding the nature of the Explora-Andenes Escarpment, Weddell Sea, in *Geological History of Polar Oceans: Arctic Versus Antarctic, NATO/ASI Ser. C*, edited by U. Bleil and J. Thiede, pp. 163–169, Kluwer Academic Press, Dordrecht, Netherlands, 1990.

Hinz, K., and W. Krause, The continental margin of Queen Maud Land/Antarctica: Seismic sequences, structural elements, and geological development, *Geol. Jahrb.*, *E23*, 17–41, 1982.

Hodell, D. A., and P. F. Ciesielski, Southern Ocean response to the intensification of northern hemisphere glaciation at 2.4 Ma, in *Geological History of Polar Oceans: Arctic Versus Antarctic, NATO/ASI Ser. C*, edited by U. Bleil and J. Thiede, pp. 707–728, Kluwer Academic Press, Dordrecht, Netherlands, 1990.

Hollister, C. D., and R. B. Elder, Contour currents in the Weddell Sea, *Deep Sea Res.*, *16*, 99–101, 1969.

Huybrechts, P., The Antarctic ice sheet and environmental change: A three-dimensional modelling study, *Rep. Polar Res.*, *99*, 241 pp., 1992.

Imbrie, J., J. D. Hays, D. G. Martinson, A. MacIntyre, A. C. Mix, J. Morley, N. G. Pisias, W. L. Prell, and N. J. Shackleton, The orbital theory of Pleistocene climate: Support from a revised chronology of the marine $d^{18}O$ record, in *Milankovitch and Climate*, edited by A. Berger, J. Imbrie, J. Hays, G. Kukla, and B. Saltzmann, pp. 269–305, D. Reidel, Hingham, Mass., 1984.

Juckes, L. M., The geology of north-eastern Heimefrontfjella, Dronning Maud Land, *Br. Antarct. Surv. Sci. Rep.*, *65*, 1–44, 1972.

Kaharoeddin, F. A., M. R. Eggers, E. H. Goldstein, R. S. Graves, D. K. Watkins, J. A. Bergen, and S. C. Jones, ARA Islas Orcadas cruise 1578 sediment descriptions, *Contrib. 48*, Sediment. Res. Lab., Fla. State Univ., Tallahassee, 1980.

Kaul, N., Detaillierte seismische Untersuchungen am östlichen Kontinentalrand des Weddell-Meeres vor Kapp Norvegia, Antarktis, *Rep. Polar Res.*, *89*, 120 pp., 1991.

Kellogg, T. B., and D. E. Kellogg, Antarctic cryogenic sediments: Biotic and inorganic facies of ice shelf and marine-based ice sheet environments, *Palaeogeogr. Palaeoclimatol. Palaeoecol.*, *67*, 51–74, 1988.

Kellogg, T. B., R. S. Truesdale, and L. E. Ostermann, Late Quaternary extent of the West Antarctic Ice Sheet: New evidence from the Ross Sea cores, *Geology*, *7*, 249–253, 1979.

Keir, R. S., On the late Pleistocene ocean geochemistry and circulation, *Paleoceanography*, *3*, 413–445, 1988.

Kipfstuhl, J., Zur Entstehung von Unterwassereis und das Wachstum und die Energiebilanz des Meereises in der Atka Bucht, Antarktis, *Rep. Polar Res.*, *85*, 88 pp., 1991.

Kroopnick, P., The distribution of ^{13}C of ΣCO_2 in the world oceans, *Deep Sea Res.*, *32*, 57–84, 1985.

Kuhn, G., W. Ehrmann, M. Hambrey, M. Melles, and G. Schmiedl, Glacio-marine sedimentary processes in the Weddell Sea and Lazarev Sea, in Die Expeditionen ANTARKTIS IX/1-4 des Forschungsschiffes 'Polarstern' 1990/91, *Rep. Polar Res.*, *100*, 223–244, 1992.

Labeyrie, L. D., and J. C. Duplessy, Changes in the oceanic $^{13}C/^{12}C$ ratio during the last 140 000 years: high-latitude surface water records, *Palaeogeogr. Palaeoclimatol. Palaeoecol.*, *50*, 217–240, 1985.

Ledbetter, M. T., and P. F. Ciesielski, Post-Miocene disconformities and paleoceanography in the Atlantic sector of the Southern Ocean, *Palaeogeogr. Palaeoclimatol. Palaeoecol.*, *52*, 185–214, 1986.

Mackensen, A., H. Grobe, H.-W. Hubberten, V. Spiess, and D. K. Fütterer, Stable isotope stratigraphy from the Antarctic continental margin during the last one million years, *Mar. Geol.*, *87*, 315–321, 1989.

Mackensen, A., H. Grobe, G. Kuhn, and D. K. Fütterer, Benthic foraminiferal assemblages from the eastern Weddell Sea between 68 and 73°S: Distribution, ecology and fossilization potential, *Mar. Micropaleontol.*, *16*, 241–283, 1990.

Martin, J. H., Glacial-interglacial CO_2 change: The iron hypothesis, *Paleoceanography*, *5*, 1–13, 1990.

Martinson, D. G., N. G. Pisias, J. D. Hays, J. Imbrie, T. C. Moore, Jr., and N. J. Shackleton, Age dating and the orbital theory of the Ice Ages: Development of a high-resolution 0 to 300,000-year chronostratigraphy, *Quat. Res.*, *27*, 1–29, 1987.

Melles, M., Paläoglaziologie und Paläozeanographie im Spätquartär am Kontinentalrand des südlichen Weddellmeeres, Antarktis, *Rep. Polar Res.*, *81*, 190 pp., 1991.

Miller, H., and H. Oerter, Die Expedition Antarktis V mit FS Polarstern 1986/87, Bericht von den Fahrtabschnitten ANT V/4-5, *Rep. Polar Res.*, *57*, 1–57, 1990.

Miller, H., J. P. Henriet, N. Kaul, and A. Moons, A fine-scale seismic stratigraphy of the eastern margin of the Weddell Sea, in *Geological History of Polar Oceans: Arctic Versus Antarctic, NATO/ASI Ser. C*, edited by U. Bleil and J. Thiede, pp. 131–161, Kluwer Academic Press, Dordrecht, Netherlands, 1990.

Mix, A. C., and R. G. Fairbanks, North Atlantic surface-ocean control for Pleistocene deep-ocean circulation, *Earth Planet. Sci. Lett.*, *73*, 231–243, 1985.

Mortlock, R. A., C. D. Charles, P. N. Froelich, P. Zibello, J. Saltzman, J. D. Hays, and L. H. Burckle, Evidence for lower productivity in the Antarctic Ocean during the last glaciation, *Nature*, *351*, 220–222, 1991.

Oppo, D. W., and R. G. Fairbanks, Variability in the deep and intermediate water circulation of the Atlantic Ocean during the past 25,000 years: Northern Hemisphere modulation of the Southern Ocean, *Earth Planet. Sci. Lett.*, *86*, 1–15, 1987.

Oppo, D. W., R. G. Fairbanks, A. L. Gordon, and N. J. Shackleton, Late Pleistocene Southern Ocean d^{13}C variability, *Paleoceanography*, *5*, 43–54, 1990.

Orheim, O., and A. Elverhøi, Model for submarine glacial deposition, *Ann. Glaciol.*, *2*, 123–128, 1981.

Oskierski, W., Verteilung und Herkunft glazial-mariner Gerölle am Antarktischen Kontinentalrand des östlichen Weddellmeeres, *Rep. Polar Res.*, *47*, 132 pp., 1988.

Peters, M., Die Vulkanite im westlichen und mittleren Neuschwabenland, Vestfjella und Ahlmannryggen, Antarktika, *Rep. Polar Res.*, *61*, 186 pp., 1989.

Potter, J. R., and J. G. Paren, Interaction between ice shelf and ocean in George VI sound, Antarctica, in *Oceanology of the Antarctic Continental Shelf, Antarct. Res. Ser.*, vol. 43, edited by S. S. Jacobs, pp. 35–58, AGU, Washington, D. C., 1985.

Prell, W. L., J. Imbrie, D. G. Martinson, J. J. Morley, N. G. Pisias, N. J. Shackleton, and H. F. Streeter, Graphic correlation of oxygen isotope stratigraphy application to the late Quaternary, *Paleoceanography*, *1*, 137–162, 1986.

Pudsey, C. J., P. F. Barker, and N. Hamilton, Weddell Sea abyssal sediments: A record of Antarctic bottom water flow, *Mar. Geol.*, *81*, 289–314, 1988.

Rex, D. C., Age of dolerite from Dronning Maud Land, *Br. Antarct. Surv. Bull.*, *11*, 101, 1967.

Ruddiman, W. F., M. E. Raymo, D. G. Martinson, B. M. Clement, and J. Backman, Pleistocene evolution of northern hemisphere climate, *Paleoceanography*, *4*, 353–412, 1989.

Sarnthein, M., K. Winn, J.-C. Duplessy, and M. R. Fontugne, Global variations of surface ocean productivity in low and mid latitudes: Influence on CO_2 reservoirs of the deep ocean and atmosphere during the last 21,000 years, *Paleoceanography*, *3*, 361–399, 1988.

Savin, S. M., and R. G. Douglas, Stable isotope and magnesium geochemistry of recent planktonic foraminifera from the Southern Pacific, *Geol. Soc. Am. Bull.*, *84*, 2327–2342, 1973.

Singer, J. K., and J. B. Anderson, Use of total grain-size distributions to define bed erosion and transport for poorly sorted sediment undergoing simulated bioturbation, *Mar. Geol.*, *57*, 335–359, 1984.

Sundquist, E. T., and W. S. Broecker (Eds.), *The Carbon Cycle and Atmospheric CO_2: Natural Variations Archean to Present, Geophys. Monogr. Ser.*, vol. 32, 627 pp., AGU, Washington, D. C., 1985.

Thomas, R. H., The dynamics of the Brunt Ice Shelf, Coats Land, Antarctica, *Br. Antarct. Surv. Sci. Rep.*, *79*, 1973.

Voss, J., Zoogeographische und Gemeinschaftsanalyse des Makrozoobenthos des Weddellmeeres (Antarktis), *Rep. Polar Res.*, *45*, 130 pp., 1988.

Weaver, F., Pliocene paleoclimatic and paleoglacial history of East Antarctica recorded in deep-sea piston cores, *Contrib. 36*, Sediment. Res. Lab., Fla. State Univ., Tallahassee, 1973.

Wefer, G., G. Fischer, D. K. Fütterer, R. Gersonde, S. Honjo, and D. Ostermann, Particle sedimentation and productivity in Antarctic waters of the Atlantic sector, in *Geological History of Polar Oceans: Arctic Versus Antarctic, NATO/ ASI Ser. C*, edited by U. Bleil and J. Thiede, pp. 363–379, Kluwer Academic Press, Dordrecht, Netherlands, 1990.

Wollenburg, I., Sedimenttransport durch das arktische Meereis, Ph.D. thesis, 132 pp., Univ. of Kiel, Kiel, Germany, 1991.

Wright, R., and J. B. Anderson, The importance of sediment gravity flow to sediment transport and sorting in a glacial marine environment: Eastern Weddell Sea, Antarctica, *Geol. Soc. Am. Bull.*, *93*, 951–963, 1982.

Wu, G., J. C. Herguera, and W. H. Berger, Differential dissolution: Modification of late Pleistocene oxygen isotope records in the western equatorial Pacific, *Paleoceanography*, *5*, 581–594, 1990.

Yiou, F., G. M. Raisbeck, D. Bourles, C. Lorius, and N. I. Barkov, ^{10}Be in the ice at Vostok, Antarctica during the last climatic cycle, *Nature*, *316*, 616–617, 1985.

Zwally, H. J., J. C. Comiso, D. L. Parkinson, W. J. Campbell, F. D. Carsey, and P. Gloersen, Antarctic sea ice, 1973–1976: Satellite passive microwave observation, *NASA Spec. Publ.*, SP-459, 206 pp., 1983.

(Received August 25, 1991;
accepted June 1, 1992.)

PALEOECOLOGICAL IMPLICATIONS OF RADIOLARIAN DISTRIBUTION AND STANDING STOCKS VERSUS ACCUMULATION RATES IN THE WEDDELL SEA

DEMETRIO BOLTOVSKOY

*Departamento de Ciencias Biológicas, Facultad de Ciencias Exactas y Naturales, Universidad de Buenos Aires,
1428 Buenos Aires, Argentina
Consejo Nacional de Investigaciones Científicas y Tecnícas, Buenos Aires, Argentina*

VIVIANA A. ALDER

Instituto Antártico Argentino, 1010 Buenos Aires, Argentina

In the Scotia and Weddell seas, polycystine radiolarians dwell chiefly at depths between 200 and 300 m, their vertical patterns being strongly associated with the higher temperatures characteristic of the Warm Deep Water. At scales of approximately 400 to 2000 km and ~30 days, radiolarian horizontal quantitative distribution trends are not visibly affected by ice cover or primary production. On the other hand, comparison of polycystine standing stocks at 0 to 400 m versus their accumulation rates at 400 to 900 m indicates that >90% of the shells are lost to sedimentation. It is suggested that mechanical fragmentation by grazing (rather than dissolution) is primarily responsible for this loss. Deep habitat and high destruction rates in the water column are important factors which hinder the use of Antarctic polycystine thanatocoenoses for paleoecological reconstructions.

INTRODUCTION

As opposed to calcareous microfossils, the siliceous remains of silicoflagellates, diatoms, and radiolarians are abundant and widespread in Subantarctic and Antarctic sediments [*Lisitzyn*, 1974], especially since the Miocene [*Kennett*, 1978], and thus constitute a major tool for interpretation of past biogeographic and climatic settings [*Hays*, 1965; *Ciesielski and Weaver*, 1974; *Lozano and Hays*, 1976; *Burckle and Burak*, 1988]. However, the biogeographic patterns of these fossil assemblages are meaningful in ecological terms only as long as they reflect changes in the overlying waters, especially in the upper mixed layer, which is most affected by seasonal and long-term climatic shifts [*Boltovskoy*, 1988]. While less abundant than phytoplanktonic cells, siliceous radiolarian skeletons are usually larger and more robust than the former and are therefore thought to preserve better in the fossil record [*Milliman and Takahashi*, 1992]. Analyses of the distributional patterns of polycystine radiolarians in the plankton of the Weddell Sea and comparison with their accumulation rates in sediment trap collections suggest that associations between these deep-living protists and near-surface hydrology and primary production are weak and that radiolarian destruction rates in the upper 500–1000

m are very high. The mechanisms involved and their potential bearing on paleoceanographic surveys are discussed.

MATERIALS

Samples used for this study were collected from November 1988 through January 1989 in the area of the Weddell-Scotia Confluence (49°W, between 57°S and 61°30'S, 11 stations, 95 samples mainly from 0 to 400 m) and from January 1989 in the Weddell Sea (61°S through 78°S, 12 stations, 72 samples from 0 to 150 m) (Figure 1). About 10 L of water were retrieved from the chosen depths with Niskin and GoFlo bottles, filtered with a 15-μm mesh sieve and preserved with Lugol's acid solution. Counts were performed in 10- or 25-mL settling chambers under an inverted microscope. Detailed analyses of the distributional patterns of the dominant microplanktonic organisms recorded in these materials are reported elsewhere [*Alder and Boltovskoy*, 1991*b*; *Boltovskoy and Alder*, 1992; V. Alder and D. Boltovskoy, The ecology of microzooplankton in the Weddell-Scotia Confluence area: Horizontal and vertical distribution patterns, submitted to *Marine Ecology Progress Series*, 1992].

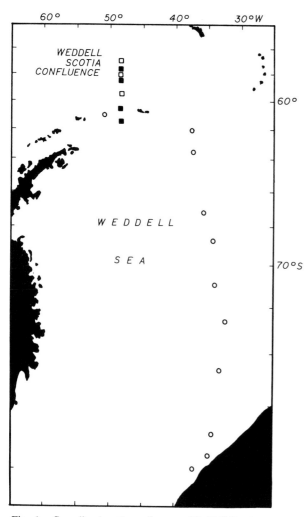

Fig. 1.   Sampling locations. Squares, Weddell-Scotia Confluence area, 0–400 m, with a few samples down to 1150 m (solid
squares denote sites with two stations). Circles, Weddell Sea,
0–150 m.

VERTICAL DISTRIBUTION OF RADIOLARIA AND RELATIONSHIPS WITH NEAR-SURFACE CONDITIONS

Figure 2 illustrates the vertical distribution of polycystines along the two cruise tracks. In the Weddell
Sea, where samplings were taken down to only 150 m,
radiolarian abundances increase consistently with
depth. In the Weddell-Scotia Confluence area, polycystine numbers increase down to 200 m, the depth of 400
m hosting somewhat lower numbers of shells. Their
average population depth is 240 m. While most previous
surveys of the vertical distribution of Antarctic Radiolaria were restricted to data from the upper 150–200 m,
our results are in general agreement with these earlier
reports insofar as sarcodine abundances are highest

away from the surface [*Morley and Stepien*, 1985;
Gowing, 1989; *Gowing and Garrison*, 1991].

Because percentages of empty radiolarian shells increase sharply with depth [*Petrushevskaya*, 1971; *Gowing*, 1986; *Gowing and Coale*, 1989], it could be argued
that the high numbers of these deep-dwelling sarcodines
recorded in our Antarctic profiles are largely represented by settling organisms, already devoid of proto-
plasm, rather than by in situ living populations. We
anticipate, however, that the concentrations recorded
down to 400 m are mainly represented by live cells,
rather than discarded skeletons. *Nöthig and Gowing*
[1991], for example, found that, on the average, only
20% of the polycystine shells retrieved from the 200- to
500-m stratum (Weddell Sea) are empty skeletons. Furthermore, high percentages of empty skeletons at
deeper levels would require high standing stocks closer
to the surface; our profiles do not show noticeable peaks
above the layers of maximum shell concentrations (Figure 2), which again suggests that the organisms retrieved represent in situ populations.

The above results indicate that in the Weddell Sea,
polycystine peak concentrations are associated with
layers deeper than 150–200 m, where the Warm Deep
Water dominates, characterized by higher temperatures
than the overlying strata [*Gordon*, 1967; *Carmack and
Foster*, 1977; *Morley and Stepien*, 1985; *Nöthig et al.*,
1991; *Nöthig and Gowing*, 1991]. This relationship,
which was also noticed during the winter [*Gowing and
Garrison*, 1991], is best reflected by the highly significant correlation between radiolarian abundances and
water temperature (Figure 3). On the other hand, as
opposed to most other microplanktonic organisms,
which react conspicuously to ice cover, at our sampling
scales radiolarians (as well as foraminifers) show less
change in association with the ice edge (Figure 4).
Integrated radiolarian and foraminiferal abundances
were not associated with chlorophyll *a* (Figure 5).

The two uncouplings (i.e., radiolarian abundance
versus ice cover and primary production) have been
noticed before in Antarctic waters [*Gowing*, 1989; *Gowing and Garrison*, 1991; *Boltovskoy and Alder*, 1992; V.
Alder and D. Boltovskoy, submitted 1992]. Lack of
association with the horizontal and vertical distribution
of chlorophyll *a* is especially interesting because in
extrapolar waters both radiolarians and foraminifers
have been reported to concentrate in areas of chlorophyll maxima [*Kling*, 1979; *Fairbanks and Wiebe*, 1980;
Dworetzky and Morley, 1987]. We speculate that the
deeper distribution of radiolarian in the Antarctic is
probably a response to lower competition with other
microzooplankters (chiefly the very abundant heterotro-
pic dinoflagellates and ciliates [*Boltovskoy et al.*, 1989;
Alder and Boltovskoy, 1991a]) and to higher water
temperatures. The near-surface primary production signal, which dissipates with depth, in turn accounts for

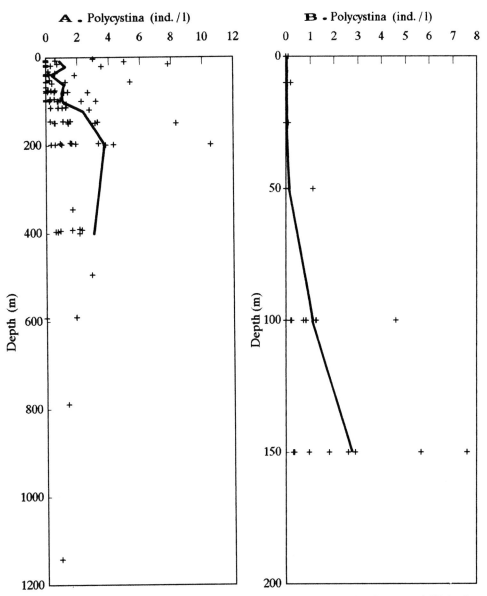

Fig. 2.   Vertical distribution of polycystines (*a*) in the Weddell-Scotia Confluence and (*b*) in the Weddell Sea (refer to Figure 1). Actual data points and average values (line).

blurred relationships with phytoplanktonic patches and the ice edge.

Given the depth at which radiolarians dwell in the Weddell Sea and their lack of association (at these short-term temporal scales) with ice coverage and primary production at the surface, it seems reasonable to assume that the environmental signal of their resulting thanatocoenoses does not lend itself to simple interpretations. For one thing, this signal is probably warmer than the corresponding upper mixed layer, not only because of the temperature regime at their preferred

depth, but also because the Warm Deep Water originates at lower latitudes and can laterally advect subtropical radiolarian shells to the Antarctic zone. *Abelmann and Gersonde* [1991] and *Abelmann* [1992] recorded up to 80% of transitional to subtropical polycystines in sediment trap samples deployed at 700 m in the area of the Polar Front (approximately 50°S, 6°E) and significant numbers in traps from several locations as far south as 65°S. Interestingly, all these traps were located below 360 m, and none of those deployed closer to the surface collected extrapolar species. The degree to which this

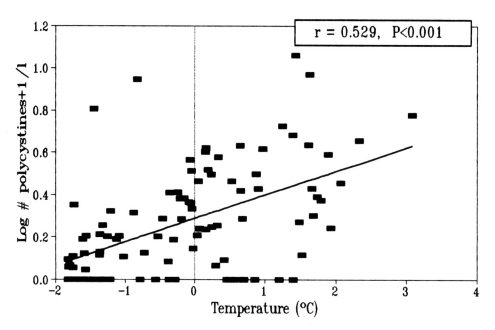

Fig. 3.   Linear regression and actual data points for polycystine abundances versus temperature in the Weddell-Scotia Confluence area (data from 0 to 400 m).

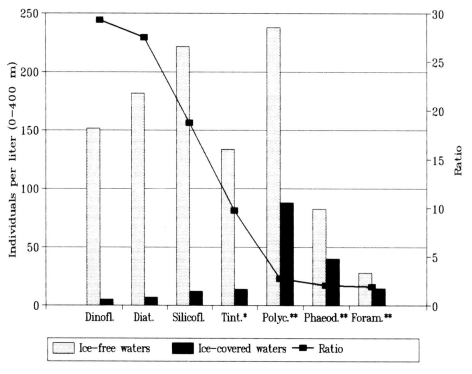

Fig. 4.   Average abundances of selected microplanktonic groups in ice-free versus ice-covered (>70%) waters in the Weddell-Scotia Confluence area. Diatom data are for 0–100 m. Asterisk, ×10; double asterisk, ×100.

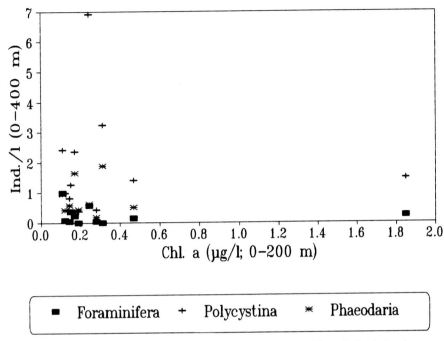

Fig. 5.   Relationships between chlorophyll *a* and radiolarian and foraminiferal abundances.

"contamination" can bias sedimentary assemblages is largely unknown because of our lack of adequate data on radiolarian distributions in the plankton.

RADIOLARIAN STANDING STOCKS VERSUS ACCUMULATION RATES

Data illustrated in Table 1 summarize some recent Weddell Sea estimates of polycystine abundances in the

TABLE 1.   Estimate of Radiolarian Loss to Sedimentation in the Weddell Sea

	Summer	Winter
Polycystines/m^2 in O to 400-m water column	611,000	7600
Radiolarians/m^2/day in sediment trap at 400–900 m	144–1111	7–8
Estimated polycystine turnover rates, years		
Maximum	11.6	2.9
Minimum	1.5	2.6
Estimated polycystine loss in the water column assuming a life span of 1.5 months, %	92–99	95

Plankton data are from expeditions shown in Figure 1 (summer) and from *Gowing and Garrison* [1991] (winter; approximately 57°S to 62°S, 35°W to 49°W, June–August 1988, 17 stations sampled from 5 to 210 m). Sediment trap data are from *Abelmann* [1992] and *Abelmann and Gersonde* [1991] (year-round time series traps deployed at 62°26.5'S, 34°45.5'W, 863 m (WS1) and 64°53.1'S, 2°33.7'W, 360 m (WS3)).

water column and their accumulation rates in sediment trap at 360 and 863 m. During the summer, average concentrations down to 400 m are over 600,000 individuals/m^2, while in the winter the figure drops to about 8000 shells/m^2 (the latter value includes only individuals >50 μm and is therefore underestimated). Sediment trap data from depths of 400 to 900 m yield approximately 100–1000 shells/m^2/day in the summer and 7–8 shells/m^2/day in the winter. According to these numbers, assuming a 100% retrieval of polycystine shells in the sediment traps, it would take 1.5 to 11.6 years for the entire summer population to double [see *Takahashi*, 1983]. A similar analysis for the winter yields values of 2.6 to 2.9 years. In other words, these figures indicate that the average turnover rate of a radiolarian specimen would vary between 1.5 and 11.6 years. Such life spans are clearly far too large for the radiolarians, which most probably take less than 1 month to reach maturity and reproduce [*Casey et al.*, 1970; *Anderson*, 1983; *Caron and Swanberg*, 1990]. A similar estimate for the silicoflagellates also yields unrealistically high life spans (average concentration in 0–100 m of 40,000,000 cells/m^2, peak sedimentation rate of 400,000 cells/m^2/day; average turnover rate of over 3 months). Thus assuming a life span of 1.5 months for the polycystines, these destruction rates would range around 92–99%, which means that only 1 to 8% of the shells produced in the upper 300 m reach 400–900 m in an identifiable fashion (Table 1).

Admittedly, these data are preliminary and spotty, and the production versus accumulation ratios derived might be somewhat biased. For example, counts of planktonic materials were mostly carried out on samples filtered through 15- or 20-μm mesh sieves, while sediment trap ones were filtered through 44 μm, thus missing the smaller species and earlier developmental stages. One of the sediment trap deployments (WS3, see Table 1) was located rather far away from the area covered by the plankton samplings, which can account for regional differences in radiolarian production and for lateral subsurface advection of radiolarian skeletons [*Boltovskoy*, 1988]. Other potential sources of error are inclusion of dead individuals in the radiolarian standing stocks in the water column and spotty planktonic coverage inadequately representing long-term averages. In addition, estimates of polycystine life spans are indirect and rough and might be higher at the low Antarctic temperatures. Nevertheless, the imbalance seems too high to be readily dismissed as an artifact. If an average life span of 1–2 months is correct, for 0% destruction/ dissolution in the upper 400 m the ratio standing stock in the plankton (in individuals per cubic meter) versus accumulation rate at 400 m (in individuals/m^2/yr) should be around 1:5000 to 1:10,000, while in the data presented here it is roughly 1:90, that is, 50 to 100 times higher than expected. Interestingly, a similar analysis for the radiolarians of the California Current (based on planktonic data from *Boltovskoy and Riedel* [1987] and sediment trap results from *Gowing and Coale* [1989]) yields doubling times of approximately 15 days, which is remarkably close to figures calculated by different methods [*Caron and Swanberg*, 1990].

Several previous investigations concluded that up to 80% of the biogenic silica produced in surface Antarctic waters dissolves above 100–200 m [*Kozlova*, 1961; *Nelson and Gordon*, 1982; *Leventer and Dunbar*, 1987]. These high rates have been chiefly attributed to the diatoms, whose frustules are lighter and more susceptible to dissolution than radiolarian skeletons. On the other hand, comparisons of trapped biogenic silica at various depths in the water column and on the seafloor suggest that, with the exception of the Phaeodaria, most siliceous remains do reach the bottom, and dissolution, which affects over 90% of total siliceous flux, occurs at the water-sediment interface (*Gersonde and Wefer* [1987]; see the review of *Milliman and Takahashi* [1992]).

The most plausible explanation for this inconsistency is probably destruction of radiolarian shells by zooplanktonic grazing, rather than dissolution in the water column; although the skeletal fragments are not identified in sample counts, they do contribute to total biogenic silica as measured chemically. It is probable that krill, which feeds chiefly on microplankton-sized particles [*Quetin and Ross*, 1985], plays a major role in this process and is responsible for these higher proportions of destructive grazing in the Antarctic than elsewhere [*Peinert et al.*, 1989]. *Gersonde and Wefer* [1987] and

Abelmann and Gersonde [1991] noticed that mechanical breakdown affects the largest diatom frustules very strongly, grazers' fecal pellets containing almost exclusively unidentifiable fragments. *Leventer and Dunbar*'s [1987] data also support this conclusion: in their sediment trap samples from McMurdo Sound they recorded a 47–79% drop in the number of diatom frustules between 34 and 220 m, while opal flux decreased only 13–40% in the same interval. Although sediment trap results from the Antarctic [e.g., *Abelmann*, 1992] and elsewhere (see the reviews of *Wefer* [1989] and *Milliman and Takahashi* [1992]) do not seem to endorse our conclusion, over 90% of all trap deployments carried out to date have been located at depths below 300 m, where mesozooplanktonic and macrozooplanktonic abundances drop sharply [e.g., *Vinogradov*, 1968]. Thus steepest changes in the degree of preservation of the microplanktonic remains are expected to take place in the upper 100 to 400 m, where most of the grazers are concentrated; farther down mechanical fragmentation of the shells would be much less important, which could account for the lack of clear differences between levels below approximately 0.5 km.

While the argument for destructive radiolarian grazing, most probably by krill, is supported by much indirect evidence, the scarcity of data at hand does not allow dismissing alternative or complementary hypotheses which could contribute to explain the standing stock versus sedimentation rate imbalance reviewed. As was noticed above, unmatching plankton versus sediment trap locations and questionable radiolarian doubling rates are probably important points which have to be addressed in order to confirm our conclusions. On the other hand, vertical stability of the water column, especially near the ice edge (where, incidentally, most planktonic collections were carried out), could slow down radiolarian sedimentation, thus contributing to the enhancement of dissolution in the upper water column.

An important implication of high destruction rates by grazing in the upper part of the water column is that, because it is a biologically controlled phenomenon, its incidence will vary greatly taxonomically, geographically, vertically, and seasonally, thus restricting our capability of drawing broad generalizations regarding its influence on plankton-sediment relationships. As was noted above, destruction of shells by grazing is highly selective, larger and more fragile shells being affected more than the smaller and robust ones [*Honjo*, 1976; *Gersonde and Wefer*, 1987]. Furthermore, different radiolarian depth habitats can favor the destruction of the shallower-living taxa (where zooplanktonic grazing is stronger), thus enhancing the contribution of the deeper (and, most probably, less environment sensitive) ones in the sediments. Destruction of shells by grazing would also be subject to large seasonal fluctuations due to differences in the physiological state of the consumers and to their seasonal vertical migrations [e.g., *Lancraft et al.*, 1991].

Acknowledgments. The Instituto Antártico Argentino, the Alfred Wegener Institut (Germany), and the European Science Foundation financed and operated the cruises on which materials were collected. Irene Schloss and Andrea Abelmann put at our disposal unpublished data. Joseph Morley and Kozo Takahashi offered valuable suggestions for improving an earlier draft of the manuscript. The senior author's participation in the International Conference on the Role of the Southern Ocean in Global Change: An Ocean Drilling Perspective was made possible thanks to financial aid from the U.S. National Science Foundation, channeled through J. P. Kennett.

REFERENCES

Abelmann, A., Radiolarian flux in Antarctic waters (Drake Passage, Powell Basin, Bransfield Strait), *Polar Biol.*, in press, 1992.

Abelmann, A., and R. Gersonde, Biosiliceous particle flux in the Southern Ocean, *Mar. Chem.*, 35, 503–536, 1991.

Alder, V. A., and D. Boltovskoy, Microplanktonic distributional patterns west of the Antarctic Peninsula, with special emphasis on the tintinnids, *Polar Biol.*, 11, 103–112, 1991a.

Alder, V. A., and D. Boltovskoy, The ecology and biogeography of tintinnid ciliates in the Atlantic sector of the Southern Ocean, *Mar. Chem.*, 35, 337–346, 1991b.

Anderson, O. R., *Radiolaria*, 355 pp., Springer-Verlag, New York, 1983.

Boltovskoy, D., Equatorward sedimentary shadows of near-surface oceanographic patterns, *Speculations Sci. Technol.*, 11, 219–232, 1988.

Boltovskoy, D., and V. A. Alder, Microzooplankton and tintinnid species-specific assemblage structures: Patterns of distribution and year-to-year variations in the Weddell Sea (Antarctica), *J. Plankton Res.*, 14(10), 1992.

Boltovskoy, D., and W. R. Riedel, Polycystine Radiolaria of the California Current area: Seasonal and geographic patterns, *Mar. Micropaleontol.*, 12, 65–104, 1987.

Boltovskoy, D., V. A. Alder, and F. Spinelli, Summer Weddell Sea microplankton: Assemblage structure, distribution and abundance, with special emphasis on the Tintinnina, *Polar Biol.*, 9, 447–456, 1989.

Burckle, L. H., and R. W. Burak, Fluctuations in late Quaternary diatom abundances: Stratigraphic and paleoclimatic implications from Subantarctic deep-sea cores, *Palaeogeogr. Palaeoclimatol. Palaeoecol.*, 67, 147–156, 1988.

Carmack, E. C., and T. D. Foster, Water masses and circulation in the Weddell Sea, in *Polar Oceans*, edited by M. J. Dunbar, pp. 151–166, Arctic Institute of North America, 1977.

Caron, D. A., and N. R. Swanberg, The ecology of planktonic sarcodines, *Rev. Aquat. Sci.*, 3, 147–180, 1990.

Casey, R. E., T. M. Partridge, and J. R. Sloan, Radiolarian life spans, mortality rates, and seasonality gained from Recent sediment and plankton samples, in *Proceedings of the Second Planktonic Conference*, edited by A. Farinacci, pp. 159–165, Tecnoscienza, Rome, 1970.

Ciesielski, P. F., and F. M. Weaver, Early Pliocene temperature changes in the Antarctic seas, *Geology*, 2, 511–515, 1974.

Dworetzky, B. A., and J. J. Morley, Vertical distribution of Radiolaria in the eastern equatorial Atlantic: Analysis of a multiple series of closely spaced plankton tows, *Mar. Micropaleontol.*, 12, 1–19, 1987.

Fairbanks, R. G., and P. H. Wiebe, Foraminifera and chloro-phyll maximum: Vertical distribution, seasonal succession, and paleoceanographic significance, *Science*, 209, 1524–1526, 1980.

Gersonde, R., and G. Wefer, Sedimentation of biogenic siliceous particles in Antarctic waters from the Atlantic sector, *Mar. Micropaleontol.*, 11, 311–332, 1987.

Gordon, A. L., Structure of Antarctic waters between 20°W and 170°W, *Antarct. Map Folio Ser.*, folio 6, 10 pp., Am. Geogr. Soc., New York, 1967.

Gowing, M. M., Trophic biology of phaeodarian radiolarians and flux of living radiolarians in the upper 2000 m of the North Pacific central gyre, *Deep Sea Res.*, 33, 655–674, 1986.

Gowing, M. M., Abundance and feeding ecology of Antarctic phaeodarian radiolarians, *Mar. Biol.*, 103, 107–118, 1989.

Gowing, M. M., and S. L. Coale, Fluxes of living radiolarians and their skeletons along a northeast Pacific transect from coastal upwelling to open ocean waters, *Deep Sea Res.*, 36, 561–576, 1989.

Gowing, M. M., and D. L. Garrison, Austral winter distributions of large tintinnid and large sarcodinid protozooplankton in the ice-edge zone of the Weddell/Scotia seas, *J. Mar. Syst.*, 2, 131–141, 1991.

Hays, J. D., Radiolaria and late Tertiary and Quaternary history of Antarctic seas, in *Biology of the Antarctic Seas II*, *Antarct. Res. Ser.*, vol. 5, edited by G. Llano, pp. 125–184, AGU, Washington, D. C., 1965.

Honjo, S., Coccoliths: Production, transportation and sedimentation, *Mar. Micropaleontol.*, 1, 65–79, 1976.

Kennett, J. P., The development of planktonic biogeography in the Southern Ocean during the Cenozoic, *Mar. Micropaleontol.*, 3, 301–345, 1978.

Kling, S. A., Vertical distribution of Polycystine radiolarians in the central North Pacific, *Mar. Micropaleontol.*, 4, 295–318, 1979.

Kozlova, O. G., Quantitative contents of diatoms in waters of the Indian sector of the Antarctic, *Dokl. Akad. Nauk SSSR*, 138, 107–210, 1961.

Lancraft, T. M., T. L. Hopkins, J. J. Torres, and J. Donnelly, Oceanic micronektonic/macrozooplanktonic community structure and feeding in ice covered Antarctic waters during the winter (AMERIEZ 1988), *Polar Biol.*, 11, 157–167, 1991.

Leventer, A., and R. B. Dunbar, Diatom flux in McMurdo Sound, Antarctica, *Mar. Micropaleontol.*, 12, 49–64, 1987.

Lisitzyn, A. P., *Sediment Formation in the Oceans*, 438 pp., Nauka, Moscow, 1974.

Lozano, J. A., and J. D. Hays, Relationship of radiolarian assemblages to sediment types and physical oceanography in the Atlantic and western Indian sectors of the Antarctic Ocean, *Mem. Geol. Soc. Am.*, 145, 303–336, 1976.

Milliman, J. D., and K. Takahashi, Carbonate and opal production and accumulation in the ocean, in *Global Surficial Geofluxes: Modern to Glacial*, edited by T. M. Usselman, W. Hay, and M. Meybeck, National Academy Press, Washington, D. C., in press, 1992.

Morley, J. J., and J. C. Stepien, Antarctic Radiolaria in late winter/early spring Weddell Sea waters, *Micropaleontology*, 31, 365–371, 1985.

Nelson, D. M., and L. I. Gordon, Production and pelagic dissolution of biogenic silica in the Southern Ocean, *Geochim. Cosmochim. Acta*, 46, 491–501, 1982.

Nöthig, E. M., and M. M. Gowing, Abundance and distribution of large protozooplankton and nauplii in the Weddell Sea, Antarctica in late winter, with special emphasis on phaeodarian radiolarians and their feeding ecology, *Mar. Biol.*, 111, 473–484, 1991.

Nöthig, E. M., U. Bathmann, E. Fahrbach, R. Gradinger, J. C. Jennings, L. I. Gordon, and R. Makarov, Regional relationships between biological and hydrographical properties in the Weddell Gyre in late austral winter 1989, *Mar. Chem.*, *35*, 325–336, 1991.

Peinert, R., B. von Bodungen, and V. Smetacek, Food web structure and loss rate, in *Productivity of the Ocean: Present and Past*, edited by W. H. Berger, V. S. Smetacek, and G. Wefer, pp. 35–48, John Wiley, New York, 1989.

Petrushevskaya, M. G., Spumellarian and Nassellarian Radiolaria in the plankton and bottom sediments of the central Pacific, in *The Micropaleontology of Oceans*, edited by B. M. Funnell and W. R. Riedel, pp. 309–318, Cambridge University Press, New York, 1971.

Quetin, L. B., and R. M. Ross, Feeding by Antarctic krill, *Euphausia superba*: Does size matter?, in *Antarctic Nutrient Cycles and Food Webs*, edited by W. Siegfried, P. Condy,

and R. Laws, pp. 199–204, Springer-Verlag, New York, 1985.

Takahashi, K., Radiolaria: Sinking population, standing stock, and production rate, *Mar. Micropaleontol.*, *8*, 171–181, 1983.

Vinogradov, M. E., *Vertical Distribution of Oceanic Zooplankton*, 320 pp., Nauka, Moscow, 1968.

Wefer, G., Particle flux in the ocean: Effects of episodic production, in *Productivity of the Ocean: Present and Past*, edited by W. H. Berger, V. S. Smetacek, and G. Wefer, pp. 139–153, John Wiley, New York, 1989.

(Received October 26, 1991;
accepted March 25, 1992.)

DEEP SEA DRILLING PROJECT AND OCEAN DRILLING PROGRAM CRUISES SHOWING THE TWO OR THREE CO-CHIEF SCIENTISTS ON EACH LEG

Initial Reports of the Deep Sea Drilling Project are available from the U.S. Government Printing Office, Washington, D. C.

Initial Reports and Scientific Results of the Ocean Drilling Program are available from The Ocean Drilling Program, College Station, Texas.

Barker, P. F., R. L. Carlson, D. A. Johnson, et al., *Initial Rep. Deep Sea Drill. Proj.*, 72, 1024 pp., 1983.

Barker, P. F., I. W. D. Dalziel, et al., *Initial Rep. Deep Sea Drill. Proj.*, 36, 1079 pp., 1977.

Barker, P. F., J. P. Kennett, et al., *Proc. Ocean Drill. Program Initial Rep.*, 113, 785 pp., 1988.

Barker, P. F., J. P. Kennett, et al., *Proc. Ocean Drill. Program Sci. Results*, 113, 1033 pp., 1990.

Barron, J. A., B. Larsen, et al., *Proc. Ocean Drill. Program Initial Rep.*, 119, 42 pp., 1989.

Barron, J. A., B. Larsen, et al., *Proc. Ocean Drill. Program Sci. Results*, 119, 1003 pp., 1991.

Bolli, H. M., W. B. F. Ryan, et al., *Initial Rep. Deep Sea Drill. Proj.*, 40, 1079 pp., 1978.

Bougault, H., S. C. Cande, et al., *Initial Rep. Deep Sea Drill. Proj.*, 82, 667 pp., 1985.

Burns, R. E., J. E. Andrews, et al., *Initial Rep. Deep Sea Drill. Proj.*, 21, 931 pp., 1973.

Ciesielski, P. F., Y. Kristoffersen, et al., *Proc. Ocean Drill. Program Initial Rep.*, 114, 815 pp., 1988.

Ciesielski, P. F., Y. Kristoffersen, *Proc. Ocean Drill. Program Sci. Results*, 114, 826 pp., 1991.

Davies, P. J., J. A. McKenzie, A. Palmer-Julson, et al., *Proc. Ocean Drill. Program Sci. Results*, 133, in press, 1992.

Davies, T. A., B. P. Luyendyk, et al., *Initial Rep. Deep Sea Drill. Proj.*, 26, 1129 pp., 1974.

De Graciansky, P. C., C. W. Poag, et al., *Initial Rep. Deep Sea Drill. Proj.*, 80, 1258 pp., 1985.

Eldholm, O., J. Thiede, E. Taylor, et al., *Proc. Ocean Drill. Program Sci. Results*, 104, 1141 pp., 1989.

Gradstein, F. M., J. N. Ludden, et al., *Proc. Ocean Drill. Program Initial Rep.*, 123, 716 pp., 1990.

Haq, B. U., U. von Rad, et al., *Proc. Ocean Drill. Program Initial Rep.*, 122, 826 pp., 1990.

Hay, W. W., J.-C. Sibuet, et al., *Initial Rep. Deep Sea Drill. Proj.*, 75, 1303 pp., 1984.

Hayes, D. E., L. A. Frakes, et al., *Initial Rep. Deep Sea Drill. Proj.*, 28, 1017 pp., 1975.

Heath, G. R., L. H. Burckle, et al., *Initial Rep. Deep Sea Drill. Proj.*, 86, 804 pp., 1985.

Hollister, C. D., C. Craddock, et al., *Initial Rep. Deep Sea Drill. Proj.*, 35, 129 pp., 1976.

Hsü, K. J., J. L. La Breque et al., *Initial Rep. Deep Sea Drill. Proj.*, 73, 798 pp., 1984.

Kennett, J. P., R. E. Houtz, et al., *Initial Rep. Deep Sea Drill. Proj.*, 29, 1197 pp., 1975.

Lancelot, Y., E. Seibold, et al., *Initial Rep. Deep Sea Drill. Proj.*, 41, 1259 pp., 1978.

Larson, R. L., R. Moberly, et al., *Initial Rep. Deep Sea Drill. Proj.*, 32, 980 pp., 1975.

Laughton, A. S., W. A. Berggren, et al., *Initial Rep. Deep Sea Drill. Proj.*, 12, 1243 pp., 1972.

Ludwig, W. J., V. A. Krasheninnikov, et al., *Initial Rep. Deep Sea Drill. Proj.*, 71, 1187 pp., 1983.

Maxwell, A. E., P. P. von Herzen, et al., *Initial Rep. Deep Sea Drill. Proj.*, 3, 806 pp., 1970.

Mayer, L. A., F. Theyer, et al., *Initial Rep. Deep Sea Drill. Proj.*, 85, 1021 pp., 1985.

Montadert, L., D. G. Roberts, et al., *Initial Rep. Deep Sea Drill. Proj.*, 48, 1183 pp., 1979.

Moore, T. C., Jr., P. D. Rabinowitz, et al., *Initial Rep. Deep Sea Drill. Proj.*, 74, 894 pp., 1984.

Perch-Nielsen, K., P. R. Supko, et al., *Initial Rep. Deep Sea Drill. Proj.*, 39, 1139 pp., 1977.

Pierce, J., J. Weissel, et al., *Proc. Ocean Drill. Program Initial Rep.*, 121, 1000 pp., 1989.

Prell, W. L., J. V. Gardner, et al., *Initial Rep. Deep Sea Drill. Proj.*, 68, 495 pp., 1982.

Roberts, D. G., D. Schnitker, et al., *Initial Rep. Deep Sea Drill. Proj.*, 81, 923 pp., 1984.

Ruddiman, W. F., R. B. Kidd, E. Thomas, et al., *Initial Rep. Deep Sea Drill. Proj.*, 94, 1261 pp., 1987.

Ruddiman, W. F., M. Sarnthein, et al., *Proc. Ocean Drill. Program Sci. Results*, 108, 519 pp., 1989.

Schlanger, S. O., E. D. Jackson, et al., *Initial Rep. Deep Sea Drill. Proj.*, 33, 973 pp., 1976.

Schlich, R., S. W. Wise, Jr., et al., *Proc. Ocean Drill. Program Initial Rep.*, 120, 648 pp., 1989.

Simpson, E. S. W., R. Schlich, et al., *Initial Rep. Deep Sea Drill. Proj.*, 25, 884 pp., 1974.

Suess, E., R. von Huene, et al., *Proc. Ocean Drill. Program Sci. Results*, 112, 738 pp., 1990.

Theide, J., T. L. Vallier, et al., *Initial Rep. Deep Sea Drill. Proj.*, 62, 1120 pp., 1981.

Veevers, J. J., J. R. Heirtzler, et al., *Initial Rep. Deep Sea Drill. Proj.*, 27, 1060 pp., 1974.

Von der Borch, C. C., J. G. Sclater, et al., *Initial Rep. Deep Sea Drill. Proj.*, 22, 890 pp., 1974.

von Rad, U., B. U. Haq, et al., *Proc. Ocean Drill. Program Sci. Results*, 122, 934 pp., 1992.

Wise, S. W., Jr., R. Schlich, et al., *Proc. Ocean Drill. Program Sci. Results*, 120, 1155 pp., 1992.